A Century of U.S. Naval Intelligence

Lieutenant Theodorus B. M. Mason organized the Office of Intelligence, Bureau of Navigation, and served as its first Chief Intelligence Officer, June 1882–April 1885.

A CENTURY OF U.S. NAVAL INTELLIGENCE

Captain Wyman H. Packard, USN (Retired)

A joint publication
of the
Office of Naval Intelligence and the Naval Historical Center

Department of the Navy
Washington
1996

∞ The paper used in this publication meets the requirements for permanence established by the American National Standard for Information Sciences "Permanence of Paper for Printed Library Materials" (ANSI Z39–1984).

Library of Congress Cataloging-in-Publication Data

Packard, Wyman H.
A century of U.S. naval intelligence / Wyman H. Packard.
p. cm.
Includes bibliographical references (p.) and index.
ISBN: 978-1-907521-78-2
1. United States. Office of Naval Intelligence—History—20th century. I. Title. II. Title: Century of U.S. naval intelligence.
VB231.U54P33 1994 94–35197
359.3'432'0973—dc20

For sale by the U.S. Government Printing Office
Superintendent of Documents, Mail Stop: SSOP, Washington, DC 20402–9328

To all those who
prepared me for, encouraged me in, and helped me with,
this extensive research and writing effort

Contents

List of Tables

FOREWORD

For three decades since his retirement from the U.S. Navy, Captain Wyman H. Packard has worked diligently to compile his monumental history of U.S. Naval Intelligence. Two previous editions appeared in classified, limited distribution form, but former Director of Naval Intelligence RAdm. Thomas A. Brooks foresaw a wider audience for Capt. Packard's labors and directed that a new, revised edition be prepared for open publication.

The resulting work is the product of a gratifying cooperation between the Office of Naval Intelligence and the Naval Historical Center, which throughout the project has provided major support to Capt. Packard's researches and which saw this volume through the publication process. The joint effort is intended to provide intelligence professionals, scholars, and the general public with a detailed, topical accounting of the long and varied activities of U.S. Naval Intelligence on behalf of the nation. Equally important, it is hoped that the book's detailed references to resources for further research will spark more work in a field that has not been adequately explored by historians in the past.

The role of naval intelligence in the success of the U.S. armed forces in time of war and in periods of often precarious peace deserves wider appreciation; Capt. Packard has indeed performed a magnificent service to the Office of Naval Intelligence through his painstaking labors.

M. W. CRAMER
Rear Admiral, U.S. Navy
Director of Naval Intelligence

THE AUTHOR

Capt. Wyman H. Packard, USN (Ret.), served in active duty intelligence billets from 1946 to 1965, including one or more tours as Assistant Naval Attaché, District Intelligence Officer, and Fleet Intelligence Officer. He also served in the Office of Naval Intelligence as Liaison Officer to other agencies, head of the Collection and Dissemination Branch, head of the Foreign Intelligence Division, and Deputy Director of Naval Intelligence.

His initial experiences in general line billets in the Navy following graduation from the U.S. Naval Academy in 1935 were in *Tennessee* (BB 43) in the Pacific, in *Goff* (DD 247) in the Atlantic and Caribbean, and in *Hornet* (CV 8) during its delivery of the Doolittle B-25 bomber group to its launching site in the Western Pacific for the attack on Japan. Capt. Packard continued to serve on board *Hornet* during the Battle of Midway, the air support to the Solomon Islands occupation forces, and the Battle of Santa Cruz where the carrier was sunk.

His first opportunity to use the output from intelligence production organizations was in 1942–1944, after the loss of *Hornet* when he was required to remain in the South Pacific as Assistant Operations Officer on the newly assembled staff of Adm. William F. Halsey, Commander South Pacific, in Noumea, New Caledonia. In 1945–1946, he worked on his first historical document, a highly classified summary of COMINCH participation in joint planning of special amphibious landing operations in World War II.

After retirement in October 1965, he started his research for this book on a volunteer basis and as time from his periodic consultant tasks permitted. Early classified editions issued by ONI had limited distribution in 1974 and 1978. When previously classified material was cleared for unclassified research, both the Director of Naval Intelligence and the Director of Naval History induced the author to work on this unclassified version.

He has also written numerous articles on various aspects of naval intelligence for the U.S. Naval Institute *Proceedings* and a thesis for its 1968 *Naval Review*. Since 1970, he has been corresponding secretary of his Naval Academy Class of 1935, writing a class news column for the Alumni Association monthly magazine *Shipmate*.

Capt. Packard resides in Arlington, Virginia.

PREFACE

This book is designed for use as a reference work, a topical chronology. Although those who have been involved in the events and situations described may find the book of interest, it has not been written primarily for that purpose.

Nearly all naval commands have been participants in the work of naval intelligence. Some have been involved regularly in the collection of intelligence information, and most operating commands have been required to collect intelligence information under certain situations. Almost all commands, afloat and ashore, have been users of intelligence to varying degrees, depending on their missions and foreign involvement. Most major operating commands have had intelligence-processing capabilities for adapting available intelligence information to meet their own specific needs.

To produce a comprehensive history of intelligence in the Navy, it would have been necessary to research the history of all past and present elements of the Navy, not just of specific intelligence organizations. This task, of course, would have required much more than the part-time effort that this product represents. However, it is hoped that this book, although far from complete and containing many gaps, will be of value to future researchers not only of U.S. Naval Intelligence but also of the Navy's history in which naval intelligence has played such an important part.

The administrative histories of the Office of Naval Intelligence (ONI), various other intelligence organizations and major commands in World War II, and the command histories of the Naval Intelligence Command (NIC) and its field activities that have been produced since 1967 are voluminous and very detailed. To carry subjects through the time periods involved, many of the World War II histories have been summarized in this work. The individual command histories have been exploited to a limited extent as time permitted. To facilitate follow-on research, sources, which for the most part are located in the Operational Archives of the Naval Historical Center in Washington, D.C., are provided at the end of each chapter.

This writer has been repeatedly impressed by the similarities in procedures in the research of history and intelligence. Neither history nor intelligence should be considered either factually complete or completely factual in spite of the best and most earnest efforts of the historian or intelligence analyst. There are too many gaps and inconsistencies in the information available. Intelligence is the continuous processing of current information for its implications for the future, in the light of the past. History is, in effect, the compilation of intelligence after the fact.

After several atttempts to produce an expression of gratitude to specific individuals who helped me in the researching for this history during the past 27 years, I have concluded that producing such a list would be repetitious of the many names that appear in the endnotes. I am quite certain that those individuals would approve any effort to avoid such duplication and thus to keep from adding unnecessary pages to this book.

I must mention, however, some of those who helped in other ways, performing special services along the way:

RAdm. Earl F. Rectanus, Director of Naval Intelligence (DNI) (July 1971–September 1974), encouraged me to "get serious" with my research on the history of naval intelligence, and suggested on 24 April 1972 that it be an all source reference work for future use by the Office of Naval Intelligence. He approved my use of desk space, classified storage, and typing assistance in and by the Naval Intelligence Command.

Capt. Barney Garbow of NIC arranged for the above support service in the command's spaces in the Hoffman Building in Alexandria. He also approved the use of Intelligence Reserve Units to take their annual two-week active duty for training doing historical research of ONI files held not only in ONI and the Naval Historical Center but also in the Federal Record Center in Suitland, Maryland.

Capt. Glenn Fugate, USNR, head of Reserve Programs Department of NIC, administered this helpful effort, with the effective assistance by his technical assistant, Martha Nelson.

Some of the early participants in the Reserve effort included: Cdr. S. J. O'Neill, LCdr. H.J.T. Powell, LCdr. M. T. Buford, LCdr. K. Trough, Capt. Ed F. Russell, Jr., LCdr. John Sherwood III, Cdr. David D. Gilboe, LCdr. L. S. Newman, Jr., LCdr. Henry E. Hovland, Lt. George Reynolds, Cdr. Curt W. Hibbard, Cdr. Willaim H. Moberger, LCdr. Robert F. Wolff, LCdr. Kenneth Hagen, Cdr. Willaim C. Cook, and Lt. Carl VerSteeg.

An early visit to see Capt. Paul Ryan, Deputy Director of Naval History, and VAdm. Edwin B. Hooper, Director of Naval History, led to a meeting with Dr. Dean C. Allard, head of the Operational Archives Branch in the Washington Navy Yard. All gave strong encouragement, and Dr. Allard provided desk space for the reservists in their file research efforts. The Operational Archives staff were all helpful in locating and making available the many files and records containing information pertinent to the history of naval intelligence.

Dr. Allard continued to provide guidance and support as he advanced to Senior Historian of the Naval Historical Center and then to Director of Naval History, especially after NIC moved to Suitland and I moved to the Operational Archives.

Those in ONI who provided early help by turning over some of their historical files to the Operational Archives for my use include: Vincent D. Engels, Capt. Bill Hatch, L. E. Gingell, Capt. C. J. Oleniacz, Robert Hilbish, Frank Moran, Dr. George Kidd, Roy Adolfson, Capt. C. Dale Everheart, and B. L. Willard.

Of recent support, RAdm. Thomas A. Brooks gave the early push to have this unclassified version published, and RAdms. Edward D. Sheafer, Jr., and Michael W. Cramer carried it through. A. D. Baker III of ONI performed the time-consuming task of doing the initial editing and filling in ship and aircraft designations and many of the first names of individuals. Shirley Fickett, also of ONI, typed and computerized the initial manuscript and helped with the proofing. The manuscript was copy-edited by SSR, Incorporated and wordprocessed and indexed by Frank Services, Incorporated. The Naval Historical Center's editor, Sandra J. Doyle, and editorial assistant, Wendy Karppi, prepared the work for publication. Deborah Rhode of the U.S. Government Printing Office's Typography and Design Section designed the layout of the book.

The opinions and conclusions expressed herein are mine alone and do not necessarily reflect the views of the Department of the Navy or any other agency of the U.S. Government.

W. H. PACKARD

INTRODUCTION

Naval intelligence is the accumulated knowledge on the naval science and developments in all maritime countries; the naval capabilities, activities, and intentions of all potentially hostile and friendly countries; and the characteristics of all possible areas of naval operations. It has been a requirement within the U.S. Navy ever since intelligence was used to justify the procurement of the Navy's first ships.

Additionally, naval intelligence includes the Navy's contribution to joint military and national intelligence efforts.

Naval intelligence, and the craft of military intelligence in general, has also been aptly defined by a number of authorities during this century. A few of the more noteworthy examples follow:

1923

It is impossible for us to predict in what part of the globe our armed forces may be called into action in support of diplomacy. If our commanders are to be able to brush aside the "fog of war" and make decisions that will lead them to success, the power to do so can only come from an intelligence service. Such a service can hardly be built in a day, a month, or a year. It must have been functioning quietly and steadily for many years if all the voluminous but vital information is to be collected, arranged for ready use, and kept up-to-date. The brilliant outcome of a campaign springs inevitably from years of quiet preparation. . . . Peace time is the golden time for intelligence work, for not only is the information easier to collect, but time is available for its proper collating, compilation, distribution and study.[1]

1936

Naval intelligence comprises an evaluated knowledge of nations, primarily their war-making capacity, and especially of naval and maritime factors. It is the most complete and authentic information on a probable or actual enemy or theatre of operations, critically analyzed, and incorporating the strategical and tactical conclusions to be drawn therefrom. *Evaluation* is the most critical and systematic analysis of information to determine its probable credibility and accuracy, its significance, its relevance and importance, and its conclusions. Information subjected to this process becomes "intelligence."[2]

1940

The *Naval Intelligence Service* is an organization under the primary control and direction of the Chief of Naval Operations. It comprises, as a whole, the Office of Naval Intelligence (ONI), a division of the Office of the Chief of Naval Operations (OPNAV); the intelligence organizations in the Naval Districts and outlying Naval Stations; and in the Forces Afloat. Naval Attaches and intelligence officers abroad are an integral part of ONI.[3]

1946

It is axiomatic that all reasoned decisions must be based on information that is correct, timely and properly interpreted. Also it is axiomatic that good intelligence is good economy because it makes possible the most effective disposition of our own forces. Reduction in quality of intelligence is not good economy. If anything, the weaker our operating forces, the greater will be the necessity for good intelligence.[4]

1951

You can never really become an Intelligence Officer of the inspired class unless you happen to be born with that delicate touch which produces a reasonable and measurable evaluation without full knowledge of all the facts; but there are characteristics which you can develop even if they do not come as part of your standard equipment at birth. The first is an attitude of constant suspicion—an unwillingness to take anything for granted; the second is a form of scientific mindedness—the ability to approach all things with a sense of analytical inquiry; and finally, perhaps more important than anything else, a restraint which enables one to remain silent.[5]

1953

The day has long since passed when the intelligence responsibility of our Naval establishment could safely be entrusted to officers as a collateral duty to other assignment. The intelligence officer's job is too big to be successfully carried on by just anybody, however well intentioned. The good intelligence officer today must be a person with specialized training. This does not mean, of course, that he should set himself up in some ivory tower away from all contact with the operating forces. Quite the contrary. No one can be a good intelligence officer without a detailed and intimate knowledge of the needs and capabilities of the operating forces. There is a place, to be sure, both in the regular establishment and in the reserve, for people who perform intelligence duties exclusively. But, it is equally important that the Navy also maintain a substantial reservoir of regular line officers who understand intelligence work. Such officers as these can carry out to the Fleet the full story of what intelligence is able to do to assist command.[6]

1955

In the broadest sense, intelligence underlies our estimate of the enemy and thus helps to guide our political strategy.[7]

---

[1] ONI-8: *Instructions for Intelligence Officers*, May 1923, 3–4.

[2] ONI-19 (1933), *Naval Intelligence Manual*, 1936 revision, Articles 103 and 106.

[3] ONI-19 (1933), 1940 change, article 110.

[4] Rear Admiral Thomas B. Inglis, USN, Chief of Naval Intelligence, letter to the Secretary of the Navy, ser 02204P32, accession 3770, box 1, 8 Aug 1946, Day File, Naval Historical Center, Operational Archives (AO).

[5] General Walter Bedell Smith, quoted by Rear Admiral Felix L. Johnson, Director of Naval Intelligence, in an address at the Army War College, 19 Feb 1952.

[6] Secretary of the Navy Robert B. Anderson, in a speech at the graduating exercises, Naval Intelligence School, 26 Jun 1953, quoted in *ONI Review,* 1953, 431.

[7] *Killian Report*, 14 Feb 1955, 25.

TITLES FOR THE HEAD OF THE OFFICE OF NAVAL INTELLIGENCE

June 1882–February 1973

Date of Origin or Change	**Title**
15 June 1882	Chief Intelligence Officer
1911	Director of Naval Intelligence
30 October 1945	Chief of Naval Intelligence
1 November 1948	Director of Naval Intelligence
1 June 1954	Director of Naval Intelligence and Assistant Chief of Naval Operations for Intelligence
1 July 1967	Assistant Chief of Naval Operations for Intelligence and Commander, Naval Intelligence Command
15 March 1971	Director, Intelligence Division (OP-942) and Assistant Chief of Naval Operations for Intelligence (OP-009); COMNAVINTCOM
February 1973	Director of Naval Intelligence (OP-009) and COMNAVINTCOM; additional duty Assistant for Intelligence Support (OP-094Q) in Command Support Programs

Note: Except for the pre-1911 period, the title Director of Naval Intelligence, or DNI, is used throughout this book when referring to the head of the Office of Naval Intelligence.

CHAPTER 1

General Summary

Before ONI

In the period between the end of the Civil War and the establishment of the Office of Naval Intelligence (ONI) in 1882, the U.S. Navy gradually deteriorated into a force in name only. The voters of the country had no interest in foreign affairs or in maintaining a strong navy, and Congress accordingly found no incentive to budget funds to construct new warships and little justification to operate and maintain old, ineffective ones. Research to advance the innovations in naval science and technology introduced during the Civil War received no official encouragement.

In Europe, however, the opposite situation prevailed. There, the maritime powers, particularly those that had had naval attachés in Washington during the U.S. Civil War, advanced the research ball, introduced their own ingenuity, and came up with improvements in ship design, ship construction techniques, propulsion, and weapons. At the same time, new concepts for the employment and support of navies were developed.

Although the U.S. Navy was not allowed to compete with this maritime progress, it could and did observe it in anticipation of the time when new ships might be authorized. Beginning shortly after the Civil War, ships cruising to European ports were expected to try to keep abreast of developments abroad. For example, in May 1866, when Assistant Secretary of the Navy Gustavus V. Fox was sent aboard the monitor *Miantonomoh* as emissary to Czar Alexander of Russia, Secretary of the Navy Gideon Welles instructed Fox to collect details on the important naval situations in the various countries of Europe that he would visit en route to and from Russia. Fox was to observe foreign methods of building, repairing, and laying up naval vessels and to compare European naval vessels with those of the United States. Fox visited ports in England, France, Denmark, and Finland on his way to Russia; the return trip included stops in Sweden, Germany, France again, and at various Mediterranean ports during the winter before arriving in the United States in May 1867.[1]

Officers were also sent individually and in groups by the various Navy bureaus to observe installations and techniques according to the interests of the sponsoring bureau. Chief Engineer James W. King made four trips to Europe, the first in 1867 to visit dockyards in England and France. On King's second trip to Europe, in 1869, soon after his appointment as Chief of the Bureau of Steam Engineering, he investigated the progress being made in compound steam engine design. In 1873, King collected information about various naval appliances. Finally, in 1875–1876, he was directed by the U.S. Senate to examine and report on ships of war and the mercantile marine. The data that King collected on his last trip were published under the title *European Ships of War*. He later authored *The Warships and Navies of the World*, which was recognized as the authority on the subject at the time of its publication in 1883.[2]

In 1870, Commo. Christopher R. P. Rodgers was sent to England, France, and Russia to obtain information on naval administration and logistic matters. Also in 1870, Capt. Edward Simpson, assisted by LCdr. Joseph D. Marvin, was provided with letters of introduction to U.S. diplomatic representatives in all major European countries and was instructed to collect information on naval developments, particularly those relating to ordnance.

In 1872, Capt. Simpson was relieved by Cdr. Francis M. Ramsay. Cdr. Ramsay was additionally ordered to report to the U.S. minister in England "as Naval Attaché to his legation." This appears to be the first time that the title "naval attaché" was accorded to a U.S. naval officer. The U.S. naval attaché system, as part of a naval intelligence organization, was not established until 1882, but Ramsay's orders and duties entitle him to be recognized *de facto* as the first U.S. naval attaché.[3]

U.S. Navy observers followed the naval operations of the war in South America between Chile and the bordering countries of Peru and Bolivia in 1879–1881 and the fighting in Egypt during the 1882 British Tel-el-Kebir campaign. LCdr. Dennis W. Mullan, executive officer of the frigate *Adams*, was detailed to accompany the staff of Gen. Baquedano, the Chilean commander in chief. Mullan's reports to the Navy Department later provided part of the material for one of the early ONI War Series reports, *War on the Pacific Coast of South America, 1879–81*. LCdr. Casper F. Goodrich was assigned to the staff of LtGen. Sir Garnet Wolseley during the Tel-el-Kebir campaign, and his reports were also published in the ONI War Series as *Report of the British Naval and Military Operations in Egypt, 1882*.[4]

Origin of ONI: 1882

As the reports from overseas observers accumulated in the various Navy bureaus and with little or no coordination between the bureaus of the information obtained, the Secretary of the Navy found it increasingly difficult to get unanimity in the opinions of the bureau chiefs as to which developments in Europe were important to the future needs of the U.S. Navy. Congressmen were unhappy about the conflicting views and theories presented to them by Navy spokesmen on the specifications for new ships. To correct the situation and to provide an authoritative official source on what the new Navy should be, Secretary of the Navy William H. Hunt created an advisory board to establish positions on new ship construction requirements that could be uniformly voiced by the Secretary and the bureau chiefs. The board began its work on 11 July 1881, but it needed factual, objective information upon which to base its advice to the Secretary.[5]

Lt. Theodorus B. M. Mason, having completed an extended leave of absence from 1877 to 1879, during which he had traveled throughout Europe to study the progress in naval science, had assembled an impressive bundle of reports. He was a qualified naval observer, knew what information was available, and knew how to get it. Consequently, in late 1881, when Mason had his first chance to start advancing his views on the need for collecting naval information abroad, he had a receptive, although selective, audience. If advantage was to be derived from progress in naval science elsewhere, as he advocated, the U.S. Navy should assign naval attachés to embassies and legations throughout the world to collect intelligence on the subject. He also strongly recommended that a section be set up in the Secretary of the Navy's office where intelligence reports could be assembled, correlated, and made available for professional use in the movement that was getting underway to rebuild the Navy.[6]

Lt. Mason soon gained access to Secretary Hunt, who apparently agreed with Mason's idea for setting up an intelligence office; on 23 March 1882, the Secretary issued General Order No. 292 establishing an Office of Intelligence:

General Order No. 292

NAVY DEPARTMENT,
Washington, March 23, 1882

An "Office of Intelligence" is hereby established in the Bureau of Navigation for the purpose of collecting and recording such information as may be useful to the Department in time of war, as well as in peace.

To facilitate this work, the Department Library will be combined with the "Office of Intelligence" and placed under the direction of the Chief of the Bureau of Navigation.

Commanding and all other Officers are directed to avail themselves of all opportunities which may arise to collect and forward to the "Office of Intelligence" professional matter likely to serve the object in view.

William H. Hunt
Secretary of the Navy

Instead of making the intelligence office part of his office, as Mason had recommended, Secretary Hunt placed it in the Bureau of Navigation, the Navy bureau responsible for various matters including personnel. There were two probable reasons: Secretary Hunt was a "short timer" and wanted to put the new office where he knew it would receive support (Commo. John G. Walker, Chief of the Bureau of Navigation, was aggressively interested in foreign development and favored Mason's proposals), and he could tie the intelligence organization to the Navy Department Library, where the intelligence office would immediately have access to the professional books and reports on foreign developments already being collected and assembled by the Bureau of Navigation.

William E. Chandler replaced Hunt as Secretary of the Navy on 17 April 1882. Mason, following up to ensure that the general order would be implemented, found Secretary Chandler receptive. On 15 June 1882, Mason found himself assigned to the Bureau of Navigation to organize the new Office of Intelligence and to serve as its first "Chief Intelligence Officer."

Mason's problems were many, several of them stemming from the fact that the new office, established by a Navy Department general order and not by Congress, functioned more as a board and, therefore, could receive no direct funding until it had been authorized by Congress. Three officers were assigned to assist Mason: Lt. M. Fisher Wright, who reported to the Bureau of Navigation for special duty on 1 July 1882; Lt. A. G. Berry, who was on duty as the Signal Officer, Washington; and Ens. Templin M. Potts, who had reported to the Navy Department in Washington but was not yet otherwise assigned. Clerks were borrowed from other offices as needed and as available, and office space was assigned in the State, War, and Navy Building.[7]

After studying the filing systems used by other U.S. Government departments and bureaus, Mason selected the one used by the State Department. From that system, he devised a card catalogue method suitable to the needs of his office as interpreted from the guidance that Secretary Chandler had given him in a letter dated 25 July 1882 (and probably written by Mason):

NAVY DEPARTMENT
Washington, D.C. July 25, 1882

To Lieutenant Theodorus B. M. Mason In Charge of the Office of Naval Intelligence, Bureau of Navigation, Navy Department:

The following rules and regulations are established for the Office of Naval Intelligence:

The object of the Office will be to collect, compile, record and correct information on the following subjects:

1st: The cruising fleets of foreign powers.

2nd: The war material of foreign powers.

3rd: The nautical personnel of foreign powers.

4th: The armament of foreign ports including their lines of communication.

5th: The facilities of foreign governments for transporting troops and material.

6th: The facilities of foreign governments for improving torpedo boats and torpedo defenses.

7th: The facilities on foreign coasts and in foreign ports for landing men and supplies.

8th: The facilities for obtaining coals and supplies in all quarters of the globe.

9th: The actual capabilities of foreign merchant steamers and the true routes followed by regular steamship lines.

10th: Information in regard to our own Navy.

11th: Information in regard to our mercantile marine.

12th: Information in regard to our coast defense.

13th: Information which may be of use to our officers in their professional studies.

14th: Information which may be of use to our mercantile marine.

In order to publish information and to assist officers in their studies, a monthly bulletin will be published containing reprints, original articles and reports, not of a confidential nature, which contain information of general interest; a bibliography of current service literature, a reprint of general orders, and the movements of ships and officers when proper to be published together with such other information as it may be necessary or convenient to promulgate through this channel.

In order that the objects of the office may be fully realized, only such officers as have shown an aptitude for intelligence staff work or who by their intelligence and knowledge of foreign languages and drawing give promise of such aptitude, should be employed.

In order to collect information a corps of correspondents, in the persons of Naval Attachés to our foreign legations [and of] special aids to our commanding officers aboard will be organized.

The Bureaus of the Navy Department, having in their archives information which it will be necessary to collate, are directed to give the office full access to the same. The younger officers of the service will be encouraged in collecting and reporting intelligence and in writing articles on naval subjects.

A record will be kept of those officers who are specially qualified in languages, drawing and as Intelligence Staff Officers.

The Department Library will furnish access to all the current service literature. The United States Naval Institute, a voluntary organization of the officers of the Navy for the purpose of facilitating study, will be encouraged by a contribution to its journal of such matter as may be thought proper from time to time.

William E. Chandler
Secretary of the Navy

These original rules and regulations for ONI are of particular and continuing interest. They list information requirements that, with a few updates and additions, still provide an acceptable definition of the scope of naval intelligence. The rules and regulations anticipated that ONI's products would be of value and interest to all naval officers in their professional advancement and duties. Chandler's letter recognized the need to have officers of the Navy collect and report on intelligence relating to naval subjects and also recognized a similiarity in mission between the U.S. Naval Institute and ONI.

Lt. Mason organized his new office along functional rather than geographic lines in order to facilitate the correlation of intelligence material according to its usefulness to his expected primary customers, the Secretary of the Navy and the bureaus of the naval establishment. At that time, the bureaus with an interest in foreign conditions and developments were Navigation, Steam Engineering, Ordnance, Construction and Repair, Yards and Docks, and Medicine and Surgery. Mason set up an index of subjects pertinent to the interests and intelligence requirements of each bureau customer.[8]

At first, data relating to army warfare matters were also listed on cards by ONI, but it soon became apparent that trying to keep informed on foreign navies would keep the small ONI staff fully employed. Therefore, when the War Department established its own intelligence office in 1885, ONI's files of military information were turned over to the Army.[9]

Apparently, Mason also initially had difficulty gaining access to the foreign information that the other Navy bureaus held. On 7 October 1882, Secretary Chandler issued a circular to all bureau chiefs stating that

> access to the records and information of each Bureau should ordinarily be freely given, without formality, to the other Bureaus, their officers and clerks. If access to records, or information deemed confidential by the Chief of the Bureau containing them, is sought, such access may be refused until the question is submitted to the Secretary. . . . Good judgement and good nature will likely always determine any differences, without raising questions for the decision of the Secretary.

Finally, the new office sent its first overseas observer, LCdr. French E. Chadwick, to be Naval Attaché, London. Chadwick, who reported to the American legation on 15 November 1882, was an excellent choice. He exhibited an exceptional talent as an observer not only on naval matters, but also on scientific and technical subjects relevant to the reconstruction of the U.S. Navy.[10]

At the same time, orders were issued to Lt. John C. Soley that were similiar to Chadwick's and had the same date (28 October 1882), sending him to temporary duty in France for the general purpose "of obtaining full and accurate information in reference to the organization of certain branches of the naval administration in that country." Soley was required to make periodic reports and was to obtain assistance from the U.S. legation, but he was not granted the title of naval attaché.[11]

Thus, ONI was established on an austere basis. It had the start of a collection and processing system, sound rules for its guidance, and enthusiastic leadership, but it had an inexperienced staff and no funds.

Early Years

Three more officers, Lt. William H. Driggs, Lt. Sidney A. Staunton, and Passed Assistant Engineer J.P.S. Lawrance, were assigned in 1883 to the Bureau of Navigation for special duty, and two officers departed, but only Lt. Mason was overtly identified as being affiliated with the Office of Naval Intelligence. The Navy Department Library, as of 1 January 1883, was still a separate organization under Professor James R. Soley, USN; it served ONI, however, as a procurer of foreign publications and technical literature.[12]

ONI's problems with the Navy technical bureaus gradually subsided as the latter found the work of ONI of value to their planning and design work. ONI's output also proved effective in justifying to Congress the need for appropriations for Navy expansion.

There was no request for appropriations for ONI by name in 1883, but the Chief of the Bureau of Navigation did request a limited appropriation of $12,000, "to publish in convenient form such information obtained from leading professional periodicals and from other sources as will be of value to naval officers, to the Merchant Marine and to the shipbuilding interests of the country."[13]

Secretary Chandler announced the establishment and purpose of ONI in his *SECNAV Annual Report, 1883*:

> In order that the Department may be supplied with the fullest and most accurate information as to the progress of naval science in this and in other countries, and the condition and resources of foreign navies, an Office of Naval Intelligence has been established for the collection and classification of such information, and for its publication, as far as may be advantageous and suitable.[14]

The first requests for funds for the "Collection of Naval Intelligence" appeared in the Secretary's *Annual Report* for 1884 under the heading, "For Support of the Bureau of Navigation for Fiscal Year 1886": "for the publication of current technical information for the instruction of the personnel of the Navy and mercantile marine—$10,000. For salary of one copyist $900.00. For salary for one laborer $660.00, total $11,560.00." Congress did not approve the request, nor would it do so knowingly for any subsequent similar request until Fiscal Year 1900.[15]

Another four officers reported to ONI during 1884: Lt. William H. Beehler, Lt.(jg) Washington I. Chambers, Ens. Frank R. Heath, and Ens. William L. Rodgers. No officers were detached, leaving nine assigned at the end of the year.

Numerous events and changes in 1885 influenced the development of ONI. Five naval officers reported to ONI for duty: Lt. Wainwright Kellogg, Lt.(jg) John C. Colwell, Lt.(jg) Alexander Sharp, Jr., Ens. George H. Stafford, and Lt. Raymond P. Rodgers, who was the relief for Lt. Mason as Chief Intelligence Officer. Three officers were detached in addition to Mason, making a net gain of one, for a total of ten officers at the end of the year.[16]

ONI's naval attaché collection capabilities were doubled by the assignment of Lt. Benjamin H. Buckingham as Naval Attaché, Paris, on 11 November 1885. Buckingham was also accredited to St. Petersburg and Berlin. It seems probable that one of the inducements to the Navy for Buckingham's assignment was a letter to the Chief Intelligence Officer from Edward W. Very, former Assistant Chief of the Bureau of Ordnance (and inventor of the Very Pistol signal gun), who had resigned from the Navy as a lieutenant and was employed by the Hotchkiss Company, an armaments manufacturer in Paris. Very had been forwarding to ONI a great volume of intelligence on small caliber weapons, for whose manufacture Hotchkiss could claim preeminence. Very's letter of 6 May 1885 indicated that he would have to drop most of his "little private intelligence department" because of the increase in the volume of his design work for Hotchkiss.[17]

Collection capabilities were also expanding through the cooperation of Hydrographic Branch Offices then existing at Boston, New York, Philadelphia, Baltimore, New Orleans, and San Francisco. The branch offices provided charts, sailing directions, light lists, etc., to commanding officers of naval ships and masters of merchant ships, and solicited from them information on their experiences and observations, particularly information that would update the Hydrographic Office publications or help future visitors to foreign ports. The latter information was passed to ONI.[18]

A new ONI customer was created with the convening of the first class of the Naval War College at Newport, Rhode Island, in September 1885. The new college and ONI had much in common: they were both concerned with the study of worldwide progress in naval science and with conveying that knowledge to naval officers to improve their development and employment of ships and weapons. Both were also newly established and found their support and their opposition coming from the same groups within the Navy. Additionally, much of what ONI produced was needed by the War College, not only for its courses but also to add factual conditions, and thus realism, to considerations of strategic and tactical classroom problems. Secretary Chandler's response to a Senate inquiry about the advanced course to be set up at the Naval War College reads like an intelligence requirements list that would be ONI's responsibility to fulfill:

> The constant change in the methods of conducting naval warfare, imposed by the introduction of the armored ships, swift cruisers, rams, sea-going torpedo boats, and high-power guns, together with the more rigid methods of treating the various subjects belonging to naval science, render imperative the establishment of a school where officers may be enabled to keep abreast of the improvements going on in every navy of the world.[19]

Additional details on the ONI-Naval War College relationship are discussed in Chapter 31.

Another change in 1885 was the assignment of the Navy's war planning function to ONI. An order of 31 March 1885 (now lost but repeated in part in a general order issued in 1892) stated:

> The duty of the office [ONI] shall be to collect and classify information upon all subjects connected with war, or which can have a bearing upon naval action, and to prepare detailed plans of campaigns covering all contingencies of active naval operations.

The assignment of war planning to ONI was a logical step, considering that the guidance to ONI from the time it had been established required the organization to assemble information not only about foreign naval forces but also about our own. Having war planning done by the office that had the responsibility for gathering the information needed for such planning, no doubt, seemed sensibly simple at the time. War planning remained an ONI function until 1904, when it was made the joint responsibility of ONI and the General Board; the arrangement continued until establishment of the post of Chief of Naval Operations (CNO) in 1915, when a separate division of that office was set up for War Plans.[20]

From 1886 to 1897, ONI continued to operate with an average staff of ten officers, one or two borrowed clerks, and no appropriation except for Fiscal Years 1889 and 1890, when $1,600 was appropriated for one stenographer transferred to ONI from the Secretary of the Navy's office. In Fiscal Year 1891, an appropriation for the stenographer was requested but was denied by Congress.

Even the annual appropriation for professional and technical reference books for the Navy Department Library was reduced in 1886 from $2,500 to $1,000. The reduction was indirectly prejudicial to ONI's collection efforts, for the organization depended to some extent on the library's funds for the purchase of foreign reference books.

The Secretary of the Navy's *Annual Reports* for the years 1886–1897 contained lengthy discussions on foreign naval developments that were obviously supplied by ONI. The office was also frequently involved in supplying studies and reports to the naval committees of Congress. Branch Hydrographic Offices reported a considerable demand for any ONI reports on hand.

In 1888, Congress sanctioned an appropriation for "Pay, Miscellaneous" that included the "cost of special instruction at home or abroad in maintenance of students and attachés, and information from abroad and the collection . . . thereof." The appropriation gave the naval attaché system its first legislative recognition. It also gave the attachés an expense account for funds needed for the collection of information. Up to that time, an attaché was ordered to his post "without pay," which meant no pay beyond that of his rank. Cdr. Chadwick, however, could draw on an account with a London banking house. From its inception, ONI was closely tied to the decisions made by both the Navy and Congress relating to the rebuilding of the Navy. It was therefore considered proper to support ONI and the naval attachés from the appropriation fund titled "Increase of Navy" when no other funds were available; this may have been the source of the funds for Chadwick's account. Also in 1888, a third naval attaché was added to the team of intelligence collectors with the arrival of Lt. Nathan Sargent at the U.S. Embassy in Rome, on 20 November 1888. Sargent was also accredited to Vienna.[21]

The Secretary of the Navy in 1889 evaluated ONI and its naval attachés:

> At the very time when the first cruisers were being designed, the Department took steps to supply its wants of experience by the systematic acquisition of information as to naval progress abroad. The establishment of the office of Naval Intelligence and the assignment of Naval Attachés to duty in Europe, both of which measures date from 1882, have been of incalculable assistance in the work of reconstruction; and it is proper to refer especially to the untiring and successful efforts of Commander F. E. Chadwick, the first attaché sent out, whose extraordinary ability and judgment during six years of difficult service in England and on the Continent have had a lasting influence upon naval development in this country. The results subsequently obtained have shown the wisdom of the policy adopted at the outset.[22]

Lt. Rodgers was detached as Chief Intelligence Officer on 22 July 1889, and Cdr. Charles H. Davis reported as his relief on 16 September.

On 30 September 1889, Capt. Alfred Thayer Mahan, having been relieved by Cdr. Casper F. Goodrich as president of the Naval War College, reported for special duty to the Bureau of Navigation to serve as an advisor to the Secretary of the Navy and to prepare outline plans in case of war with various foreign nations. In performing his duties, Mahan had access to ONI files and conferred with Chief Intelligence Officer Cdr. Davis. From time to time, Mahan would even use ONI stationery.[23]

On 21 October 1889, the Navy Department Library and War Records Office under Professor Soley was transferred from the Bureau of Navigation to the Office of the Secretary of the Navy. The reason for the change has not been determined, but it may have been in anticipation of the arrival of Commo. Francis Ramsay as Chief of the Bureau of Navigation on 1 November 1889. Soley had been a frequent lecturer at the Naval War College, and a strong supporter; Commo. Ramsay, on the other hand, had been against the establishment of the War College. Soley and Ramsay, therefore, were in opposing camps relative to the college and, thus, also toward ONI.[24]

Professor Soley performed well in the Secretary's office, and, on 18 July 1890, he was appointed by Secretary Benjamin F. Tracy to fill the long vacant billet of Assistant Secretary. The Library and War Records Office remained under his supervision, but LCdr. Frederick M. Wise succeeded Soley as librarian and officer in charge of the War Records Office.

On 3 October 1890, Secretary Tracy advised the Chief of the Bureau of Navigation that ONI "including all correspondence with Naval Attachés, is hereby transferred from the Bureau of Navigation to the Office of the Secretary." The office was placed under the Assistant Secretary by a 17 October 1890 circular to chiefs of the Navy bureaus:

> The Office of Naval Intelligence and the Naval Attachés having been transferred to the Secretary's Office are hereby placed under the supervision of the Assistant Secretary, who will conduct the correspondence in relation thereto. Chiefs of Bureaus desiring information from abroad will furnish the Assistant Secretary with memoranda of the information required.

The move of ONI to the office of the Secretary placed it in the position that Lt. Mason had originally recommended when he was advocating the establishment of an Office of Naval Intelligence.[25]

In requesting an appropriation for ONI in his *Annual Report* for 1890, Secretary Tracy stated:

> The work of the Office of Naval Intelligence is constantly increasing; with the increase in construction and the growing work of arming and equipping the new ships, the importance of this office is felt by every bureau and office of the Department and in the service at large. With this steady increase of work, the office labors under the enormous disadvantage of having no clerical force authorized, although prior to the current fiscal year one stenographer had been allowed by the appropriation bill.

During a visit of the armored cruiser *Baltimore* to Chile in 1891, a crew member was attacked. When the two governments were unable to reach a settlement, the Navy Department prepared for war. Capt. Alfred Thayer Mahan was again called to Washington from the Naval War College to head a naval board consisting of himself, the Assistant Secretary of the Navy, and an officer from ONI. President Benjamin Harrison sent a war message to Congress, but Chile apologized and agreed to pay indemnities. The establishment of the Navy special board indicated the need for a permanent planning staff and served as another argument for those pushing for the establishment of a General Staff. ONI, however, was available on a continuing basis to provide the information and advice needed by temporary planners.[26]

Secretary of the Navy Tracy, by General Order No. 372 of 25 June 1889, had directed a major reorganization of the Navy Department effective 1 July 1889. The reorganization set up what became known as the Board of Construction, composed of the Chiefs of the Bureaus of Ordnance, Equipment, Construction, Steam Engineering, and Yards and Docks, plus the Chief Intelligence Officer. The board served as "a standing committee of advice in reference to questions arising in the design and construction of new ships." By 1892, the board had become indispensable to the Department in its efforts towards the reconstruction of the Navy, according to Tracy. The deliberations of the board were influenced in large measure by the information supplied by the Chief Intelligence Officer.[27]

On 2 September 1892, Cdr. French Chadwick, who had previously served so capably as Naval Attaché, London, became the Chief Intelligence Officer (Cdr. Davis having been detached on 31 August.) ONI's analytical capabilities were expanded in January 1893 when the Marine Corps assigned 1stLt. Lincoln Karmany to the ONI staff as its first Marine Corps officer.[28]

By 1893, ONI was maintaining records on the navies of Argentina, Austria, Brazil, Chile, China, Denmark, England, France, Germany, Greece, Holland, Italy, Japan, Portugal, Russia, Spain, Sweden and Norway (then combined), and Turkey. Also, between 1893 and 1896, ONI was keeping the Navy Department "thoroughly informed of everything from abroad likely to affect the construction or equipment of our new battleships, cruisers, or torpedo boats, building or about to be built." Various reports and studies on those matters were furnished to the Secretary of the Navy and to several boards formed to advise him.[29]

Cdr. Chadwick was detached from the Office of Naval Intelligence on 20 June 1893 and was relieved as Chief Intelligence Officer by Lt. Frederick Singer, who reported on 30 June. Lt. Singer had previously been an ONI staff officer from 1887 to 1890.

With the outbreak of the Cuban Revolution in 1895, much of ONI's effort was shifted to accumulating information on the Spanish Navy and on Spanish coastal defenses, both at home and in the Spanish colonies. The possibility that the United States might become involved influenced its war planning considerations, especially the decision to keep sufficient forces in the Atlantic to protect the East Coast. In 1897, for example, even when Japan appeared to be preparing to contest American efforts to protect U.S. citizens and interests during unsettled conditions in Hawaii, it was decided not to shift any battleships from the Atlantic to the Pacific.[30]

An act of Congress, approved on 12 January 1895, reduced to 1,000 the number of copies of any government publication that could be printed after 31 December 1899. The act stimulated a review of necessary reductions in the liberal (3,000 copies) distribution of ONI publications. It had been normal ONI practice to provide copies to all officers on the active list of the Navy (it was felt that ONI publications were intended primarily for their instruction), large public libraries, and constituents of senators and congressional representatives in response to individual requests. Exchanges of publications were also made with selected publishers and, through the naval attachés, with foreign naval officials.[31]

LCdr. Richard Wainwright, who had relieved Lt. Singer as Chief Intelligence Officer on 4 April 1896, took steps in 1896–1897 to overhaul the indexing and filing of ONI records and to remove obsolete matter from the working files. With the increase in the number of active ships in the Navy, the number of commissioned officers that could be spared for shore assignments, such as to ONI, had been reduced accordingly. The number of officers assigned to ONI, in fact, dropped from ten in 1896 to five in 1897. Consequently, Wainwright found it necessary to have information in the working files up to date, in immediately usable form, and easily retrievable by the on-loan clerical staff at the very first sign of possible conflict.

Under ONI, and reporting to it, were the three previously mentioned naval attachés accredited to various countries in Europe, with each attached to one or more legations or embassies. From those sources, ONI was in constant receipt of valuable information of scientific and naval character and was kept fully posted on the developments in the naval world abroad.

When he took over ONI in 1896, Wainwright determined the need for an authorized and trained clerical force having some degree of permanence.

He requested Fiscal Year 1898 appropriations to pay the salaries of one clerk, one assistant draftsman, and one laborer, for a total of $3,260. The request was disapproved (or ignored) by Congress, and ONI continued to borrow civilian personnel from other sources: one clerk from the Bureau of Supplies and Accounts, one assistant draftsman from the Bureau of Construction and Repair, and one laborer from the Nautical Almanac Office.[32] In addition, in 1897, ONI obtained the services of four copyists, paying them under the Naval Appropriations Act fund titled "Increase of the Navy." The augmentation helped to alleviate the load on those who remained at ONI's Washington office as the naval officers moved to sea.[33]

On 19 April 1897, Theodore Roosevelt became Assistant Secretary of the Navy and immediately indicated that his supervisory function over ONI was going to involve more than signing a few papers. He established close contact with the Chief Intelligence Officer, Richard Wainwright. Roosevelt wanted not only all the information ONI could provide on the world's navies, but he also sought Wainwright's advice on many matters not directly involving intelligence. When LCdr. Wainwright was transferred to sea duty in November 1897, Roosevelt took the occasion to express his views in a letter to the Secretary of the Navy on the functions that the Chief Intelligence Officer should fulfill: "They should be more than merely heads of such offices; they should provide advice and assistance to the civilian heads of the Department in a manner as would a Board of Admiralty member."[34]

In 1895, when tension with Spain increased over the Cuban situation, LCdr. Raymond P. Rodgers, Naval Attaché, Paris, was additionally accredited to Madrid. Lt. William S. Sims relieved Rodgers on 5 March 1897 as Naval Attaché, Paris, St. Petersburg, and Madrid. On 1 July 1897, Lt. George L. Dyer was ordered to Madrid to serve exclusively as naval attaché at the U.S. legation, taking over on 10 August 1897. However, Lt. Sims had officiated long enough in Madrid to establish contacts and set up lines of communication from his Spanish contacts to his normal post in Paris. These contacts continued to function effectively after Lt. Dyer had to close shop in April 1898 at the start of the Spanish-American War. On 15 November 1987, Cdr. Richardson Clover relieved LCdr. Wainwright as Chief Intelligence Officer.

Spanish-American War

After the *Maine* blew up at Havana on 15 February 1898, the Navy Department began negotiations for the purchase of ships and munitions of war from abroad. The negotiations were conducted through the naval attachés in Europe and through commercial agents in Washington.[35]

The activities of the naval attachés, in addition to the purchase of war material, were devoted principally to reporting the movements of the Spanish squadrons of Admirals Cervera and Camara. The attachés also established and maintained a corps of special agents. The attaché reports were forwarded to ONI and formed a special information report series. The series was destroyed after the war because the reports contained much data that would have been compromising to the special agents (some of whom were in high positions in Spain) had the contents—and thus the agents' identities—ever been made public.[36]

At the time of the sinking of the *Maine*, several ONI staff officers had already been ordered to sea duty without replacement. When Congress declared war on Spain on 25 April, there remained in ONI four officers plus a retired ensign, Edward E. Hayden, who had been recalled to active duty on 23 April. Capt. John R. Bartlett, USN (Ret.), reported on 1 May as the relief for the Chief Intelligence Officer, Cdr. Richardson Clover.[37]

On the day after the declaration of war with Spain, ONI was transferred from the office of the Assistant Secretary back to the Bureau of Navigation by order of the Secretary.[38]

On 5 April, Ens. Arthur B. Hoff was assigned as an Assistant Naval Attaché, London, in order to aid in the extensive arrangements being undertaken by the attachés in both London and Paris to purchase ships and material in anticipation of war with Spain.

The Naval War Board was established on the day war was declared. Capt. Alfred Thayer Mahan was ordered back to active duty on the same day and reported to the board on 9 May. Assistant Secretary Theodore Roosevelt was a member of the board until he severed his connection with it on 9 May and resigned as Assistant Secretary on the tenth. RAdm. Montgomery Sicard was president of the board. The other member was Capt. Arent S. Crowninshield, Chief of the Bureau of Navigation; Lt.(jg) Alphonso H. Cobb, USN (Ret.), was secretary.[39]

The Naval War Board was the outgrowth of an informal advisory board that had existed for some time. It was the duty of the board to advise the Secretary of the Navy about the Navy Department's strategic policy. ONI's responsibility to the board was to make available information from its files about the defenses of Cuba, Puerto Rico, and the Philippines, as well as the location of the principal ships of the Spanish fleet. During the approximately five months of war, the amount of intelligence information furnished by ONI to the Naval War Board was reported by Capt. Bartlett to equal that furnished to all the bureaus of the Department of the Navy combined.[40]

As of 1 July 1898, following successive detachments of officers to sea duty, there remained in ONI only one retired captain and one retired ensign, plus the usual borrowed clerks and messengers. Retired Capt. John R. Bartlett on 9 July was also ordered to additional duty to relieve RAdm. Henry Erben, USN (Ret.), as Chief of the United States Auxiliary Naval Forces. Bartlett, with Lt. H. L. Satterlee as his assistant, operated the Auxiliary Naval Forces headquarters from ONI spaces in the State, War, and Navy Building. Capt. Bartlett also continued to serve as superintendant of the Coast Signal Service, a position he had held prior to becoming Chief Intelligence Officer.[41] For more details on intelligence activities during the Spanish-American War, see Chapter 32.

During the Spanish-American War, it became apparent that the Navy had a critical need for a qualified observer of naval affairs in the Far East. This situation was rectified on 10 September 1898 when Lt. Albert L. Key was ordered as naval attaché to the U.S. legations at Tokyo and Peking.[42]

Although Capt. Bartlett had only served as Chief Intelligence Officer during the Spanish-American War (1 May to 15 October 1898), he wrote a report on the operations of ONI that may be considered to be the first annual report of the Chief Intelligence Officer. In the absence of any previous such report, he included much historical data about ONI for the sixteen years since its establishment. In his summary of the work of ONI, Bartlett mentioned many of the subjects of its numerous publications and stated: "In addition to the labor incident to the preparation of these volumes, the office has furnished to Senators and Representatives and Congressional committees, in response to requests from them and under the authority of the Department, an amount of information which in the aggregate is very great."[43]

At the conclusion of the hostilities with Spain, Cdr. Clover returned on 15 October to complete his interrupted tour as Chief Intelligence Officer, and LCdr. William H. Driggs also returned from sea duty to resume his staff intelligence duties. Capt. Bartlett and Ens. Hayden were detached, and Lt. Horace W. Harrison was ordered to ONI on 1 November, raising the office complement to three officers. In a supplementary report to Bartlett's annual report, Cdr. Clover stated that the office, as then staffed and organized, could not meet its requirements. He pointed out that all foreign naval powers had offices corresponding to ONI that were established on a permanent and liberal basis. One European power was said to have had an official annual budget of at least $200,000 for obtaining and preserving the type of information that ONI should have been collecting and that would have given the United States a great advantage in case of war. Clover stressed that all foreign nations were kept fully posted about our ships and fortifications and our improvements in the arts of war and that for this country to fail to do the same was a gross error.

Because of the scarcity of officers in the U.S. Navy, and with only three on duty in ONI, four department-head billets in ONI were vacant at the end of 1898. In the past, there had always been between seven and twelve officers for whom there had been plenty of work that was, in itself, an education. Cdr. Clover concluded that as the Navy's strength increased, "We should not neglect to use every effort to keep posted as to the improvements abroad in the machinery and science of war, and to obtain such other information as is necessary in the hour of national peril."[44]

Although the Navy Library was still part of the Office of the Secretary of the Navy and was thus separated organizationally from ONI, a request for sufficient appropriated funds was made in 1898 "to obtain the most recent treatises on professional subjects, expensive books of scientific and technical value, and such periodicals, foreign and domestic, as contain data of value to naval officers, and especially to supply the needs of the Office of Naval Intelligence, which depends mainly upon this library for the technical and professional information used in its publications."[45]

Official Recognition

The existence of ONI was finally offically recognized by Congress in 1899 when it appropriated funds for Fiscal Year 1900 to allow ONI to employ five clerks, one translator, one assistant draftsman, and one laborer.

During the Spanish-American War, the ONI staff had been reduced to the point where it could only handle urgent war reqests for information from its files. The routine work of processing newly received information and filing it where it would be ready for future emergencies could not be done. Subsequent efforts to reduce the backlog occupied the entire staff, which by 30 June 1899 had been increased to seven officers and the borrowed civilian force mentioned above (and which ONI acquired as permanent staff the next day).

Under a joint resolution adopted by Congress on 12 January 1899, ONI was permitted to print 3,000 copies of its publications for calendar year 1899, thus temporarily setting aside for ONI the 1,000-copy printing limitation then imposed on all other government publications.

There was also a change in policy in 1899 relative to the release of information to foreign representatives in Washington. Prior to that time, foreign representatives had been able to make requests for information directly to any U.S. Navy officer or activity, and they usually got what they wanted. As a result, our naval attachés, who were attempting to collect information on a basis of reciprocity, were having difficulties because they didn't know what information had or had not been given out in Washington to the foreign representatives. To correct the situation, Department Order 22 was issued requiring that official information for attachés or other foreign officials in Washington or abroad be furnished only through ONI. The order brought about an improved bartering position for the U.S. naval attachés, but it also increased the volume of classified material. Much information that had previously been published in the General Information Series was shifted to a Confidential Series so that its dissemination to foreign representatives could be controlled.

At this time, ONI was organized on a functional basis, with an administration department that dealt with attachés abroad and foreign attachés in Washington plus five technical desks covering ships, ordnance, personnel, communications, and steam engineering and electricity. The steam engineering desk was also responsible for intelligence on medicine and surgery.[46]

Capt. Charles D. Sigsbee relieved Cdr. Richardson Clover as the Chief Intelligence Officer on 1 February 1900. As of 30 June, the number of officers assigned to ONI was down to five. In that year, however, lack of office space became an additional problem. The Chief of the Bureau of Navigation recommended that an additional office be made available for the exclusive use of the Chief Intelligence Officer so that he might have the required privacy when conducting discussions on confidential matters. Clover had been sharing an office with two subordinates who could not be moved to other rooms without causing overcrowded conditions elsewhere. As a temporary solution, a nearby storeroom was rearranged to permit part of the clerical force to occupy it.[47]

By a joint resolution of 12 January 1900, Congress allowed ONI to print 2,000 copies of its publications, 1,000 less than for 1899 but a number deemed sufficient to meet official needs. The demand for ONI's *Notes on the Spanish-American War* had far exceeded the supply, and Congress authorized a reprinting of the entire series as Senate Document No. 388, Fifty-sixth Congress, First Session.[48]

Efforts to establish a General Staff in the Navy culiminated on 13 March 1900. By General Order No. 544, Secretary of the Navy John D. Long established the General Board to be composed of Admiral of the Navy George Dewey as senior member and eight other members, including the President of the Naval War College and his principal assistant, the Chief Intelligence Officer and his principal assistant, and the Chief of the Bureau of Navigation, with the last serving as chairman of the Executive Committee of the board.[49]

By now the clerical force of ONI was adequate and highly efficient, and it was recommended that the principal clerk of the office, whose duties were of a highly confidential character, should have his salary raised from $1,600 to $1,800 per year and that the principal stenographer and typist should receive a salary of $1,200 per year, an increase of $200. The office was unable to retain the services of highly efficient clerks at a salary of only $1,000. Two clerks who had been receiving that salary had recently transferred out of ONI for better paying jobs elsewhere in the government. It was not deemed wise to transfer ONI clerks, who had a knowledge of the confidential work of the office, to other departments of the government.[50]

The Chief Intelligence Officer again invited attention to the desirability of maintaining ONI's civilian employees permanently, instead of losing them from time to time by reason of promotion. The circumstances that governed transfer cases were applied to the naval officers serving in ONI. It was the duty of the ONI staff officers to read thoroughly all available literature on the whole scope of the world's naval information. The need for experience suggested the longest possible tours of duty in the office, especially in view of the small number of officers then assigned to ONI.[51]

The policy on the assignment of personnel to ONI was changed when the Chief of the Bureau of Navigation shifted some of the intelligence-processing work to civilians, and reduced the number of officers from seven to five. Work was impeded while the new system was still under trial, but Capt. Sigsbee, still the Chief Intelligence Officer in 1902, expressed the expectation that, when the new staff arrangement had been in operation for a while, satisfactory results would be obtained.[52]

When the 1903 Naval Appropriation Bill was under discussion, both before the congressional naval committees and in the Senate and House of Representatives, much information about foreign naval progress was furnished by ONI to senators and representatives at their request.[53]

Translation of foreign language documents were regularly performed by ONI for the Office of the Secretary of the Navy, for the Bureau of Navigation, and, when required, for other bureaus of the Navy Department. The volume of the work was reported, in the ONI annual report for 1902, to be considerable. The same report stated, "It is the aim of the office [ONI] to carry on the work according to the present policy of the Department, that is to say, to have as much work as possible done by civilian employees."

In 1903, the Navy Department's order of October 28, 1892, "Instructions in Regard to Intelligence Duty," was superseded by a modified order that stated the duties of ONI to be "to collect and classify information upon all subjects connected with war or which can have a bearing upon naval action or plans of campaign and to cooperate with the General Board in the preparation of detailed plans covering all contingencies of active operations by naval forces afloat or on shore."[54]

Radios were first installed on board naval ships in 1903 and revolutionized the potential for near real-time distribution of intelligence. Although it probably wasn't realized at the time, the introduction of radio to naval operating forces established the basic means for providing or exchanging operational intelligence to or between ships at sea. That radio communications could become a source of information as well as a security weakness was apparently not appreciated until about a decade later.

After two decades in the State, War, and Navy Building, ONI moved across the street in 1903 to the Mills Building at the corner of Seventeenth Street and Pennsylvania Avenue.[55] Cdr. Seaton Schroeder relieved Capt. Sigsbee on 1 May 1903 as Chief Intelligence Officer and continued in the position until 18 April 1906, when he was relieved by Capt. Raymond Rodgers.[56]

Only three staff intelligence officers were on board as of 30 June 1903. One of the assigned officers was on temporary duty at the Naval War College until September, and another was due to report for duty on 15 July. The number of officers assigned to ONI was increased to eight in 1906.[57]

The Russo-Japanese War in 1904 brought about the assignment of Lt. Newton A. McCully to Russia as an assistant naval attaché in the hope that he would be permitted to visit the war zone and observe the operation of Russian naval forces. Although he finally reached the Far East in April, McCully received unfavorable treatment because the Russians believed that President Roosevelt supported the Japanese. The bulk of McCully's reports covered the ground defenses of Port Arthur because he was not allowed to visit any ship or repair facilities. For more details on Lt. McCully's experiences, see Chapter 3.[58]

In the major reorganization of the Navy Department effected by Secretary of the Navy George von L. Meyer in December 1909, the Office of Naval Intelligence was put under the cognizance of the Aid for Operations, one of four aids established to advise the Secretary of the Navy and the heads of the four Navy administrative divisions: Personnel, Material, Inspections, and Operations of the Fleet. The senior division, by reason of its duties, was the Division of Operations of the Fleet, which was headed by RAdm. Richard Wainwright, a former Chief Intelligence Officer. As Aid for Operations, he was an *ex officio* member of the General Board and served as its executive head under Adm. Dewey, its president. The aids had daily access to the Secretary of the Navy and met with him weekly as a council.[59]

On 11 May 1909, Capt. Charles Vreeland relieved Capt. Rodgers as head of ONI. This had been Rodgers's second tour as Chief Intelligence Officer, and it gave him an accumulated total of seven years in that billet. On the other hand, Capt. Vreeland remained for only seven months, and on 17 December he was relieved by Capt. Templin Potts, who had previously served in ONI as a junior officer from 1883 to 1885.

A program of Japanese language training was initiated in 1910 by assigning junior officers under training to the naval attaché's office in Tokyo. The program was interrupted in 1913 when Secretary of the Navy Josephus Daniels, with President Wilson's approval, established a policy of having as few naval officers on shore duty as possible. The language student billets were among those eliminated.

On 20 November 1911, the title Chief Intelligence Officer was changed to Director of Naval Intelligence (DNI) by General Order No. 132. This was the first change in title for the head of ONI since the office had been founded. The Assistant Chief became the Assistant Director, and all other officers continued to be staff intelligence officers.

In 1912, as a result of General Order No. 139 of 16 December 1911, ONI became responsible for censorship of articles written by persons in the Navy or employed by the Navy Department. In the 1913 *Navy Regulations*, ONI was also made responsible for the censorship of photographs. For additional details on ONI involvement in censorship, see Chapter 23.

On 25 January 1912, Capt. Thomas S. Rodgers relieved Capt. Potts as DNI, remaining in the position until 15 December 1913. He, in turn, was relieved by Capt. Henry F. Bryan, who filled in until the arrival one month later (20 January 1914) of Capt. James H. Oliver.[60]

World War I

In April 1914, ONI made its second move, this time from the Mills Building to the Navy Building at 1734 New York Avenue, NW, Washington, D.C.

At the outbreak of World War I in August 1914, the Director of Naval Intelligence was in general charge of the Office of Naval Intelligence and was responsible for all official correspondence, including correspondence with the U.S. naval attachés abroad and with foreign naval attachés in the United States. The other officers on duty in ONI were in charge of the functional desks for ships, communications, ordnance, personnel, and engineering.[61]

ONI policy was to limit collection to technical information that would assist the Navy in improving the capabilities of the fleet rather than to gather intelligence of an operational nature. In 1914, ONI had eight officers and eight civilians, and 75 percent of their time was spent in clipping and filing newspaper articles, a duplication to some extent of the clipping services being performed by various Navy bureaus.[62]

Information collection by naval attachés was expanded in 1914 by ending double accreditation of the naval attaché in Paris to both France and Russia. Capt. Newton McCully was assigned to Petrograd in hope of improving the flow of information about Russia's participation in World War I.[63] Full-time naval attachés were also assigned to Tokyo and Peking, thus discontinuing the practice of using one officer to cover both posts. Also in 1914, in order to obtain better observations of the wartime use of aircraft, two aeronautical specialists, Lt. John H. Towers and 1stLt. Bernard L. Smith, USMC, were assigned as assistant naval attachés at London and Paris, respectively.[64]

With the creation of the post of Chief of Naval Operations on 3 March 1915, the Office of Naval Intelligence was designated the Division of Naval Intelligence (OP-16), one of nine divisions of the Office of Naval Operations (OPNAV). Incident to the change was a comprehensive reorganization of the naval intelligence service intended to provide the machinery for obtaining all possible information from available foreign sources and for processing and filing it in ONI.[65] Secretary of the Navy Josephus Daniels expressed the view that "complete and correct information is the first requisite for judicious decision and intelligent action."[66]

One of the new divisions of Naval Operations was responsible for the development of war plans, thus relieving ONI of its long-standing direct involvement with that function. Of course, ONI continued to produce the intelligence needed for war planning and provided it to the War Plans Division as required.

The sinking of the liner *Lusitania* on 7 May 1915 shifted U.S. public opinion about the war in Europe from neutral to pro-Allies. For ONI, preparations for possible participation in the war included taking a greater interest in the activities of potential German agents in the United States. There were only eight officers and eight civilians in ONI in 1915. So, little was done other than planning a major reorganization to add counterintelligence to ONI's functions.

With World War I underway in Europe, the need for improved intelligence collection became apparent, and in October 1915, ONI set up the War Information Service in response to a recommendation by the General Board. Funds for the service were provided by Congress in the Naval Appropriation Bill of 29 August 1916. The Naval Information Service was the overseas version of the War Information Service. Both services were undercover operations reporting directly to ONI. Branch offices were set up for domestic collection, separate from the Aid for Information organizations at the naval district headquarters. The overseas augmentation was to operate from neutral countries after the outbreak of hostilities.[67]

One of the agents placed by the Naval Information Service was Edward Breck, who had served as a Navy agent in Spain during the Spanish-American War. Breck reported on German activities in Brazil, Chile, and Argentina and worked to influence public opinion in those countries in favor of the Allies and against Germany.[68] John Held, Jr., (later to become famous as a cartoonist) conducted land reconnaissance of the eastern coast of Central America for the Naval Information Service, sketching coastal areas and ports and keeping lookout for potential hiding places for German submarines.[69]

In 1915, the Navy took a step toward the development of aerial photo reconnaissance when it requested the Eastman Kodak Company to develop an aerial camera. Specifications for the camera had been determined from experiments with various hand-held cameras conducted at Guantanamo and Pensacola, as well as from combat experience during the Vera Cruz incident. The first production order for aerial photo equipment was placed by the Naval Observatory on 10 January 1917 with Eastman Kodak for twenty "aero cameras and accessories."

Beginning in 1916, the war in Europe induced a rapid expansion in ONI. Counterintelligence, in which ONI had not previously been involved, received the greatest emphasis. It was reasoned that the Allies were already producing intelligence for the support of operating forces, but that the U.S. Navy was very vulnerable internally to German acts of sabotage. Strong security measures were needed quickly.[70]

In testimony before the House Committee on Naval Affairs, Sixty-fourth Congress, First Session, on 16 March 1916, CNO RAdm. William S. Benson stated that he wanted the Office of Naval Intelligence expanded. At the same hearings, Secretary of the Navy Josephus Daniels on 31 March stated that he wished to add $50,000 to the Navy appropriation for collecting information "at home," in addition to abroad, without having to specify how the funds were spent. As finally written and passed, the appropriation bill gave ONI the new authority for confidential collection of information at home but limited the funds to $30,000.[71]

With the funds made available for confidential use, an undercover ONI branch office was established in New York City, and others followed in major U.S. industrial cities.[72] See Chapters 21 and 22 for more details on domestic intelligence activities during World War I.

The United States entered the World War on 6 April 1917 with six naval attachés and two assistant naval attachés accredited to ten countries, including Germany and Austria. The attachés were withdrawn from these two countries upon the severing of diplomatic relations. During and after the war, naval attachés were accredited to the following additional capitals: Madrid, Spain; Christiania, Norway; Stockholm, Sweden; Copenhagen, Denmark; Rio de Janeiro, Brazil; Buenos Aires, Argentina; Santiago, Chile; Lima, Peru; Quito, Ecuador; Caracas, Venezuela; Montevideo, Uruguay; Mexico City, Mexico; Havanna, Cuba; Lisbon, Portugal; Warsaw, Poland; Constantinople, Turkey; Sofia, Bulgaria; and Bucharest, Romania.[73] See Chapter 3 for additional details.

RAdm. William Sims arrived in England in early April 1917, just prior to the U.S. declaration of war, to act as Commander U.S. Naval Forces Operating in European Waters. At a meeting on 10 April with the British First Sea Lord, Adm. Sir John Jellicoe, Sims learned for the first time the extent of the losses that the Allies were suffering from German unrestricted submarine warfare.

On 14 April, Sims cabled the Navy Department to explain the seriousness of the situation. Sinkings were reaching a scale that would have England militarily impotent in a few months if the trend was not soon reversed. The destruction of German submarines was being greatly exaggerated in the open press (ONI's principal source for war loss information). For morale purposes, the British government had not been disclosing the facts concerning Britain's plight. Furthermore, until the United States entered the war, the naval attaché in London was not given the true statistics either. After RAdm. Sims was designated commander of U.S. naval forces in Europe, he also took over the title and duties of Naval Attaché, London.[74]

Capt. Roger Welles, Jr., relieved Capt. Oliver as Director of Naval Intelligence on 16 April 1917 and served in the position throughout World War I. He was promoted to rear admiral during his tenure, becoming the first flag officer to fill the post of Director of Naval Intelligence.[75]

Secretary of the Navy Daniels, in his *Annual Report* for 1918, stated:

> The exigencies of war have imposed new and important duties upon Naval Intelligence. . . . Its duties abroad have increased many fold; at home it has been in touch with the 15 Naval Districts and branch offices—done a most important work in protecting naval and other plants making war materials, preventing sabotage, and in keeping an eye on alien enemies or others with a destructive propensity. A staff of vigilant and discrete confidential officers and civilians have been on alert to ferret out spies and other dangerous characters and secure their arrest. Too much commendation cannot be given to the zealous, discrete [*sic*], and patriotic men, the character of whose services was necessarily unknown to the public. To Rear Admiral Roger Welles, and his exceptionally fine assistants, the country owes more than can ever be known.
>
> A detailed account of the work of the office may not be stated as it is of a highly confidenital character, but, generally speaking, the scope of its activities include observation, investigation, and report of all subjects affecting the Navy and the prosecution of the war from a naval point of view. It includes naval operations at sea and on land, the status, changes, and progress of the material and personnel of foreign navies, and a close counterespionage watch at home. This latter includes the investigation of unauthorized radio stations, of alien enemies and suspects, of matters connected with the cable and mail censorship which affect the Navy, the protection of waterfronts and vessels, and of plants having contracts with the Navy Department with a view of safeguarding those against sabotage.
>
> The guarding of our ships while in port and the guarding against the danger from enemy agents among the passengers and crews on both our trans-Atlantic and coastwise ships have been largely performed by the Office of Naval Intelligence, and the results achieved bear eloquent witness to the efficiency of the service rendered.[76]

When the Armistice was declared on 11 November 1918, there were 306 naval reservists plus 18 civil service clerks and messengers serving in ONI.[77]

ONI moved twice in 1918. In February, it shifted quarters from the Navy Building to Corcoran Court, a temporary building next door on New York Avenue. Then, in September, it went to the newly built Main Navy Building in Potomac Park on the south side of Constitution Avenue, between Seventeenth and Nineteenth Streets.[78]

Between Two World Wars

ONI activities diminished at the end of World War I, but its policy did not change. It dealt "primarily with strategic subjects and to a lesser extent with tactics and logistics." Strategic intelligence was needed to prepare the annual reports submitted by the Director of War Plans to the Chief of Naval Operations. The reports were known as the "Estimates of the Situation and Base Development Plans." They had two sections, "Political Situation," which reflected world conditions, and "Information," which summarized the needs of the Naval Intelligence service.[79]

RAdm. Albert P. Niblack reported as the relief for RAdm. Welles as DNI on 1 May 1919, Welles having been detached on 31 January 1919. The Assistant Director had probably served as Acting DNI during the three-month interim. RAdm. Niblack had previously served as naval attaché at Berlin and Rome in 1897 and early 1898 and as the first naval attaché to Argentina, Brazil, and Chile in 1910 and 1911. One of his lasting contributions to ONI, which he performed while in office, was the production of *The Office of Naval Intelligence: Its History and Aims*, published in 1920 by the Government Printing Office.

When the U.S. Government failed to ratify the Treaty of Versailles, it lost the right to have representation on the Naval Inter-Allied Commission of Control, to which the German government was required to turn over technical information on the design of its naval ships, weapons, and equipment. Consequently, an American delegation, headed by Capt. Walter R. Gherardi of ONI, was sent to Germany in January and February to gather such information on its own.[80]

In July 1919, Secretary of the Navy Daniels directed that the Library and Office of Naval Records in the Secretary of the Navy's office be combined with the Historical Section of the CNO's staff creating the Office of Naval Records and Library in the Naval Intelligence Division of OPNAV.[81]

Following the war, ONI dropped back to an office force of forty-two in 1920 and returned to its prewar interests in all maritime countries. Most of its counterintelligence responsibilities were terminated. The monograph system of filing data on foreign countries was implemented, and with limited staff and little supervision ONI started to accumulate a mass of undigested, unclassified material on many non-naval subjects. (The monographs were post-bound, looseleaf folios of related papers on a given topic and, as time went on, often grew to almost unmanageable size.) The Foreign Branch was organized geographically, except for the section on foreign merchant shipping. The branch consequently lost interest in what its customers wanted, and the customers lost interest in what ONI should have been doing for them.

Through its intelligence officers in the naval districts, ONI rendered valuable assistance to the district commandants in their apprehension of military deserters. Officers were detailed for intelligence duty at New York, San Francisco, and Honolulu. The work of the intelligence officer on the staff of the Commander in Chief, Asiatic Fleet (CINCAF) proved so successful that the assignment of officers for similar duty to all fleets was considered.

The general policy of ONI was to make itself a center for the collection of information of value to the Navy. Information was compiled for the Planning Division of the Office of the Chief of Naval Operations, the General Board of the Navy, and the Naval War College. Dissemination of information was also continued to other government departments and to the bureaus of the Navy, and information was published and disseminated for the use of officers afloat.

RAdm. Andrew T. Long took over as Director of Naval Intelligence on 24 September 1920, RAdm. Niblack having been detached from that post on 17 September to report to London as naval attaché.[82]

All communications between the Office of Naval Intelligence and other U.S. Government departments and agencies had to be carried on between the Secretary of the Navy and the head of the organization involved. Furthermore, communications between ONI and the Office of the Attorney General (which included all correspondence to and from the Federal Bureau of Investigation) had to be routed via the Navy's Judge Advocate General.[83]

The primary mission of the Foreign Branch of ONI was to produce evaluated information about all foreign navies. But, particularly during the period 1920 to 1939, the Foreign Branch apparently concerned itself more with secondary objectives, such as military, political, economic, and sociological intelligence, which could have been provided by the Army's Military Intelligence Service, the Department of State, the Department of Commerce, and other government agencies.

There was a lack of any stable, comprehensive plan of organization for the best use of the personnel and facilities available to ONI. The lack of adequate funds and personnel, and the lack of continuity of personnel, contributed to the problem. The bulky, undigested, and unevaluated monograph materials accumulated by ONI made effective fulfillment of the Navy's intelligence requirements virtually impossible.[84]

Capt. Luke McNamee reported as the next Director of Naval Intelligence on 27 September 1921, RAdm. Long having been detached on 29 June. (RAdm. Nathan C. Twining is listed as the DNI in the *Navy Directory* of 1 July 1921; actually, he was on temporary duty at ONI, preparing for assignment as Naval Attaché, London, where he reported on 27 August 1921.)

On 12 January 1922, the Director of the War Plans Division, in a memorandum to the CNO, recommended the establishment of a Press Relations Office to be located within ONI. Its purpose would be

> to furnish correct information; to actively and definitely contradict incorrect public statements; to familiarize the people of the United States with the work and needs of the naval service; and to promote interest in the Navy. . . . It is but necessary to refer to the recent [Billy Mitchell] bombing tests to show the power of propaganda and the weakness of the lack of propaganda.

ONI concurred in the recommendation on 14 January 1922 and stated: "The closest liaison would be necessary between the publicity office and the head of the Navy Department." The ONI endorsement also requested an officer "to maintain close contact with the Secretary of the Navy, the Chief of Naval Operations, ONI, and the Bureaus and to be responsible for all information and news releases." The Secretary of the Navy approved, and he issued a directive to all bureaus and offices of the Navy Department, dated 21 February 1922 and entitled "Navy Department Information Section under the Office of Naval Intelligence" (for details, see Chapter 33).[85]

In 1922, the one officer responsible for the Japanese area was additionally assigned the general supervision of all intelligence work and all correspondence with U.S. naval attachés in foreign countries.[86]

According to the organization chart for ONI in 1923, the Director of Naval Intelligence had general supervision over all intelligence work, handled the correspondence with U.S. naval attachés in foreign countries (the task that in 1922 had been assigned to the Japanese desk), and maintained liaison with foreign naval attachés in the United States.

Section C of ONI, the geographical analysis organization, in 1923 comprised seven desks. Desk A had cognizance over South and Central America and the West Indies, compiling the monographs on those areas. Desk B gathered information from other government departments, prepared statistics on foreign navies, and tried to gather material for eight different monographs. Desk C collected information on the Near East, Southern Europe, and the new Baltic states. Desk D supervised the Japanese monograph and tried to bring compiled data concerning the Japanese navy up to date. Desk E collated information about foreign shipping and accumulated data for monographs about U.S. overseas possessions. Desk F was interested in Western Europe and also gathered statistics on the comparative naval strengths of the Washington Treaty nations (other than Japan). Desk G gathered material for the monographs dealing with China, India, Siberia, and other Far East areas.[87]

ONI continued censorship of photographs and motion pictures of naval subjects, but by 1923 it discontinued its censorship of publications and articles written by naval authors.[88]

Capt. Henry J. Hough replaced Capt. McNamee as Director of Naval Intelligence on 20 December 1923, the latter having been detached on 1 November. In the same year, the District Intelligence Officer, 3rd Naval District (DIO-3ND) successfully acquired the Japanese naval code then in effect from the office of the Japanese Naval Inspector of Material in New York, beginning the U.S. Navy's effort to exploit Japanese naval communications.[89]

On 1 September 1925, Capt. Hough was detached as DNI, and on 9 October Capt. William W. Galbraith reported as his replacement.

Naval attaché posts continued to be maintained during 1925 at London, Paris, Rome, Berlin, The Hague, Tokyo, Peking, Buenos Aires, Rio de Janeiro, and Santiago.[90]

The Naval Intelligence Volunteer Service was created by the Naval Reserve Act of 28 February 1925 (Public Law 512, Sixty-eighth Congress, First Session). Initially, very little effort was devoted to procuring reserve officers for the service because of the predominantly pacifist outlook of the general public and press at that time. For more information on the Intelligence Reserves, see Chapter 29.

An article about the Office of Naval Records and Library published in the U.S. Naval Institute *Proceedings* of January 1926, entitled "Our Vanishing History and Traditions," by Capt. Dudley W. Knox of ONI, pleaded the cause of preserving the vanishing naval archives. Knox appealed especially to former officers, their descendants, and their families to make available any documents in "family papers." The article sparked widespread interest and ultimately resulted in the establishment of the Naval Historical Foundation, a nonprofit organization that accepts and retains gifts of documents, relics, etc., for the Office of Naval Records and Library (now the Naval Historical Center).[91]

On 1 July 1926, Capt. Arthur J. Hepburn reported as DNI, relieving Capt. Galbraith who had been detached on 11 June. The billet was gapped for approximately three months when Capt. Hepburn left on 20 September 1927 and Capt. Alfred W. Johnson reported on 12 December.

Naval Intelligence policy during the late 1920s was expressed in the "Estimate of the Situation and Base Development Plans" dated 19 April 1927:

> All preparation for war and operations in war must be premised upon the best available information. This information is of three general classes, (a) information concerning our own forces, numbers, conditions, rate of mobilization; (b) similar information on enemy forces; (c) information regarding probable theatres of operation. The [Office of Naval Intelligence] is the principal agency of the Department in the gathering and dissemination of information in regard to the forces of possible enemy navies.[92]

Activities continued by ONI in 1927 included dissemination of information to the several executive departments and bureaus of the Navy Department; publication of secret information for the use of the Navy; maintenance of liaison with foreign officers in the United States, particularly the foreign naval attachés; public relations duties; and the collection, classification, and filing of old records.[93]

In 1928, ONI continued to collect information pertaining to naval matters and to disseminate it to interested parties, including Congress. Censoring photographs and motion pictures about naval subjects continued. The Public Relations and Records and Library Sections also continued to operate as before. Naval attachés were located in London, Paris, Rome, Berlin, The Hague, Tokyo, Peking, Rio de Janeiro, Buenos Aires, Santiago, and Mexico City.[94]

The evaluation of information was considered by some not to be a function of ONI. In a lecture on "Naval Intelligence" at the Army War College on 6 June 1928, Capt. David McD. LeBreton, a former Assistant Director of Naval Intelligence, stated that the Navy Department had two agencies where such evaluations should be made, the Naval War College and the War Plans Division of the Office of the Chief of Naval Operations. He also declared that, as that held true for political and general military information, it should also be the case with technical information. LeBreton continued that there were five large bureaus in the Navy Department, each concerned with a special branch of engineering. It was the business of ONI (F-Branch), he said, only to obtain information from abroad for the bureaus and to distribute it to them for interpretation and evaluation. The continuing general bias in the Navy against the evaluation of information by ONI would haunt the organization in 1941.[95]

The prejudice against ONI's evaluating the information it collected was corrected in a Statement of Functions for the Intelligence Division approved by the Chief of Naval Operations on 15 April 1929 that specified the primary duty of naval intelligence to be "the collection of all classes of information concerning foreign countries . . . the evaluation of this information, and its dissemination as intelligence." This was the first documented Navy recognition that information must be evaluated before it can become "intelligence."

Through its naval attachés abroad and intelligence officers afloat and ashore during the late 1920s, ONI followed foreign naval and military progress and developments in technical fields of special interest to the Navy and the U.S. merchant marine. In cooperation with the Military Intelligence Divison of the Army and other executive departments of the government, ONI acquired useful information about the national and military policies of foreign powers and their political, social, economic, and industrial conditions, and information relating to the strength and disposition of foreign armed forces. The information was supplied to the President and Congress for use in considering appropriation bills and also for use by the American delegates to the various conferences on the limitation of armaments. Public Relations and the Naval Records and Library Section continued to function as before.[96]

A proposal to centralize all United States intelligence efforts was received at ONI from the DIO-3ND in April 1929. The idea had been initiated by John A. Gade, a New York businessman who had been the Naval Attaché, Copenhagen, during World War I. Gade proposed that "the various intelligence units of the government be left exactly as they are now but that they may be considered as spokes of a wheel, the hub of which is a Central Intelligence organization. Into this Central Intelligence pours all information from the various spokes—Naval Intelligence, Military Intelligence, Secret Service, Department of Justice, and Department of Commerce." The Army received the same proposal, but, after joint discussion with the Navy, no action was taken. Gade was nearly two decades ahead of his time.[97]

During 1930, ONI continued collecting information about foreign navies and air forces and sent it to Congress for use in determining the annual naval appropriation bill. Naval public relations and the accumulation of documents and reports was also continued. A U.S. naval attaché, with headquarters at Tegucigalpa, Honduras, was accredited for the first time to the Central American countries.[98]

Another collateral duty was added for ONI's Office of Naval Records and Library on 28 April 1930 when a Secretary of the Navy Order appointed retired Capt. Dudley Knox as curator for the Navy Department. The curator was responsible for the collection and preservation of objects, trophies, and relics of historical and inspirational value to the

Navy, except for materials permanently assigned to the Naval Academy and other naval stations. Knox also had cognizance over matters connected with a proposed naval museum in Washington, D.C.[99]

On 20 June 1930, Capt. Harry A. Baldridge reported as DNI, Capt. Johnson having been detached on 18 June. Upon his detachment, Alfred Johnson was nominated by President Herbert Hoover to represent him in Nicaragua "to carry on to a further point of advancement the cooperation of the Government of the United States in electoral matters which was extended during the Presidential elections in 1928, and which the Government of Nicaragua has requested shall likewise be extended in connection with the impending Congressional elections and the later Presidential elections." As a result of the nomination, Capt. Johnson was appointed by the Nicaraguan Supreme Court as chairman of the National Board of Elections. He served with the rank of Envoy Extraordinary and Minister Plenipotentiary.[100]

Navy policy toward Naval Intelligence was stated in the "Estimate of the Situation for 1933," dated 27 April 1931:

> The importance of providing ample funds and adequate personnel for the efficient functioning of information agencies of the Navy Department can hardly be overemphasized. Any curtailment of their needs in this direction will adversely affect the preparation of the Navy for national defense and hence the value of the Navy in providing its share towards the national security. During the period covered by the estimate [1933], due to probable extensive readjustments, it is particularly important that constant and intimate contact be maintained with trends of thought and events throughout the world as regards naval forces. This adds to the peacetime activities of the ONI, and is of vital importance in the consideration of our own estimates and plans.[101]

Capt. Hayne Ellis reported as DNI on 1 June 1931 as the relief for Capt. Baldridge, who was detached on 29 May.

By 1931, the work involved in the production and updating of intelligence monographs had become so great that the limited number of personnel assigned to ONI could not cope with it. A decision was, therefore, reached to limit the scope of the work on monographs to those sections that represented essential naval, political, and economic information required by the War Plans Division and by the commanders of forces afloat for their "Estimates of the Situation."[102]

The 1931 policy was commendable in concept, but, because of resource constraints, its execution was less than successful. A DNI secret memo of 31 July 1934 mentioned a task to watch movements of Japanese merchant shipping because any interruption, change, or cessation in shipping operations would be one of the earliest indicators of hostile intent. That important job had to be given to a temporary duty officer in the Far East Section of ONI.

Insufficient personnel, both officer and clerical, was a constant handicap. The Foreign Branch had ten permanent personnel and one temporary staff member, compared to twenty-one Army officers and clerks in the Foreign Intelligence Section of the War Department (G-2). By way of further contrast, the U.S. section of the Imperial Japanese Navy's equivalent of ONI was headed by a captain and had eleven commanders and lieutenant commanders assigned, plus an undetermined number of clerical assistants. Obviously, the Imperial Japanese Navy was more interested in the United States than the U.S. Navy was in Japan, in spite of Japan's occupation of Manchuria in 1931.[103]

Data on the navies and air forces of foreign nations were prepared by ONI throughout the 1930s for use by Congress in preparation of the annual naval appropriation bill and for naval technical advisors of the U.S. delegation at the International Conference for the Reduction and Limitation of Armaments at Geneva in 1932. ONI also continued its public relations and historical records operations. ONI's *Quarterly Information Bulletin* had to be discontinued in 1932 for lack of funds; it had been published monthly or quarterly since 1919.[104]

Starting in 1932, the Far East Section (OP-16-B-12, the old Desks D and G of Section C) became very active during the Sino-Japanese conflict. In addition to keeping informed on the Japanese Empire, China, Siam, Manchukuo, the Philippine Islands, Guam, Hawaii, Samoa, the Aleutians, and the strategic harbors of the Pacific, OP-16-B-12 prepared periodic summaries of the Sino-Japanese situation, estimates on the Japanese internal situation for the War Plans Division, and articles for ONI publications.[105]

In 1933, as funding continued to dwindle, the naval attachés at Rio de Janeiro, Tegucigalpa, and The Hague were withdrawn. The Netherlands accreditation was assigned to the Naval Attaché, Berlin. Liaison with foreign attachés, the War Department, and the State Department was handled by the Assistant DNI. A roster of the foreign attachés was updated and posted in the ONI Office Orders by the Chief Clerk. Foreign requests for visits to naval and industrial establishments were handled by the Security Section.[106]

To provide up-to-date guidance on intelligence to the Naval Establishment, ONI produced the *Naval Intelligence Manual* (ONI-19), a confidential registered publication submitted on 5 October 1933 by DNI Capt. Hayne Ellis and approved by CNO Adm. William H. Standley.

The need for intelligence in the Navy, and therefore the major objective of Naval Intelligence, was expressed in 1933 as follows:

> The intimate relations now existing between the nations of the world, caused by the close interlocking of political and social interests and their economic interdependence bring about rapid changes in the sentiments of the nations of the world; and furthermore, the political opinion of the masses of nations may be rapidly swayed by the utilization of the press, radio and similar means.
>
> The mobility and destructive power of army, navy and air arms are such that a nation surprised by war is on the verge of defeat. Therefore, it is vital to national security that plans for possible wars be made in advance, that possible wars be foreseen in time to permit the development of forces which will be required to enforce threatened national policies, or if not this, at least to permit the disposition of available forces to affect favorably the international situation or to facilitate the desired strategical deployment.
>
> The Naval Limitations Treaties, together with the rapidity of technical developments in ship and aircraft construction, propulsion machinery, all weapons, and methods of attack and defense, necessitate technical superiority; therefore, the closest observation for (and the collection of) information of such developments in foreign navies is essential to prevent "surprise" and maintain if possible "technical superiority."
>
> The increased number, types and mobility of naval units, [and] the increased range, rapidity and efficiency of naval communications has so increased the area, rate of development, and complexity of naval operations and naval warfare that during war timely intelligence of the enemy forces and the existing conditions within the theatre of operations is a vital factor of success in naval war.
>
> To successfully present a timely and up-to-date picture of the above situation and other pertinent factors, is a major objective of Naval Intelligence.[107]

All normal ONI activities were continued in 1934, including the collection of information, the conduct of public relations activities, and the accumulation of records. ONI handled arrangements for the Navy Department for the cruises of foreign ships and aircraft in U.S. territory and the visits of foreign naval officers to Washington. The naval attaché in Santiago and the assistant attaché for air in Rome were both brought home due to lack of funds. After serving for almost three years as Director of Naval Intelligence, Capt. Hayne Ellis was detached on 21 May 1934, being replaced on 4 June by Capt. William D. Puleston.[108]

When the United States established diplomatic relations with the USSR in 1934, the U.S. Embassy staff that arrived in Moscow on 7 March included Marine Capt. David R. Nimmer as the assistant naval attaché. A Navy captain was supposed to have been assigned as naval attaché, but ONI's interest in Soviet naval affairs at that time was minimal. No Navy captain was assigned, and Capt. Nimmer served as the naval attaché for almost one year. The office was closed officially on 16 February 1935, ostensibly because of the Soviet government's refusal to settle debts owed to the United States by the previous regime; actually, however, the closing was due to the lack of courtesies extended to Nimmer, and the restriction on contacts imposed on him, compared to those granted the Soviet naval attaché in the United States. The value of the information being obtained in the USSR, it was judged, did not justify the cost of maintaining the office in Moscow.[109]

In 1935, a serious threat to the morale and efficiency of U.S. naval personnel developed from the subversive efforts of radical groups in the United States. It became essential to uncover the sources of harm and to seek the means to counter them. In addition, the activities of foreign secret agents, both in the United States and in neighboring countries, threatened the safety of the fleet and the naval shore establishment. In light of the subversive activities, the ONI-produced "Estimate of the Situation for 1937," dated 30 March 1935, pointed out the great need for funds and adequate personnel for the various intelligence agencies, stating that "agencies are now inadequate to accomplish their mission." Specific mention was made of the need for more clerical personnel to properly evaluate and disseminate information and to administer more effectively the field services of both foreign and domestic intelligence. Also mentioned was a plan to strengthen fleet intelligence and to provide measures for combating foreign agents and subversive organizations in the United States.[110]

The officer allowance for ONI in 1935 was sixteen active duty U.S. Navy line officers, two retired officers recalled to active duty, and three Marines. The disturbed conditions throughout the world placed ever-increasing demands upon the Intelligence Division, and the shortage of officers was rendering it more and more difficult to carry out the division's mission properly and efficiently. The absolute minimum of officers on permanent duty to handle adequately the conditions existing in 1935 was deemed to be twenty-four, distributed within the organization as follows: Director and Assistant Director, two; Administrative Branch, two; Intelligence Branch, one; British Empire Unit, one; Far East Unit, three; European Unit, two; Latin America Unit, two; Dissemination Unit, one; Investigating Unit, one; Security Unit, two; Public Relations Branch, three; Planning Section, one; and Historical Branch, three.

The permanent civilian force in OP-16 in 1935 totaled twenty. The number was so inadequate and had such serious adverse effects on accomplishing the Intelligence Division's mission that three clerks and two translators were made available under National Industrial Recovery Act (NIRA) funds. The steadily increasing demands on ONI made it essential that the temporary increases be made permanent. Due to representations made about the personnel shortages in ONI, Congress appropriated funds for Fiscal Year 1936 for an additional seven research clerks and two clerk stenographers. The increase was over and above the temporary help furnished under NIRA and did not lessen the need for making the temporary help permanent.[111]

As of March 1935, OP-16-B remained the "OP-Code" designator for ONI's Intelligence Branch in the Office of the Chief of Naval Operations. B-1 was responsible for dissemination, and B-2 through B-7 constituted the Domestic Intelligence Section. The Foreign Intelligence Section was made up as follows: B-10, the British Empire Unit; B-11, the Far East Unit; B-12, the Western Europe Unit; B-13, the Central Europe Unit; B-14, the Eastern Europe Unit; B-15, the Balkans and Near East Unit; B-16, the Latin American Unit; and B-17, the Foreign Commerce Unit (which was inactive in peacetime).[112]

The permanent civilian force was reduced to nineteen in 1936. In addition, two translators and three clerks were employed under NIRA funding, and, from other available moneys, six more clerks were employed for an average period of three months each on temporary appointments. Even with the eleven additional employees, it was not possible to meet the output demands on the Intelligence Division, although their presence was most helpful.

Naval attaché offices were established during 1936 in Rio de Janeiro and Lima in addition to the attaché offices that were continued in Berlin, Brussels, Buenos Aires, London, Paris, Peking, Rome, and Tokyo.[113]

The United States naval mission to Brazil had been so successful that it was expanded to eight officers in 1936. At that time, ONI was made responsible for naval missions to foreign countries, and the personnel assigned to the missions received a six-week briefing period before proceeding to their assigned countries. Reports were made to ONI by the naval mission staffs on the proficiency and personnel of the navies they assisted in training. Close and cordial contact was maintained between the Naval Attaché, Rio de Janeiro, and the head of the naval mission to Brazil.[114]

District intelligence offices were in urgent need of adequate clerical assistance in 1936. In all naval districts, the organizations for collecting information vital to security were being impeded by a lack of personnel. District intelligence office work was far in arrears. It was recommended that one additional position be established in each district, except the 7th and 8th, to provide proper support for the district intelligence officers.[115]

In the ONI "Estimate of the Situation for 1939" dated 16 April 1937, particular emphasis was placed on the need for counterintelligence. Rapidly changing political and military situations had resulted in a marked increase in international espionage. Damage to the fleet and to naval shore installations and industrial plants was anticipated.[116]

In 1937, naval attaché offices were added at Santiago, Chile, and Bogota, Colombia. Funds were obtained for the establishment of nine additional clerks for the naval district intelligence offices under the Fiscal Year 1938 congressional appropriation. Funds were to be requested for three additional clerks under the Fiscal Year 1939 estimate in order to complete staffing an intelligence office for each naval district.[117]

In June 1937, the Chief of Naval Operations assigned the following duties to the foreign intelligence sections of ONI:

> 1. Collection of all classes of pertinent information especially affecting naval and maritime matters, with particular attention to the strength, disposition, and probable intentions of foreign naval forces;
> 2. Dissemination of the above;
> 3. Direction of the activities of the U.S. Naval Attachés;
> 4. Cognizance over all communications with U.S. Naval Missions abroad and also with foreign naval attachés accredited to the United States; and
> 5. Maintenance of liaison with other government departments for the exchange of information from abroad.[118]

Capt. Puleston was detached on 30 April 1937 after nearly three years as DNI, and RAdm. Ralston S. Holmes replaced him on 1 May. Work continued throughout 1938 on collecting, evaluating, and disseminating information, and on keeping records. Naval attachés were established at Guatemala City and Lisbon, and the naval attaché office was disestablished at Brussels.[119]

Preparations for War

On 11 June 1939, RAdm. Walter S. Anderson relieved RAdm. Holmes as Director of Naval Intelligence. Anderson believed strongly that the United States was going to become involved in the impending war, and he began to try to get the naval intelligence organization ready. He sent out additional naval attachés to capitals that had not previously had them, including all member countries of the British Commonwealth. Anderson established a section in ONI to keep track of the merchant shipping routes of the world, and he initiated the training of officers for censoring duties. Approximately 225 Naval Reserve officers were trained in groups of twenty-five. Anderson established a Strategic Information Section to gather information on request and furnish it to the requestor, and a Secret Intelligence Section to handle confidential agents.

With the start of World War II in Europe, the Navy was ordered on 6 September 1939 to establish a Neutrality Patrol in the western Atlantic to observe and report the movements of "combatant vessels of nations in a state of war." Before the end of 1939, the types of ships to be reported were expanded to include German merchant vessels.[120]

A reorganization for ONI was approved on 5 December 1939. Immediately under the Assistant DNI was the Foreign Intelligence Branch (OP-16-F) with the following sections: F-1, British Empire; F-2, Far East; F-3, Western Europe; F-4, Central Europe; F-5, Eastern Europe; F-6, the Balkans and the Near East; F-7, Latin America; and F-8, Enemy Trade (inactive in peacetime).[121]

Naval attaché offices were reestablished at Brussels, Mexico City, Havana, and The Hague in 1939.[122]

The chairman of the Navy General Board, in a letter to the Secretary of the Navy dated 31 August 1939 titled "Are We Ready?" discussed the Navy's deficiencies and included the following comments on intelligence:

> Generally speaking, the Naval Intelligence Service is approaching adequacy as deficiencies in funds and personnel are being remedied. The need [for] additional personnel and facilities increases with deteriorating world conditions. Additional civilian personnel will require additional funds.
>
> The Domestic Intelligence Service comprises the Intelligence Organizations in the Districts, including regular Naval Officers, Reserves, and a small number of agents. Expansion is dependent upon the use of Reserves. . . . Disregarding allocations for censorship duties, the Reserves now enrolled are more than half of those required for intelligence duties.
>
> The Foreign Intelligence Service comprises Naval Attachés. Information and reports are also received from State, War, Commerce, commercial firms, and individuals. The network of information is good as far as it goes, but the information obtained consists primarily of that which foreign countries are willing to release. More adequate coverage is considered essential.[123]

DNI Anderson, in his memo of 21 May 1940 on "the Readiness of the Naval Establishment to Meet a Serious Emergency," stated that "the Naval Intelligence Service Operating Plans are considered sufficient and effective, at home and aboard, to execute the task assigned Naval Intelligence in Basic War Plans"; and that, if the requirements for personnel and material necessary to carry out the war plans on M-day were provided, "the Organization of the Naval Intelligence Service will be sufficient and effective," but that "the present organization of the ONI is not considered sufficient for effective operation. Additional officer and civilian personnel, additional office space, and additional funds are required to meet present conditions."[124]

On 10 June 1940, in a personal memo from Anderson to RAdm. Ernest J. King, then a member of the General Board, the above statements were reiterated, and Anderson added:

> We have at present no intelligence network abroad other than Naval Attachés. When and if the need for agents appears, I believe we can handle the situation. Our plan and organization for combating espionage and subversive activities at home are progressing continuously. We are constantly considering new "fields" and methods for "tightening up."[125]

During 1940, naval attaché offices were opened in Venezuela, Sweden, the Dominican Republic, and Turkey.[126]

According to regulations issued on 23 October 1940, the duties of the Foreign Intelligence Branch of ONI included securing "all classes of . . . information concerning foreign countries, especially that affecting naval and maritime matters, with particular attention to strength, disposition and probable intentions of foreign naval forces," and evaluating the information collected and disseminating it "as advisable."

There were rumors in 1940 that German submarines were operating from bases in the West Indies. At the specific direction of President Roosevelt, the Navy sent an American civilian yacht on a cruise to the area, manned and officered by Navy personnel in civilian attire. They examined various ports in the West Indies and along the northern and eastern coasts of South America. The Navy personnel found no evidence that any of the ports were being used by German submarines.[127]

When the main battle fleet moved from its West Coast ports to Pearl Harbor in the spring of 1940, the Commander in Chief, U.S. Fleet (CINCUS) directed the preparation and execution of plans for the security of the fleet while it was berthed in the Hawaiian area. The plans included daylight-to-dusk naval air patrols, seven days a week, by Patrol Wing (PATWING) Two, using twelve aircraft daily to search the sea around the islands to a radius of 180 miles. When the return of the fleet to its West Coast home ports was deferred indefinitely, the commander of PATWING-2 became concerned about the engine time being accumulated on his aircraft and the resultant overhaul workload. In July 1940, he reduced the patrol effort to six aircraft daily. CINCUS efforts to get the Navy Department to fill the additional patrol-plane needs in the Pacific were unsuccessful.

On 1 November 1940, Capt. Patrick N. L. Bellinger took command of PATWING-2. He advised CINCUS that the 300-mile patrol of only the western sector, then being flown by six to twelve planes each day, was inadequate protection. Bellinger estimated the need for an 800-mile radius search 360 degrees around the Hawaiian islands, with Pearl Harbor as the center. This patrol would require fifty ready aircraft each day; Bellinger had only sixty, each of which could be flown only every second or third day.

On 28 November 1940, CINCUS Adm. James O. Richardson sent CNO Adm. Harold R. Stark, a draft of a proposed revision of a CINCUS directive to improve the security of the fleet in the Pacific. The plan required a long-range air reconnaissance from Pearl Harbor by fleet patrol planes. Adm. Stark replied that wartime measures, such as the continuous air patrols, were not necessary As a result, the requirement for additional patrol aircraft was not included in the CINCUS directive issued on 5 December 1940, a deficiency that continued until the Japanese Pearl Harbor attack and contributed to its success.[128]

In December 1940, the Director of Naval Intelligence hired an American businessman to develop a covert intelligence collection organization. The private citizen was authorized to establish an office in New York as a "representative of the DNI in matters relating to [ONI's] Foreign Intelligence Service." He selected and hired undercover agents for foreign placement. After the Office of the Coordinator of Information (OCI, subsequently OSS, Office of Strategic Services) was established in the summer of 1941, the ONI covert organization was shifted to OCI on 15 October 1941. At the time of the transfer, thirteen agents had been recruited. The Public Relations Branch was removed from ONI and was set up directly under the Secretary of the Navy as the Office of Public Relations by a SECNAV directive dated 28 April 1941.[129]

In January 1941, the Foreign Intelligence Branch (OP-16-F) was expanded by the addition of three new sections: F-9, Special Intelligence; F-10, Statistical; and F-11, Strategic Information. F-8 was renamed Foreign Trade. The new organizational format was in effect at the time of the Pearl Harbor attack.[130]

During 1941, naval attaché offices were opened in the Union of South Africa, Australia, Thailand, Canada, Uruguay, and Argentina.[131]

During the year before the attack on Pearl Harbor, there were four different directors of Naval Intelligence. RAdm. Walter Anderson finished a regular two-year tour in January 1941, and he left for a battleship division command. Anderson was replaced on a temporary basis by the Assistant Director of Naval Intelligence, Capt. Jules James, who served as Acting Director until March 1941. James was replaced by Capt. Alan G. Kirk, who served as DNI from 1 March to 15 October and then was detached for a command at sea. RAdm. Theodore S. Wilkinson then assumed the post and was in charge of ONI at the time of the Pearl Harbor attack. Wilkinson had had no previous experience in an intelligence billet.[132]

On 27 January 1941, Ambassador Joseph C. Grew in Tokyo reported receiving rumors that the Peruvian minister had heard from several sources that, in the event of trouble between the United States and Japan, the Japanese intended to make a surprise attack against Pearl Harbor with all of their forces and equipment. On 1 February, the same day that Adm. Husband E. Kimmel relieved Adm. Richardson as CINCUS, ONI passed Grew's report to the Commander in Chief, U.S. Fleet. ONI advised that it placed no credence in the rumors and that, based on the current disposition and employment of Japanese Navy and Army forces, no move against Pearl Harbor appeared imminent or planned for the foreseeable future.[133]

In April 1941, Capt. Kirk, in a discussion on the scope of his duties with Adm. Royal E. Ingersoll, Vice Chief of Naval Operations, and Capt. Richmond Kelly Turner, head of the War Plans Division, stated rather strongly that his DNI job should include "interpreting possible enemy intentions, and that ONI should prepare the section of the formal estimate known as 'Enemy Intentions.'" Turner felt the War Plans Division "should prepare this section and should interpret and evaluate all information concerning possible hostile nations, from whatever source received." Turner also believed ONI "was not charged with sending out any information that would initiate any operations on the part of the fleet, or fleets anywhere." Adm. Stark approved the position taken by Turner, but the written instructions of 23 October 1940 requiring ONI to evaluate and disseminate information remained unchanged.[134]

RAdm. Anderson, Director of Naval Intelligence from June 1939 to January 1941, stated that there had been no restrictions placed on his dissemination of intelligence to the operating forces, and that if there had been, he would have requested detachment. He had had no problems with Capt. Turner when the latter took over as head of War Plans Division in 1940.[135]

In connection with the dissension between Capt. Kirk and Capt. Turner over the evaluation of intelligence information, it is worth noting that Turner had been commanding officer of the heavy cruiser *Astoria* (CA 34) when that ship returned the body of the deceased Japanese ambassador to Japan in April 1939. As a result of the approximately ten days he had spent in Japan on the mission, Turner felt that he knew a great deal about the Japanese. His views were constantly at variance with those of ONI on the subject.[136]

Another endemic problem at the outset of World War II was the dissension between the Director of Naval Intelligence and the Director of Naval Communications (DNC) over the control of the dissemination of communications-derived intelligence. The transcript of the Japanese navy operational code had been broken by the Office of Naval Communications, and the translation was done by Japanese linguists supplied by ONI. The DNC maintained that he had a prior right to present the translations of the more interesting items at the Secretary of the Navy's morning conference. The Director of Naval Intelligence countered that the translation should be evaluated and interpreted by the experts in ONI and that he should make the presentation to the Secretary. The problem was resolved in favor of the Director of Naval Intelligence, but the loss of the responsibility continued to rankle the communications organization.[137]

In response to the need for a general government agency to collect and analyze strategic information, as well as to develop a secret undercover intelligence service, the Office of the Coordinator of Information was established by a presidential order on 11 July 1941. The order directed that the departments and agencies of the government make information relating to national security available to OCI. It was noted that nothing in the duties and responsibilities of OCI should interfere with the duties and responsibilities of the regular military and naval advisors to the President.[138]

Liaison with OCI was established initially under the Administrative Branch of ONI, and Cdr. Richard E. Webb was appointed as administrative liaison officer. All requests from OCI and DNI for information or data were to be sent through Webb, although it was desired that direct contact also be maintained between personnel in ONI and OCI. In September 1941, LCdr. Alvin D. Chandler became liaison officer, and by 9 October 1941 eleven Navy and two Marine Corps officers had been detailed by ONI to serve in the coordinator's office at OCI. LCdr. John L. Riheldaffer, USN (Ret.), replaced LCdr. Chandler in October when liaison with OCI was placed under the Special Intelligence Branch (OP-16-Z), where it remained throughout the war.[139]

The Director of Naval Intelligence, in a 10 October 1941 letter to all branch and section heads, directed that, when providing information to representatives of OCI, they should ensure that "the Navy Department's evaluation and interpretation of the data is made clear . . . so that the compilation of similar data, and the preparation of the Coordinator's report, may not suffer from a lack of full appreciation of the Navy Department's evaluation."[140]

In the summer of 1941, in anticipation of U.S. involvement in a world conflict, ONI began positioning naval observers, naval liaison officers, and consular shipping advisors in principal ports and potential "hot spots" throughout the world. Naval observers had been sent to various ports in Brazil earlier in the year in connection with support to U.S. Navy ships operating in the Neutrality Patrol.[141]

In August 1941, the U.S. Navy language students in Japan were withdrawn, reaching Shanghai on Labor Day. The Naval Attaché, Tokyo, had recommended the precautionary move to ensure against the students being interned when and if Japan initiated hostilities.

Cdr. Arthur H. McCollum, head of the Far East Desk of ONI, left for London on 25 August 1941 and returned in October. When he returned, McCollum found that the *Intelligence Digest* had projected an attack on Siberia by the Japanese. The interpretation had been published at the direction of the Director of War Plans, Kelly Turner. McCollum saw no change in the situation to warrant the projection. Accordingly, the next *Digest* went out without the Siberian fairy tale. The omission did not sit well with Turner, and soon a directive came out that ONI could not send out any evaluations; it could only report facts.[142]

A CNO message of 16 October 1941 reported the resignation of the Japanese cabinet and directed due precautions and preparatory deployments. Adm. Kimmel's response included putting submarines on "war patrol" and sending twelve patrol planes to Midway to carry out daily patrols within 100 miles of the island.[143]

Another CNO message dated 24 November alerted CINCPACFLT Adm. Kimmel, CINCAF Adm. Thomas C. Hart, and the commandants of the 12th, 13th, and 14th Naval Districts about the possibility of Japanese hostile action, stating that "a surprise aggressive movement in any direction including an attack on Philippines or Guam is a possibility."[144]

The 27 November 1941 warning message directed the entire naval establishment to take the defensive dispositions required by the effective war plan, WPL-46 (RAINBOW). The message required that contributory (supporting) plans be put into operation. The contributory plans for the defense of Pearl Harbor required the establishment of aerial reconnaissance against possible attack. The war-warning message went not only to the Pacific Fleet but also to the Atlantic Fleet, which was practically at war already in carrying out the neutrality patrols, and it went to the Asiatic Fleet, where Adm. Hart took all of the necessary defensive steps. Hart got his ships and aircraft out of the Philippines. The only ships left in Manila Bay were those that couldn't be used in operations at sea. The message was also addressed to every naval district in the United States and to the Panama Canal Zone.

The only forces that did not carry out their contributory plans were the Pacific Fleet and the 14th Naval District. The latter, which covered the Hawaiian Islands, had no aircraft under its command because all planes in the Hawaiian Islands were under the control of Adm. Kimmel's Pacific Fleet.[145]

The so-called Stark 27 November 1941 warning message was released by Adm. Ingersoll and was addressed to CINCAF and CINCPAC for action, and to CINCLANT (Commander in Chief, Atlantic) and SPENAVO (Special Naval Observer), London, for information:

> This is considered a war warning message.
>
> Negotiations with Japan looking toward stabilization of conditions in the Pacific have ceased, and an aggressive move by Japan is expected within the next few days. The number and equipment of Japanese troops and the organization of naval task forces indicates an amphibious expedition against either the Philippines, Thai, or Kra Peninsula or possibly Borneo. Execute an appropriate defensive deployment prepatory to carrying out the tasks assigned in WPL 46. Inform District and Army authorities. A similar warning is being sent by War Department. SPENAVO [London] inform British. Continental Districts, Guam, Samoa directed to take appropriate measures against sabotage.

In the first week of December 1941, intercepted diplomatic coded messages from the Japanese foreign office directed their offices in London, Paris, and other cities to burn communications codes and confidential papers. A check confirmed that the Japanese were actually carrying out the order in Washington and New York. On 3 December, ONI sent messages to CINCAF, CINCPACFLT, and the commandants of the 14th and 16th Naval Districts informing them of the Japanese action. At the same time, ONI sent instructions to all endangered outposts to destroy their codes and papers and to signal back when these actions had been accomplished. Information copies went to CINCPACFLT and CINCAF. The replies came back from all of the various posts by 5 December 1941, confirming that the prewar steps were being taken. The military commanders were expected to recognize that the precautions were a serious preparation for war. Also at that time, Maj. Gregon A. Willams, USMC, Assistant Naval Attaché, Shanghai, was ordered to go to his war post at Foochow.

The naval aide to the President, Capt. John R. Beardall, was informed of ONI's message when it was sent out, and the State Department was also informed. ONI's message raised no alarm because the U.S. Government was set for war; it just was not known when or where the conflict would begin.[146]

Two other intercepted diplomatic messages had been received that would have been indicators that Pearl Harbor was the possible target, if they had been available at CINCPACFLT. On 2 December, Tokyo asked its Honolulu consul "whether or not there are any observation balloons above Pearl Harbor or if there are any indications they will be sent up. Also advise whether the warships are provided with anti-mine nets." On 6 December, the Japanese consul at Honolulu responded to Tokyo, "In my opinion battleships do not have torpedo nets."[147]

The attempt to make ONI into a mere collection agency had serious consequences that the OPNAV's War Plans Division refused to acknowledge in 1941. Without an organization devoted exclusively to collecting, correlating, and evaluating all available naval intelligence information received from all sources, a serious gap existed in the Navy's readiness. The Navy was ripe to become a victim of surprise.

Although disagreement between War Plans and ONI had been noisy during the early stages of Kirk's administration of ONI, by the time Wilkinson took charge, the subordinate position of Naval Intelligence relative to War Plans was a fait accompli. Turner, now a rear admiral, was keeping to himself the job of evaluating intelligence information and the results of his work. He had daily strategic estimates made up in his own division, but he did not show them to ONI, and Adm. Stark did not require them to be "chopped" by ONI.[148]

RAdm. Wilkinson had ready access to Adm. Stark, although he met most frequently with Stark's assistant, Adm. Ingersoll. The relations between Wilkinson and Turner, on the other hand, were something less than cordial. This was not a personal matter. Adm. Turner always distrusted Naval Intelligence, no matter who headed the organization.

Turner could not distinguish between types of information, particularly later during congressional hearings, when he tried to blame ONI for failing to send out information that he had previously claimed as the prerogative of his office but which his office had failed to send.[149]

There were many investigations to determine who in Washington and Honolulu was to blame for the Japanese success in achieving surprise at Pearl Harbor. The investigations all seem to have failed to recognize that, prior to its attack on Pearl Harbor, Japan's other operations and actions had created a successful deception that covered the preparations and movements of forces positioning for the Pearl Harbor attack. The movement of Japanese forces toward targets in Southeast Asia had not been covered up and had been reported from Japan and China and by air patrols from the Philippines. On the other hand, complete secrecy had been achieved on the movements by the Pearl Harbor striking force.

Voluminous information had been received to indicate that a Japanese attack was coming somewhere, and for several months alerts had been sent out to the operating forces, culminating in Adm. Stark's war-warning message to Adm. Kimmel on 27 November 1941 identifying the targets as "either the Philippines, Thai, Kra Peninsula or possibly Borneo." The most significant clue—the requirement for a report on the berthing locations of ships in Pearl Harbor from the Japanese consul in Honolulu—had not been available to Kimmel; it had been lost in Washington in the mass of other intercepts, including messages to Japanese consuls for similar information about other U.S. and world ports. The U.S. high command could not believe that the Japanese would make such a strategic mistake as to attack Pearl Harbor, and all intelligence reports were evaluated in that context. See Chapter 35 for more discussion on this controversial subject and subsequent events in the Pacific.

World War II

Soon after the United States entered World War II, submarines were ordered to conduct reconnaissance of selected Japanese-held islands in the central Pacific. Little was known in the United States about the defenses and support installations in the Japanese-controlled islands. Based on the information collected by the submarine missions, RAdm. William F. Halsey's carrier task force attacked the Marshall Islands on 1 February 1942.[150]

In June 1942, the Office of the Coordinator of Information was abolished by the President, and its intelligence functions were turned over to the Office of Strategic Services, an agency established directly under the Joint Chiefs of Staff. The liaison between the Division of Naval Intelligence and the OSS continued as it had previously been carried on between Naval Intelligence and OCI.[151]

A special intelligence officer, already charged with passing on the qualifications of applicants for intelligence appointments in the Division of Naval Intelligence, was assigned duty at the Bureau of Personnel to process the applications. The officer was given additional duties as liaison officer with the Division of Naval Intelligence for planning procurement of Class I-V(S) Naval Reserve intelligence specialist personnel.[152]

On 20 July 1942, RAdm. Harold C. Train relieved RAdm. Wilkinson as Director of Naval Intelligence.

In September 1942, reflecting on the experiences of the war to date, the Vice Chief of Naval Operations commented on the Naval Intelligence service as a whole:

> Decentralization has been carried out in the past. A minimum of positive directives and control has been exerted by the Office of Naval Intelligence. However, actual experience under war conditions has demonstrated that these policies have not fulfilled the purpose of the Office of the Chief of Naval Operations.[153]

A major reorganization of ONI took place in March 1943, and the title of the second in command was changed from Assistant Director to Deputy Director (see Chapter 28).

As part of the March 1943 reorganization, a North American Desk was established in the Intelligence Branch and was given cognizance over the collection of intelligence (as opposed to counterintelligence) within the continental United States and Alaska. Valuable information on foreign countries was available within the United States, and each naval district intelligence office set up a foreign intelligence section to exploit domestic resources and to collect intelligence information of value to the operating forces. To improve administration of the domestic collection program, a contact register, containing the names of sources in each naval district, was begun in September 1943.[154]

The Operational Intelligence Section was established for a short time (April to August) in 1943. Its brief existence was due mainly to the strong view of most of the senior officers in ONI that no part of that organization should be devoted exclusively to the production of intelligence for one type of customer (see Chapter 18).

On 25 September 1943, RAdm. Roscoe E. Schuirmann, Intelligence Officer to the Commander in Chief, U.S. Fleet (COMINCH), relieved

RAdm. Train as Director of Naval Intelligence, serving thereafter in both billets (see Chapter 16).

The first successful U.S. Navy wartime periscope photo reconnaissance mission was conducted by *Nautilus* (SS 168) in September 1943 to obtain information for the landings by U.S. forces on Tarawa. The panoramic photographs obtained by the submarine provided information on gun installations and beach defenses. The photographs also confirmed beach contours and the locations of exposed reefs. On 19 November, which was D-Day minus one for the Tarawa landings and the end of five days of air attack and preliminary bombardment by surface ships, *Nautilus* again entered Tarawa lagoon to update previously obtained information and to determine the success of the softening-up effort. The submarine found a new, still-undamaged, six-to-eight-foot wall of heavy logs built on the beaches. *Nautilus* also observed large coastal-defense guns and still-operable small guns on the beaches. This information, plus information on surf conditions, was reported to the amphibious task force commander.[155]

The use of submarines to support coastwatchers and guerrilla forces in the Philippines was inaugurated on 14 January 1943 when *Gudgeon* (SS 211) landed six men and one ton of equipment and supplies on the island of Negros. A second such mission was carried out by *Tambor* (SS 198) on 5 March at Mindanao. Thereafter, small landing parties and supplies were landed at about five-week intervals in the central and southern Philippines by selected Seventh Fleet submarines in conjunction with their regular war patrols. Supply operations continued until 23 January 1945, for a total of forty-one missions. *Seawolf* (SS 197) did not make its 6 October 1944 landing and was listed overdue as of that date—the only submarine lost on support operations.[156]

Between January 1943 and April 1945, as "Cdr. Robert E. Norden, USN," LCdr. Ralph G. Albrecht, USNR, of the Special Activities Branch (OP-16-Z), made 309 radio broadcasts directed to officers of the German navy, particularly submarine officers, to undermine the morale of the enemy and to lower German combat efficiency. The success attained in Albrecht's psychological warfare effort was confirmed by evidence from German naval prisoners of war and other sources that the broadcasts were consistently listened to. "Norden's" reports of submarine losses and other facts, prior to official disclosure by the German Ministry of Marine, won him a following of interested listeners and served to discredit the German leaders who, on numerous occasions, found it advisable to refute "Norden's" statements.[157]

In May 1944, the Special Activities Branch was made responsible for determining escape and evasion methods that could be used by captured United States personnel. An officer from the Air Intelligence Group (OP-16-V) was temporarily detailed to OP-16-Z for the purpose of determining to what extent existing Army arrangements for escape and evasion might be applicable to naval personnel.[158] The OP-16-V officer on loan to OP-16-Z handled liaison with the Army Military Intelligence Service's X-Division (MIS-X), which was concerned with escape and evasion matters. Selected Navy and Marine personnel were given MIS-X indoctrination prior to assignment to fleet or field duty so they might, in turn, brief combat personnel as required. Material, both physical aids and intelligence, was distributed to fleet and other commands through MIS-X or OP-16-V in accordance with requirements.[159]

The capture of the German submarine *U-505* by a U.S. carrier task group on 4 June 1944, although netting a major haul for technical intelligence from captured hardware, could have caused a loss of critical intelligence in other areas had German naval headquarters learned too soon of the submarine's capture. The codes that *U-505* carried had previously been broken by the Allies. Coming on the eve of the Normandy landings, the capture of the codes might have induced the Germans to change to another code just when the ability to read their communications was most urgently needed. The disclosure of the capture would also have temporarily reduced the very valuable support of communications-derived intelligence to the Allied antisubmarine effort.

Captured documents of naval interest were becoming available in increasing quantities in 1944 as the invasion of Europe expanded and as more islands were captured in the Pacific. Japanese ships sunk at island atolls and in the harbors of the Philippines were found to be fruitful sources of information for future operations. Intelligence teams went ashore right behind the early landings to ensure that documents and equipment of intelligence value were acquired and exploited before the souvenir hunters started their collection efforts. Information in the captured documents was often of immediate tactical value, showing strengths and weaknesses and characteristics of defensive installations yet to be attacked or that could be circumvented. Enemy documents were sometimes found in the pockets of U.S. dead or wounded. Had the documents been turned in by the souvenir hunters and been properly exploited, such action might have saved them from becoming casualties.

On 3 October 1944, the Technical Intelligence Center of ONI was established and was designated OP-16-PT (see Chapter 11).

RAdm. Leo H. Thebaud relieved RAdm. Schuirmann on 24 October 1944 as Director of Naval Intelligence and Intelligence Officer for COMINCH.

The Washington Document Center, a central agency for handling captured Japanese documents, was made a part of ONI on 14 February 1945 and was designated OP-16-WDC. It was located at the Steuart Building at Fifth and K Streets, NW, Washington, D.C.[160]

In April 1945, the title of the OPNAVs' intelligence organization was changed from the Division of Naval Intelligence to the Office of Naval Intelligence, the name it had had throughout its early years until 1915. On 6 September 1945, Commo. Thomas B. Inglis relieved RAdm. Thebaud as DNI and soon thereafter he was promoted to rear admiral.

Post–World War II Period

Following the conclusion of World War II, the COMINCH organization was dissolved and the Office of the Chief of Naval Operations (OPNAV) was reorganized on 10 October 1945. The part of COMINCH that had been responsible for operations and was to continue as part of OPNAV became OP-03, while COMINCH's intelligence organization became part of ONI. In the new OPNAV organization, ONI was assigned to OP-02 (Administration) and was designated OP-23. The title of Director of Naval Intelligence was changed to Chief of Naval Intelligence.

As a result of the valuable technical information developed from the interrogation of German scientists captured as Germany was overrun, Project Paperclip was established in July 1945 to procure and exploit foreign technical personnel. The Navy technical bureaus were particularly interested in acquiring the services of German naval construction and ordnance specialists. See Chapter 11 for further details on Paperclip.

The first area conference of naval attachés was held in London on 26 November 1945 under the auspices of the Commander Naval Forces, Europe (COMNAVEU). Such a conference had been proposed by the naval attaché in London, Commo. Tully Shelley, to the Chief of Naval Intelligence and COMNAVEU; both supported the idea. Twenty-five naval attachés and naval observers from Europe and the Mediterranean area attended.

In December 1945, ONI designated Cdr. Rufus L. Taylor as the U.S. member of a so-called Five Power (U.S., Britain, France, the USSR, and China) Committee to exploit the abandoned offices and material of the Japanese military attachés in Europe. It was uncertain what might be found, but the United States chiefly hoped that some cryptologic information might be recovered. Cdr. Taylor was selected for the job because of his Japanese language qualifications and for his ability to recognize cryptologic material and information. Little of intelligence value was discovered, however. A Belgian cryptologic system was found by Taylor in the Japanese naval attaché's office in Paris; he recovered a complete description, wiring diagrams, etc., but not the machine itself. Taylor was able to extract the documentary material without Soviet and Chinese members knowing about it (the French and British members had dropped out).[161]

Although ONI had been in existence since 1882, no specific delineation of its duties or its relation to the rest of the Navy had been incorporated into *Navy Regulations* until the publication on 20 June 1946 of the *Advanced Changes to U.S. Navy Regulations, 1920*, was approved by President Truman on 14 June. The revision inserted a new Section 9, Article 425, in Chapter 6, stating that Naval Intelligence, under the CNO, was "the organization charged with the execution of the intelligence and counterintelligence mission of the Naval establishment." Paragraph 3 of Section 9 declared that the Chief of Naval Intelligence shall have cognizance over all phases of collection, evaluation, and dissemination of all types of intelligence in the Naval establishment, except as provided in Article 421, which covered communications intelligence. Paragraph 6 stated that the Chief of Naval Intelligence was to have cognizance over the security of classified information and control over the disclosures of naval classified information to foreign governments. Paragaph 7 stated that although naval intelligence was under the CNO, the Chief of Naval Intelligence "shall disseminate immediately to appropriate parts of the Naval establishment intelligence within their cognizances which does not relate to matters under the jurisdiction of OPNAV as set forth in Article 433." Paragraph 7 also stated, "Activities of the Naval establishment shall coordinate all intelligence matters with the Chief of Naval Intelligence."[162]

To achieve the most effective implementation of the new provisions in *Navy Regulations*, the Chief of Naval Operations expressed to the Commandant, U.S. Marine Corps the desire to have Marine Corps intelligence activities in the operating forces and in the Navy Department fully integrated with naval intelligence. The integration was to include all aspects of naval intelligence, especially the development of plans and doctrine for the use of amphibious operational intelligence, the assignment and training of regular Marine Corps officers in intelligence work, and the procurement and training of Marine Corps Reserve officers as specialists for intelligence duties in anticipation of mobilization.[163]

On 1 August 1946, ONI was shifted from the Administration Division to the Operations Division of OPNAV, and its designator became OP-32. Concurrently, the Office of Naval Records and Library was removed from ONI and was combined with the Office of Naval History under the Deputy Chief of Naval Operations (DCNO) for Administration (OP-02).

Article 12-006 of the *Naval Intelligence Manual* (ONI-19[A]) stated:

> Upon Executive Order of the President, the U.S. Coast Guard becomes a part of the Naval establishment in time of emergency or war. For this reason, a close relationship must exist between Naval Intelligence and the intelligence organization of the U.S. Coast Guard. This relationship will be delineated, for guidance of Naval Intelligence, in accordance with agreements reached between the Chief of Naval Operations and the Commandant of the Coast Guard.

A joint ONI–Coast Guard committee was set up in 1946 to review the intelligence relations between the Navy and the Coast Guard. The ONI representatives were Cdr. Thomas R. Mackie and Capt. Herman E. Keisker, USNR (Inactive). They were instructed to pay special, but not exclusive, attention to espionage, sabotage and subversion; security of classified naval information; cooperation in Washington, the naval districts, and the operating forces; and cooperation in foreign and operational intelligence aspects. The committee was to recommend, in detail, agreements that would accomplish the above. Since the interests of all ONI branches were to be considered and included in the agreements, ONI branch heads were directed to render all possible assistance to the ONI–Coast Guard committee.[164]

In October 1946, the Kilgore Committee of Congress requested "information regarding the amount of money expended annually by the Navy Department on all its intelligence operations from 1936 to 1946." It amplified the request to include "total amounts of money expended annually by ONI, including all direct and indirect charges such as the salaries of civilian and military personnel on duty with Naval intelligence."

To indicate to the congressional committee the scope of the research necessary to produce the requested figures, it was pointed out that (a) during the period from 1936 to 1946, naval intelligence had received no single appropriation to cover all its expenditures; (b) the only appropriation made directly to, and administered by, naval intelligence was "Salaries, Office of Naval Intelligence," covering salaries of civilian personnel working in ONI in Washington, and, in addition, a portion of the appropriation "Miscellaneous Expenses, Navy" (maintenance of naval attachés and collection of information) was administered and accounted for by Navy intelligence; (c) as an integral part of the Navy, Naval Intelligence received salaries, services, equipment, supplies, etc., from various appropriations under the cognizance of the Office of the Secretary of the Navy, the Bureau of Supplies and Accounts, the Bureau of Yards and Docks, and the Bureau of Aeronautics; (d) in the case of naval districts, expenditures from the above allotments were made by the district commandants, who, in turn, furnished civilian personnel and services to the district intelligence officers; and (e) all military salaries were paid from the appropriation "Pay and Subsistence of Naval Personnel."[165]

In the years immediately following World War II, there was a drastic reduction in the size of the Navy's operating forces and in the number of its personnel, but the need for intelligence expanded as the scope of the subjects requiring coverage escalated under the pressures of the Cold War. To retain and provide career opportunities for a selected number of World War II Naval Reserve intelligence personnel who had acquired significant expertise, a series of "ALNAVs" (messages addressed to the entire naval establishment) was issued in 1945 inviting the reserve personnel to request transfer to the regular Navy. The restricted line (Special Duty Intelligence) personnel designator 163X, was created and incorporated in the Officer Procurement Act of 1947.[166]

In 1948, a selection board was convened to select, from unrestricted line officer applicants, ten regular officers to be designated as 1630 (Special Duty Intelligence) officers. The number was in addition to the transferees from among the naval reservists. Subsequent annual selections were to build the 1630 community to thirty. The input into the 1630 community was almost exclusively from among former unrestricted line intelligence subspecialists, plus an additional limited number of Naval Reserve graduates from the Naval Intelligence School.

The *Naval Intelligence Manual* was superseded on 1 May 1947 by ONI-19(A), *Naval Intelligence Manual–1947*, a confidential, registered publication incorporating many of the lessons learned in World War II. It was signed by Chief of Naval Intelligence RAdm. Thomas B. Inglis and was approved by Secretary of the Navy James V. Forrestal.

The National Security Act of 1947 provided, among other things, for the coordination of the intelligence activities of the U.S. Government. As a result, the responsibilities of the Chief of Naval Intelligence were broadened to satisfy the requirement to participate in the production of intelligence to meet national level requirements.[167]

In 1948 (and for several years thereafter), the Pentagon suffered the pangs of the service unification required by the National Security Act of 1947. The new Air Force, including its intelligence organization, was heavily involved in efforts to justify an expanded mission and thus obtain a bigger slice of the defense budget at the expense of the other services, particularly the Navy. Nuclear strategic bombing by B-36 intercontinental bombers was said to render all other strategies obsolete. Secretary of Defense Louis B. Johnson, who was intent on cutting any defense costs that he viewed as duplicative, seemed to favor Air Force proposals that would eliminate activities by the other services. For a while, it appeared that naval aviation and perhaps even Marine Corps aviation would be eliminated. Air Force intelligence produced studies to support Air Force claims and proposals. The Navy's air intelligence production, which often refuted the Air Force intelligence output, stimulated the Air Force to claim that ONI's organization was inadequately and improperly staffed to process air intelligence and that it should relinquish the function. The Navy's reaction was to assign more aviators to ONI.[168]

On 1 November 1948, RAdm. Inglis, in his memo serial 13601P32, requested that the title Chief of Naval Intelligence be changed to Director of Naval Intelligence, stating, "This change in title would effect a further step in standardizing the nomenclature used in various intelligence agencies in that the Intelligence Divisions of the Army and Air Force and the CIA are now headed by a 'Director'." The request was approved when Chief of Naval Operations Adm. Louis E. Denfeld initialed the memo.

In December 1948, ONI moved from the Main Navy Building to the Pentagon.

On 22 April 1949, the Navy's General Board (an advisory group of retired senior officers that was soon to be disestablished) held a formal hearing on aspects of the "Five Major Problems Reported by Naval Activities" related to internal security and intelligence. The agenda quoted extracts of comments about internal security and intelligence made by various naval activities and offices: "Internal security is very weak. All internal security organizations, including ONI, need increased funds and personnel. The Navy needs a central agency to enforce security in the Navy."[169]

In 1949, a Scientific and Technical Unit was established in Heidelberg, Germany, under the technical direction of ONI and the Commander Naval Forces, Germany (COMNAVFORGER) to collect scientific and technical information available in Europe. The unit was placed under the COMNAVFORGER intelligence officer for administrative purposes.

On 15 September 1949, RAdm. Inglis was relieved by RAdm. Felix L. Johnson as Director of Naval Intelligence.

Korean War

The overt and covert facilities for acquiring intelligence information in peacetime, such as the CIA, military missions, attachés, and naval observers, were unsuccessful in providing operational commands with any warning of impending hostilities in Korea in 1950. Factors contributing to the intelligence failure included the state of mind induced by high-level policy statements; a lack of appreciation of the capabilities of North and South Korean forces that resulted in faulty estimates by the intelligence agency exercising theater intelligence coordination and direction; and, within the Naval Establishment, a shortage of competent personnel.[170]

With the outbreak of hostilities in Korea, it became immediately apparent that the workload of naval intelligence would require additional manpower in ONI, in the district intelligence offices, at the naval attaché posts, and on fleet intelligence staffs. After requirements for increases in personnel allowances were determined and approved, there was a considerable time lag in getting the new billets filled with qualified people. Naval Reserve intelligence officers were ordered to active duty, and the training of reserve officers at the Naval Intelligence School was expanded. The processing of recalled officers by the Bureau of Personnel and the Bureau of Medicine, however, and delays in reporting, resulted in up to a four-month lag between an officer's nomination for a particular billet and his reporting date. The requirement for trained air intelligence officers was a separate problem that was met to some extent by providing air intelligence training for reserve officers at the Naval Intelligence School and at fleet activities. The filling of civilian billets was complicated by competition among the various agencies of the government for qualified people.[171]

Only one intelligence officer was assigned to the Seventh Fleet when the Korean War broke out on 25 June 1950. About sixteen intelligence officers were rounded up from the fleet in Pearl Harbor and flown out to augment the Seventh Fleet intelligence staff. DNI Felix Johnson proceeded to the Pacific the next day to get first-hand information on the situation. Johnson visited Commander Naval Forces, Far East, VAdm. C. Turner Joy in Tokyo and spent about a week in Korea and another in Pearl Harbor. Officers were sent from ONI to Pearl Harbor to replace those who went to the Seventh Fleet.[172] On 27

June 1950, in connection with charges that the intelligence community had been caught by suprise by the North Korean invasion of South Korea, RAdm. Roscoe H. Hillenkoetter, Director of Central Intelligence, gave testimony to a Senate committee in closed session on the concentration of North Korean forces along the 38th parallel.

CIA document ORE 3-49, dated 28 February 1949, stated:

> Withdrawal of U.S. forces from Korea in the spring of 1949 would probably in time be followed by an invasion, timed to coincide with Communist-led South Korean revolts, by the North Korean People's Army possibly assisted by small battle-trained units from Communist Manchuria. Although it can be assumed that South Korean Security forces will eventually develop sufficient strength to resist such an invasion, they will not have achieved that capability by the spring of 1949. It is unlikely that such strength will be achieved before January 1950.

A similar CIA publication, ORE 18-50, dated 19 June 1950, stated:

> Trained and equipped units of the Communist People's Army are being deployed southward in the area of the 38th Parallel. "People's Army" and Border Constabulary units there equal or surpass the strength of southern Korean Army units similarly deployed. Tanks and heavy artillery have also been moved close to the Parallel in recent months.
>
> The ultimate local objective of the Soviet Union and of the North Korean regime is the elimination of the southern Republic of Korea and the unification of the Korean peninsula under Communist domination. To this end, an open invasion of the Republic by northern Korean military forces has thus far been delayed in favor of a coordinated campaign involving political pressure within southern Korea, subversion, propaganda, intimidation, economic pressure, and military actions by infiltration of guerrilla forces.

Thus, the capabilities and ultimate intention to invade had been ascertained, but the timing had not.[173]

Airborne electronics intelligence (ELINT) collection became more sophisticated in 1951 when the Martin P4M Mercator came into the Ferret aircraft inventory, and the aircraft were home-based at Port Lyautey, Morocco, and at Sangley Point in the Philippines to cover Europe and the Far East, respectively. The intelligence collection aircraft detachments would eventually become known as fleet air reconnaissance squadrons (VQ). The crews of ELINT collection aircraft were on the front lines of the intelligence war during the Cold War period and suffered casualties accordingly.

The Korean War demonstrated the Navy's need for an adequate number of carrier and heavy photographic aircraft capable of performing day and night photo reconnaissance. Of equal importance was the need for adequately equipped facilities, staffed by properly qualified personnel, to process and interpret the photographs obtained.[174]

United Nations forces in the Korean War exercised control of both the air and sea, but the opportunities thus available for collecting intelligence were not fully exploited. During the first six months of hostilities, the enemy was able to move large numbers of ground troops over considerable distances without detection. In the initial phases of the war, information obtained from sightings, photography, prisoner-of-war interrogations, and covert sources failed to provide the intelligence support needed on enemy movements, dispositions, and intentions.[175]

The difficulties experienced by most naval commands were caused by an initial shortage of intelligence personnel and by delays in the arrival of additional personnel from the rear. As late as 30 April 1951, shortages of qualified photo interpreters and linguists still existed.[176]

The Cold War

A large part of the area characteristics data used in producing annexes to Operation Orders and Op-Plans in the 1950s was obtained from Navy Hydrographic Office publications and charts. In that respect, "Hydro" was an intelligence source, but it was a collector, producer, and disseminator of intelligence of a specialized type. Therefore, it was later put under the supervision of the Defense Intelligence Agency (DIA) when the latter organization was established in the early 1960s.

The Navy Hydrographic Office collected data from many sources, including the organization's own survey ships and other government agencies. The office also had reciprocal exchange arrangements with hydrographic offices of foreign countries and a corps of more than 1,500 observers in the merchant services of almost all maritime nations who regularly sent in hydrographic observations. The U.S. Navy's operating forces also forwarded soundings and other data when the data appeared to be inconsistent with information found in office publications and charts. The office's publications included *Sailing Directions*, *Light Lists*, *Weekly Bulletins*, *Notices to Mariners*, and *Daily Memoranda*. The Hydrographic Office also broadcasted urgent information to all merchant and naval shipping through its Hydrolant and Hydropac shortwave radio systems.[177]

Negative views about having civilian personnel in charge of the administrative elements of ONI were expressed in 1952 by a Board for the Review of Functions and Workload of ONI composed of Capt. W. R. Wilson, Capt. S. E. Jones, Cdr. N. D. Zimmerman, and Cdr. C. F. Pick. Their views reflected the general Navy view of its Civil Service employees:

> Civilian supervisory personnel have considerable influence and, in effect, provide the "continuity" in many areas in ONI. In some cases, this results in their participation in administration. Such participation, however, should be limited by certain considerations. It is not believed sound that any integrated (civilian and military) component of ONI, such as a section or unit, should function under the direct command and administrative control of a civilian. This is particularly the case where military personnel are involved as subordinates. In the opinion of the Board, moreover, intelligent self interest on the part of the civilian administrators causes them occasionally to give undue weight to the "size of organization," as this is an important factor in [the] justification of their respective positions.

In 1952, the Technical Unit, the Coast and Landing Beaches Section, and the Naval Facilities Section of ONI were moved from the Pentagon to building 166 at the Naval Gun Factory, Washington Navy Yard. This move began the physical breakup of ONI's production organization, making it more difficult for the analysts in the various components to collaborate in their research and correlation of information on related subjects.

When the National Security Council assigned to the CIA the responsibility for all collection of foreign intelligence information from domestic sources in 1952, ONI and CIA worked out a mutually satisfactory "salt water" agreement whereby ONI would continue to collect intelligence from waterborne sources through the district intelligence offices.[178]

On 1 July 1952, RAdm. Johnson was relieved as Director of Naval Intelligence by his deputy, RAdm. Richard F. Stout, until the arrival of RAdm. Carl F. Espe on 1 December 1952.

National policy concerning publicity about the intelligence agencies of the U.S. Government was expressed in 1952 in part as follows:

> . . . Since any publicity, factual or fictional, concerning intelligence is potentially detrimental to the effectiveness of intelligence and to the national security . . . departments and agencies represented by membership on the Intelligence Advisory Committee [Depts. of State, Defense (Army, Navy, Air Force, JCS), Justice, the Atomic Energy Commission, and CIA], shall take steps to prevent the unauthorized disclosure for written or oral publication of any information concerning intelligence or intelligence activities. The head of each department or agency will determine his channel for granting such authorization as may be necessary.

To implement the nondisclosure policy in the Navy, the Secretary of the Navy issued the following guidance:

> *Naval Intelligence:* The most rigorous adherence to the spirit and intent of the quoted national policy is necessary if adequate protection is to be accorded clandestine and intelligence activities. Accordingly, the open publication of material on these subjects will normally not be authorized. The Director of Naval Intelligence is responsible, within the Department of the Navy, for implementing the terms of this policy, and all proposed news and publicity releases on Naval Intelligence matters will be referred to the Director for decision or recommendation. Any recommendation for release must include an unqualified statement that the material desired for release will not compromise any intelligence aims, *modus operandi*, personalities, or accomplishments, and will not jeopardize any current or projected operations. Further, since intelligence operations and material are normally classified, recommendations for release will also indicate the grounds upon which declassification can be justified.
>
> *Intelligence of a Joint Nature:* The Department of the Navy will not release, or approve the release of, any information on the intelligence or clandestine activities of other government agencies. All requests for such release will be referred to the department or agency concerned. Proposed releases of Navy participation in joint intelligence activities will be coordinated with the other agency involved, by the Director of Naval Intelligence.[179]

The first Fleet Intelligence Center was established in May 1953 by Commander in Chief, U.S. Naval Forces, Eastern Atlantic and Mediterranean (CINCNELM) and was activated at Port Lyautey, Morocco, in March 1954. For details about the development of all of the Fleet Intelligence Centers, see Chapter 40.

ONI's first venture into the use of automation was compiling the data on merchant ship characteristics being received in response to National Intelligence Requirements Memorandum (NIRM) 5 (see Chapter 20).

Instability in the Guatemalan government prompted the United States to take an interest in the possibility that arms were being covertly imported into that country. On 20 May 1954, surveillance of shipping in the Gulf of Honduras was initiated by the U.S. Navy to determine the identity of ships suspected of transporting arms. ONI provided and kept current a list of ships that should be watched. The surveillance effort lasted until the end of June 1954, when a new government was established and stability returned to Guatemala.[180]

On 1 June 1954, as part of a major OPNAV reorganization, ONI was placed directly under the Vice Chief of Naval Operations, and the Director of Naval Intelligence was designated OP-92. For ONI, the reorganization corrected the previous weakness of having to handle intelligence support and administrative matters through the DCNO (Operations). The change also put ONI at the same organizational level as its counterparts in the Army and Air Force.[181]

As of April 1954, there were forty special duty intelligence (1630) officers on active duty, including one rear admiral, nine captains, fifteen commanders, eight lieutenant commanders, five lieutenants, and two lieutenants (junior grade).

In the summer of 1956, the Acoustic Intercept Data Analysis Program was established, with ONI's Technical Section serving as the control point and acoustic intelligence production center. Two Bureau of Ships laboratories (Naval Electronics Laboratory, San Diego, and Data Processing Unit, Brooklyn) performed technical processing of the recordings of acoustic signatures. The use of essentially indifferent outside activities of inadequate capacity to perform the initial processing of the acoustic intelligence recordings soon proved unsatisfactory (see Chapter 14).

RAdm. Laurence H. Frost relieved RAdm. Espe as Director of Naval Intelligence on 1 June 1956.

On 9 August 1956, the ONI Planning Board was established to increase high-level support and guidance for the production of all types of ONI plans. The responsibilities of the Planning Board were to formulate planning directives for all types of planning required by ONI; provide broad guidance for the detailed development of ONI plans; resolve all divergencies involving ONI plans; review final drafts of ONI plans prior to their submission to the Director of Naval Intelligence; and develop and maintain a master priority list of approved major ONI program objectives, which was to be reviewed as necessary to ensure that available ONI resources were used in a manner best designed to discharge the missions of the Director of Naval Intelligence.

An ONI Plans Coordination Committee was concurrently established as a subordinate group to implement the direction developed by the Planning Board. The head of OP-92B1C, the ONI Plans Coordinator, served as chairman of the committee.[182]

On 31 August 1956, British Joint Services member RAdm. Barnard, RN, advised U.S. Director of Naval Intelligence RAdm. Frost that British Director of Naval Intelligence RAdm. Inglis would again propose, during Frost's forthcoming trip to London, that the U.S. and British navies fully integrate two Royal Navy officers into the Soviet satellite sections of ONI and two U.S. Navy officers into the same section of the Naval Intelligence Division (NID) in London. Both Adm. Harry D. Felt (VCNO) and Adm. Arleigh Burke (CNO) penciled their disapproval of the idea on Frost's memorandum of information about the conversation.

On 7 September 1956, Inglis sent a follow-up letter to Frost asking that the officer exchange be an item for discussion during Frost's forthcoming London visit. Inglis commented favorably on the current arrangement of having two CINCNELM officers fully integrated into the NID merchant shipping section. Inglis considered the arrangement beneficial to both CINCNELM and NID. Inglis, however, also considered the link between ONI and NID, through the liaison officer from Naval Attaché, London, to be unsatisfactory. The liaison officer, although permitted to go where he wished and see whomever and whatever he wanted, only visited NID occasionally. The U.S. Navy officer was not being fully employed in the intelligence processing business, and RAdm. Inglis was of the opinion that the U.S. officer couldn't evaluate the material he saw and couldn't give a studied opinion to the NID officers concerning the reports he read. In any event, the proposal died with the Anglo-French invasion of Suez in 1956 and the consequent cooling of U.S.-British relations.[183] For further information, see Chapter 39.

The Hungarian revolt on 23 October 1956, and the resultant flow of escapees into Austria and thence to the United States, offered intelligence collection opportunities that were exploited by the Army and Air Force. The Navy, however, had no personnel pool from which to draw for the exploitation of such spur-of-the-moment intelligence collection opportunities, and the low potential of the refugee sources to supply information of a strictly naval concern did not justify temporarily taking people away from fulltime billets to participate in this collection effort. Furthermore, the Israeli attack on Egypt on 29 October, followed by the British and French landings in Suez on 6 November, kept ONI fully occupied. When the United States sided with the USSR in the United Nations and called for a cease-fire in Egypt, the traditionally favorable relations between ONI and the British Naval Intelligence Division had to be nurtured unofficially while the U.S.-British climate was officially frosty.

ONI did, however, participate in a limited fashion in the interrogation of the escapees from the Hungarian revolt. William E. W. Howe of ONI's Scientific and Technical Intelligence Center (OP-

922F2) responded to a request for volunteer interrogators, was provided with a Hungarian linguist from the 3rd Naval District, and spent two weeks at Camp Kilmer, New Jersey, participating in the screening and brief interrogation of a sizable group of technically-oriented refugees. The cooperation between the CIA and the armed services intelligence organizations was excellent.[184]

Cold War requirements led to the inauguration on 1 July 1957 of a barrier patrol between Newfoundland and the Azores. WV-2 Warning Star radar aircraft and destroyer escort radar pickets (DERs) conducted air and surface reconnaissance and reported their contacts to Commander in Chief, Atlantic Fleet for evaluation.

Crisis situations that developed in 1958 put special requirements on intelligence-producing organizations. The unusual events included the U.S. landing in Lebanon in July in response to a request for help by the local government against an Egyptian-inspired coup attempt and the Chinese Communist shelling of Kinmen and the Matsu Islands from August to November; the latter led to the establishment of the U.S. Taiwan Defense Command on 10 September, with Commander Task Force 72 as the Navy component commander.

By August 1959, there were 229 special duty intelligence officers on active duty; 104 were 1630s and 125 were 1635-designated reservists. There were also 501 unrestricted line subspecialists. The Keith Board, which was convened to study the need for subspecialists in the Navy, identified and recommended a minimum of 379 intelligence subspecialist billets. It also recommended that intelligence subspecialists be given greater consideration for selection for the Armed Forces Staff College, for courses at the Naval War College, and for attendance at the National War College.[185]

In 1960, OP-922G, the Basic and Technical Intelligence Branch, was broken up when the Naval Scientific and Technical Intelligence Center (NAVSTIC) was set up at the Naval Observatory and the Coast and Landing Beach Section and the Ports and Naval Facilities Section were shifted to the Navy Service Center in South Arlington. The latter was redesignated as OP-922G, the Basic Intelligence Branch.

In the belief that "a single host Service acting as executive agent for administration and logistics at each [attaché] station could do the job more simply and economically," the Secretary of Defense ordered the "executive agency" principle put into effect at all attaché posts, effective 1 July 1960.[186]

On 15 September 1960, RAdm. Frost was relieved as Director of Naval Intelligence by RAdm. Vernon L. Lowrance.

The Cuban Missile Crisis and the Creation of the Defense Intelligence Agency

The Bay of Pigs landing operation in Cuba in April 1961 involved ONI only as the channel (and cut-out) between the Navy and the Central Intelligence Agency for arranging the Navy's support to the operations. The CIA's rules on security of information on the operation were so tight that only three or four officers in ONI other than the director were aware of the preparatory support arrangements.

The Navy's member of the Watch Committee, Capt. Rufus Taylor, learned of the general nature of the Bay of Pigs operation when the committee was briefed by its CIA representative. Capt. Taylor also sat in when Deputy Director of Central Intelligence Gen. Cabel briefed CNO Adm. Arleigh Burke and DCNO (Operations) VAdm. Wallace M. Beakley. The Navy was committed to protecting the landing force on the high seas but wasn't to participate in the landing. Beakley was concerned that, if the landing were to go wrong, there would be no way to recover the troops, which is exactly what happened.

One of the reasons for the failure of the Bay of Pigs operation was the lack of any opportunity for the people of the Joint Staff or subordinate levels of the Army, Navy, and Air Force to work on the details of the operation and spot the flaws that would have been more readily apparent to those accustomed to working on plans for similar operations. Only the top service chiefs were informed of the operation, and they were not given its details.[187]

The Defense Intelligence Agency (DIA) was activated on 1 October 1961, under the following guidelines:

> 1. The purpose for creating DIA was to improve the effectiveness and responsiveness of Department of Defense intelligence products and activities, and to bring about efficiencies and economies in the use of resources.
>
> 2. DIA was to evolve on a carefully planned and phased schedule so as not to disrupt or degrade intelligence processes throughout the entire military structure.
>
> 3. The Secretary of Defense reserved for later consideration the transfer of selected intelligence functions that in the meantime would remain with the military services.
>
> 4. Any DIA planning that involved taking over functions and resources of the military departments was to be reviewed and approved by the Joint Chiefs of Staff and the Secretary of Defense.

Approximately 350 people and many of their functions were transferred to DIA from ONI. All processing responsibilities were removed, except in the areas of merchant marine, scientific and technical, and special (related to National Security Agency) intelligence. The joint Army-Navy-Air Force effort in air intelligence that had existed since World War II was almost completely transferred to DIA.

On 6 June 1962, ONI's Acoustic Intelligence Analysis Facility became operational as part of NAVSTIC in Building 52 at the U.S. Naval Observatory.

As a preamble to the Cuban missile crisis in 1962, there had been numerous reports, mostly at the rumor level, that missile sites were being built and missiles being received in Cuba. Photos of ship deck cargo tended to confirm the latter, and overhead photographs by U-2 aircraft confirmed the existence of the sites themselves and the fact that the missiles were of a strategic type. That in turn led President Kennedy to initiate a quarantine action against Soviet shipments to Cuba.

The dramatic reversal of course by Soviet merchant shipping that had been bound for Cuba at first wasn't accepted by Secretary of Defense Robert S. McNamara, but it was given a good deal of credence by CNO Adm. W. George Anderson. McNamara was perceived as having a tendency to reject any intelligence information that didn't fit his preconceived notions of what ought to be taking place. Another point of friction between the Secretary and the CNO was McNamara's taking operational play away from Adm. Anderson. McNamara would go into Operations Plot, call Navy ships directly by radiophone, and direct the commanding officers on where to go and what to do. In some cases, he personally determined which U.S. ships would board Soviet ships.[188]

The Naval Photographic Interpretation Center was very much involved in the Cuban missile crisis, first in the detection from photographs of the Soviet shipments of missiles into Cuba in late summer of 1962, and then in the interpretation of low overflight photography in October and November to ensure that the Soviet missile bases were being dismantled. Light Photographic Squadron (VFP) 62, which flew RF-8 Crusaders, supplied much of the latter photography (see Chapters 4 and 38).

ONI also kept the top policy levels of the government informed about the increase in merchant ship traffic to Cuba, in both Communist bloc and non-Communist flag vessels. In October 1962, ONI provided a study for that purpose on *The Pattern of Shipping to Cuba from January 1962 to 26 October 1962*.[189]

On 25 June 1963, RAdm. Lowrance was relieved by RAdm. Rufus Taylor, the first intelligence specialist (1630 designator) to occupy the billet of Director of Naval Intelligence. The previous Navy policy of filling the DNI billet with an unrestricted line officer had been considered desirable because, in theory, it assured that the products of ONI would be scrutinized and guided with the customers' requirements receiving top consideration. It had also been felt that ONI's image as part of the Navy could best be projected to the executive offices ("E-ring") of the Pentagon and to the senior officers of the Navy by a senior unrestricted line officer. RAdm. Taylor, although a restricted line specialist, was an acceptable exception because of his earlier fifteen or more years of experience as an unrestricted line officer.

Secretary of the Navy Notice 5450 of 25 February 1964 changed the title of the U.S. Naval Photographic Interpretation Center to the Naval Reconnaissance and Technical Support Center, a field activity of the Chief of Naval Operations, administered by ONI.

War in Vietnam

The reported attacks by North Vietnamese motor torpedo boats against U.S. Navy ships patrolling in the Gulf of Tonkin in August 1964 led to reprisal air strikes and the requirement for air target and antiaircraft defense intelligence on North Vietnam. U.S. military involvement in support of South Vietnam proliferated thereafter, as did the requirements for intelligence on the military capabilities of both the North and South Vietnamese. In connection with the latter, the conflicting information being received from Military Assistance Command, Vietnam (MACV), and from the attachés at Saigon precipitated Secretary McNamara into disestablishing the attaché organization in Vietnam. The naval attaché in particular had not been enchanted by the expertise or combat readiness of the South Vietnamese navy, but McNamara evidently did not wish to receive such negative information.

On 12 December 1964, McNamara issued DOD Directive C-5105.32 to provide general policy and guidance for the establishment, maintenance, and direction of a single Defense Attaché System (DAS) as an organizational function of the Defense Intelligence Agency. The Director of DIA assumed operational control of DAS on 1 July 1965. On 17 August 1965, the Secretary of the Navy designated the Director of Naval Intelligence as the point of contact within the Department of the Navy for defense attaché matters. All naval attachés and assistant naval attachés accredited to foreign governments and "other DOD personnel assigned to attaché posts" became part of the Defense Attaché System. The control and administrative functions relating to the naval attaché system were assumed by DIA.[190]

The Combs Board reviewed the specialist and subspecialist programs in the Navy in 1964 and, in its report of 17 December 1964, recommended that the rank structure for officer intelligence specialists (1630s) in the grades of lieutenant commander through captain should be increased from 60 percent to 75 percent. It also called for the Director of Naval Intelligence to be designated as the "code sponsor" for the 135X (Air Intelligence) officers and recommended that a quota of ten 135X officers be established at the Defense Intelligence School. It was recognized that the requirements for intelligence special duty officers had increased substantially in the past four years, primarily due to the establishment of the Defense Intelligence Agency and its requests for more 1630s.[191]

RAdm. Taylor was relieved as Director of Naval Intelligence on 27 May 1966 by his deputy, Capt. Maurice H. Rindskopf, who served as Acting Director of Naval Intelligence until RAdm. Eugene B. Fluckey, a well-decorated World War II submariner, arrived on 22 July 1966. Capt. Rindskopf continued as Fluckey's deputy.

The Israeli-Egyptian crisis of late May 1967 and the Israeli attack on the *Liberty* (AGTR 5) on 8 June stimulated the updating of available intelligence on the Middle East. Many Naval Reserve intelligence officers were used on two weeks' active duty under training to assist in the effort, particularly at the Fleet Intelligence Center, Atlantic (FICLANT).

When the Naval Intelligence Command (NAVINTCOM) was established on 1 July 1967, RAdm. Fluckey dropped the title of Director of Naval Intelligence and became Commander Naval Intelligence Command. He retained the title of Assistant Chief of Naval Operations (ACNO) for Intelligence, OP-92. See Chapter 28 for more details about the establishment of the Naval Intelligence Command.

Following the establishment of NAVINTCOM, pressure became strong to move its headquarters out of the Pentagon. As far as OP-09B, ACNO (Administration), was concerned, it was more important that the headquarters leave than that a suitable new location be found. Under such pressure, a temporary, less-than-satisfactory location was the only acceptable solution, and a move to the privately owned Hoffman Building, then under construction in Alexandria, Virginia, was to be carried out as soon as possible. The cost of establishing secure spaces and communications at the new location assured that the so-called temporary location would be essentially permanent. Advantages to the move were the availability of space for expansion and the chance for needed personnel force growth outside the stringent OPNAV personnel ceiling.

The capture by North Korea of the euphemistically designated Environmental Research Ship *Pueblo* (AGER 2) on 23 January 1968 put a temporary damper on all sensitive collection operations peripheral to Communist coastal areas. For more details on the event, see Chapter 5.

On 10 June 1968, Capt. Frank M. Murphy, Deputy ACNO (Intelligence), relieved RAdm. Fluckey temporarily until RAdm. Frederick J. Harlfinger II, another submariner, reported on 12 August as OP-92 and Commander Naval Intelligence Command.

In 1968, a change to Title 10, U.S. Code, allowed for the expansion of the special duty officer community, and many air intelligence officers changed from the 1350 to the 1630 designator, causing an increase in the 1630 community to over 1,100 officers.[192]

On 23 June 1969, major elements of the Naval Intelligence Command commenced moving into the Hoffman Building from the fifth floor of the Pentagon, and the Translation Division moved to the Hoffman Building from the Naval Security Station on Nebraska Avenue, NW, Washington, D.C. On 15 September, the Naval Investigative Service Headquarters started its move to the Hoffman Building from the Fairmont Building in North Arlington, Virginia. Portions of the Naval Scientific and Technical Intelligence Center and the Naval Intelligence Processing System Support Activity (NIPSSA), another ONI subsidiary, started moving to the Hoffman Building from various locations. By 30 October, all the moves had been completed.

On 15 March 1971, a change in the OPNAV organization dropped the Director of Naval Intelligence one command level below that which the office had maintained since June 1954. The Office of Command Support Programs (OP-094) was established on the same date, and the Office of ACNO (Intelligence), OP-092, was disestablished and made the Intelligence Division of OP-094. Its new designation was OP-942. The development of a detailed organization for OP-094 and a proposed mission and functions statement for the various staff officers, divisions, and branches was to be submitted for approval by 1 May 1971.[193]

Capt. Earl F. Rectanus, who had already been selected for promotion to rear admiral, was "frocked" and relieved RAdm. Harlfinger on 22 July 1971 as Commander Naval Intelligence Command. Harlfinger was promoted to vice admiral as OP-094.[194] RAdm. Rectanus became OP-942 and OP-009; the latter was a double-hat for OP-942 according to OPNAV Notice 5430 of 5 March 1971.

In February 1973, the designation of the Office of the Director of Naval Intelligence changed from OP-942 to OP-009, and the director returned to an organizational position directly under the CNO. The director was also assigned additional duty as OP-094Q, Assistant for Intelligence Support, Command Support Programs. The resumption of the traditional title, Office of Naval Intelligence, was also approved at that time.[195]

Chapter Notes

1. James A. Knowles, Jr., "Blue Water Monitor," U.S. Naval Institute *Proceedings*, Mar 1973, 79, hereafter *USNIP*. Fox was Assistant Secretary of the Navy until 26 November 1866.

2. *ONI Review*, Jul 1956, 301; and Lt. William A. Sachse, "Our Naval Attaché System," *USNIP*, May 1946, 662.

3. *ONI Review*, Jul 1956, 302.

4. Ibid., 303.

5. *Annual Report of the Secretary of the Navy, November 28, 1881* (Washington: GPO, 1881), 5–6 (hereafter *SECNAV Annual Report* with date).

6. Capt. John M. Ellicott, "T.B.M. Mason," *USNIP*, Mar 1952.

7. Radm. Albert G. Berry, USN (Ret.), "The Beginning of the ONI," *USNIP*, Jan 1937.

8. *ONI Review*, Apr–May 1957, 172.

9. Berry, "The Beginning of the ONI."

10. Sachse, "Our Naval Attaché System," May 1946, 662–70.

11. *ONI Review*, Jul 1956, 303.

12. *Navy Directory*, 1883.

13. *SECNAV Annual Report*, *1883*, 244.

14. Ibid., 26.

15. *SECNAV Annual Report*, *1884*, 164.

16. *Navy Directories* for 1884, 1885.

17. *ONI Review*, Apr–May 1957, 173.

18. *SECNAV Annual Report*, *1885*, 87, 150–53.

19. Ltr to U.S. Senate, 11 Feb 1885, cited in Adm. Austin M. Knight and Lt. William D. Puleston, *History of the Naval War College* (Newport, RI: U.S. Naval War College, 1916), 17.

20. *ONI Review*, Apr–May 1957, 177.

21. *SECNAV Annual Report*, *1888*, 2; 50th Cong., *U.S. Statutes*, 459; and Sachse, "Our Naval Attaché System," 666–67.

22. *SECNAV Annual Report*, *1889*, 7.

23. Mahan ltr to RAdm. Stephen B. Luce, 3 Sep 1901; and *Naval Directory*, 1889.

24. Knight and Puleston, *History of the Naval War College*, 17; and Henry Beers, "Development of the Office of CNO," 14.

25. SECNAV ltr, 3 Oct 1890.

26. Beers, "Development of the Office of Naval Intelligence," 23–24.

27. *SECNAV Annual Report*, *1892*, 48.

28. *Navy Directory*, 1 Jul 1893.

29. *SECNAV Annual Report*, *1893*, 6–12.

30. Capt. Damon E. Cummings, *Admiral Wainwright and the United States Fleet* (Washington: GPO, 1962), 177.

31. *SECNAV Annual Report*, *1898*, 376–77.

32. *SECNAV Annual Report*, *1896*, 48–49.

33. *SECNAV Annual Report*, *1897*, 45.

34. Cummings, *Admiral Wainwright*, 73–78.

35. Capt. Herbert E. Cocke, USN, "History of ONI," MS, Office of Naval Intelligence, 1931, 9.

36. Ibid., 10.

37. *Navy Directory*, 1 Jul 1898, 77.

38. *SECNAV Annual Report*, *1898*, 2:32.

39. Ibid. 2:33; Beer's "Development of the Office of the CNO" states that Cdr. Richardson Clover, Chief Intelligence Officer, was a member of the board until he left for sea duty.

40. *SECNAV Annual Report, 1898*, 1:33.

41. Ibid., 113.

42. *ONI Annual Report*, *1899*, 466.

43. *SECNAV Annual Report*, *1898*, 1:374–77.

44. Ibid., 379.

45. Ibid., 63.

46. *ONI Annual Report, 1899*, 464–66.

47. *SECNAV Annual Report*, *1900*, 456.

48. Ibid., 525–27.

49. Vernon E. Davis, *History of the Joint Chiefs of Staff in World War II*, vol. 1, *Organization and Development* (Washington: Joint Chiefs of Staff, 1972), 3.

50. *SECNAV Annual Report*, *1900*, 525–27.

51. *SECNAV Annual Report*, *1902*, 96.

52. Ibid., 20.

53. Ibid., 25.

54. Cocke, "History of ONI," 15–16.

55. RAdm. Albert P. Niblack, USN, *The History and Aims of the Office of Naval Intelligence* (Washington: GPO, 1920).

56. *Navy Directories*, 1903–1906; Cocke, "History of ONI," passim.

57. *SECNAV Annual Report*, *1903–1906*, passim.

58. Capt. Henry S. Shields, USAF, "A Historical Survey of U.S. Naval Attachés in Russia: 1904–1941," MS, Defense Intelligence School, Mar 1970, 6–13.

59. Cummings, *Admiral Wainwright*, 213–16. The term "aid" was used in official correspondence until the end of the war.

60. *Navy Directories*, passim.

61. Cocke, "History of ONI," 17.

62. Department of the Navy, "Administrative History of the Office of Naval Intelligence in World War II," 10 Jul 1946, unpublished MS, 9, hereafter *ONI WWII Admin History*.

63. Shields, "Naval Attachés in Russia," 21.

64. Sachse, "Our Naval Attaché System," 670.

65. *ONI WWII Admin History*, 4.

66. *SECNAV Annual Report*, *1915*, 10.

67. James R. Green, "The First Sixty Years of the Office of Naval Intelligence" (M.A. thesis, The American University, 1963), 36; and Cocke, "History of ONI," 18.

68. Green, "First Sixty Years," 65.

69. Press release, U.S. Navy Memorial Museum, Jul 1975.

70. *ONI WWII Admin History*, 149.

71. House Committee on Naval Affairs, *Estimates Submitted by the Secretary of the Navy: Hearings*, 64th Cong., 1st sess., 16 Mar 1916, 3:3133, 3635.

72. Edward Macauley, "Naval Intelligence Before and During the World War," 19 Nov 1943, interview transcript, in author's files, Operational Archives (OA).

73. Cocke, "History of ONI," 28–29.

74. Elting E. Morison, *Admiral Sims and the Modern American Navy* (Boston: Houghton Mifflin, 1942), 342–45.

75. *Naval Directories*, 1917–1918.

76. *SECNAV Annual Report, 1918*, 24–25.

77. Niblack, *History and Aims of ONI*, 23.

78. Ibid., 2; Cocke, "History of ONI," 2.

79. *ONI WWII Admin History,* 15–16, quoting ONI-19, Art. 105.

80. Department of the Navy, *U.S. Naval Activity Relative to the Armistice of 1918 and the Peace Conference of 1919* (Washington: GPO, 1919), 17, 26

81. *ONI WWII Admin History,* 1147.

82. *SECNAV Annual Report, 1920*, 156–57.

83. *ONI WWII Admin History,* 185–89.

84. Ibid., 822.

85. F. Donald Scovel, "Helms A'Lee: History of the Development of the Public Affairs Function in the U.S. Navy, 1861–1941" (M.A. thesis, University of Wisconsin, 1968), 83, 85.

86. *ONI WWII Admin History*, 592.

87. Ibid., 506–7.

88. *Annual Report of the Chief of Naval Operations, FY 1923*, hereafter *CNO Annual Report*.

89. Adm. Royal E. Ingersoll, Oral History, Columbia University, New York, NY, 1965, 45, 102–3, copy in OA.

90. *CNO Annual Report, FY 1925*.

91. *ONI WWII Admin History,* 1165.

92. Ibid., 19.

93. *CNO Annual Report, FY 1927*.

94. *CNO Annual Report, FY 1928*.

95. *ONI WWII Admin History,* 511–12.

96. *CNO Annual Report, FY 1929*.

97. *ONI WWII Admin History,* 67.

98. *CNO Annual Report, FY 1930*.

99. *ONI WWII Admin History,* 1160.

100. *New York Herald Tribune*, 18 Jun 1930.

101. *ONI WWII Admin History,* 21.

102. Ibid., 507–8.

103. Ibid., 509.

104. *CNO Annual Report, FY 1932*; and *Annual Report of the Director of Naval Intelligence, FY 1935*, 1, hereafter *DNI Annual Report*.

105. *ONI WWII Admin History,* 594.

106. *CNO Annual Report, FY 1933*; and Office of Naval Intelligence, Administrative Branch (OP-16-A), *Annual Report, FY 1933*, 2, hereafter ONI Admin Branch, *Annual Report*.

107. Office of Naval Intelligence, *Naval Intelligence Manual* (ONI-19) (Washington: GPO, 1933), para. 2010.

108. *CNO Annual Report, FY 1934*.

109. Shields, "Naval Attachés in Russia, 36–53.

110. *ONI WWII Admin History,* 23.

111. *DNI Annual Report, FY 1935,* 5–6.

112. *ONI WWII Admin History,* 536.

113. *DNI Annual Report, FY 1936*.

114. VAdm. Felix Johnson, interview by author, 14 May 1975. Johnson, Director of Naval Intelligence (1949–1952), had served in the Naval Mission to Brazil from August 1936 to December 1938.

115. *DNI Annual Report, FY 1936*, 6.

116. *ONI WWII Admin History,* 24.

117. *DNI Annual Report, FY 1937*; and ONI Admin Branch, *Annual Report, 1937*.

118. *ONI WWII Admin History,* 511.

119. *CNO Annual Report, FY 1938*.

120. VAdm. Walter S. Anderson, Oral History, Columbia University, NY, 1963, 224–29, copy in OA; and James O. Richardson, *On the Treadmill to Pearl Harbor: The Memoirs of Admiral J. O. Richardson as Told to Admiral George C. Dyer, USN (Retired)* (Washington: GPO, 1973), 156–57.

121. *ONI WWII Admin History,* 540.

122. *CNO Annual Report, FY 1939*.

123. *ONI WWII Admin History,* 31, Appendix C.

124. Ibid., 47, Appendix N.

125. Ibid., 48, Appendix O.

126. *CNO Annual Report, FY 1940*.

127. Anderson oral history, Columbia University, 231.

128. Richardson, *Treadmill to Pearl Harbor*, 337–38, 353–60.

129. Scovel, "Helms A'Lee," 129.

130. *ONI WWII Admin History,* 540.

131. *CNO Annual Report, FY 1941*.

132. Green, "The First Sixty Years of ONI," 96; and U.S. Congress, *Report of the Joint Committee on the Investigation of the Pearl Harbor Attack*, 79th Cong., 2d sess. (Washington: GPO, 1946), 1807.

133. *Pearl Harbor Investigation*, 77.

134. Roberta Wohlstetter, *Pearl Harbor: Warning and Decision* (Stanford: Stanford University Press, 1962), 317.

135. VAdm. Walter S. Anderson ltr to author, 20 Sep 1974.

136. Adm. Alan G. Kirk, Oral History, Columbia University, NY, 1965, 180, copy in OA.

137. Ibid., 179.

138. DNI (OP-16-1) memo, ser 1933416, 16 Jul 1941.

139. DNI (OP-16-A) ltr, ser 2104916, 19 Aug 1941.

140. OP-16 ltr, ser 2726116, 10 Oct 1941.

141. *ONI Review*, Jun 1947, 3; and "History of the Fourth Fleet," MS history, 9.

142. RAdm. Arthur H. McCollum, Oral History, U.S. Naval Institute (USNI), Annapolis, MD, 2 vols., 1971, passim, copy in OA.

143. *Pearl Harbor Investigation*, 104.

144. Ibid., 98.

145. Ingersoll oral history, Columbia University, 78–80.

146. McCollum oral history, USNI.

147. *Pearl Harbor Investigation*, 148.

148. Wohlstetter, *Pearl Harbor*, 322.

149. Ibid., 315, 318.

150. *ONI Review*, Nov 1946, 14.

151. "History of the Special Activities Branch, 17 Jun 1940–31 Oct 1943," MS history, 11, box 16, Job 3679, Federal Records Center, Washington National Records Center (FRC/WNRC), Suitland, MD.

152. Vice Chief of Naval Operations (VCNO) ltr OP-16-A-6 ser 2723016, 26 Jun 1942, cited in *ONI WWII Admin History*, 55.

153. VCNO ltr, ser 01815916, 5 Sep 1942, cited in *ONI WWII Admin History*, 2.

154. *ONI WWII Admin History*, 739–40.

155. *ONI Review*, Nov 1946, 15–17.

156. Capt. Arthur H. McCollum, "Submarine Activities with Guerrillas," MS (on microfilm), Job No. 420, OA.

157. Norden material WNRC, file P15, box 16, Job 3679, Historical files, FRC/WNRC.

158. OP-16-Z memo, 1 Jun 1944, ibid.

159. OP-16-Z memo, 2 Feb 1945, ibid.

160. OP-16-Z ltr, ser 197916, 14 Feb 1945, ibid.

161. VAdm. Rufus L. Taylor, interview by author, 23 Nov 1975, tape recording.

162. OP-32, "Command Narrative and Quarterly Summary Report, 1 Jul 1946–30 Sep 1946"; and *ONI Review*, Sep 1946, 35.

163. OP-32X ltr to Commandant, Marine Corps, ser 05950P32, 3 Sep 1946, ONI Day File.

164. OP-32X memo, ser 05946P32, 28 Aug 1946, ONI Day File.

165. Chief of Naval Intelligence (OP-32C) memo, ser 17376P32, 8 Nov 1946, ONI Day File.

166. Pamphlet, *Career Planning Info for Intel SDOs*, 1947; also ALNAVs 202–45, 207–45, 213–45, 271–45, 335–45, and 338–45.

167. ONI-70-1, chap. 1, 4 Nov 1957, 25.

168. Capt. Warren J. Bettens ltr to author, 17 Aug 1974.

169. Chairman General Board, ser 00061EN2, 8 Apr 1949, General Board Records, OA.

170. Commander in Chief, Pacific Fleet (CINCPACFLT), *Interim Evaluation Report No. 2: Korean War Naval Operations*, 355.

171. "History of Administrative Problems in the Korean War," 2: 18.

172. VAdm. Felix Johnson, MS memoir, 31 Dec 1973.

173. DNI memo to CNO, OP-322Y2 ser 000327P32, 12 May 1951.

174. CINCPACFLT, *Interim Evaluation No. 2: Korean War Naval Operations*, 213.

175. Ibid., 355

176. Ibid., 1743

177. *ONI Review*, Oct 1952, 408.

178. ONI Instruction 003820.36A of 3 Jun 1952.

179. SECNAV Instruction 03800.4 of 19 Jul 1952.

180. Commander in Chief, Atlantic Fleet (CINCLANTFLT), Supplement to *Annual Report, 15 Apr 1954 to 30 Jun 1954*, 5–6.

181. VAdm. Carl F. Espe ltr to author, 25 Sep 1974.

182. ONI Internal Instruction 5203.1, 9 Aug 1956.

183. OP-92 memo, ser 0005P92, 31 Aug 1956; and British Admiralty ltr to RAdm. Frost (DNI), 7 Sep 1956.

184. William E.W. Howe ltr to author, 1983, 3.

185. Keith Board Report, 21 Aug 1959, Tab D.

186. JANAF (Joint Army/Navy/Air Force) *Attaché Letter*, No. 3, 1 Jul 1959.

187. VAdm. Rufus L. Taylor ltr to author, Oct 1975.

188. Ibid.

189. OP-922N ltr, ser SSO/00323.

190. SECNAV Instruction 05400.11 of 17 Aug 1965.

191. Combs Board Report, 17 Dec 1964, Appendix G.

192. Pamphlet, *Career Planning for Intel SDOs*, 1968.

193. OPNAV Notice 5430 of 2 Mar 1971.

194. NDW Notice 5060 of 9 Jul 1971.

195. OPNAV Notice 5430 of 1 Feb 1973.

CHAPTER 2

Intelligence Collection

Information collected for intelligence production purposes includes documents, sketches, photographs, sensor data, and hardware. The means of collection are varied and are categorized in this chapter according to the types of sources available to the Navy. This chapter and Chapter 3, concerning naval attachés, contain most of the information on collection up to World War II. Chapters 4 through 8 also contain information on intelligence collection. Chapter 32 on the Spanish-American War includes details on collection efforts which took place during that conflict. Domestic collection was mostly performed in support of counterintelligence, but information was also collected domestically in support of intelligence, and the domestic collection effort is covered in Chapter 22.

The word reconnaissance, as used in this book, encompasses surveillance, since most search operations start as reconnaissance of specific areas for the purpose of collecting both positive and negative information. If search operations detect targets deserving further collection effort, the targets may be kept under surveillance for varying periods of time until the required information is obtained. Then reconnaissance is resumed to find new targets.

The Navy's in-house collection capabilities originally were almost totally in the area that is now termed "human intelligence" (HUMINT). Most intelligence information in the early years of the Navy was collected by human senses, either on board ship or at naval attaché posts. In about 1886, the first sensor was introduced into the fleet when cameras were supplied to help people retain the images they were seeing. Of course, spy glasses had been in use as long as there had been navies, but they were an aid to seeing, not a recording sensor like the camera. When a useful air reconnaissance capability became available to the fleet in about 1914, it was still dependent on human observation, occasionally supplemented by hand-held cameras.

The introduction of radios aboard ship in 1903 unknowingly opened up a new type of intelligence source that was first exploited during World War I. The radio was a device that shifted radio frequencies to audio frequencies, and it needed the human ear and brain to transpose the signals into information of intelligence interest.

Up to and even during the initial phases of World War II, HUMINT was still the primary source of intelligence information, augmented by the more sophisticated cameras and radio interception. Consequently, all supplementary chapters on collection in this book, except Chapter 3, begin for the most part during or after World War II.

Before ONI

Collection of information on foreign naval subjects began long before there was an Office of Naval Intelligence. Benjamin Franklin, while serving as minister to the French court, operated a sizable naval section in addition to his diplomatic staff. Franklin's naval officers were involved in the procurement of naval vessels, crews, supplies, and information. Their efforts, however, to collect information on the British and the British navy, although good enough to help make John Paul Jones a success, were frustrated in some respects because one of their sources, Edward Bancroft, a friend of Franklin and supposedly a reliable American secret agent, was actually a double agent in the employ of the British. Bancroft made numerous trips to England, where he furnished the British with information about American supplies, negotiations, and ship movements. The information he brought back to Franklin, although authentic, was so out of date that, by the time it was received, it had little value. The true nature of Bancroft's activities did not become known until many years later.[1]

One of the earliest requests for intelligence support made to the Navy was received in 1800 from

President John Adams, who requested that Secretary of the Navy Benjamin Stoddert

> employ some of his clerks in preparing a catalogue of books for the use of his office. It ought to consist of all the best writings in Dutch, Spanish, French and especially in English, upon the theory and practice of naval architecture, navigation, gunnery, hydraulics, and all branches of mathematics subservient to the profession of the sea. The lives of all the admirals, English, French, Dutch or any other nation, who have distinguished themselves by the boldness and success of their navigation, or their gallantry and skill in naval combat. If there are no funds which can be legally applied by the Secretary to purchase of such a library, application ought to be made to Congress for assistance.[2]

Another early intelligence collection request was made by Secretary of the Navy Samuel L. Southard in 1827 to U.S. ambassadors and ministers in all the capitals of the maritime powers:

> May I therefore beg the favor of you, from time to time to communicate to me such information respecting the naval forces of Great Britain, or other nations, as you may be able to procure without inconvenience; especially respecting the number, situation, use and employment of their vessels; the number, character, etc., of their navy and dock yards; the number and mode of furnishing their seamen; the means of educating their officers; the amount and character of the expenditures; and, generally, anything which will enable this department completely to comprehend the extent and character of the naval means of the nation.[3]

When Commo. Matthew C. Perry was given command of the expedition to Japan on 24 March 1852, he used all available means to learn everything he could about Japan while he waited for the ships of his squadron to be readied. Charts of Japanese waters were secured from the Netherlands at a cost of $30,000, and, through book collectors in New York and London, Perry gathered all the available, useful literature relating to Japanese history, customs, and manners.[4]

Perry had previously been involved in the collection of information from abroad, and he appears to have been the first U.S. naval officer to be sent out to do so. In 1838, Secretary of the Navy Mahlon Dickerson sent Perry, then commanding officer of the steam sloop *Fulton*, to Europe to collect information on optics and "the use and construction of steamships for naval purposes." Perry obtained a wealth of technical information and new ideas as a result of his inquiries in England and France, where a "revolution in naval science was in progress."[5]

Similar assignments were given to other officers from time to time. In 1854, for example, Engineer in Chief Daniel B. Martin went to Europe to collect information on screw propellers and steam machinery.[6] During the Civil War, William H. Aspinwall and John M. Forbes (neither of them naval officers) were in England to report on the construction of warships in private shipyards.[7] During the practice cruise of 1863, Stephen B. Luce, Commandant of Midshipmen at the Naval Academy, prescribed what the midshipmen were to note in their journals:

> To be useful in after years, and to induce a habit of acquiring useful knowledge, it shall contain certain notes on the pilotage of each place, on the facilities for water, provision, wood and stores; facilities for docking and repairs, on the anchorage and tides, on the military defenses and etiquette peculiar to the place, the name and brief description of every man-of-war in port and such other items as may prove useful for future reference.[8]

Secretary of the Navy George M. Robeson, in an effort to encourage naval officers to collect information, put out a circular on 17 March 1874:

> Officers of the several branches of the naval service are requested to collect and forward to the Navy Department all information on professional subjects, and also in any branch of natural science which may come under their observation.
>
> The opportunities are excellent, and it will be gratifying to the Department to see that they are availed of, and that the results obtained be forwarded to the respective Bureaus for publication when thought of advantage, either professionally or to men of science.[9]

In 1877, at Luce's urging, Secretary of the Navy Richard W. Thompson had an officer ordered abroad to "study the systems of naval education in foreign countries with a view to establishing a post-graduate course." The report was buried and not used, to Luce's disgust.[10]

Early Years

The establishment of the Office of Naval Intelligence was partly justified by the need to correlate the information being collected independently by the various Navy bureaus. Obtaining information for its own files was one of ONI's first official collection tasks.

The translation and processing of data available in foreign books and publications in the stacks of the Navy Department Library was another ONI collection effort. It was initiated by Lt. Theodorus B. M. Mason, the first head of ONI and a linguist of renown. Mason was often called upon throughout his naval career to serve as foreign observer, translator, and interpreter for various Secretaries of the Navy.

The use of naval attachés for the collection of intelligence was prescribed in the first Secretary of the Navy instructions to Mason issued on 25 July 1882. For details on the activities of naval attachés, see Chapter 3.

Requests for intelligence collection were sent in the form of letters from the Chief of the Bureau of Navigation (BUNAV) to the commander in chief of the fleet or the commander of the station concerned. For example, on 24 November 1886, Commo. J. G. Walker wrote RAdm. Stephen Luce requesting that Lt. Charles C. Rogers of the *Galena* do some work for ONI while his ship visited Cuba. Walker requested that Rogers visit ONI for instructions while *Galena* coaled at Norfolk prior to the ship's departure for the Caribbean.

In a trip report of 9 October 1888 that was sent to the superintendent of the Naval Observatory, Lt. Albert G. Winterhalter reported on visits to various observatories and establishments of scientific character in Meudon, France; Greenwich, Richmond, Ealing, Oxford, and Cambridge, England; Leyden, Holland; Duesseldorf, Bonn, Strasbow, Berlin, Potsdam, Dresden, Leipzig, and Munich, Germany; and Vienna, Austria. He brought back numerous pieces of optical equipment in the form of purchases and gifts from manufacturers. One, a marine spy glass presented by M.R.R.&C. Avizard of Paris, upon comparison with the standard U.S. Navy glass, showed optical superiority. By the direction of the BUNAV Chief, it was made the U.S. Navy standard to which future optical performance of glasses had to aspire.[11]

The Hydrographic Branch Office in Boston, in connection with its collection efforts, commented:

> The importance of boarding every arrival cannot be overestimated. It stimulates the seamen observers to better work, gives them an idea that the Government appreciates their voluntary labors, and turns them into active agents to induce others to help.[12]

The first "General Instruction in Regard to Intelligence Duty" appeared in a Navy Department order of 31 March 1885. The order unfortunately cannot be located now, but in 1892 it was issued in a modified form, and some interesting portions are quoted below:

> Commanders-in-Chief are directed to appoint an officer, preferably of their personal staff, to perform the duty of Fleet Intelligence Officer, and commanders of vessels are directed to appoint an officer of their command to perform the duty of Intelligence Officer; in each case the name and rank of the officer so appointed shall be reported to the department.
>
> In order that there may be no duplication of purchases [of information], the Intelligence Officer, before leaving a foreign port, shall prepare a list of purchases made by him. This list shall be enclosed in a package sealed with the ship's seal, and addressed to "Commanding Officer of the next U.S. ship-of-war visiting the port of . . . ." This package shall be left with the consular officer of the U.S. with a request that it shall be delivered to the order of the commanding officer of the next U.S. ship-of-war visiting that port.
>
> The Commanding Officer shall, on his arrival, request from the consul such package. The Intelligence Officer shall make such additions, over his signature, as may be necessary, and the package shall be returned to the consul before leaving port.[13]

Lt. William S. Sims was the Intelligence Officer on the staff of Commander in Chief, Asiatic Fleet RAdm. Frederick (Turk) Rodgers during 1901–1902. Sims submitted a steady flow of reports to ONI, many as a result of his contact with the British navy's top gunnery expert, Sir Percy Scott. Scott had developed what he called "continuous aim," using telescopic sights that he had developed. Sims's reports were seen by President Theodore Roosevelt, stimulating Roosevelt's interest and providing the impetus for the improvement of gunnery in the U.S. Navy.[14]

In 1903, the Navy Department's order of 28 October 1892, "Instructions in Regard to Intelligence Duty," was modified by an order that included instructions regarding intelligence duty in cruising ships. The instructions were similar to those in the prior order, except that there was no mention of the practice of leaving a sealed package at each foreign port visited by a U.S. man-of-war.

World War I

To increase its sources of information on the conflict in Europe, both at home and abroad, ONI set up the War Information Service in October 1915. The general plan for the service was to secure information on the state of preparation of the navies of probable enemies, the movements of probable enemy forces and the progress in their preparations for war, and the political and financial conditions of the probable enemy country. The plan could be used in time of war and provided for obtaining information through neutral countries and from secret agents in enemy countries.

To put the War Information Service into effective operation, funds were required. Congress was appealed to, and the Navy Appropriation Bill, approved 29 August 1916, contained a provision by which funds could be expended by ONI "for obtaining information from abroad and at home." The process for expending money under the provision was as follows: The Director of Naval Intelligence (DNI) would draw a requisition for the amount required. After being approved by the Secretary of the Navy, the requisition form was taken to the Bureau of Supplies and Accounts, and a check was drawn to the order of the DNI. The proceeds of the check were deposited in one or more of the three banks with which ONI maintained accounts: National City Bank of New York; Riggs National Bank of Washington, D.C.; and National Metropolitan Bank of Washington, D.C.[15]

The Naval Information Service was the overseas version of the War Information Service. It was intended to operate from neutral countries after the outbreak of hostilities. Posts in England, Germany, Japan, China, Argentina, and Chile were to be augmented by five officers and six clerks each. The plan, strong on secrecy and weak on details, was apparently put together rapidly in response to a recommendation from the General Board of the Navy, a group of very senior officers who advised the Secretary of the Navy and the newly created post of Chief of Naval Operations (CNO).[16]

As part of the War Information Service plan, arrangements were made to take advantage of the collection potential of other government departments and agencies, as well as of private activities. The potential sources included the War Department and its military attachés, the State Department with its Diplomatic and Consular Service, the Justice Department with its Secret Service, the Treasury Department with its Secret Service and Coast Guard, the Post Office Department with its Secret Service, the Commerce Department and its agents, retired officers living abroad, commercial firms with representatives abroad, news-gathering organizations, neutral but friendly officials at home and abroad, and foreign neutral secret service organizations.[17]

Early in the spring of 1916, DNI Capt. James H. Oliver instructed Maj. John H. Russell, USMC, to get in touch with large commercial firms throughout the country and obtain their assistance in collecting information of potential value to the Navy. Many heads of such firms were visited, and a system of information collection was developed. In early 1917, Maj. Russell was placed in charge of Section A of ONI, where he put into effect a system of obtaining and sending out agents. The agents were first tried out on work close at hand; their spheres of action were gradually extended as their proficiency increased.[18]

The Office of Naval Intelligence maintained name lists on 4-by-6-inch and 5-by-8-inch cards of prospective and hired agents and informants, arranged by foreign country. Most of the employed agents were Americans residing abroad as businessmen, missionaries, or teachers. They reported on enemy or potential enemy activity, economic conditions, and military preparations. The cards contained some biographical data for each agent, including an evaluation of the agent's reliability, occupation, background experiences, salary and expenses paid, and whether the agent was a volunteer.

Another card file (on 3-by-5-inch cards) was maintained on potential sources of information, arranged alphabetically by geographic location, on consular agents and representatives of other government departments. Names of officials were usually shown, arranged alphabetically by location. Other subseries within the file were "War Slate, Agents," "Paid Agents," and "Volunteer Agents," arranged alphabetically by name of person.[19]

When the United States broke relations with Germany in 1917, German espionage and propaganda activities in South America were redoubled. The large German population in Brazil not only planned to keep that country from joining the Allies but talked boldly of "uprisings" and joining in action with Germans in neighboring countries. The United States urgently needed to find out more about German activities in South America.

Separate from the attaché system was the system of agents who operated especially in South America. One such agent was Edward Breck, an American of varied accomplishments, who spoke Portuguese and Spanish as well as German. Breck, who had served the Navy as an agent in Spain during the Spanish-American War, again offered his services to the Navy. In February 1917, he was accepted by Naval Intelligence, and on 3 March, a month before the United States declared war, he sailed for Brazil as "Dr. Ernst Brecht."

"Dr. Brecht" became active both in Brazil and Argentina and later in Chile. He reported on German activities in those countries and influenced public opinion there in favor of the Allies and against Germany. For his services, and because of his fluency in Portuguese, Breck was appointed attaché in Lisbon in May 1918.[20]

Another agent was John Held, Jr., later famous as the creator of cartoons about the "roaring twenties," who served for fourteen months in 1917–1918 as a member of a naval intelligence mission in Central America. Under the guise of an archeological research expedition, he and three colleagues sought out potential hiding places for German submarines. Held also made a number of archeological drawings, but most of his sketches were of coastal scenes at such places as Belize, Campeche, Vera Cruz, and Cozumel.[21]

Capt. Walter R. Gherardi, headed an American fact-finding delegation to Germany in January and

February 1919. Several similar missions were sent to Germany and Central Europe following World War I, most of which included naval officers as members.

Article 209 of the Treaty of Versailles required the German government to turn over to the Naval Inter-Allied Commission of Control all information and documents on the design of warships, including details about their guns, munitions, torpedoes, mines, explosives, wireless telegraphic apparatus, and, in general, everything relating to naval war material. The U.S. Government, having failed to ratify the Treaty of Versailles, was not represented on any of the Inter-Allied Commissions of Control, and most of the technical information obtained under authority of Article 209 of the treaty was consequently not turned over to the United States.[22]

Between Two World Wars

In time of peace, the Navy's sources of intelligence information were naval attachés; consular reports; State Department dispatches (reports from ambassadors and ministers); the files of ONI; reports from Navy shipboard intelligence officers, tourists, businessmen, newspaper correspondents, and commercial travelers; and commercial reports from business firms, newspapers, periodicals, and agents. In time of war, information was to be obtained through agents, code breakers, travelers, and spies. In time of peace, an intelligence officer was designated on board each U.S. Navy ship, and it was his duty to forward reports on every port visited and to act as an agent of ONI. A source that became increasingly important was code breaking, which, in time of peace, had to be studied and mastered. A potentially fruitful, but never properly exploited, source of information was U.S. citizen travelers, who should have been questioned at once upon their return from an area of interest.[23]

To provide guidance to the naval service on the information to be collected on foreign areas, *Instructions for Intelligence Officers* (ONI-8) was published on 1 May 1923. ONI-8 was the first formal publication on collection known to have been published by ONI. It was approved by CNO Adm. Robert E. Coontz and Secretary of the Navy Edwin Denby and was designed primarily for use by the operating forces. In its introduction, ONI-8 stated:

> Our first thought on entering a foreign harbor must be, "What information can I collect which would be of value to me as the commander of an attacking force? What information must I have after having taken this place in order that I may intelligently administer its affairs? What information must I have if I am to hold this place against attack from land and sea?"

ONI-8 began with an alphabetical list of subjects for which information was required, giving monograph section numbers for each subject, followed by a numerical arrangement of subjects by monograph section numbers. ("Monograph" was the term applied in ONI to the continuously updated dossiers maintained on various countries.) The final section was an arrangement of subjects by order of military importance, with detailed guidance on the elements of information needed under each subject and, in some cases, how to collect it.

Reports were to be made on the Intelligence Report Form, with the title sheet giving the subject, subheading, and corresponding monograph index numbers as shown in ONI-8. Thus, reports received by ONI could be readily collated with information on the same subject that had been received from other sources. Ships' intelligence officers were encouraged to submit reports to ONI on persons in foreign areas who were pro-American and who might serve as channels of communication in peace or as sources of information in time of war without incurring suspicion.[24]

When Japan received a mandate over the former German islands in the Pacific after World War I, it was agreed that the United States could send men-of-war to visit the islands at any time. When the United States indicated an intention to do so, however, Japan refused permission. The United States did not strenuously insist upon its rights, but finally, in 1923, permission was obtained for the light cruiser *Milwaukee* (CL 5) to visit the Marshall and Caroline Islands. The ship made excellent surveys of several islands, and the photographs, soundings, and other information obtained were of great value. Stops were made at Truk, Jaluit, Eniwetok, and other islands of particular interest to ONI.[25]

During 1923, other light cruisers in the Marshall and Caroline Islands added materially to ONI's supply of important information. There were serious gaps in the information on the Aleutian Islands and practically no charts, although it was believed that the Japanese had made good surveys of the U.S.-owned islands. Correction of these deficiencies was needed as soon as possible and was therefore prescribed by the Director of War Plans in his "Estimate of the Situation and Base Development Plan" of 17 March 1924.[26]

In October 1927, the light cruiser *Marblehead* (CL 12) proceeded from Shanghai to Kobe, Japan, while the Japanese fleet was conducting major exercises in the area of the cruiser's track. The Japanese fleet's tactical communications circuits were monitored in what is believed to be the first use of a ship by the U.S. Navy to gather intelligence by radio intercept.

Lt. Ellis M. Zacharias, the Asiatic Fleet Intelligence Officer at the time, was on board *Marblehead* during the collection cruise. The Naval Attaché, Tokyo, met *Marblehead* at Kobe and brought Zacharias back to Tokyo. From there Zacharias went with Lt. Edward S. Pearce out into Tokyo Bay to look at the Japanese fleet when it returned from maneuvers.[27]

Beginning in 1927, the Latin American Desk (OP-16-B-16) of ONI sought to keep its monograph material on Latin America current by requesting various units of the U.S. Fleet on cruises to correct or enlarge the monograph information in ONI. Monographs and reports of pertinent information were furnished to naval operating units in Latin American waters. From 1935 on, the effort was particularly fruitful and apparently well organized, judging by the frequent reports from the Special Service Squadron.[28]

When Lt.(jg) Henri H. Smith-Hutton was a Japanese language student in Japan in the late 1920s, he was given the task of accompanying Lt. J. J. Ballantine, an aviator on the staff of Commander in Chief, Asiatic Fleet, while Ballantine, who didn't speak Japanese, inspected Japanese naval air stations over a two-month period. Before the pair visited an air station, they studied previous reports to see what information might have been omitted. Then, after the visit, they would compare the reports with what they had seen and update them accordingly. Ballantine was well liked by the Japanese naval aviators, and he reported that they were good pilots. As was proven later, such reports by experienced observers did not receive appropriate distribution. The aviators of the U.S. Navy had to learn in actual combat that the Japanese naval aviators were very good, as were their aircraft and weapons.[29]

To take advantage of the collection potential of officers visiting foreign countries, the following procedures prescribed in *Navy Regulations* were repeated in the *ONI Intelligence Manual* (ONI-19) in 1933:

> No officer of the Navy or of the Marine Corps shall proceed to a foreign country on special duty connected with the service except under orders proposed by the Bureau of Navigation or by the Major General Commandant of the Marine Corps as the case may be. A copy of each such order shall immediately be filed in the Bureau of Navigation and in the Office of Naval Intelligence. The Office of Naval Intelligence shall in each case prepare a letter for the Secretary of the Navy's signature informing the Department of State of the intended visit and the general nature only of the duty on which the officer is to be sent, in order that the diplomatic representatives of the U.S. in the countries to be visited may be informed in regard thereto. The written official report made by such officer with respect to this mission shall be transmitted by him to the Office of Naval Intelligence for further reference and ultimate file.[30]

Navy Regulations also required that "the Naval Constructor shall examine all foreign ships and naval establishments he may be permitted to visit, and shall make detailed reports thereon to the Office of Naval Intelligence of all matters that may be of interest in his particular branch."[31]

In 1933, the sources available for collection of intelligence information for the Navy Department were considered to be personal observation and studies by naval attachés, intelligence officers of naval units, and individuals or unofficial agencies cooperating with ONI; intelligence furnished by the Army's Military Intelligence Division; intelligence furnished by other government departments, primarily State, Commerce, Justice, and Treasury (Coast Guard); foreign and U.S. Government documents; foreign and domestic professional periodicals; foreign and domestic press; foreign and domestic charts, maps, etc.; and merchant marine officers and personnel.[32]

In the mid-1930s, some of the principal sources for ONI's Far East (OP-16-B-11) were reports from Marine Corps intelligence officers stationed in China. Pertinent reports on Japanese-controlled islands in the Pacific were also submitted by overseas units of the Marine Corps.

The results of reconnaissance during cruises by units or squadrons of the Asiatic Fleet provided monograph data on China, Japanese ports, Russian Pacific ports, and other points of importance in the Far East. The Far East Desk requested extensive photo and hydrographic intelligence covering the approaches, harbors, beaches, and installations at Japanese Mandate Islands, such as Truk, Wotje, and the Palaus, from destroyer *Alden* (DD 211) during the ship's passage from the United States to Asiatic waters, and the resulting reports revealed an early grasp of the extent of intelligence information needed for amphibious operations.[33]

In 1936, the Japanese Combined Fleet anchored in Tokyo Bay, and hourly trips by launches were made around the fleet to permit the public to get a close look. The Naval Attaché, Tokyo, did not miss the opportunity to collect information on the Japanese navy and sent all his officers, singly, to make the launch trip. Cameras were not allowed, but a refreshment stand on the dock had packets of photo postcards of Japanese navy ships for sale to the public. A complete set, which included several ship pictures not previously held by the naval attaché, was purchased.[34]

A request was sent by the Western European Desk of ONI to light cruiser *Boise* (CL 47) to supplement monograph material during the ship's visit to Monrovia, Liberia, during the fall of 1938; a similar request was sent to light cruiser *Honolulu* (CL 48) to obtain information on the possible use of Madeira and the Azores for aviation purposes.[35]

Reports were also received from individual officers attached to other bureaus of the Navy Department. The reports indicated that the gathering of intelligence often depended upon the initiative and foresight of these officers. Awareness of the value of intelligence reports was also displayed by individual Americans traveling or residing in Japanese territories.[36]

As commander of a naval task force that visited Trinidad in 1938, RAdm. Walter S. Anderson drew up an intelligence report on Trinidad and its harbor and submitted it to ONI. Such reports were supposed to be made by officers of ships whenever the opportunity arose.[37]

In 1939, a secret CNO memo stated:

> A real undercover foreign intelligence service, equipped and able to carry on espionage and counter-espionage, does not exist. Compared with the organization and activities of foreign nations, this lack on the part of ONI is recognized as a distinct weakness. Naval Intelligence is spending nothing. The amount of intelligence received is in direct ratio to the amount of money made available, and spent. . . . The lack of a real undercover intelligence service, in the foreign field, is considered a serious defect that should be remedied.[38]

On 6 September 1939, the CNO ordered Commander Atlantic Squadron RAdm. Alfred W. Johnson (a former Director of Naval Intelligence) to set up a neutrality patrol for the Atlantic Ocean. The initial orders stated, "At earliest practicable date, establish combined air and ship outer patrol for observation approximately along the line east from Boston to latitude 42°30′N, longitude 65°W, thence south to latitude 19°N, then around eastward outline of Leeward and Windward Islands to Trinidad. Observe and report in code, movements of combatant vessels of nations in state of war."

The first major breach in "impartial" neutrality was made when the movements of German merchant ships were required to be reported and those of the Allied nations were not. In mid-December 1939, heavy cruiser *Tuscaloosa* (CA 37) and other ships trailed the German merchant vessel SS *Columbus* out of Vera Cruz and picked up survivors when the German ship was scuttled upon being intercepted by a British destroyer.[39]

A worldwide network for the surveillance of Japanese merchant ships was in effect in 1939. Included in the network was a coast-watcher service in China under the direction of the Assistant Naval Attaché, Shanghai. The worldwide collection effort was under the supervision of Cdr. Arthur H. McCollum, head of ONI's Far East Section.[40]

As Director of Naval Intelligence in 1940, RAdm. Walter Anderson was very much aware of the need for a secret intelligence service, especially in Mexico and South America. In December 1940, he hired Warren B. Phillips, an American businessman, to develop a covert intelligence collection organization. Phillips was authorized to establish an office in New York as a "representative of the DNI in matters relating to its Foreign Intelligence Service." Anderson felt it would be desirable for Phillips to be in close contact with Maj. F. D. Sharp, head of the U.S. Army Military Intelligence Division's New York office. Phillips learned that Sharp had office space and would be willing to have Phillips move in. On 11 December 1940, BGen. Sherman Miles, Army Assistant Chief of Staff for Intelligence, agreed to a request from Anderson to permit the arrangement.

By 6 February 1941, Phillips had two offices, one with Sharp at 1260 Sixth Avenue and the other at District Intelligence Office, 3rd Naval District (DIO-3ND). Phillips now had the position of Representative of the Special Intelligence Service of the Division of Naval Intelligence. Initially, Phillips believed that his tasks, like Sharp's, included interviewing persons newly arrived from foreign countries and that he and Sharp should coordinate their efforts to avoid overlapping. When Capt. Alan G. Kirk took over as Director of Naval Intelligence, however, he made it the responsibility of the DIO-3ND to perform the debriefing task, and Phillips was instructed to work "solely with special agents in the field."

Next, Phillips ran into problems from Cdr. W. Vincent Astor, USNR, who had been designated by President Roosevelt in June 1940 "to coordinate the intelligence work in the New York area." Astor learned that Phillips had been in touch with a former British intelligence operator in New York, that he was selecting agents to be sent abroad whom he had hired at what Astor considered exorbitant salaries ($4,000–$6,000 per year and $10 per day plus travel expenses), and that Phillips had entire charge of expenditures from the Navy's "Secret" fund of about $100,000. Astor felt that Phillips was indiscreet, unreliable, and a social climber. He so informed the commandant of the 3rd Naval District, RAdm. Adolphus Andrews, and then the President. Andrews took the problem to CNO Adm. Harold R. Stark, who on 20 May 1941 informed Andrews: "You should know that ONI, with approval of the Secretary and others, is attempting to create what we call here a Special Intelligence Service." Stark defended Phillips's loyalty, patriotism, and integrity and concluded with the statement that Secretary of the Navy Frank Knox had determined to continue Phillips in the job. After the Office of the Coordinator of Information (OCI) (subsequently Office of Strategic Services [OSS]) was established in the summer of 1941 and the military services had decided that the new organization should take over undercover intelligence operations, Phillips and the thirteen agents he had recruited were reassigned to OCI on 15 October 1941.[41]

World War II

To expand foreign collection capabilities, naval observers, naval liaison officers, and consular shipping advisors were assigned to various foreign port cities and other focal points. Naval observers were similar to naval attachés but were stationed in selected locations for the performance of specific duties. They were not under cover, and they were accredited through the usual diplomatic channels.

Consular shipping advisors were stationed in countries that, for diplomatic, political, or other reasons, could not appropriately receive naval observers. Consular shipping advisors could be naval officers or civilians, but in accordance with an agreement between the Department of State and the Navy Department, they were required to be attached in a civilian capacity to the staffs of consulates. Consular shipping advisors were available to perform some of the duties outlined for naval observers but, to protect the position of their consulates, it was necessary that they come under the control of the senior foreign service officer at the post.[42]

The sources of intelligence information being exploited by ONI in 1941 were naval attachés; government departments (State, Treasury, Justice, Agriculture, and Commerce); journals, newspapers, and other publications; observations and information obtained from naval units visiting foreign ports; specially appointed naval missions; naval observers stationed overseas; and overt and covert agents. In the Western Hemisphere, agents were used only on special occasions, the collection of intelligence at that time being primarily an FBI responsibility. Outside the Western Hemisphere agents were employed for covert activities, and, as previously mentioned, an office was operating in New York to maintain contact with private individuals and firms.[43]

Following the commencement of hostilities in December 1941, "target squads" in the continental United States began the collection and evaluation of enemy technological developments and personnel. These squads were organized by certain district intelligence offices to obtain information in the files of subdivisions of federal, state, and local government offices.[44]

A standard procedure was set up to interrogate Army, Navy, and Marine Corps officers returning from areas about which information was either limited or inadequate.[45] Excellent intelligence information was also obtained from individuals in the United States who had been to or had lived in foreign countries of interest. The collection effort was also carried out by each district intelligence office, and the results were forwarded to ONI.[46]

In August 1942, two submarines carried Marine Corps Col. Evans F. Carlson, LtCol. James Roosevelt, and the 2d Marine Raider Battalion, a small raiding force, to Makin Island to harass the Japanese garrison and conduct reconnaissance. "Carlson's Raiders" returned to Pearl Harbor with many captured documents, including plans, charts, battle orders, and one top secret map that provided the air defense capabilities of all Japanese-held Pacific islands, the strength of the air forces on them, and the forces' radius of operations, methods of alert, types of aircraft, and operation plans for future emergencies. It was assumed that the Japanese would immediately change their plans, since they would know the plans had been compromised, but they did not do so until the Iwo Jima landings in 1945.[47]

In 1941, ONI had decided to establish the post of U.S. Naval Observer, Suez, to keep the Navy Department informed of happenings in that port and to act in a service capacity to American merchant ships discharging there. The Red Sea had been opened to American ships by a presidential Executive Order of 11 April 1941, and the first ship arrived at Suez on 4 July. Orders were issued on 8 October for a lieutenant (jg), U.S. Naval Reserve, and a second lieutenant, U.S. Marine Corps Reserve, to proceed to Suez. The naval officer arrived on 7 December.

The activities of the Suez office mainly involved compiling reports on current happenings of naval and military interest and rendering service to American merchant ships. After the establishment of the U.S. Naval Armed Guard on board merchant vessels, the naval observer was required to make salary payments to the Navy personnel.[48]

On 27 March 1942, the title of the Suez post was changed to U.S. Naval Liaison Office. All the intelligence that emanated from the office was obtained from the British authorities or with their assistance. Close contact was maintained with other U.S. naval activities in the Middle East. In that area, the Joint Intelligence Committee, Africa/Middle East (JICA/ME), acted as a clearinghouse and exercised authority over intelligence matters. Intelligence reports to ONI were sent via JICA/ME. At least once a fortnight an officer carrying mail from Suez went by car to Cairo, 90 miles away. On those trips, visits were made to JICA/ME, the naval attaché office, and many other offices in Cairo, both American and British.[49]

On 14 April 1942, the destroyer *Roper* (DD 147) sank the German submarine *U-85* just north of Cape Hatteras. There were no survivors, but the recovery of the notebook of an engine room rating gave particulars on the submarine's construction, its operations, and the arrangement of the engineering spaces. During the salvage efforts, an officer from the Special Intelligence Section of ONI and a British officer temporarily assigned to the section noticed certain discharges from the hull that were one of the first clues to what was subsequently identified as the "submarine bubble target"; it was a device used by the Germans to confuse sonar operators by creating a turbulence to give a false submarine echo. Details of the device, learned while it was still in an experimental stage, were subsequently obtained through prisoner-of-war (POW) interrogations.[50]

On 16 June 1942, the ONI Special Intelligence Section (OP-16-Z) was designated as being responsible for the dissemination of information derived from examination and analysis of captured enemy naval equipment, as well as for the control and disposition of the equipment itself. On 25 June 1942, a directive establishing the procedure for processing such material was issued by the VCNO to the naval districts and forces afloat. Working relationships were developed with the technical bureaus of the Navy Department, the Army, the Marine Corps, and the Office of Economic Warfare. Subsequently, the Joint Chiefs of Staff (JCS) directed the establishment of a centralized enemy equipment control organization, similar to that set up by the Navy, for the Army and the Navy. On that basis, a complete and rapid exchange of information between the services in the field of technical intelligence exploitation was effected.[51]

The Japanese repatriation transport *Asama Maru*, with North and South American diplomats and other repatriates from Japan, Hong Kong, Saigon, and Singapore, arrived in Lourenco Marques in Portuguese East Africa on 23 July 1942 to exchange passengers with the Swedish liner *Gripsholm*, which brought Japanese repatriates from the United States.

Upon his arrival at Lourenco Marques, former U.S. Ambassador to Japan Joseph C. Grew, a repatriation passenger on *Asama Maru*, received a secret State Department message from the local American consul stating that one of the Chilean newsmen arriving on the ship who was to leave on the *Gripsholm* was known to be carrying a roll of 35mm film. The film had been given to him by the Japanese military and contained pictures of Japanese triumphs in Saigon, Singapore, the Philippines, and the Dutch East Indies. The pictures were to be used to illustrate Japanese propaganda stories in South America. The message further stated that the Navy Department had agreed that Cdr. Henri Smith-Hutton, the former Naval Attaché, Tokyo, who was also among those being repatriated, was to be directed to search for the Japanese film and bring it back to the United States. Smith-Hutton was authorized to select only one officer to help him. He suggested Marine Corps Maj. Gregon Williams, who had been Assistant Naval Attaché, Shanghai.

After the *Gripsholm* departed Lourenco Marques en route to Rio de Janeiro, the Chilean newsmen were identified and found to be occupying a stateroom well below decks. Their daily routine was observed, and plans were made for an undetected search of their room and luggage. Fortunately, it was hot below decks, and the newsmen spent most of their time topside and were seldom in their stateroom during the day. Smith-Hutton's wife was detailed to keep an eye on the men while her husband and Maj. Williams took turns searching and standing guard in the passageway. On the second attempt, Williams found the roll of film, and Smith-Hutton took it to the United States as directed.[52]

The North American Desk (OP-16-FN) was established in the Intelligence Branch of ONI on 20 March 1943 and was given cognizance over intelligence activities (as opposed to counterintelligence) within the continental United States and Alaska. Every naval district contained valuable information on foreign countries. Each district intelligence officer set up a Foreign Intelligence Section to coordinate the collection of all intelligence of value to naval operating forces.

In September 1943, work on a "Contact Register" was begun. A record of all sources of information was received from the naval districts and filed in OP-16-FN; each source was listed on an 8-by-10.5-inch form that showed his identification and gave data on area(s) and/or subject(s) of knowledgeability, with one copy filed by name and one or more by area, as appropriate.[53]

A group operating under VAdm. H. Kent Hewitt, Commander Naval Forces, Northwest African Waters (COMNAVNAW), had a specific mission in Italy: "to use investigative experience amplified by language background to obtain information of immediate technical and strategic importance which was not currently available through established sources."[54]

Two teams of naval intelligence officers landed near Gela in Sicily with advance Army combat troops on 10 July 1943. They moved west along the coast, surveying the ports of Gela and Porto Empedocle. At Agrigento, they located the Italian naval headquarters building. The find was reported to the commander of the Advanced Bases Group at Licata. The senior naval intelligence officer urged that a thorough search of the Italian naval headquarters building be made. Permission was granted to send out a reconnaissance party, but the group was instructed to open up the port of Porto Empedocle first.

The reconnaissance party, consisting of LCdr. S.A.D. Hunter and Lts.(jg) George G. Brownell and Paul Alfieri, finally reached their objective. They found the headquarters to be the nerve center for the entire southern coast of Sicily. The quantity of classified documents, both in cabinets and desks, was tremendous. It took two days and three nights to screen the material. Three safes, when finally opened, contained safe routes through Italian minefields for various ports and the Straits of Messina, secret code books, recognition signals, and demolition plans for ports and cities, including Porto Empedocle, Messina, and Palermo.

After the group's efforts to get transportation for the mass of material were unsuccessful, they carried it on foot in several trips to Porto Empedocle, a distance of about 4 miles. Although the commander of the Advance Bases Group still didn't recognize the importance of the documents, he arranged to have those considered by others to be most important put aboard the amphibious force flagship *Biscayne* (AGC 18), lying off Gela. The chief of staff of the task force commander, RAdm. Richard L. Conolly, was so impressed by the sampling of documents that he ordered a destroyer into Licata to pick up the rest and take the material to VAdm. Hewitt at Algiers. Instead, however, Commander in Chief, Mediterranean (CINCMED) Adm. Cunningham, RN, had the destroyer put in at Malta. From there, the documents were flown to London for processing by the British Admiralty.

Fortunately, some of the material of immediate importance to the U.S. Army had been delivered directly to Gen. George S. Patton's advance intelligence officer. He was quick to realize its value and had the documents translated, duplicated, and distributed to U.S. Seventh Army and British Eighth Army units.[55]

A considerable number of German naval documents were also captured by U.S. Navy intelligence officers at the headquarters of the Third *Schnellboots* Flotilla at Porto Empedocle and Agrigento, Sicily, between 10 and 14 July 1943. The documents formed part of a collection of German and Italian papers that was taken directly to Malta aboard a U.S. warship on orders of CINCMED. Some materials were returned to U.S. naval authorities and were used as the basis for reports by COMNAVNAW on S-boat (from the German word for "fast boat," or motor torpedo boat) characteristics, organization, operations, personnel, discipline, lack of Nazi indoctrination, tactics under air attack, etc.[56]

The German submarine *U-505* was captured by the destroyer escort *Pillsbury* (DE 133) on 4 June 1944 with the help of *Chatelain* (DE 149) and aircraft from the escort carrier *Guadalcanal* (CVE 60). Coming on the eve of OVERLORD (the invasion of mainland Europe through Normandy), the capture of the *U-505*, which had everything from acoustic torpedoes to the most secret German code books and tactical publications aboard, proved one of the war's major windfalls for Allied intelligence. Fortunately, the capture did not cause the Germans to change their codes at that critical time.[57]

The increased number of sunken Japanese ships made it desirable to equip teams for light salvage work and diving. One of the most successful hauls from a sunken ship was made by a Seventh Fleet team exploiting the Japanese heavy cruiser *Nachi*, which had been sunk in Manila Bay by Third Fleet carrier aircraft in November 1944. Carrier pilots, Philippine guerrillas, and Japanese prisoners who saw the vessel go down furnished rough fixes on the *Nachi*'s position, which was finally located by echo ranging in 93 feet of water. Many hydrographic charts found aboard the ship carried annotated locations of minefields and defenses. There were also secret plots, diaries, ship's logs, blueprints, technical documents, and volumes on Japanese doctrine and tactics. Most important among the captured papers were fleet operation plans and orders dating back to before Pearl Harbor.

Intelligence centers commenced early in the war to organize teams to follow up all landings and important operations. The teams normally included specialists in naval and aviation material, bomb and mine disposal, and the appropriate languages. The invasion of Saipan alone produced 27 tons of Japanese documents, and the Crash Intelligence Section of the Saipan intelligence exploitation team found 23 Zeke fighters, most of them in flyable condition, as well as 30 aircraft engines and 300 boxes of spare parts.[58]

During the first six months of 1944, approximately 130 large cases of Japanese documents were received by ONI from the Joint Intelligence Center, Pacific Ocean Areas (JICPOA). In addition, documents were received by OP-16-FE (Far East) from other sources, such as the Naval Research Laboratory, the Hydrographic Office, and the Air Intelligence Group (OP-16-V), for translation from Japanese into English. The documents were from areas such as the Marshalls, Gilberts, Kiska, and Guadalcanal and included blueprints of Japanese equipment such as the latest airplanes, carburetors, and radars; also examined were Japanese charts, ship logs, war diaries, field manuals, and a code book.

The backlog of untranslated material accumulated so rapidly that, in May 1944, approximately twenty recently arrived graduates of the Navy School of Oriental Languages were brought into ONI's Washington offices on temporary duty to work solely on the accumulated documents.[59]

In September 1944, thirty additional language officers, mostly WAVES (Women Accepted for Volunteer Emergency Service), were assigned permanently to the OP-16-FE translation unit. By February 1945, eighty-one officers (including WAVES), nine enlisted personnel, and five civilians were assigned to processing, translating, evaluating, and disseminating captured Japanese documents. Even these personnel were insufficient in number to handle the task.[60]

A directive issued by ONI on 23 June 1945 advised of the establishment of the Captured German Document Center, run by the Army, and assigned eight officers, nine yeomen, one civilian analyst, and two clerk-stenographers to help process the sudden influx of large quantities of German documents taken prior to the official German surrender on 8 May 1945. The processing of documents by the center included receiving, recording, summarizing, indexing, and disseminating. The index proved to be a valuable aid to the Bureaus of Ships, Ordnance, and Aeronautics and to the Hydrographic Office, as well as to various other groups and agencies interested in research and historical projects.[61]

Representatives of the Navy and War Departments, the British War Office, the Australian and Canadian armies, and the various Pacific theaters of operation attended the Japanese Document Conference held from 28 December 1944 to 15 January 1945. This group proposed that the Washington Document Center (WDC), although a joint-service section, be placed under the Director of Naval Intelligence, who would be "solely responsible for its operation." Upon official approval of the plan, the Director of Naval Intelligence established the WDC (OP-16-WDC) by his letter OP-16-Z serial 197916 of 14 February 1945. The WDC was located at the Steuart Building at Fifth and K Streets, NW, Washington, D.C.

The mission of the Washington Document Center was to serve as a central agency for the handling of captured Japanese documents. Its duties were to receive from theater document sections all Japanese documents of intelligence value after theater exploitation, to pre-scan and sort documents by general subject categories, to assign documents to Army and Navy translation agencies, to maintain the minimum records necessary for efficient and expeditious receipt and distribution, and to maintain close working liaison with the Pacific Military Research Section (PACMIRS) and the Far East Section of ONI.

On 29 August 1945, a second Japanese Document Conference was convened at which the representatives proposed that the translation sections of PACMIRS and ONI's Far East Section be consolidated with the Washington Document Center and that an advance echelon of WDC be established in Japan. A library of seized Japanese documents was to be established at the WDC. The first component of the advance echelon arrived in Japan in November 1945, composed of Army and Navy specialists "trained and briefed in current Washington interests to insure high quality of documents to be evacuated." During the period 4 March 1945 to 21 October 1945, the WDC received, processed, and disseminated 146,324 Japanese documents ranging from calling cards to encyclopedia sets.[62]

In the postwar ONI plan of October 1945, the North American Desk, with its files, functions, and personnel, was taken out of the Foreign Branch and shifted to the Domestic Branch as OP-23D3, the Contact Register. Arrangements were completed whereby the officer detail sections of the Bureau of Personnel (BUPERS) would advise OP-23D3 of the names of all naval officers applying for permission to travel abroad (in accordance with *BUPERS Manual*, Article H-1804) so that they could be contacted and briefed about needed information.[63]

The clearing of approaches to landing beaches was the primary function of underwater demolition teams (UDT). Beginning in 1943 in World War II, beach reconnaissance prior to the execution of amphibious landings was a secondary function of great importance to the success of the subsequent landings. UDT observations afforded the best means of substantiating or disproving prior intelligence on an area, as well as providing new information to be used in charts by the attacking forces.

In the European theater, where it was essential not to disclose in advance the location of prospective landing beaches, the demolition phases of operations coincided closely with the first assault waves. In the Pacific theater, however, demolition and extensive reconnaissance could be, and was, conducted up to seven days before a landing, there being little chance for the enemy to move in reserves from other islands or areas to reinforce threatened objectives.

Each UDT was briefed on conditions to be expected at the landing area, based on pre-assault information that was often sketchy and inadequate. An initial reconnaissance was usually necessary to determine the presence of mines, obstructions, and natural breaks in the offshore reefs before demolition work could begin. Each UDT or reconnaissance unit was debriefed as soon as possible upon return to its ship to record conditions it had observed. These included the surf at the beach, the depth of the water where the surf broke, the distance from the beach where the surf broke, the set and speed of the observed currents, the extent and type of obstacles to be found in the approaches to the beach, the gradient (slope) of the beach out to the 3-fathom line, the location of any antiboat mines, and the extent and type of any obstacles found on the beaches. An estimate was made of the team's ability to clear the approaches to the beach, and to neutralize any defensive positions that had been observed. A master chart of all beaches, incorporating the above information, was produced and provided to landing force commanders.[64]

The U.S. Navy, as part of its Atlantic antisubmarine warfare effort, established radio direction finder (D/F) stations at Amagansett, Long Island; Bahia and Belem, Brazil; Curacao, Dutch West Indies; Dupont, South Carolina; Cape Farewell, Greenland; Guantanamo Bay, Cuba; Jan Mayen Island; Jupiter, Florida; Poyners' Hill, North Carolina; San Juan, Puerto Rico; Toro Point, Canal Zone; Trinidad; and Winter Harbor, Maine. The British and Canadians also had D/F stations on the shores of the Atlantic. All of these stations were grouped into nets; each net possessed its own internal communications system with external radio links from the net control station to the main plotting centers in London, Ottawa, and Washington. All stations were able to tune immediately to any enemy submarine transmission heard by any other station. Tip-offs on enemy transmissions were also received by Navy ships that had radio D/F equipment. The main plotting center in Washington plotted bearings on each enemy submarine transmission from all receiving stations and would send a fix to the hunter-killer groups at sea.[65]

The Cold War Era

In 1949, a *Naval Intelligence Requirements-Periodic Summary* (NIRPS) was first published by ONI to furnish a secret, comprehensive, and detailed statement of the information required for the production of intelligence necessary to fulfill the "Primary Intelligence Objectives" of the Navy.[66] It was to be revised periodically; the first change was issued on 11 June 1952 as ONI Instruction 003820.21, Change 1.

To improve Navy intelligence on Soviet undersea warfare, it was proposed in early 1950 to strengthen field collection agencies in countries contiguous to the USSR that offered good "listening post" prospects; expand activities concerned with the interrogation of returning POWs and defectors, including provision for additional naval interrogators; give more guidance to naval attachés on undersea warfare (USW) intelligence requirements; furnish the Office of Special Operations, Central Intelligence Agency (CIA), with precise collection requirements pertaining to USW intelligence; increase merchant ship and commercial aircraft reporting on submarines and mines; increase the flow of Russian periodicals and documents; and train additional Russian linguists and employ more effectively those who had received language training at the Naval Academy and at the Naval Intelligence School.

As a follow-up to its presentation to the Low Board (headed by RAdm. Francis S. Low), ONI canceled its previous collection guidance on information required on foreign submarines (Chief of Naval Intelligence letter, serial 00234P32 of 30 September 1948) and issued Naval Intelligence Requirements Memorandum No. 13, *Submarine Warfare Intelligence*, which listed those items that ONI considered to be the most important to collect on foreign submarines, their ability to conduct combat operations and their vulnerability to countermeasures.

The collection of intelligence information by merchant ships was handled in 1950 by selected merchant marine personnel who were briefed prior to departure from U.S. ports by district intelligence officers and other interested agencies. When additional requirements or guidance developed after their departures, further briefings were given by naval attachés, if possible. After ports of interest were visited, merchant marine personnel were debriefed by naval attachés or observers at their next port of call or at the earliest opportunity. Upon their return to the United States, further debriefing was accomplished by district intelligence officers and any other agencies that had participated in the predeparture briefings. Liaison with the home offices of the owners of the merchant ships was maintained by the district intelligence offices.[67]

Special collection organizations that were active during the Korean War included the Field Research Unit, Far East Command, which operated covert collection agencies covering all phases of enemy activity; the Combined Command for Reconnaissance Activity in Korea, which coordinated all unconventional warfare operations in the Korean theater; and Task Force Kirkland, a guerrilla agent unit operating on the east coast of the Korean peninsula from the bomb-line to Wonsan. By 1952, however, the concentrated efforts of the North Koreans to tighten their security considerably reduced the effectiveness of friendly agents, and photo reconnaissance became the primary source of intelligence information for the operating forces of the Navy.[68]

Directives issued by Army Intelligence (GHQ-G2) controlled the handling and distribution of captured North Korean documents. They made no provision for the distribution of documents of naval interest to naval commands. The lack of linguists in the Navy, however, and the scarcity of documents of value to the Navy made the prescribed procedures satisfactory as long as such conditions continued.[69]

The emphasis in intelligence collection in the 1950s shifted from a general effort to a primary concern for information on Sino-Soviet-Satellite bloc naval subjects, although information on other areas was not precluded. This change is apparent in the collection guidance contained in ONI Instruction 003820.5 of 1950 and the guidance in ONI 49-2A of January 1960.

While military assistance advisory groups (MAAG) and naval missions were recognized as being in favorable positions for collecting intelligence information, their personnel were not intended to make such reports officially. Based on a JCS directive, however, unified commanders were to ensure cooperation between MAAG personnel and appropriate attachés on an informal basis, with the attachés taking particular care to protect MAAG personnel from being disclosed as the sources for any information thus provided. The same guidance was given to naval mission personnel in OPNAV Instruction 003810.3.

In mid-July 1951, information was received that a Soviet MiG-15 fighter aircraft had been shot down and was on a mud flat in the Yellow Sea off the west coast of Korea, awash at low tide. On 15 July, Commander U.N. Blockading and Escort Force (Commander Task Force 95) was directed to recover the MiG. A task group under RAdm. Alan K. Scott-Moncrieff, RN (CTG 95.1), succeeded in salvaging the major portions of the aircraft, furnishing much valuable intelligence on the characteristics of the fighter.[70]

Night amphibious landings by Republic of Korea troops were conducted from time to time on the northeast coast of Korea behind enemy lines under the operational control of CTF 95. The primary mission of the landings was the capture of prisoners and documents for intelligence exploitation purposes. Commander Amphibious Forces, Far East (CTF 90) made a fast attack transport available to CTF 95 on request to train, transport, and land the raiding parties. Four such landings were made in the first half of 1952, two by raiders from the fast transport *Wantuck* (APD 125) and two by commando units transported by *Horace A. Bass* (APD 124).

Although some landings failed completely in their missions and several had to be canceled due to heavy surf or low visibility conditions, the commando ability of the landing parties steadily improved, and each landing provided experience that increased the effectiveness of the operations. Actual accomplishments achieved by raiding party landings, however, had not been impressive as of mid-1952, considering the time and effort expended. Enemy vigilance, early detection, and opposition forced early withdrawal of the raiding parties in some cases. The more successful raids were attributed to better pre-raid intelligence, improved planning and coordination, greater self-confidence and spirit in the Republic of Korea Army landing party, and last, but not least, luck.[71]

The National Security Council placed the responsibility for the collection of foreign intelligence from domestic sources with the CIA. Because of the extensive naval interest in maritime matters and the capabilities of the Navy district intelligence offices, a mutually satisfactory agreement (known as the Salt Water Agreement) was reached in 1952 by ONI and the CIA: Exploitation of maritime sources for counterintelligence and security purposes was the responsibility of the district intelligence offices, without prior clearance from, or coordination with, the CIA. Exploitation of owners, operators, and agents of shipping companies for purposes of collecting intelligence was made the responsibility of the CIA. The district intelligence office could exploit such sources for intelligence purposes, subject to prior coordination with the local CIA field office, the latter arranging the interview and having the option to participate. Exploitation of masters, officers, and crews of merchant ships for intelligence purposes was made the responsibility of the district intelligence offices. Access to DIO sources by the CIA in exceptional cases was not precluded, but such special interviews had to be arranged by the district intelligence offices.[72]

In connection with the collection of information through the boarding of merchant ships, "special category" merchant ships were considered to be of three types: (a) Soviet and Soviet Satellite-registered merchant ships, (b) Non-Communist registered merchant ships under Soviet or Soviet-Satellite control (owned or chartered), and (c) Non-Communist registered or controlled merchant ships currently involved in trade with the Soviet Bloc. "Charlie" (category C) ships were considered good potential sources for information of naval interest and were given special attention by naval district boarding officers.[73]

The basic concepts on the collection of information needed to produce beach intelligence for the purpose of amphibious landings were enunciated in

1953, based largely on lessons learned in World War II and the Korean War. The "direct method" for obtaining beach intelligence was by making on-the-spot surveys of the objective beach area. Specially formed, equipped, and trained Marine Corps teams, usually working in conjunction with Navy hydrographic survey teams, performed triangulation surveys laid upon strict ground control nets. Underwater demolition teams were also used. Both UDTs and amphibious reconnaissance patrols were employed surreptitiously to get last minute answers about a beach area prior to actual landing. The "indirect method" involved analyzing aerial photographs, submarine soundings, large scale hydrographic charts, reports of interrogations of former local inhabitants and prisoners of war, and reports of visits by attachés.

In general, for the collection of beach data the beach was understood to be that area lying between the low-water line (usually, mean low-water spring tides are used by compilers of hydrographic charts as the datum plane) and the maximum point or line inland reached by the water under storm conditions. The important features of any given beach for which information was required by strategic and operational planners were as follows: its geographic location, length (both overall and usable), width (at both low tide and high water), gradient or degree of slope, material and firmness, trafficability (ability to support the movement of personnel, materials, and vehicles without special measures), obstacles (both natural and fabricated, on the beach and offshore), exits (current and potential), and beach defenses.[74]

During the 1950s and 1960s, beach survey teams from the Marine Corps Topographic Battalion conducted surveys of coastal areas in various countries including Turkey, Morocco, and Thailand. In most cases, the survey operations were conducted from survey ships with the permission of the country being surveyed.

On 20 May 1954, the CNO initiated action requiring surveillance of the Gulf of Honduras and the adjacent Caribbean area to determine the identities of ships transporting munitions to Guatemala. ONI provided and kept current a list of suspect ships. The CNO directed that any of the suspect ships attempting to enter the Gulf of Honduras were to be escorted to Panama for inspection.

Merchant ship reporting was instituted in the Caribbean, and the commander of the Caribbean Sea Frontier established air patrols that covered all entrances to the Caribbean and established surface patrols for close surveillance in the Gulf of Honduras. Five patrol aircraft squadrons, four destroyers, one attack transport, and two submarines were assigned to the surveillance duties, but the submarine patrols were soon discontinued. Numerous suspect ships were kept under close surveillance, but no ship was actually diverted to Panama for inspection. On 23 June 1954, an anti-Communist government was established in Guatemala, bringing stability to that country and relieving the danger to U.S. nationals living there. On 30 June, the surveillance was discontinued.[75]

In 1955, under guidance from the commanders of the various naval sea frontiers, merchant ships and fishing vessels were enlisted to assist in intelligence collection efforts in a program designated merchant intelligence (MERINT). Reports of aircraft formations flying toward the United States and of sightings of military vessels, submarines, unidentified flying objects (UFOs), and guided missiles were to be made to the nearest military or civilian radio station for eventual relay to the appropriate U.S. sea frontier commander.[76]

Six Sound Surveillance Stations (SOSUS) were placed in operation in Fiscal Year 1956, and a Sound Surveillance Evaluation Center was activated at the Commander Eastern Sea Frontier headquarters. A confirmation of the effectiveness of the southern segment of SOSUS was accomplished in November 1955.

On 1 February 1956, a Sound Surveillance Control Center (SSCC) was activated in the Commander in Chief, Atlantic Fleet (CINCLANTFLT) Operational Intelligence Plot where SOSUS and high-frequency radio direction-finder (HF/DF) submarine contact reports could be quickly coordinated. Oceanographic Unit operations watch personnel staffed the SSCC and assumed control of the Atlantic SOSUS on 1 February 1956. Oceanographic Unit cryptologic watch officers handled SOSUS communications through the CINCLANTFLT staff communications center.[77]

In 1956, the Hoover Commission Report on Intelligence Activities recommended "that the Navy expand its collection effort." As of March 1956, the Navy had implemented the recommendation by increasing the use of the intelligence collection potential of naval ships visiting foreign ports, improving the cooperation between military assistance advisory groups and naval attachés, and improving the guidance to collectors.[78]

By the end of Fiscal Year 1957, the Atlantic SOSUS consisted of several stations and two evaluation centers. Two additional stations were nearing completion. The Oceanographic Unit, Norfolk, personnel had been integrated into the CINCLANTFLT staff.[79]

During the 1956 Hungarian revolt, the Army and Air Force units from the Intelligence Staff (G-2) U.S. Army, Europe, and an Air Force interrogation team from Berchtesgaden were concentrated along the Austro-Hungarian border in order to spot potential intelligence sources among the escapees. Also, the Army attaché at Vienna advised the Army intelligence staff in Heidelberg on the knowledgeability of any sources passing through his area. Based on the initial alerts, military intelligence teams on board ships carrying Hungarian refugees to the United States made out preliminary interrogation reports that were delivered to the local military intelligence officer at Camp Kilmer, New Jersey. Phase III interrogations took place in the United States on the basis of inter-agency requirements. ONI had one yeoman at Camp Kilmer to review and obtain copies of interrogation reports of interest to the Navy.[80]

The Defector Committee (Germany) included representatives from the European Command, U.S. Army, Europe, U.S. Air Force, Europe, the Department of State, Commander Naval Forces, Germany (COMNAVGER), various covert organizations, and the Defector Reception Center (DRC). At its meeting of 25 July 1957, the COMNAVGER representatives were Capt. Richard H. Tenney and Cdr. A. R. Czerwonky.

The Defector Reception Center was the only interrogation center in Western Europe handling high-level sources who qualified as defectors. In 1958, the Navy had one officer interrogator permanently assigned to the DRC. He was administratively under the Commander in Chief, U.S. Naval Forces, Eastern Atlantic and Mediterranean Representative, Germany. His duties involved participation on committees and panels connected with the acceptance of, and interrogation-scheduling for, sources believed to be knowledgeable in matters of naval interest, the preparation of interrogation briefs based on Navy requirements and the source's knowledgeability, the interrogation of sources, and the preparation of the resultant reports.[81]

On 1 July 1957, Commander Barrier Force, Atlantic inaugurated a full-length barrier between Argentia, Newfoundland, and Lajes in the Azores, using destroyer escort radar pickets (DER) and WV-2 Warning Star aircraft. All air contacts made by the barrier patrol forces were reported by "Flash" precedence to CINCLANTFLT for evaluation.[82]

Guidance for monthly informal letters to naval attachés and other Naval Intelligence collection activities was issued on 14 February 1958 by ONI Instruction 5200.1B. The purpose of the letters, which had been sent out at least since November 1952, was to provide intelligence collection and administrative guidance. The collection guidance in the letters included a periodic and constructive overall evaluation of the activity's collection effort; such general collection guidance as would enable the collection activity to redirect its planned collection effort to subjects, trends, or developments of current interest to ONI; a current evaluation of message and "Joint Weekly Attaché Message" ("Weeka") reporting; a recapitulation of outstanding Specific Requests for Information (SRI) not yet fulfilled; and information about changes in Naval Intelligence Collection Instructions and in other general collection guidance publications.[83]

In January 1960, the Navy, together with the other military services, engaged in an unprecedented collection effort that produced a vast quantity of new data on Soviet ballistic missile firings into the Pacific. A unique portion of the information was collected by submarines. No other collection platform at that time was able to observe the detachment of the data capsule from the nose cone, its descent by parachute, and its recovery by one of the Soviet missile range instrumentation ships (SMRIS). All potential electronic intelligence (ELINT) and other forms of intelligence data on the intercontinental ballistic missile (ICBM) and the SMRIS were collected by the various units prior to, during, and after the missile's surface impact.[84]

The Joint Reconnaissance Center (JRC) was established within the Joint Staff by JCS secret memo 1107-60 (Appendix to Enclosure A to JCS 2150/51) of 26 October 1960. The JRC became operational in March 1961 as part of J-3 and was responsible for keeping cognizant of all sensitive peacetime reconnaissance for the JCS. For additional information on the JRC, see the chapters on the various types of reconnaissance.

In early 1962, the director of the Defense Intelligence Agency established policies and procedures for the submission, validation, and control of Specific Intelligence Collection Requirements (SICR). Thereafter, the use of ONI's Specific Request for Information (collection directives) was terminated. The Attaché Collateral Support Section of ONI (OP-922H1), which had levied Navy-generated requirements directly upon appropriate Navy collectors, continued to give guidance to and monitor Navy collection operations and continued to serve as the screening section for requirements generated within ONI and the Navy Hydrographic Office. Requests for information from any other agency or office within the Navy Department were treated as requests for research and were routed to the appropriate analytical section of ONI or to the Navy Hydrographic Office. A SICR was originated by the sections if they were unable to supply the requested information.[85]

The Cuban Missile Crisis

The Joint Chiefs of Staff established a blockade of Cuba on 23 October 1962. Subsequently, the JCS substituted the word "quarantine" for "blockade" to avoid the possible legal connotation of the latter as an act of war. To carry out the quarantine, Commander Second Fleet was designated by CINCLANTFLT as Commander Task Force 136, the Quarantine Force Commander.

On 23 October, destroyer-type units of Task Group 136.1 were positioned at twelve stations, 47 miles apart, on an arc of a 500-mile radius from Cape Maisi (the easternmost extremity of Cuba) from 27°30′N, 70°W to 20°N, 65°W. The stations were given the prefix code word WALNUT. The original picket line was outside the operational range of Cuban aircraft. When it was determined by air reconnaissance that the Cuban Air Force was in a poor state of readiness to launch an attack, the quarantine arc, with the approval of JCS and CINCLANT, was moved closer to Cuba but still to seaward of the Bahama Islands. The move was accomplished on 30–31 October, and the new stations were designated by the code name CHESTNUT.

An estimated average of 46 ships, 240 aircraft, and 30,000 personnel were directly engaged in the effort to locate and identify ships involved in the Cuban trade. The majority of the 200 ships identified as being of interest to quarantine control were initially intercepted by aircraft. Once it was determined that a ship was of interest, a surface unit was vectored to intercept it.

In addition to Task Force 136 surface units, five destroyer-type ships assigned to Commander U.S. Naval Forces, Key West intermittently participated in quarantine operations near Havana, in the Old Bahama Channel, and in the southern approaches to the Florida Straits. Commander Caribbean Sea Frontier conducted similar operations with destroyers in the Windward Passage and along the southern coast of Cuba.

Between 24 October and 4 November, many suspicious Soviet ships turned back when intercepted and never did get to Cuba. Others with nonsuspicious cargo slowed or stopped, apparently waiting guidance from the Kremlin. In due course, the latter ships proceeded to Cuban ports.

On 26 October, the Lebanese cargo vessel SS *Marucla* was boarded by a party from the destroyers *John R. Pierce* (DD 753) and *Joseph P. Kennedy Jr.* (DD 850) to inspect cargo and papers. No prohibited material was found, and the ship was cleared to proceed.

Between 5 and 11 November, the quarantine force was involved in checking on nine Soviet ships that the Soviet delegation in the United States had identified as being the ones that would carry Soviet missiles out of Cuba. The Soviet delegation had provided the dates of departure and the number of missiles to be carried on each of the nine ships. The United States, in turn, designated three rendezvous points where U.S. Navy ships would carry out the agreed-on inspection of the Soviet merchant ships involved. The Soviet ships, however, made no apparent effort to pass through the rendezvous points, nor did they depart from Cuban ports on the dates that had been specified. Consequently, the air and surface search effort continued in order to find and identify the designated ships. Aerial and surface photography and visual observation were used to verify the number of missiles being withdrawn from Cuba. Eventually, all nine Soviet ships were located, and the forty-two Soviet missiles to be removed from Cuba were counted.

Between 11 and 21 November, several ships were trailed and six others were designated as being of special interest, but no offensive weapons were detected on the ships intercepted and photographed during the period. The quarantine was lifted and Task Force 136 was dissolved on 21 November.

Concurrent with the lifting of the quarantine, President John F. Kennedy announced that the Soviet Union had indicated that all forty-two Il-28 Beagle light bombers in Cuba would be removed within thirty days. Surveillance operations were carried out during 1–6 December to verify their removal by ship.

The appearance of the Soviet naval auxiliary *Terek* in the Western Atlantic on 18 October 1962 became a matter of prime antisubmarine warfare concern because it was thought that the ship might be supporting submarines. Soviet fishing trawlers were also active in the Western Atlantic and were kept under surveillance throughout the Cuban crisis. The Argentia Submarine-Air Barrier was established on 24 October 1962 to detect any Soviet submarine activity in the area and consisted of seventeen patrol (VP) aircraft and ten U.S. Navy submarines, assisted by Canadian antisubmarine forces. Flight operations reached 120 hours per day. The barrier was terminated on 13 November. During the crisis, two Soviet Foxtrot-class submarines were caught on the surface, and one of them was forced to surface after thirty-five hours of continuous sonar contact. Naval Air Reserve units logged about 350 hours of surveillance on a voluntary basis, and reported over 190 different surface and subsurface contacts.[86]

The Vietnam War Era

On 14 February 1963, information was received from the U.S. naval attaché at Santo Domingo in the

Dominican Republic and the U.S. Coast Guard that members of the leftist FALN organization had seized the Venezuelan motor ship *Anzoategui*, which had departed from La Guaira, Venezuela on 12 February 1963. After the government of Venezuela requested assistance, the U.S. Joint Chiefs of Staff directed that a search be conducted for the ship. The estimated position, south of the Dominican Republic, indicated that the *Anzoategui* might be headed for Cuba. A surface and air barrier was established from Puerto Rico west to the Yucatan Channel by two aircraft carriers and twelve destroyer-type ships. Search aircraft from land stations and the aircraft carriers covered the water areas from the Lesser Antilles west to the Yucatan Peninsula. The *Anzoategui* was sighted by aircraft on 16 February east of Trinidad and was thereafter kept under surveillance by aircraft, the Navy-owned, civilian-manned oceanographic research ship *Josiah Willard Gibbs* (T-AGOR 1), and two U.S. Navy destroyers from San Juan until the *Anzoategui* was under control of the Brazilian navy on 20 February.[87]

The Monthly Informal Letters to Navy intelligence collection activities were changed to Quarterly Informal Letters in February 1963. The change was necessitated by the establishment of the Defense Intelligence Agency and the resultant reduction in personnel in the production and management sections of ONI. Additionally, to provide timely guidance, a monthly newsletter series was started that contained items applicable to the entire Navy collection system and not just to individual collectors.[88]

Navy participation in the activities of the Defector Reception Center at Frankfurt in 1963 was managed by the CINCUSNAVEUR Representative, Germany, who served on the nominating panel and voted on the acceptance of sources as being of primary interest to the Navy. The representative would then assign a case officer to exploit the accepted source. The case officer prepared the Information Reports (IR) based on the collected information and submitted the reports to the DRC for publication as a DRC report. An advance copy of an IR was also to be sent to ONI by the case officer if the content justified independent circulation.

Because of its high selectivity and worldwide operation, the DRC exploited a fairly constant flow of defectors. The quality of information yielded by the DRC operations was maintained at a high level. Although REPGER used the center without directly contributing material, funds, or personnel, there were many disadvantages to not having people assigned exclusively to the DRC. Deficiencies in foreign language resources precluded direct exploitation of the more valuable defectors (unless they spoke German) and required the use of interpreters unfamiliar with naval or maritime subjects and terminology, a situation that usually introduced errors or omissions in the translations.

Defense Intelligence Agency officers who visited the DRC during an attaché conference in Frankfurt in September 1963 reported, "The continuous failure of the Service Intelligence Agencies in Europe to recognize the extreme importance of this very small but highly qualitative effort at the DRC is most discouraging, but not new."

The DIA report also stated that the Navy effort was even more sporadic and perfunctory. The Navy had a total of five officers in Frankfurt to handle all intelligence, counterintelligence, security, clandestine, and interrogation functions. None of the five, or their staff, spoke Russian, and whenever a source with knowledge on naval subjects appeared in the DRC, the Navy had to borrow personnel from other agencies to assist in the exploitation effort.[89]

In 1964, the critical situation developing in Southeast Asia resulted in a significant shift of both air and surface intelligence collection activity to that area. On 17 December 1964, the Navy commenced armed reconnaissance missions over Laos on a continuing basis. Also in December, a significant number of air reconnaissance operations were conducted by Seventh Fleet air units over the Republic of Vietnam to satisfy Commander U.S. Military Assistance Command, Vietnam requirements. For more details, see Chapters 4 and 5.[90]

During 1966–1968, the South Vietnamese navy conducted an intelligence collection effort in Cambodia. Early in 1967, agents of the collection net reported that Sihanoukville was the transshipment point through which supplies were being infiltrated to the Viet Cong in South Vietnam. The material was off-loaded in Sihanoukville and shipped by truck to Viet Cong areas along the Vietnamese-Cambodian Border. The information was reported to Washington by Commander Naval Forces, Vietnam in numerous Intelligence Information Reports (IIR), but intelligence reports on Viet Cong activities in Cambodia were discredited in Washington in deference to Prince Sihanhouk's claim that Cambodia was neutral and not involved with the Viet Cong or the North Vietnamese.

In the spring of 1968, a submarine was stationed off Sihanoukville to check on merchant ships entering port. The rules of engagement, however, required the submarine to remain 15 miles off the coast. A number of Chinese Communist ships were identified; from that distance, however, it was not possible to determine that the ships had actually entered port at Sihanoukville.

Thus, "directed intelligence" for political purposes hindered collection and exploitation of intelligence. In Vietnam, there was a tendency to exaggerate the results of offensive operations. There was also great pressure from Washington to provide statistics on trucks destroyed, facilities hit, board-and-search operations conducted, body counts, etc. The data were often inflated to include speculated rather than confirmed results.[91]

The Navy supplied gunfire support on a lavish scale to friendly forces in South Vietnam, but there wasn't any proof of the effectiveness of the bombardments since there was no feedback from the "customers." Targets were also hit in North Vietnam, but again there was no post-strike photographic reconnaissance on gunfire targets to show the results. Spotting aircraft, usually an A-1 Skyraider whose pilot was equipped with binoculars, could sometimes provide damage assessment, but based on Korean War experience, the results were probably overenthusiastically reported.[92]

On 22 August 1969, a staff study on the advisability of the establishment of a collection operations management plot (COMP) as a part of the Operations Coordination Branch of the Intelligence Collection Division of the Naval Intelligence Command was initiated. Capt. Earl F. Rectanus was head of the division, and LCdr. Charles A. Peterson was head of the Operations Coordination Branch. Their study was completed and forwarded on 5 September 1969. On 19 September the COMP was established in Room 5D718 in the Pentagon. On 10 October 1969, the Intelligence Collection Division and the Operations Coordination Branch moved to Rooms 1200 and 1280 in the Hoffman Building in Alexandria, Virginia.

Collection by Navy operating forces in 1970 included visits to foreign ports by U.S. Navy and Coast Guard ships, and observations of Soviet ships at sea such as the helicopter-carrying cruiser *Moskva* in the Mediterranean; Soviet Pacific Fleet units in the Indian Ocean; transits by Soviet Northern Fleet units to west of Gibraltar to rendezvous with the *Moskva* group; units embarked on the Northern Sea Route transit (an annual observation in the Bering Strait in late August–early September); the Soviet worldwide exercise OKEAN; and task groups en route to and operating in the Caribbean Sea.

Excessively slow processing of Navy Specific Intelligence Collection Requirements (SICR) for electronic intelligence by the Defense Intelligence Agency resulted in missed opportunities to collect against requirements that might otherwise have been satisfied during Soviet naval operation OKEAN in 1970. Discussions on the collection problems were held with BGen. Daniel Graham, U.S. Army (DIACO), on 23 April 1970 and with Capt. Carr (DIACO-1) and Col. Hezlep (DIACO-1D) on 29 April, resulting in DIA actions to expedite the processing of ELINT SICRs.[93]

In October 1971, Commander Naval Intelligence Command (COMNAVINTCOM) hosted the 3rd Inter-American Naval Intelligence Conference. Eleven Latin American navies were represented by heads of their naval intelligence organizations or other senior intelligence personnel. A revised collection agreement was approved that was subsequently ratified by nine of the countries. The conference greatly stimulated the level of active intelligence collection cooperation, particularly by Brazil, Uruguay, Argentina, and Venezuela, and it improved U.S. bilateral relationships with each of the countries.[94]

Many friendly foreign navies maintained surveillance operations in areas where there were few, if any, U.S. Navy intelligence collection resources. In an effort to optimize the worldwide database of the Ocean Surveillance Information System (OSIS), numerous initiatives were made with foreign navies to exploit their ocean surveillance capabilities and to enhance the quality of their reporting. In addition, scientific and technical intelligence information was exchanged with a number of countries on a bilateral basis.[95]

A separate, more secure room was authorized for the Naval Intelligence Command's Collection Operations Management Plot in Fiscal Year 1971. Early in Fiscal Year 1972, construction of spaces, displays, and communications for the plot were completed. The COMP thereupon became a 24-hour alert center serving as COMNAVINTCOM's point of contact for coordinating and supporting timely collection operations worldwide.

During Fiscal Year 1972, a collection opportunity (COLOP) management system that had been initiated the previous year was further developed and expanded. The system was designed to correlate collection requirements with resources and to coordinate resource applications against specifically identified opportunities.[96]

A *Shipboard Intelligence Officer's Collection Guide* was developed during Fiscal Year 1972 to permit collateral duty intelligence officers aboard ships to identify specific collection requirements quickly relative to broad collection targets and to provide guidance on the appropriate collection resources to be applied. During Fiscal Year 1973, the *Guide* was in use on all Pacific Fleet units and on some Sixth Fleet units.

Other guidance for shipboard collateral duty intelligence officers was published by the Naval Intelligence Command in the *Fleet Intelligence Newsletter*. It provided informal guidance on intelligence procedures and applicable changes, gave guidance for reporting, and provided general intelligence of interest to individual ships or units.

To assist fleet collection managers in identifying the high priority collection requirements applicable to specific COLOPs, the Naval Intelligence Command started a Requirements Advisory Program during fiscal year 1973 whereby the requirements applicable to each COLOP target could be identified relative to collection resources and priorities. The advisories varied in length from a single page to thirty or forty pages and covered a wide variety of subjects from port visits to major Soviet fleet exercises.

A six-month test of a Navy concept to expedite the feedback of evaluations to collectors was completed in February 1973. The value of using the new message evaluation format over the old Department of Defense evaluation form, which was sent by mail, was proven. In addition to providing timely advice and evaluation to collectors, the message format expedited cancellation of satisfied collection requirements, thus permitting earlier redirection of collection resources to other collection requirements. Based on the results of the test, the Defense Intelligence Agency took steps to modify *DIA Manual 58-2* accordingly.[97]

Collaboration with other navies in the collection of information continued during fiscal year 1973. The Naval Ocean Surveillance Information Center (NOSIC) at Suitland, Maryland, and the Fleet Ocean Surveillance Intelligence Center (FOSIC) at Norfolk, Virginia, received ship location data and other reconnaissance information. In the Pacific, the FOSIC at Commander in Chief, Pacific Fleet headquarters in Oahu, Hawaii, received HF/DF information.[98]

Chapter Notes

1. *ONI Review*, Jul 1956, 300.

2. John Adams, *The Works of John Adams, Second President of the United States* (Boston: Little, Brown & Co., 1854), 9:47.

3. "Notes on Early History of Naval Intelligence in the U.S.," *ONI Review*, Apr–May 1957, 169.

4. George R. Clarke, William O. Stevens, Carroll S. Alden, and Herman F. Krafft, rev. ed., *A Short History of the United States Navy* (Philadelphia: J. B. Lippincott, 1939), 227.

5. *ONI Review*, Jul 1956, 301; see also letters to Officers, Ships of War, 25:303, in Naval Records Collection (RG 45), National Archives (NA).

6. General Letter Book, no. 50, 286, RG 45, NA.

7. "William Henry Aspinwall," 1:396–97, and "John Murray Forbes," 3:507–8, in *Dictionary of American Biography* (Charles Scribner's Sons, 1927–1964).

8. RAdm. Albert Gleaves, *Life and Letters of Rear Admiral Stephen B. Luce, U.S. Navy* (New York: Putnam, 1925), 83.

9. *General Orders and Circulars Issued by the Navy Department From 1863 to 1887* (Washington: GPO, 1887), 126.

10. Capt. John M. Ellicott, "Three Navy Cranks and What They Turned," *USNIP*, Oct 1924, 1615–28.

11. *SECNAV Annual Report, 1888*, 175.

12. *SECNAV Annual Report, 1889*, 342.

13. Capt. Herbert E. Cocke, USN, "History of ONI," MS, Office of Naval Intelligence, 1931," 6.

14. VAdm. Alfred W. Johnson, Oral History, Columbia University, New York, NY, 1964, 36, copy in OA.

15. Cocke, "History of ONI," 17–20.

16. James R. Green, "The First Sixty years of the Office of Naval Intelligence" (M.A. thesis, The American University, 1963), 66.

17. Cocke, "History of ONI," 18.

18. Maj. John H. Russell, USMC, ltr to Director of Naval Intelligence (DNI), 7 Aug 1919.

19. Inventory, 1966, 50–51, in Records of the Office of the Chief of Naval Operations (RG 38), NA.

20. Green, "First Sixty Years," 65; and Josephus Daniels, *Our Navy at War* (New York: Doran, 1922), 259–60.

21. Navy Museum, press release, Jul 1975, on the occasion of an exhibit of Held's work.

22. ONI, "USN Activity re Armistice of 1918 and Peace Conference of 1919," Section 103 of "German Monograph," 17, 26, box 1, Job 7631, FRC/WNRC.

23. RAdm. Albert P. Niblack, USN, *The History and Aims of the Office of Naval Intelligence* (Washington: GPO, 1920), 12.

24. ONI-8, 1923, 59 et seq.

25. Adm. Royal E. Ingersoll, Oral History, Columbia University, New York, NY, 1965, 48, copy in OA.

26. Department of the Navy, "Administrative History of the Office of Naval Intelligence in World War II," 10 Jul 1946, unpublished MS, 18, hereafter *ONI WWII Admin History*.

27. Capt. Ellis M. Zacharias, *Secret Missions, The Story of an Intelligence Officer* (New York: Putnam, 1946), 104.

28. *ONI WWII Admin History*, 582.

29. Capt. Henri H. Smith-Hutton, Oral History, USNI, Annapolis, MD, 1979, 1:73–78, copy in OA.

30. ONI-19, *Naval Intelligence Manual*, 1933, para. 1011.

31. Ibid., para. 1009.

32. Ibid., para. 2008.

33. *ONI WWII Admin History*, 595–96.

34. Capt. Joseph Finnegan, MS narrative, OA.

35. *ONI WWII Admin History*, 587.

36. Ibid., 596.

37. VAdm. Walter S. Anderson, Oral History, Columbia University, New York, NY, 1963, 220, copy in OA.

38. Lecture at Naval Intelligence School, "The History of Naval Intelligence," 2 Aug 1946, 16.

39. James O. Richardson, *On the Treadmill to Pearl Harbor: The Memoirs of Admiral J. O. Richardson as told to Vice Admiral George C. Dyer, USN (Retired)* (Washington: GPO, 1973), 156–58.

40. RAdm. Arthur H. McCollum, Oral History, USNI, Annapolis, MD, 1971, 1:245–50, copy in OA.

41. Thomas F. Troy, "The Coordination of Information and British Intelligence," *Studies in Intelligence*, vol. 18, no. 1–5, Spring 1974, 82–86. Phillips was referred to as "Wallace B. Phillips" in a Col. William J. Donovan memo to President Roosevelt, dated 5 May 1942.

42. *ONI WWII Admin History*, 529.

43. Capt. Howard D. Bode, USN, memo, 10 Feb 1941, file A8-2, box 16, Job 3679, FRC/WNRC.

44. *ONI Quarterly*, vol. 2, no. 3, 1958, 11–13.

45. *ONI WWII Admin History*, 500.

46. Testimony of RAdm. Thomas B. Inglis 27 Mar 1946, in House, *Navy Department Appropriations Bill for 1947: Hearings Before the Subcommittee of the Committee on Appropriations*, 79th Cong., 2d sess., 1946, 1590.

47. Zacharias, *Secret Missions*, 318.

48. *ONI Review*, Dec 1946, 24.

49. Ibid., 27.

50. OP-16-Z, "War History," MS history, 8–9, box 16, Job 3679, FRC/WNRC.

51. Ibid., 12.

52. Smith-Hutton oral history, USNI, 1:377–78.

53. *ONI WWII Admin History*, 739–40.

54. *ONI Quarterly*, vol. 2, no. 3, 1958, 11–13.

55. *ONI Review*, May 1946, 30–35.

56. Commander Naval Forces, North African Waters ltr, 25 Sep 1943, box 7, Job 7631, FRC/WNRC.

57. Philip K. Lundeburg, "American Antisubmarine Operations in the Atlantic, May 1943–May 1945" (Ph.D. diss., Harvard University, 1953), chap. 10, 29–32.

58. *ONI Review*, Nov 1945, 41.

59. *ONI WWII Admin History*, 716.

60. Ibid., 717.

61. Ibid., 683–84, 686.

62. Ibid., 894–98.

63. Ibid., 745–46.

64. ONI, *Operational Intelligence (OPINTEL) Notes*, Aug 1945, 6–9.

65. Navy Dept. press release on "Radio Direction-Finding and the U-boat," 13 Jan 1946.

66. OP-322H (Collection and Dissemination Branch) NIRPS ltr of transmittal, 1949 (precise date missing).

67. VAdm. Francis F. Low, *Study of Undersea Warfare* (Low Report), ser 001P003, 22 Apr 1950, Annex, 2, 3, Post 1 Jan 1946 Command File, OA; and DNI ltr to Commander in Chief, U.S. Naval Forces, Eastern Atlantic and Mediterranean, 16 Jan 1950.

68. Commander in Chief, Pacific Fleet (CINCPACFLT), *Interim Evaluation Report No. 5: Korean War Naval Operations*, 5-72.

69. CINCPACFLT, *Interim Evaluation Report No. 2: Korean War Naval Operations*, 1745.

70. Commander Seventh Fleet, Report of Operations, 28 Mar 1951–3 Mar 1952, 15.

71. CINCPACFLT, *Interim Evaluation Report No. 4: Korean War Naval Operations*, 7-19.

72. ONI Instruction 003820.36A of 3 Jun 1952.

73. ONI Instruction 05550.6 of 15 May 1953.

74. *ONI Review*, Oct 1953, 450–51.

75. Commander in Chief, Atlantic Fleet (CINCLANTFLT), supplement to *Annual Report, 15 Apr 1954–30 Jun 1954*, 5–6.

76. *ONI Quarterly*, vol. 8, no. 4, 1955, 3–4.

77. *CINCLANTFLT Annual Report, FY 1956*.

78. OP-922 ltr, ser 00063P92, 12 Mar 1956.

79. *CINCLANTFLT Annual Report, FY 1957*, 30.

80. S&T Subcommittee of Intelligence Coordination Committee(G), Minutes of meeting of 13 May 1957, OP-922 H4 files.

81. Defector Files, OP-922H4 files.

82. *CINCLANTFLT Annual Report, FY 1958*, 36–37.

83. ONI Instruction 5200.2A of 8 Feb 1963.

84. OP-92B1 memo, ser 000907P92, 4 May 1960, "Annual Report to the President's Board of Consultants on Foreign Intelligence Activities."

85. ONI Instruction 3821.2 of 22 Mar 1962.

86. *CINCLANT Annual Report, 1962*, 14–19.

87. *CINCLANTFLT Annual Report, 1963*, 25.

88. ONI Instruction 5200.2A of 8 Feb 1963.

89. Defector File, OP-922H4 files.

90. *CINCPACFLT Annual Report, 1965*, 18, 53.

91. Task Force 168, Project LLIVN (Lessons Learned in Vietnam), Annex C, 3–4, 19.

92. Ibid., 23–24.

93. Naval Intelligence Command NIC-35 memos, 23 Apr 1970 and 7 May 1970.

94. DNI, "Report to the Chief of Naval Operations for Fiscal Year 1972," 45, hereafter *DNI Annual Report to CNO*.

95. Ibid., 36, 47.

96. Ibid., 18–19.

97. *DNI Annual Report to CNO, 1973*, III-2–III-4.

98. Ibid., V-6–V-7.

CHAPTER 3

Naval Attachés

Establishment of the Naval Attaché System

When the Office of Naval Intelligence was established in 1882, it was Lt. Theodorus B. M. Mason's basic concept that naval attachés had to be assigned to U.S. embassies and legations in the naval and maritime countries throughout the world and that ONI would be the office to support and coordinate their activities. Consistent with Mason's concept is the following statement in the first written directive from the Secretary of the Navy to Lt. Mason dated 25 July 1882: "In order to collect information, a corps of correspondents, in the persons of naval attachés to our foreign legations, will be organized."

To start the intelligence collection system, orders were issued to LCdr. French E. Chadwick, on 28 October 1882, assigning him to London as the naval attaché for the purpose of obtaining full and accurate information on British progress in naval science.[1] In addition to information on British naval matters, he was to retain from his prior orders of 12 July 1882, which had assigned him to Europe, instructions to collect information of use to the Coast Guard Service, the administration of the Lighthouse Service, the Coast Survey, the Hospital Service, the Meteorological Service, and the Board of Trade.[2] Information was also desired on inspection systems for steam vessels and for chain and cable. By Chadwick's prior orders, he was also authorized to contact the Russian, Swedish, Norwegian, Danish, German, Dutch, and Belgian embassies in London for details on their lighthouse services, and he was also to visit Holland, Belgium, Germany, and Sweden if the visits might yield the desired information.[3]

The choice of Chadwick as the first actual U.S. naval attaché was a fortunate one and showed not only good judgment on the part of the selector but also an appreciation of Chadwick's competence in naval science. Belonging to the small group of well-educated and scientifically learned naval officers that included such men as Capt. Alfred T. Mahan and Cdr. William T. Sampson, Chadwick exhibited a fine capability for observing foreign navies and acquiring and reporting information about them, as well as for recognizing information potentially useful to U.S. planning for naval reconstruction.

Orders similar to Chadwick's and carrying the same date were issued to Lt. John C. Soley, except that his assignment was for temporary duty in France, and he was not designated a naval attaché.[4]

As indicated in Chapter 1, the first designated U.S. naval attaché was Cdr. Francis M. Ramsay, who was given that title in 1872 while representing the Bureau of Ordnance at the U.S. legation in England. Ramsay, however, did not collect information for the Navy as a whole or in response to requirements from any elements of the Navy other than the Bureau of Ordnance.

Another pre-ONI officer who carried the title but not the functions of naval attaché was Commo. Robert W. Shufeldt, assigned as naval attaché to the U.S. Legation, Peking, from 1881 to 1882. Shufeldt's title was merely a cover for his true mission, the making of a treaty with Korea. In the latter effort, Shufeldt obtained the backing of the Chinese viceroy by aiding him in the reorganization of the Chinese navy.[5]

Ens. George C. Foulk was another naval officer who was given the title of naval attaché but who did not carry out the normal functions associated with that title. During two cruises on the Asiatic Station, he had studied Japanese, Chinese, and, to a small extent, Korean. Consequently, when members of a Korean special mission arrived in Washington in mid-September 1883, he was directed by President Chester A. Arthur to assist in rendering suitable attentions to the mission. Lt. Mason of ONI was similarly assigned, and the two officers accompanied the mission during its trips through the eastern part of the United States. At the conclusion of the visit, the mission returned to Korea on board the steam frigate *Trenton* via the Mediterranean and the Suez Canal. Ens. Foulk was appointed by the President as naval attaché to the new American legation in Seoul and accompanied the Korean mission on its return, arriving in Chemulpo (Inchon) Harbor, Korea, on 31 May 1884. At that time, Korea had no navy, and Foulk became U.S. chargé d'affaires *ad interim* in December 1884, and in that capacity continued, dealing expertly with problems of extreme delicacy and complexity, for nearly three years.[6] He was not considered a naval attaché in the strict sense of the word, even though he did have the title and diplomatic privileges associated with the position.[7]

The second official naval attaché sent out by ONI was Lt. Benjamin H. Buckingham, who served at the U.S. Embassy, Paris, from 11 November 1885 to 30 March 1889. Buckingham was additionally accredited to St. Petersburg and Berlin. Lt. Nathan Sargent was the third naval attaché and was the first to be posted to Rome. Sargent was also accredited to Vienna and served in both the Italian and the Austro-Hungarian Empire positions from 20 November 1888 to 31 October 1895.[8]

Naval Attachés and the Spanish-American War

Covering six European countries with three officers continued until just prior to the Spanish-American War. In 1892, probably in recognition of changing political alignments, accreditation to Germany was shifted from the attaché in Paris to the one in Rome.

When U.S.-Spanish relations started to cool in 1895, LCdr. Raymond Perry Rodgers, Naval Attaché, Paris, was additionally accredited to Madrid. His relief on 30 April 1897, Lt. William S. Sims, was similarly accredited until July 1897, when Lt. George L. Dyer was ordered as Naval Attaché, Madrid. Dyer held the assignment until the commencement of the Spanish-American War in April 1898. The number of naval attachés assigned to Europe then went back to three and remained so until just prior to World War I.

In the Far East, Cdr. Francis M. Barber was ordered as Naval Attaché, Tokyo, in 1895, following the victory of Japan over China in 1894. Barber was also accredited to Peking during the same period, but, upon his detachment without relief, both posts were left vacant until 10 September 1898, when Lt. Albert L. Key was ordered to Tokyo with double accreditation to Japan and China. The need for a naval attaché stationed in the Far East had been demonstrated in the Spanish-American War.

The industry demonstrated by Lt. Sims in his intelligence duties in the cruiser *Charleston* in the Far East led to his appointment as Naval Attaché, Paris, where he arrived in early March 1897. (In 1888, he had asked for and been given a year's leave to study French in France.)[9]

With his predecessor, Lt. Rodgers, Sims made a long trip to Russia and an extended tour of the principal naval bases of Europe. Sims relieved Rodgers on 30 April. His reports soon caught the eye of Assistant Secretary of the Navy Theodore Roosevelt. The reports sent by Sims dealt chiefly with methods used by European countries in ship construction and target practice. After six months of observations, Sims was convinced that American gunnery was far less effective than that of the great foreign navies, and he so reported in January 1898.

Initially, Sims was accredited to Spain in addition to France and Russia. He was informed before leaving Washington, however, that his Spanish assignment would soon be withdrawn and that a new naval attaché would be appointed to the post. Consequently, Sims was further advised that it would not be necessary for him to concern himself with the Spanish area, as all necessary information could be obtained in the interim from American consuls in Spain. Sims felt that the idea of obtaining information that way was wishful thinking because some of the U.S. consuls were actually Spanish citizens. Accordingly, Sims asked to be supplied with the funds necessary to establish his own agents. Secretary of the Navy John D. Long at first opposed Sims's request, but Capt. Mahan eventually convinced Long of its value. Money was supplied, and Sims commenced building up an agent network in Spain. The agents recruited were of many nationalities and professions, and in due course Sims had observers in most of the larger Spanish cities and ports. Many of the agents were quite effective, but one, a doctor whose practice was among the aristocracy of the Spanish capital, proved in time to have relied on a fertile imagination for most of his information.

Sims's early reports to ONI, based on information gathered by his agents, included details on the defenses of Cartagena and the availability of 200,000 tons of coal in the Canary Islands, defended by only a few guns with limited ammunition. In early June, he passed on information that when the Spanish cruiser *Cristobol Colon* had arrived in Santiago, Cuba, the ship's boilers were in need of repair and her speed had been reduced to one-third normal. A few days later, Sims reported on the Spanish plans to send Adm. Camara and his ships to threaten the American forces under Dewey at Manila. Camara and his fleet did indeed depart Spain on 16 June.

When the War Strategy Board was informed of Sims's report, Capt. Mahan wrote to Sims that he should leak information that a U.S. fleet under Commo. John C. Watson was to be sent to blockade Spain and molest its commerce. The defeat of Cervera at Santiago, plus possibly the above threat, induced the Spanish to recall Camara to protect the Iberian Peninsula.

With the end of the Spanish-American War, Sims returned to his normal but prolific reporting on the characteristics of new ships launched by France and Russia, the location and facilities of new bases, changes in naval regulations, and operational casualties. One report in a lighter vein explained the French method of training carrier pigeons; someone added to the report that the United States should crossbreed pigeons with parrots so that the birds could vocalize their messages, for which comment Sims was reproved by the Navy Department.

When Sims was detached on 1 June 1900, his reports as naval attaché in Paris filled twenty-two letterpress books containing 11,000 pages, a record that probably still stands.[10]

After the coastal battleship *Maine* blew up in Havana Harbor on 15 February 1898, the Navy Department began negotiations for the purchase abroad of ships and munitions of war. The negotiations were conducted partly through U.S. naval attachés in Europe.

The activities of the naval attachés, in addition to purchasing war material, were devoted principally to reporting the movements of the Spanish squadrons of Cervera and Camara. The attachés maintained a corps of agents whose reports were forwarded to ONI and formed a special information series. The reports were destroyed after the war because they contained much data that, if made public, would have been compromising to the agents, some of whom were in high positions in Spain.[11]

Ens. Arthur B. Hoff was sent to London as an assistant naval attaché during the period of the Spanish-American War. Hoff was the first officer to be designated an assistant naval attaché.

Pre–World War I Period

In Latin America, the first naval attaché was Lt. Marbury Johnston, who was ordered to Caracas in 1902, first as an assistant to the U.S. minister and then as naval attaché. Britain and Germany had imposed a joint naval blockade on Venezuela and had seized some of that country's gunboats in an effort to force the payment of past debts. The United States considered the situation to be a threat to the Monroe Doctrine, and President Theodore Roosevelt moved U.S. warships to a position of readiness in Puerto Rican waters until arbitration was accepted. Lt. Johnston remained in Venezuela only from 8 January 1903 to 20 September 1903; then the post was closed.[12]

Increased U.S. interest in and concern about Russo-Japanese relations and the possibility of armed conflict was reflected in a letter of 23 December 1903 from the ambassador to Russia, Robert McCormick, to Secretary of State John Hay. McCormick reported that the Russian government had agreed to a request to allow LCdr. Charles C. Marsh, the naval attaché at Tokyo, to be responsible for "watching and visiting Russian possessions and naval activities in the Orient." The Russians specified that Marsh would not be allowed to establish himself permanently in either Port Arthur or Vladivostok since they wanted no regularly assigned foreign attaché in either place. The outbreak of war between Russia and Japan put an end to the project at that time.[13]

On 8 February 1904, the State Department asked Ambassador McCormick to seek permission for Lt. Newton A. McCully to accompany Russian naval forces in case of war (that same day the Japanese had launched a surprise torpedo boat attack on the Russian Port Arthur squadron). On 19 February, McCormick replied that the American request would be answered along with similar ones from other nations, but that the Russians had stated that no quarters or commissariat were available. By a dispatch of 11 March 1904 from the Secretary of State to McCormick, he was instructed to make McCully the assistant naval attaché in St. Petersburg pending receipt of permission for him to join Russian forces in the war zone.[14]

By 15 March 1904, Lt. McCully was in St. Petersburg with the title of assistant naval attaché. Officially, he had been given permission by the Russian foreign office to join the Far Eastern forces. After some delays, he reached Moscow on 2 April and Manchuria some time later in April.

McCully reported his journey on the Trans-Siberian Railroad completely, giving details on the route, roadbed, bridges, locomotives and rolling stock, and the Russian proficiency in handling the large quantities of war supplies and personnel being moved eastward. McCully's problems, however, really began when he tried to get to Port Arthur. A Russian officer in charge of all foreign attachés in the war zone requested that McCully surrender the permit that he had received from the minister for foreign affairs in St. Petersburg in order to have it checked. The permit, officially sealed, stated that McCully was permitted to accompany either the Russian army or navy and directed that no one was to interfere with his efforts to do so. Of course, this permit was "lost" by the Russian officer, and after repeated inquiries McCully was given a replacement that was unsealed and that gave him permission to visit Port Arthur only.

The closer McCully came to Port Arthur, the more difficulties he had with the Russian authorities. As he awaited permission to proceed from Liaoyang to Port Arthur, McCully witnessed French and German attachés going through with practically no delay. The attachés of other nations were being given free access to Port Arthur and had few restrictions on their regular inspections of Russian warships, shore defenses, and repair facilities.

The deliberate affront to the United States was a result of the view throughout Russia that the United States, particularly President Roosevelt, was strongly pro-Japanese. In addition, in the first week of the war before McCully had arrived in the area, two Russian ships had been sunk in the Korean harbor of Chemulpo by a Japanese squadron. The U.S. gunboat *Vicksburg* was nearby and witnessed the violation of a neutral harbor, but the ship's commanding officer, Cdr. William A. Marshall, refused to join the commanders of British, French, and Italian warships present at the time of the attack in a protest of the incident. The *Vicksburg*'s captain also refused to take Russian wounded aboard either his ship or two American transports also present in the harbor. As a result, the wounded were landed ashore and taken captive by the Japanese. Marshall felt that his instructions from the Navy Department justified his actions. Furthermore, his ship was small and his facilities inadequate. The resulting uproar in Russia, however, strained relations with the United States even further. An article in the Russian newspaper *Novoye Vremya* (*New Times*) on 11 April 1905 accused the United States of encouraging Japan to go to war with Russia in order to stop the spread of Russian influence into Manchuria and ultimately China. It quoted a U.S. senator as maintaining that the U.S. sphere of influence should include Russia's Pacific coast and should stretch inland along the Amur River to the Transbaikal and Yakutsk areas.

All these incidents and anti-Russian statements were undoubtedly responsible for generating the anti-U.S. attitudes of the Russians and contributed to the slights and difficulties that impeded McCully's accomplishment of his mission throughout the period of his assignment in the Russian Far East.

Much of the information on Russian ships that McCully was able to report was obtained from conversations with other foreign attachés. The bulk of his other reports were about the ground defenses of Port Arthur and on the state of morale, sanitation, and medical needs of the Russian troops. At the end of hostilities in 1905, McCully was withdrawn without a relief, and the accreditation to Russia was returned to the Naval Attaché, Paris.[15]

By 1907, the increasing importance of Germany as a naval power was recognized, and LCdr. William L. Howard was divested of his responsibilities in Austria-Hungary and Italy to devote his time exclusively to observing the German navy and its development.

The next naval attaché assigned in Latin America was Cdr. Albert P. Niblack who was accredited to Argentina, Brazil, and Chile to mark the hundredth anniversary of independence for Argentina and Chile. Niblack served from 1 July 1910 to 24 November 1911. In 1912, Argentina and Chile received individual attachés, and in 1914 Brazil was also similarly covered.[16]

In 1910, junior officers were sent to Tokyo to learn the Japanese language, where they were assigned under the naval attaché. The Japanese language study program was interrupted in 1913 when Secretary of the Navy Josephus Daniels instituted a policy of having as few naval officers as possible on shore duty and the student naval officers were brought home.[17]

World War I

Russia

The start of World War I revived U.S. interest in Russian military affairs. On 10 August 1914, the Navy again designated then-Capt. Newton McCully as the new naval attaché in Russia, terminating the double accreditation of ALUSNA (American Legation, U.S. Naval Attaché) Paris, LCdr. Samuel I. M. Major. McCully's orders instructed him to leave his command, the stores ship *Glacier* (AF 4), at San Diego and proceed to St. Petersburg by way of the Pacific and the Trans-Siberian Railroad. He arrived on post on 6 October and for the next three years was in Petrograd (as St. Petersburg had been renamed), making full use of acquaintances from his previous tour and of his fluency in the Russian language.[18]

During 1914, full-time attachés were assigned to Tokyo and Peking, discontinuing the previous practice of sharing one officer between both posts. Also in 1914, the practice of assigning assistant naval attachés was expanded with Lt. John H. Towers and 1stLt. Bernard L. Smith, USMC, both aeronautical specialists, added to London and Paris, respectively, in October.[19]

In May 1917, the Navy Department nominated Cdr. Clater S. Crosley to succeed Capt. McCully as Naval Attaché, Petrograd. But by February 1918, Cdr. Crosley was requesting a transfer back to active-duty status since he believed that there was no naval activity in Russia at that time, and that none was likely. Crosley had also found life in Russia unfavorable to his health. By March, Crosley's transfer request had been approved.

In February 1918, having had enough of the delaying tactics of the new Soviet regime at the Brest Litovsk peace talks, Germany reopened its offensive into the Russian heartland. As a result, the foreign diplomatic missions evacuated Petrograd, and by the end of July the American legation finally reestablished itself at Archangel in north Russia.

Again the problem of finding a naval attaché to send to Russia arose, and again Newton McCully, now a rear admiral, was proposed for the assignment. Ambassador to Russia David R. Francis agreed, but when McCully arrived, it was revealed that his primary duty was as Commander U.S. Naval Forces, Murmansk Region, rather than as the naval attaché on the ambassador's staff at Archangel. The ambassador insisted upon the appointment of a regular attaché, complaining that he had been without one since Crosley left in February 1918. When Lt. Sergius Riis arrived early in September at Archangel, he was appointed as the acting naval attaché. McCully gave his approval to the arrangement, which continued until the ambassador departed in November 1918.[20]

The United States Enters the War

With the entry of the United States into the war in April 1917, the posts at Berlin and Vienna were closed, and naval attachés or assistants were assigned for the first time in Lisbon, Copenhagen, Stockholm, and Christiania (Oslo). The latter three posts were initially assigned to LtCol. James C. Breckinridge, the first Marine officer to serve as an attaché. In Latin America, naval attachés were assigned for the first time in 1917–1918 to Peru, Ecuador, Uruguay, and Cuba. In the latter post, LCdr. Carlos V. Cusachs went as naval advisor to the Cuban government as well as naval attaché.[21]

The U.S. naval attaché office in Paris became a liaison office, an information office, a housing office, and a purchasing office, and kept contact with the U.S. legation in Zurich. With operations, communications, logistics, graves registration, intelligence, counterintelligence, and staff representation in France, the office was extremely busy. A lengthy report on the activities of the U.S. naval attaché's office at Paris during World War I is contained in "History of the Office of the U.S. Naval Attaché, American Embassy, Paris, France, during the Period Embraced by the Participation of the U.S. in the War of 1914–1918," a paper by LCdr. Charles O. Maas, USNRF, the assistant naval attaché in Paris during the war.[22]

The United States entered the war with six naval attachés and two assistant naval attachés on station.[23] They were accredited to ten countries, including Germany and Austria. At the end of hostilities in November 1918, there were fifteen naval attaché posts covering eighteen countries.[24] In addition to their intelligence activities, the attaché offices had jurisdiction over the investigation of officers, crews, and passengers in ships bound for or arriving from America; the senders and receivers of cablegrams; the inspection of cargoes and shipments; firms suspected of trading with the enemy; and coastal watches for enemy vessels and for persons who might be giving aid or information to such vessels.[25]

Naval Attaché, Scandinavia

LtCol. Breckinridge was first made naval attaché for all of Scandinavia, resident in Christiania (Oslo). After some time, Mr. E. B. Robinette was made his civilian assistant, and the brothers H. U. and John A. Gade were enrolled as junior and senior lieutenants, United States Naval Reserve Force (USNRF), respectively. Later, John Gade was promoted to lieutenant commander and ordered to Denmark as naval attaché, his brother going as his assistant. Robinette was enrolled as a lieutenant, USNRF, and was ordered to duty in Sweden.

The Naval Attaché, Scandinavia, reported on enemy ship movements, locations and types of enemy minefields, and enemy troop strength and movements. The office also collected examples of enemy propaganda and endeavored to expose its inaccuracies to gain the sympathy of the Scandinavian press.

Passport control was one of the best weapons in counterespionage work, but only during a few months of the war did the naval and military attachés have the right arbitrarily to refuse anyone the privilege of traveling to the United States. Gaining the friendship and confidence of the local police officials was found to be important, as their friendly tips of possible enemy agents and couriers were most helpful and usually reliable.

When officers on merchant ships were discovered serving as enemy couriers, they were induced to identify their enemy contacts and in other ways to betray the enemy's trust in them, or they were exposed for other counteraction. Cooperative commercial travelers were given guidance for collecting information and for developing sources from among dissatisfied citizens in the enemy country.[26]

Naval Attaché, Copenhagen, John Gade reported on the movements of all German vessels and aircraft through or from Danish waters and the Heligoland Bight, activities in German harbors and shipyards, German minefields in Danish and adjacent German harbors, the movements of German merchantmen in those harbors, the personnel of the German navy, and the movements of German raiders. Gade kept informed on activities of the German intelligence and espionage services in Denmark, and he procured information from Germany through agents sent to and living in Germany and through deserters from the German army and navy. The attaché was informed about political and economic conditions in Germany and the Scandinavian countries by keeping in close touch with the Danish army and navy intelligence services, the Danish foreign office, politicians and businessmen familiar with Danish-German relations, and Scandinavian military and naval attachés accredited to the German government.[27]

Through the United States Military Control Office, the naval attaché was able to ensure the security checking and blocking of any suspects attempting to pass to and from America.

The qualifications for an effective naval attaché were described over a decade later by former naval attaché LtCol. Breckinridge:

> We need two things, and we need them badly. These are a knowledge of languages away and beyond the usual American ability to stutter in something or other; we simply must cultivate the study of languages because we are a joke in any international gathering, and helplessly tied to the apron strings of some translator. The other thing is to have a small class in which to teach what intelligence duty is. Our officers are shot out to different countries for every reason except that they are fitted to go there. To begin with, they should know the language fluently, know the history of the people and the country, something about their social conditions and persuasions, their national ambitions and prejudices, their music and arts, and their literature. They then will be at home and not otherwise. A competent attaché should be first fitted to be an attaché, over and beyond knowing how to reply to an invitation, what to wear, where to expect his seat at table to be, and fill in a monograph when he feels like making an effort at doing work. If he is prepared for that sort of work, there is no limit to what he can do. If not prepared for it, he will do routine, perform the ritual, and might just as well be "popular" at home as anywhere else.[28]

Naval Attaché, Rome

The Naval Attaché, Rome, during World War I was LCdr. Charles R. Train, who was later promoted on post to commander and then to captain.

Upon the declaration of war by the United States against Germany, Capt. Harry M. Hodges, USN (Ret.), a resident in Italy, was ordered to the attaché's office where he assumed the duty of communications officer. By the fall of 1917, the work had become too extensive for the naval attaché to handle alone. It was evident that the U.S. would soon declare war against Austria-Hungary, which would mean active participation by United States forces in the Mediterranean and probably in the Adriatic. Accordingly, LCdr. Roland Riggs, USN (Ret.), was sent as Assistant Naval Attaché, Rome, in October 1917.

A courier service for handling American confidential correspondence was established between Rome and Paris in November 1917.

The naval attaché office in Rome was divided into five departments: intelligence, operations, aviation, pay, and communications. It was planned to have an officer in charge of each department, but that could not be done due to the shortage of personnel. The naval attaché took charge of operations and acted as liaison officer to the Italian Ministry of Marine.

In February 1918, the Italian Aviation School at Bolsena began operations with about fifty American officers and students under the command of LCdr. Calderon, RIN. In April, LCdr. John L. Callan, USNRF, reported for duty and took charge of the aviation department of the naval attaché's office. LCdr. Riggs was placed in charge of office management and the intelligence department, with the naval attaché retaining responsibility for operations.

In May 1918, Naval Attaché, Rome, placed LCdr. C. M. Peck, USNRF, as port officer and assistant naval attaché at Genoa, where the U.S. consul general had offered Peck the use of space in his office. The port officer's mission was to assist in the berthing, discharge, and prompt sailing of all American ships. His duties included interviewing masters or officers of all U.S. vessels; inspecting naval armed guard crews; investigating offenses committed by U.S. naval personnel on vessels other than those regularly commissioned in the Navy; transmitting sailing orders from the naval attaché; and briefing masters of vessels about necessary precautions in designated danger zones and about convoy doctrine. Prior to their sailing, Peck would furnish convoy and escort commanders with the latest information on submarine and mine activities. He also examined crew lists furnished by the consul general to determine whether there were any suspects who should be removed.

In October 1918, Capt. Harold C. Train was officially appointed by the Navy Department as staff representative of the force commander in London. At that time, Train commanded all U.S. Navy activities in Italy and had the titles of naval attaché, staff representative of the force commander, and commander of the U.S. naval aviation forces in Italy.[29]

Naval Attaché, Stockholm

The naval attaché office at Stockholm was established during August 1917, with LtCol. Breckinridge in charge. Almost the first work of the office was to attempt to change public opinion in favor of the Allies. Breckinridge found that American movie pictures were better and more popular with the Swedish public than German films. By allowing films to come into the country under control of the naval attaché, the films were shown only in theaters where a contract had been signed stipulating that German films were not to be shown and requiring American and Allied educational and propaganda films to be shown twice a day. As a result, German films were excluded from most of the theaters. With the assistance of Liberal and Socialist leaders in the Swedish government and of certain strong financial groups whose interests lay with the Entente powers, a news bureau was formed to take over the distribution of Allied and world news to the Swedish press. Breckinridge reported:

> Thus, we finally, after many months of work, in the face of hostile public opinion and of an organized press campaign against the scheme, took from the hands of the German-controlled bureau the distribution of our own news. The result was that all American and Allied news reached the press and a certain kind of Central Power and Russian news which had hitherto been suppressed because it was not in the interests of the Central Powers to have it published was also given to the press of Sweden.[30]

Advantage was taken of visits to Germany by Swedish businessmen and by officers of the Swedish army and navy to obtain information. Information was also obtained from Germans visiting Sweden, from Swedish and German sailors from vessels trading with Germany, and from German deserters.

Movements of German war vessels through the Kattegat were learned about by a system of coast watchers established along the western coast of Sweden. Considerable information on Germany was also secured from Swedish government officials with whom the naval attaché's office had formed a close relationship.

Information on Russia, Finland, and the Baltic provinces was secured from Allied and neutral representatives and others coming out of those countries, as well as from agents sent there. Information about Sweden was largely secured from official sources. In addition to the agents employed exclusively by the attaché's office, other agents were employed jointly with the U.S. military attaché and with the British and French attachés. Working with the British, the U.S. attaché's office set up an office in Finland to get information across the border from Moscow.[31]

Naval Attaché, The Hague

The Naval Attaché, The Hague, during World War I was Lt.(jg) Eugene D. McCormick. He secured his information about the German navy from German naval deserters, refugees, neutral travelers, and workers in enemy countries and obtained news from agents with lines of communications into enemy territory, along the Dutch coast, and along the Dutch-German and the Dutch-Belgian frontiers.

The German intelligence service took all possible steps to prevent the U.S. naval attaché in The Hague from gaining access to German deserters and attempted to involve the naval attaché's office in trouble with the Dutch police authorities on account of its activities. As a rule, it was found that deserters who had just crossed the frontier gave accurate information but that, after a few weeks in Holland, their information became unreliable.

Some of the best information came from Belgian refugees who had been at work on dredges in Zeebrugge, Blankenberghe, Knocke, and Ostend or had been involved in constructing gun platforms, coastal railways, aviation fields, and other installations of military value. Getting naval information from Dutch commercial travelers who had been in Germany for business purposes met with little success; they were usually either pro-German or else afraid of being convicted in Germany for espionage.

McCormick's office in The Hague was not successful in establishing a permanent line of communication into enemy country. U.S. agents established along frontiers were able to obtain information about the movements of German ships going in or out of Zeebrugge Harbor and on the Ems River. They obtained a number of German deserters and interviewed travelers going to and from Germany.

The counterespionage efforts directed by McCormick were carried out by two groups of agents, the head of each reporting directly to the naval attaché. The agent groups attempted to keep enemy agents off ships plying between America and Holland, tried to prevent smuggling on those ships, investigated suspects, and confiscated German propaganda material.[32]

McCormick's office at The Hague also handled agent reports from Ostend about German ships, searchlights, and gun emplacements, and made extracts from the German press. One report stated that the Netherlands was full of unreliable Belgian agents who were more bother than the German agents.[33]

Naval Attaché, Madrid

The office of Naval Attaché, Madrid, was reopened 26 May 1917 with Capt. Benton C. Decker as naval attaché. To begin the intelligence work in Spain, a general call was sent out through all the U.S. consulates in Spain requesting Americans in Spain with a knowledge of the Spanish language and character to communicate with the naval attaché. The majority of the American agents in Spain were obtained as a result of this request.

Capt. Decker was advised that his principal duty would be to prevent enemy agents from embarking secretly for America and that U.S. agents should only be of sufficient number to accomplish the desired results. A complete surveillance of the Spanish coast was not deemed necessary.

The cooperation among the Allied representatives in Spain was excellent. In April 1918, the first combined meeting of the naval attachés from Allied countries was held in the office of then-RAdm. Decker, who was still serving as Naval Attaché, Madrid. Afterwards, the conferences were regular weekly affairs.

In May 1918, RAdm. Decker was relieved by Capt. Walter S. Crosley. Crosley endeavored to obtain more money for extending the work of his office but was unable to do so.

The conditions in Spain were not strictly neutral. There were good grounds for believing that German refugees and crews from interned vessels were being used for war purposes. Spanish officials were believed to be working for the Germans and against the Allies. Propaganda against the Allies was permitted, but Allied propaganda was restricted. Signal stations on Spanish territory were used to communicate with German submarines at sea; there were frequent communications by boat between Spanish territory and German submarines; and Allied ships had been torpedoed in Spanish territorial waters. The U.S. naval attaché was constantly trying to obtain data to enable Allied forces to capture enemy submarines and to prove that Germany and Austria were violating the neutrality of Spain.[34]

Naval Attaché, Lisbon

The Naval Attaché, Lisbon, during the latter part of World War I was LCdr. Edward Breck, USNRF, of Spanish-American War fame. Breck devised a scheme whereby the Portuguese government itself took over the task of watching the frontiers. Two agents of the Portuguese Preventive Police were assigned to the office of the naval attaché, placing Breck in the position of a police commissioner with the power to arrest anyone.

In addition to the regular work of the office at Lisbon, the naval attaché had to arrange for the repairs of the smaller types of Allied warships at facilities in the area as well as for the payment of repairs through the force commander in London. All arrangements for docking, provisioning, patrolling, unloading, and storing ammunition were made by Breck.[35]

Naval Attaché, London

When the United States entered World War I, Capt. William D. MacDougall was Naval Attaché, London. On 9 April 1917, RAdm. William Sims, with his aide, Cdr. John V. Babcock, arrived in England as a representative of the U.S. Navy. Sims's mission was to study the naval situation and learn how the U.S. Navy could best and most quickly cooperate in the naval war. The naval attaché and his two assistants served as Sims's staff until August. After Sims was designated Commander U.S. Naval Forces Operating in European Waters, he relieved Capt. MacDougall, taking over the duties of naval attaché in addition to his other duties. Cdr. Babcock became the head of the Intelligence Section of the European force commander's headquarters.[36]

Retrenchment Post–World War I

At the conclusion of hostilities many wartime attaché posts were discontinued or consolidated. By 1921, the office in Copenhagen was again responsible for all Scandinavia, a good area from which to observe the new Communist regime in Russia. By 1925, the naval attaché in Berlin was also accredited to the Scandinavian countries.[37]

The method of obtaining information in foreign countries was principally through the exchange of information of equal importance. The acquisition of information by any questionable method was strictly frowned upon. It was ONI's policy that naval attachés should use only reputable business methods and that they were to avoid anything savoring of "gumshoe" techniques in their collection of information. One of the difficulties in finding suitable officers to serve as naval attachés had been that maintenance allowances were ridiculously small, and officers who volunteered for the duty did so with the full knowledge that their pay and allowance would not be sufficient to meet their expenses. Seldom was a naval attaché able to live on his pay and satisfactorily perform the duties expected of him, which led to the deduction that only officers with private means could afford to take the position. Therefore, the pay situation created automatic prejudice and a gratuitous assumption that the position of naval attaché was something of a sinecure.[38]

RAdm. Roger Welles, Jr., Director of Naval Intelligence during World War I, also found great difficulty in selecting officers for the duty of naval attaché:

> The mere fact that an officer knew a foreign language was not positive proof that he would make a good attaché. . . . He should be a man with a keen imagination, able to draw correct conclusions from very scanty evidence, courteous in manner, a man of the world (but not too worldly) and, in general, with sufficient intelligence to be a good mixer in all classes of society.[39]

The prospective attachés, Welles believed, should not only study foreign languages, but also diplomacy, international law, the constitutional law of the countries to which they were assigned, the foreign policy of the countries, and modern international relations in general.

As the likelihood dimmed for a responsible government emerging in Russia that the United States would approve and recognize, the need for assigning an attaché there lessened. Ambassador Francis left for health reasons in November 1918, and the other members of the legation and the American military forces in northern Russia departed a year later. RAdm. McCully departed in November 1919 and reported for duty with Allied forces operating in the Black Sea and southern Russia. There, at the request of the State Department, he engaged in various kinds of intelligence work and, on several occasions, traveled inland to observe conditions firsthand.[40] But since the United States did not recognize any Russian government at that time, McCully was not accredited or given the privileges or title of naval attaché.[41]

In 1920, naval attachés were maintained at London, Paris, Rome, Madrid, Lisbon, The Hague, Copenhagen, Tokyo, Peking, Rio de Janeiro, Buenos Aires, and Santiago, Chile. There were also assistant naval attachés at London, Paris, Rome and Tokyo. The naval attaché at Copenhagen was also accredited to Norway and Sweden. A naval officer was stationed in Berlin as an unofficial advisor to the U.S. Commissioner of Control, who oversaw U.S. interests during the occupation of Germany by the victorious Allied forces. In addition to naval data, the naval officers furnished information concerning political changes, social disturbances, and conditions in the former Central Powers countries.[42]

There was no limitation on freedom of movement by U.S. naval attachés in Japan in the 1920s and 1930s. Any non-Japanese person entering "fortified zones," such as the areas around Yokosuka, Kure, and Sasebo, was quickly spotted by the police, however, and followed while in that area. Photographing and sketching were forbidden, and the Japanese police were firm but courteous in their enforcement of the constraints.

ONI guidance to the Naval Attaché, Tokyo, was in the form of requesting reports that were as complete as possible on specific subjects, usually subjects mentioned in previous reports. The requests were usually contained in personal letters from the head of the ONI Far East Section. Personal letters were exchanged frequently and conveyed ONI's feelings about the work the attaché was doing.

In Tokyo, it was known that any information on the Japanese Mandated Islands was of great interest and importance, so every scrap of relevant information obtainable from any source was sent to ONI. Unfortunately, the few Japanese who could be contacted and had ever been to the former German-controlled islands were simple merchants who had made no unique observations, and they added little to the information on hand. Efforts to arrange for ship visits or to get visas to go to the Mandates were unsuccessful.[43]

In April 1921, the American ambassador to Poland wrote to Secretary of State Charles Evans Hughes about the excellent opportunities for securing information about the Soviet Union at his post because Polish missions would soon be going into Russia. Soviet delegations were also arriving in Poland, and trade posts were being established along the frontier. The U.S. Embassy in Poland also recommended LCdr. Hugh W. Koehler because of his experience in Russian affairs and his fluency in the language. Koehler had assisted Adm. McCully for over a year in the Crimea in 1919–1920 and had traveled in disguise "all through the Ukraine." LCdr. Koehler was appointed to the naval attaché post at Warsaw on 25 May 1921.

A further discussion in connection with Koehler's assignment was whether he would be given a roving commission as naval observer in the newly established Baltic republics of Latvia, Estonia, and Lithuania. The State Department questioned the advisability of the roving commission, believing that the U.S. commissioner for the area and the consuls in Latvia and Estonia covered it adequately. Any of Koehler's activities in Lithuania, the State Department felt, should be coordinated with the commissioner. It seems likely that, in the end, Koehler received some sort of permission to operate at will in the three countries, although there are no formal letters of appointment to give specific dates. Koehler completed his tour in July 1922.[44]

In the years after World War I, the number of attachés assigned to foreign countries was never great; at times they numbered only eight or ten. It was difficult to keep even that small number in the field. For example, in 1922 the naval attaché in Warsaw was paid with Department of State funds and was, for all intents and purposes, an officer of that department. A similar situation happened at Havana in 1923. In some cases, the practice of early ONI days was resorted to, and an attaché would be assigned to more than one post. The attaché accredited to Paris in 1923 also was accredited to Madrid and Lisbon; the naval attaché at Berlin covered Copenhagen and Oslo.[45]

In 1922, Capt. Edward H. Watson, Naval Attaché, Tokyo, received publicity concerning his return to the United States as the result of an entrapment effort by a Japanese naval officer who tried to sell him secret Japanese publications. Watson had also clandestinely inspected Japanese naval ports, according to the Japanese press.[46]

In early 1923, Naval Attaché, Berlin, Cdr. William F. Halsey, Jr., wrote to Robert Murphy, Vice Consul, Munich, that he had picked up rumors that Bavarian factories were turning out diesel engines especially designed for Japanese submarines in violation of the Treaty of Versailles. Murphy was able to confirm the rumors.[47]

While serving in Berlin, Halsey was also responsible for obtaining a newly invented stereoscopic range finder and for sending it back to the United States, where it was tried out by the Navy and adopted to take the place of the coincidence range finder then in use in both the British and U.S. navies.[48]

From 1926 through 1928, naval attachés were located in London, Paris, Rome, Berlin, The Hague, Tokyo, Peking, Rio de Janeiro, Buenos Aires, Santiago, and Mexico City. The latter post was established in 1926.[49]

The Naval Attaché System in the 1930s

In the 1930s, under standard procedures, a naval attaché was ordered to report to the American ambassador or minister in the country to which he was accredited and to consider the State Department official to be his superior officer. The attaché was, by courtesy, a member of the diplomatic corps but was not a diplomat. He was the direct representative of the Navy Department and an official agent for gathering information. The attaché forwarded his reports directly to ONI but was expected to keep the ambassador or minister informed as to their content, except for contents of reports of a purely technical nature. The attaché had a dual responsibility, first to the Navy and secondly to the ambassador or minister. His role was a delicate one, requiring tact and judgment. The Navy Department laid down the principle that a fleet commander in chief on his own station could not issue an order to a naval attaché.

After an officer had been selected for assignment to duty as an attaché, it was the practice to order him to Washington for temporary duty in ONI for the purpose of having him review the intelligence reports on the country to which he was accredited. The prospective naval attaché also visited the technical bureaus of the Navy Department to see if they had any information that would be of value to him or if the bureaus desired any particular information from the country to which he was going. The naval attaché was also a special disbursing officer, and during his briefing period he was given instruction in keeping his accounts. When there was a change of attachés, it was normal practice to have the officer report at his post two to four weeks before taking over the duties of the office so that he could be thoroughly instructed by the officer to be relieved. The officer being relieved was usually ordered to duty in ONI so that he could review and update all of his reports.[50]

Commencing in 1930, a naval attaché for air was attachéd to a foreign legation when a separate air ministry had been established in the country to which he was accredited. Usually, the one officer would perform the dual roles of naval attaché and naval attaché for air.

On 11 March 1931, all naval attachés were especially instructed by a Director of Naval Intelligence letter to collect special items of intelligence. These included information on U.S. overseas commercial interests, the overseas commercial interests of foreign countries, combat intelligence material, data for limitations-of-armament studies and congressional hearings, and advances in technical naval science in foreign countries.[51]

The naval attachés at Santo Domingo and Mexico City were withdrawn on 24 July 1931 and 1 January 1932, respectively.[52] The naval attachés at Rio de Janeiro, Tegucigalpa, and The Hague were withdrawn on 21 December 1932, 31 May 1933, and 30 June 1933 respectively, and on the latter date the Naval Attaché, Berlin, was additionally accredited to The Hague. Establishment of the Naval Mission to Brazil on 15 November 1932 was the reason for the decision that the naval attaché could be withdrawn there. Naval Missions in Latin America came under ONI's OP-16-FL until January 1942, when they were transferred to the Pan-American Division of the Office of the Chief of Naval Operations (OPNAV), OP-17.[53]

The office of Naval Attaché, Brussels, was established on 21 August 1933, and LCdr. John Gade (former Naval Attaché, Copenhagen, during World War I) was detailed to the station with the express understanding that the U.S. Government was to be subjected to no expense incidental to his office. The Communications Division of OPNAV, however, did allot $150 to cover the cost of his official communications.[54]

The Navy Department made available to the naval attaché at Tokyo a "maintenance allowance" of $300 per month. The assistant naval attaché similarly received $200 per month. The allowances were provided for entertaining and bettering the acquaintance of officials and other knowledgeable sources. The funds permitted the attachés to widen their circle of friends and contacts and thus to improve the reporting capabilities of their office.

Great attention was paid to newspapers, magazines, and all official publications issued by the Japanese navy department and government. One valuable member of the attaché's staff was a retired chief yeoman, Leonard Wagner, who had been in the office since 1920. He had become an expert on the Japanese budget, among other things, and each year he prepared a detailed breakdown of the naval budget as it appeared in the Japanese *Official Gazette*, a publication that corresponded to the U.S. *Congressional Record*.[55]

When the U.S. Navy sustained a 15 percent pay cut in 1933, the Naval Attaché, Tokyo's maintenance allowance was cut about 25 percent to $225 per month, and the assistant naval attaché's allowance was cut to $150.[56]

Special collection instructions were seldom received from ONI, but the assistant naval attaché at Tokyo frequently exchanged letters with the officers of the Far East Section of ONI, particularly with Lt. Arthur H. McCollum. While the letters were not the same as official word from the Director of Naval Intelligence, they did provide helpful guidance, and there was no feeling of working in the dark on the part of the staff in Tokyo.[57]

The naval attaché office in Japan concluded that about 95 percent of the information it sought was readily available in open sources if one knew where to look and could read Japanese; only 5 percent was secret and obtainable only with luck or by clandestine means. The Japanese were justifiably proud of their merchant marine and published excellent pictures in maritime magazines whenever a new ship went into service. The photographs proved valuable in World War II for use by U.S. Navy submarines in identifying the ships that they attacked.

The naval attaché's office in Tokyo in 1933–1934 had an allowance of $300 per year to pay for information obtained through informants, but the fund was seldom touched.

Japanese police made each naval attaché pass a test within a few months after arrival in Japan. A phone call would be made by an anonymous caller requesting an appointment. When met by the attaché, the "informant" offered to sell plans for a naval base or proposed that he be hired as an undercover agent. It became part of the turnover routine to warn one's relief that such an approach would be made and to decline it.[58]

U.S. naval attachés were required to conduct themselves in a spirit of entire frankness and to be careful to show willingness to observe all the local rules regarding forbidden zones and police regulations. Every government knew that the naval attaché was detailed to get information concerning the local naval establishment and recognized that the attaché was entitled to take every legitimate means to procure that information. It was held that resorting to dubious methods would not bring results that could compensate for the loss of prestige in the eyes of foreign officials. The Navy Department directed that a naval attaché in the performance of his duties would employ "only such means as are consonant with his official position and the diplomatic relations that he bears to the government which receives him as naval attaché."[59]

The tasks of the naval attaché were stated in ONI's official manual in 1933 to be:

> In time of peace, to collect information on the naval strength and power for waging war of the country to which he was accredited and to cooperate with other U.S. Goverrnment agencies in the collection of information.
>
> In time of war, to collect information on the composition of enemy naval forces, their movements, and probable intentions; to cooperate with other government agencies in the collection of war information; to cooperate with the Chief of Mission in the performance of his duties; to evaluate and supply the Navy Department, and other governmental agencies concerned, with the information collected.[60]

The same manual indicated that the standard sources and methods available to naval attachés for collecting information were

> the host country navy department and air service; visits to ships, dockyards, and other host government establishments; associations with naval and military officers; visits to industrial establishments; associations with industrialists and other civilians; and the press correspondents, press clipping bureaus, and U.S. foreign service and commercial officers.[61]

A Joint Senate House Naval Affairs Committee indicated in one of its reports in early 1934 that maintenance allowances for naval attachés should be cut in half. Capt. Walter S. Anderson explained to the naval committee that this change would make it possible for only rich officers or men with rich wives to take jobs as naval attachés. The funds were restored.[62]

The office of naval attaché in Santiago, Chile, was closed on 30 June 1934 and the assistant naval attaché for air in Rome was ordered home without relief because of lack of funds.[63] Naval Attaché, Rome, was additionally accredited as naval attaché for air, and within a few years all naval attachés were accredited as naval attaché and naval attaché for air in order to facilitate their contacts with foreign air arms.[64]

The assistant naval attaché at Paris and his counterpart at Rome served during Fiscal Year 1934 without maintenance allowances for the extraordinary expenses incidental to their official positions.[65]

When the United States established diplomatic relations with the USSR and sent an initial embassy staff to Moscow in March 1934, Capt. David R. Nimmer, USMC, accompanied the group as the assistant naval attaché. Capt. Nimmer had previously been a Russian language student in Harbin, Manchuria, just prior to the time that Japan established the Kingdom of Manchukuo. A Navy captain was supposed to have had the attaché post, but he turned down the assignment at the last minute out of a desire to obtain command experience at sea. Consequently, a Marine officer gained the distinction of being the first U.S. naval attaché officially accredited to the USSR.

According to Nimmer, there was no one in ONI in 1934 who was interested in Soviet naval affairs per se. The original staffing plan for the naval attaché's office included three officer attachés, plus a dentist, surgeon, and paymaster, and twenty-nine enlisted men to support them and perform general Embassy duties as guards, messengers, chauffeurs, pharmacists, and electricians. In actuality, Capt. Nimmer and two Navy and six Marine enlisted men arrived in the Soviet Union on 7 March, and no more Navy or Marine personnel were assigned at that time.[66]

The specific information that ONI had instructed Nimmer to attempt to obtain included particulars on aircraft armament, cannon, and projectiles larger than .50-caliber; data on gunsights for flexible aircraft machine guns; methods of mounting bombs and torpedoes on large seaplanes; and chemical notes on diesel fuels. Nimmer's letter of 31 October 1934 to ONI attests to his lack of success in fulfilling the requests:

> Both the Chancery and the Consulate General, as well as the military and naval attachés, are experiencing the greatest difficulties in obtaining replies to communications, or unevasive answers to verbal queries. This conduct on the part of the Russians is not confined solely to dealings with Americans, but all diplomatic missions seem to be having like troubles.[67]

The difficulties that Capt. Nimmer had encountered in obtaining meaningful and useful information from the Soviet government, especially concerning publications and requests for specific information, were summed up in a letter to ONI written toward the end of his tour on 14 December 1934:

> As to the difficulties with the Russians, their fanatical secretiveness over the most trivial matter and their abject terror to make decisions without first referring, through the chain of command, the business in hand to the Minister of Defense [Voroshilov], makes the complete accomplishment of any single piece of business a major and generally unsuccessful undertaking; and of the officials all along the line, evaders and liars.

The Navy Department was becoming increasingly unhappy with the lack of cooperation being extended to Capt. Nimmer by the Soviets in comparison with the cooperation that the United States gave to the Soviet naval attaché in Washington, Adm. Paul Oras. In November 1934, ONI registered a complaint with Oras during one of his visits to the office. Oras immediately cabled Minister of Defense Klimenty Voroshilov. The latter arranged a meeting with Nimmer and claimed that his orders had been for the American to be shown "everything," implying that any shortcomings were the result of unofficial acts by individual commanders and not a reflection of Soviet government policy.

By December 1934, Capt. Nimmer was becoming increasingly pessimistic about what the Moscow naval attaché office was accomplishing. Thus, when a crisis in Soviet-American relations arose in early 1935, a convenient way of showing Washington's displeasure was to withdraw the military and naval attachés. The crisis concerned the Soviet refusal to make any settlement of the debts owed to the U. S. Government and private U.S. companies by previous Russian regimes. On 6 February 1935, Secretary of State Cordell Hull officially informed Ambassador William Bullitt that the naval attaché office would be closed and its personnel withdrawn immediately. The office was closed on 16 February 1935.

The lack of ONI protest at Nimmer's recall indicates that the cost of operating the Moscow office far outweighed the value of the information being received from it. Even the increased efforts of the Soviets after the Oras incident failed to make the post more attractive.[68]

Other naval attachés during Fiscal Year 1935 were stationed at London, Paris, Rome, Berlin, Brussels-Lisbon, Tokyo, Peiping, and Buenos Aires.[69]

The uncertain conditions and technical developments in Europe, on which the Navy Department needed to be kept as fully informed as possible, made necessary a strengthening of the naval attaché offices in certain countries. An additional assistant naval attaché (intended to be a Construction Corps lieutenant commander who would also be qualified as a naval aviator), was ordered to London, and plans were made to send an assistant naval attaché to Rome. Lack of funds, however, precluded the employment of additional clerical help, the need for which was being acutely felt in London, Paris, Berlin, and Rome. It was recommended that additional funds be provided for the employment of four additional clerks or that four chief yeomen be ordered to attaché office duty.[70]

Naval attaché offices were established in Rio de Janeiro and Lima in August 1935. Other attaché offices were continued during Fiscal Year 1936 in Berlin, Brussels, Buenos Aires, London, Paris, Peiping, Rome, and Tokyo.[71] In 1937, new offices were established at Santiago, Chile, and Bogota, Columbia.[72] During 1938, naval attaché offices were continued in London, Paris, Rome, Berlin, Tokyo, Peiping, Rio de Janeiro, Buenos Aires, Lima, Santiago, and Bogota. The office in Brussels was closed.[73]

The naval attaché organization abroad consisted of attachés and, in some cases, assistant attachés, each accredited to one or more foreign countries in Europe, Asia, and South America. In addition, there were a small number of officers attending schools or engaged in the study of foreign languages whose activities came under the Office of Naval Intelligence. Naval missions in Brazil, Argentina, and Peru, each consisting of a few officers and enlisted personnel, were also in part under ONI but were not included in the Naval Intelligence Service. Altogether, in 1938, there were twenty-seven officers assigned as attachés or assistant attachés, assisted by approximately thirty enlisted personnel or civilians; in addition, there were twenty-two officer students under ONI control or sponsorship.

Each naval attaché had a unit in ONI directly concerned with the activities of the attaché's organization and to which matters concerning his activities, whether originating within or outside the organization, were referred for consideration or recommendation. In addition, under the Assistant Director of Naval Intelligence, there was an active Foreign Intelligence Section to coordinate and administer the entire naval attaché system's activities, both within ONI and in the field.[74]

In January 1938, Naval Attaché, London, Capt. Russell Willson participated with Capt. Royal E. Ingersoll, who was assigned to the Office of the CNO, in conversations with the British about removing the limitations on the size of naval ships established by the Second London Naval Limitation of Arms Conference of 1935. Ingersoll represented President Roosevelt, and with Willson also initiated arrangements for developing joint codes, joint radio call letters, and the means for distributing these items prior to war conditions. Planning for collaboration against Japan in the Pacific was also discussed with the British.[75]

Additional naval attaché accreditation by post in 1939 was as follows:

NA Office	Also Accredited to
London	Naval constructor was assistant to France, Italy, Germany, Netherlands
Paris	Spain (until May); assistants to Spain and Portugal; Supply Corps assistant to Italy, Netherlands, England, Germany
Rome	Yugoslavia
Berlin	Norway, Sweden, Denmark, Finland
Buenos Aires	Uruguay
Bogota	Venezuela, Panama, Ecuador
Guatemala City	El Salvador, Honduras, Costa Rica, Nicaragua

Offices also continued at Tokyo, Peiping, Rio de Janeiro, Lima, Santiago (Chile), and Lisbon. New offices were reestablished at Mexico City in August 1938, at Brussels on 15 April 1939, and at Havana in May 1939; the naval attaché at Havana was also accredited to the Dominican Republic. The Hague office was reopened in August 1938.[76]

Prelude to World War II

On Thursday, 24 August 1939, the naval attaché at Paris, in a message to the Chief of Naval Operations, estimated that all German forces were in position to enter Poland not later than Friday night. He also expressed the opinion that England and France would fight. Germany invaded Poland on 3 September, and Great Britain and France did declare war on Germany.[77]

The Naval Attaché, Berlin, Cdr. Albert E. Schrader, maintained a war diary as a daily record of events from 1 September 1939 to 24 March 1941 that was submitted to ONI as a series of reports. Schrader's sources of information, in addition to the various U.S. press and radio broadcasting representatives in Germany, included the daily (but often thrice-weekly or even less) briefings of foreign attachés by the German naval ministry (attaché group), initially by a LCdr. von Davidson, then by a Capt. Mirow; news from local press and radio; German navy head Adm. Raeder (who, when he had special items to be passed to the United States, would call in Cdr. Schrader); the naval attaché community, particularly the Swedish, Greek, Italian, and Soviet attachés (there was considerable sparring with the latter two); limited travel outside Berlin, usually on diplomatic courier trips to external neutral cities; other members of the U.S. Embassy staff; and foreign attaché trips to English Channel ports, Belgium, the Netherlands, and Paris in July 1940 after the French surrender. As an example of Schrader's liaison with the press, George Kidd of the United Press was interviewed on 13 September 1939 following his return from witnessing the start of the war from Danzig; Kidd also called on Schrader on 10 March 1940 after making a two-day tour of the Upper Rhine front, bringing information on gun emplacements manned by the German navy east of Strasbourg.[78]

LCdr. Henri H. Smith-Hutton returned to Tokyo as naval attaché on 28 April 1939, relieving Capt. Harold M. Bemis. In reviewing the files of the Tokyo office, Smith-Hutton found that the best reports submitted to ONI had been those prepared in 1936–1937 by naval aviator LCdr. Ralph A. Ofstie, Assistant Naval Attaché for Air. Ofstie had inspected the Japanese naval air installations thoroughly and had learned a great deal about them.[79]

During the last few months of Capt. Bemis's tour, however, the Japanese navy department declined to allow visits to some of the Imperial Japanese Navy yards and bases that Bemis had requested, informing Bemis that the Japanese naval attaché in Washington had not been given permission to visit similar facilities in the United States. Since attaché visits were based on reciprocity, Bemis could not visit bases and airfields as he had previously been allowed to do. In turning over the post, Bemis advised Smith-Hutton not to make similar visit requests; it might induce the Japanese to try again at ONI, and he understood informally from Washington that they would prefer not to give the Japanese permission to visit U.S. yards and bases because of the big naval building program then in progress. Thus, the mutual inspection arrangements that had existed for many years were terminated, and thereafter there was no chance to visit Yokosuka, Kure, or Sasebo, the Imperial Japanese Navy's main shipbuilding yards.

Although many contacts with Japanese friends were lost because of the increasing practice by the police of interrogating visitors to U.S. residences, the Naval Attaché, Tokyo's reports continued to be voluminous. The *Official Gazette* was followed very closely for every mention of the Japanese navy in debates in the Diet or the House of Peers. There were a number of popular naval publications and magazines, and there was quite a lot published on the merchant marine, with many pictures. Publications provided much information worthy of reporting. All pictures of Japanese merchant ships appearing in magazines and newspapers, and the detailed plans available from unclassified sources, were forwarded to ONI for use in preparing recognition manuals. In addition, every week for a year and a half, a few detailed maps of different sections of the country had been purchased, until a complete topographic map of all Japan was in hand. It was then bound and sent to ONI. Although the maps didn't show fortified areas or military and naval bases, they did show cities, towns, railroads, and terrain features, all of which were of value in planning air targeting operations in wartime.[80]

Most of the above reports contained no startling information but were full of nonclassified data of potential wartime interest and value to the Navy. Very few reports were sent by cable because of their lack of urgency. Furthermore, cables were expensive.[81]

One covert source provided information that the torpedoes carried by Japanese destroyers appeared to be larger than 21 inches in diameter and were probably nearer to 25 inches (they were in fact of 24-inch diameter). The source also reported that the torpedoes used oxygen for fuel. No comment was received from Washington when this new information was sent to ONI. Another report from the same source noted that the cruiser *Mogami*'s main battery weapons were not as shown in *Jane's Fighting Ships*. Instead of five turrets with three 6-inch guns each, *Mogami* had five turrets with two 8-inch guns each. The Bureau of Ordnance commented on the report that it was impossible to reconfigure a ship designed for 6-inch guns to one for 8-inch guns. Both reports, however, were later found to be entirely correct.[82]

One of the pre–World War II intelligence collectors in Japan was Lt.(jg) Stephen Jurika, Jr., the assistant naval attaché in Tokyo from June 1939 to September 1941. Under LCdr. Smith-Hutton, Jurika participated in most of the collection effort conducted outside of the office. As naval attaché, Donald J. Smith-Hutton handled the policy, protocol, and entertainment activities of the office; LCdr. McCallum, the senior assistant naval attaché, handled the office work.

Jurika attempted to witness all ship launchings. In Kobe, he would reserve a room overlooking the shipyard, either at the Tor Hotel up on a hill, or on the fourth floor of the Oriental Hotel. Fairly good pictures were possible with a telephoto lens. At Yokohama, it was easier. Some of the Standard Oil representatives lived on a bluff overlooking the Mitsubishi yards.

In 1940, a Japanese Zero fighter aircraft was on display at the Haneda International Airport. Jurika, a naval aviator, went to see it and was allowed to sit in the cockpit, where he found the nameplate written in English. He noted that the weight of the Zero was about half that of the U.S. Navy's F4F Wildcat, but that the horsepower was the same, giving the Zero better speed, climbing, and maneuvering capabilities. About three months after he submitted his report, ONI chided him that he should be more careful in reporting the characteristics and estimated weight of Japanese aircraft.

Permission was never given to visit Japanese naval installations, but seaplanes were operated from a naval air station on the Chiba peninsula on the east side of Tokyo Bay, where there was also a good golf course nearby from which Jurika could observe their activities.

Jurika went once a quarter to the Philippines to get in his required flight time. He made the trip from Tokyo to Manila and back on regularly scheduled President Lines ships, and he got to know the masters of the ships quite well. They were all U.S. Naval Reserve officers and glad to cooperate. The first stop was Kobe, Japan, and, on leaving port, the ship would pass as close as possible to the Mitsubishi shipyards, where naval ships were being built. On the landward side of the shipyard, observation of construction progress was blocked by matting. Jurika always had a stateroom on the starboard side of the ship and would take a series of pictures from the porthole, with his Leica mounted on a tripod, as the liner passed the shipyard. Comparing the pictures on successive trips gave a good appreciation of progress of the construction.

If, on Jurika's trips to the Philippines, the ship was passing southeast of the island of Kyushu when any Japanese aircraft carriers were operating out of Ariake Wan, the President Lines ship would slow down so that Jurika could check the timing of their landing or launching operations and note the characteristics of other maneuvering evolutions.

Jurika made a special effort to collect target intelligence. There were many commercial pamphlets available, and whenever he drove from Tokyo to Yokohama, Jurika would travel via a different route, noting all the industrial complexes that stretched without interruption between the two cities. Jurika was also able to get a complete series of land-use maps of Japan and, beginning in July 1940, he worked to fill in on the maps what were considered primary targets and the points of identification needed for aerial approaches to the targets.

Jurika obtained a lot of the information for the target maps from the Soviet assistant naval attaché. Jurika met him on the tennis court and, after they had lunched together, the Russian wanted to know what information on Japan he and Jurika could exchange. The Soviets had been collecting information on Japanese industrial establishments for years, and their attaché supplied Jurika with information on the locations of factories in Tokyo and its suburbs. The U.S. Navy officer responded with information from Japanese newspapers and magazines. The Soviet-supplied information saved Jurika three or four months of driving around, and he found the data to be accurate. The Russians had the best espionage collection net in Japan, using Japanese Communists as sources.

Jurika made many attempts to reach the Mandated Islands. There was a four-engined flying boat that went each week from Tokyo to Ponape and Truk. For six months he tried to buy a ticket on the plane, but each request was refused for various reasons. Finally, Jurika obtained a written statement from the Japanese that he was not allowed to visit the Mandated Islands. He did get some third-hand information, however, from the Soviet assistant naval attaché who had contacts with Japanese fishermen who had been to the islands. The main information of value was the identification of those islands that were prohibited to the fishermen and were thus assumed to be military bases.

The American Club, located in the commercial district of Tokyo, was the meeting place for news correspondents and industrial representatives, both those based in Japan and those passing through. They would often hold forth and analyze conditions and situations in the Far East, and they often served as good sources and provided leads to anticipated events. Japanese newspapers and magazines became better information sources when, in mid-1940, the Japanese government found it desirable to bring the war in China home to its people and to enlist their support. More and more articles, photos, and news accounts of the war in China appeared in the public media.[83]

On 20 March 1940, the Naval Attaché, Berlin, was relieved of his Scandinavian accreditations by LCdr. Ole O. Hagen, USN (Ret.). On the 31st, the assistant naval attaché in Berlin, LCdr. Edward R. Durgin, was detached prior to the arrival of his relief, LCdr. Arthur H. Graubart, who reported on 24 May. LCdr. Hagen was stationed in Sweden, with additional accreditation to Norway, Denmark, and Finland.[84]

Other new posts in the naval attaché system opened during Fiscal Year 1940 in Venezuela, the Dominican Republic, and Turkey.[85]

Transportation available from Berlin to the United States in July 1940 was a daily train to Geneva (which took one day), followed by a once-a-week bus from Geneva to Barcelona (which took two days), then a train from Barcelona to Lisbon. From Lisbon one could proceed by Pan American Clipper or by ship to the United States. The same route was used for official diplomatic pouch mail.

The sale of fifty overaged destroyers to the British by the United States in August 1940 upset the German navy ministry and made it less cooperative in handling the naval attaché's requests for information.

Hitler was apparently a believer in astrology. Foreign attachés in Berlin, therefore, were usually influenced in their forecasts on the timing of possible big events, such as the invasion of England, by checking to see if the right planets were in alignment. Two such days viewed with apprehension were 24 July and 15 August 1940, but other anticipated preparatory steps failed to materialize.[86]

After the invasion of Belgium and northern France in July 1940, Naval Attaché, Paris, was transferred to Vichy, France. After the German occupation of all of France on 11 November 1942, all personnel at the naval attaché's office at Vichy were transferred to Baden Baden under German custody.[87]

Cdr. Roscoe H. Hillenkoetter, the naval attaché to Vichy France, made a quick trip to Morocco and Algeria in 1940. He reported to the U.S. Embassy at Vichy that he was agreeably surprised and encouraged by what he had observed: contrary to rumors (planted by the Germans), Hillenkoetter found that the Nazis had left French Africa almost completely to its own devices. He stated that "if France is going to fight again anywhere in this war, I believe North Africa will be the place."[88]

In April 1940, President Roosevelt decided to send a senior naval officer to London for informal discussions with the British Admiralty. The orders, first issued in July 1940, provided that RAdm. Robert L. Ghormley would serve as naval attaché, with his assistants to be designated as assistant naval attachés. Since this was to be a special (and probably temporary) assignment, however, Ghormley's title was changed to "Special Observer." Ghormley was attached to the U.S. Embassy but accredited directly to the British Admiralty.

Members of the office of Naval Attaché, London, were invited to attend special-observer joint meetings with the British when British proposals for Anglo-American naval cooperation pertaining to their specialties were to be discussed.

In September 1940, the British Committee for Joint Cooperation with the United States under Sir Sidney Bailey had been designated by the Admiralty to receive all requests from the U.S. naval attaché for technical information. Between 12 September 1940 and 2 July 1941, 395 such requests were made. In addition, the naval attaché continued to use Section V of the (British) Naval Intelligence Division (NID) as the normal liaison channel with the Admiralty for naval matters relating to the British Empire. Officers sent by the various Navy bureaus to observe British tactics, inventions, etc. came under the U.S. naval attaché as did officers who were assigned aboard British ships as observers. On 22 October 1940, there were thirty-two officers designated as Assistant Naval Attaché, London.[89]

In early 1941, ONI began sending naval observers to various key Brazilian ports. The first was LCdr. William A. Hodgman, USN (Ret.), who arrived at Recife on 26 February. At first Hodgman obtained office space in the U.S. consulate, but he later moved to the third floor of the Bank of London building, close to the waterfront, where he could overlook the harbor.[90]

When Commander Cruiser Division Two, RAdm. Jonas H. Ingram, arrived on 10 May 1941 to check out Recife and Bahia as replenishment ports for the Neutrality Patrol, LCdr. Hodgman was able to advise Ingram on the facilities of each port and to point out the superiority of Recife over Bahia for naval patrol purposes. Hodgman also had made the initial contacts with local Brazilian authorities that led to Ingram's later favorable relationship with them.

Recife became the center of U.S. naval activity in the South Atlantic, and in August, 2ndLt. D. J. Kendall, USMCR, arrived as assistant naval observer to help provide services to the increasing number of U.S. ships visiting Recife.

Other naval observers assigned in Brazil in 1941 were Lt. M. B. Saben, USN (Ret.), to Bahia, arriving 1 October; LCdr. Hugh C. Frazer, USN (Ret.), to Natal on 14 October (relieved on 28 October 1941 by Lt. L. K. Winans, USNR); and LCdr. Edward Breed, USNR, to Belem, on 17 November.[91]

Cdr. Schrader was relieved by Capt. Adolph von Pickhardt as Naval Attaché, Berlin, 1 April 1941, shortly after passage by Congress in March of the Lend-Lease bill, an event that made the Germans even less friendly than before.[92]

New naval attachés were assigned during Fiscal Year 1941 to the Union of South Africa, Australia, Thailand, Canada, Uruguay, and Argentina.[93] In anticipation of an approaching world conflict, ONI, in the summer of 1941, began sending naval observers, naval liaison officers, and consular shipping advisors to all the principal ports and hot spots in the world.[94]

The U.S. naval attaché system at the time of the Pearl Harbor attack employed 133 officers, 200 enlisted men, and no civilians.[95]

The Navy Department in 1938 had started planning for the reestablishment of the naval attaché office in Moscow, but the plan had been vetoed by the State Department. The question was again opened in May 1941, and again the State Department opposed the idea because of strained Russo-American relations and because of American disapproval of Soviet foreign policy. Inadequate living quarters and the difficulty in obtaining information were again cited. After discussions on 15 May 1941 between Capt. Alan G. Kirk, Director of Naval Intelligence, and Ray Atherton and Loy Henderson, Acting Chief and Assistant Chief of the Division of European Affairs, State Department, respectively, the matter was referred to the U.S. Embassy, Moscow. The Embassy, in turn, sounded out the Soviet government.

On 22 June 1941, the Germans invaded the USSR, and the next day the Soviet ambassador to the United States informed Henderson that the Soviet government agreed to the stationing of a U.S. naval attaché and his staff in Moscow. The Soviet government questioned the need for four officers, the number requested by ONI. The explanation that two would be concerned with naval aviation partially satisfied the Soviets, but the request to assign one to live in Vladivostok caused further delay.

On 7 August 1941, the Soviets agreed to four officers (the German armies were getting close to Moscow), but they still opposed stationing a naval observer or U.S. shipping advisor at Vladivostok. Navy Lts. Samuel B. Frankel and George D. Roullard arrived in Moscow at the end of September.[96]

When George Roullard finally got approval to go on to Vladivostok, he went as an assistant naval attaché with the same privileges that he would have had in Moscow, but he was not permitted to wear a uniform. Roullard, together with a yeoman first class, set up an office in February 1942 at the U.S. consulate, where he was to act as if he were a member of the Vladivostok consulate staff. Roullard was not to reveal his U.S. Navy identity except to personnel of the Soviet navy who were permitted liaison with him. The Soviets didn't want to give the Japanese consulate in Vladivostok any justification to request similar privileges. Roullard's primary duties were to report ship movements and information of interest concerning the movements of Lend-Lease supplies to the USSR via the Pacific route.[97]

Naval Attachés During World War II

Japan, 1941

When military extremists took control of the Japanese government on 16 October 1941 and Gen. Hideki Tojo became prime minister, all foreign naval attachés in Japan were informed that the navy ministry had to be advised one week in advance of any plans to travel more than 15 miles from Tokyo, and the exact itinerary had to be provided. Capt. Henri Smith-Hutton, the naval attaché, made a test run west to the tourist resort of Miyajima accompanied by his wife, going by train through Osaka and Kobe. At stops in Hiroshima (an important military center) and Kure (a naval base), the train attendant pulled down the shades and told them to remain in their compartment. At Miyajima, they were escorted by a policeman and a detective to their hotel and whenever they left their rooms. Following the trip, Smith-Hutton notified Ambassador Joseph Grew and ONI that he could not be counted on to give advance warning of Japanese naval moves.

On 8 December 1941, Capt. Smith-Hutton was alerted that something serious had happened when he tuned in the American radio station at Shanghai shortly after he awoke at 0630. The announcer was reading a directive from the American consul general advising Americans to remain calm and to stay off the streets. There was no clarifying announcement, and the station signed off.

Smith-Hutton called his wife and told her something had happened in Shanghai, and he suggested that they walk over to the U.S. Embassy chancery in the adjacent garden. At his office, he learned that Radio San Francisco had announced the Japanese attack on Pearl Harbor. Smith-Hutton notified Ambassador Grew, who instructed Smith-Hutton to go to the Japanese navy department, four blocks away, to find out if the broadcast from San Francisco was correct. Smith-Hutton went to the office of RAdm. Nakamura, senior aide to the navy minister, Adm. Shigetaro Shimada. Nakamura confirmed that the broadcast was true. Smith-Hutton returned to the embassy and so advised the ambassador.

Except for one cipher for use in emergency communications, the naval attaché's office had burned its classified papers and codes four days before. The destruction of the code machine was accomplished with a small hammer. The small bits of metal were placed in about twenty envelopes. Late that night, Smith-Hutton and his assistant, LCdr. Martin R. Stone, drove towards Yokohama, dropping one envelope into the water at each bridge they crossed. The final envelope was thrown into the moat at the Imperial Palace.[98]

The embassy staff in Tokyo was locked up in the large embassy compound on 8 December 1941, and all those who lived outside the compound were allowed to visit their homes to bring in clothing, personal belongings, etc. Ambassador Grew's residence had three bedrooms, and he took in a number of the senior embassy officials and their wives, including the Smith-Huttons. The Ambassador's mess included the Smith-Huttons plus four other Americans, and other messes were established in the apartments near the chancery. After about six days, the Americans were allowed to send one of the chauffeurs out for fresh vegetables and fish.

After about ten days, the Swiss minister was allowed to come into the U.S. Embassy. After that, he visited fairly regularly, bringing news received via short wave radio. No mail was permitted, but staff members were allowed to receive the *Japan Advertiser*, an English-language newspaper published in Tokyo.

On 18 April 1942, the day of the Doolittle raid, the air raid sirens started about 1100, but since there had been a series of air defense drills, they were not taken to be unusual. Japanese aircraft, however, usually took part in the air attack drills, giving a rare chance to see them from the ground. Capt. Smith-Hutton and his assistant, LCdr. Stone, among others, frequently took advantage of the excellent view from the top of one of the Embassy compound apartment buildings just to see the show. On the morning of the Doolittle raid, there were several fighters in the air, and the interned Americans watched them with binoculars as usual. At 1130, they saw a plane flying low over the northern part of the city, about 6 miles from the chancery. The observers thought they heard faint gunfire but couldn't be sure, and then they saw a cloud of black smoke coming up from an area that looked to be about under the path of the low-flying aircraft. The plane had disappeared to the west, and they decided that the Japanese were making the drills more realistic.

At about noon, they returned to the Embassy for lunch when a plane flying low swooped over the Embassy and quickly disappeared behind the large trees to the west of the compound. It was seen only by Mrs. Smith-Hutton and Crocker, the first secretary, who was very nearsighted, but he agreed with Mrs. Smith-Hutton that the aircraft had had American markings, not Japanese *hinomaru* (sun) insignia. Also, the engines seemed to have a different sound. Neither, however, could describe the plane well enough to permit identification. Later, the Japanese did announce that the aircraft were American. Thereafter, the police at the gates seemed to pay more attention to blackout curtains at the chancery.[99]

During his internment in Tokyo, Smith-Hutton collected a complete file of the English-language newspaper, *Japan Times*, and the Japanese newspaper, *Tokyo Nichi Nichi*. He also made a card file on every Japanese naval officer of the rank of commander and above, with the duty stations of the officer and remarks on his personality. When the naval attaché staff was evacuated, they were told not to take any written material out of the country. Smith-Hutton, however, felt the Japanese would be too busy to bother with a thorough inspection of the passengers' luggage. So, although the file of newspapers he had was quite bulky, he divided it up among numerous boxes and suitcases. In addition, Smith-Hutton carried on his person a diary of important events and observations that he had kept during internment. Fortunately, there was no inspection, and Smith-Hutton turned over all the material to the Far East Section of ONI upon his return to Washington.[100]

The Embassy staff in Tokyo remained interned in the Embassy compound until 17 June 1942, when they embarked on the first diplomatic exchange ship, *Asama Maru*, which departed Yokohama on 25 June. After stops at Hong Kong, Saigon, and Singapore, the Japanese liner rendezvoused with the neutral Swedish passenger ship *Gripsholm* at Lourenco Marques in the Portuguese East African colony of Mozambique. Here an exchange was made with the Japanese Embassy staff internees from the United States. Cdr. Ethelbert Watts was sent there from ONI to assure a body-for-body accounting.[101]

Naval Attaché, London

The post of Naval Attaché, London, was held by the following officers during World War II:

Name	Dates
Capt. A. G. Kirk	1 Oct 1939–Nov 1940
RAdm. R. L. Ghormley (also Special Naval Observer [SPENAVO])	Nov 1940–Mar 1941
Capt. C. A. Lockwood, Jr. (also SPENAVO)	9 Mar 1941–12 Mar 1942
Cdr. E. W. Litch (Acting NA)	12 Mar 1942–12 May 1942
RAdm. A. G. Kirk (also Chief of Staff, Commander Naval Forces Europe [COMNAVEU])	12 May 1942–9 Feb 1943
Capt. P. H. Bastedo (also Chief of Staff COMNAVEU, 13 Feb–25 May 1943)	9 Feb 1943–17 Oct 1943
RAdm. G. B. Wilson (also Chief of Staff COMNAVEU, 25 May 1943–13 Apr 1945)	17 Oct 1943–13 Apr 1945

Name	Dates
RAdm. L. T. Dubose (also Chief of Staff COMNAVEU)	13 Apr 1945–25 Jul 1945
RAdm. S. S. Lewis (also Chief of Staff COMNAVEU)	25 Jul 1945–1 Aug 1946

Prior to 25 July 1941, the Naval Attaché, London, had the additional title of naval attaché for air. Thereafter, a naval aviator was designated as naval attaché for air and assistant naval attaché. The officers holding the position were:

Name	Dates
Cdr. R. A. Ofstie	25 Jul 1941–7 Mar 1942
Cdr. E. W. Litch	7 Mar 1942–7 Aug 1942
Cdr. F. B. Kaufman	7 Aug 1942–15 Jan 1943
Capt. A. I. Malstrom	15 Jan 1943–1 Dec 1943
Capt. H. B. Miller	1 Dec 1943–15 Aug 1944
Capt. W. F. Kline	15 Aug 1944–8 Nov 1945[102]

On 27 May 1941, following President Roosevelt's proclamation of a state of unlimited emergency, arrangements were completed for a working amalgamation of the staff of the naval attaché and special naval observer in London. Capt. Charles A. Lockwood, Jr., became chief of staff to RAdm. Ghormley but retained his primary designation and duties as Naval Attaché, London. The two organizations were not merged, but their activities were coordinated by Lockwood, and some of the other key personnel served as members of both staffs.[103]

On 28 May, with the increased staffing of the office, the naval attaché organization was expanded from four sections to seven departments: Administration and Security, Communications, Supply and Disbursing, Operations, Fleet Observers, Technical, and Aviation. The Technical Department was the most significant and in due course was combined with the Material Section of the staff of the special naval observer.

In July 1941, the senior assistant naval attaché for air was raised to the position of Naval Air Attaché and at the same time was assigned as Air Officer on the SPENAVO staff. Cdr. Ralph Ofstie assumed the position on 25 July 1941.

Attempts were occasionally made by various Navy bureaus and offices to communicate directly with the special naval observer on nonintelligence matters, but the practice was discouraged. To assure coordinated action by Navy Department offices and reduce administrative duplication by the special naval observer, all correspondence, regardless of subject, was sent via ONI.

The large number of officers carried as assistant naval attachés became a concern. Many, in fact, were observers in function rather than attachés. Accordingly, those officers who were observers of wartime operations had their titles changed to naval observer, lost their attaché status, and served primarily on the staff of the Special Naval Observer.[104]

When the United States entered World War II, the Special Naval Observer became Commander Naval Forces, Europe.

When Adm. Harold R. Stark relieved RAdm. Ghormley as COMNAVEU, he took steps to unify all U.S. naval activities in the United Kingdom under his staff. Stark's Chief of Staff, RAdm. Alan Kirk, was designated Naval Attaché, London. The arrangement assured the continued coordination of naval intelligence reporting that had previously been achieved by double-hatting the special naval observer's chief of staff and the naval attaché.[105]

The naval attaché reported on the enemy's organization, operations, and plans and on Allied and enemy material and weapons. Operational intelligence was increasingly the function of COMNAVEU. Thus, both organizations were involved in the gathering of intelligence. Direct liaison was maintained not only with the Admiralty but also with other British and Allied intelligence services in London. Officers assigned to the naval attaché were also assigned for additional duty to COMNAVEU.

On 28 April 1942, the Technical Section of the office of the naval attaché was consolidated with the Maintenance Division of COMNAVEU and became the Material Division of COMNAVEU. Capt. Thorwald A. Solberg remained as head of the Material Division, which was concerned with general engineering and naval construction, scientific research, and standardization of components.

Initially, liaison with governments in exile in London was maintained by the Intelligence Division of Commander Naval Forces, Europe. In June 1942, however, COMNAVEU recommended that a naval attaché be assigned to the ambassador for liaison, and Capt. John L. Callan was assigned the duty in September 1942.[106]

On 9 February 1943, Capt. Paul H. Bastedo relieved Adm. Kirk as naval attaché and as Acting Chief of Staff COMNAVEU in London. On 25 May 1943 Capt. (soon to be RAdm.) George B. Wilson reported as Chief of Staff, and thus, for the first time since May 1941, the functions of the chief of staff and naval attaché were organizationally separated. It was decided, however, that some of the intelligence functions should be placed under an assistant chief of staff for intelligence, and Capt. Bastedo, while still naval attaché, was assigned to that billet with responsibility for censorship, travel control, security, combat intelligence, political warfare, prisoner-of-war matters, shipping statistics, and historical files. The Technical Section, however, reported directly to the chief of staff. On 17 October 1943 when Capt. Bastedo left, RAdm. Wilson took on additional duties as naval attaché.[107]

The Allies' Combined Chiefs of Staff determined that attachés should be appointed to countries as they were liberated, even though Supreme Commander Allied Expeditionary Forces (SCAEF) still had the responsibility for military and administrative matters. The U.S. Joint Chiefs of Staff concluded that matters of long-term policy and matters outside the area of SCAEF responsibility required handling by attachés and that the latter's functions would not interfere with those of the SCAEF mission. Where a conflict arose, SCAEF would have the authority to decide the matter. Accordingly, on 27 March 1945, Capt. Dallas D. Dupre, the intelligence officer on the staff of Adm. Kirk, Commander Naval Forces, France, was detached and ordered to report to the U.S. Ambassador to France as the naval attaché and naval attaché for air.[108]

Naval Attachés in South America

In early 1942, VAdm. Jonas Ingram inspected the various U.S. naval facilities in Brazil at Recife, Bahia, Maceio, and Natal. As a result, he had Lt. L. K. Winans relieved as naval observer at Natal by LCdr. Charles B. Gary, who had previously been assigned as an observer in Maceio. The latter port was useless because of its shallow water and scarcity of supplies, but Natal was the location of a commercial seaplane base superior to any in the Caribbean or elsewhere in Brazil. The work of the naval observer was becoming more oriented toward logistic support of operating forces, and only infrequently was the naval observer staff able to work on ONI's requirements for the collection of intelligence.[109]

In April 1942, Assistant Naval Attaché, Rio de Janeiro, Capt. E. Edward Brady, Jr., accompanied VAdm. Ingram on his visit to see Brazilian President Vargas. It was the diplomatic highlight of Ingram's experience in Brazil and led to the "unfreezing" of Brazilian commercial shipping and the naming of Ingram as Vargas's Sea Lord and Naval Advisor.[110]

Brazil declared war on Germany and Italy in August 1942, and on 12 September, the Naval Attaché, Rio, informed Ingram that President Vargas had ordered Brazilian navy forces to be placed under Commander South Atlantic Forces (COMSOLANTFOR), Ingram's new title, effective 15 September.[111]

RAdm. Augustin T. Beauregard was made Naval Attaché, Rio de Janeiro, in 1942. In mid-November, Commander in Chief, U.S. Fleet Adm. Ernest J. King directed Ingram and Beauregard to reorganize U.S. naval activities in Brazil to provide better collaboration in support of the operating forces. As a result, a naval operating base (NOB) was established at Rio with RAdm. Beauregard as the commanding officer. All assistant naval attachés in Rio were shifted to operational duties in the new NOB organization. A new naval attaché was ordered to Rio. In addition, all naval observers on the east coast of South America were placed directly under VAdm. Ingram as COMSOLANTFOR, but in each of the observer's offices there were to be one or more intelligence officers working directly under ONI. These intelligence officers were also to maintain liaison with the naval attaché at Rio de Janeiro and with COMSOLANTFOR.[112]

During 1942, the naval observer's staff in Recife was increased to seven officers. Lt. H. A. Richey, USNR, was the executive officer and assistant for intelligence, and Ens. (later Lt.) W. F. McKenna was the assistant for communications intelligence. LCdr. Hodgman was promoted to commander. His responsibilities varied widely, from ensuring rapid replenishment of ships (including convoys) to quick delivery of mail to U.S. military personnel in the area. He also handled the leasing of a building for use as the headquarters for COMSOLANTFOR, and he moved his own offices into the same building.[113]

The number of ship sinkings in the South Atlantic increased in late 1942, and the Naval Observer, Natal, was kept busy with the debriefing of survivors and in arranging for their air transportation back to the United States.[114]

The missions assigned to naval observers in Brazil by ONI included the collection and classification of information and the establishment of proper liaison between the U.S. Navy and the Brazilian armed services and civil government, as well as with other U.S. Government agencies. A naval observer office was established in Santos, Brazil, in March 1942, with LCdr. W. M. Kilcoin, USNR, in charge. Until 17 April, his first and sole duty was to aid in the negotiations for the commissioning of transports *Monticello* (AP 61) and *Lejeune* (AP 74), the former Italian *Conte Grande* and the German *Windhuk*, respectively, which had been interned in Santos at the outbreak of the war in 1939.[115]

In May and June, connections were established to provide sources of information on Axis sympathizers and groups in Brazil, as well as surveillance of suspicious persons. As a result, it was possible for the naval observer in Santos to report suspected persons to the police and to effect their detention or removal to Brazil's interior. Sources of information developed by the Santos office included the American consul, the local British intelligence staff, local representatives of the U.S. Federal Bureau of Investigation (FBI), the Brazilian army and navy, police and customs officers, representatives of steamship companies, ship chandlers, and others having information on cargo movements, purchases of stores and supplies, the activities of dock employees, the crews of neutral merchant ships, and persons involved in waterfront incidents.

Reports and information were exchanged at weekly or semiweekly meetings of the naval observer at Santos, the American consul, FBI representatives, and the Brazilian security and intelligence staff held either at the office of the American consul or the office of the naval observer. Requirements of the naval observer for police investigations were handled through the local FBI office, but when the assistance of the maritime police, dock authorities, or Brazilian army or navy was required, the naval observer made the requests directly. Close contact was maintained with the local port captain, a Brazilian naval officer, in matters relating to surveillance of crews of suspect ships, the denial of certain supplies to selected neutral merchant ships, and the control of access to dock areas.

When, in June 1942, the Naval Observer, Santos, was assigned the added responsibility of ship reporting and routing, with its attendant heavy increase in communications, most intelligence functions had to be handled by LCdr. Kilcoin's assistant, Lt. James H. Redington, USNR. Ens. John P. Fitzpatrick, USNR, was the assistant intelligence officer. As time permitted, ONI was supplied with detailed reporting on subjects of naval interest such as landing beaches, naval installations, harbor defenses, and supplies, and on heavy industry, transportation, commerce, communications facilities, public utilities, hospitals, political and subversive activities, and current events.

On 11 November 1942, the naval observer office at Santos received a Director of Naval Intelligence memorandum to all naval attachés and observers that defined the counterintelligence information needed by ONI as that relating to the enemy's plans and capacity for espionage or sabotage, the type of information the enemy agents desired and what they were actually getting, and the methods used by the enemy to transmit espionage reports.[116]

In 1943, the various naval observer posts in Brazil had expanded into Naval Operating Facilities (NOF) and were designated as such on 1 June. At Recife, Cdr. Hodgman had been promoted to captain and had the title Commandant NOF Recife, reporting directly to Commander South Atlantic Forces. Similarly, at Bahia, the original naval observer, former Lt. Saben, was now a commander and was Commandant NOF Bahia. Although Bahia had expanded substantially, it remained smaller than the NOF at Recife. In the first half of 1943, Bahia was a major convoy assembly point.[117]

Naval Observer, Belem, LCdr. H. V. Whelan, USNR, became Commanding Officer NOF on 30 April 1943. NOF Belem became the center of all U.S. naval activity around the mouth of the Amazon. Being approximately 1,000 miles from Recife, Belem was the stopping-off point for high-ranking officials and other visitors passing through. Included among the outlying facilities at Belem was a naval supplementary radio station. The intelligence officers assigned at NOF Belem during the war included Lt. J. W. Meehan, Jr., USNR; Lt.(jg) R. L. Ramsey, USNR; and Lt. R. T. Davis, Jr., USNR; successively. By 31 December 1943, some time was available for work on intelligence reporting in support of ONI's monograph-updating effort by the intelligence officer.[118]

At NOF Natal, a Joint Intelligence Committee was organized in early 1943 and was composed of local representatives from the various U.S. Navy activities at Natal, the U.S. Army, the U.S. vice consul, and the Army provost marshal. Meetings were held in the NOF offices on a biweekly basis to discuss intelligence and security problems of mutual interest.[119]

The Naval Attaché, Montevideo, served as Commander South Atlantic, VAdm. Ingram's contact with the Uruguayan government. He obtained a favorable response from the Uruguayans to Commander Fourth Fleet's request for unwritten permission to use Uruguayan territory for sea and land aircraft operations in the event of U-boat activity off the Rio de la Plata.[120]

Naval Attachés in the USSR

During 1942–1943, attaché posts in the USSR were located at Moscow, Murmansk, Archangel, and Vladivostok. The ports were all expanded throughout the period, especially in Murmansk, where a lack of American civilian personnel and the complications of handling the traffic of the Murmansk convoys placed a heavy burden on the staff.

Information obtained by the Naval Attaché, Moscow, from the Russians included material on the Imperial Japanese Navy and on Japanese-controlled ports and installations, as well as a large amount of technical data on German mines, ships, etc.[121]

The U.S. Military Mission to the USSR in World War II operated directly under the ambassador in order to permit access to Stalin. It was established on 18 October 1943, commanded by MGen. John R. Deane, U.S. Army, and divided into three divisions: Military, Navy, and Supply. An Air Division was added later. The functions of the Navy Division were typical of those of a naval attaché office and included:

1. Advising the head of mission and the ambassador on naval matters.

2. Placing personnel in Soviet-controlled ports where U.S. merchant ships called, primarily at Murmansk, Archangel, and Vladivostok, but also at Odessa, Constanta, Novorossisk, Gdynia, and Danzig.

3. Acting for the War Shipping Administration as representatives of U.S. shipping interests in port.

4. Assisting U.S. personnel ashore, helping survivors, and helping to acquire survivors' stores.

5. Collecting data on port facilities, convoys and their routing, and repair facilities.

6. Providing communication services for the exchange of weather information, Lend-Lease matters, and traffic to outlying stations.

In general, the Navy Division received more cooperation from the Soviets than did the U.S. mission as a whole. The Red Navy staff was more approachable than the Red Army staff, apparently because the Red Navy believed there was much to be gained by collaborating with representatives of the much larger U.S. Navy. On the other hand, the Red Army proudly considered itself superior to any other army and could see very little to be gained from collaboration with the U.S. Army.[122]

In February 1944, an agreement was reached to exchange intelligence information on Japan between the chief of the naval section of the U.S. Military Mission to the USSR, RAdm. Charles E. Olsen, and the acting chief of the Soviet naval general staff.[123]

The office of the Assistant Naval Attaché, Archangel, USSR, was closed as of 3 August 1946.[124]

Naval Attachés in Africa

LtCol. William A. Eddy, USMCR, the naval attaché at Tangier from before World War II to October 1943, had grown up in the Middle East and spoke Arabic fluently. He was invaluable throughout the various stages of the North African operations. "No American knew more Arabs or more about power politics in Africa. He was one of a kind, unique; we could have used a hundred like him."[125]

On the recommendation of the naval liaison officer in Durban, South Africa, a naval liaison/observer post was established at Lourenco Marques, Mozambique (Portuguese East Africa) in February 1943. Since Portugal was a neutral country, the post was designated as a civilian billet with the title of Assistant to the Consul General. Lt. John W. Scott, Jr., was the first officer assigned to the post, and he was assisted by CY Charles W. Adam, USNR.

Personnel assigned to the Lourenco Marques post wore civilian clothes and had special passports. Naval fiscal support for the post was handled between the Navy and State Departments, funds being transferred as the consul general's reports were received. The Collection and Classification of Information fund was handled separately at a local bank. Logistics, medical services, and discreet communications were handled through the American Legation, U.S. Liaison Office, Durban.

The second officer assigned to Lorenco Marques was Lt. Donald H. Scott, USNR (no relation to John W.). He took over in August 1944 and remained until the post was officially closed on 15 December 1945. In January 1945, CY Adam was relieved by two yeomen first class, Henry E. Mezger and George Morales.

The office of the assistant to the consul general was located in an apartment overlooking the main square of Lourenco Marques about one block from the waterfront. It also looked down the street that ran parallel to the waterfront and housed all the bars. Merchant seamen and armed guard personnel couldn't walk into the rest of the town without crossing the line of bars. One of the functions of the post was to help keep the personnel out of trouble while they were on shore liberty. At times, there were as many as six U.S. merchant ships in port, mainly to pick up iron and chrome ore for the United States. One of the best preventatives against disturbances was to deliver U.S. mail as soon as a ship arrived, and particularly before liberty started. (U.S. mail arrived once a week via a British Overseas Airways aircraft from Cairo.)

In collaboration with the British routing officer, located next door, the U.S. naval observer's office hired a Jewish refugee from Germany named Karl to be the "shore patrol." He would circulate among the bars at night and would encourage seamen who had had one too many to return to their ships. When a ship was getting ready to sail and a seaman was missing, Karl could usually find him. Karl also kept seamen out of blacklisted bars (those owned and operated by Germans or German sympathizers), and in carrying out his other duties, he would, on occasion, pick up bits of information that would be worthy of reporting back to ONI.

Another function of the officer assigned to Lourenco Marques was to resolve difficulties between shipboard U.S. Navy Armed Guard officers and the masters of merchant ships. One such case involved an Armed Guard officer who had persisted in holding target practice at sea with the weapons on the bridge every afternoon while the master was trying to get his much needed siesta. It took about four days of discussions with each of them separately before Lt. Scott could get the disputants to meet together with him and resolve their difficulties.

In addition to delivering mail to each newly arrived ship, the naval observer briefed the crew on the nature of the port, its regulations and facilities for liberty, and which establishments were to be avoided. Baseballs, bats, and gloves were also loaned. Once every two or three months a seaman would spend his liberty in the local Portuguese jail, which was not a luxury hostelry. The word would get around, and it had a sobering influence on subsequent liberty parties for the next couple of months.[126]

Carrying out the above nonintelligence functions facilitated the maintenance of contacts and visits to other parts of the Portuguese colony and neighboring territories, which in turn permitted the gathering of information of value to naval intelligence. About nine or ten reports were submitted each month on the average, most of them in response to ONI requests.[127]

The means of communication between Lourenco Marques and ONI included pouch mail via the previously mentioned British aircraft and cables sent via the U.S. Navy liaison office at Durban, South Africa. Classified cables were encrypted in strip cipher for transmitting shipping information, and a "one-time pad" was available for more sensitive cable traffic during the last year of World War II. Portuguese cable facilities were adequate, but the rates to the United States were higher than from South Africa to the United States. There were no known German sympathizers in the Portuguese cable office, but the possibility was recognized.

One request from ONI in early September 1945 was for all publications on Northern and Southern Rhodesia and Nyasaland, including maps, telephone directories, business directories, and *Who's Who*-type publications. After checking with the U.S. Consul in Johannesburg for such publications that might be available there, Lt. Scott made a twelve-day shopping trip to Beira, Lusaka, Livingstone, Bulawayo, and Salisbury, contacting both government and commercial outlets. The resultant reports forwarded about fifty pounds of publications to ONI.[128]

Naval Attachés in Scandinavia

Cdr. Francis A. Klaveness was senior Assistant Naval Attaché, Stockholm, from June 1944 to August 1945, when he relieved Capt. Walter Heiberg as Naval Attaché, Stockholm. During that period, the Germans, Italians, and Japanese were also in Sweden, and Klaveness reported, "We were watching each other, and the Swedish Secret Service was watching all of us—very efficiently."

Among other tasks, Klaveness maintained contact with a Norwegian Navy underground organization known as the RMO through the Norwegian naval attaché's office. The RMO furnished information on German commerce to and from Norway; on arrivals and departures of naval ships, transports, and merchant ships; and on VIPs, etc., and also provided a certain amount of political information and the names of collaborators.

RMO sources kept a close but covert watch on the German battleship *Tirpitz*, moored in Trondheim, and sent in reports on its status once or twice a week. Most of the information came from Norway by means of couriers, although some of it apparently was received by clandestine radio. The information was screened by the U.S. naval attaché and sent by coded message to London and ONI if its timeliness warranted such handling; otherwise it was sent by diplomatic pouch, twice a week. After the war, Cdr. Klaveness received the Norwegian Cross of Freedom from King Haakon VII for his work in collaboration with the Norwegian underground.[129]

Naval Attachés in the Post–World War II Period

On 26 November 1945, under the auspices of Commander Naval Forces, Europe, a conference was held in London by twenty-five naval attachés and naval observers from the European and Mediterranean area. Commo. Tully Shelley, the naval attaché at London, had discussed such a possibility with RAdm. Leo H. Thebaud (Director of Naval Intelligence) and then with Adm. H. Kent Hewitt (COMNAVEU), both of whom supported the idea. The purposes of the conference were to assist in defining and establishing the procedures for accomplishing the attaché mission by exchanging information and to analyze the operation of foreign naval forces without reference to geographic or national boundaries that might be inimical to U.S. interests. The conference produced an "Estimate of the Situation" and a "List of Subjects Requiring Research and Study." A second conference was held 25–29 June 1946 in London, and a third 18–24 June 1947, the latter under the auspices of Naval Attaché, London.[130]

In connection with his attendance at the attaché conference in London in June 1946, Chief of Naval Intelligence RAdm. Thomas B. Inglis inspected European naval attaché posts. He was impressed by the universal plea for additional personnel to gather the quantities of useful naval intelligence information available. Administrative overhead work, such as meeting and entertaining VIPs, obligations to the Ambassador as a member of his staff, arranging for visits of U.S. Navy ships, acting as disbursing agents, and changing currency for ships' companies, could be accomplished by assigned personnel, but little time was available for collecting intelligence, ostensibly the primary mission for the attachés.[131]

Foreign nations were becoming more and more security conscious. Consequently, it was becoming more difficult for naval attachés to find the information that they were in the foreign countries to obtain. Naval attachés had access to all legitimate sources of information, and about 80 percent of the information that they collected was obtained from overt sources. Many of their contacts were social and required a large budget for entertaining. Such social contacts proved very valuable both prior to and during World War II. Before the war, the attachés had been underfunded, but in the postwar period it was desirable to improve that situation. Accordingly, an increase in attaché maintenance funds was requested of Congress.[132]

In January 1946, Cdr. Klaveness and Lt. H. Hove shifted from Stockholm to Helsinki to become the first U.S. naval attaché and assistant naval attaché to be resident in Finland. All previously accredited naval attachés to Finland had been resident in Stockholm, Berlin, or Copenhagen. Klaveness considered the work in Finland more satisfying than in Sweden because of the wholehearted manner in which he was received and the ready cooperation offered by the Finnish defense staff. His other sources of information included retired Finnish army officers, Finnish businessmen, government officials from civilian branches of the Finnish government, other diplomats, social contacts, and a couple of paid informants.[133]

A new element was added to the Navy's intelligence organization in London when Commo. Robert E. Robinson, Jr., reported on 6 February 1946 as head of the London branch of the Office of Naval Research. Robinson was designated an assistant naval attaché, but at first he had additional duty on COMNAVEU's staff. Commo. Robinson had two junior officers who performed administrative functions. The collection of scientific and research information in Europe was conducted by civilian scientists under the direction of Dr. H. M. MacNeille.

On 1 June 1946, Chief of Naval Operations Adm. Chester W. Nimitz informed COMNAVEU that the arrangement by which his Chief of Staff also served as Naval Attaché, London, would have to be terminated, due to budgetary and reduction-in-force requirements. On 1 August, Commo. Tully Shelley became naval attaché in succession to RAdm. Spencer S. Lewis but retained his assignment as COMNAVEU intelligence officer. Nimitz pointed out that naval attachés were subject to the orders and wishes of the chiefs of the diplomatic missions to which they were accredited, and coordination control was to be exercised by the Chief of Naval Intelligence. However, the CNO did agree that COMNAVEU could task the Naval Attaché, London, for intelligence information needed by his command. COMNAVEU had desired a separate but coordinated intelligence organization in Europe, but the CNO announced his intention of ordering Shelley's relief to duty as naval attaché only when the time came.

On 27 January 1947, Capt. Gill M. Richardson reported as COMNAVEU's intelligence officer. His arrival completed the formal separation of COMNAVEU and the naval attaché, but the practical interworking of the two organizations continued.[134]

In August 1947, RAdm. W. E. Parry, RN, proposed in a letter to the Naval Attaché, London, that a U.S. naval officer be assigned to the Naval Intelligence Division of the Admiralty to help process the increased volume of material becoming available on Russia. Chief of Naval Intelligence Inglis responded that ONI was also short of qualified intelligence officers and could not establish a new billet at, or assign an officer full time to, the NID. He would not object, however, if the Naval Attaché, London, assigned an officer from his staff for the duty, but it wasn't to be considered a permanent detail.[135]

In February 1947, ONI used its monthly publication, the *ONI Review*, to solicit applicants for naval attaché posts at the forty-three countries where they were then being maintained. "Officers interested in assignment as naval attaché[s], now, or in the future, should submit a letter to the CNO giving summary of their previous assignments and listing foreign language qualifications, if any. Inquiries are invited from officers in the ranks of commander, lieutenant commander, and lieutenant, and Marine officers of corresponding ranks." In May 1948, there were a total of 120 naval officers on attaché duty at the forty-three posts.[136]

The *Naval Intelligence Manual* (ONI-19[A]), issued in 1947, prescribed the mission of naval attachés, observers, and liaison officers as being

> To represent the U.S. in a thorough and creditable manner, and to procure and report all information and intelligence obtainable and of value to the U.S. They are to report particularly on matters of naval interest, both in peace and war, to the Chief of Naval Intelligence. In execution of their mission, they will perform the following tasks:
>
> a. The collection, preliminary evaluation and forwarding to CNI of all information and intelligence on the strength and war potential, and naval strength and disposition of forces of the country or countries to which they are accredited or assigned.

b. The preparation of plans for the expansion of their activities in the event of war in which the U.S. is a belligerent.

c. Cooperation with other U.S. agencies abroad in the collection, evaluation and dissemination of intelligence of interest to other government agencies.

d. In time of war, collection of information on the composition of enemy naval forces, their movements and probable intentions, and cooperation with other government agencies in the collection of war information.[137]

To ensure the most efficient utilization of personnel and to achieve the maximum effectiveness in the execution of their missions, the intelligence activities of the military and naval attachés were to be closely integrated in each country by the following measures:

a. Procurement of adjacent office spaces.

b. Maintenance of a joint intelligence library, accessible to both attachés and their staffs.

c. Exchange of pertinent data and allocation of assignments for intelligence work to avoid duplication of effort.

d. Such other measures as directed or found to be expedient.[138]

Capt. Smith-Hutton, who served as Naval Attaché, Paris, from May 1947 to May 1952, commented on long tours for naval attachés:

Within limits, it seems to me that long tours have many advantages from an intelligence point of view, but much depends on the attitude of the individual attaché. Our service conditions officers to "move on" after three years. So, frequently officers complain that they get "stale" after three years. If the attaché feels that he is stagnating in a post, he should not be required to stay.

Also, there could be technical reasons for changing officers frequently because, in these days of new weapons and techniques, an officer can get quickly out of date. Finally, thought might be given to the character of the people of the host country. As an example, I think that any competent U.S. Attaché would be well received in London (or any English speaking country) and be effective after a very short time. The French are very different, more difficult to get to know, and do not welcome new personnel readily. It usually takes much longer for a U.S. officer to do well in Paris than it does in London. Thus, there are so many variables that it is difficult to be positive on this. Generally speaking, if an attaché's career is not going to be harmed by a long stay, I think ONI will benefit more if the tours are made a bit on the long side.[139]

Relations with the French navy were so good that French naval personnel would supply virtually any information requested. For example, LCdr. Joseph V. Meigs, Jr., Assistant Naval Attaché for Air, spent two months photographing all French naval ports and bases, working with the French navy from both the air and the ground. The French navy received copies of all the photographs for their records. When the Military Assistance Advisory Group (MAAG), France, was organized, the naval attaché's office spent several months working with it to get the newly arrived organization off to a good start.[140]

Lt. Robert H. Dreher, USNR, Assistant Naval Attaché, Moscow, from 21 June 1946 to 29 April 1948, was the victim of an MVD spy frame-up six days before his scheduled date of departure. On 23 April he received a phone call from a Soviet Customs official who claimed that he wished to discuss "official business" concerning the shipment of American goods. Initially, Dreher had been resident in Odessa, with the assigned task of helping expedite shipments from the United States for the embassy in Moscow. The customs official had been officially designated by the Soviet government as Dreher's contact in carrying out the task. When Dreher reached the customs office in response to the phone call, the official began talking volubly about other matters, and then the Soviet secret police rushed in and placed both under arrest.

Ambassador Walter Bedell Smith was called to the Kremlin on 26 April and told by Deputy Foreign Minister Andrei Vishinsky that Dreher must leave the USSR at once. Dreher had applied for an exit visa ten days before the incident. He left on schedule.[141]

Beginning in 1948, a weekly interpretative summary, entitled the "Joint Weeka," was prepared by virtually all U.S. embassies and legations, based on an agreement between the Departments of State, Army, Navy, and Air Force. It was the product of a joint effort by Embassy officers and the Army, Navy, and Air Force attachés, and it was forwarded to Washington by expeditious means. Its purpose was to provide Washington, appropriate foreign service posts, and major U.S. military commands overseas with timely, integrated, and interpretive commentary on political, military, economic, and psychological events and developments of importance to the United States.

In 1960, "Joint Weekas" were still being produced in accordance with joint instructions, concurred in by the Office of Naval Intelligence, that had been promulgated by Department of State Instruction CA-29 of 1 July 1957. The naval attaché at each station prepared the Navy section of the "Joint Weeka," reporting and commenting on changes in naval policy and capabilities, significant changes in naval commands, flag officer assignments, organizations, equipment and strength, significant movements of naval units, and noteworthy incidents and operations. He also provided items on naval aviation for use in the Air section.[142]

In a briefing prepared by Naval Attaché, Rome, on 13 April 1949 for the Director of Naval Intelligence, problems were presented, and in some cases solutions were suggested that in varying degrees were typical of situations at other naval attaché posts in 1949. The briefing reported that an excessive proportion of the available time was being spent on fleet business, retarding the exploitation of intelligence resources. Indicative of the preponderance of the workload for the fleet was the communications load for a typical month: 469 fleet messages compared to twenty-five to and from CNO/ONI. Just the processing of the communications traffic required many extra work hours. No solution was suggested, but the Director of Naval Intelligence tried unsuccessfully to have Commander in Chief, U.S. Naval Forces, Eastern Atlantic and Mediterranean (CINCNELM) supply two junior communications officers to Naval Attaché, Rome.

The Rome briefing also mentioned the time-consuming tasks of making travel arrangements for visiting VIPs. The attaché office made hotel, travel, and theater reservations; provided guide services for both sightseeing and shopping; and arranged for visas, sojourn permits, audiences with the Pope, and in some cases entertainment. All these extra services reduced the time available for performing the naval attaché's primary intelligence duties.

The briefing went on to point out that the Atlantic Pact Military Aid Program was about to be applied to Italy. Although the State Department was expected to administer the program, it was anticipated that the ambassador would require the three service attachés to provide him with advice on Italian proposals for aid. Although this situation would add further to the naval attaché's workload, it was recognized that it would offer increased possibilities for expanding contacts with potential for intelligence exploitation.

The briefing concluded by pointing out that, although the Marine Security Guard at the Embassy was not under the naval attaché for command or administrative purposes (except for paying assigned personnel), the naval attaché did find that any adverse comment on the military bearing or performance of any of the Marine personnel became his responsibility to correct. It was recommended that an officer be assigned by the Marine Corps to each Security Guard Unit to serve additionally in the naval attaché's office.[143]

An effort to consolidate the attaché networks of the three Services was tried in 1949 when Secretary of Defense Louis A. Johnson directed a 30-percent reduction in personnel assigned to attaché duty. The primary purpose of the consolidation, of course, was to reduce costs by eliminating certain duplications of facilities. Under the new arrangement, one armed forces attaché in smaller countries represented all three services. In countries where conditions justified an attaché from each service, one was designated as the senior military attaché. This attaché was responsible for coordinating the administration of attaché functions, including consolidation of communications, libraries, transportation, tools, supplies, office personnel, and finances. The senior military attaché also served as principal advisor to the chief of mission on matters that concerned more than one service. The other attachés, however, retained the right to deal directly with the Ambassador on matters that concerned their own special fields of interest. The senior military attaché was in a position similar to the chairman of a joint committee but was not in military command.

The senior military attaché initially was determined by the relative rank of the officers present at a given post, but each service, through normal rotation, was to select future attachés to achieve senior representation in accordance with the following schedule:

> *Army*: Twenty-one (Afghanistan, Belgium, Bulgaria, Burma, Columbia, El Salvador, Ethiopia, France, Hong Kong, Hungary, India, Korea, Lebanon, Mexico, New Zealand, Rumania, Spain, Switzerland, Syria, Uruguay, and the USSR)
>
> *Navy*: Seventeen (Australia, Ceylon, Chile, China, Denmark, Dominican Republic, Egypt, Great Britain (to rotate with Air Force, Navy first), Greece, Indonesia, Iraq, Italy, Morocco, the Netherlands, Norway, the Philippines, and Singapore)
>
> *Air Force*: Twenty (Argentina, Brazil, Canada, Cuba, Czechoslovakia, Finland, Guatemala, Iran, Ireland, Israel, Pakistan, Peru, Poland, Portugal, Sweden, Turkey, Thailand, Union of South Africa, Venezuela, and Yugoslavia)

As a result of the Johnson consolidation, the Army closed sixteen attaché posts and the Air Force, twelve. The Navy closed eight: in Columbia, Iran, Israel, Lebanon, Peru, the Union of South Africa, Uruguay, and Venezuela. In all, the Navy reduced the number of its attaché personnel by 30 percent; the Army, 39 percent; and the Air Force, 37 percent. The naval attaché offices lost five aircraft and the Air Force thirty; net reductions in vehicles were ninety-nine for the Army, thirty-five for the Navy, and thirty-nine for the Air Force.

In addition to closing these posts, the Navy took drastic personnel reductions in Argentina, Brazil, Canada, Chile, China, Cuba, and Thailand and minor reductions in Egypt, Great Britain, Greece,

Indonesia, Korea, Mexico, Morocco, the Netherlands, Portugal, and Sweden. The stations least affected were in the areas of operations of the Sixth Fleet and CINCNELM.[144]

The Third Joint Pacific and Far East Attaché Conference was hosted by Commander in Chief, Pacific in Honolulu, 12–16 September 1949. Attending were six naval attachés, two naval liaison officers, and five ONI representatives. Committees were formed and reports were prepared on military, air, and naval attaché administrative problems, the coordination of intelligence activities in the field, and miscellaneous problems not within the purview of the other committees.[145]

A typical example of the ongoing communication among the naval attachés was the Fourth Latin American Attaché Conference, held at Fort Amador, Canal Zone, 23–28 January 1950. RAdm. Felix I. Johnson, Director of Naval Intelligence, headed the ONI delegation to the conference. Capt. Arthur L. Maher and Capt. James G. Lang also attended from ONI and gave presentations. In addition to all U.S. naval attachés from Central and South America and the Caribbean Islands, Capt. John B. Cleland, Intelligence Officer, 15th Naval District (DIO-15ND); Col. E. L. Lyman, Intelligence Officer (G-2), Headquarters, U.S. Marine Corps; and Capt. Roger B. Nickerson, Intelligence Officer, Caribbean Sea Frontier, also attended, as did Capt. Jack C. Renard, Intelligence Officer (N-2), Atlantic Fleet, who gave a presentation about the Atlantic Fleet Intelligence Center.[146]

For guidance on the relationship between naval attachés and Navy members of MAAGs, the following instruction was sent to naval attachés accredited to Western European countries in 1950:

> ONI has been informed that the Naval members of the Military Assistance Advisory Groups have been directed to submit all information which becomes available to them of intelligence value to the naval attaché as the proper channel for the coordination and collection of intelligence information. It is the belief . . . that this should be the chief and proper channel for the collection of such information.[147]

As a result of the reduction in personnel assigned to naval attaché posts, intelligence reporting dropped in volume in the first half of 1950. From Argentina, 33 reports were received, compared to 172 for the same period in 1949; the post had been reduced from three officers to one. For Sweden, 58 reports were received in 1950, compared to 191 for the same period in 1949. For the Netherlands and Belgium, there was a 69-percent decrease in the number of reports compared to 1949, and only one officer and two civilians were assigned to cover both countries. Reporting from Portugal suffered a 40-percent decrease and from Spain a 35-percent decrease. Decreases in reporting from other reduced posts were proportional.[148]

Naval Attachés During the 1950s

Korea

With the commencement of hostilities in Korea in June 1950, the naval attaché, Cdr. John P. Seifert, assisted in the evacuation of U.S. nationals from Seoul. He remained in Korea with ten other U.S. Embassy personnel, joining with a small nucleus of top South Korean government officials to provide continuity of government during the initial advance of the North Korean Army. He also acted as assistant intelligence officer on the Army staff and was responsible for planning air missions for Navy carrier air strikes.[149]

Cdr. Seifert had been the first naval attaché assigned to South Korea, reporting on 20 April 1950. Naval liaison officers had staffed the post when it was established on 21 January 1949.[150]

Attaché Conferences

Joint attaché conferences among the three services were held periodically during the 1950s to discuss common problems and to exchange ideas on procedures and practices. Representatives from the three service intelligence agencies attended and held discussions with their individual attachés. In 1950, the conference was held in Frankfurt, Germany, and Director of Naval Intelligence Felix Johnson combined attendance at the conference with an inspection tour of naval attaché posts in Western Europe.[151]

Organization of the Attaché System

At the request of the Joint Chiefs of Staff (JCS), Secretary of Defense Gen. George C. Marshall, USA, abolished the senior military attaché concept on 17 July 1951 and established a limited executive agency for logistical and administrative functions. The latter concept was abandoned on 4 March 1952, with each military department again assuming responsibility for providing its own logistic and administrative support.[152]

A Department of Defense letter of 18 February 1952, signed by the heads of the three service intelligence agencies, reaffirmed the practice of having a joint service attaché communications and coding office at each station to meet the needs of all service attachés at each post, a system that had been established on 13 August 1949 and further described by an earlier DOD letter of 12 October 1951. The service responsible for providing communications and coding was also to provide the necessary personnel and material.

The 1952 letter also emphasized the need for each service attaché to coordinate reporting with the other attachés when the information being reported was of interest to one or more of the other services. The coordination, however, was to be at the discretion of the originator if the material reported was of such nature as to require special handling. Furthermore, no report was to be unnecessarily delayed in order to achieve coordination. If there was nonconcurrence with any intelligence report by an attaché of another service, however, the nonconcurrence was to be noted in the report. Nonconcurring attachés were to report their disagreements to their directors of intelligence if the subject matter of the report was of major concern to their services. The DOD letter also recognized the general principle that each service attaché would establish and maintain his own contacts for the collection of intelligence information pertaining to his own service.

An exception was to be made by mutual agreement between the service attachés involved in the same case, when one attaché had established and developed a contact source that provided intelligence vital to another service whose attaché was unable to achieve the same degree of rapprochement. In such a case, the maximum possible coordination was to be maintained between the two service attachés involved in order to achieve full exploitation of the source.[153]

Organizational Relationships in Europe

A policy statement, issued by the Office of the Chief of Naval Operations when CINCNELM and Mediterranean moved back to London from Naples in June 1952, prescribed the relationship between CINCNELM and the U.S. naval attachés in its area of naval responsibility in matters that affected its operations. Specifically,

> 1. Naval attachés were to provide CINCNELM with whatever strategic, operational, and political intelligence CINCNELM might require. CINCNELM was to advise all naval attachés of the general nature of the reports desired from them and, on occasion, of the specific subjects to be investigated.
>
> 2. Naval attachés were to maintain up-to-date rosters of inactive U.S. Naval Reserve personnel living in their areas and were to act as local representatives of CINCNELM in connection with administrative matters concerning such personnel.
>
> 3. Naval attachés were to deal directly with CINCNELM, or with subordinate commanders designated by CINCNELM, in connection with visits or transits of naval vessels, flights of aircraft, or naval parties on authorized missions. The attachés were to render appropriate assistance on such occasions, and, if practicable, to attend at the ports of visit or to send suitable representatives.
>
> 4. Upon request by CINCNELM or subordinate commanders designated by CINCNELM, and in accordance with existing instructions, naval attachés were to take any necessary action to arrange clearances for visits of U.S. naval ships and to advise the requesting authority when clearance had been received.
>
> 5. Ordinarily, CINCNELM would request clearance directly from the naval attachés for flights of aircraft under CINCNELM's operational command. In connection with such flights, the naval attachés were to endeavor to ensure speedy clearances, to arrange, where possible, for automatic clearances to become effective upon receipt of a flight plan, and to keep CINCNELM fully informed of special requirements in the foreign country concerned and of appropriate operational data about aircraft operations in the countries to which they were accredited.

Additionally, CINCNELM was to be the Chief of Naval Operation's representative in dealing with the British Admiralty, except in the exchange of intelligence, which was a function of the Naval Attaché, London.[154]

Naval Attaché Aircraft

In 1948, the following naval attaché posts had Beech JRB Expediter aircraft assigned: Buenos Aires, Rio de Janeiro, Ottawa, Bogota, Havana, Cairo, London, Paris, Guatemala City, Rome, Mexico City, Batavia, Lima, Madrid, Stockholm, and Caracas. In addition, Capetown and Nanking had Douglas R4D Skytrains assigned, and Melbourne had a Beech SNB Kansan, a trainer version of the JRB twin-engined utility transport.[155]

In 1957, the following naval attaché posts had Douglas R4D Skytrain aircraft: Melbourne, Copenhagen, Cairo, New Delhi, Tehran, Baghdad, Seoul, and Mexico City. In addition, Athens, Djakarta, and Oslo operated Grumman UF amphibians.

The assignment of an aircraft to a particular service attaché was determined by interservice agreement. Naval attaché aircraft were needed in areas of potential naval or amphibious operations where there were one or more of the following factors: an extensive coastline; an extremely large area of accreditation (Naval Attaché, Cairo, also covered Libya, Sudan, and Ethiopia); poor transportation facilities; an otherwise inaccessible area of accreditation (Naval Attaché, New Delhi, was also assigned to Nepal); and/or a strategically important location.[156]

The policy statement on the use of U.S. naval attaché aircraft issued in August 1955 required that

> to the extent feasible, at posts where Navy maintains and operates U.S. naval attaché aircraft, the U.S. naval attaché shall furnish air support to U.S. service attachés and State Department representatives where such support is necessary to the accomplishment of their missions.
>
> The use of attaché aircraft for all purposes, intelligence as well as others, will at all times remain the responsibility and prerogative of the attaché maintaining the aircraft.[157]

Naval Attaché Procedures and Duties

After being detailed to a foreign post, but before proceeding to it, the prescribed procedure in 1956 was for the prospective naval attaché to report to ONI for at least one month's temporary duty for briefing and indoctrination. During the temporary duty period, the new attaché was to spend as much time as possible with the geographic desk responsible for the assigned country. The prospective attaché was to become familiar with the current situation and important personalities at the new post by means of briefings and by reading appropriate intelligence reports and their evaluations. The new attaché was also to become conversant with collection and reporting procedures, problems, and facilities. Indoctrination was to include a briefing on security by the Security Division of ONI, at which time the prospective attaché was informed of any particular security problems known to exist at that post. The newly detailed attaché was also to be thoroughly instructed as to the administrative and fiscal procedures applicable at the prospective post.[158]

As of April 1956, U.S. naval attachés were located in the following countries: Argentina (also accredited to Paraguay), Australia (also to New Zealand), Brazil, Burma, Canada, Ceylon, Chile, Cuba (also to Haiti), Denmark, Dominican Republic, Egypt (also to Libya and Ethiopia), Finland, France (also to Switzerland), Germany, Great Britain, Greece, Hong Kong (headed by an assistant naval attaché officially accredited to Great Britain), India (also to Nepal), Indonesia, Iran, Iraq, Italy, Japan, Korea, Lebanon (also to Syria), Mexico (also to Guatemala, Honduras, Costa Rica, El Salvador, and Nicaragua), Morocco, the Netherlands (also to Belgium), Norway, Pakistan, Peru, the Philippines, Poland, Portugal, Singapore (headed by an assistant naval attaché officially accredited to Great Britain), Spain, Sweden, Taiwan, Thailand, Turkey (at Ankara, but with a suboffice at Istanbul), Uruguay, USSR, Venezuela, Vietnam (also to Cambodia and Laos), and Yugoslavia.[159]

In 1956, the Naval Attaché, Bonn, Capt. Frederick J. Harlfinger, had been designated by Ambassador Conant as his coordinator of research and development. Because of that, Harlfinger recommended to the Director of Naval Intelligence that his relief should be briefed long and well not only by appropriate sections of the Office of the Chief of Naval Operations but also by the Assistant Secretary of the Navy's office, the Office of Naval Research, and the various Navy engineering bureaus. Harlfinger emphasized that with his small staff and the myriad of Navy problems, both U.S. and German, he would be hard pressed to find time to perform his primary job of intelligence collection, particularly if the Commander Naval Forces, Germany organization were to be disestablished (which it was, not long after).[160]

In 1957, the forty-six U.S. naval attaché posts, including four suboffices, were staffed by 116 naval officers, 20 Marine officers, 171 Navy enlisted, 6 Marine enlisted, 63 U.S. civilians, and 158 foreign citizen civilians.[161]

In 1957, the primary task of the naval attaché was to collect and report all information of naval interest, accenting that which might affect the plans and strategy of the Navy and Marine Corps. The task was categorized as being so important that it should never be relegated to a secondary consideration in favor of other duties. Other attaché duties included handling protocol matters, requests from the chief of the diplomatic mission, assistance to visiting U.S. Navy ships (including recreation arrangements for visiting personnel), procurement of supplies and provisions, communications and cryptographic problems, disbursing and supply responsibilities, and general administration duties and reports.[162]

The qualifications for naval attachés, as described in 1957, were:

> The naval attaché should have at least a working knowledge of the language of the country to which he is to be assigned. If he lacks this qualification upon assignment to post, he should take immediate steps to acquire it. The naval attaché should also be familiar with the customs, culture, history, domestic and foreign political trends, and general background of the country to which assigned. He should be particularly familiar with its naval and air organizations.[163]

The policy on placement of naval attaché posts in 1957 required consideration of the size and complexity of a country's navy and its facilities; the role of its navy in the political and economic affairs of the country; the magnitude of its seagoing trade; its strategic location as an observation post; its contribution to the naval strength of NATO (North Atlantic Treaty Organization), SEATO (Southeast Atlantic Treaty Organization), and other alliances in which the United States was a participant; and the scope and frequency of U.S. Navy visits to that country.

A statement on the justification for the naval attaché system, prepared and approved in 1957 by the Office of the Chief of Naval Operations, declared:

> The policy of the United States Government is to achieve the maximum national welfare. Achievement of this objective is constantly affected in varying degrees by the activities of other nations. Consequently, an intimate knowledge of the objectives and capabilities of foreign nations becomes a necessity. . . . Where a foreign government possesses significant naval or maritime capability, a professionally qualified naval officer is [to be] included among the official observers as naval attaché.
>
> Knowledge of the naval and maritime capabilities of foreign nations is important for both the strategic planner and the operational commander. The Navy also has a vital interest in all areas in which land, sea, and air forces may be employed. Obviously, intelligence of the military geography, hydrography, and oceanography of coastal areas is obtainable in time of peace at far less cost in man-hours, lives and money than in time of war. . . . From its collected data, the Department of the Navy acquires the knowledge necessary to plan adequately the training, readiness and preparation for war (required by Article 201, *U.S. Navy Regulations*) of its operating forces. Long experience has shown that this data can be obtained most proficiently and economically by the naval attaché.[164]

A procedure was adopted in May 1957 to provide selected naval attachés with Briefs of Naval Interest that could be discussed with naval officials of the host countries as a form of quid pro quo for the information that they supplied to the attachés. The program was implemented by the Collection and Dissemination Branch (OP-922H) of ONI and proved to be an effective way for naval attachés to develop and maintain favorable contacts with selected foreign officials.[165]

The requirement for naval attachés and liaison officers to submit "Post Reports" on a regular basis was established on 30 December 1957. Prior to that time, Post Reports were required at irregular periods to supplement information on living conditions and other personal matters contained in State Department Post Reports and Army attaché Station Reports. The new reports were not to duplicate the State and Army reports but were to cover matters particularly pertinent to Navy personnel and pertaining to special and unusual requirements of the post. The first report was to be submitted on 1 February 1958 and thereafter on 1 February of even-numbered years. Supplemental reports or reports of no change were required on 1 February of odd-numbered years. Subjects to be covered were a "post description" and "requirements and information" on addresses, animals (pets), automobiles, beverages, clothing, food, household effects, housing, insurance, language, medical, money, offices, recreation, servants, schools, social practices, telephone numbers, and transportation and shipping. The use of additional paragraphs and/or amplifying information was encouraged.[166]

Through the medium of the "collection plan," in 1958 ONI had each naval attaché assess both the potential for intelligence that existed at his post and his capability for collecting it. An assessment was to be made against each standing intelligence collection requirement. The survey reports enabled ONI to know more accurately the actual strengths and weaknesses of its attaché system.

The two-week special attaché course presented at the Naval Intelligence School was reviewed during 1958 and was then improved to give a maximum of technical training in intelligence collection and report writing as well as instruction in photography for intelligence use.[167]

In March 1959, Dr. Herbert York of the Office of the Secretary of the Defense sent representatives to London to seek space for an office for a "scientific liaison person" who would coordinate the Weapons Systems Evaluation Group of the Operations Research Office representatives and assist the Combined Policy Committee groups in the area of scientific and technical research and development. Dr. York had previously addressed a letter on the same subject to the Army attaché in London. The creation of the scientific liaison post was looked on by the service attachés in London as the start of a move by the Department of Defense to establish itself as the point of coordination between the services and the British as well as to move into the embassy organization and push the attachés out.[168]

To institute economies in the administration and logistic aspects of the attaché system, on 11 May 1959 the Secretary of Defense issued a memorandum to the three service secretaries in which he set forth the general principle that "it is inescapable that a single host Service acting as executive agent for administration and logistics at each station could do the job more simply and economically." The memo called for the implementation of that principle, with the provision that operational relationships between attachés and their parent services would not be abridged. To test the concept and to identify and eliminate any problem areas, nine "pilot stations" were organized under the executive agency principle on 1 July 1959. In essence, the responsibility of the service attaché designated as the executive agent was to provide the administrative and logistic support necessary for the efficient operation of the offices of all U.S. military attachés at the same post. The basic directive for the executive agency responsibilities was JANAF (Joint Army/Navy/Air Force) Attaché Letter No. 3 of 1 July 1959.

Eight of the nine pilot stations at which the executive agency principle was to be tried out, beginning 1 July 1959, were Tokyo, Mexico City, Rio de Janeiro, Ottawa, Bonn, Hong Kong, Ankara, and Pretoria. ONI issued its Instruction 05410.2 on 26 June 1959 to the naval attachés at the affected stations to provide additional detailed information to supplement the general instructions contained in the JANAF Attaché Letter. The Executive Agency principle became effective for all stations on 1 July 1960.[169]

Farewell to Cuba, 1961

The last naval attaché accredited to Havana, Cuba, prior to the closing of the U.S. Embassy there in March 1961 was Cdr. Joseph H. Floyd, who reported on 2 August 1960. He had been briefed at ONI for duty as Naval Attaché, Chile, but at the last moment was redirected to Havana because of the premature departure of Capt. Charles R. Clark, Jr., in July 1960. As a result, Floyd arrived at his post without the benefit of adequate briefing on the sensitive situation that had developed since the Castro-led Communist takeover of Cuba.

Floyd's collection activities were aimed primarily at Soviet bloc shipping entering Cuba and the types and tonnage of cargo being imported. He also reported on the activities of known Communists and extreme nationalists. Normal contacts included numerous Cuban nationals employed by various U.S. Government agencies and U.S. business offices, and U.S. citizens residing in Cuba. Most of the contacts were those of the previous naval attaché and could be picked up without the usual introduction during a turnover period. The contacts were used to provide information on the destinations of cargoes delivered by Soviet bloc ships.

Other sources for Cdr. Floyd included Cuban national "walk-ins" disillusioned with the Castro regime, attachés of other foreign countries, and prostitutes who had had close contact with Cuban government officials.

In November 1960, an Army attaché source reported that Soviet missiles were being moved from the docks to the interior of Cuba. The Army attaché could not get away to confirm the report and asked Floyd to do so. Floyd sighted what appeared to be missiles, but it was dark and they were still crated; the type of missile could not be determined. He reported his observations to the U.S. ambassador and to ONI but obtained no reaction from either.

Another noteworthy report received by Floyd was information on the full spectrum of radio frequencies used by Cuba provided by a Cuban radioman. Some duplicate, but not necessarily undesirable, reporting was discovered during a conversation between Floyd and the CIA representative when they found that both had been tasked to determine Castro's health and the prescription drugs he was using.

The naval attaché post at Havana was disestablished on 31 March 1961, and Floyd and his assistant, LCdr. P. H. Klepak, were ordered to the naval base at Key West to await further assignment by the Bureau of Personnel. When in Key West, Floyd aided in the recruitment of Cuban nationals for a forthcoming operation that, in due course, was identified as the Bay of Pigs invasion. Guidance in this effort was received from Cuban nationals who were known to be sympathetic to the U.S. operation and who had been highly effective in the overthrow of Fulgencio Batista but had been expelled by Castro.[170]

Attaché Organizational Relationships in the 1960s

In a visit report of 3 July 1961 by two OP-922H2 (Collection Division) representatives, the Naval Attaché, Bonn, Capt. J. B. Thro, was quoted as stating that his primary functions were to serve as the official channel for navy to navy contact, to work to help strengthen the German navy, to maintain official liaison with the West German government agencies, and to report on matters of naval interest with respect to West Germany. Capt. Thro reportedly believed that the mission, duties, and functions of Commander in Chief, U.S. Naval Forces, Europe Representative, Germany (CINCUSNAVEUR REPGER) did not conflict with or duplicate the efforts of Naval Attaché, Bonn, but rather that the two organizations complemented each other since REPGER covered an area not permitted to the naval attaché.

Relationships between naval attachés and naval missions in the same country varied, usually as a result of the personalities and backgrounds of those involved. In some cases, they maintained frequent and favorable contact, but in many, the attachés and the naval mission personnel operated as strangers or even antagonists. To try to improve the situation, an OPNAV instruction was issued in September 1961 to establish policy and to provide guidance to U.S. naval missions and naval attachés concerning their relationship in intelligence matters:

> While MAAGs, JUSMAAGs [Joint U.S. Military Assistance Advisory Groups], and Naval Missions are not established as intelligence agencies, U.S. military personnel assigned to them, by nature of their assignments, have access to a volume of intelligence information of significant value to the United States. In some instances the acquisition and timely transmission of this information may contribute as

> much to the accomplishment of United States national objectives as the pursuits of their normal functions. In the national interest, it is essential that intelligence information acquired by U.S. nationals be made available to United States intelligence agencies. Therefore, U.S. personnel attached to the above or similar activities are required to effect the maximum cooperation with service attachés to assure that such intelligence information is made available to intelligence agencies.[171]

At the end of 1963, the Secretary of Defense desired that the military attachés in Vietnam be incorporated under Commander U.S. Military Assistance Command, Vietnam (COMUSMACV). The concept was also discussed at a meeting between the Joint Chiefs of Staff and the Secretary of Defense. That meeting led to a JCS message of 26 November, requiring COMUSMACV to work with the U.S. embassy in Saigon to develop and submit a proposal placing the attachés under the operational control of COMUSMACV in such a way that their military activities would be coordinated by that headquarters while not changing their diplomatic status. The Office of Naval Intelligence was disturbed about the development, which would later prove to degrade the amount and quality of the factual reporting from the Vietnam area.[172]

One of the problems that the Naval Attaché, Saigon, had in 1963 relative to his reporting of information on the South Vietnamese navy was that the chief of the Navy Section, MAAG claimed that any reports that tended to cast the South Vietnamese navy (VNN) in a derogatory light were criticisms of MAAG efforts and would not be tolerated. The naval attaché was thus frustrated in his efforts to make objective reports since the VNN capabilities were patently not up to the standards set by the MAAG. Secretary of Defense Robert S. McNamara also objected to factual independent reporting by the naval attaché in Saigon and closed the office in January 1965.[173]

Creation of the Defense Attaché System

On 12 December 1964, McNamara issued Department of Defense Directive C5105.32, establishing the Defense Attaché System (DAS) "to improve the collection of intelligence information and the management of the total attaché effort." The DAS was placed under the supervision of the Defense Intelligence Agency (DIA).

Since 1882, when the practice of assigning naval attachés to selected foreign countries became routine procedure, the composition and tasks of the system had been adjusted to meet the varying requirements of each country and situation. Consistently throughout the period, however, the chief asset of the naval attaché system had been its recognized status as a part of the U.S. Navy. Each naval attaché had a full-time Navy job and was able to do it better because the attaché was known as a bonafide representative of the U.S. Navy. People abroad, both foreign and American, would seek out naval attachés because they were such representatives.[174]

On 1 July 1965, the DIA assumed operational control of the Defense Attaché System, and the Director of Naval Intelligence was designated as the point of contact for defense attaché matters within the Department of the Navy. By the above-mentioned DOD directive, the military departments were to provide the director of DIA with assistance and logistic and administrative support, as well as the specialized training required to establish, maintain, and administer the system. The services were also to nominate to the DIA director qualified personnel for assignment to the DAS. All naval attachés and assistant naval attachés accredited to foreign governments and all other DOD personnel assigned to attaché posts were assigned to the DAS. Staff authorizations required by DIA to establish, maintain, and administer the DAS were to be provided on an approximately equal basis from the military departments, including civilian personnel as needed.[175]

The U.S. Navy liaison officer in Hamburg was transferred from the Naval Attaché, Bonn, to the jurisdiction of the CINCUSNAVEUR REPGER organization effective 1 July 1965. The move was motivated partly to save the Hamburg office for the Navy at a time when all service attachés were being shifted to the DAS. The reorganization in Germany was also motivated by a State Department reciprocity consideration; it was U.S. policy not to permit foreign attachés in the United States to be based at any location other than in the capital city, Washington, D.C.[176]

The office of the Assistant Naval Attaché, Istanbul, was far larger than its parent office in Ankara in 1965, and it generally performed a specific mission—surveillance of the Turkish Straits—that bore little relation to the function of the naval attaché. Consideration was being given to changing the Istanbul office to give it a CINCUSNAVEUR representative status, but that was not done, and the office was eventually closed.[177]

A review by the Deputy Assistant Secretary of Defense for Intelligence in 1970 concluded that the DAS was not working satisfactorily. The attachés were serving too many masters; they were responsible to the Defense Intelligence Agency for intelligence collection and to their respective services for representational matters. The problem was further exacerbated by DIA's ineffective management of the system. DIA refused to recognize representation requirements as co-equal to collection requirements, even though representation contacts were usually needed first in order to establish contacts for collection. The Navy recommended that control of the attachés be returned to the respective services.[178]

This recommendation was disapproved on 31 March 1970, but the Joint Chiefs concurred that the operation of the Defense Attaché System was unsatisfactory and requested increased access to the attachés by the services and greater emphasis on representational matters.[179]

The military services, particularly the Navy, had a compelling need for authoritative and responsive representation in many foreign countries. The service attachés, as elements of the DAS, an identified component of the Defense Intelligence Agency and under its administrative and operational control, were too obviously intelligence-oriented, to the detriment of service representation needs (and collection capabilities). As field representatives of an intelligence agency, their accessibility as service representatives was inhibited. Furthermore, the title worn by the defense attaché often preempted the other service attachés from involvement in issues and functions of exclusive concern to the respective services or in which their knowledge and experience made them most competent to take action or to provide authoritative guidance. Also, the designation of the representative of one of the services as *the* defense attaché degraded the position of the representatives of the other services in the eyes of foreign officials.[180]

In April 1973, the responsibility in the Naval Intelligence Command (NIC) for overseeing the naval attachés was transferred to the Foreign Operations Division (NIC-32). As a result, NIC-32 became the focal point for all overt foreign naval intelligence cooperation programs. The Attaché Programs Section within NIC-32 was responsible for reviewing and endorsing personnel nominations for attaché billets, monitoring attaché training, arranging briefings and debriefings for attaché personnel, staffing Joint Chiefs of Staff papers, preparing Director of Naval Intelligence correspondence on attaché programs, and coordinating actions within the DIA on all matters pertaining to the Defense Attaché System.[181]

Table 3.1.
Naval Attachés Accredited Under Defense Attaché System, 1974

Country	NA Rank	ANA Rank	DATT	AAT*
Australia	Capt.	LtCol., USMC	Air Force	—
Belgium	Cdr.	LCdr.	Army	—
Brazil	Capt.	LCdr.	Air Force	—
Canada	Capt.	LCdr.	Air Force	—
Chile	Capt.	Lt.	Navy	—
Denmark	Capt.	Maj., USMC	Navy	—
Dominican Republic	LtCol., USMC	—	Navy	—
Ecuador	Capt.	—	Army	—
Finland	Cdr.	LCdr.	Air Force	—
France	Capt.	LCdr. & LtCol., USMC	Army	—
Germany	Capt.	LCdr.	Army	—
Greece	Capt.	LCdr. & Maj., USMC	Navy	—
Hong Kong	Capt.	LCdr. & LtCol, USMC	Navy	Diplomatically assigned to UK
India	Capt.	LCdr.	Air Force	Nepal
Indonesia	Col., USMC	LCdr./Lt.	Army	—
Iran	Cdr.	—	Air Force	—
Israel	Cdr.	LCdr.	Air Force	—
Italy	Capt.	Cdr. & Maj. USMC	Navy	—
Jamaica	LtCol., USMC	—	Navy	Haiti
Japan	Capt.	Cdr.	Navy	—
Khmer Republic	LCdr.	LCdr.	Army	—
Lebanon	Col., USMC	—	Navy	Cyprus
Liberia	Cdr.	—	Navy	Ivory Coast, Sierra Leone, Ghana
Malagsay	LCdr.	—	Air Force	—
Malta	Cdr.	—	Navy	—
Mexico	Capt.	Maj., USMC	Army	El Savador, Guatemala, Honduras, Nicaragua, Costa Rica
Morocco	Capt.	Maj., USMC	Navy	—
Netherlands	Capt.	Maj., USMC	Air Force	W. Samoa
New Zealand	Capt.	—	Navy	—
Norway	Capt.	Cdr.	Air Force	—
Pakistan	Capt.	(Vacant)	Air Force	—

Country	NA Rank	ANA Rank	DATT	AAT*
Peru	Capt.	—	Navy	—
Philippines	Cdr.	—	Air Force	—
Poland	Cdr.	—	Army	—
Portugal	Capt.	Lt.	Air Force	—
Senegal	LtCol., USMC	—	Navy	Gambia, Mali
Singapore	Capt.	LCdr.	Army	—
South Africa	Cdr.	Maj., USMC	Air Force	—
Spain	Capt.	LCdr.	Army	—
Sri Lanka	Cdr.	—	Navy	—
Sweden	Capt.	LCdr.	Air Force	—
Taiwan	Capt.	(Vacant)	Air Force	—
Thailand	Capt.	Maj., USMC	Air Force	Laos, Burma
Turkey	Capt.	Maj., USMC	Air Force	—
United Kingdom	RAdm.	Capt.; LtCol., USMC; Capt. and Cdr. resident at Bath	Navy	—
Uruguay	Capt.	—	Navy	—
USSR	RAdm.	Cdr., LCdr., LtCol., USMC	Navy	—
Venezuela	Capt.	LCdr.	Air Force	Trinidad and Tobago
Vietnam	Capt.	—	Army	—

Chapter Notes

1. *Navy Register*, 1884.

2. ONI ltrs to naval attachés, commanding officers, and other officers, no. 12, 32–33, Naval Records Collection (RG 45), NA.

3. Lt. William A. Sachse, "Our Naval Attaché System," USNIP, May 1946, 663.

4. Ibid., 666.

5. Ibid., 663.

6. "The Korean Mission to the United States in 1883," *Royal Asiatic Society Journal*, vol. 18, 1929.

7. *ONI Review*, Jul 1956, 303.

8. *Navy Register*, various editions.

9. Elting E. Morison, *Admiral Sims and the Modern American Navy* (Boston: Houghton Mifflin, 1942), 48–49.

10. Ibid., 54–56.

11. Capt. Herbert E. Cocke, USN, "History of ONI," MS, Office of Naval Intelligence, 1931, 9–10.

12. Sachse, "Our Naval Attaché System," 668–70.

13. Capt. Henry S. Shields, USAF, "A Historical Survey of U.S. Naval Attachés in Russia: 1904–1941," MS, Defense Intelligence School, Mar 1970, 6.

14. Ibid., 7. McCully's own reports of his travels in Russia have been published as *The McCully Report: The Russo-Japanese War, 1904–1905,* Richard A. von Doenhoff, ed. (Annapolis: Naval Institute Press, 1977).

15. Ibid., 8–13.

16. Sachse, "Our Naval Attaché System," 668–69.

17. Ibid., 670; and *New York Times*, 25 May 1913, sec. 3, 8.

18. Shields, "U.S. Naval Attaché System," 21.

19. Sachse, 670.

20. Shields, 22, 24–25.

21. Sachse, 668.

22. LCdr. Charles O. Maas, USNRF, MS report, Individual Reports, World War II Command File, OA.

23. Cocke, "History of ONI," 28.

24. *ONI Review*, Aug 1956, 346.

25. Department of the Navy, "Administrative History of the Office of Naval Intelligence in World War II," 10 Jul 1946, unpublished MS, 12, hereafter *ONI WWII Admin History.*

26. Cocke, "History of ONI," App., "NA Christiana."

27. Ibid., chap. 4, 2.

28. MajGen. James C. Breckenridge, USMC, ltr to Director of Naval Intelligence (DNI), 4 Oct 1933, Intelligence Files, OA.

29. Cocke, "History of ONI," App., "NA Rome."

30. Ibid., App., "NA Stockholm."

31. Ibid.

32. Ibid., App., "NA, The Hague."

33. James R. Green, "The First Sixty Years of the Office of Naval Intelligence" (M.A. thesis, The American University, 1963), 61.

34. Cocke, "History of ONI," App., "Madrid."

35. Ibid., App., "Lisbon."

36. Ibid., App., "London"; and Morison, Sims, 367.

37. *Navy Directory,* various editions.

38. RAdm. Albert P. Niblack, USN, *The History and Aims of the Office of Naval Intelligence* (Washington: GPO, 1920), 6.

39. Welles ltr to Head, Naval Post Graduate School, 17 Dec 1919.

40. Shields, "U.S. Navy Attachés in Russia," 26–27.

41. *SECNAV Annual Report, 1920,* 39.

42. Ibid., 156.

43. Capt. Henri H. Smith-Hutton, ltr response to author's inquiry, Apr 1974.

44. Shields, 28–29.

45. *ONI WWII Admin History,* 523.

46. ONI, *Monthly Information Bulletin,* vol. 7, 15 Jul 1922, 52–53.

47. Robert Murphy, *Diplomat Among Warriors* (Garden City: Doubleday, 1964), 35.

48. Adm. William F. Halsey, USN, *Admiral Halsey's Story* (New York: McGraw-Hill, 1947), 48.

49. *CNO Annual Reports, 1926* and *1928.*

50. Cocke, "History of ONI," 3–4.

51. *ONI WWII Admin History,* 27, 1205.

52. *CNO Annual Report,* FY 1932, 36.

53. *CNO Annual Report,* FY 1933, 20, 37.

54. OP-16-A, "Annual Report, Fiscal Year 1934," 1.

55. Capt. Henri H. Smith-Hutton, Oral History, USNI, Annapolis, MD, 1979, 1:138–39, copy in OA.

56. Ibid., 141.

57. Ibid., 163.

58. Capt. Henri H. Smith-Hutton, USN, "Tokyo, Dec. 1941," *Shipmate,* Dec 1973, 13–15.

59. ONI-19, *Naval Intelligence Manual,* 1933, para. 6018.

60. Ibid., para. 6035.

61. Ibid., para. 6041.

62. VAdm. Walter S. Anderson, Oral History, Columbia University, New York, NY, 1963, 191, copy in OA.

63. *CNO Annual Report, FY 1934,* 27.

64. Library of Congress, Federal Research Division, *The Military Attaché System: Its History and Present Status,* ca. 1971, 24–25.

65. *CNO Annual Report, FY 1934,* 2.

66. Shields, "U.S. Naval Attachés in Russia," 36–38.

67. Ibid., 42.

68. Ibid., 50–53.

69. *DNI Annual Report, FY 1935,* 1.

70. Ibid., 6.

71. *DNI Annual Report, FY 1936,* 1.

72. Ibid., 1.

73. Ibid., 1.

74. *ONI WWII Admin Report,* App. H.

75. Adm. Royal E. Ingersoll, Oral History, Columbia University, New York, NY, 1965, 72–74, copy in OA.

76. *DNI Annual Report, FY 1939,* passim.

77. James O. Richardson, *On the Treadmill to Pearl Harbor: The Memoirs of Admiral J. O. Richardson as told to Vice Admiral George C. Dyer, USN (Retired)* (Washington: GPO, 1973), 153.

78. Cdr. Albert E. Schrader, war diary, passim.

79. Smith-Hutton oral history, USNI, 1:265, 268.

80. Ibid., 1:280–83.

81. Ibid., 1:287–88.

82. Ibid., 1:301–2.

83. Capt. Stephen Jurika, Jr., USN, Oral History, USNI, Annapolis, MD, 1975–76; and additional Jurika tapes in author's possession.

84. Schrader war diary.

85. *CNO Annual Report, FY 1940.*

86. Schrader war diary, passim.

87. *ONI WWII Admin History,* 666.

88. Murphy, *Diplomat Among Warriors,* 82.

89. "Administrative History of Commander Naval Forces, Europe (COMNAVEU) in World War II," MS, 2–5, 21–23, hereafter *COMNAVEU WWII Admin History.*

90. "Administrative History of Commander Fourth Fleet (COM4THFLT) in World War II," 9, hereafter *COM4THFLT WWII Admin History.*

91. Ibid., 20–25.

92. Schrader war diary, passim.

93. *CNO Annual Report, 1941,* 10–11.

94. *ONI Review,* Jun 1947, 3.

95. Green, "Sixty Years of ONI," 102.

96. Shields, "U.S. Naval Attachés in Russia," 56–60.

97. *ONI Review,* Jun 1947, 3–7; Shields, 61.

98. Smith-Hutton, "Tokyo, Dec. 1941," 13–15.

99. Smith-Hutton ltr response, Apr 1974.

100. Smith-Hutton oral history, USNI, 1:375–76.

101. Smith-Hutton, "Tokyo, Dec. 1941," 13–15; and RAdm. Arthur H. McCollum, Oral History, USNI, Annapolis, MD, 1971, 2:453, copy in OA.

102. "History of Naval Attaché, London, 1939–1946," MS, 9–12, OA.

103. *COMNAVEU WWII Admin History,* 20–21, 27.

104. Ibid., 23–35.

105. Ibid., 87–88.

106. Ibid., 108–13, 125–28.

107. Ibid., 113–14.

108. Ibid., 346–48.

109. *COM4THFLT WWII Admin History,* 50–51.

110. Ibid., 56.

111. Ibid., 76–79.

112. Ibid., 105–8.

113. Ibid., 110–19.

114. Ibid., 122–23.

115. "Administrative History of NOF, Belem, Brazil, in World War II," folder 24, passim, hereafter *NOF Belem WWII Admin History.*

116. "Administrative History of NOF, Santos, Brazil, in World War II," folder 24(19), passim, hereafter *NOF Santos WWII Admin History.*

117. *COM4THFLT WWII Admin History,* 190–210; "Administrative History of COMSOLANT (Commander Naval Forces, South Atlantic) in World War II," 7.

118. *NOF Belem WWII Admin History,* passim.

119. "Administrative History of NOF, Natal, Brazil, in World War II," folder 24, passim, OA.

120. Lundeberg, "Antisubmarine Operations in the Atlantic," chap. 6, 19.

121. *ONI WWII Admin History,* 693–95.

122. *ONI Review,* Aug 1946, 10–11.

123. Radm. Charles E. Olsen memo, 25 Apr 1944, box 1, TSC 6192, RG 10430, OA.

124. CNO ltr, ser 14946P32, 8 Aug 1946, Accession 3770, box 1, ONI Day File, OA.

125. Murphy, *Diplomat Among Warriors,* 110.

126. Capt. Donald H. Scott, "Notes on Mozambique," forwarded to author, 13 Apr 1976.

127. Capt. Donald H. Scott, lecture outline of 25 Apr 1949.

128. Scott ltr to author, 5 Sep 1976.

129. Capt. Francis A. Klaveness ltrs to author, 27 Feb 1975 and 19 Mar 1975.

130. *COMNAVEU Admin History,* Oct 1945–Mar 1947, 77–79.

131. CNI ltr, ser 1520P32, 23 Aug 1946, Accession 3770, box 1, OA.

132. Testimony of RAdm. Thomas B. Inglis, *Navy Department Appropriations Bill for 1947: Hearings Before the Subcommittee of the Committee on Appropriations,* 79th Cong., 2d sess., 1946, 1587–1603.

133. Klaveness ltrs to author, 27 Feb 1975 and 19 Mar 1975.

134. *COMNAVEU Admin History,* Oct 1945 to Mar 1947, 70–76.

135. CNO ltr, ser 0003337P32, 4 Sep 1947.

136. *ONI Review,* Aug 1956, 348.

137. ONI-19(A), May 1947, para. 5004.

138. Ibid., para. 5012.

139. Smith-Hutton ltr response, Apr 1974.

140. Ibid.

141. *Washington Post,* 16 Aug 1948, 1–2.

142. ONI-49-2A, Navy Intelligence Collection Instructions of 1 Jun 1960, IV-8-IV-9.

143. Naval Attaché, Rome, ltr to DNI, ser 00035, 11 May 1949, accession 10430, box 5, TSC 6799, OA.

144. *ONI Review,* Dec 1949, 29–31.

145. CNO (OP-322H2) ltr, ser 06490P32, 2 Aug 1949, file A19/EN3, box 5, Job 12876, OA.

146. CNO (OP-322H2) ltr, ser 002459P32, 9 Dec 1949, file A19/EN3, box 5, Job 12876, OA.

147. DNI ltr, ser 000117P32, 13 Feb 1950.

148. CNO (OP-322H2B3) memo, 24 Aug 1950, Naval Attaché file, box 1, Job 14236, OA.

149. John P. Seifert, interview by author, 2 Dec 1975.

150. Naval Attaché Rosters, 1949, in author's possession.

151. Felix Johnson, interview by author, 14 May 1975. The author also attended the conference and accompanied RAdm. Johnson on part of his inspection tour.

152. Library of Congress, *Military Attaché System,* 28.

153. ONI Instruction 005201.3 of 21 Feb 1952.

154. CNO (OP-302C) ltr, ser 081P52, 22 May 1952, "Incoming Correspondence 1952–53" file, box 1, Job 12876, OA.

155. DCNO (Operations) memo, ser 542P32, 25 Feb 1948, box 1, Job 14236, OP-32F Day File 1947–1949.

156. *Naval Intelligence Quarterly,* vol. 10, no. 2, 1957, 3–11.

157. ONI Instruction 04630.2 of 19 Aug 1955.

158. ONI-70-1, 20 Jun 1956, 62.

159. *ONI Review,* Aug 1956, 349.

160. Naval Attaché Bonn personal ltr to DNI, 7 Dec 1956, "Admiral Frost" files, OA.

161. Naval Intelligence Collection Instructions of 1 Jun 1960, IV-8.

162. ONI-70-1, 4 Nov 1957, chap. 1, 58.

163. Ibid., 61.

164. *Naval Intelligence Quarterly*, vol. 10, no. 2, 1957, 3–11.

165. CNO msg, 24 May 1957, OP-922H4 files, OA.

166. ONI Instruction 5213.6 of 30 Dec 1957.

167. CNO (OP-92B1B) memo, ser 000601P92, 8 May 1958.

168. Naval Attaché, London ltr, 6 Mar 1959, "Admiral Frost" files, Nov 1958–May 1959.

169. ONI Instruction 05410.2 of 26 Jun 1959; and Library of Congress, *Military Attaché System,* 30.

170. Joseph H. Floyd, interviews by author, 7 Jun 1975 and 10 Aug 1975.

171. OPNAV Instruction 003800.8A of 27 Sep 1961.

172. CNO (OP-922H) personal ltr to Naval Attaché, Saigon and Pacific Fleet, Fleet Intelligence Officer, 5 Nov 1963, OA.

173. Commander in Chief, Pacific Fleet [Capt Donald M.] Showers personal msg to DNI [RAdm. Rufus L.] Taylor, H4 files, OA.

174. Capt. Wyman H. Packard, USN, "The Naval Attaché," *USNIP*, Apr 1965, 130–33.

175. SECNAV Instruction 05400.11 of 17 Aug 1965.

176. Naval Intelligence Command (NIC-321F) ltr, ser 0055–68, 3 Jun 1968, H4 files, OA.

177. OP-922H ltr, ser 00449–65, 15 Mar 1965.

178. NIC-3531 memo, 19 Jan 1970.

179. NIC-35 memo, 10 Apr 1970.

180. *DNI Annual Report to CNO, FY 1972,* 33

181. *DNI Annual Report to CNO, FY 1973,* V-1.

CHAPTER 4

Air Reconnaissance

This chapter includes a sampling of the information contained in the files of the Aviation History Section of the Air Warfare Division of the Office of the Chief of Naval Operations about the various types of intelligence collection performed by naval air squadrons. This sampling, in turn, is based on histories submitted by the participating squadrons and is arranged chronologically by types of collection, such as photographic, patrol, and multisensor.

Much of the data collection by naval air reconnaissance squadrons was in direct support of the operating forces of the Navy. As a consequence, the initial processing of data was performed by fleet resources within the areas where the collection was performed. Final processing and the correlation of information from other sources has been at the fleet or Washington level. Therefore, there is a close relationship between this chapter and Chapters 12, 13, 15, 18, and 34.

Development of Aerial Photographic Intelligence Capabilities (VD, VC, and VJ Squadrons)

Aerial photography in the U.S. Navy had its origin about the time that naval officers were first learning to fly. Snapshots from handheld cameras were taken by many of the early naval aviators. Officially, some experiments in taking aerial photographs were conducted at Guantanamo in 1913 when the small naval air arm first operated with fleet units. During the Vera Cruz incident in April 1914, Lt. Patrick N. L. Bellinger made flights to photograph the harbor. Such pioneering efforts were undertaken using personal cameras rather than government supplied equipment.

A more formal program was developed at Pensacola during the winter of 1914–1915 when SC3 W. L. Richardson, whose hobby was photography, introduced improvements on earlier efforts. He was later made an officer and designated the head of the Navy's first photography school.

In 1915, the Navy requested the Eastman Kodak Company to develop an aerial camera to meet certain specifications, and during that year and the next, tests were made at Pensacola of this and other makes of camera. On 10 January 1917, the Navy placed its first production order for aerial photo equipment when the Naval Observatory requisitioned twenty "aero cameras and accessories" from Eastman Kodak.

A few aerial photographs were taken of enemy bases and of combat action during World War I, but there appears to have been little operational application made of the results of these pioneer efforts, probably because photo interpretation had not developed very far toward its current sophisticated state. In the postwar period, aerial photo equipment and techniques slowly progressed as experience was gained when filling requests from various civil agencies of the government. For example, in the summer of 1926, a Navy photographic unit, equipped with three Loening OL-1 amphibious aircraft, made the first aerial mapping photographs of Alaska at the request of the Department of the Interior. Other units made aerial surveys of different parts of the United States, Central America, and the Caribbean islands.

In the 1920s and 1930s utility squadrons and sometimes patrol squadrons generally carried out photographic work for the fleet. The missions included some mapping and photographing of ships, stations, and torpedo and gunnery practices.

One exception was the search for Amelia Earhart Putnam after she disappeared on 3 July 1937 on the Lae, New Guinea, to Howland Island leg of her around-the-world trip. Aircraft carrier *Lexington* (CV 2) was sent from Long Beach to search the probable area of Earhart's flight track and her last-known position near the Gilbert Islands. No documentary evidence has been found that the Navy took advantage of the opportunity to collect photo intelligence on Japanese activities in the islands in proximity to the search area. Assistant Chief of Naval Operations RAdm. James O. Richardson labeled the reconnaissance effort a headache for the Navy. He stated that the tremendous expenditures for gasoline for *Lexington*'s search aircraft had put a severe strain on aviation funds.[1]

The establishment of special photographic units in both the Atlantic and Pacific Fleets took place only a short time before the United States became an active participant in World War II. The Fleet Air Photographic Units were set up in May 1941 to receive and assimilate photographic personnel and material and to coordinate and conduct advanced training in all aerial photo services as required. The formation of the Atlantic unit was complete by 9 June 1941, and a photo lab was set up in the Administration Building at Naval Air Station (NAS), Norfolk. Its first aircraft was received in September 1941. In the Pacific, although Commander Scouting Force had recommended the establishment of a photo unit in each patrol wing to be composed of photographic and interpretation sections able to move to any area as required, only one unit, set up in May 1941, was in existence when the war began.[2]

On 15 July 1942, less than a month before the Marines landed on Guadalcanal, photo interpretation officers and photographer's mates making up the first photo interpretation unit reported for duty in the South Pacific. The first reconnaissance mission over Guadalcanal was flown from Australia on 17 July by Army B-17s using Navy photo equipment. The cameras were operated by LtCol. Merrill B. Twining and Maj. William B. McKean, 1st Marine Division operations officers who were acting as observers on this flight.

Until November 1942, when Marine Photographic Squadron (VMD) 154 reported, the only airplanes available in the South Pacific for long-range photo missions were Army B-17s, although some local missions were made by Navy PBY Catalinas. In April 1943, the Navy's Photographic Squadron (VD) One arrived in the South Pacific. In August 1943, photographic interpretation and photographic squadrons in the South Pacific were combined as Fleet Air Photographic Group One.

In the Central Pacific, Photographic Squadrons Three and Four were commissioned on 15 February and 15 July 1943, respectively, to become Photographic Group Two on 4 January 1944. On 1 June 1944, VD-5 was commissioned and was added to Photographic Group Two in November. The four VD squadrons covered the entire combat area of the war in the Pacific, operating long-range aircraft from land bases.

In the Atlantic, VD-2 was primarily involved in training carrier photo pilots for both the Atlantic and Pacific Fleets. A squadron detachment was set up in May 1944 at the Naval Air Facility, New Cumberland, Pennsylvania, and its Photographic Reconnaissance Training School provided photo pilots for carrier duty in all areas until the end of the war.

Post–World War II Navy plans provided for the retention of two long-range photographic squadrons, VD-5 in the Pacific and VD-2 in the Atlantic, under the new designations VPP-1 and VPP-2, respectively. In 1948, the squadrons were redesignated VP-61 and VP-62; they were decommissioned in 1950 after two composite squadrons had been organized in early 1949.

Composite Squadrons (VC) 61 and 62 were assigned carrier-type aircraft and were the first formally organized carrier photo units. The need for a long-range shipboard photo reconnaissance capability soon became apparent, and Utility Squadrons (VJ) 61 and 62 were commissioned in 1951 to fulfill the requirement. The subsequent light photographic squadrons (VFP) and heavy photographic squadrons (VAP) were the direct descendants of the VC and VJ squadrons.[3]

The first propeller-driven aircraft complement of Composite Squadron 62 included ten F8F-2P Bearcats, and four F4U-5P and two F4U-4P Corsairs. VC-62 was commissioned on 3 January 1949; its mission was "to train and maintain the readiness of units for carrier-based photographic reconnaissance of designated targets in areas of Naval Operations." VC-62 was to provide detachments of two photo planes and specially trained pilots to each East Coast air group.

The F8F-2P Bearcats and F4U-4P/5P Corsairs were essentially standard fighters with a camera mounted in the fuselage. In many respects, the pilots found these airplanes poorly suited to their photo reconnaissance missions. Neither aircraft type was equipped with a trimetragon camera installation, but the problem was solved in the F8F-2P by attaching a camera capsule to the centerline bomb rack. The arrangement did not work for the F4Us because their inverted gull wings blocked the field of view of the oblique cameras. Another problem was the lack of a photo viewfinder. With the restricted visibility that both aircraft types afforded the pilot, getting the right object in the center of the picture was largely a matter of luck until professional skill was gained through experience.

The first jet-powered photo reconnaissance aircraft was the F2H-2P Banshee. It carried three cameras, all of which could be aimed from the cockpit, and was received by VC-62 in mid-1951. A photo viewfinder allowed the pilot to see what was under his plane and to center his picture on the target. The F2H cruised at 400 knots, was capable of altitudes in excess of 45,000 feet, and solved most of the previous problems.

The faster jet-powered airplanes, however, required faster cameras. The K-17, 6-inch focal-length cameras, with their six-tenths of a second recycle, gave the proper overlap at low altitudes, but the pictures were blurred because of the aircraft's speed. The problem was solved by using a magazine that kept the film moving continuously at a speed related to the aircraft's ground speed.[4]

Light Photographic Squadron 63, originally designated Composite Squadron 61, was commissioned on 20 January 1949 at NAS Miramar, San Diego. On 2 July 1956, it was designated VFP-61, but on 1 July 1959 its designation was changed again to Composite Photographic Squadron (VCP) 63. Finally, on 1 July 1961, it became totally carrier-based, the designation changed to Light Photographic Squadron 63, and its shore-based A3D-2P Skywarrior aircraft transferred to Heavy Photographic Squadron (VAP) 61 at Guam.

In July 1950, the first Korean War photo detachment, led by Lt. L. W. Moffit, was deployed on board *Philippine Sea* (CV 47). In the fall of 1950, VC-61 acquired its first photo-modified F9F-2P Panther jet, making possible high-speed aerial photo reconnaissance over North Korean forces. In June 1952, the F2H-P Banshee joined the squadron, and three of the aircraft were placed on board *Valley Forge* (CV 45) in the forward area. Lt. Charles Hooper, assigned to *Valley Forge*, was the first Navy pilot to complete 100 photographic missions over Korea. A later photo detachment operated from *Essex* (CV 9).

As the Banshee became obsolete, it was succeeded by the F9F-8P Cougar. Then, in September 1957, VFP-61 received its first F8U-1P Crusader, giving longer range, more time over target, higher altitudes, and a top speed in excess of 1,000 miles per hour. In July 1959, the A3D-2P Skywarrior, a photo version of the A3D (now A-3) bomber, was added to the inventory. The A3D-2P was primarily intended for mapping and charting, and, except for speed and altitude, it had roughly the same photo capabilities as those of the F8U-1P. In addition, the A3D-2F had an inflight refueling capability, and its cameras and magazines could be changed in-flight, giving it a greater data-gathering capability.[5]

The Korean War demonstrated to the Navy the need for an adequate number of carrier- and heavy-photographic aircraft capable of performing day and night photo reconnaissance to meet naval requirements. Of equal importance was the need for adequately equipped facilities, staffed by properly qualified personnel, to process and interpret the photographs obtained.[6]

Shipboard Aerial Reconnaissance (VFP and VAP Squadrons)

Task Force (TF) 77, the heavy aircraft carrier group operating in Korean waters, initially lacked a photo aircraft capable of obtaining at high altitudes the large-scale photographs required for accurately pinpointing enemy antiaircraft gun positions. The 12-inch focal length of the K-17 camera of the F9F-2P required runs at low altitudes.[7]

On 13 January 1952, TF 77 acquired two F2H-2P Banshee aircraft on temporary loan from the Marines and two K-38, 36-inch focal length cameras from the Air Force. A single run at 15,000 feet with the borrowed camera equipment provided coverage equivalent to three runs with the K-17, 12-inch camera at 5,000 feet. The resultant workload for the photo interpreters was halved, and they had to handle only one-third the number of prints. With the improved equipment, the photo interpreters provided the task force with flak analysis and flak mosaics within twenty-four hours of receiving the processed film.

On 3 February 1952, Commander Task Force (CTF) 77 recommended that the F2H-2P aircraft with the K-38 camera be adopted as standard carrier equipment as soon as practicable. As of 1 July 1952, there were seven F2H-2Ps deployed in the Western Pacific; three on board *Philippine Sea* and four on board *Essex*, two of the four carriers that were rotating through TF 77.[8]

Many of the photo reconnaissance missions flown by carrier aircraft searched for targets before a carrier strike was launched so that a more satisfactory selection of targets could be made than had been possible from photographs received from the Fifth Air Force.[9]

In July 1952, most TF 77 carrier photo planes were F9F-2P Panthers. In November, the F9F-5P arrived in the Korean theater on board *Oriskany* (CV 34), and, by February 1953, there were two carriers with F9F-5P and two with the markedly superior F2H-2P Banshee. The Banshee had a more flexible camera installation than the Panther, with three cameras rotatable to right or left, an excellent viewfinder, and ready accessibility of cameras for loading, removal, or adjustment. Its speed and endurance were superior to the F9F-2P. Most F2H-2P photography was taken with the K-38, a 36-inch focal length camera operated at 15,000 feet. Slow recycling of the camera, the lack of image-motion compensation, and slow shutter speeds, however, made it necessary for the F2H-2P to slow to 220–260 knots in order to obtain satisfactory results.[10]

The F9F-5P was considerably superior to the F9F-2P but was inferior in speed and endurance to the F2H-2P. The main advantages of the F9F-5P over the F9F-2P were its provision for mounting the K-38, with its 36-inch lens, and the convenience of being able to be escorted by fighter aircraft with the same general flight characteristics.[11]

In April 1953, Lt. H. D. Williams and CMM Raines of *Philippine Sea* perfected a modification to the standard K-38 camera equipment that effectively compensated for image motion at high speed and low altitude. Thus, it became possible to spot many productive targets that had gone previously undetected.[12]

With the establishment of the Taiwan Defense Command in 1955 and the requirement to collect intelligence that might indicate Chinese Communist preparation or intentions to attack or invade Taiwan or the Pescadores Islands, periodic photo reconnaissance flights were flown by Task Force 72 patrol squadrons along the Chinese coast to determine any changes in military readiness or capabilities to support offshore operations. Direct contact was maintained with the Chinese Nationalist Air Force headquarters to provide guidance on the flight plans and to assure a coordinated effort. The reconnaissance flights became a high-volume business and quickly outgrew the initial arrangement of having the photo lab work done by the Commander Seventh Fleet's amphibious warfare command ship (AGC) at Keelung (Chi-lung). When the Taiwan Defense Command's photo lab became operational, the command completed photo interpretation work, sent out daily dispatches based on the work, and prepared targeting sheets on mainland installations.[13]

As related above, the Banshee was replaced by the F9F-8P Cougar, which served well until it was replaced by the supersonic F8U-1P Crusader between 1957 and 1960. The Crusaders brought a whole new concept to aerial photography; they were equipped with miniaturized cameras and the CAX-12 camera system, which had focal lengths ranging from 1½ to 12 inches and a negative format of 70mm and larger. The CAX-12 system provided for a recycle rate of six pictures per second. Sharp pictures could be obtained at aircraft speeds in excess of Mach 1.1. The unarmed, photo-configured Crusaders later provided the intelligence data that made possible more effective attacks against hostile installations in Southeast Asia.[14]

Beginning in March 1960, Navy photo squadrons operated over Cuba to fulfill photo intelligence requirements for the armed services, the Central Intelligence Agency, and the National Security Agency. The photography resulted in complete coverage of the Cuban coastline, as well as coverage of areas pertinent to the evaluation of collateral intelligence information. Quarterly coverage was made of the Cuban south coast, including Largo Cay and the Isle of Pines. The Naval Photographic Interpretation Center processed the film and provided photo interpretation reports to the above agencies and the armed services.[15]

Light Photographic Squadron 62 became very much involved in the Cuban missile crisis in 1962. On the evening of 22 October, President John F. Kennedy announced to the free world the opposition of the United States to the Soviet placement of nuclear missiles on Cuban soil. VFP-62 was among the first reconnaissance assets to be called upon to conduct low-level photographic flights over Cuban territory the next morning.

VFP-62 detachments operated both from NAS Key West and from aircraft carriers throughout the critical period until 26 November; it was awarded the Navy Unit Commendation for the effort, the first time such an award had been made in peacetime and also the first time that a President had personally made the presentation.[16]

On 9 February and 5 June 1963, in accordance with Joint Chiefs of Staff (JCS) directives, Commander in Chief, Atlantic (CINCLANT) subordinated aircraft made two low-altitude photo reconnaissance passes over targets in Cuba. There were no hostile reactions to either mission, both of which were conducted by two aircraft.[17]

The increased need for photo reconnaissance over the Republic of Vietnam in 1963–1964, the initiation of low-altitude reconnaissance over Laos, and the decision by higher authority to employ Pacific Fleet photographic processing and interpretation facilities for the exploitation of the major portion of the material obtained on the missions created a requirement for expanded Navy photo facilities in the Pacific Fleet area.[18]

During the period 21 May–30 September 1964, VFP-63 detachments participated in special operations over Laos. On 6 June 1964, Lt. C. F. Klusmann, flying from *Kitty Hawk* (CVA 63), was shot down while on a mission over Route 7 in central Laos and was captured by the Pathet Lao. On 29 August, however, he was able to escape confinement and, with the help of guerrilla forces, made his way to Udorn, Thailand, on 31 August 1964. On 5 August 1964, the same VFP-63 detachment also participated in raids against North Vietnamese motor torpedo-boat bases in the Gulf of Tonkin.[19]

Heavy Photographic Squadron 61, equipped with RA-3B Skywarrior aircraft and based at NAS Agana, Guam, had the mission to provide photographic services and intelligence for naval operations as directed by Commander Seventh Fleet or higher authority. Its tasks included aerial photographic reconnaissance, including radar photographic reconnaissance; preparing and disseminating initial photographic intelligence reports and studies; operating assigned photographic and photo intelligence facilities (at the Fleet Air Photographic Lab); obtaining cartographic photography and preparing charts based on the photos (VAP-61 had the Navy's only cartographic photo capability); and furnishing photographic services to ships, shore-based fleet aircraft squadrons, and other naval activities as required.[20]

From May 1964 to 31 January 1970, VAP-61 provided photographic reconnaissance support to Task Force 77 (Yankee Team) operations by maintaining a detachment, usually on board attack carriers operating in the South China Sea and the Gulf of Tonkin, to provide reconnaissance flights over hostile Southeast Asian areas, locating truck convoys and vectoring attack aircraft against them. In January 1968, the detachment was shifted to the air base at Danang in the Republic of Vietnam. With the bombing halt against North Vietnam in November 1968, the VAP-61 detachment concentrated its efforts against infiltration routes.

After five and one-half years of continuous combat support in Southeast Asia, VAP-61's commitment to CTF 77 was terminated on 31 January 1970. It shifted its major emphasis to Thailand mapping projects, which it completed in March 1970. In April and September 1970, a detachment from VAP-61 was sent to Osan Air Base, Republic of Korea, to conduct an environmental survey under the auspices of the Naval Oceanographic Office. While based at Osan, the VAP-61 detachment also fulfilled numerous photographic requirements for various agencies in Korea on a not-to-interfere basis.

Also in April 1970, a two-plane VAP-61 detachment was sent to Australia for a seven-month cartographic operation over New Guinea and Australia. From its home base at Agana, the detachment additionally accomplished a significant amount of cartographic and reconnaissance photography over the U.S. Pacific Trust Territories, Okinawa, Japan, Midway, and the Philippines to satisfy fleet and national tasking.

In January 1971, a detachment from VAP-61 commenced operations in the Republic of Vietnam to update existing maps and to obtain cartographic photography of selected Cambodian cities. Operating from the Royal Thai Navy Air Base at Utapao, Thailand, it was able to fulfill urgently needed photographic requirements for several commands in South Vietnam and Thailand, in addition to its primary cartographic task.

On 1 July 1971, VAP-61 (less the Fleet Air Photo Lab) was consolidated with Fleet Air Reconnaissance Squadron (VQ) One at Agana.[21]

Light Photographic Squadron 62 was based at NAS Cecil Field, Florida, in the mid 1960s and deployed detachments aboard attack carriers of the Atlantic Fleet. The mission was to conduct aerial photographic reconnaissance in support of fleet carrier operations in the Atlantic and Mediterranean and to support CINCLANT special operations in the Caribbean area.

Most Atlantic Fleet attack carriers were normally deployed to the Mediterranean, but, in June 1966, *Franklin D. Roosevelt* (CV 42), with VFP-62 Detachment 42 embarked, was deployed to Southeast Asia. While on Yankee Station, the VFP-62 detachment fulfilled special photo-reconnaissance requirements, including obtaining aerial photo coverage of lines of communication, cities, airfields, and petroleum storage areas. Special targets covered by the VFP-62 detachment were mostly pinpoint targets for which photography was needed by embarked staffs and the ship's strike targeting personnel. The photo coverage could be either pre- or post-strike.

There were usually three or four missions flown each day, and the photo lab and interpretation work included taking a quick look at the film as soon as it arrived to find significant targets of a transient nature that might be worthy of a flash report; titling the film and making duplicate positive prints, which were in demand by the targeting and air intelligence staff for locating new targets; preparing an Initial Photo-Interpretation Report (IPIR); preparing an Operational Report after the last mission each day, to be sent by message giving the photo coverage and times; and packaging the film and IPIRs for the first COD (Carrier-On-Board Delivery) aircraft departure the next morning. Detachment 42's two officer photo interpreters and two enlisted photographic technicians could not perform the thorough processing job required. Two more technicians were requested but did not arrive during the deployment.[22]

Patrol Reconnaissance (VP and VPB Squadrons)

World War II

Among the tasks assigned to the Navy prior to World War II by *Joint Action of the Army and Navy* was to provide and operate a system of offshore scouting and patrol to give timely warning of an attack. Shipborne aircraft could be counted on to search out to about 150–250 miles and still have a reasonable fuel reserve. The shore-based patrol aircraft could search to a 700–800-mile radius with safety. In 1940, all aircraft built for maritime patrol and scouting were assigned to the fleet, and the responsibility for offshore scouting and patrol to give timely warning of an attack, therefore, rested with the fleet commanders.

In early March 1940, before the Battle Fleet sailed from West Coast ports to Hawaii, Commander Base Force (COMBASEFOR) was directed by Commander in Chief, U.S. Fleet (CINCUS) to prepare and execute plans for the security of the U.S. Fleet while it was berthed in the general Pearl Harbor area. The resulting COMBASEFOR order required a seven-day-a-week, daylight-to-dusk naval air patrol of the outer sea area around the Hawaiian Islands. Specifically, the area to be covered was a circle having a 180-mile radius with its center at Ahua Point, close by the entrance to Pearl Harbor. Commander Patrol Wing Two Capt. Aubrey W. Fitch was to use twelve aircraft daily for the search.[23]

When the return of the fleet to its West Coast home ports was deferred indefinitely, the aerial reconnaissance patrols were continued, but not without considerable strain on the pilot training program, which both the Chief of Naval Operations and the Bureau of Aeronautics had given high priority. Thus, in July 1940, Patrol Wing Two reduced its security patrol to six planes daily, each flying an average of 8.5 hours. This still totaled in excess of 1,500 hours per month for the wing and increased by over 50 percent the normal monthly flying time. Capt. Fitch was concerned about the amount of engine time being accumulated and the resultant overhaul load.

The Hawaiian area was well down on the Navy Department's priority list for additional patrol squadrons. Higher priority had been given throughout 1940 to assignments in the Atlantic or to the sale of patrol planes to prospective allies. CINCUS efforts to get the Navy Department to fill the needs for patrol planes in the Pacific were unsuccessful.

When Capt. Patrick N. L. Bellinger took command of Patrol Wing Two on 1 November 1940, he was unhappy about the daily reconnaissance burden and advised CINCUS that it would require fifty ready patrol planes each day to search an 800-mile, 360-degree area around Pearl Harbor. He had only sixty-odd patrol aircraft. Planes and crews could fly only every second or third day. The 300-mile patrols being flown by six to twelve planes each day were an inadequate protective search. Furthermore, they only covered the western sector.

On 28 November 1940, Adm. James O. Richardson (CINCUS) wrote Adm. Harold R. Stark (CNO) sending him a draft of a proposed revision of a CINCUS directive to tighten up external security. The directive prescribed a long-range air reconnaissance patrol to be maintained from Pearl by fleet patrol planes. Stark responded that wartime security measures were not yet required and that continuous air patrols were not necessary. Consequently, Richardson's reconnaissance requirement was not included in the final directive when it was issued on 5 December 1940.[24]

In 1941, in connection with the Neutrality Patrol, Patrol Squadron 54, flying PBY-2 Catalinas, was based in Norfolk, with a detachment at Bermuda. The squadron's instructions on what to do in case of a sighting were ambiguous. No submarines or other unusual activity were detected by the Bermuda-based patrol during 1941.

VP-92 was one of four squadrons commissioned in 1941 to be equipped with the new PBY-5A amphibious patrol aircraft. Originally, the aircraft were intended for use on the Neutrality Patrol out of Iceland, but the Pearl Harbor attack forced a change, and the amphibians were deployed to the Pacific. The aircraft were delivered at San Diego. Again the orders were changed, sending VP-92 to operate under Commander 10th Naval District out of San Juan, Puerto Rico, over the Caribbean where German submarines were sinking merchant vessels with little or no interference. Squadron detachments were dispatched to critical locations around the Caribbean wherever and whenever sinkings indicated submarine activity. One three-plane detachment operated from Guantanamo, and others were deployed to Trinidad and Windward Island bases. Operations were generally locally conceived and controlled, with the heaviest concentrations placed mainly on high-density shipping routes such as the Windward and Bahama Channel passages. In the summer of 1942, VP-92 command moved to Guantanamo, and a convoy control center was established to provide operational and intelligence information to the aircraft crews. VP-92 devoted its main efforts thereafter to escorting convoys. On the night of 27 August 1942, a VP-92 plane caught German submarine *U-94* on the surface inside a convoy, bracketing it with four 600-pound depth bombs and damaging the submarine so that it couldn't operate submerged; *U-94* was finished off by a Canadian convoy escort, HMCS *Oakville*.

Shortly after their arrival in the Caribbean, the PBY-5As were equipped with radar, which improved their submarine detection capabilities tremendously. This was particularly true at night and in low visibility, when a submarine had much less time to submerge after visually sighting the approach of the very slow aircraft.

In September 1942, VP-92 terminated operations in the Caribbean and flew to Freetown, Sierra Leone. Here it operated with British forces until the invasion of North Africa, when it shifted to Morocco.

There, with VP-71 from Iceland, VP-92 made up Fleet Air Wing (FAW) 15, based at Port Lyautey (now Kenitra), Morocco. Its patrol areas covered the Atlantic approaches to Gibraltar and the offshore coastal routes south past the Canary Islands.[25]

A daily mid-ocean barrier sweep designed by Dr. Jacinto Steinhardt of the Navy's Antisubmarine Warfare Operational Research Group to intercept blockade runners was inaugurated in April 1943 by VP-83 and VP-94, based at Natal, Brazil. Dr. Steinhardt also developed a "long-gambit" search plan to improve the South Atlantic airmen's low average in regaining lost radar contacts on submarines. When the German submarines began operating in loosely coordinated pairs, the aerial gambit plans became almost doubly effective.[26]

The Navy assigned three PB4Y Liberator land-based patrol bomber squadrons (VPB-103, VPB-105 and VPB-110) to the British Coastal Command in late summer 1943. They were based first at St. Eval, Cornwall, and later at Dunkeswell, Devon, about 15 miles northeast of Exeter. In the winter of 1944–1945, the three Navy squadrons were joined by VPB-112 and VPB-117, based at Upottery, Devon. The five VPB squadrons made up Fleet Air Wing Seven. Their mission was to conduct antisubmarine warfare in the English Channel, the Bay of Biscay, and the southwest approaches to both. The Liberators were equipped with radar, a low-level bombsight, sonobuoys, cameras, night lights, depth bombs, and torpedoes. All operational direction was received from the Plymouth headquarters of the Coastal Command, where there was an extensive control room fully equipped and staffed by U.S. Navy air intelligence officers and a few Royal Air Force officers. Preflight briefings were thorough and were conducted like clockwork in a professional, business-like atmosphere. Patrol areas were small (50 miles on a side) and very specific as to location. Postflight debriefings were equally thorough, with all crew members participating, and the gist of the mission narrative was put on a scrambler teletype to other control rooms in the Plymouth sector.[27]

Post–World War II

Up to 1 February 1949, VP-26 maintained a three-aircraft unit of PB4Y-2 Privateers at Port Lyautey, Morocco, for "electronics work" and search-and-rescue operations. In due course, the unit was expanded to nine aircraft and accomplished several electronic intelligence (ELINT) collection tasks.[28]

Aerial reconnaissance of the Taiwan Straits area between 22°N and 27°N was commenced on 20 July 1950 by U.S. naval aircraft. Reports were made of sightings of ships transiting the area.[29]

Korean War

Aerial reconnaissance at the start of the Korean War in July and August 1950 was of little value due to the inaccuracies and lack of timeliness of the reports of ship sightings from units of the Far East Air Force. Fishing boats were reported as combatant merchant ships or ships of larger size, ship locations were incorrectly given (the ships were either underway in water too shallow for navigation or at anchor in water too deep for anchoring), friendly combatants were identified as merchant ships, and sunken ships were repeatedly reported as underway. All reports were investigated by blockade commanders and usually represented a waste of limited resources and the transmission of numerous messages over already overcrowded circuits.

Fleet Air Wing Six was commissioned on 4 August 1950, and Fleet Air Japan was commissioned on 9 August. Direct communication between VP-6 aircraft and elements of the blockade force was directed on 5 October and produced a workable coordination between naval air reconnaissance and surface units.

By 1 September, the intelligence officer at Commander Naval Forces, Far East (COMNAVFE) was checking air reconnaissance reports before passing them on to the operating forces. By early November, COMNAVFE was operating a Theater Shipping Surveillance Center that afforded a means of sifting out reports of known friendly shipping and obviously erroneous reports. It also kept air commands informed about the location and movement of United Nations' surface forces and units in areas over which the reconnaissance aircraft were operating.[30]

Daily reconnaissance by patrol aircraft contributed considerably to the blockade of North Korea, obtaining intelligence information on merchant shipping, providing weather information for United Nations forces, providing flare illumination for Marine night intruders, locating enemy radar installations, and patroling the Taiwan Strait. Sorties were flown by Navy PBM-5 Mariners, PB4Y-2 Privateers, and SP-2 Neptunes and by Royal Air Force Sunderlands covering the Yellow Sea, Sea of Japan, Tsushima Strait, and the waters off the China coast and Taiwan.[31]

Chinese Communist vessels and aircraft several times fired upon U.S. Navy patrol aircraft in international waters during the early 1950s. During 1951, aircraft were shot at on 30 September off Tsingtao by two frigates, on 4 October off Lien Yun by a frigate, and on 6 November off Swatow by a small combatant craft. During 1952, U.S. Navy aircraft received fire on 20 January off Tsingtao from two frigates, on 12 July off the Kiangsu coast from frigates in two separate incidents, on 27 July off the mouth of the Yangtze from two frigates, and on 20 September from two MiG-15 fighters. The incidents continued during the following year: on 16 March off Tsingtao, a frigate shot at Navy aircraft; and, on 7 April, a landing ship off the coast 130 nautical miles north of Shanghai opened fire.[32]

Naval patrol squadrons were deployed as follows in support of the Korean War:

- Atsugi, Japan: one land-based squadron.
- Iwakuni, Japan: one land-based squadron, one seaplane squadron, and one RAF Sunderland flying boat wing.
- Okinawa: one land-based squadron and a detachment of four to six seaplanes from Sangley Point, the Philippines.
- Sangley Point, Luzon, Philippines: one seaplane squadron, less detachments.
- Boko Ko in the Pescadores: a detachment of two seaplanes from Sangley Point when weather permitted.

The briefings and debriefings of these squadrons and detachments were accomplished by the squadron air intelligence officers, augmented by air intelligence officers assigned to Fleet Air Wing staffs and seaplane tenders.[33]

The Cold War Era and the Vietnam War

In early July 1957, two Soviet destroyers and a tanker were observed traveling southbound in the Red Sea and were believed to be headed for the Soviet Far East. Task Force 72, augmented by five airborne early-warning aircraft, and additionally using those of its own patrol aircraft not needed in Taiwan Straits patrol, conducted reconnaissance to intercept and track the Soviet ships. Contact was made in the Indian Ocean on 10 July 1957, and daily observations were made thereafter through the Strait of Malacca (now The Straits), the South China Sea, and up to 31°29'N, 126°30'E, where FAW-6 assumed surveillance responsibility on 21 July. The trailing operation demonstrated the feasibility of using long-range, early-warning aircraft for shipping surveillance in the Indian Ocean.[34]

Patrol Squadron Two deployed to Marine Corps Air Station Iwakuni, Japan, in March 1966 and began flying daily submarine and shipping surveillance-reconnaissance in the Sea of Japan and random coverage of the North Pacific along the Kurile Island chain, the Yellow Sea, and over the La Perouse and Taiwan Straits.[35]

In February 1966, VP-1 deployed to Iwakuni under the operational control of Commander Task Group 72.4 (CFAW-6). The squadron was equipped with twelve SP-2H Neptunes.

A seven-aircraft VP-1 detachment was established at Tan Son Nhut Air Base in Saigon on 17 February 1966 as Task Group 115.2, Air Patrol Group, Market Time Operations. Its mission was to prevent the infiltration by sea of men, arms, and supplies to the Viet Cong. The detachment worked with the Coastal Surveillance Force, Vietnam, which consisted of radar picket destroyer escorts, minesweepers, Coast Guard cutters, and the new 50-foot Swift boats. The surveillance area started at the 17th parallel and extended south around the Mekong Delta to the Cambodian border. Two Vietnamese navy observers accompanied each flight to help identify local coastal craft, and they were in radio contact with the Vietnamese navy surface craft to receive intercepts of suspicious contacts developed by their operations. The aircraft were also in radio contact with the Vietnamese navy commander of surface patrol forces in each patrol area along the coast as it was overflown.

Ocean surveillance air patrols were flown over the Yellow Sea, the Sea of Japan, and the South China Sea-Gulf of Siam areas. In "rigging" merchant-ship contacts (passing close-aboard at low altitude for identification purposes), photos were taken from the bow and after stations of the aircraft. The craft flew at 100-foot altitudes while each member of the aircrew carried out specific sighting assignments.

Electronic intelligence was collected on the electronic environment in the Sea of Japan, the Yellow Sea, and the Sea of Okhotsk. Special requirements were received from the Pacific Command ELINT Center at Fuchu, Japan, and ELINT search plans were established by CFAW-6, placing special emphasis on shipborne radars. This resulted in numerous signals being obtained from destroyers, frigates, submarines, and auxiliaries. Visual and photographic correlation was obtained in most cases.[36]

VP-40, equipped with P-5M Marlin seaplanes, was deployed to Sangley Point in the Philippines from March through September 1966 and was under the operational control of Commander Task Group 72.3 (CFAW-8). The squadron maintained a detachment at Camranh Bay, South Vietnam, supported by *Pine Island* (AV 12), to participate in Market Time coastal patrol operations, and also at Buckner Bay, Okinawa, supported by *Salisbury Sound* (AV 13), to conduct ocean surveillance in the Straits of Taiwan. A similar and final deployment with P-5M seaplanes was made by VP-40 from May to November 1967. During the extended period of seaplane operations by VP-40 in the Far East dating back to the Korean War, the seaplane had proven its value and effectiveness in ocean reconnaissance. Particularly during its period of support in the South Vietnam conflict, the seaplane's capability to operate from tenders in open ocean waters immediately adjacent to combat areas provided intelligence collection services that were not otherwise available.[37]

In November 1967, VP-40, with its P-5M Marlin seaplanes, was relieved at Sangley Point by VP-1, marking the end of the Navy's use of seaplanes for patrol in the South China Sea and concluding a long and effective service by seaplanes in air reconnaissance that dated to World War I. VP-1 was equipped with the more sophisticated Lockheed P-2H Neptunes, which had greater speed and endurance than the P-5M. As soon as VP-1 arrived, its squadron was required to assign a detachment of six planes to the Air Force base at Camranh Bay in South Vietnam to support Market Time coastal patrol operations. The detachment was the first Navy patrol to operate full-time from the Air Force base. VP-1 was assigned to the operational control of Commander Task Group 72.3 (CFAW-8 until August, when CFAW-10 took over).

VP-1's mission from Naval Station, Sangley Point, in the Philippines was to conduct shipping surveillance over the South China Sea to within close proximity of Hainan and the Chinese mainland. Flight tracks were named after wristwatch brands. All shipping contacts of over 1,000 tons were "rigged" and photographed, and their position, course, and speed recorded. Radio contact reports were made on any merchant ships of interest, and particular note was made of those visibly carrying military cargo to Haiphong. ELINT recordings were made of any emitters, and the recordings were forwarded to the Pacific Command ELINT Center in Japan. Photographs were also taken of Lincoln and Woody Islands in the Chinese Communist-controlled Paracel Island group whenever flight tracks passed nearby. All photographic, electronic, and visual information collected by the squadron was evaluated by squadron intelligence personnel before being submitted to Commander Task Group 72.3's air intelligence staff and to the antisubmarine classification center for forwarding to higher authority.

The Camranh Bay detachment operated under Commander Coastal Surveillance Forces, Vietnam (CTF 115). It conducted 24-hour radar and visual surveillance of sea areas along the coast of the Republic of Vietnam in order to detect and prevent seaborne infiltration of insurgent forces.[38]

VP-9 deployed to Adak, Alaska, on 1 December 1967, relieving VP-28. The squadron was assigned to the operational control of Commander Alaskan Sea Frontier (CTF 33) and was equipped with nine P-3B Orions. To eliminate the need to maintain a detachment at Shemya, VP-9 rotated one plane and crew through the facility for a one-night stopover each day.

Daily patrols were flown along prescribed tracks (given feminine names) covering an area between 40°N latitude and the Bering Straits and between 155°W longitude and the Soviet coastal waters along the Kamchatka Peninsula. In addition, one daily flight was made for antisubmarine sound surveillance system (ASW SOSUS) coordination and correlation with Naval Facility, Adak. The ASW flights were suspended temporarily following the North Korean capture of *Pueblo* (AGER 2) on 23 January until 30 April 1968, because it was feared that classified information about SOSUS on board the ship had been compromised.

The primary mission of VP-9 at Adak was to localize and prosecute submarine contacts believed to be other than friendly and to collect photographic intelligence (PHOTINT), electronic intelligence, and acoustic intelligence (ACINT) on the contacts. Secondary missions included obtaining ELINT and photographic, electronic, and acoustic intelligence on Soviet naval surface ships; photos of Soviet bloc merchant ships; and photos of Japanese and Soviet fishing vessels.

The Alaskan area, with numerous Soviet submarines unwittingly providing target services, was an excellent ASW training station for patrol squadrons. When VP-9 arrived at Adak, VP-28 was in the process of working a contact evaluated as being a Soviet conventional submarine that had been detected by SOSUS on 27 November 1967. VP-9 joined VP-28 in the operations on 30 November while still in the process of relieving. SOSUS had lost contact after 27 November, and VP-28's efforts were negative on the 27th, 28th and 29th. On the 30th, VP-9 held low-frequency acquisition and ranging (LOFAR) sonobuoy contact and evaluated the target as a Soviet nuclear submarine. Naval Facility, Adak, through SOSUS, generated another contact on 30 November that was evaluated as a Soviet nuclear submarine. VP-9's contact was held simultaneously by SOSUS at 1033 Zulu hours on 30 November. The same contact was retained by VP-9 on 2 December, and it was determined by the contact's track that the submarine was moving into Commander Task Force 32, Commander Hawaiian Sea Frontier's area. LOFAR sonobuoys successfully held the same contact on 20 and 21 January 1968 during its return to port.

Another contact of particular interest, detected first by SOSUS on 7 December 1967, was prosecuted from 7 to 13 December by VP-9 along a track of nearly 2,000 miles following the great circle course from the middle of the West Coast of the United States to its home port. It was determined that such transiting submarines passed through the Aleutian Island chain at Amukta Pass.

During the period December 1967 to March 1968, the Soviet missile range instrumentation ships (SMRIS) *Chazhma*, *Chumikan*, *Sakhalin*, *Sibir*, and *Suchan* (later renamed *Spassk*) operated in the CTF 33 area. On 16 March, during a SMRIS relocation flight, a series of radar contacts was investigated and visually identified as being seven Soviet naval ships, including a *Don*-class submarine tender, a *Skoryy*-class destroyer, two *Riga*-class frigates, and three small craft.

On 17 March, a *Kotlin*-class destroyer was identified, and on the 18th, six submarine contacts were developed in the same general area, indicating that a significant Soviet naval exercise was in progress. The operations continued in the CTF 33 and 32 areas until May 1968. One of the submarines was identified from VP-9 photos as a unit of the *Golf* class, capable of launching the SS-N-5 ballistic missile. When two range instrumentation ships were found 700 nautical miles east of the submarines during a SMRIS relocation flight, it was speculated that the *Golf* was to fire a practice missile. But the event, if it took place, was not observed while the VP-9 aircraft was in the area.

During VP-9's Adak deployment, ELINT intercepts were reconfirmed on twenty-four land-based radars and were made on submarines and surface ships. Intercepts relating to surface ships and some of the submarines were visually correlated to specific units. All ELINT data were sent to the Pacific Command ELINT Center for processing.[39]

The operations of the VP-1 detachment at Camranh Bay consisted of night flights in the Gulf of Tonkin to detect by radar any high-speed surface targets that might be a threat to U.S. Fleet units operating in the area. Through February of 1969, no detections were made. In the third quarter of 1969, VP-1 turned in its P-2H Neptunes and received nine new P-3B Orion aircraft.[40]

On 1 April 1969, VP-9 relieved VP-47 as Task Unit 72.3.6 at Naval Station, Sangley Point, under the operational control of Commander Task Group 72.3. VP-9 maintained a detachment at Naval Air Facility, Camranh Bay, for Market Time coastal patrol operations. An aircraft from the squadron successfully located a trawler believed to be attempting an infiltration of the South Vietnamese coast in the first known infiltration attempt since February 1968. Otherwise, the deployment followed the usual pattern.[41]

CFAW-6 was based at Marine Corps Air Station, Iwakuni, in 1970 under the operational control of Commander Patrol Force, Seventh Fleet (CTF 72). CFAW-6 had a primary mission of antisubmarine warfare and secondary missions of ocean surveillance, search and rescue, and aerial mining. Its designation was CTG 72.4, and under its control were various patrol squadrons based in Japan.[42]

In February 1970, VP-1, with its new P-3 Orion aircraft, was deployed to Marine Corps Air Station at Iwakuni, Japan, and operated under the operational control of CFAW-6. VP-1 also maintained a detachment at the Royal Thai Navy Air Base at Utapao to support Market Time patrol and to conduct interdiction operations off the Vietnamese coast.

In March, VP-1 aircraft searched for and located the hijacked freighter *Columbian Eagle* and kept the ship under surveillance until it entered Cambodian territorial waters. In April and early May, VP-1 participated in the observation of the Soviet Pacific Fleet's portion of the worldwide naval exercise OKEAN, filing seven intelligence reports on the operating characteristics of fourteen Soviet surface units that participated. Other air reconnaissance efforts by VP-1 aircraft flying from Naha, Okinawa, included photographing Soviet submarines, merchant ships, warships, auxiliaries, and oceanographic research ships, and observing Soviet antisubmarine operations.[43]

In April 1971, VP-1 deployed to Naval Station, Sangley Point. Due to the impending closure of that facility, it moved in the same month to NAS Cubi Point, Philippines, becoming the first patrol squadron to be assigned there. VP-1 was designated Task Unit 72.3.1 (Philippine Air Patrol Unit) and Task Unit 72.3.6 (Vietnam Air Patrol Unit and Philippine Air Patrol Group). VP-1's duties included Market Time air patrols, ocean surveillance air patrols, Communist bloc ship relocations, and special operations as tasked. A permanent detachment of four aircraft and five crews was maintained at Camranh Bay in South Vietnam in order to conduct daily Market Time air patrols to detect infiltration of arms and supplies into South Vietnam by sea.[44]

VP-1 deployed in November 1972 to NAS Cubi Point and operated under the operational control of CTG 72.3. The squadron's duties included flying numerous tracks through the South China Sea and the Indian Ocean in support of Market Time operations and in gathering intelligence information on shipping traffic in those areas.[45]

VP-5, home based at NAS Jacksonville, Florida, maintained surveillance of ship transits from ports in Florida to ports in the Caribbean in order to alert U.S. forces to any Cuban attempts to seize ships.

In March 1972, VP-5 deployed to Naval Air Facility, Sigonella, Sicily. At the end of May, the squadron established a two-plane detachment at Rota, Spain, to support surveillance operations against a group of Soviet submarines in the North Atlantic that was moving south toward the Mediterranean. The operations continued into June and included five days of barrier operations south of Sardinia and in the straits of Sicily, as well as continuous on-top coverage of a Soviet *Juliett*-class cruise missile submarine as it transited east to Alexandria, Egypt. Surfaced Soviet *Foxtrot*-class diesel attack submarines were photographed, and each submarine's acoustic signature was correlated with its unit identity as determined by photographs.[46]

Multisensor Aerial Reconnaissance (VW, VAH, and RVAH Squadrons)

In early February 1953, Airborne Early Warning Squadron (VW) One deployed a detachment of three PB-1W Flying Fortresses (the Navy version of the B-17) to operate from NAS Atsugi for approximately six weeks. The task was to demonstrate whether or not land-based AEW (Airborne Early Warning) aircraft might contribute to, and improve the effectiveness of, fast carrier task force operations in wartime against enemy air attack. During the period 24 February to 23 March 1953, PB-1W aircraft equipped with AN/APS-20 radar operated as airborne combat information centers, stationed on a barrier patrol 50 miles long, centered 30 to 40 miles from the task force on a line normal to the line of bearing of expected enemy air attack. An analysis of the Flying Fortress operations indicated that the PB-1W did increase Task Force 77's early warning ranges against aircraft targets approaching the task force and that an AEW aircraft equipped with height-finding radar and IFF (Identification Friend-or-Foe) interrogator-responder would improve the overall force's airborne early warning capability.[47]

The Greenland-Iceland-United Kingdom Early Warning Barrier had been activated on 1 August 1961. At that time, Airborne Early Warning Barrier operations along the Argentia-Azores axis were discontinued. The new barrier was joined on the western end by the DEW (Defense Early Warning) Line Extension, which was operated by Commander in Chief, North American Air Defense Command (CINCNORAD), and extended across Greenland. At the eastern end, in the United Kingdom, the barrier was joined with the Allied Command Europe Early Warning System.

Commander Barrier Force, Atlantic controlled barrier operations from the Operations Control Center at Keflavik, Iceland. Two aircraft stations were maintained, one to the east and one to the west of Iceland. The patrol stations were flown at random, unpredictable intervals. The aircraft were WV-1 Warning Stars (military Lockheed Constellations, later retyped EC-121) homeported at Argentia, Newfoundland, and they were rotated to Keflavik for barrier operations.[48]

Air reconnaissance missions over Cuba were conducted daily during 1963. The missions included photo coverage of the entire island on a routine basis and against specific targets when required; ELINT missions to maintain a current electronic order-of-battle (EOB) and for special intelligence in support of designated objectives; shipping surveillance to maintain accurate information on location, loading, and destinations of shipping; weather reconnaissance in support of high-altitude photo missions; and special support operations for Cuban overflights as directed by the Joint Chiefs of Staff. CINCLANT prepared reconnaissance schedules for JCS approval and provided fighter, search and rescue, and communications support to all reconnaissance flights conducted in the Cuban area.

An average of 361 reconnaissance flights was flown over Cuba per month in 1963. A large majority was in direct support to Strategic Air Command U-2 flights over Cuba. Next in magnitude of effort were the daily shipping surveillance missions to detect ships approaching and departing Cuba and to provide intelligence information concerning the types of cargo being carried.[49]

In 1964, with JCS approval, CINCLANT undertook aerial photographic coverage of Haiti and the Dominican Republic on a limited scale. Patrol squadrons in the Caribbean area continued collecting photographs of ships and deck cargo inbound to Cuba.

The average number of flights per month against Cuba dropped from 361 in 1963 to 242 in 1964. One shipping surveillance route out of Guantanamo was flown daily by the Military Air Transport Service until 14 April 1964, but thereafter in 1964 such flights were made only on special occasions for specific purposes.[50]

Air reconnaissance flights over Communist infiltration routes in Laos were approved in principle by President Lyndon Johnson on 17 March 1964 and became part of Commander in Chief, Pacific's (CINCPAC) OPLAN 37-64 of 21 April for expanded U.S. participation in the Vietnam War. Up to that time, Laotian Premier Souvanna Phouma had refused to sanction U.S. flights over his country. However, faced with the annual spring offensive by the North Vietnamese-supported Pathet Lao, which threatened to overrun the Plain of Jars in northeast Laos, he approved U.S. flights on 19 May.

Kitty Hawk launched the Navy's first Yankee Team flight over Laos on 21 May. Because the Navy was able to report Pathet Lao movements and strongholds to the American embassy in Laos, the Royal Laotians were able to check the Pathet Lao at the edge of the Plain of Jars. The Navy reconnaissance flights also collected intelligence information on North Vietnamese bases and infiltration routes supporting the Viet Cong in South Vietnam.

The Joint Chiefs of Staff limited the number of Laos missions to nine per day and required CINCPAC to forward plans for low-level reconnaissance missions for their approval at least 36 hours in advance of each flight.

In the first combat employment of the new ELINT aircraft, on 30 May 1964, an EA-3B Skywarrior launched from *Kitty Hawk* pinpointed several radar-controlled antiaircraft guns in Laos.

Yankee Team flights completed updating U.S. intelligence on the Plain of Jars by the end of June 1964. Air reconnaissance then concentrated on detecting new Communist movements and dispositions in support of a Royal Laotian counteroffensive against the Pathet Lao. In mid-July, Navy aircraft collected intelligence information for Operation TRIANGLE, a Royal Laotian army campaign to regain control of Routes 7 and 13.[51]

On 1 September 1964, Heavy Attack Squadron (VAH) One was designated Reconnaissance Attack Squadron (RVAH) One, having received six supersonic RA-5C Vigilante aircraft in July 1964. RVAH-1's new mission was to conduct all-weather, multisensor tactical reconnaissance flights to obtain current, integrated intelligence information and, secondarily, to conduct all-weather, air-to-surface attack operations. On 20 March 1965, RVAH-1 embarked in *Independence* (CVA 62) for its first deployment with RA-5Cs and its first participation in combat operations in Southeast Asia.[52]

VAH-11 was similarly designated RVAH–11 on 1 July 1964 and received its RA-5C aircraft on 10 July 1964. The squadron's first combat flights were made from *Forrestal* (CVA 59) at Yankee Station on 25 July 1967, but a fire on the flight and hangar decks of *Forrestal* on 29 July destroyed three of the RA-5Cs and severely damaged two others. The squadron's sixth plane escaped involvement by being at NAS Cubi Point on that day.

RVAH-11 returned to Mayport, Florida, on *Forrestal*, arriving 12 September 1967 and, upon being re-equipped with replacement aircraft, embarked in *Kitty Hawk* on 10 October for return to Southeast Asia. It resumed combat operations on 23 December. Poor weather conditions over North Vietnam in early 1968 limited optical reconnaissance, and emphasis was placed on collection with side-looking radar (SLAR) and passive electronic countermeasures (PECM) sensors, with generally good results. Flight crews became proficient in low-altitude, high-speed, day-and-night, all-weather SLAR techniques.

In mid-February 1968, three of the RVAH-11 RA-5C Vigilantes were equipped with a new infrared (IR) imaging system. During the squadron's second tour on Yankee Station commencing 24 February, a pioneer effort was begun to develop combat infrared sensor reconnaissance tactics over North Vietnam. Many of the techniques that the squadrons had been using during nighttime, side-looking radar flights were found to be readily adaptable to the infrared missions.

The bombing halt in March 1968 restricted combat operations, but the need for reconnaissance missions increased to ensure coverage of the large number of North Vietnamese supply routes and traffic control areas. Weather improved markedly during the bombing halt, and maximum advantage was taken of the side-area coverage capability available from the RA-5C's 18-inch focal length pan camera. The improvement in infrared sensors increased the amount and quality of the intelligence gathered by that sensor. On one mission, twenty-five motor vehicles and over fifty waterborne logistics craft were detected by IR.[53]

During the 1960s, Airborne Early Warning Squadron One was permanently based at Agana, Guam, with a four-plane detachment maintained at Naval Station, Sangley Point. VW-1's mission at Sangley Point was to provide airborne early warning in the Gulf of Tonkin for Seventh Fleet units operating in the South China Sea by providing a 12-hour, on-station coverage each night from 1900 local time to 0700 the next morning.

In February 1967, the Sangley Point detachment was reduced to two aircraft and moved to the Marine Corps Air Station, Chu Lai, Republic of Vietnam.

In 1967, two secondary missions were assigned to the VW-1 detachment at Chu Lai to provide North Vietnamese coastal surveillance between 19°N and 20°40'N latitude to detect high-speed targets (motor torpedo boats) exiting the coast in that area and to control carrier aircraft launched to investigate or attack the surface contacts. The VW-1 detachment was also expected to provide in-flight tracking of, and communication link with, RA-3B Skywarrior infrared photo reconnaissance aircraft during their low-altitude reconnaissance missions over North Vietnam. The RA-3Bs were attached to Heavy Photographic Squadron (VAP) 61 detachment at NAS Cubi Point. The RA-3Bs were normally in contact with Yankee Team aircraft carriers during their transits of the South China Sea but out of contact during their low-altitude flights over land. To provide the communications relay service, the VW-1 aircraft would shift its station to about latitude 18°20'N or as necessary to maintain radar contact with the RA-3B. An RA-3B was lost on 25 August 1967; a VW-1 aircraft was tracking it and saw the RA-3B disappear from the radar screen.[54]

The Tactical Airborne Reconnaissance System (TARS) was to have been the follow-on, carrier-based, multisensor reconnaissance system to replace the RA-5C. Original planning called for TARS to reach operational status in the 1975–1976 time frame. The Office of Naval Intelligence (ONI) worked with the Office of the Chief of Naval Operations and Naval Material Command representatives on concept formulation and development planning for the new system.

Proposal requests were to be issued in Fiscal Year 1969, but due to the tight money situation in 1968 and the reduction in research and development dollars for Fiscal Years 1969 and 1970, operational development was expected to slip to 1978–1980.[55]

ONI was also working with the Assistant Chief of Naval Operations for Air (OP-506) and the Naval Material Command on White Dove II (GROSBEAK), a tactical ELINT-COMINT airborne collection platform originally conceived by another government agency and later turned over to the Navy. It was to be used by VQ-1 in the Vietnam conflict and consisted of modular collection equipment in EP-3 aircraft, along with associated automatic ground readout equipment. The ground system was to be shipped to the Pacific Command ELINT Center at Fuchu, Japan, in the early summer of 1968, with three EP-3s to follow in January or February 1969. The system was expected to augment and improve VQ-1's "Big Look" capability, which at that time was represented by elderly EC-121 SIGINT aircraft used for collection and broadcast of Vietnamese surface-to-air missile activity on a real-time basis to support strike aircraft.[56]

In January 1969, RVAH-11 was again at Yankee Station on board *Kitty Hawk*, and again seasonal monsoon weather interfered with reconnaissance missions over North Vietnam. Alternate missions were flown over Laos, which helped to improve squadron proficiency and maintain its readiness in night IR and PECM sensor combat techniques. By April, the weather had improved, and top priority shifted to Blue Tree missions over North Vietnam to observe enemy activity while the bombing lull continued.

Following the shooting down of an EC-121 reconnaissance aircraft by North Korea in April, *Kitty Hawk* moved to Sasebo, Japan, in May for brief deployments to "Defender" Station near South Korea, and RVAH-11 conducted surveillance of the Korean area.

With the return of *Kitty Hawk* to Yankee Station at the end of June, good weather reappeared over North Vietnam and the Gulf of Tonkin, but visibility over Laos was marginal. Camouflage detection infrared film was used to help in the Ho Chi Minh Trail interdiction effort.

On 2 October 1969, the Chief of Naval Operations deleted the nuclear attack capability from the RVAH squadron's mission. All-weather, multisensor tactical reconnaissance thereafter remained the RA-5C's only mission.[57]

A plan was developed in 1969 for consolidating most, if not all, airborne intelligence collection training at Albany, Georgia. It was anticipated that the principles developed in the airborne intelligence collection training facility could be extended to cover surface and submarine collection training and that ultimately the latter two could be consolidated with the airborne training facility.[58]

As an example of a typical Mediterranean deployment, RVAH-11 spent seven months on board *Independence* in that area, arriving in July 1970. An arresting hook failure on 23 July caused the loss of one aircraft and its crew, and RA-5C carrier operations were suspended during the ensuing investigation. On 4 August, limited flight operations were resumed to fly Bonnie Blue missions in response to national ELINT tasking. Excellent results were obtained.

Between 21 and 25 August, RVAH-11 participated in an exercise to locate and identify all ships within a 250-mile radius of the *Independence* Task Group. The RA-5C was particularly well suited for that type of effort, for it possessed long range, high speed, and a capability to photograph all contacts.

In September, developments in Jordan prompted the movement of the bulk of the Sixth Fleet's forces to an area south of Cyprus. The squadron flew over 200 flight hours to conduct surveillance of Soviet ships in the eastern Mediterranean and to collect ELINT. After twenty days of continuous flight operations, all RA-5C carrier operations were suspended throughout the Navy when another arresting hook failed on an RA-5C on board *Ranger* (CVA 61) off the U.S. West Coast. The restrictions were lifted on 29 October, but the Jordan situation had eased, and the remainder of RVAH-11's deployment was uneventful.[59]

Circuit Gold aircraft were specially configured, multisensor EP-3A Orions. Five of the aircraft were assigned to CINCLANTFLT and CINCPACFLT and were supported by ONI's Task Force 168 for use in a wide variety of collection programs. The most significant contribution to intelligence collection made by Circuit Gold aircraft during Fiscal Year 1973 was in the reconnaissance of the May 1973 major Soviet fleet exercise in the Norwegian Sea.[60]

ELINT Aircraft Operations (VQ Squadrons)

The P4M-1 Mercator came into the "ferret" aircraft inventory in 1951. The aircraft were configured for five ELINT sensor-operator positions, which were occupied by aviation and Naval Security Group operator teams headed by an aviation electronics officer. The teams were assigned to Naval Communications Unit (NCU) 32 under Commander in Chief, U.S. Naval Forces, Eastern Atlantic and Mediterranean and to NCU-38 under Commander in Chief, Pacific Fleet and operated out of Port Lyautey, Morocco, and Sangley Point, Luzon, the Philippines, respectively.

Port Lyautey and Sangley Point served only as home bases. The planes and the Navy communications units based at Port Lyautey were on continuous deployment to such staging points as Naples and Malta to cover the western Mediterranean and Adriatic; to Athens and Nicosia, Cyprus, to cover the eastern Mediterranean, Black Sea and Middle Eastern countries; to the London area to cover the Baltic and part of the North Sea; to various West German airfields to cover the East-West German border and part of the Baltic; and to Keflavik, Iceland, to cover the far North Atlantic and Norwegian Sea. The Sangley Point planes and NCU deployed to Taiwan and to Iwakuni and Atsugi, Japan, to cover the China Sea, Yellow Sea, Sea of Japan, and Sea of Okhotsk; and to Shemya, Alaska, to cover missile and satellite re-entry telemetry at the impact areas near Kamchatka.[61]

The first designation for the Port Lyautey group was as Detachment Able (Det Able) of Airborne Early Warning Squadron Two, with Cdr. R. R. Sparks as officer in charge of VW-2 and Lt. John W. Douglas as officer in charge of the associated NCU-32. At Sangley Point, the airborne ELINT effort started in October 1951 as a special projects division of the air operations department. The division was formed into an electronics countermeasure group on 12 May 1953 and was assigned as a detachment of VW-1.

The next step taken toward the establishment of the eventual fleet air reconnaissance squadrons (VQ) was the disestablishment of the "Dets" and their redesignation as Electronics Countermeasures Squadrons (ECMRON) One and Two, which were commissioned in June and September 1955, respectively. The aircraft then being used for ELINT collection were still the P4M-1 Mercator and the P2V Neptune, but in the spring of 1956, the P2Vs were returned to the United States, as they were beyond their overhaul limitations. In September 1956, the first A3D-1Q Skywarrior (later EA-3A) aircraft were delivered, and in March 1958, P2V-5F Neptune aircraft, powered by two piston engines and two turbojets, were added. The P2V-5F continued in VQ service until 1960, and in the same year the last of the P4M-1Q Mercators was withdrawn from service.

In November 1958, the first detachment of ECMRON-2 moved to Rota, Spain, from Port Lyautey, and the remainder made the shift in January 1959. On 1 January 1960, ECMRON-1 moved from Sangley Point to Atsugi, and the squadrons were renamed Fleet Air Reconnaissance Squadrons One and Two.[62]

The VQ squadrons, the Navy's primary airborne electronic reconnaissance assets in the 1950s and 1960s devoted to support of the operating forces, were particularly effective in ELINT collection in the Far East, the Mediterranean, and sometimes in the Black Sea. One early successful mission into the Black Sea ended with the P4M being pursued by Soviet aircraft across Turkey to a landing at Cyprus. That incident was written up in the *Saturday Evening Post* by Hanson W. Baldwin, but fortunately the article didn't compromise the fact that two airborne signals of great interest (an early type air-intercept radar and an airborne altimeter) were recovered by the mission. The VQ squadron crews, operating on the front lines of the intelligence war, were unsung heroes of the Cold War during that period.[63]

Navy patrol and Marine Corps reconnaissance squadrons also collected ELINT data, but only incident to their visual search missions.

In February 1960, the WV-2 Warning Star (later designated EC-121M) began to arrive in the VQ squadrons to replace aging P4Ms and P2Vs. The A3D-2Qs (later designated EA-3B) were also replacing the A3D-1Qs.

In October 1962, VQ-2 sent a three-plane detachment to Florida for temporary deployment for the duration of the Cuban crisis.

In January 1965, the first permanent detachment of VQ-2 aircraft was deployed on board Sixth Fleet carrier *Saratoga* (CV 60) in the Mediterranean to operate under the operational control of the carrier air group commander. On 1 March 1965, operational control of VQ-2 officially shifted from Commander Fleet Air, Mediterranean (COMFAIRMED) to Commander Sixth Fleet; COMFAIRMED retained administrative control.

In December 1965, the first VQ-2 detachment of two aircraft, three crews, and their ground support personnel was sent to Southeast Asia to augment VQ-1. The deployment continued until 1969.[64]

The arrival of the EC-121Ms and the EA-3Bs represented an increase in the capabilities of the fleet air reconnaissance squadrons for direct fleet support and for intelligence collection activities against naval and national targets.[65]

CINCLANTFLT's increased requirements in connection with the Cuban crisis during October and November 1962 led to a proposal that four VQ aircraft be assigned permanently to the Atlantic

Fleet to expand its collection capability. ONI's Electronic Intelligence Section (OP-922Y4) prescribed the technical equipment requirements for the aircraft and staffed the request through the Office of the Chief of Naval Operations. As a result, three EC-121Ms and one EA-3B were deployed from Rota, Spain, to Florida. In February 1963, one of the EC-121Ms was returned to Rota due to Commander in Chief, U.S. Naval Forces, Europe (CINCUSNAVEUR) requirements for its services.[66]

Because of Soviet advances in technology, standard ELINT collection equipment and recorders gradually became insufficiently sensitive to intercept data-link and missile-associated telemetry. To cope with the problem, special equipment was designed and installed on several VQ aircraft. The data collected proved to be of high quality and, in some areas, were the only data collected.

Some specific collection equipment, developed by Sanders, Inc., under the guidance of the Bureau of Weapons and ONI's ELINT analysts, obtained information that aided in the production of a defensive countermeasures system later installed in Navy A-3 Skywarrior heavy attack aircraft.[67]

As of 1963, the mission of the two VQ squadrons was "to conduct electronic reconnaissance missions in support of Fleet operations in order to obtain essential information on the electronic warfare capabilities of targets of naval interest." Home bases in 1963 continued to be Rota and Atsugi, but operations were conducted by VQ-2 from Aviano, Italy; Brize Norton, England; Incirlik, Turkey; Sigonella, Sicily; Wheelus Air Force Base, Libya; and Wiesbaden, West Germany. VQ-1 conducted its operations using Shemya and Adak, Alaska; Okinawa; Cubi Point, the Philippines; Bangkok, Thailand; and Darwin, Australia, as bases.[68]

The appearance in Cuba in mid-1961 of Soviet-made early warning radars and associated weapon control systems required increased surveillance and peripheral reconnaissance during 1962. The limited ELINT collection capability of VAW-33 and VMCJ-2 aircraft was augmented in May 1962 by the temporary assignment of a WV-2Q (EC-121M) Warning Star SIGINT platform. Numerous new Cuban radar installations located by the WV-2Q confirmed the suspected rapid buildup of Soviet forces on the island and pointed up the requirement for intensified collection efforts to maintain current radar order-of-battle information. In October 1962, three WV-2Q (EC-121M) and one A3D-2Q (EA-3A) Skywarrior aircraft were assigned to the operational control of Commander in Chief, Atlantic, for the purpose of keeping track of developments in Cuba. At the end of 1962, two WV-2Qs were continuing the SIGINT collection effort.[69]

Through machine-processing equipment acquired by VQ-1 during 1962, the processing of SIGINT data was accomplished within 72 hours after the collecting aircraft had landed; no further processing was necessary at the Pacific Command ELINT Center, other than to add the resultant data to its data bank.[70]

VQ-1, based at Atsugi, continued its outstanding contribution to the many requirements for ELINT during 1962. Flying a variety of tracks, the squadron conducted routine and special collection operations along the Sino-Soviet periphery, in Southeast Asia, and off Indonesia. Using specially configured aircraft, a detachment of VQ-1 participated in the collection of intercontinental ballistic missile (ICBM) intelligence in the vicinity of Soviet impact areas on the Kamchatka Peninsula and in the mid-Pacific. The squadron was the largest contributor of ELINT in the Pacific Command area, outproducing all other air collection efforts combined.[71]

VQ-1 continued to collect ELINT data from Soviet radars in the Far East and against Soviet ICBMs launched to Pacific missile test impact areas. The squadron originated and successfully tested a new concept in ELINT collection techniques known as Brigand, which permitted passive detection of a transmitting radar and more accurate pinpointing of its location. Initial results from Brigand were impressive, and the system underwent additional research and development at squadron headquarters and in Washington.[72]

Although the aircraft assigned to the fleet air reconnaissance squadrons were used for ELINT collection in a direct support role, they were also intended to perform many other functions for the naval operating forces, such as monitoring fleet tactical communications, monitoring fleet noncommunication emissions, and conducting tests and exercises of fleet electronic warfare capability and readiness.[73]

In addition to continuing its reconnaissance flights in execution of the national ELINT plan, VQ-1 flew extensive sorties in conjunction with Yankee Team carrier aviation and associated operations in Southeast Asia. Throughout the summer and until late October 1964, VQ-1 EA-3B Skywarrior aircraft operations from Shemya continued successful collection against Soviet space operations and missile activity in the Kamchatka Peninsula area.[74]

In 1968, VQ-1 was based at NAS Atsugi and was still equipped with EC-121M Warning Stars and EA-3B Skywarrior aircraft. Its mission was to conduct electronic reconnaissance in support of fleet operations in order to obtain information and intelligence on areas and targets of naval interest. Its area of operations extended from the Aleutian Islands to the Indian Ocean. VQ-1 maintained a task element (TE 70.2.3.1) at Danang Air Base in South

Vietnam to conduct electronic warfare operations in direct support of U.S. combat forces in Southeast Asia. Primarily, TE 70.2.3.1 was engaged in electronic surveillance of the Demilitarized Zone and in providing surface-to-air missile and fire-control radar warnings in response to CTF 77 tasking. As an example, in December 1968, VQ-1 flew a series of missions in search of fire control radars in Laos.[75]

In 1971, VQ-1 was based at NAS Agana, Guam, as Task Unit 70.2.3 under the direct control of Commander Seventh Fleet. During July 1971, Heavy Photographic Squadron 61 and Airborne Early Warning Squadron One were combined with VQ-1.

To perform its electronic warfare mission, VQ-1 operated out of Singapore, Alaska, Wake Island, Midway, Korea, Japan, Taiwan, the Philippines, and Vietnam, locating the positions of a long list of hostile emitters. Cartographic mapping and photographic reconnaissance missions were flown over Hawaii, the U.S. Trust Territories, Okinawa, Korea, Japan, Vietnam, the Philippines, and Cambodia. A permanent detachment was also maintained at Naval Air Facility, Atsugi, to conduct reconnaissance missions along the Soviet coast.

Southeast Asian operations by VQ-1 provided nearly 1,000 warnings of impending MiG fighter and surface-to-air missile attacks against endangered U.S. aircraft. VQ-1 also provided warning services in the case of surface-to-surface missile threats and other surface attack threats. In addition to tactical warning services, VQ-1 updated electronic order-of-battle data for the Southeast Asian area and thus contributed to CTF 77 combat unit readiness. MiG fighter aircraft were deployed to airfields in North Vietnam, south of 20°N latitude, almost continuously throughout 1971. Surface-to-air missile activity increased and moved south and west during the year. Toward the end of 1971, Soviet-made Fan Song missile fire-control radar intercepts decreased and reported missile firings increased, lending credence to the postulation that the SA-2 missile system had a backup optical fire-control capability.

An exceptional collection opportunity offered itself between August and October 1971 when the Soviet Pacific fleet conducted extensive out-of-area operations. VQ-1 obtained data on many Soviet naval emitters and "fingerprinted" a majority of the participating combatants. The collection operations were unique because they included the first known employment of a coordinated effort by RA-3B Skywarrior photo aircraft and EP-3B Orion Big Look aircraft. The active photo collection efforts by the RA-3B generated many signals not normally elicited by the EP-3B in its normal passive-active role. Four antiship cruise missile exercises were monitored by VQ-1 during the operations; all of the signals normally associated with cruise missile operations were intercepted and recorded.

A significant development in the VQ-1 intelligence department was the activation of a new special intelligence communications (SPINTCOM) facility at Guam and the indoctrination of former VAP-61 and VW-1 personnel in the use of the system. The new facility enabled VQ-1 at Guam to achieve near real-time communications with aircraft airborne in the Sea of Japan, the north and central Pacific, and other shore-supporting stations. The size of the associated special security officer (SSO) facility provided, for the first time, space adequate for briefing and administering SSO materials.

During 1971, VQ-1 maintained detachments at Danang; Atsugi (established in July); Cubi Point (photo aircraft); Utapao, Thailand (photo aircraft); and a Seawing Det in both *Enterprise* (CVAN 65) and *Coral Sea* (CV 43).

At the end of 1971, VQ-1 had the following aircraft: one TA-3B for training, ten EA-3Bs, three RA-3Bs, one C-121J, one WC-121N, three EC-131Ms, and two EP-3Bs.[76]

Commanding officers of VQ-1 between 1954 and 1976 were as follows:

Name	Dates
LCdr. E. R. Hall	Jun 1954–Jun 1956
Cdr. W. H. Hoff	Jun 1956–Aug 1956
Cdr. H. Larson	Aug 1956–Nov 1957
Cdr. N. P. Byrd, Jr.	Nov 1957–Nov 1958
Cdr. R. C. James	Nov 1958–Oct 1959
Cdr. R. Knopke	Oct 1959–Jan 1961
Cdr. T. E. Moore	Jan 1961–Dec 1961
Cdr. J. W. Jenkins	Dec 1961–Nov 1962
Cdr. W. J. Wacker	Nov 1962–Oct 1963
Cdr. A. T. Holt	Oct 1963–Nov 1964
Cdr. F. Carment, Jr.	Nov 1964–Nov 1965
Cdr. M. E. Klein	Nov 1965–Nov 1966
Cdr. R. F. Dreesen	Nov 1966–Dec 1967
Capt. R. M. Delorenzi	Dec 1967–Feb 1970
Capt. C. L. Chute	Feb 1970–Jul 1971
Capt. J. Akins	Jul 1971–Sep 1974
Capt. T. W. Connolly	Sep 1974–Aug 1976

In 1972, VQ-2 conducted electronic warfare and countermeasures support for the Sixth Fleet and collected electromagnetic intelligence on Mediterranean littoral countries, the Soviet Mediterranean fleet, and other naval theater interests. In addition, the squadron was tasked with conducting reconnaissance missions as determined by the United States Intelligence Board (USIB) and as directed by the Joint Chiefs of Staff and CINCUSNAVEUR for contribution toward the national collection effort.[77]

During 1972, VQ-2 maintained detachments on board Sixth Fleet carriers operating in the Mediterranean as well as detachments on a regular basis at Athens, Greece; Ramstein, West Germany; and Key West, Florida. In support of special operations, squadron aircraft and flight crews were deployed to other areas such as Sola and Bodo, Norway; Sigonella, Sicily; Incirlik, Turkey; Lajes, in the Azores; Souda Bay, Crete; and Aviano and Decimomannu, Italy.

Mediterranean Floor Door missions were flown from Sixth Fleet aircraft carriers and shore facilities at Rota, Aviano, and Athens. Frequently, missions flown from carriers were performed in conjunction with RA-5C Vigilante shipboard aircraft flying nationally scheduled Floor Sponge missions. The coordinated missions proved effective, with the EA-3B providing tactical evaluations of the target environment while the RA-5C readout offered a strategic picture of the total environment.

Baltic Floor Door missions were flown from Ramstein Air Force Base in West Germany by EP-3E aircraft of VQ-2, with the EA-3B as the alternate aircraft. Three missions per month were scheduled. VQ-2's primary tasks were concerned with Soviet naval activity and with monitoring Soviet RDT&E (research, development, test, and evaluation) from the Soviet fleet testing and weapons range areas.

Splinter Arm missions were flown from Key West to provide SIGINT support for CINCLANT contingency plans concerning Cuba. Operational control for the Splinter Arm missions was under the Commander in Chief, Atlantic Fleet.

Floor Leader missions were flown by VQ-2 only in the Mediterranean, where their primary task was to collect ELINT in support of national-level requirements promulgated by the Defense Intelligence Agency. A secondary task was to collect SIGINT in support of national and direct fleet support requirements. Floor Show missions were scheduled by CINCUSNAVEUR, in support of peacetime COMINT and ELINT collection programs and also provided direct support to Sixth Fleet units in the Mediterranean.[78]

Commanding officers of VQ-2 between 1954 and 1976 were as follows:

Name	Dates
Cdr. M. L. Kalin	Sep 1955–Jul 1957
Cdr. R. R. Sparks	Jul 1957–Oct 1958
Cdr. C. H. Sigley	Oct 1958–Oct 1959
Cdr. P. D. Halpin	Oct 1959–Apr 1961
Cdr. A. G. Elder	Apr 1961–Apr 1962
Cdr. H. E. Fitzwater	Apr 1962–May 1963
Cdr. R. M. Davis	May 1963–May 1964
Cdr. C. A. Kiser	May 1964–May 1965
Cdr. J. H. McConnell	May 1965–Jun 1966
Cdr. A. D. Burkett	Jun 1966–May 1967
Cdr. E. V. Laney	May 1967–May 1968
Cdr. T. E. Daum*	May 1968–Jun 1968
Cdr. R. W. Arn	Jun 1968–Jul 1969
Cdr. H. G. Hatch	Jul 1969–Jul 1970
Cdr. A. A. Gallotta	Jul 1970–Jun 1971
Cdr. J. E. Taylor	Jun 1971–Jul 1972
Cdr. J. F. McRae	Jul 1972–Jul 1973
Cdr. J. D. Meyer	Jul 1973–Jul 1974
Cdr. D. J. Alberg	Jul 1974–Jul 1975
Cdr. D. N. Hagen	Jul 1975–Jul 1976

*Killed on active duty while serving as commanding officer.

Table 4.1. Primary U.S. Navy Electronic Reconnaissance Aircraft

Designation	Mfg.	Approx. Yrs. of Service	Description
1. PBY-5 (Catalina)	Consolidated	1944–1945	Twin reciprocating engine amphibian; eight-man crew
2. PB4Y-1 (Liberator)	Consolidated	1943–1947	Four-engine reciprocating; twin tail; 10-man crew
3. PB4Y-2 (Privateer)	Consolidated	1945–1950	Four-engine reciprocating; single tail; 10-man crew
4. P2V (Neptune)	Lockheed	1947–1960	Twin reciprocating; 10-man crew
5. P4M-1Q (Mercator)	Martin	1950–1960	Twin reciprocating and twin jet; 16-man crew
6. A3D-1Q (Skywarrior)	Douglas	1956–1959	Twin jet; four-man crew
7. A3D-2Q (Skywarrior) redesignated EA-3B in 1962	Douglas	1959–	Twin jet; present seven-man crew
8. WV-2Q (Warning Star) redesignated EC-121M in 1962	Lockheed	1960–1974	Four-engine reciprocating; 31-man crew
9. EP-3B (Batrack*)	Lockheed	1969	Four-engine turboprop; 28-man crew
10. EP-3A (Aries*)	Lockheed	1971–present	Four-engine turboprop; 28-man crew

*Unofficial nickname

Source: Capt. Don Charles East, USN, "History of U.S. Navy Fleet Air Reconnaissance Squadrons One and Two (VQ-1 and VQ-2)," Newport, RI, 1986, privately printed monograph.

Table 4.2. Incidents of U.S. Navy Electronic Reconnaissance Aircraft Facing Hostile Fire

Date	Parent Unit	Type of Aircraft	Location	Casualties	Country
8 Apr 1950	VP-26, Det A	PB4Y-2	Baltic Sea	10 Missing	USSR
6 Jun 1951	VP-6	P2V	Sea of Japan	10 Missing	USSR
19 Jan 1953	VP-22	P2V	Formosa Strait	11 Dead	PRC
4 Sep 1954	VP-19	P2V	Sea of Japan	1 Dead 9 Rescued	USSR
22 Jun 1955	VP-9	P2V-5	Bering Strait	7 Wounded	USSR
22 Aug 1956	VQ-1	P4M-1Q	Off PRC Coast	16 Missing	PRC
16 Jun 1959	VQ-1	P4M-1Q	Sea of Japan	1 Wounded	N. Korea
14 Apr 1969	VQ-1	EC-121M	Sea of Japan	31 Dead	N. Korea
20 Sep 1969	VQ-2	EA-3B	Danang, RVN	None	N. Vietnam

Chapter Notes

1. James O. Richardson, *On the Treadmill to Pearl Harbor: The Memoirs of Admiral J. O. Richardson as told to Vice Admiral George C. Dyer, USN (Retired)* (Washington: GPO, 1973), 121.

2. Navy Photographic Reconnaissance Squadron (VFP) 63, "History of VFP-63," 1968. Naval Aviation History Branch, Naval Historical Center, holds all command histories for Navy aircraft squadron.

3. "History of VFP-63," passim.

4. VFP-62, "History of VFP-62," 1968, passim.

5. "History of VFP-63," passim.

6. Commander in Chief, Pacific Fleet (CINCPACFLT), *Interim Evaluation Report No. 2: Korean War Naval Operations*, 213, OA.

7. CINCPACFLT, *Interim Evaluation Report No. 3, Korean War Naval Operations*, 5-48.

8. CINCPACFLT, *Interim Evaluation Report No. 4, Korean War Naval Operations*, 3-46.

9. CINCPACFLT, *Interim Evaluation Report No. 6, Korean War Naval Operations*, 3-77.

10. CINCPACFLT, *Interim Evaluation Report No. 5, Korean War Naval Operations*, 3-58, 3-59.

11. Ibid., 3-59.

12. Commander Seventh Fleet, Report of Operations, 20 May 1952–27 Jul 1953, 13.

13. Capt. Rudolf J. Fabian, taped interview for author, 21 Jan 1976.

14. "History of VFP-62."

15. Director of Naval Intelligence (DNI) ltr to Director of Central Intelligence (DCI), ser 000972P92, 29 Aug 1960.

16. *VFP-62 Newsletter*, Feb 1963.

17. Commander in Chief, Atlantic (CINCLANT), *Annual Report, FY 1963*, 16.

18. *CINCPACFLT Annual Report, 1 Oct 1963–26 Jun 1964*, 51.

19. VFP-63, "History Summary," 1964.

20. Heavy Photographic Squadron (VAP) 61, *Command History, 1969.*

21. VAP-61, *Command History, 1970–Jul 1971.*

22. VFP-62, *Command History, 1966.*

23. Richardson, *Treadmill to Pearl Harbor*, 337–38.

24. Ibid., 353–60.

25. Capt. Warren J. Bettens, USN, ltrs to author, 31 Aug 1974 and 1 Oct 1974.

26. Philip K. Lundeburg, "American Antisubmarine Operations in the Atlantic, May 1943–May 1945" (Ph.D. diss., Harvard University, 1953), 1, 3, 18.

27. Bettens ltrs, 31 Aug 1974 and 1 Oct 1974.

28. Commander in Chief, U.S. Naval Forces, Eastern Atlantic and Mediterranean (CINCNELM), *Report of Operations, FY 1949*, 3–4.

29. CNO (OP-322Y2E) memo ser 0064P32, 25 Jan 1951, box 1, Job 12876, OA.

30. CINCPACFLT, *Interim Evaluation Report No. 1: Korean War Naval Operations,* 1008–9.

31. CINCPACFLT, *Interim Evaluation Report No. 6*, 1–6.

32. CNO (OP-322N) ltr ser 32B1/00204, 13 Jul 1953.

33. CINCPACFLT, *Interim Evaluation Report No. 5*, 4-36–4-37.

34. Commander Taiwan Patrol Force, *Command History,* 2 Mar 1960, 6.

35. Patrol Squadron (VP) 2, *Command History, 1966*, passim.

36. VP-1, *Command History, 1966.*

37. VP-40 history summaries.

38. VP-1, *Command History, 1967*

39. VP-9, *Command History, 1968*, passim.

40. VP-1, *Command History, 1969*, passim.

41. VP-9, *Command History, 1969*, passim.

42. Commander Fleet Air Wing (CFW) 6, *Command History, 1970*, passim.

43. VP-1, *Command History, 1970*, passim.

44. VP-1, *Command History, 1971*, passim.

45. VP-1, *Command History, 1972*, passim.

46. VP-5, *Command History, 1972*, passim.

47. CINCPACFLT, *Interim Evaluation Report No. 6*, 3-90.

48. *CINCLANT Annual Report, FY 1962*, 42.

49. *CINCLANT Annual Report, FY 1963*, 15.

50. *CINCLANT Annual Report, FY 1964*, 9, 16.

51. VAdm. Edwin B. Hooper, Dean C. Allard, and Oscar P. Fitzgerald, *The U.S. Navy and the Vietnam Conflict*, vol. 1, *Setting the Stage to 1959* (Washington: Naval History Division, 1976), 45–50, 55–56.

52. Navy Reconnaissance Heavy Attack Squadron (RVAH) 1, Aviation History Summary, 1 Oct 1965, passim.

53. RVAH-11, *Command History, 1967*, passim

54. Airborne Early Warning Squadron (VW) 1, *Command History, 1967*.

55. The RA-5C was retired without a dedicated, full-capability shipboard reconnaissance aircraft replacement. The later TARPS (Tactical Airborne Reconnaissance Photographic System) pods for F-14 Tomcat interceptors were originally intended as an interim solution but have since become the standard carrier air group reconnaissance system.

56. OP-92U memo to CNO, "Status of Major ONI Programs," 10 Jun 1968.

57. RVAH-11, *Command History, 1969*, passim.

58. *DNI Report to CNO, FY 1969*, passim.

59. RVAH-11, *Command History, 1970*, passim.

60. *DNI Report to CNO, FY 1973*, III-6.

61. John Douglas, ltr response to author query, 4 Oct 1973.

62. Air Reconnaissance Squadron (VQ) 2, *Yearbook*, 1969–70, passim; and LCdr. J. D. Ward (EW officer of VQ-1), ltr response to author query, 27 Jul 1973.

63. William E.W. Howe, notes on ELINT, 20 Jun 1973, 10, response to author query.

64. VQ-2 *Yearbook*, 1969–1970, passim.

65. Operations Intelligence (OPINTEL) Briefing, 1 Jun 1963, 3, author's files.

66. *CINCLANT Annual Report, FY 1963*, 55–56.

67. OPINTEL Briefing, 1 Jun 1963, 3.

68. ONI Intelligence Briefing, Feb 1963 (precise date not given), author's files, OA.

69. *CINCLANT Annual Report, FY 1962*, 51.

70. *CINCPACFLT Annual Report, FY 1962*, 39.

71. *CINCPACFLT Annual Report, FY 1963*, 43.

72. *CINCPACFLT Annual Report, 1 Oct 1963–26 Jun 1964*, 51.

73. OPNAV Instruction 003430.2B of 7 Feb 1964.

74. *CINCPACFLT Annual Report, 26 Jun 1964–30 Mar 1965*, 54.

75. VQ-1, *Command History, 1968*, 1–3, 10.

76. VQ-1, *Command History, 1971*, passim.

77. VQ-2, Annual Operations Summary (OpSum), 1 Mar 1973, chap. 1, para. A.

78. VQ-2, *Command History, 1972*.

CHAPTER 5

Surface Reconnaissance

Surface reconnaissance is one of the earliest forms of naval intelligence collection. As mentioned frequently throughout Chapter 1 and again in Chapters 34 through 39 on operating forces, as well as in Chapter 32, personnel on board U S. Navy ships have long been used to observe conditions and identify ships present in various ports and to seek out hostile forces in various ocean areas. Observers have also been recruited from among the officers and crews of merchant ships, including foreign-flag vessels, to observe and report on ports visited and ships sighted. Cameras have been supplied by the Office of Naval Intelligence in selected cases, and, in some instances, film has been supplied for use in privately owned cameras.

Naval attachés and district intelligence officers were the normal contacts with merchant ship observers, particularly following World War II. During that war, as well as in World War I, merchant ship observers were found to be of marginal value as a reconnaissance device. Most Allied merchant ships in wartime traveled in convoys with naval escorts, and the latter would report any sightings of operational value through their chains of command.

In the late 1940s and early 1950s, greater emphasis was placed on the collection of information by merchant ship observers, and a special code designation was instituted to protect source identities. The sighting, photographing, and reporting of Communist ships was given top priority, information on their movements not being available through other normal overt sources. One indirect value of surface reconnaissance was that it provided a relatively open source to which to attribute information gathered by more sensitive systems of collection.

By the early 1950s, the Navy was using combatant ships in intelligence collection roles, and, a decade later, specialized intelligence collection ships, converted from various types of naval auxiliaries, began to appear in the fleet. This chapter deals primarily with the use of U.S. Navy surface ships in reconnaissance and electronics intelligence collection roles from the early 1950s to 1973. In World War II, U.S. Navy use of specialized intelligence collection ships was surprisingly limited; perhaps the best known example was the abortive cruise of the schooner *Lanakai* under then-Lt. Kemp Tolley.[1]

Development of the Surface Reconnaissance Ship Concept

In 1952 the Taiwan Patrol Force, Task Force 72, began conducting surface reconnaissance in the Taiwan Strait using destroyers of the Seventh Fleet. Their purpose was to prevent any attack on, or invasion of, Taiwan (then Formosa) and the Pescadores Islands and to ensure that these islands were not used as a base of operations against the Chinese mainland by the Chinese Nationalists, except in the event of a Chinese Communist attack. In early February 1953, as the result of a major U.S. policy change, the latter part of the above mission was deleted.[2]

In the early 1960s, the Soviet navy began operating more extensively outside its coastal waters, allowing for closer scrutiny of its units. To take advantage of the opportunity to learn more about Soviet operations and the defenses along the Soviet and People's Republic of China coasts, a series of reconnaissance and surveillance operations was conducted in the Far East, using destroyers or submarine rescue ships (ASR). A small Naval Security Group detachment, with jury-rigged receivers and antennas, was placed in each of the assigned U.S. Navy ships for use while cruising along the Chinese and Soviet coasts. The operations proved so useful and productive that Commander in Chief, Pacific Fleet (CINCPACFLT), proposed to the Chief of Naval Operations (CNO) in February 1965 that a small ship be outfitted to perform intelligence collection as a permanent task. In this way the crews involved would develop expertise in such operations.

The Far Eastern coastal surveillance effort was designed to be completely overt. CINCPACFLT explained the mission of the operations in a September 1965 message as "to determine Soviet reaction to a small unarmed naval vessel which is overtly a naval surveillance ship deployed in Soviet Navy operating areas." Thus, the surveillance ship was to be perceived as an observation platform wearing the "uniform" of the U.S. Navy, operating in places where it was entitled to be, and in no way disguising its nationality, appearance, or function.[3]

In early April 1962, the destroyer *DeHaven* (DD 727) made the first Desoto Patrol along the northern coast of China from the Taiwan Strait to the vicinity of Tsingtao and back. Thereafter, the patrols were made on a random basis every month or two into such areas as the Sea of Japan (including the Gulf of Tartary), the Yellow Sea, South China Sea, and Java Sea. In December 1962, *Agerholm* (DD 826) extended the patrols into the Gulf of Tonkin. The presence of Desoto Patrols in strategically important areas stimulated reactions, which in turn provided an opportunity for intelligence collection. Particularly noteworthy reactions occurred when patrols entered an area for the first time. Highly useful signal intelligence (SIGINT) and photographic intelligence were thus obtained.[4]

Intelligence coverage of the Soviet naval convoy during its annual transit of the northern sea route along the northern coast of the USSR was performed for the first time in 1963. An icebreaker (AGB) supplied by Commander Alaskan Sea Frontier, with an embarked helicopter, collected valuable intelligence concerning the number, class, and characteristics of the Soviet naval vessels entering the Pacific with the convoy.[5]

A radar picket destroyer escort (DER) was kept in position to maintain close surveillance of the Soviet missile-range instrumentation ships throughout their intercontinental ballistic missile (ICBM) tracking operations in the mid-Pacific from November 1963 to April 1964, inaugurating a policy of maintaining surveillance of such ships by surface craft.

The two peripheral reconnaissance missions conducted by destroyers from the Taiwan Strait Patrol Force along the Chinese and North Vietnamese coasts were performed for the purpose of stimulating military reaction, and they afforded unique opportunities to collect electronic intelligence and photographic intelligence.

During April 1964, the submarine rescue vessel *Chanticleer* (ASR 7) conducted "snooper" operations in the vicinity of Soviet naval operating areas off Vladivostok and the southern Maritime Province to collect intelligence information on the Soviet spring naval exercises. *Chanticleer*'s operating procedures were not unlike those of the Soviet SIGINT trawlers that had been shadowing U.S. naval operations since the beginning of the 1960s.[6]

Commander Seventh Fleet (COM7THFLT), in January 1964, prescribed the mission of the Desoto Patrols as being intended "to probe peripheral areas of concern to COM7THFLT and to collect all-source intelligence in order to increase both the COM7THFLT and national fund of information concerning both military and civil activity of the Asiatic Communist Bloc." VAdm. Thomas H. Moorer, COM7THFLT, ordered random cruises to be conducted about once every three months in order to gather intelligence information on Communist air and sea defenses and to make hydrographic and weather observations. Moorer specified the general track for the patrols but gave commanders flexibility to change course as desirable to collect additional intelligence. To derive maximum collection on Communist reaction to the patrols, flights by Seventh Fleet reconnaissance aircraft were scheduled to coincide with them.[7]

In mid-February, in response to an urgent request from Gen. Paul D. Harkins, Commander U.S. Military Assistance Command, Vietnam (COMUSMACV), for radar photography of the North Vietnamese coasts, the destroyer *John R. Craig* (DD 885), with Commander Destroyer Squadron (COMDESRON) One, Capt. Daniel S. Appleton, embarked, began the first Desoto Patrol in the Gulf of Tonkin under the rules specified by COM7THFLT. *Craig*'s orders directed the ship to remain 15 nautical miles off the Chinese coast and 4 nautical miles off the North Vietnamese mainland. Several days of fog limited visual collection, but much valuable electronic intelligence that fulfilled the ship's basic mission was collected. A Chinese *Kronstadt*-class patrol boat and an unidentified plane shadowed *Craig* during part of the patrol. The destroyer completed its patrol on 8 March 1964 and returned to Taiwan as the Chinese Communists issued their 280th "serious warning" of an alleged violation of their territorial waters.[8]

The Desoto Patrols in the Gulf of Tonkin led to the incidents of the same name when torpedo boats attacked the destroyer *Maddox* (DD 731) on the night of 2 August 1964, and *Maddox* and *Turner Joy* (DD 951) were believed to have been similarly attacked on 4 August.[9]

On 12 September 1964, CINCPACFLT Adm. Thomas Moorer recommended another Desoto Patrol into the Gulf of Tonkin. CINCPAC Adm. Ulysses S. G. Sharp approved the plan, and the Joint Chiefs of

Staff authorized a three-day cruise to approach no closer than 20 nautical miles to the North Vietnamese mainland and no closer than 12 miles to Communist-held islands, except that the destroyers could approach the 3-mile limit if in hot pursuit of attackers. At dawn on 17 September, the destroyers *Morton* (DD 948) and *Richard S. Edwards* (DD 950), with COMDESRON 52 (Capt. Edward E. Hollyfield) embarked in *Morton*, entered the Gulf of Tonkin. On the night of 18 September, the destroyers were closed by two high-speed targets, which were taken under fire by both U.S. Navy ships and were believed to have been sunk.[10]

Specialized Surface Reconnaissance Ships

Surface collectors of intelligence of indirect support to the Navy—but under the technical direction and control of the National Security Agency—were the technical research ships *Belmont* (AGTR 4), *Georgetown* (AGTR 2, designated AG 165 prior to 1 April 1964), *Jamestown* (AGTR 3), *Liberty* (AGTR 5), and *Oxford* (AGTR 1) and the miscellaneous auxiliaries *Pvt. Jose Valdez* (T-AG 169) and *Sgt. Joseph Muller* (T-AG 171). Their primary mission was to collect SIGINT to satisfy national requirements. They also collected data for electromagnetic propagation studies and to check out advanced communications systems.[11]

Georgetown conducted collection operations off the north coast of Cuba from April through June 1964 and again from March to May 1965. During January to March 1965, the ship operated off the west coast of South America.[12]

Jamestown conducted technical research operations along the coast of Africa in the summer of 1964 and during the winter of 1964–1965 in support of U.S. Navy electronic research projects. Visits were made to Capetown, South Africa, and Freetown, Sierra Leone, during the summer operations and to Dakar, Capetown, and Lagos in the winter.[13]

From 2 February to 29 May 1965, *Oxford* cruised from the Canary Islands to Nigeria, South Africa, and Sierra Leone conducting "technical operations" en route. *Oxfoxd* continued on to the South China Sea and operated there and in the Gulf of Siam for the remainder of the year. The converted C1-AM-V1 class cargo ship USNS *Pvt. Jose Valdez* conducted similar operations along the Atlantic and Indian Ocean coasts of Africa from June 1964 to May 1965.[14]

During the summer of 1964, two reconnaissance patrols, using submarine rescue ships, were conducted in the Vladivostok area to collect general, photo, and electronic intelligence data during Soviet fleet maneuvers. Also, one fleet ocean tug (ATF) conducted a surveillance mission in the vicinity of the Kurile Islands and in the Petropavlovsk area to observe a Soviet fleet exercise.[15]

The environmental research ship *Banner* (AGER 1) carried out reconnaissance operations in the Western Pacific from August 1965 to December 1967 under the code name CLICKBEETLE. During ten of the missions, *Banner* was the target of Soviet and Chinese Communist harassment. The hostile actions included shouldering, closing to short range and maneuvering dangerously, one minor collision, closing with guns trained on *Banner*, surrounding the ship with trawlers, and signaling "Heave to or I will open fire." In two of the instances, Commander Naval Forces, Japan (COMNAVFORJAP) directed *Banner* to move to another area, and in two other incidents involving Chinese harassment, destroyer or destroyer escort assistance was dispatched from the Taiwan Defense Patrol.[16]

In 1967, the commanding officer of *Banner*, Cdr. Charles R. Clark, perceived no threat from the Soviets because they had many surveillance ships operating and would have much to lose by seizing his ship. "The North Koreans and the Red Chinese were a very doubtful factor, because nobody knew what they would do, but the fact that we were in international waters, doing legal operations, was our greatest protection."[17]

The success of *Banner*'s CLICKBEETLE operations led to the fitting out of two additional small surface collectors: *Pueblo* (AGER 2) and *Palm Beach* (AGER 2).[18]

Banner's operations had accomplished the following objectives in support of the operational concept of the CLICKBEETLE program: (1) intelligence had been collected in fulfillment of high-priority national intelligence objectives; (2) Soviet reaction to a small overt naval surveillance ship deployed in Soviet naval operating areas had been determined; (3) the effectiveness of a small ship acting singly and primarily as a naval surveillance and intelligence collection unit had been tested; (4) SIGINT signals that were not interceptable from shore-based sites or were not intercepted in sufficient depth by other mobile or shore facilities to allow adequate exploitation had been collected; (5) photographic, acoustic, hydrographic, and other intelligence material on targets of opportunity had been collected; and (6) any intercepted information of "critic" or "spot report" nature had been reported.[19]

Between 21 June and 14 September 1966, *Atakapa* (ATF 149), equipped with electronics intercept equipment, first conducted electronic surveillance of a Soviet fleet exercise off Norway and then operated briefly in the Baltic. While waiting for the exercises to begin, *Atakapa* was positioned in the Norwegian

Sea for the purpose of detecting and identifying Soviet out-of-area submarines. This surveillance effort resulted in the sighting and trailing of two Zulu-class submarines, thereby confirming the identities, actions, and intentions of the Mediterranean-bound units. During the Soviet exercises, *Atakapa* collected extensive intelligence information that no other platform or facility had collected and proved the concept of employing small ships as intelligence collectors. Despite the tug's long operating endurance, however, its slow speed proved inadequate for fleet surveillance, and the use of faster ships was recommended. In the Baltic, *Atakapa*'s operations established a U.S. presence in the area and obtained considerable SIGINT data that was, however, of limited value since most of the data were available from other collection sources. Soviet reaction to *Atakapa* being in the Baltic was negligible.[20]

In January 1966, *Jamestown* conducted surveillance operations off the coast of Cambodia and continued to operate in the Western Pacific throughout the year. *Oxford* also continued to operate in the Southeast Asian area in 1966.[21]

On 15 November 1966, *Banner*, while conducting reconnaissance in the CLICKBEETLE IX operating area, was harassed by a large number of Chinese Communist fishing boats and steel hulled trawlers. Commander Task Force 72 sent the destroyer *Everett F. Larson* (DD 830) to extricate *Banner* from the situation.[22]

On 8 June 1967, while conducting reconnaissance off the Egyptian coast, *Liberty* was attacked, allegedly by mistake, by Israeli aircraft and motor torpedo boats and received severe damage from a torpedo and gunfire. The Israelis acknowledged their mistake and offered apologies, and *Liberty* proceeded to Valletta, Malta, for immediate repairs. The ship, however, was so extensively damaged that it was not completely repaired and never operated again in its assigned role.[23]

After the *Liberty* incident, it became a practice of Commander Sixth Fleet to station a destroyer over the horizon in the vicinity of the reconnaissance ship. The Vice Chief of Naval Operations also directed that the reconnaissance ships be armed in the future. On 14 December 1967, the CNO directed that *Banner*, *Pueblo*, and *Atakapa* be armed with a minimum of two .50-caliber machine guns prior to their next missions.[24]

Atakapa conducted reconnaissance operations in the Mediterranean from June into October 1967. *Pueblo*, a converted 25-year-old Army FS (small coastal freighter), which had been inactive for several years following ten years of service in the Philippines, was refitted by the Navy as a surface reconnaissance ship in 1967. *Oxford* and *Jamestown* continued their collection operations in the Western Pacific area throughout 1967.[25]

As of 1967, Black Sea operations were being carried out on at least an annual basis. Electronic intelligence riders and combat cameramen were assigned during such operations to expand the intelligence collection resources of the Black Sea-deployed U.S. Navy ships. In the succeeding years, two such operations were conducted during 1968, three in 1969, and four in 1970.[26]

Pvt. Jose Valdez conducted collection operations along the African coast with visits to Mombasa, Lourenco Marques, Luanda, Dakar, and Monrovia during the period 17 January to 28 August 1968.[27]

Georgetown and *Sgt. Joseph Muller*, an unarmed ship with a civilian crew, conducted intelligence collection operations off Cuba from July to October 1968.[28]

The AGTRs *Oxford* and *Jamestown* continued their collection operations in the Western Pacific throughout 1968. The modified fleet tug *Atakapa* operated in the Mediterranean from June into October 1968.[29]

CLICKBEETLE Phase II was initiated by Commander Task Force 96 (COMNAVFORJAP) Operation Order (OPORDER) 301-68 of 3 January 1968. Phase I had successfully tested the operational feasibility and political implications of using one small trawler-type ship as a naval surveillance and intelligence collection unit. Phase II expanded the program to use two ships (*Banner* and *Pueblo*) to provide continuous coverage of a selected area or operation. The program's objectives also included testing Soviet reaction to the continuous presence of a U.S. intelligence collection ship in Soviet naval operating areas. It was expected that the experience gained from Phase II, and the procedures and equipment developed therefrom, would lead to the implementation of Phase III—the employment of more AGERs.[30]

During the CLICKBEETLE Phase II operations, the collection ships were specifically required to (1) remain a minimum of one mile outside the Communist-claimed territorial waters—no closer than 13 nautical miles from Communist-claimed land; (2) avoid any action that would be considered harassment and not close Communist ships to within 500 yards except for briefly closing to 200 yards as necessary for visual or photographic coverage of items of unusual interest; (3) avoid collisions with Communist ships; and (4) avoid actions that could be construed as provocative, including the display of weapons in a manner that could be misinterpreted as an intention to use them.

Code words pertinent to the CLICKBEETLE Phase II operations included BREEDER, assigned by the

Chief of Naval Operations for AGER intelligence collection operations in Atlantic and Pacific Fleets; ICHTHYIC, assigned by the Commander in Chief, Pacific for AGER intelligence collection operations in the Pacific; and GRAVY BOAT, assigned to unclassified research operations in an ocean environment in support of research programs. PINKROOT One was the code word specifically assigned to *Pueblo*'s first mission.[31]

The mission of the Naval Security Group Detachment (NAVSECGRUDET) on board an AGER during deployment was to intercept, exploit, and report on foreign electromagnetic emissions for SIGINT purposes. The detachment's paramount function was to provide direct support to the AGER commanding officer, and its secondary function was to satisfy specified fleet and national collection requirements. Routine operational control for SIGINT functions was exercised by Director Naval Security Group through Director Naval Security Group, Pacific. Technical control of SIGINT operations was exercised by the National Security Agency. Intelligence coordination with COMNAVFORJAP and Commander Naval Forces, Philippines was effected by CINCPACFLT and COM7THFLT, as required. Command of NAVSECGRUDET was exercised by the AGER commanding officer through the officer in charge of the detachment.[32]

Pueblo was captured on 23 January 1968 by the North Korean navy in international waters off Wonsan on her first mission. Documents probably acquired by the North Koreans from *Pueblo* included copies of the CLICKBEETLE II OPORDER and the instructions that were the documentary sources for several of the above paragraphs.

The decision to assign *Pueblo* to an area off North Korea was based on several considerations: (1) weather conditions in January are marginal off Petropavlovsk and Vladivostok, so those areas were rejected; (2) operations off the Chinese coast were eliminated because planners were uncertain as to how *Pueblo*, on her first mission, would react to the harassment and intimidation to be expected from the Chinese; (3) the Tsushima Strait was rejected because previous missions had obtained only marginal results; and (4) there were priority intelligence collection requirements on North Korea against which it was thought *Pueblo* could collect.[33]

The risk for *Pueblo* to operate in international waters off North Korea was considered minimal because (1) her sister AGER, *Banner*, was operating in the same area at that time and had previously operated in the Yellow Sea off the North Korean coast without problems; (2) the AGTR *Oxford* had operated in the same area as proposed for *Pueblo* in February 1967, again without problems; and (3) there had been no information indicating any sort of aggressiveness by the North Korean navy other than against fishing vessels or South Korean navy ships in North Korean territorial waters.[34]

The Foreign Broadcast Information Service had monitored North Korean broadcasts of 8 and 11 January 1968 that warned against provocative acts in territorial waters by "spy boats" disguised as fishing boats. Cdr. Bobby R. Inman, Chief of the Current Intelligence Branch, CINCPACFLT, later said that such warnings were an annual event occurring when South Korean fishing boats moved north in January and that they would cause no reevaluation of the risk. A Director National Security Agency (DIRNSA) message of 29 December 1967, which noted activity by North Korea against aerial reconnaissance, the sinking of a South Korean patrol escort (PCE), and increased border and fishing craft incidents, evaluated the North Korean threat situation as being on the increase. Cdr. Inman did not see the message until February 1968; he believed that, had he seen it prior to the *Pueblo* incident, it would have triggered an all-out review of existing intelligence. Inman doubted, however, that the DIRNSA message would have changed his risk evaluation.[35]

At the time of the *Pueblo* capture, *Banner* was at sea but was recalled because of communication problems. *Banner* was then sent into the Sea of Japan with a task force including the carriers *Enterprise* (CVAN 65) and *Ranger* (CVA 61) and their escorts, with the constraint that the AGER remain closer to the center of the task force than to any Communist-held territory. *Banner* was used as a sort of picket for early warning and to collect any other pertinent information. Subsequently, *Banner* operated in the Sea of Japan during an antisubmarine task force transit.[36]

The third AGER conversion, *Palm Beach*, patrolled in the Norwegian Sea and the eastern North Atlantic area from 27 June to 22 July 1968 to collect SIGINT and to conduct visual surveillance of Soviet naval units operating off the coast of Norway. Similar surveillance operations were carried out by *Palm Beach* during the period 1–12 August 1968.[37]

The first phase of an AGER study, providing the rationale for the procurement of three medium-endurance SIGINT collection ships and their associated equipment, was completed in June 1968. Phase II of the study was to have provided a rationale for outyear procurement of AGERs through 1970. Money had been requested for Phase II, and the draft report was scheduled for completion by 1 November 1968. The ships, however, were never acquired.[38]

In 1969, *Palm Beach* operated in the eastern Mediterranean from 4 to 25 June to maintain SIG-

INT surveillance of Soviet naval units at or near anchorages at Kithira, Alexandria, and Crete. During 11-28 August and from 30 August to 11 September, *Belmont* conducted surveillance of the Soviet helicopter-carrying cruiser *Moskva* in the eastern Mediterranean. The patrols gave the officers of the deck of *Belmont* an opportunity to train at formation-keeping with the Soviet ships.[39]

The following surface reconnaissance ships were decommissioned in 1969 and 1970:

Ship	Decommissioned
Oxford (AGTR 1)	19 Dec 1969
Georgetown (AGTR 2)	19 Dec 1969
Belmont (AGTR 4)	16 Jan 1970
Banner (AGER 1)	14 Nov 1969
Palm Beach (AGER 3)	1 Dec 1969

Intelligence-gathering ships like *Belmont* and *Palm Beach* had been withdrawn in 1969 from service with the Sixth Fleet in favor of portable SIGINT vans mounted on destroyers.[40]

The radar picket destroyer escorts *Thomas J. Gary* (DER 326) and *Calcaterra* (DER 390) were reconfigured to act as naval tactical reconnaissance ships (NTRS) in 1969–1970 by the installation of a special electronic package (OICS Van) to meet the requirements for "multi-sensor collection of intelligence on maritime targets of opportunity or interest, for on-board processing of data and its preliminary evaluation, and for transmittal of tactical and significant information in near real-time." *Calcaterra* operated in the eastern Mediterranean during the Jordan Crisis (17–25 September 1970). The decision to decommission *Palm Beach* was made when the NTRS conversion of *Thomas J. Gary* proved to be satisfactory.[41]

The improvements to the Navy's surface collection (Diamonds) capabilities, represented by *Thomas J. Gary* and *Calcaterra* in the Atlantic Fleet, were impeded by fiscal constraints when $1.9 million for three additional OICS Vans and $2.4 million for the modification of two Pacific Fleet DERs were dropped from the budget in 1969.[42]

Surface Reconnaissance Programs at the Beginning of the 1970s

Task Force 168, an element of ONI's collection organization, provided the following support to SILVER FOX collection operations into the Black Sea: prebriefing officers of staffs and ships, updating collection guidance for each operation, organizing and training an augmenting collection support team and funding its travel and per diem, shipping special equipment to the port of embarkation, installing and checking equipment prior to deployment, training ship's personnel to assist with equipment and data recording, supervising collection during deployment, returning data and equipment to appropriate facilities for analysis after deployment, collating all collected intelligence information into a comprehensive report, and publishing a Silver Fox report for each operation, incorporating the lessons learned in order to upgrade support to future operations. There were two SILVER FOX missions during Fiscal Year 1972.[43]

During Fiscal Year 1973, three more SILVER FOX operations were executed, all supported by the newly established Forward Area Support Team, Task Group 168.3. Intelligence collection targets in the Black Sea included naval cruise missile development and testing, new ship construction, antisubmarine warfare testing, command and control, tactical workups, vertical takeoff and landing (VTOL) aircraft operations, and other important naval developments.[44]

A pilot program was initiated during Fiscal Year 1972 to determine the feasibility of using Military Sealift Command (MSC) ships in ocean surveillance. The cargo ship *Col. William J. O'Brien* (T-AK 246) was provided with photo equipment and recognition guides, and the ship's civilian officers were briefed for a 90-day deployment. The results were excellent and led to an expansion of the effort. Initially, five MSC ships were to be supplied with photo equipment and all MSC ships were to be given guidance for reporting. A proposed Commander Military Sealift Command instruction was drafted to implement the program.[45]

Chapter Notes

1. See RAdm. Kemp Tolley, *The Cruise of the Lanakai* (Annapolis: Naval Institute Press, 1977).

2. Commander in Chief, Pacific Fleet (CINCPACFLT), *Interim Evaluation Report No. 5: Korean War Naval Operations*, 5–94.

3. Testimony of John L. Marocchi, *Record of Proceedings of the Pueblo Court of Inquiry*, Classified Annex, 198–246.

4. *CINCPACFLT Annual Reports, FY 1962*, 12; *FY 1963*, 43.

5. Ibid., *FY 1963*, 44.

6. *CINCPACFLT Annual Report, 1 Oct 1963–26 Jun 1964*, 52.

7. Edward J. Marolda and Oscar P. Fitzgerald, *The United States Navy and the Vietnam Conflict,* vol. 2, *From Military Assistance to Combat, 1959–1963* (Washington: Naval Historical Center, 1987), 31.

8. Ibid., 31–32.

9. Ibid., 70–82.

10. Ibid., 97–99.

11. See individual ship operations files for these units, OA, NHC.

12. Communications Satellite Relay (CSR-8) msgs 302132Z Apr

1964 and 161652Z Nov 1964.

13. CNO msg 021653Z Jul 1964, and CSR-8 msg 101357Z Dec1964.

14. CSR-8 msg 241458Z Dec 1964, and T-AG-169 Operational Files, OA.

15. *CINCPACFLT Annual Report, 26 Jun 1964–30 Mar 1965*, 54.

16. *Pueblo Inquiry*, Classified Annex, vol. 1, exhibit 18-16.

17. Ibid., 227.

18. Ibid., Classified Annex, 198–204.

19. Ibid., exhibit 18-2: CTF 96 OPORDER 301-68 of 3 Jan 1968.

20. *Atakapa* (ATF 149), *Command History, 1966*, passim.

21. *Jamestown* (AGTR 3), Operational File, OA.

22. Taiwan Patrol Force, *Command History, 1966*, Addendum, 3.

23. Commander Sixth Fleet (COM6THFLT), *Command History, 1967*.

24. *Pueblo Inquiry*, Classified Annex, 198.

25. *Atakapa, Command History, 1967*; *Pueblo Inquiry*, 10, 1858; *Oxford* (AGTR 1) and *Jamestown* Operational Files, OA.

26. COM6THFLT, *Command Histories, 1967*, II-F; *1968*, II-D; *1969*, II-C; *1970*, II-D.

27. USNS *Pvt. Jose Valdez* (T-AG-169), *Command History, 1968*.

28. *Calcaterra* (DER 390), *Command History, 1968*.

29. See individual ships in operational files, OA.

30. *Pueblo Inquiry*, exhibit 18-2.

31. Ibid.

32. *Pueblo Inquiry*, exhibit 18-39: DIRNAVSECGRU SIGINT LOI 1-68 of 2 Jan 1968.

33. *Pueblo Inquiry*, Classified Annex, 256–64: Lt. Edward A. Brooks, COMNAVFORJAP staff.

34. *Pueblo Inquiry*, Classified Annex, 198-88 and 198-86, and exhibits 18-17, 18, and 19.

35. Ibid., Classified Annex, exhibits 18-17, 18, and 19; and 714-13 and 198-181.

36. Testimony of Capts. T. L. Dwyer and F. A. Pease, *Pueblo Inquiry*, Classified Annex, 198-60, and 198-126.

37. *Palm Beach* (AGER 3), Patrol Reports, 22 Jul and 12 Aug 1968.

38. Director of Naval Intelligence (DNI) memo to CNO, 20 Jun 1968.

39. *Palm Beach* Patrol Report #2 of 25 Jun 1969; and Commanding Officer, *Belmont* (AGTR 4) ltr, 27 Sep 1969.

40. COM6THFLT, *Command History, 1969*, II-C.

41. *Calcaterra Command History, 1970*; and CNO (OP-03) memo, 11 Jul 1969.

42. DNI (OP-092Y) memo, "Significant Events in Naval Intelligence for 1969."

43. *DNI Report to CNO for FY 1972*, 22.

44. *DNI Report to CNO for FY 1973*, III-6.

45. *DNI Reports to CNO for FY 1972*, 36; and *FY 1973*, V-6.

CHAPTER 6

Submarine Reconnaissance and Support

Reconnaissance Missions

Only a few days after the United States entered World War II, submarine reconnaissance missions were ordered. *Pompano* (SS 181), for example, was ordered to patrol the Marshall Islands, sink Japanese ships, and determine what forces and equipment the Japanese had at their various island bases. Four days later, that submarine's orders were changed to make reconnaissance her primary mission, and *Pompano* proceeded to take a look at Wake, Eniwetok, Ujelang, Ponape, Rongelap, and Bikini.

A similar patrol was made by *Dolphin* (SS 169), which reconnoitered the Arno, Maloelap, Wotje, Kwajalein, and Jaluit Atolls to within 500 yards of the offshore reefs. Shortly thereafter, *Tautog* (SS 199) conducted surveillance of the Bikini, Rongelap, Kwajalein, Ujae, Utirik, and Taongi Atolls. Little was known of enemy installations on the isolated Pacific islands at that early stage of the war. Based on the information collected by the submarine reconnaissance patrols, the Navy made its first retaliatory strikes against enemy territory with the attack by RAdm. William F. Halsey's carrier task force on the Marshalls on 1 February 1942.

Periscope photography had been contemplated prior to World War II, and preliminary steps taken to obtain suitable cameras and adapt them to periscope use. No thought, however, had been given to the degree of exactness required if reconnaissance photographs were to be of intelligence value. Periscope photography was considered useful in identifying ship targets and proving that ships had been sunk.

Periscope photo reconnaissance was not fully developed until near the end of 1943, and much credit for the effective experimental work that went into perfecting it goes to the crew of *Nautilus* (SS 168). Amphibious landings were being planned for the Gilbert Islands for November 1943, and RAdm. Richmond Kelly Turner requested submarine reconnaissance of Tarawa, Makin, and Abemama. *Nautilus* was designated to perform the mission. Brackets for mounting the camera on the periscope were built at Submarine Base, Pearl Harbor, and an enlisted photographer was assigned to the submarine for temporary duty. A darkroom was fitted in the submarine's lower sound room, because it was thought necessary to process the film on board to permit retaking any pictures that didn't turn out satisfactorily.

Thus equipped, *Nautilus* departed Pearl Harbor on 16 September 1943. Photographic problems that had to be overcome included vibration in the periscope and the low light available through the optical system of the periscope. Tactical obstacles included offshore reefs and possible defensive minefields that had to be watched for and avoided, the danger of detection by Japanese lookouts and sentries, and possible chart inaccuracies. In spite of these and other problems, the mission was accomplished. It revealed that the Hydrographic Office charts for the target area were generally correct about the contours of the various islands but that their orientation was out as much as 11 degrees in some instances. The panoramic photographs provided information on gun installations and other beach defenses, particularly the wire and log barricades erected on reefs and beaches.

On 19 November, D-Day minus one, *Nautilus* again entered Tarawa lagoon. The atoll had been under air attack for the past five days, and a preliminary surface bombardment was then in progress. *Nautilus* found that new, 6- to 8-foot, heavy log walls had been built on the beaches since the submarine's first visit and that the defenses were so far undamaged. The large coastal defense guns and the small beach guns were also undamaged, and the latter were fired at *Nautilus*. The new information was reported to the task force commander, along with estimates on the height of the surf at the various beaches.

The work of *Nautilus* on this first submarine reconnaissance mission solved the most difficult of the problems inherent to periscope photography procedures and proved the value of submarine reconnaissance. As a consequence, all subsequent amphibious operations in the Pacific during World War II were preceded by submarine photo reconnaissance. Each reconnaissance mission involved taking up to 2,000 individual pictures. When the submarine had returned to port, the prints, negatives, and charts showing the locations from which the pictures had been taken were turned in to the Joint Intelligence Center, Pacific Ocean Areas (JICPOA) for further processing, interpretation, and dissemination.

The following submarine photo-reconnaissance missions were conducted in the Pacific during World War II:

Ship	Date	Commanding Officer and Area
Nautilus (SS 168)	Sep 1943	W. D. Irvin Tarawa, Makin, Abemama
Seal (SS 183)	Nov 1943	H. B. Dodge Kwajalein
Spearfish (SS 190)	Nov 1943	J. W. Williams Jaluit, Kwajalein, Wotje
Tarpon (SS 175)	Dec 1943	T. B. Oakley, Jr. Wotje, Kwajalein, Mili, Maloelap
Searaven (SS 196)	Jan 1944	H. M. Dry Eniwetok
Seal (SS 183)	Jan 1944	H. B. Dodge Ponape
Thresher (SS 200)	Mar 1944	D. C. MacMillan Oroluk, Nomi
Greenling (SS 213)	Mar 1944	J. D. Grant Saipan, Tinian, Guam
Salmon (SS 182)	Apr 1944	H. K. Nauman Ulithi, Woleai, Yap
Seawolf (SS 197)	Jun 1944	R. B. Lynch Palau
Burrfish (SS 312)	Jul 1944	W. B. Perkins Palau, Yap*
Spearfish (SS 190)	Nov 1944	C. C. Cole Iwo Jima
Swordfish (SS 193)	Dec 1944	K. E. Montross Okinawa†

*In addition to conducting photo reconnaissance, *Burrfish* made a landing party reconnaissance on the beaches of Palau and Yap using a landing party supplied by Commander, Amphibious Force, Pacific. Much valuable information was obtained on the beach and landing conditions on one Palau beach. Intense Japanese radar activity at other Palau beaches made landing party reconnaissance impracticable. Two landings were made on Yap; three men were lost on the second landing.

† Failed to return. Cause unknown.

In the closing months of the war in the Pacific, *Redfish* (SS 395) and *Runner* (SS 476) made submerged reconnaissance of coastal waters of Hokkaido and Honshu in early July 1945 to determine the existence of Japanese minefields in those waters prior to Third Fleet coastal bombardments during the period 10–20 July.[1]

Submarine Support to Coast Watchers

After the formal surrender of the Philippines to the Japanese early in 1942, the few Americans who escaped hid in remote parts of the archipelago, where there was no rapid and dependable means by which they could send Allied forces any information. Accordingly, it was decided to test the feasibility of making landings by submarine to supply the "coast watchers" with lightweight communications and surveillance equipment.

The first supply attempt was made on 14 January 1943, when the *Gudgeon* (SS 211) landed six men and 2,000 pounds of equipment and supplies on the island of Negros. This was followed by *Tambor* (SS 198), which landed a small party with about two tons of supplies at Labangan, Mindanao, on 5 March 1943. Thereafter, at about five-week intervals, small parties of personnel, each with about two tons of stores, were landed at various points in the central and southern Philippines by special missions carried out by selected submarines during their regular war patrols.

The cooperation of the natives in the southern part of the Philippines area was extremely good, and an organization of guerrilla forces, under the direction of Gen. Douglas MacArthur's Southwest Pacific command, was set up along recognized military lines.

The success of a guerrilla organization in the Philippines was due in large measure to the feasibility of supplying it with a modicum of arms, ammunition, medical supplies, radios, and funds. When the requirements of the supply effort, plus the expansion of the coast watcher and communications net, mounted to proportions that could not be handled by submarines as part of their regular war patrols, a special supply unit was organized in October 1943.

The submarines *Narwhal* (SS 167), *Nautilus*, *Seawolf* (SS 197), and *Stingray* (SS 186) were assigned the primary duty of carrying out supply and evacuation missions in the Philippines area. That the efforts were highly successful was proved by the rapid growth of an efficient net of coast watchers, weather observers, and aircraft spotters. At the time of the initial U.S. landings in Leyte, a net of 120 small radio stations completely covered the central and southern Philippines, with additional, but incomplete, coverage of Luzon. The Navy staffed and operated two control stations in Mindanao to screen the guerrilla traffic and to pass on to Seventh Fleet units those intelligence items of operational importance. The military supplies brought in by the submarines played no small part in the organization of the Filipino natives into effective combat and reconnaissance units until, at the time of the amphibious landings on Leyte, there were an estimated 65,000 organized guerrilla troops in the Philippines south of 12°N. The coast watchers in the Philippines rendered valuable service during the ensuing campaign.

Trout (SS 202), LCdr. Albert H. Clark commanding, was ordered to deliver a party of six or seven men, $10,000 in cash, and two tons of equipment and supplies to a designated spot on Basilan Island for the purpose of establishing a secret intelligence unit in the Sulu Archipelago and Zamboanga area. This unit was to become a coast watcher net to conduct a general reconnaissance survey and arrange for delivery of extra supplies to guerrilla units. *Trout*'s mission was completed on 26 May 1943.

While on patrol on 7 May 1944, *Crevalle* (SS 291), LCdr. Francis D. Walker, Jr., commanding, was ordered to pick up captured documents believed to be of high value at a site on the south coast of Negros, near Tolong, together with twenty-five U.S. evacuees. The special mission was accomplished on 11 May 1944; forty persons were evacuated, including twenty-eight women and children. The passengers were disembarked at Darwin on 19 May. Incident to the mission, a limited amount of food, ammunition, and canteen supplies was transferred to forces ashore. While in Molucca Passage during the return trip, *Crevalle* was subjected to severe depth-charging by Japanese forces.

Nautilus, LCdr. G. S. Sharp commanding—after delivering men and supplies off the mouth of Amney River, Mindoro, off Canayaon, Bohol and off Lagoma, Leyte, between 9 and 14 July 1944—was directed to make a pickup of important captured documents at Balatong Point, Negros. The additional special mission was accomplished on 16 July 1944.

Cero (SS 225), LCdr. Edward F. Dissette commanding, departed from Woendi on 17 October 1944 to discharge fourteen men and twenty tons of supplies at a pre-arranged spot on the west coast of Luzon. After arriving there, first at Darigayos Inlet on 25 October, and making no contact on the 25th, 26th, 27th, or 28th, or at Santiago Cove on the 29th or 30th, *Cero* finally made contact off the mouth of the Masanga River on the east coast of the island, and the personnel and supplies were off-loaded. Two Navy pilots and two evacuees were picked up for return to Pearl Harbor. Important, urgent Army intelligence sketches for Commander Southwest Pacific, and documents containing Philippine intelligence for Commander Seventh Fleet were brought back and disseminated to the proper authorities.

While on patrol, *Blackfin* (SS 322), under LCdr. George H. Laird, Jr., was directed on 14 November 1944 to pick up captured cryptographic and other secret documents plus technical equipment at a site west of Camuong River on the north coast of Mindoro. The mission was completed on 18 November, and the submarine rendezvoused off Morotai for transfer of the documents, which were then transported by air through the facilities of Commander Thirteenth Air Force to Seventh Fleet Intelligence Center, Hollandia (now Djajapura, Indonesia).

Seawolf, LCdr. Albert M. Bontier commanding, departed from Darwin on 1 August 1944 and on 7 August delivered a party of one officer, four radio operators, and one meteorologist, together with 14,538 pounds of supplies, to Tawitawi to reinforce an existing intelligence party. *Seawolf* then landed a party of one noncommissioned officer and five men and 7,153 pounds of supplies at a new site on northern Palawan on 9 August to set up a coast-watcher intelligence station. The missions were accomplished without undue incident, and *Seawolf* received a "Well Done" from Commander Task Force 72.

On its next mission, *Seawolf* was lost. The submarine departed Brisbane 21 September 1944 to discharge a party of seventeen men and six tons of cargo at a spot on the east coast of Samar and to pick up a Maj. Sabarre and eleven men who were to be landed on Batan Island together with nine tons of cargo. The estimated time of arrival at Samar was 6 October 1944, but the date was not met, and the submarine was listed overdue as of that date.

Gar (SS 206), LCdr. Maurice Ferrara commanding, departed Woendi, Biak, on 4 December 1944 with orders to deliver thirty-five tons of supplies to a prearranged spot off Darigayos Inlet (16°49′N, 120°19′E) and, upon completing the mission, to shift to operational control by Commander Submarine Force, Pacific for routing to Pearl Harbor. The mission was completed on 11 December with an added stop at Santiago Cove (17°17′2″N, 120°24′5″E) for a dawn pick-up of intelligence and Japanese documents of utmost secrecy to be delivered to the Army. One naval officer passenger was also picked up for the return trip.

A total of nineteen Seventh Fleet submarines were assigned to carry out supply and evacuation missions in the Philippines from 1 February 1943 to 23 January 1945. They participated in a total of forty-one missions over the two-year period. *Narwhal* participated in nine, followed by *Nautilus* with six and *Stingray* with five. *Seawolf* was the only submarine lost in the two-year period. On three other missions, the enemy attacked, but the submarines involved escaped without damage. Four missions involved delivery of important mail in addition to the delivery of personnel and cargo. Seven missions involved the pickup of "important captures or secret documents of intelligence value." During the nineteen missions, 331 persons and approximately 1,325 tons of supplies were delivered, and approximately 472 personnel were evacuated from the Philippines.[2]

Post–World War II Reconnaissance in the Pacific

Sea Dog (SS 401) conducted the first of a series of reconnaissance patrols outside Soviet territorial waters along the Siberian coast in May 1948. The primary benefit from the first patrol was the determination that there were no shoal water, pinnacles, anchored mines, or other dangers to submerged navigation in the international waters in that area. Efforts to sweep up to the 12-mile limit, however, were foiled by ice. Capt. T. A. Huckins, an intelligence officer from the Alaskan Joint Staff, was a passenger from Adak back to Kodiak during the patrol. He brought on board known Russian radio calls and frequencies, intelligence estimates on locations of Russian air bases, and a Russian dictionary. The material was passed on to *Blackfin*, the next submarine involved in the series of patrols.[3]

When the Korean War started on 25 June 1950, there were four submarines and one submarine rescue ship in the Western Pacific area. On 1 July 1950, Commander Seventh Fleet ordered all submarines to Yokosuka, Japan. It had been determined that any use of submarines would have to be confined mainly to the acquisition of intelligence information.

On 13 July 1950, Joint Zones for submarine operations were established as follows:

Zone 1: Between 24°26′N and 121°30′E to the China Coast.

Zone 2: Between 22°24′N and 116°20′E.

Zone 3: The sea area west of a line connecting 26°N, 122°42′E and 30°N, 124°E.

Zone 4: The sea area west of a line connecting 26°N, 121°30′E and 28°N, 122°42′E.

The first submarine reconnaissance patrols were started on 17 and 18 July by *Catfish* (SS 339) and *Pickerel* (SS 524) in Zones 3 and 1, respectively.

On 22 July, Joint Zones 5 and 6 were established south of 45°45′N, extending from 140°E to 145°E and bounded on the south by the island of Hokkaido. On 23 July, *Remora* (SS 487) departed Yokosuka for the first patrol of the La Perouse Strait.

On 1 August, China coast Zones 1 through 4 were taken over by surface patrols. Joint Zone 7 for periscope photo-reconnaissance was designated as the east coast of Korea between 40°N and 41°N latitude.

The primary mission of the submarine patrols in the La Perouse Strait was to obtain intelligence information for Commander Naval Forces, Far East (COMNAVFE) and other operational commanders about the movement of foreign shipping. Between July and November 1950, the main items of interest obtained by the patrols were the large amount of Soviet shipping observed, the sighting of three Soviet submarines, and the Russian use of searchlights to identify passing ships.[4]

In December 1950, *Besugo* (SS 321) was assigned to the La Perouse Strait patrol area, but the weather conditions were so bad that reconnaissance efforts were ineffective, and the submarine patrols were discontinued for the duration of the winter months.[5]

From 4 April to 6 December 1951, submarine reconnaissance was conducted continuously in the La Perouse Strait area. Both visual and photographic collection provided much needed intelligence information on Soviet ships and shipping trends.[6]

Submarine patrols of La Perouse Strait were resumed on 1 March 1952 and continued until 10 December; visual and photographic surveillance was conducted of shipping east, north, and west of Hokkaido. The submarines also made reports to COMNAVFE on Soviet and Chinese Communist sea and airborne activity. To increase the effectiveness of the patrols and to improve the security of submarine operations, patrol areas were expanded by the establishment of Joint Zone 10 in August and Joint Zone 11 in October 1952. Zone 10 was bounded by latitudes 44°30′N and 46°30′N and longitudes 138°45′E and 140°E; Zone 11 included waters north of latitude 46°N and between longitudes 143°55′E and 145°E. In December 1952, a reconnaissance patrol was conducted by *Scabbardfish* (SS 397) off the South China coast.[7]

On 22 January 1953, patrols of the La Perouse Strait were again resumed in order to maintain continuity of shipping surveillance and to provide submarine crews with experience in cold weather operations. Limited amphibious landing and raiding operations from submarines were also carried on during the period of February through July 1953. The cessation of Korean War hostilities on 27 July 1953 caused no change in the submarine reconnaissance operations. *Pomfret* (SS 391), on station at the time, remained on patrol until relieved in August by *Ronquil* (SS 396).[3]

Chapter Notes

1. *ONI Review*, Nov 1946, 14–16.
2. RAdm. Arthur H. McCollum, USN, "Submarine Activities with Guerrillas," Paper No. 420, passim, OA.
3. CO *Sea Dog* (SS 401) ser 003, 12 Jun 1948, with First Endorsement by Commander Submarines, Pacific, 23 Jun 1948.
4. Commander in Chief, Pacific Fleet (CINCPACFLT), *Interim Evaluation Report No.1: Korean War Naval Operations*, 1047–50.
5. CO *Besugo* (SS 321) ltr with encls., 23 Dec 1950.
6. CINCPACFLT, *Interim Evaluation Report No. 3: Korean War Naval Operations*, 12–4.
7. CINCPACFLT, *Interim Evaluation Report No. 4: Korean War Naval Operations*, 6-3; and *Interim Evaluation Report No. 5*, 5-9–5-100.
8. CINCPACFLT, *Interim Evaluation Report No. 6: Korean War Naval Operations*, 6-3.

CHAPTER 7

Prisoner-of-War Interrogation

POW Interrogation During World War II

The Special Intelligence Section (OP-16-F-9) of the Office of Naval Intelligence, set up in January 1941, was responsible for, among other things, prisoner-of-war (POW) interrogations. Early in the summer of 1941, members of the staff commenced drawing up plans for the interrogation of naval POWs. In September 1941, a comprehensive recommendation on the organization and its methods of operation was submitted. Instead, a joint Army-Navy committee was designated by the Secretaries of War and Navy to draft recommendations on the handling and custody of POWs. The recommendations were made to, and approved by, the Secretary of the Navy on 4 October 1941 and approved by the Secretary of War on 10 October 1941. The agreement provided that the Army would assume custody of all POWs and that persons captured by the Navy would be delivered to the Army as soon as practicable.

Other preparatory actions for POW interrogations were taken by OP-16-F-9 in October 1941. A Naval Reserve officer was assigned to London to receive training in British methods of interrogation. Also, Navy Department activities were requested to furnish questionnaires on desired subjects to be covered during the interrogation of POWs. Following OP-16-F-9's receipt of the questionnaires, they were promulgated to all ships and stations by the Chief of Naval Operations on 5 December 1941, together with instructions as to procedures for the interrogation of POWs.

Also in October, the first ONI opportunity to conduct a POW interrogation arrived when the German cargo ship M/S *Odenwald*, running the blockade from Japan and at the time of its capture masquerading as the American vessel *Willmoto*, was apprehended off Brazil by the light cruiser *Omaha* (CL 4). Although attempts were made by the German crew to scuttle the vessel, *Omaha* was successful in bringing *Odenwald* and her crew into San Juan. The crew was brought to the United States for internment. From documents captured with the vessel and through interrogations of the crew, valuable information was developed by the Special Intelligence Section on operational aspects of German blockade running, such as the routes and rendezvous points for blockade runners making French ports.

A joint Army-Navy conference on organizational needs, interrogation procedures, and internment facilities resulted in more recommendations that were approved by the Secretary of the Navy on 18 December 1941 and the Secretary of War on 6 January 1942. In accordance with the recommendations, the War Department undertook to establish two joint interrogation centers, one in the vicinity of Washington, D.C., at Fort Hunt, Virginia, on 2 August 1942, and the other on the West Coast at Byron Hot Springs, California, on 15 December 1942.

The first German naval prisoners captured by U.S. forces were from the submarine *U-352*, sunk by the U.S. Coast Guard cutter *Icarus* (WPC 110) off the Carolina Capes on 9 May 1942. Interrogation was eventually conducted when the joint interrogation center at Fort Hunt was opened in August, and a fairly complete history of the boat was obtained.[1]

On 5 August 1942, the Special Intelligence Section became the Special Activities Branch (OP-16-Z) and was organized as follows:

— Chief of Section: In charge.

— Administrative Officer: Washington Office Administrator, and in charge of security, procurement of supplies, assignment and direction of clerical personnel, indoctrination of new personnel, and acting as custodian of registered publications.

— Interrogators: Interrogated prisoners at interrogation centers and ports of arrival, participated in salvage operations and in preparation of reports.

— Censorship Officer: Responsible for the censoring of POW mail.

— Research and Analysis Unit: Reviewed all incoming intelligence material, indexed pertinent information in the master, ship name, and boat lists, and circulated reports of general and specific interest throughout the section.

— Personnel on Watch List: Available for watch standing at interrogation centers, for transcription of recordings, and for editorial work in preparation of reports.[2]

In October 1942, two Japanese language officers were detached from OP-16-Z, one to be assigned to the 14th Naval District and the other to the Allied Translator and Interpreter Section then being established in Brisbane, Australia. The head of the German Interrogation Section accompanied the officers to Australia to assist in the establishment of interrogation facilities on a combined service basis. He returned to the United States in company with the head of the British Admiralty interrogation unit with recommendations for more interrogation facilities and procedures for the Pacific theater.

When the Brisbane unit became operational, the officer at the 14th Naval District was transferred to Australia, and three additional officers were sent to Brisbane from the United States. In the summer of 1943, one of the officers in Australia was transferred to Noumea, New Caledonia, to assist in interrogations there. An officer was also ordered to Brisbane in February 1943 to handle captured enemy equipment and, after a few months, was transferred to Noumea for similar work there.[3]

Early in 1943, two officers from OP-16-Z were assigned to the collection of material derived from POW interrogations that might be of use for psychological warfare purposes. The two offices worked closely with the Special Warfare Branch (OP-16-W) in the preparation of the material.[4]

At the time of the establishment of the Joint Intelligence Collection Agency in North Africa in March 1943, the head of the Special Activities Branch went to London and worked out an arrangement with the Admiralty for combined participation by the Allies in interrogation centers at Algiers and Cairo. Three officers trained by OP-16-Z were assigned to Algiers and two were assigned to Cairo.[5]

On 6 May 1943, the Director of Naval Intelligence and the Army Assistant Chief of Staff for Intelligence (G-2) approved a joint agreement on the operation of interrogation centers. The agreement clarified and formalized existing procedures for Army administrative control, and it provided that operational, tactical, and technical information from military and naval POWs be disseminated only by the cognizant service in order to ensure proper evaluation. Under the agreement, it was possible to reduce the size of the naval interrogation unit, with the Army taking over certain functions that naval personnel had previously performed.[6]

U-118, sunk on 12 June 1943, was a new type of minelaying submarine weighing 1,600 tons. Through interrogation of prisoners from the German submarine, it was possible by mid-July to obtain scale drawings of the boat and details of its mines and their method of operation. The intelligence success was illustrative of the changed direction of POW interrogation that put emphasis on determining enemy technical developments in the fields of electronics, torpedoes, and armament. Improved interrogation techniques and the increased willingness of some prisoners to divulge information made it possible to derive much information on enemy developments in new technologies that were still in an experimental stage.[7]

With the increased emphasis on technical subjects, interrogators were able to obtain complete information and drawings on new types of German acoustic torpedoes, supplementary data on the circling torpedo used against convoys, information on the stationing of communications intelligence personnel on U-boats, information on radar and counter-radar installations and methods, details on modifications of armament, and information on German submarine tactics. Much intelligence acquired from POW interrogations during the period was regarded by operational and technical personnel as being of immediate importance for adapting Allied antisubmarine warfare equipment and methods to counter enemy devices and tactics.

Because of the increased submarine activity off Brazil, Commander Fourth Fleet requested an officer to facilitate preliminary interrogations. In August 1943, one ONI officer, Lt.(jg) J. R. Mullen, USNR, from OP-16-Z was assigned to the task.[8]

Because only a few Japanese POWs were being sent to the United States, it was decided in September 1943 to bring all Japanese language personnel from Byron Hot Springs back to Washington and assign them to the Far East Section of the Intelligence Branch of ONI. Navy personnel remaining at Byron Hot Springs after the shift were one officer, one enlisted person, and one civilian.

The large numbers of enemy naval personnel captured in July and August 1943, and the increasing volume of Army prisoners from the Mediterranean theater, proved the inadequacy of the interrogation center at Fort Hunt. As a temporary expedient, the Army developed facilities at Fort Meade, Maryland, and Pine Grove Furnace, Pennsylvania, as holding camps for POWs awaiting interrogation. The facilities became available in late September.

German naval POWs taken in the Mediterranean were sent to London or Washington after a brief preliminary interrogation. If a ship was sailing for the United Kingdom first, the prisoners went to London; if it was sailing to the United States first, the POWs were sent to Washington. Special exceptions were made for talented technical personnel, who were flown to London.

The Naval Intelligence Division (NID) 1/PW was the British counterpart to ONI's OP-16-Z, and Lt. C. L. Kuhn, USNR, was assigned as the liaison between OP-16-Z and NID 1/PW. The British asked Kuhn what plans OP-16-Z would furnish Navy interrogators if and when cross-channel operations took place. Lt. Kuhn checked with Capt. Norman S. Ives, the officer on the staff of Commander Task Force 122 responsible for port administration, and Ives said he wanted a minimum of four Navy interrogation teams to accompany the American assault forces.[9]

Naval POWs interrogated by OP-16-Z in 1942 and 1943, numbering in excess of 700, came from the following enemy ships:

— German *U-352*, sunk by USCG *Icarus* (WPC 110) on 9 May 1942 off Cape Hatteras.

— Japanese heavy cruiser *Mikuma*, sunk at the Battle of Midway, June 1942.

— German *U-210*, sunk by HMCS *Assiniboine* on 6 July 1942 in the North Atlantic, south of Cape Farewell.

— German *U-701*, sunk by an Army bomber on 7 July 1942 off Cape Hatteras.

— Japanese submarine *RO-61*, sunk by *Reid* (DD 369) in August 1942 in the Aleutians.

— German *U-94*, sunk by corvette HMCS *Oakville* and a U.S. Navy plane on 27 August 1942 in the Caribbean.

— German *U-162*, sunk by destroyers HMS *Quentin*, *Pathfinder*, and *Vimy* on 3 September 1942 off Trinidad.

— German *U-512*, sunk by U.S. Army bomber on 10 October 1942 in the Trinidad area.

— Japanese heavy cruiser *Furutaka*, sunk on 12 October 1942 in the Solomons at the Battle of Cape Esperance.

— German *U-595*, sunk by a British plane on 14 November 1942 off Cape Khamis, Algiers.

— German cargo ship *Anneliese Essberger*, scuttled on 21 November 1942 in the South Atlantic after being intercepted by *Milwaukee* (CL 5), *Cincinnati* (CL 6), and *Somers* (DD 381).

— Japanese destroyer *Takanami*, sunk 30 November 1942 in the Solomons at the Battle of Tassafaronga.

— German *U-164*, sunk by a Navy patrol plane on 6 January 1943 off Brazil.

— German *U-606*, sunk by USCG *Campbell* (WPG 32) and Polish destroyer *Burza* on 22 February 1943 in the North Atlantic.

— German merchant ship *Speybank*, sunk by probable German submarine 3 March 1943 in the South Atlantic.

— German merchant ship *Kota Nopan*, scuttled 10 March 1943 in the South Atlantic after intercept by *Savannah* (CL 42) task group.

— Italian submarine *Archimede*, sunk by a Navy patrol plane 15 April 1943 off Brazil.

— German *U-203*, sunk by planes from carrier HMS *Biter* and by destroyer HMS *Pathfinder* on 25 April 1943 in the North Atlantic.

— German *U-128*, sunk by Brazilian aircraft and *Moffett* (DD 362) and *Jouett* (DD 396) on 17 May 1943 off Brazil.

— German *U-569*, sunk by plane from *Bogue* (CVE 9) on 22 May 1943 in the North Atlantic.

— German *U-521*, sunk by *PC-565* on 2 June 1943 off the Virginia Capes.

— German *U-118*, sunk by a plane from *Bogue* on 12 June 1943 near the Canary Islands.

— German *U-409*, sunk by destroyer HMS *Inconstant* on 12 July 1943 in the Mediterranean.

— German *U-487*, sunk by a plane from *Core* (CVE 13) on 13 July 1943 off the Azores.

— German *U-67*, sunk by a plane from *Core* on 16 July 1943 off the Azores.

— German *U-513*, sunk by a Navy patrol plane on 19 July 1943 off Brazil.

— German *U-662*, sunk by a Navy patrol plane on 21 July 1943 off Brazil.

— German *U-527*, sunk by a plane from *Bogue* on 23 July 1943 off the Azores.

— German *U-598*, sunk by a Navy patrol plane on 23 July 1943 off Brazil.

— German *U-591*, sunk by a Navy patrol plane on 30 July 1943 off Brazil.

— German *U-199*, sunk by a Brazilian plane on 31 July 1943 off Brazil.

— German *U-615*, sunk by a Navy patrol plane on 7 August 1943 east of Trinidad.

— German *U-664*, sunk by a plane from *Card* (CVE 11) on 9 August 1943 north of the Azores.

— German *U-604*, scuttled after an attack by a Navy plane 11 August 1943 off Brazil.

— German *U-185*, sunk by a plane from *Core* on 24 August 1943 in the North Atlantic.

— German *U-841*, sunk by HMS *Byard* 17 October 1943 in the North Atlantic.

— German *U-848*, sunk by a Navy plane during November 1943 in the South Atlantic (only one survivor, who died).

— German *U-172*, sunk by *Bogue* task group during December 1943 in the Mid-Atlantic.[10]

Navy POWs interrogated by OP-16-Z during 1944 came from the following ships:

— Three German blockade runners intercepted in January in the South Atlantic.

— German *U-231*, sunk by British aircraft on 13 January north of the Azores.

— German *U-177*, sunk in the Atlantic in February.

— German *U-761*, sunk in the approaches to the Mediterranean on 24 February by U.S. and British aircraft and HMS destroyers *Anthony* and *Wishart*.

— German *U-575*, sunk in the mid-Atlantic by U.S. and British aircraft and ships from the *Bogue* task group on 13 March.

— Japanese submarine *I-35*, sunk by *Frazier* (DD 607) off Tarawa during the attack on Tarawa.

— German *U-801*, sunk by *Block Island* (CVE 106) task group on 17 March off the Cape Verde Islands.

— German *U-1059*, sunk by *Block Island* task group on 19 March off the Cape Verde Islands.

— German *U-856* sunk by destroyer *Champlin* (DD 601) and escort *Huse* (DE 145) off Long Island, New York.

— German *U-515*, sunk by *Guadalcanal* (CVE 60) task group on 9 April near Madeira.

— German *U-68*, sunk by *Guadalcanal* task group on 10 April northwest of Madeira.

— German *U-371*, sunk in the Mediterranean by U.S., British, and French escorts on 5 May.

— German *U-66*, sunk by *Block Island* task group on 1 May west of the Cape Verde Islands.

— German *U-616*, sunk in the Mediterranean on 17 May by U.S. destroyers and British aircraft.

— German *U-960*, sunk in the Mediterranean on 19 May by *Niblack* (DD 424) and *Ludlow* (DD 438) and British aircraft.

— German *U-505*, captured in mid-Atlantic by *Guadalcanal* task group on 4 June.

— German *U-490*, sunk in the central Atlantic by *Croatan* (CVE 25) task group on 11 June.

— German *U-860*, sunk in the south Atlantic by aircraft from *Solomons* (CVE 67) on 15 June.

— German *U-233*, rammed and sunk off Halifax by escorts *Baker* (DE 190) and *Thomas* (DE 102) on 5 July.

— German *U-1229*, sunk in the North Atlantic by aircraft from *Bogue* (CVE 9) on 20 August.

— A German destroyer sunk in the Mediterranean during 1944.[11]

On the basis of information and drawings produced from the interrogations of German naval prisoners in the United States, research agencies were able to construct a working model of a new German acoustic torpedo. This development made it possible to continue tests and experiments for the development of countermeasures.[12]

In March 1944, one officer and two civilian agents were detached from their Washington duties and returned to the West Coast interrogation center because of the increased number of Japanese POWs arriving there. The influx was the result of a change in Commander in Chief, Pacific Fleet policy intended to speed up the transfer to the United States of prisoners taken in the Pacific. In view of the amphibious character of Japanese military operations in the Pacific, arrangements were made with the Army to interrogate all Japanese prisoners jointly. This was different from the practice with German POWs where, in general, interrogations were conducted exclusively by officers of the cognizant service.[13]

Personnel from OP-16-Z were sent to Europe prior to the Normandy landings to assist in the interrogation of German prisoners of war. Some were assigned to various front-line activities for tactical debriefings. Those assigned to British port parties worked along the French coast. Similarly, some OP-16-Z personnel were attached to the Joint U.S. Navy-Royal Navy 20th Assault Unit, exploiting captured documents and hardware as well as POWs; they also visited all major coastal installations, especially submarine pens, from Le Havre to Lorient and including Brest and Cherbourg. Other OP-16-Z personnel worked with Alsos units (technical specialists sent to Europe ostensibly to recover weapons technology but primarily charged with obtaining information on the German nuclear weapons program) and Mobile Explosive Investigation Unit 3 prior to the latter's breakup to form the Naval Technical Mission in Europe (NAVTECHMISEU). Various mine and bomb disposal personnel assigned to assault units in the European, African, and Mediterranean areas also participated in the POW interrogation effort.[14] See Chapter 11 for further information on the Alsos effort and NAVTECHMISEU and Chapter 39 for a fairly detailed report on the interrogation effort in that area following the Normandy landings.

During 1945, naval POWs interrogated by OP-16-Z were from the following enemy ships and submarines:

— German *U-546*, sunk in the North Atlantic by eight U.S. escorts on 24 April.

— German *U-1228*, *234*, *805*, *873*, *548*—all surrendered 9 May in Europe at the end of hostilities.

— German *U-858*, surrendered 14 May and sent into Portsmouth, New Hampshire.

— German *U-530*, surrendered at Buenos Aires on 10 July.

— German *U-977*, surrendered at Buenos Aires on 17 August.

— Japanese submarine *I-365*.

— Japanese cruiser *Natori*.

— Japanese escort destroyer *Matsu*.

— Japanese submarine *I-24*.[15]

Obtaining "negative information" was occasionally an important factor in the interrogation of prisoners. As an example, reports received from Stockholm early in November 1944 indicated that the Germans were installing launch platforms for V-1 missiles on some of the new Type XXI submarines. Norwegian reports also concluded that the boats were being prepared for attacks on New York City and other U.S. ports. This alarming intelligence understandably aroused the concern of those responsible for the defense of the East Coast, particularly when six German submarines in Group *Seewolf* were detected heading west in the North Atlantic in the early spring of 1945. When one of the submarines, *U-546*, was sunk on 24 April and her survivors rescued, it became very important to determine, if possible, the mission of the group, perhaps thereby refuting the previous reports. Survivors from *U-546* were sent to Argentia, Newfoundland, and before many hours passed it became evident to naval interrogators that Group *Seewolf* had not been dispatched to deliver the long-heralded V-1 attack against East Coast cities but simply to conduct a vigorous diversionary campaign against shipping in North American waters.

Subsequent Turkish reports that the Germans planned to begin bombarding the Atlantic seaboard with stratospheric V-3 bombs served further to confirm suspicion that the warnings were Nazi-inspired psychological warfare.[16]

In May 1945, intensive interrogation was conducted of the German army and navy officers and technical specialists who were passengers in the *U-234* en route Japan when it surrendered to U.S. forces at the end of European hostilities. Details about the exchange of information between Germany and Japan were particularly important and proved to be invaluable not only in connection with the continuing war against Japan but also as a windfall in support of U.S. advances in electronics and other fields of research.[17]

A contemporary listing of the outstanding accomplishments of the German Interrogation Section of OP-16-Z during World War II concluded that it

— was one of the principal sources on which Commander in Chief, U.S. Fleet, Combat Intelligence, had based its appreciation of German U-boat warfare. POW interrogation provided information on U-boat tactics, new equipment, morale, and unit disposition;

— was the sole source of advance information on the German acoustic torpedo;

— was the sole source of advance information on the *schnorchel* breathing device that permitted German submarines to operate their diesel engines while running submerged at shallow depths;

— was the sole source of advance information on German high-underwater-speed U-boats;

— had provided information of great value on German communications;

— had provided continuous and detailed information on the development of a German search receiver that was built to intercept Allied radar;

— had deduced the operational use of the German submarine bubble target during salvage operations of *U-72* in May 1942 and confirmed it during interrogation of *U-701* POWs in July 1942; and

— was the principal source of information on a number of other U-boat developments such as radar decoy balloons and spar buoys, antiradar covering for *schnorchels* and U-boat superstructures, the helicopter carried by some long-range U-boats, mines and torpedoes, U-boat minelayers, and infrared sensors.

In the spring of 1944, a German espionage agent, Oskar Mantel, who was to have been landed in the United States, was picked up from a U-boat sunk in the Atlantic. He was turned over to the FBI and corroborated a great deal of information on the German intelligence service, particularly its methods of operation and the identity of the personnel in the Paris *Abwehrstelle*.[18]

POW Interrogation During the Korean War

Interrogations of POWs within the Korean theater were conducted in accordance with Army intelligence directives. The lack of trained interrogators within the Navy was partially offset by the use of Republic of Korea navy personnel, particularly in connection with intelligence teams that had been sent ashore. The contents of POW interrogation reports from the Army were disappointingly lacking in items of naval interest. As of May 1951, the 200 POW interrogation reports then completed mentioned nothing on the enemy navy or naval matters, although some of the prisoners had lived in, trained at, or passed through such ports as Songjin, Hungnam, Wonsan, Hamhung, and Yanggang.

An example of the continuing problem concerning unsatisfactory interrogation of POWs by the Army is the capture of a large, new, heavily constructed sampan on 22 May 1952 by the minesweeper *Murrelet*

(AM 372). One of the crew of five taken prisoner volunteered information about the location of a minefield but gave no details on its location. The results of later official interrogation of the prisoner, which should have expanded the information on this very important subject, were never made available to Commander Mine Squadron 3.[19]

The Interrogation Desk (OP-322Y4E) was activated in the Office of Naval Intelligence in October 1951. By March 1952, the organization was ready to inaugurate an interrogation training program. At that time, there were no trained POW interrogators in ONI to fill Navy requirements in Russian or Soviet satellite languages. To correct the situation, it was recommended that a vigorous program be initiated to procure civilian linguists as volunteer reserves to be trained in POW interrogation in OP-322Y and that a specific number of graduates from each class of the Russian language course at the Naval Intelligence School be assigned to intensive POW interrogation training in OP-322Y.[20]

POW Interrogation During the Vietnam War

Throughout the Vietnam War, the South Vietnamese were responsible for the custody and processing of prisoners of war. Frequently, knowledgeable POWs who had information of interest to the U.S. Navy were not interrogated on naval subjects by their captors or, if they were, the interrogations were inadequate. On a number of occasions, Navy personnel were unable to question high-interest POWs until after the South Vietnamese were finished. The POWs had on occasion been physically abused by the South Vietnamese and, by the time U.S. Navy interrogators were given access to them, they were no longer capable of recalling information of interest. Additionally, any time-sensitive information the POWs may have had was no longer of value. In the few instances where early access to potentially important POWs was gained by the Navy, the access had to be obtained through high-level negotiations by the Military Assistance Command, Vietnam.[21]

The author has not had the opportunity to research the files of the Defense Intelligence Agency (DIA) for information relating to the interrogation of prisoners of war during the Vietnam conflict. Such debriefings, of course, were conducted in-country and often on an ad hoc basis. Thus, unless a full report had been sent back to DIA in Washington, it is quite probable that many of the records of naval-related interrogations were not preserved.

Chapter Notes

1. OP-16-Z, "History of the Special Activities Branch, 17 Jun 1942–31 Oct 1943," MS, 4–5, 9–10, OA.

2. *ONI Roster*; Capt. John L. Riheldaffer ltr to LCdr. Ralph G. Albrecht, 8 Sep 1942, OP-16-Z misc. file, OA.

3. "History of Special Activities Branch," 14–15.

4. Ibid., 18.

5. Ibid., 16.

6. Ibid., 20–21.

7. Ibid., 18.

8. Ibid., 20–21.

9. OP-16-Z ltr to Lt. Kuhn, 29 Dec 1943, OP-16-Z misc. file.

10. "History of the Special Activities Branch," 9–20; and Erich Gröner, von Dieter Jung, and Martin Maass, *Die deutschen Kriegsschiffe, 1815–1945* (Koblenz: Bernard & Graefe Verlag, 1985) 3:82–119.

11. Ibid.

12. OP-16-Z memo, 1 Mar 1944.

13. OP-16-Z memo, 1 Apr 1944.

14. Capt. C. J. Oleniecz ltr to author, 13 Dec 1974.

15. "History of the Special Activities Branch," 9–20; and Gröner, et al., *Die deutschen Kriegsschiffe*.

16. Philip K. Lundeburg, "American Antisubmarine Operations in the Atlantic, May 1943–May 1945" (Ph.D. diss., Harvard University, 1953), chap. 12, 25–26.

17. OP-16-Z memo, 4 Jun 1945.

18. George Kidd (OP-322F114) memo to Capt. Harris W. Baltazzi (OP-32B1), 23 Dec 1948. Kidd, who later became ONI's senior civilian, had been a POW interrogator during the Normandy landings on detachment from OP-16-Z and is believed to have written much of the manuscript history cited above; in 1948 he was the naval officer in charge of the ONI German Desk.

19. Commander in Chief, Pacific Fleet, *Interim Evaluation Report No. 2: Korean War Naval Operations*, 1744–45.

20. OP-322Y4E memo, 12 Mar 1952, box 1, "Incoming Correspondence File, 1952–1953," Job 12876, FRC/WNRC.

21. Task Force 168, Lessons Learned in Vietnam (LLIVN), Annex C, 29.

CHAPTER 8

Human Intelligence

This chapter discusses the field of human intelligence (HUMINT) in a necessarily fragmentary format, and the story of the Navy's formal involvement in HUMINT is ended in 1970 because of security concerns. Many other chapters in this history discuss various aspects of human intelligence, but this chapter relates specific events not covered elsewhere. It also describes the evolution of the administration of Navy HUMINT.

World War II Experience

Recognizing the need for an undercover intelligene collection organization within the Office of Naval Intelligence in 1939, various ONI officers considered action to correct the deficiency. Cdr. Francis D. Pryor, USN (Ret.), ONI's Plans Officer (OP-16-X) since 1931, devoted considerable time to the collection and preparation of material for the publication of ONI-22, a preliminary effort toward the production of a manual on espionage. Lt. Marsden J. Perry, USNR, had also collected material on the subject. Cdr. Elliot B. Nixon, head of Domestic Intelligence in 1939, was placed in charge of ONI's espionage functions.[1]

The Special Intelligence Section (OP-16-F-9) of ONI's Foreign Intelligence Branch was established on 17 June 1940 for the purpose of obtaining, training, and administering secret agents. It was staffed by one retired officer, Pryor, and a chief yeoman.

In the spring of 1941, arrangements were made with the State Department for the assignment of control officers or vice consuls by the Army and Navy to locations in North Africa for the purpose of collecting intelligence. By the summer of 1941, two Navy representatives, selected by OP-16-F-9, had been posted to Algiers, and one each to Casablanca, Oran, and Tunis. Later a vice consul was sent to Dakar.

Also in April 1941, the OP-16-F-9 section head, Cdr. John L. Riheldaffer, USN (Ret.), made a trip to the 11th Naval District and arranged for the development and operation of an intelligence network along the west coast of Mexico under the immediate direction of an assistant to the district intelligence officer. Riheldaffer held the ultimate responsibility for the administration and direction of the net in Mexico, and he received all reports from it. In July 1941, one of Riheldaffer's agents detected Japanese smuggling of mercury from Mexican ports; the detection resulted in the smuggling being stopped.[2]

OP-16-F-9 also engaged in the cultivation of contacts within business organizations abroad that might be in a position to furnish information concerning possible enemy countries. The Special Intelligence Section also employed a number of agents under special contract; one went to the Far East, two traveled through the Far East and Middle East, three were assigned in the Middle East, one was sent to Spain, one went to West Africa, and two traveled to the Caribbean and Mexican areas. In general, the agents were individuals who traveled in the areas of interest for open purposes and who accepted the task of collecting and reporting information that might be of value to the Navy.

Upon the establishment of the Office of Coordination of Information in July 1941, all Navy special intelligence activities were transferred to that office. The civilian employee who had been participating in the work in OP-16-F-9 took over the direction of espionage in the new organization.

As of 1 April 1942, the Coordinator of Information turned the operation of the special intelligence activities conducted in Mexico back to the Navy. From the 11th Naval District, a network was maintained that covered lower California and the Pacific coast of Mexico. In addition, two fishing boats were acquired for collecting offshore intelligence. The fishing boat cover was successful, at least insofar that a substantial monetary profit was made from the boat's catch. The principal intelligence service performed by the Mexican operation, however, was to disprove the existence of Japanese activities in the area.

A special intelligence office was established in San Antonio, Texas, from which contact was maintained with a chain of informants along the Gulf of Mexico and in Mexico City. In 1942, particular attention was paid to the possibility that enemy agents were making contact with German submarines operating extensively in the Gulf of Mexico. The counterespionage effort failed to find any evidence of shore-based support for the U-boats.

On 5 August 1942, the Special Intelligence Section was removed from the Foreign Intelligence Branch and established as the Special Activities Branch (OP-16-Z).[3]

In September 1943, the North American Theater Desk (OP-16-FN) of ONI was charged with the responsibility of supervising coastal intelligence collection and the collection of strategic intelligence from sources within the United States. FN sections were set up in each naval district to contact domestic sources, both companies and individuals, that might have information on foreign subjects and places of interest to the Navy.

OP-16-FN received the information collected by its offices in the naval districts, logged it, tabulated it on cards, indexed it, and sent it on to the analysts at the cognizant ONI geographic desks for evaluation and processing. Sources were tabulated according to their knowledge, identity, naval district of contact, and the evaluation of the information they supplied. The list of sources thus developed was the start of what eventually became known as the Navy Contact Register.

Many of the reports derived from domestic sources contained excellent information. One such report obtained by the District Intelligence Office, 9th Naval District from an old copra trader who had visited many of the islands of the Western Pacific contained information, maps, charts, and photos of the island of Tarawa, including the locations of coral reefs. The value of the report was immediately recognized, and it was quickly reproduced and distributed—but not in time to reach the Marines, who were landing with heavy losses on Tarawa, due partly to their landing craft getting hung up on the coral reefs.[4]

Post–World War II and Korea

U.S. Naval Forces Western Pacific, following World War II and the disestablishment of Commander Naval Group, China, established the Intelligence Liaison Office (ILO) in Shanghai for the purpose of maintaining contact with the various factions competing for control of China and observing their activities.

In that connection, ILO Shanghai maintained discreet contact with various officials and members of the *Ching Hung Pang*, a Chinese secret society and progenitor of the *Kuomintang*. When ILO Shanghai was first established (circa 1946), officials of the society frequently and voluntarily supplied a considerable amount of early and usually accurate political, economic, military, and counterintelligence information. The *Ching Hung Pang* also provided cover, protection, and introductions when needed in connection with intelligence and security activities. In early 1948, when the Central Government's control of metropolitan centers was becoming more tenuous, several ranking officers active in the society offered to arrange for a flow of pertinent information to the ILO and for the protection of U.S. and foreign lives and property in case of a possible general breakdown of law and order, a major emergency arising from mob action, or an attack on Shanghai by dissident military forces.[5]

During the preparations for the 15 September 1950 Inchon landing, intelligence collection was performed by Lt. Eugene F. Clark who, while assigned to Gen. Douglas MacArthur's intelligence staff, had been asked to volunteer for the task. With two interpreters, some weapons, a radio, and other supplies, he was put ashore on the island of Yonghung-do at the mouth of the channel leading to Inchon. Clark set up a "command post," recruited and organized 150 South Korean youths into his "army," and sent them on intelligence gathering missions to obtain information about Inchon's harbor defenses. Clark himself reconnoitered potential landing areas in and near Inchon, sending in information on beach conditions, tides, and navigation problems in the winding channel leading to the city.

Nightly harassing attacks against Clark's station on Yonghung-do by Red soldiers from a nearby island were initially small-scale and were beaten off. A week before the Inchon invasion, the North Koreans attacked in some force but again were repulsed when Clark attacked their boats with an armed sampan.

On the night of the invasion, when the United Nations fleet approached, Clark climbed the darkened Inchon harbor lighthouse and relighted its beacon. He then sailed out to and boarded the American command ship as it entered the harbor.[6]

In March 1951, very little coordination or liaison existed between naval forces and the covert agencies that were operating along the coasts of Korea behind enemy lines. On several occasions, United Nations naval forces and aircraft fired on small craft that were found later to have been carrying friendly agents or guerrillas. Gradually, the various agencies were identified, and measures were taken to establish liaison and to coordinate their efforts with naval forces in their areas of operations. Recognition procedures and small-craft movement schedules were developed, and adequate island defenses were established.[7]

Task Group 95.2 in Wonsan harbor, Korea, was charged with the support of the Yo Do Island Intelligence Collection Center. The center was established in May 1951 and had a widespread agent network in the Wonsan-Hamhung area. Because of increasing enemy security measures along the Korean east coast, agent work from Yo Do became correspondingly more difficult. From time to time, however, operational intelligence information was obtained that added current data on the enemy situation. Task Element 95.23 was the designation for the Navy liaison officer at Yo Do.

Task Force Kirkland, an Army guerrilla agency, also on the Korean east coast, operated agents from Nan Do Island, 10 miles off the coast from Kojo. It provided valuable local target information for naval surface and aircraft units. Kirkland agents frequently acted as shore fire controllers for naval units, and, after the bombardment, they entered the target areas to assess the damage inflicted.[8]

In early 1952, the various agencies on the west coast of Korea that were engaged in the covert collection of intelligence information were integrated under Commander Covert Clandestine and Related Activities, Korea (CCRAK), who had his headquarters in Seoul.[9]

Covert collection by CCRAK-controlled agencies and prisoner-of-war interrogations provided a limited amount of information of tactical value to the fleet. Information obtained from covert sources pertained mostly to enemy troop strengths and dispositions and was usually of little value, but sometimes it provided locations of guns in coastal areas.

On the west coast, commanders of operating units received clandestinely obtained information through direct liaison. On the east coast, an intelligence officer was assigned to Yo Do Island in Wonsan harbor by Commander Naval Forces, Far East. The officer maintained close contact with the various covert intelligence collection units that were using Yo Do Island as a base of operations.[10]

The establishment of CCRAK by Commander in Chief, Far East had resulted in a more satisfactory coordination of covert collection activities, but frequent difficulties continued in the identification and clearance of friendly small craft by the naval blockading forces and in obtaining advance notice when supporting ships were needed for such operations. In March, April, and May 1953, conferences were held between CCRAK and Commander Task Force 95 that resulted in changes in CTF 95 operational orders and resolved the clearance problem. In July 1953, the assignment of a CCRAK coordinator for east coast operations, based at Sokcho, improved coordination matters.

Naval intelligence liaison officers stationed at Yo Do Island and CCRAK representatives and coordinators on other islands on both coasts of the Korean peninsula provided an adequate and continuing flow of information on enemy order of battle, supply and troop concentration, and industrial and communications installations. Most were out of range of TF 95 units on the east coast. On the west coast, however, the same types of information provided very accurate and valuable target details and locations for use by aircraft from TF 95's light aircraft carrier unit.[11]

The Cold War Era and the Creation of HUMINT

Director Central Intelligence Directive No. 5/1, an "Agreed Activities" agreement negotiated by the Joint Intelligence Committee with the Director of Central Intelligence in 1955, recognized military participation in clandestine intelligence, defined the collection activities that the military services could conduct, and set forth procedures to coordinate them with the clandestine collection activities of the Central Intelligence Agency. The agreement was distributed to Navy unified and fleet commanders on 7 February 1955. Guidance was issued at the same time on the extent and limitation of the naval effort in clandestine intelligence in peacetime.[12]

ONI-70-1, *U.S. Naval Intelligence Manual*, issued on 20 June 1956, provided useful guidance on the handling of clandestine HUMINT sources:

> The value and continued usefulness of an informant depends upon his confidence that, in providing information, he will not jeopardize his government, business or professional position, or in some cases even his life. Hence, the utmost care must be exercised not to reveal the identity of an informant or to compromise him in any way. In order to protect, and at the same time identify a source in correspondence (or in phone conversations), a source symbol or designator shall be used in accordance with current ONI instructions.[13]

By 1960, the use of human intelligence sources by naval intelligence had become routine, and standard operating procedures had been worked out. Collection activities (naval attachés and district intelligence offices) maintained lists of persons located in their respective areas who were actual or potential sources of information of naval interest. Such lists gave the name, address, telephone number and/or usual means for contacting the individual, the type of information obtainable by the source, the source's particular qualifications, and a general estimate of reliability and cooperativeness. Separate lists were maintained for overt sources and for confidential (sensitive) sources. Lists of confidential sources developed exclusively for use in connection with counterintelligence were maintained separately.

Source code designators were assigned to confidential sources for use in information reports, correspondence, messages, and telephone conversations in order to protect their true identities. The designators were made up from the name of the place of the reporting activity, followed by the last two digits of the calendar year in which source identity was furnished to ONI, followed by a dash and the sequence number assigned to the source. A new sequence of numbers was started at the beginning of each calendar year.

An Interagency Source Register was maintained within the U.S. intelligence community. All intelligence agencies contributed to the register in order to detect the use of a source by more than one agency, as well as to minimize the possibility of being victimized by "papermills," fabricators, and other intelligence collection nuisances. Source information supplied to the register included only the source's name, date and place of birth, status, and country from which the source was controlled.[14]

An extensive study of the Department of Defense (DOD) Personnel Security Program during the spring of 1965 by a special survey team produced twenty-two recommendations for changes in the investigative procedures and organizations of the three military services.

Two of the recommendations related to the involvement of the Navy's investigative resources in other than investigative work. On 27 May 1965, Secretary of Defense Robert S. McNamara approved the recommendations, which removed HUMINT collection as a responsibility of the Navy's field investigative offices. Action was to be completed not later than one year from the date of McNamara's memorandum. On 18 June 1965, Director of Naval Intelligence (DNI) RAdm. Rufus L. Taylor recommended to Secretary of the Navy Paul H. Nitze that the decision to end the use of district intelligence offices in HUMINT collection be canceled, since the action would lead to the requirement to establish twelve to fifteen new activities to perform domestic intelligence collection.[15]

On 7 December 1965, Secretary Nitze assigned to DNI the responsibility for implementing a clandestine intelligence collection program and provided guidance for its coordination and control. ONI followed up on Nitze's order with its instruction 005430.12 of 11 January 1966, which established within ONI the control and management of the Navy's clandestine intelligence collection program.

The acronym HUMINT was introduced in 1966 to identify the intelligence collection function that uses human beings as both sources and collectors. HUMINT included overt, sensitive, and clandestine activities defined as follows:

Overt: Those activities that are conducted in such a manner as to allow acknowledgment, if necessary, without significant embarrassment to the United States.

Sensitive: Those collection activities that fall within the "gray" area between Overt and Clandestine, in that their exposure would be detrimental to the best interests of the United States and therefore must be carried out in a manner that will minimize the chances of compromise.

Clandestine: Those activities that must be conducted under maximum security constraints, on a basis of plausible denial, and with full regard to their complexity and risk.[16]

The DOD *Human Resource Intelligence Collection Implementation Plan–1966* was prepared by the Defense Intelligence Agency (DIA) in coordination with the military departments and the unified commands at the direction of the Joint Chiefs of Staff and was issued on 8 July 1966. The plan provided the basis for mutual coordination and support between the Defense Attáche System and the Human Resource Intelligence Collection System. Both operated under the management of the Director of DIA, the former under his operational control and the latter under his technical direction and coordination through the military departments and unified commands.

The DOD plan also directed the military departments to exercise command, including operational command and control, of HUMINT collection units and activities not assigned to the unified commands; provide administrative, budgetary and logistical support to DOD-controlled HUMINT collection units and activities; task DOD-controlled units and activities; conduct collection operations to satisfy validated collection requirements; and advise the Defense Intelligence Agency of those clandestine counterintelligence activities that had a positive intelligence collection potential.

On 1 April 1966, the Naval Field Operations Support Group (NFOSG) was established under an officer in charge as a field activity under the command of the Chief of Naval Operations. Its unclassified mission was to support assigned naval personnel on detached duty and develop the necessary communications and logistics channels to those personnel. Its support of HUMINT operations was classified secret. LCdr. Thomas J. Saunders was the first officer in charge.[17] DNI letter, serial 006522P92 of 28 July 1966, assigned to the NFOSG the mission of establishing a worldwide intelligence organization while preserving nonintelligence attributability of the clandestine collection operations.

The Department of Defense HUMINT plan called specifically for "the co-location of departmentally-controlled HUMINT collection elements in the Washington area and the provision of services of common concern." This concept was amplified in subsequent DIA correspondence, which included the "Plan for the Establishment of the Washington Human Resource Collection Center" prescribed in a Defense Intelligence letter of 15 May 1967.[18]

When the Naval Intelligence Command (NIC) was established on 1 July 1967, the Collateral Support Section (OP-922H1) became the Human Resources Branch (NIC-321). It was composed of seven military and civilian personnel and was headed by Capt. C. J. Oleniacz, USN (Ret.).[19]

As a result of DIA's intent to establish the Washington Human Resource Collection Center, Commander Naval Intelligence Command, in letters, serial N008 and N009, both of 9 October 1967, directed the establishment of a task force to study the Navy's HUMINT effort. The task force was composed of Capt. Wyman H. Packard, USN (Ret.), as senior member, with Capt. Howard W. Holschuh as the Naval Intelligence Command representative, Cdr. James G. Brady as the Scientific and Technical Intelligence Center representative, Cdr. D. E. Nielsen as the Naval Field Operations Support Group representative, and R. P. Ray as the Naval Investigative Service representative.[20]

The Naval Investigative Service Offices (NISO) were located in about 200 places in the United States and overseas. The information collected in support of their counterintelligence and investigative missions was often of equal value to intelligence. Furthermore, NISO resources, particularly those in foreign countries, had the potential to collect information of intelligence value on a not-to-interfere with their primary missions basis. The Naval Intelligence Command kept informed on the NISO collection potential and provided collection requirements to NISOs as appropriate.

Much of the Navy's HUMINT potential in the 1960s was similarly derived from its various commands and was incidental to their primary functions, which were not related to intelligence. So that potential collection opportunities would not be lost, the Naval Intelligence Command provided centralized direction and technical guidance to assure selective and effective exploitation.[21]

The Office of Naval Intelligence's clandestine intelligence collection program, conducted by the Naval Field Operations Support Group, had grown by 1968 to a strength of 168 officers, enlisted, and civilian specialists located at the Washington headquarters and in fifteen domestic and foreign detachments. Plans called for continuing expansion to a Fiscal Year 1972 strength of 262 personnel in twenty locations. NFOSG's clandestine mission and its connection with ONI were classified.[22]

On 15 January 1968, in preparation for the transfer of Navy personnel to a Navy element of the proposed Washington Field Activities Support Center (WFASC), the Human Resources Branch (NIC-321) was restructured and augmented with four NFOSG personnel. In the 8 July 1968 reorganization, NIC-321 became NIC-352, headed by Cdr. J. W. Wilson, and was retitled the Resources, Programs, and Coordination Branch. NIC-352's office staff was further augmented to a total of eighteen by transferring personnel from NFOSG and the Naval Intelligence Command.

When the DOD Resources Management and Coordination Branch moved to the Washington Field Activities Support Center at Fort Belvoir, Virginia, its mission and functions were stated as follows:

Mission:

> To be cognizant over Navy Department HUMINT Collection effort and to implement the responsibilities of the Division Head on all matters related to the execution, support and evaluation of human resources intelligence collection programs.

Functions:

> (1) Maintain cognizance of and manage all Navy departmental HUMINT collection activities.
>
> (2) Keep NAVINTCOM Intelligence Collection Division informed of the Navy's HUMINT collection capabilities.
>
> (3) Receive validated collection requirements from NAVINTCOM Intelligence Collection Division and serve them on the Navy's HUMINT collection resources.
>
> (4) Review reports from the Navy's HUMINT collection activities for purposes of evaluating the effectiveness of those activities and maintaining qualitative and quantitative statistics on the fulfillment of requirements.
>
> (5) Assist in the development of mid-range plans for HUMINT collection operations and programs to fulfill future requirements and assigned objectives, coordinating with the Army and Air Force elements at the WFASC as necessary to avoid duplication of effort and assure mutual support.
>
> (6) Collaborate as necessary in the planning for, and documentation of, manpower, facilities,

equipment and services requirements for Human Resources collection activities in the CIP [Consolidated Intelligence Program].

(7) Maintain the Navy HUMINT Source Registry and introduce Navy HUMINT source data into the WFASC source register.

(8) Obtain analytical aid to improve collection guidance and support to HUMINT resources.

(9) Originate and maintain up-to-date all necessary directives concerning HUMINT collection and implement provisions of the DOD Human Resources Intelligence Collection Implementation Plan, 1966, and any subsequently approved DIA or Navy Department directives concerning HUMINT collection.[23]

The Washington Field Activities Support Center, after considerable opposition from the Office of Naval Intelligence, was finally established at Fort Belvoir on 15 November 1968 as a centralized Department of Defense HUMINT activity. The plan for establishing the support center had been issued by a DIA letter of 15 May 1967, with a target date for commencing operations of 1 January 1968. The move of Naval Intelligence Command's HUMINT Resources, Programs, and Coordination Branch to Fort Belvoir took on the dual role of Naval Element, WFASC. In such a detached location, with poor communication and transportation facilities by which to maintain necessary contact with production analysts, and with marginal support from, and no real collaboration between, the services, the support center was ineffective and was disestablished on 9 January 1970.[24]

Chapter Notes

1. Pryor memo to Director of Naval Intelligence (DNI), 23 Jun 1939.

2. OP-16-Z, "History of OP-16-Z, 17 Jun 1940–31 Oct 1943," MS, 1–3, OA.

3. Ibid., 10–11.

4. Capt. Frank Klaveness memo, 27 Jan 1975, OP-16-FN papers in author's files.

5. Intelligence Liaison Officer, Shanghai ltr 48–77, 23 Nov 1948, box 1, Record Group 10430, OA.

6. *Washington Post*, 26 Apr 1953, 1, 4.

7. Commander Seventh Fleet (COM7THFLT), *Report of Korean War Operations, 28 Mar 1951–3 Mar 1952*, 24.

8. Ibid., B4, B8.

9. Commander in Chief, Pacific Fleet (CINCPACFLT), *Interim Evaluation Report No. 4: Korean War Naval Operations*, 5–24.

10. COM7THFLT, *Report of Korean War Operations, 20 May 1952–27 Jul 1953*, 29.

11. CINCPACFLT, *Interim Evaluation Report No. 6: Korean War Naval Operations*, 5-128, 5-129.

12. CNO (OP-922H4E) ltrs, ser 00078P92 and ser 00079P927, Feb 1955, in author's files, OA.

13. ONI-70-A, *U.S. Navy Intelligence Manual*, 20 Jun 1956, 75.

14. *Naval Intelligence Collection Instructions*, ONI-49-2A, 1 Jun 1960, IV-35.

15. DNI memo, 18 Jun 1965. Taylor must have convinced Secretary Nitze, for the necessary activities were not established.

16. Defense Intelligence Agency (DIA), Human Intelligence (HUMINT) Plan, 1966.

17. SECNAV Notice 5450, ser 249, 15 Mar 1966.

18. Report of HUMINT Task Force, 1 Dec 1967, encl. (1), 4, in author's files.

19. Naval Intelligence Command (NIC), NIC-352, *Command History*, 16 Feb 1970.

20. Report of HUMINT Task Force ltr of transmittal.

21. Ibid., encl. (1), 4–6.

22. DNI, "Report to CNO on Major ONI Programs," 20 Jun 1968.

23. NIC-352 ltr, ser 0126-68, 9 Nov 1968.

24. NIC memo, 19 Jan 1970.

CHAPTER 9

Translation Services

From the Origin of ONI to World War II

In 1800, President John Adams wrote to the first Secretary of the Navy, Benjamin Stoddert, showing an early recognition of the importance to the Navy of information contained in foreign documents. He advised his Secretary to estabish a library that would include the best writings on all aspects of naval science and theory as well as biographies of those foreign admirals most skilled in naval combat.[1]

The importance of foreign documents to the Navy was still recognized when the Office of Naval Intelligence was established in 1882. Department of the Navy General Order 292 directed that the new office be combined with the Navy Department Library to facilitate its work in collecting and recording naval information.[2]

One of the earliest forms of collection used by ONI was the translation of foreign books and periodicals obtained by the Navy Department Library and the various ships, bureaus, and offices of the Navy. Lt. Theodorus B. M. Mason, the first head of the Office of Naval Intelligence, was an accomplished linguist and well aware of the wealth of information on foreign naval developments available in open literature published in foreign languages. The officers initially assigned to assist Lt. Mason, however, were not necessarily proficient translators and had to exploit the publications, word by word, using foreign language dictionaries.[3]

The first appropriation bill passed by Congress to specifically mention funds for the Office of Naval Intelligence was for Fiscal Year 1900 and authorized the employment of one translator at $1,400 per year.[4]

By 1902, translations of foreign language documents were being regularly made by ONI for the Office of the Secretary of the Navy, the Bureau of Navigation, and, when required, for other bureaus of the Department of the Navy. The volume of the work was reportedly considerable.[5]

The ONI organization in 1918 contained a Translating Section (OP-16-E) that was charged with translating intelligence documents from French, Italian, Spanish, Portuguese, Russian, Dutch, Japanese, Chinese, and German. It also had to clip, file, and distribute certain foreign newspapers and periodicals received by ONI.[6]

As a result of U.S. failure to ratify the Treaty of Versailles, captured documents on the German design of warships, armaments, munitions, torpedoes, mines, wireless apparatus, and related naval war material were not made available to the United States by the Allied powers. In consequence, Capt. Walter R. Gherardi headed an American delegation sent to Germany in January–February 1919 to gather whatever information and documents it could find. Several similar missions were sent to Germany and Central Europe following World War I, and most of them included naval officers. The foreign-language material had to be translated later at the Office of Naval Intelligence.[7]

In 1930, it was recorded that the Translating Section was unable to keep up with the amount of translation work requested.[8]

By 1932, the Translating Section had three translators: Eva M. Smith, principal translator, in charge; Mary P. Stevens, senior translator, stenographer, and typist; and Johanna Boernsen, senior translator. The work of the section consisted entirely of making translations into English from French, German, Spanish, Italian, Portuguese, and Dutch, and occasionally from Swedish, Danish, and Norwegian. The translations were made to meet the needs of not only ONI but also the various bureaus of the Navy, the Naval Research Laboratory, the Naval Observatory, and other government departments.

Much of the translation work was in highly technical areas and required the translators to consult with knowledgeable technicians and research technical reference books in order to achieve correct terminology. Lack of standardization of terms among various countries added to the difficulty in translating technical subjects. An additional German language translator was requested in 1932. In 1933, however, the Translating Section (OP-16-A-6) was reduced to two translators; a request was made for another translator, this time for one qualified in French.[9]

The need for additional translators was again expressed in 1934. A total of 1,619 pages had been translated between 1 July 1933 and 30 June 1934, but the backlog was 284 printed pages, plus a number of books, pamphlets, and other records awaiting translation when higher priority workload circumstances would allow.[10]

As of 1 December 1939, the Translating Section consisted of Johanna Boernsen as chief translator and Bluma Karp, Mildred Mervine, Edwin Niggli, and a Miss Grande as translators. The section was part of the Administrative Branch and was still located in the Main Navy Building on Constitution Avenue.

World War II

During reorganization of ONI in mid-1941, the Translating Section became the Translations Unit of the Services Section of the Administrative Branch, and its designator was changed to OP-16-A-4-d. Other translators added since 1939 included Mary Masser, Beatrice Dillon, Mildry Sluth, and H. Pearson Hopper. Language translation capabilities included French, Italian, Spanish, Portuguese, Rumanian, Latin, Greek, Dutch, German, Danish, Norwegian, Swedish, Russian and other Slavic languages, Lithuanian, Finnish, Hungarian, and Hebrew. The Far East Section (OP-16-FE-2) of ONI, essentially an analytical organization, handled Japanese translations. In late 1941 or early 1942, the Translations Unit moved to the new temporary "L-Building," across the reflecting pool from the Main Navy Building.[11]

The mission of the ONI Translations Unit during World War II was to supervise the preparation of all translation work and the servicing of all language problems originating in the Navy Department. Its task was to plan, assign, and supervise the work of twelve translators and three stenographer-typists; maintain liaison with outside agencies that could handle material the section could not; and do occasional interpreting and translating that required an officer to meet prescribed security requirements.

On 4 April 1942, James N. Mosel, a civilian, came to the Translations Unit on a six-month contract to survey its work and to assist in translating. In September, he was commissioned as an ensign, and he assumed the duties of officer in charge in November. Mosel attained the rank of lieutenant by the end of the war.[12]

The first major collection of captured Japanese documents was made in August 1942 when two submarines carried Marine Col. Evans Carlson, Lt. Col. James Roosevelt, and the 1st Marine Raider Battalion to Makin Island to harass the Japanese garrison. The documents, which were brought back to Pearl Harbor, included plans, charts, air defense details on all Japanese-held Pacific islands, and battle orders.[13]

During 1942, Sluth (who had become Mrs. H. Pearson Hopper in 1941) retired and Dillon died, but the Translations Unit acquired two new translators, G. E. Hyde and A. M. Wilson. In 1943, Mervine resigned, and another translator, Gertrude W. Holinger, joined the section, as did a typist, Caroline Crichlow.[14]

The voluminous receipt of German and Italian naval documents started with the occupation of Sicily in 1943. The headquarters of the Italian navy in Sicily was captured before its files could be destroyed and yielded information on the entire disposition of the Italian and German naval forces in the Mediterranean, along with charts of minefields and safe conduct routes.[15]

The capture of the German submarine *U-505* on 4 June 1944 provided code books, logs, and tactical publications to be translated in addition to the hardware and weapons of a complete submarine.[16]

Sunken Japanese ships provided large quantities of documents, many of them of immediate as well as historic value. The heavy cruiser *Nachi*, which was sunk in Manila Bay in November 1944, provided a major haul of annotated charts of minefields and defenses, diaries, logs, blueprints, fleet operation plans and orders dating back to before the Pearl Harbor attack, and numerous books on Japanese naval tactics and doctrine.[17]

During the period that the Translations Unit was located in L-Building (1942–1944), about one-third of the office production consisted of translations of letters to and from Navy personnel for the Censorship Branch. It was an intolerably heavy load, and Ens. Mosel, the officer in charge of the unit, after repeated attempts, managed to have the task diverted to the General Censorship office in New York. Another major task, which took up about 10 percent of the unit's time between 1942 and 1945, was a translation of the German War Law that had been requested by the Foreign Intelligence Branch (OP-16-F). The task was never completed because of the pressure of more important work.[18]

In April 1945, the name and designator of the office were changed back to the Translating Section (OP-16-A-6).[19]

While the Translating Section was in the Steuart Building at 5th & K Streets, N.W. (where it had moved in November 1944), Dr. Francis R. Preveden joined the staff. Toward the end of World War II, he was also giving instruction to a rotating group of graduates of the Navy Russian language course at Boulder, Colorado, who were assigned to the section for temporary duty under instruction. The training continued into 1946 until the Boulder school closed.[20]

During the first six months of 1944, approximately 130 large cases of Japanese documents had been received by ONI from the Joint Intelligence Center, Pacific Ocean Area. In addition, ONI's Far East Section (OP-16-FE) received many documents for translation from Japanese into English from the Hydrographic Office, the Naval Research Laboratory, the various Navy bureaus, and other offices. The documents had been picked up on the captured islands of the Pacific and included blueprints of Japanese equipment, charts, logs, war diaries, field manuals, and code books. The backlog of untranslated material accumulated so rapidly that it was necessary to have approximately twenty recent graduates of the Navy School of Oriental Languages ordered to the Office of Naval Intelligence in May 1944 for temporary duty to work on translating the materials.

In September 1944, thirty more language officers, mostly WAVES (Women Accepted for Volunteer Emergency Service), were assigned to permanent duty in the Translation Unit of OP-16-FE. By February 1945, the unit consisted of eighty-one officers, nine enlisted personnel, and five civilians. Even these personnel were insufficient in number to keep up with the task of processing, translating, evaluating, and disseminating captured Japanese documents.[21]

The Washington Document Center (WDC), a joint service center for processing Japanese documents, was placed under the Director of Naval Intelligence as the result of a proposal made by the Japanese Document Conference commencing 28 December 1944. Upon official approval, the Director of Naval Intelligence established the WDC as OP-16-WDC in a letter dated 14 February 1945. The WDC office was located on the fifth floor of the Steuart Building.[22]

Although the Translation Unit of OP-16-FE and the WDC were concurrent occupants of the Steuart Building in the latter part of World War II and both were involved in the translation of foreign documents, they were not combined organizationally. Apparently, there was some effort by the WDC to do so, but when the Director of Naval Intelligence took over WDC, the effort ceased.[23]

During the period between 4 March and 21 October 1945, the WDC received, processed, and disseminated 146,324 Japanese documents ranging from calling cards to encyclopedia sets.[24]

On 23 June 1945, the Director of Naval Intelligence assigned eight officers, nine yeomen, one analyst, and two clerk-stenographers to the Captured German Document Center, run by the Army, to help with the sudden influx of German documents captured prior to the official German surrender on 8 May 1945.[25]

A second Japanese Document Conference, convened on 29 August 1945, proposed the consolidation of the translation sections of the Pacific Military Intelligence Research Section and ONI's Far East Section with the WDC and the establishment of an advanced echelon of the WDC in Japan. The first component of the advanced echelon arrived in Japan in November 1945 and was composed of Army and Navy specialists familiar with Washington interests to ensure that the documents had significant intelligence value.[26]

During World War II, the translation units of ONI prepared translations from twenty-two foreign languages into English on a variety of naval and technical subjects. Sixty percent of the work was for bureaus and offices of the Navy other than the Office of Naval Intelligence. It was estimated that the translating of correspondence alone could have kept three translators fully occupied. For additional information on World War II foreign document collection and exploitation, see Chapters 11 and 17.[27]

Korean War to the 1970s

Shortly after the start of the Korean War, Lt. James Mosel, who had been head of the Translations Unit during most of World War II, was recalled to active duty to serve again as its head. He immediately instituted a change in policy, making each translator responsible for the quality and correctness of his or her own products. Prior to the change, the chief translator had reviewed each product; this practice had had an adverse effect on productivity and morale. Mosel's change remedied the problem. In May 1951, the Translations Unit moved from the Pentagon to Building 52 at the Naval Observatory on Massachusetts Avenue, N.W.[28]

The handling and distribution of captured documents during the Korean War was controlled by directives issued by the Army's General Headquarters, Intelligence Section (GHQ-G2). No provision was made for the distribution of documents of naval interest to naval commands. The lack of Korean linguists in the Navy and the scarcity of documents of value to the Navy, however, made this deficiency in procedures unimportant.[29]

In 1955, much of the work of the Translations Section (by then OP-923M4) was in the translation of technical documents. The section was co-located with the Technical Intelligence Section (OP-922F2) at the Naval Observatory.[30]

Dr. Preveden, the senior special translator in the mid-1950s, worked in approximately twenty-seven languages and was learning others. Preveden introduced the dictation system of translating, using a dictaphone with a wax cylinder. Each cylinder was capable of taking only about two minutes of dictation and had to be scraped down between each use. Even that crude equipment permitted a dramatic increase in the productivity of each translator.[31]

In December 1957, the Translations Section of the ONI completed the translation of *The Soviet Russians as Opponents at Sea*, an analysis of German and Russian naval operations in World War II, prepared for the U.S. Navy by a group of former German naval officers under German VAdm. Friedrich Ruge. The 300,000-word study provided a baseline for all subsequent studies on the operational developments of the Soviet navy since World War II and was later published, in a greatly condensed form, by the U.S. Naval Institute.[32]

The Naval Reserve Translation Program was inaugurated in October 1959 for the purpose of using the foreign language skills of Naval Reserve personnel. Originally, eligibility for the program was limited to Intelligence Reserves, but the scarcity of linguists led to authorizing eligibility for all Naval Reserves for the translation program about two years later. The program permitted reserves to earn retirement points by doing translation assignments at home. As of 1976, there were sixty-five officers and five enlisted personnel participating.

In November 1959, one of the ONI translators, P. Thomas Koines, completed a two-year project for the Civil Service Commission (CSC) in helping to revise the CSC standards for translators and in preparing examinations in the Greek and German languages.

When the Naval Intelligence Command was established on 1 July 1967, the Translations Section was renamed the Translation Division (NIC-15). The elevation to division status and the concurrent authorization to hire four additional translators permitted setting up a more rational organization that was also more responsive to the Navy's needs. It enabled the division to hire persons with the experience needed to conduct an active, coordinated foreign document exploitation program.

Beginning in about 1968, the Translation Division started maintaining a list of private individuals with unique language capabilities who were willing and had the spare time to perform translation tasks and interpreter assignments. Individuals on the list were contacted to perform tasks or assignments that were beyond the linguistic or staff capabilities of the division. For example, in July 1970, one served as an interpreter for the meeting of the U.S. and Spanish negotiators of the Spanish Base Rights Agreements.

In January 1970, the Translation Division initiated regular publication of translations of selected excerpts from the monthly Soviet *Naval Digest (Morskoy Sbornik)*.

When the Office of the Chief of Naval Operations received messages in a foreign language, usually for the Secretary of the Navy or the Chief of Naval Operations and consequently requiring prompt distribution, the Communications Duty Office would phone the Translation Division for an oral translation. The prompt response of the division on each and every occasion induced the Assistant Chief of Naval Operations for Communications and Cryptology on 25 March 1970 to express his "most sincere appreciation" to the Commander Naval Intelligence Command for the division's proficient assistance.

Although not strictly a function in the intelligence field, the Translation Division was called on during the Vietnam War to produce manuals in Vietnamese to go with the naval equipment and weapon systems turned over to the South Vietnamese navy. For example, in 1972, several such manuals were produced for the PADD Sonar System at the request of the Naval Ordnance Laboratory at White Oak, Maryland.

In another example of assistance to other bureaus of the Navy in 1972, the Bureau of Personnel (BUPERS) requested and received prompt translation services in the production of questionnaires for use in surveying Icelandic nationals who worked at U.S. Navy facilities at Keflavik. The survey was needed for the BUPERS Intercultural Relations Program in order to determine how to improve interpersonal relationships between host nationals and U.S. naval personnel at overseas bases.

On 8 May 1972, at 1300, the Translation Division was tasked to prepare, in camera-ready copy, a warning to shipping regarding the mining of Haiphong harbor in North Vietnam. The text of the warning was to be translated into twelve languages and was to be delivered to the pilot of a plane departing at 1100 on 9 May for the West Coast and thence to Vietnam. A frantic several hours by the Translation Division, particularly its division head, Thomas Koines, plus assistance from Voice of America and CIA personnel solicited in the President's name, enabled the deadline to be met.

In September 1972, the Translation Division initiated a regular publication containing selected translations from the Soviet monthly periodical, *Shipbuilding (Sudostroinye)*.

The first translation workshop for Naval Reserve personnel proficient in Russian was conducted by William Cramer of the Translation Division at Naval Air Station, New Orleans, on 20–31 October 1975. Cramer had become the head of the Naval Reserve Translation Program in December 1970. Funding problems kept participation down to four. The enthusiasm of those attending, however, made the workshop a success. Cramer taught translation techniques that would help make the participants more productive and competent in their translation program efforts. A second workshop was conducted by Cramer on 14–25 June 1976 at Naval Air Station, South Weymouth, Massachusetts, again for reservists proficient in Russian; there were seven participants.[33]

The following officers and civilians headed ONI's Translation Section from 1942 to 1976:

Name	**Date***
Ens. James N. Mosel	Nov 1942–Mar 1946
Lt. S. Frank	Mar 1946–May 1946
Johanna Hensoldt (Chief Translator)†	May 1946–Sep 1950
Lt. James N. Mosel	Sep 1950–1952
LCdr. P. A. Wadsworth	1952–1954
Lt. H. Feeney	1954–Jun 1957
Lt. Robert B. Bathurst	Jun 1957–Jun 1960
Lt./LCdr. R. L. Muros	Jun 1960–Apr 1963
Lt.(jg) G. A. Lillquist	Jun 1963–Jun 1964
P. Thomas Koines	Jun 1964–[34]

*Approximate. Based on available rosters.
† Formerly known as Johanna Boernsen; she had been in the Translating Section since at least 1932.

Table 9.1.
Title and Organizational Designator of ONI's Translation Sections

Title	**Designator**	**Date***
Translating Section	OP-16-E	World War I
Translating Section	OP-16-A-6	1933
Translations Unit†	OP-16-A-4-d	Jul 1941
Translating Section	OP-16-A-6	Apr 1945
Translating Section	OP-23C4	Oct 1945
Translating Section	OP-32C4	1 Aug 1946
Translations Unit	OP-323M4	1 Oct 1948
Translations Section	OP-923M4	1 Jun 1954
Translations Section	OP-923M2	Oct 1964
Translation Division	NIC-15	Jul 1967
Translation Division	NIC-15/STIC-034	Jul 1971
Translation	STIC-34/NIC-15	1 Jan 1972
Translation Services Division	NISC-62	1 Jul 1972
Foreign Languages Services Office	NIC-00S3	—

*Dates in most cases are approximate and are based on when changes were first noted in available rosters.
† The Far East Section (OP-16-FE) also worked on Japanese translations.
Source: ONI personnel rosters

Table 9.2.
Locations of the Translation Section

Locations	**Date***
Corcoran Court†	1917
Main Navy Building, Constitution Ave.†	Sep 1918
L Building, Mall†	1942
Steuart Building, 5th & F Streets, NW†	Nov 1944
Main Navy Building†	Sep 1946
Pentagon†	Dec 1948
Naval Observatory†	May 1951
Malvern Building, Alexandria, VA	Mar 1967
Naval Security Station, Nebraska Ave., NW†	Jun 1968
Hoffman Building, Alexandria, VA	Jun 1969
NIC Building 1, Suitland, MD	——

*Dates in most cases are approximate, based on where changes were first noted in available rosters.
† Washington, DC.
Source: ONI personnel rosters.

Chapter Notes

1. John Adams, *The Works of John Adams* (Boston: Little Brown, 1854), 9:47.

2. General Order 292, 23 Mar 1882.

3. RAdm. Albert G. Berry, USN, "The Beginning of the Office of Naval Intelligence," *USNIP*, Jan 1937, 102–3.

4. *SECNAV Annual Report, 1899*, 464.

5. *SECNAV Annual Report, 1902*, 95.

6. ONI Office Organization Chart, 1 Sep 1918, 24.

7. Department of the Navy, *U.S. Naval Activity Relative to the Armistice of 1918 and the Peace Conference of 1919* (Washington: GPO, 1919), 17, 26.

8. Capt. Herbert E. Cocke, USN, "History of ONI," MS, Office of Naval Intelligence, 1931, 16.

9. OP-16-A, *Annual Reports, 1932*, 13–14; *1933*, 15.

10. OP-16-A, *Annual Report, FY 1934*, 13.

11. ONI rosters; and H. Pearson Hopper memo to author, Jan 1978.

12. Department of the Navy, "Administrative History of the Office of Naval Intelligence in World War II," 10 Jul 1946, unpublished MS, 137–38, hereafter *ONI WWII Admin History*.

13. Capt. Ellis N. Zacharias, *Secret Missions: The Story of an Intelligence Officer* (New York: Putnam, 1946), 318.

14. Hopper memo, Jan 1978.

15. *ONI Quarterly*, vol. 2, no. 3, 1958, 11–13.

16. Philip K.Lundeburg, "American Antisubmarine Operations in the Atlantic, May 1943–May 1945" (Ph.D. diss., Harvard University, 1953) chap. 10, 29–32.

17. *ONI Review*, Nov 1945, 41.

18. Hopper memo, Jan 1978.

19. ONI Organizational Diagram, 6 Apr 1945.

20. Ibid.

21. *ONI WWII Admin History*, 716–17.

22. Ibid., 894–95.

23. Hopper memo, Jan 1978.

24. *ONI WWII Admin History*, 897–98.

25. Ibid., 683–84.

26. Ibid., 897.

27. Ibid., 138.

28. P. Thomas Koines to author, notes on Translating Section, 3 Jan 1977.

29. Commander in Chief, U.S. Pacific Fleet (CINCPACFLT), *Interim Evaluation Report No. 2: Korean War Naval Operations*, 1745.

30. W.E.W. Howe to author, notes on the development of the Scientific and Technical Intelligence Center (STIC), undated.

31. Koines and Howe notes.

32. Koines ltr to author, Nov 1976; and Friedrich Ruge, *The Soviets as Opponents at Sea* (Annapolis: U.S. Naval Institute, 1977).

33. Information on translating services, 1957–1976 in Koines ltr, Nov 1976.

34. ONI personnel rosters.

Chapter 10

Production

This chapter deals mainly with the Navy's processing and production of intelligence, including participation in joint service production efforts. The emphasis is on the products rather than on the processes of production. More specific discussions of processing appear in Chapters 11 through 20.

Beginnings

The first steps taken by Lt. Theodorus B. M. Mason in 1882 to give the Office of Naval Intelligence a capability to produce intelligence included a determination of customers' needs and the development of a filing and indexing system that would most readily meet those needs. ONI's primary clients were the Secretary of the Navy, the Navy Department bureaus, and the Congress of the United States. Their intelligence needs were almost exclusively for information from abroad that would provide guidance in rebuilding the fleet.

Many of the Navy bureaus had been collecting intelligence information from world naval powers to meet their own technical requirements. Lt. Mason, with the Secretary of the Navy's concurrence, assembled the needed information in the Office of Naval Intelligence and correlated it with that held by the Navy Department Library. Many of the foreign books and periodicals in the library were in foreign languages, and one of the earliest collection and production techniques was the selection and translation of foreign publications judged to be authoritative and containing information on foreign navies.

Early in the history of ONI, the balance between collection and production capabilities became a problem—one that has remained to this day. The mass of uncorrelated material already on hand was beyond the processing capability of ONI's limited staff and probably was the chief deterrent to the deployment of more than one naval attaché in 1882 to collect more information.

The first publications produced by the Office of Naval Intelligence were a War Series called *Information From Abroad*, which ran through only four numbers:

No. I, 1883 *Operations of the French Navy During the Recent Wars in Tunis*

No. II, 1883 *The War on the Pacific Coast of South America Between Chile and the Allied Republics of Peru and Bolivia, 1879–1881*

No. III, 1885 *Report of the British Naval and Military Operations in Egypt, 1882*

No. IV, 1893 *The Chilean Revolution of 1891*[1]

The General Information Series (unclassified) ran through 21 numbers from 1883 to 1902 and included the highly regarded and much-used annual *Notes on Naval Progress*. (See the end of this chapter for a complete list.) In January 1902, the *New York Sun* reported praise for *Notes on Naval Progress* by a correspondent of the London *Times*:

> The Admiralty conceals its knowledge even from the House of Commons. . . . When the Parliament insists on obtaining a return on the fleets of the Powers, the bare return is given without any attempt at summarizing the results, or any endeavor to make the information of practical use for purposes of discussion. We have to go to the American Naval Intelligence to obtain a summary on this information.[2]

Publication of the General Information Series was discontinued in 1903 so that the ONI staff might occupy their time in more important work. It was the only unclassified publication produced by ONI at that time. The last issue was No. XXI, *Notes on Naval Progress in 1902*, which was mainly of interest to members of Congress. On 15 April 1902, the Senate had passed a joint resolution providing for the printing of an edition of future volumes of the General Information Series for use by Congress. Apparently, the resolution did not pass the

House Joint Committee on Printing, resulting in the termination of the series.[3]

President Theodore Roosevelt, in a note dated 2 October 1902 to the Secretary of the Navy, had stated: "The naval intelligence report is very interesting, but in my opinion altogether too bulky. The mere bulk of any document of that nature is against it." Roosevelt's Secretary amplified the complaint in a note of 6 October:

> He (the President) thinks there is much useless matter and a large number of unnecessary and expensive illustrations included in many of the reports and documents published; that many are issued at great expense which accomplish no practical good and that there is too much public printing generally.

The comments referred to the latest *Notes on Naval Progress*. Chief Intelligence Officer Capt. Charles D. Sigsbee replied to the Secretary of the Navy on 4 October that the next report would be trimmed "to the narrowest limit." On 10 October, in another memo to the Secretary, Sigsbee further explained that the annual report was made up mainly of extracts from foreign publications. Further reports would be condensed, but the extra work required to do so would "be a great tax on the time of the Staff Intelligence Officers." This was probably another factor that contributed to the decision to cease publication of the General Information Series.[4]

In 1888, the office published the first edition of *Coaling, Docking, and Repair Facilities of the Ports of the World with Analyses of Different Kinds of Coal*. The series ran through four editions and a supplement. In 1911, the publication was reissued under the title *Port Directory of the Principal Foreign Ports*. A series of Spanish-American War publications called *War Notes*, in eight volumes, was in great demand, and, in 1900, Congress authorized them to be published in one volume with the title *Notes on the Spanish-American War*.

In 1915, the office began to issue a periodic *Information Bulletin* series. In 1918, the series was retitled the *Semi-Monthly Compilation* and was subsequently issued in mimeograph form and, later in the same year, in printed form. In January 1919, the publication was superseded by the *Monthly Information Bulletin*, which was issued through 1941 and then replaced by the *ONI Weekly* in January 1942.[5]

Material prepared by the ONI for publication in reports of other divisions of the Office of the Chief of Naval Operations after World War I included strategic intelligence needed for the Political Situation section of the annual reports submitted by the Director of War Plans to the Chief of Naval Operations (CNO), which were known as Estimates of the Situation and Base Development Plans. The Political Situation section reported on world conditions.[6]

Between the World Wars—Establishing Series Publications

From 1920 to 1939, the Foreign Intelligence Branch of ONI was weak, and personnel shortages hamstrung its efforts. The desks of Section C and the sections (units) of B Branch, which later became F Branch, were largely depositories for information. They received and filed but did not collate or evaluate, and dissemination was intermittent and inconsistent. One officer, with or without clerical help, could not effectively process information on up to twenty-two countries.[7]

Cdr. George McD. Courts, in a confidential letter of 11 April 1931 to the Director of Naval Intelligence (DNI), offered some underlying causes for Navy apathy and the inadequate staffing, and thus ineffective functioning, of the Foreign Intelligence Branch:

> (1) Our geographic isolation and consequent difficulty of visualizing a national menace; (2) the fact that in World War I the problems of wartime intelligence were handled largely by the British; and (3) the non-aggressive character of our national policies toward other countries.[8]

Following World War I, ONI initiated a series of publications, which continued until 1956. The series included reference publications on foreign subjects or areas and guidance publications on intelligence procedures and techniques. Some of the early publications in the series follow.

ONI-1 *British Monograph—Dominions and Colonies*

ONI-2 *Monograph of Brazil*

ONI-3 *Monograph of Mexico*

ONI-4 *Monograph of Japan, 1920*

ONI-5 *British Monograph—British Isles*

ONI-7 *Nicaraguan Monograph*

ONI-8 *Instructions for Intelligence Officers, 1923*

ONI-9 *Chinese Monograph*

ONI-11 *Naval Estimate of Japan*

ONI-12 *Strategic Harbours of the Pacific, 1921*

ONI-12 *Strategic Harbours of the Pacific, 1929*

ONI-13 *Monograph of Japan, 1931*

ONI-16 *Instructions and Orders for Port Guards and Naval Ship Inspectors, 1932*

ONI0-18 *Pamphlet of Information on Cuba*

ONI-19 *ONI Intelligence Manual, 1933*

ONI-19 *Naval Intelligence Manual, 1936 Revision*

ONI-19(A) *Naval Intelligence Manual, 1947*

ONI-19(B) *Naval Intelligence Manual, 1949*

ONI-20 *Monograph Index Guide*

ONI-21 *Extracts from Chinese Monograph, 1934, Change #1*

ONI-22 *Notes on Espionage, Counter-Espionage and Passport Control, 1935*

ONI-23(A) *Outer Hawaiian, Wake and Marcus Islands*

ONI-27 *Monograph of Luzon and Islands off the North Coast*

ONI-29 *Geographic Monograph, Palau and Marianas Islands*

ONI-34 *Monograph of Zamboanga Peninsula, Gulf of Davao, North and South Coasts of Mindanao*

ONI-35 *Cable and Radio Censorship*

ONI-37 *Monograph of Netherlands East Indies, General, Volume I* (1935)

ONI-38 *Monograph of Netherlands East Indies, Volume II*

ONI-39 *Monograph of Aleutian Islands, Volume I*

ONI-40 to -99 More monographs

As of 1923, ONI was engaged in the preparation of monographs on the various countries of the world, concentrating on those in which U.S. national policy had the most immediate interest. The monographs were divided into subject sections as follows:

Section	Subject
100	State
200	Social Conditions
300	Finance
400	Industrial
500	Commerce
600	Communications
700	Army
800	Navy
1000	Port Directory [9]

Concept of Intelligence Production Prior to World War II

In a lecture at the Naval Postgraduate School on 16 February 1929, DNI Capt. Alfred W. Johnson discussed some of the intelligence processing pitfalls:

> Information is inherently such an exceedingly broad thing that it is obviously very necessary that we should exercise the greatest judgement [sic] and discrimination if the value of our output is to be in proportion to the size of our personnel and to the needs of the services. It is a simple matter to collect great quantities of information which no one will ever want. It is also easy to fail to collate valuable items from a mass on its way to cold storage. And after collation has been done, it is not hard to let the "chip-munk" instinct work and hide it away under lock and key where it will never do anyone any good.[10]

Naval intelligence in 1933 included Navy Department (or strategic) intelligence and combat (or operational) intelligence. Navy Department intelligence was defined as

> that produced by ONI in peace and in war. It is the product of a scientific and systematic collection and evaluation of information on the Political, Economic, Social and Psychologic, Military, Air and Naval Forces; and the Geographic Situation of a specified nation, for the purpose of arriving at a definite conception of its naval strength and effort, and an estimate of the probable initial intentions of its naval forces in case of war.
>
> It deals primarily with subjects that are strategical in nature and to a less extent with others that have to do with tactics and logistics.
>
> This intelligence is the knowledge required by the CNO to formulate an Estimate of the Situation from which may be derived basic War Plans; it is likewise essential to the Commander in Chief and subordinate commanders concerned in the formulation of basic campaign plans, or such plans that a particular situation may demand.[11]

Naval combat intelligence was described as

> that produced after the outbreak of hostilities primarily obtained by the naval forces afloat and intelligence agencies operating under orders of the commander responsible for the conduct of naval operations within the designated theatre (Fleet Zone), and secondarily such pertinent information that may be furnished by the intelligence agencies without the limits of this area and the Navy Department.
>
> This intelligence is the evaluated information required by a commander regarding the enemy forces within or approaching the Fleet Zone which will enable him to make timely distribution and employment of the forces under his command.
>
> As a general rule, this intelligence is confined to the location, strength, composition, disposition, movement, tactics, probable intentions, and condition of the enemy forces opposing or likely to oppose our own forces; together with the weather and meteorological conditions in the area of probable operations.
>
> It constitutes a vital element of the Commander's Estimate of the Situation and is essential to the preparation and execution of strategic and tactical plans.[12]

Navy Department intelligence, in time of peace, was divided as follows:

> A. Statistical—for use in Congressional Hearings, making studies on Limitation of Armament and other requirements, and for making comparative studies and estimates of naval strength.

B. Technical—for use of the Bureaus concerned, to prevent "technical surprise" and maintain if possible "technical superiority."

C. Domestic—for familiarity with the domestic situation and to permit the necessary expansion for war requirements in a rapid and systematic manner.

D. Foreign—for use of the CNO and other Naval Commanders in estimating the war capacity and naval power of the various nations, together with their probable intentions, their alliances, treaties and pacts that may effect our own naval policies and plans both in peace and in war.[13]

Evaluation was considered by ONI in 1933 to be the "critical and systematic analysis of enemy information for the purpose of determining its probable accuracy, significance and importance."

Information subjected to the evaluation process became intelligence. The officer who merely transmitted to his commander the information that he had received performed only part of his duty. Naval intelligence was to be concise, free from irrelevant matter, and ready for immediate use. It had to convey the facts and their significance and the deductions to be drawn from a consideration of the facts in connection with other intelligence already at hand.[14]

Lt. Arthur H. McCollum, who had ONI's Japanese Desk in 1933–1934, gathered photographs of Japanese ships and had scale models built from which identification data could be produced in what was apparently a "first" for that technique. Information was also obtained from pilots and engineers by which speed-power curves were devised for many Japanese warships and merchant ships. Information thus derived was published and distributed in classified publications.[15]

Overlapping jurisdiction among ONI desks, units, and sections resulted from the assignment of colonies to desks responsible for the countries to which the colonies belonged. For example, the Central European Desk (B-13) had the Dutch East Indies and Aruba, and the Western European Desk (B-12) had French colonies in the Far East. In the latter case, both B-11 (Far East Section) and B-12 were writing independently to Commander in Chief, Asiatic Fleet (CINCAF) requesting the same information on French colonies in the Far East.[16]

Desk E, British Empire, had to prepare monographs on Great Britain and its dominions, protectorates, mandates, and colonies, which encompassed the whole world and overlapped virtually all the other desks in the Foreign Intelligence Division.[17]

Peacetime chores were assigned to ONI during the late prewar days of the 1930s. For example, when President Franklin Roosevelt appointed Adm. William D. Leahy as Governor of Puerto Rico in June 1939, ONI produced a background study on politics and personalities for Leahy's use. Eunice Willson, one of the civilian analysts in the Latin American Section (OP-16-B-16), prepared the study.[18]

Shortly after then-LCdr. Arthur McCollum returned to ONI in 1939, he took on the project of updating the Japanese ship recognition publications that he had produced during his earlier tour in 1933. The work had been so secret that it took six months to locate the original ship models, and McCollum found that the previous modelmaker had died in the interim. McCollum nonetheless persevered, and his new recognition publications, with pictures and updated speed-curve data, were issued in 1941 and proved to be of considerable value to submarine operations after the United States entered the war.[19]

Each foreign section prepared a Daily Information Memorandum for the DNI to meet the demands for information on all aspects of the war in Europe. The procedure started on 5 September 1939 and continued until 29 May 1941, when such production was taken over by a special section.[20]

The *Daily Summary of World Events*, still being produced in 1952, contained items of a timely nature, consolidated and published by ONI's Foreign Intelligence (OP-322F1). No authority could be discovered at that time for the publication of the summaries, but a format had been established for them in 1945 at the request of the Chief of Naval Intelligence. (The daily summaries were continued until December 1955.)[21]

World War II Intelligence Production

Organization and Concepts

In March 1940, a CNO letter to "All Ships and Stations" explained the dissemination of intelligence information by the Division of Naval Intelligence and how the intelligence should be used. There were two general classes of documents: Class A consisted of individual reports, usually on standard report forms but also in letters and tabulations, to meet specific requirements. Class B consisted of Naval Intelligence Bulletins on subjects of general interest to the naval service that were compiled periodically. Class A reports were distributed to heads of bureaus and offices of the Navy Department, to fleets and subdivisions thereof, to naval districts and activities of the shore establishment, and to other government departments and subdivisions. Class B reports were given a wider distribution and sent to all important subdivisions, particularly when documents or pamphlets were classified "restricted" or were unclassified. For Class A reports, recipients were expected to disseminate information within the recipient organization and to other units or activities to which the recipient's interest extended to assure full use and application of the information, as well as to obtain the evaluation of technical information over which they had cognizance. Both the Library of Congress and the U.S. Naval Institute were included in the distribution of the CNO letter.[22]

In 1940, shortly after Germany had precipitated World War II, the German Desk in ONI had one officer, one civilian analyst, and one clerk. There were no functional sections; in other words, no desks devoted to processing intelligence relating to a specific subject or function, such as ships, aircraft, or amphibious warfare. Foreign naval intelligence was organized geographically. Consequently, the three workers on the German Desk had to be ready to answer spot questions and to furnish any studies required about geographic, political, economic, technical, or naval matters involving Germany, Austria, and Scandinavia.[23]

With ONI's limited personnel, contacts between all production levels were direct and personal; when the Director of Naval Intelligence wanted something, almost everyone knew of it at once, and the entire organization at the working level experienced a minor crisis. The simple structure of the intelligence organization tended to accelerate the process of intelligence production.

In June 1940, CINCAF Adm. Thomas C. Hart wrote the Chief of Naval Operations:

> About intelligence etc., we seem never to receive from ONI or other divisions of your office anything in the way of an estimate or evaluation of intelligence concerning the Far East. In fact, about a year ago, when I was in your office trying to inform myself about the situation out here, I was given nothing and told nothing except what was contained in the regular incoming reports. I thought at the time that something in the way of an evaluation, or . . . a distillation, might be a regular function of that part of ONI and might be extremely valuable. . . . I do at least directly request that we be informed of those respects in which our own estimates are disagreed with by your people.[24]

The *Fortnightly Summary of Current National Situations* was started by ONI in December 1940. The summary was to present condensed, broad-view reports about the diplomatic situation in Japan, Germany, France, Italy, Russia, and Latin America; the Japanese military situation; the Japanese naval situation; the Chinese military situation; German military, naval, and air statistics; and Italian naval and air statistics. As problems became greater and the workload heavier, the production sections of ONI were not able to contribute adequate information for the summary, and, in due course, a new section was set up that published the *Weekly Summary*.[25]

The Foreign Intelligence Branch was divided into eleven sections, seven for geographic and political areas, and four topical. Each geographic and political section maintained a monograph on each foreign country assigned to its section. The monographs, in 1940, were divided into eleven main sections: Political Forces, Social Forces, Economic Forces (Finance), Economic Forces (Industry), Economic Forces (Commerce), Cities and Towns Geography, Communications, Army, Navy, Air, and General Summary. The sections were further broken down into subtitles when the volume of material warranted.

Generally, each monograph consisted of one loose-leaf binder for each main title. Theoretically, the eleven binders making up a monograph about a foreign country contained all the evaluated material that ONI had about that country. Secret material required safe storage; material of a lower classification could be kept in a locked file. Some sections set up secret monographs that contained only items classified secret in order to reduce the volume of material requiring safe storage. Under the stress of an increased wartime workload, and suffering from inadequate numbers of personnel, the geographic sections, almost without exception, fell far behind in keeping up their monographs, and material was entered without editing, collating, or summarizing. Thus, the monographs became bulky and unwieldy. Frequently, more than one binder was needed for some of the principal titles.

The four ONI sections handling information and intelligence by topic were Foreign Trade, Special Intelligence, Statistical, and Strategic. The Foreign Trade Section collected and maintained information about cargo movements everywhere, with the exception of Japanese cargoes, which were followed in the Far East Section and in Domestic Intelligence. The Statistical Section compiled information on the strengths of navies and air forces, especially of the United States and Great Britain. The Statistical Section also compiled information on the aircraft production capacities of foreign countries. The Strategic Section gathered data on cities, towns, geographic characteristics, rail centers, communications, industrial developments, etc., and coordinated its work with the geographic sections.[26]

Identification and Characteristics Publications

Before the start of World War II, intelligence about the disposition, characteristics, and appearance of foreign naval vessels, merchant ships, and aircraft was being received and evaluated largely by ONI's foreign intelligence desks. Little information on these subjects had been distributed to the fleets. A few publications (in most cases highly inaccurate, incomplete, and elementary) had been issued about Japanese naval vessels, merchant ships, and aircraft. Two recognition manuals depicting U.S. warships had been distributed by the Bureau of Ships, but nothing was available to the fleets on the units of other nations.

In 1940, the Statistical Section, OP-16-Z, ceased to exist as a separate branch and became OP-16-F-10 of the F Branch. It took over, in addition to an increased volume of statistical work, the preparation of certain elementary publications that would provide U.S. forces with data on the appearance and characteristics of foreign ships, weapons, and aircraft.

By the fall of 1941, requests from the fleet for more information of increased scope, both in the number of countries and in the types of information covered, made creation of a separate section necessary. On 31 December 1941, the establishment of the Identification and Characteristics (I&C) Section, OP-16-F-20, was approved by the Director of Naval Intelligence. Its functions were to collect, evaluate, codify, correlate, and disseminate all available information on the characteristics and appearance of all foreign naval and merchant ships, and to translate design characteristics of U.S. and foreign ships into tables, line drawings, and models from which identification studies could be produced for use by all U.S. armed services.

The concept of a master file drawing of every ship was developed, and it proved of great value as the war progressed. The preparation of the drawings required translating photos, general arrangement plans, inboard profiles, and even prisoner-of-war sketches into highly accurate, carefully delineated plan and profile drawings. From the drawings were developed silhouettes, models, fields-of-fire diagrams, and other devices of tactical value to the operating forces.

Carefully constructed models, as accurate as possible in every detail, were built by the David Taylor Model Basin, professional model builder Van Ryper at Martha's Vineyard, and an expert model maker in the I&C Section. Photos of the models were taken from the various target angles that a submarine, surface ship, or aircraft might find of use in making an approach on the enemy. Photos of the models were provided to Time, Inc., which had a contract to produce identification manuals for the Navy.[27]

The responsibility for the preparation of complete statistical information on aircraft was less clearly defined. The Aviation Intelligence Branch of the Bureau of Aeronautics (BUAER) was preparing performance and characteristics data on foreign aircraft, and the Special Devices Section of BUAER had initiated the drafting of preliminary drawings as a basis for mass-producing training models. In the fall of 1942, an informal agreement on workload distribution was arrived at whereby the I&C Section would prepare basic master file drawings of all foreign aircraft and maintain complete photo files while the Aviation Intelligence Branch of BUAER would be responsible for characteristics data.

It was not until early 1943 that much effort could be expended on the technical aspects of ship equipment, and files were started on enemy guns, fire-control equipment, radar, and similar subjects. That aspect of the I&C Section's activities increased in importance until the Technical Intelligence Center was established in October 1944.

The various Navy technical bureaus (Ships, Ordnance, Aeronautics, etc.) and organizations such as the Office of Scientific Research and Development and the Naval Research Laboratory were all vitally interested in foreign technical development and had collected varying amounts of intelligence data.

Publication of information was undertaken jointly by ONI with other activities—initially with the Bureau of Ordnance (BUORD)—to make information on enemy ordnance collected by ONI and evaluated by BUORD available to all interested activities.

In the March 1943 reorganization of ONI that consolidated the five sections dealing with the preparation of publications, the I&C Section became OP-16-P-2. Its name, functions, and duties remained unchanged. A large percentage of the work of the section depended on the interpretation of photos. For more details on the interrelationship of the I&C Section and the Photo Interpretation Center, see Chapter 13.

Late in 1943, a close tie between personnel in the I&C Section, who were preparing basic drawings and therefore were interpreting photos of aircraft, and people in BUAER, who were preparing statistics on aircraft performance data, was so essential as to require consolidation of the two organizations. The BUAER activity had become a part of the Deputy Chief of Naval Operations (DCNO) for Air (see Chapter 12) and was organizing field teams to investigate and analyze captured and crashed Japanese aircraft. The air element of the I&C Section was transferred to the Air Information Branch of DCNO (Air) on 19 October 1943.

By a Commander in Chief, U.S. Fleet (COMINCH) directive of 13 July 1943, Adm. Ernest J. King directed initiation "of suitable measures to effect close coordination in the Division of Naval Intelligence" of all recognition publications produced by the Navy Department. This action made Naval

Intelligence directly responsible for the coordination of all information, including information from BUAER and the Army, published in the restricted monthly periodical *Recognition Journal*, prepared under contract by Time, Inc.

By 1944, the I&C Section was able to concentrate its attention on the processing and evaluation phases of its mission and to divorce itself from the details of actual publication. The art and layout staff of the section were transferred to the Publications Section.[28]

Recognition material being produced by ONI and available in 1943 included the following:

JAN #1 *Uniforms and Insignia*

ONI-41-42 *Japanese Naval Vessels*

ONI-41-42 *Recognition Supplement: Aerial Views of Japanese Naval Vessels*

ONI-54 Series *U.S. Naval Vessels*

ONI-201 *Naval Vessels of the British Commonwealth*

ONI-202 *Italian Naval Vessels*

ONI-203 *French Naval Vessels*

ONI-204 *German Naval Vessels*

ONI-205 and 235 *Russian Naval Vessels and Military Aircraft*

ONI-206 *Minor European Navies*

ONI-208J *Japanese Merchant Vessels (Revised)*

ONI-208R *Russian Merchant Vessels (Revised)*

ONI-220M *Axis Submarines*

ONI-222 *Statistical Data on Foreign Navies*

ONI-223 *Ship Shapes—Types and Anatomy of Naval Vessels*

ONI-223 K *Warships in Code*

ONI-223 M *Merchant Ship Shapes*

ONI-225 J *Japanese Landing Operations and Equipment*

ONI-226 *Allied Landing Craft*

ONI-232 *Japanese Military Aircraft*

ONI-233 *Italian Military Aircraft*

ONI-234 *German Military Aircraft*

FM-30-30 *Recognition Pictorial Manual, etc.*

FM-30-50 *Recognition Pictorial Manual, Naval Vessels*[29]

Air Intelligence Production

The establishment of an Aviation Intelligence Branch in the Bureau of Aeronautics in September 1941 (see Chapter 12) was the first of several actions taken by individual customers of intelligence to correct deficiencies in ONI's policy of producing general intelligence without regard to the specific needs of specific customers. ONI did not and could not have as complete an understanding of the needs of each customer as an in-house intelligence organization could. Soon after the United States entered World War II, COMINCH and individual operating forces set up or extensively expanded their own intelligence organizations to tailor the intelligence received from ONI and their own resources, according to their specific wartime requirements.

Foreign Intelligence Branch Production

The nature of the activities of the Foreign Intelligence Branch is indicated by the type of information in summaries and statistical tables sent to the Director of the War Plans Division in 1941. The information dealt with existing naval situations for the United States, the British Commonwealth, Germany, Japan, Italy, Turkey, France, and the Netherlands East Indies. The finished intelligence studies included (1) the strength, type, and general distribution of naval forces with expected increases of strength every six months for the next two years; (2) brief estimates of the political, economic, and financial situations insofar as they might indicate an ability to sustain military operations; and (3) statistics about British shipping losses, the amount of shipping available, and the merchant shipbuilding programs of the United States and Great Britain.[30]

Prior to the Allied landings in North Africa in November 1942, one of the big jobs and major accomplishments of ONI's French Desk (headed by Ens. Charles A. Rocheleau) was the continuing analysis of which French naval personalities were located at which bases and aboard which ships, the pro- or anti-Allied views of those personalities, and their anticipated reactions when confronted with the landings. Dr. F. McKechnie of the French Desk spent a lot of time and effort on the study; during the landings it proved to be highly accurate, according to LtCol. Homer L. Litzenberg, Jr., USMC, an intelligence officer at one of the landings.

Other activities of the French Desk at the time of the North African landings included briefing personnel who were about to depart for assignment to naval billets that would be in contact with the French; debriefing personnel returning from the North African landings; and maintaining contact with "Giraud French" and "Free French" naval representatives in Washington.

In connection with the last function, a French naval mission, representing Adm. Henri Giraud and headed by Adm. Raymond Fenard, had been established so that the French could be involved in the rehabilitation of French ships in the United States that had turned themselves in to the Allied forces following the North African landings. The U.S. Navy's liaison officer with the French mission, Lt. Cedric Worth, recognized the intelligence potential of his job and made contact with ONI's French

Desk, thus keeping himself briefed on ONI's requirements. The French Desk was part of the Western European Section (OP-16-F-3), and the chief of the section at that time was LCdr. Norman T. Ball.[31]

On 28 January 1942, the first *ONI Weekly* was issued "for the confidential information of the officers of the United States Fleet." In due course, the organization of the *ONI Weekly* stabilized into two sections: Progress of the War, consisting of events arranged by combat theaters, and Special Articles, covering strategic and tactical subjects, combat information, and historical data. The *ONI Weekly* was published throughout the war, and the sections on the progress of the war provide a excellent reference material. However, because the highest classification used was confidential, some significant events of higher classification were not mentioned.[32]

Combat Narratives

ONI during World War II produced the following Combat Narratives:

Published:

The Aleutians Campaign, June 1942–August 1943

The Battle of the Coral Sea, 4–8 May 1942

The Java Sea Campaign

The Assault on Kwajalein and Majuro, Part I

The Battle of Midway, 3–6 June 1942

The Landings in North Africa, November 1942

Early Raids in the Pacific Ocean, 1 February–10 May 1942 (Marshall & Gilberts, Rabaul, Wake & Marcus, Lae)

Solomon Islands Campaign

Miscellaneous Actions in the South Pacific, 8 August 1942–22 January 1943

Unpublished manuscripts:

"The Navy's Share in the Tokyo Raid"

"Anti-Aircraft Action, 7 April 1943, Guadalcanal-Tulagi"

"The Anzio-Nettuno Landings, January 1944"

"The Capture of the Gilberts"

"Convoy to Gaeta, 1944"

"Guadalcanal & Tulagi Bases"

"Japanese Attacks on Shipping in Guadalcanal-Tulagi Area, 1943"

"The Movement of Supplies into the Guadalcanal-Tulagi Area"

"Operations in the Marianas, Phase I: The Conquest of Saipan"

"Operations in New Guinea Waters"

"The Salerno Landings, September 1943"

"The Mediterranean Convoys"

"Pearl Harbor, 1942"

"Submarine Encounters, 31 August–15 September 1942"

"The Solomon Islands Campaign, Part XIII, Bougainville Operations 1943"

Post–World War II Intelligence Production

Operational Notes

In May 1945, *ONI Operational Notes*, Volume I, Number 1, published as the first issue of an official monthly magazine, was produced by the Operational Intelligence Branch of ONI "for the confidential information and instruction of operational intelligence officers." Publication ceased after the August 1945 issue (Number 4). Many of the articles had been prepared by operational intelligence (OPINTEL) officers recently returned from combat duty, and in the articles they related their experiences in carrying out their OPINTEL responsibilities.[33]

ONI Review

The *ONI Review* was published monthly, commencing with the November 1945 issue. It took the place of the *ONI Weekly*, which was discontinued with the 26 September 1945 issue. The *ONI Review* was to "concentrate on intelligence relating to the armed forces of foreign nations, particularly their naval forces" and from time to time would report "on such diplomatic, political or economic trends abroad as may potentially affect the security of the United States."[34]

The *ONI Review* was published regularly through April 1963. The magazine was classified confidential, and each issue contained six to ten articles on foreign naval subjects or on intelligence activities and experiences. It also carried a section entitled Intelligence Briefs that summarized recently received reports on events in various specific countries. *ONI Review* was published for the information and guidance of officers of the U.S. Navy, Coast Guard, and Marine Corps so they could have the background necessary to interpret intelligence of higher classification when required to do so. In May 1963, the *ONI Review* was combined with the *Army Intelligence Review* and the *Aerospace Intelligence Digest* and was issued by the Defense Intelligence Agency under the title *Defense Intelligence Digest*. It was to "provide all components of the Department of Defense and other U.S. agencies with timely intelligence of wide professional interest on significant developments and trends in military capabilities and vulnerabilities of foreign nations."[35]

In February 1952, because of a trend toward higher classification in the material that might be used, the *ONI Review* requested reader reaction to receiving a more interesting secret-level publication.

The November *ONI Review* reported that, based on 445 survey replies, the decision had been made to retain confidential classification, mainly because a secret classification would inevitably restrict the circulation, and, in particular, would make the magazine less available to junior officers on board ships. Secret supplements were subsequently issued on occasions when the material available warranted doing so.[36]

Effective with the January 1958 issue, the classification of the *ONI Review* was raised from confidential to secret. The action was taken because the increased volume of secret intelligence being received required more timely dissemination. Furthermore, raising the classification eliminated the need to publish the *Secret Supplement* to the *ONI Review*, which had previously been issued quarterly. The objective of the *ONI Review* was to continue as before: "To provide naval intelligence of a general nature for the information and guidance of the officers of the U.S. Navy and Marine Corps."

Completed articles for the *ONI Review* were checked to ascertain that they were releasable to the intelligence activities of the British, Canadian, Australian, and New Zealand navies. Articles determined not to be releasable were to be promulgated to U.S. Navy commanders on a "not releasable to foreign nationals" basis. No other foreign navies were to receive copies of the *ONI Review*.[37]

International Developments of Naval Interest

The *International Developments of Naval Interest (IDNI)* contained brief items prepared by the geographic desks and passed to the Maritime Branch (OP-322N) for editing and publishing. Occasionally, a request was made by OP-322N to other sections for special articles of a timely nature, but usually the items were developed from material in State Department cables. *IDNI* was published from March 1945 to December 1954, but no authority had been found as of 1952 for its publication.[38]

Naval Intelligence Quarterly

The *Naval Intelligence Quarterly* was first published in May 1948, primarily

> for distribution only to those Intelligence Reserve officers, both 1635 and 1355 AI (plus officers of other designators cleared and approved for Telecommunications Censorship billets) who are currently active in the Reserve training program through affiliation with a regularly drilling unit, company or division of the U.S. Naval Reserve. It is for use in the training of such officers and should be given no further distribution or promiscuous display.

It was classified "for official use only" and was published until 1959.[39]

Scientific and Technical Abstracts and Reports

The *Scientific and Technical Abstracts and Reports* originated in 1953, was issued monthly by the Naval Scientific and Technical Intelligence Center until 1967, when it was replaced by the *Naval Scientific and Technical Intelligence Bulletin*. It later became the *Naval Scientific and Technical Intelligence Review*.[40]

Air Intelligence Reports

Air Intelligence Reports were studies produced, starting in 1946, by the Air Intelligence Division (DI/USAF-ONI), a joint Air Force-Navy organization. The division was responsible to the Directorate of Intelligence, USAF for the production of intelligence material of all types. Since it constituted the Air Branch of ONI, it was also responsible to the Chief of Naval Intelligence for intelligence of an aeronautical nature only. Consequently, material contained in the issues of the *Air Intelligence Reports* that dealt with other matters did not necessarily express the views of ONI.[41]

Amphibious Intelligence Studies

In the fall of 1947, the Commandant of the Marine Corps, in a letter to the Secretary of the Navy, emphasized the importance of amphibious intelligence and recommended expanding ONI's organization for the purpose of producing it. In 1948, ONI established an Amphibious Intelligence Section in the F Branch with the mission "to prepare and maintain up-to-date Amphibious Intelligence studies on such areas of the world as OP-32 may direct; and to assemble current information on the amphibious forces of the world and to prepare such reports on these forces as may be directed."[42]

The Amphibious Objective Study (AOS) Program was initiated in 1954 to satisfy the need for collection and publication, in compact form, of the basic intelligence required for planning an amphibious operation in a given objective area. A format was designed, and a prototype study using Vieques Island was distributed to major fleet and Fleet Marine Force commanders for study and recommendation. After receipt of recommendations from the major commands, production of actual studies was undertaken on a limited basis. Two studies were programmed for production during Fiscal Year 1954, and an additional six studies were programmed for the following year.

Selection of objective areas for the Amphibious Objective Studies was made on the basis of existing naval and joint strategic plans and in accordance with the recommendations of the major fleet commands having cognizance over the areas under consideration. Priority of production of AOSs was maintained in an Objective Priority List, subject to modification. Changes in priorities, deletions, and additions were made as influenced by changes in strategic planning or factors affecting the rate of production. AOS numbers were assigned for area identification purposes, and they had no relation to priority. ONI Series 41 numbers were assigned when the study was published.

Because of the lack of funds and personnel and a necessary diversion of facilities to other more important projects, only two AOS studies had been completed by the end of Fiscal Year 1955. These were AOS 1 (ONI-41-1 on the Stavanger area of Norway) and AOS 2 (ONI-41-2 on the Esbjerg area of Denmark). The distribution of AOS 3 (the Do San area of Indochina) was to have been completed soon after (as of 23 August 1955).[43]

Airfields and Seaplane Stations of the World

The *Airfields and Seaplane Stations of the World*, a series of forty-one volumes, was published through a joint effort of ONI and the Directorate of Intelligence, USAF. The first volume was distributed in April 1951. When completed, it covered all areas of the world except the continental United States and contained detailed information on approximately 10,000 airfields and seaplane stations, plus the geographic coordinates and names of nearly 5,000 former airfields and stations. The series was designed primarily for long-range planning and was not distributed to aviation activities at the squadron level or below.[44]

Office of Naval Intelligence Comments

In order to provide area and fleet commanders with an ONI evaluation of significant international developments needed for planning or conducting naval operations, ONI initiated in June 1954 a message series known as *Office of Naval Intelligence Comments*. Proposed items for *ONIC* distribution were to be considered at each meeting of the Daily Summary of World Events Working Group and evaluated continually by the ONI duty officer. Each *ONIC* contained one or more items, and each item consisted of a statement on an important international development followed by ONI's comment or assessment of its significance. *ONIC* dissemination went to Commander in Chief, Pacific; Commander in Chief, Pacific Fleet; Commander in Chief, Atlantic; Commander in Chief, Atlantic Fleet; Commander in Chief, U.S. Naval Forces, Eastern Atlantic and Mediterranean; Commander Naval Forces, Far East; the four commanders of the numbered fleets, First, Second, Sixth, and Seventh Fleets; and the four sea frontier commanders.[45]

Office of Naval Intelligence Bulletins

The *Office of Naval Intelligence Bulletin (ONIB)* was a compilation of intelligence items of current naval interest, with ONI comments thereon, disseminated on a daily basis, Monday through Friday. The *ONIB* editorial board met daily at 0930 to consider items and comments submitted by the various ONI divisions, which were encouraged to prepare appropriate items within their fields of interest. The first instruction concerning the production of the *ONIB* was issued on 13 November 1958. It was revised on 10 November 1959. Items for the *ONIB* were to cover all significant international events and trends that might affect naval policy or operations. Comments were to include remarks as to the significance of the information, an assessment of the validity of the report, and what reactions to expect as a result of the event. *ONIB*s were delivered by hand within the Pentagon and mailed to addressees outside the Pentagon.[46]

National Intelligence Surveys and Estimates

The National Security Council in 1948 issued a directive setting up the *National Intelligence Survey (NIS)* program. The survey program was divided into 103 geographic areas and included all land areas of the globe. In addition, there were five basic oceanographic areas. The NIS program had two phases: (1) initial production according to Joint Chiefs of Staff priorities and intelligence agency capabilities, and (2) continuous maintenance.[47]

The production responsibilities of the Office of Naval Intelligence in the *NIS* program included the following sections: Coastal Defenses, Coastal Weather, Coasts and Landing Beaches, Electronics, Marine Climate and Oceanography, Merchant Marine, Naval Biographies, Naval Forces, Naval Map and Chart Appraisal, Naval Weapons and Science, Ports and Naval Facilities, Sea Approaches, and Shipbuilding. In addition, ONI collaborated with Air Force Intelligence on Civil Aviation and Air Forces. The U.S. Navy Hydrographic Office and the Naval Weather Service supported ONI, as appropriate, in meeting ONI's production responsibilities.[48]

The Director of Central Intelligence was responsible for coordinating *NIS* production and maintenance and for accomplishing the editing, publication, and dissemination of the surveys.[49]

The main *NIS* sections for which ONI was responsible follow:

Section 22, Coasts and Landing Beaches. A study of the near-shore oceanography of the coastline areas with data on the water depth, type of bottom, current direction and velocities, tide and swell, bioluminescence, icing conditions, etc. was produced by the Hydrographic Office for inclusion in this section. The Coast and Landing Beach Unit of ONI made up a detailed topographic study of the entire coastline and indicated the physically most suitable landing areas. The areas were then analyzed as to conditions of soil, width and length of beach, adjacent coastal topography, land communications, relationship to strategic areas, etc. Contributions to the studies, sometimes substantial, were received from the Beach Erosion Board of the Army Corps of Engineers. The Navy's Photo Interpretation Center also contributed valuable details to the Coasts and Landing Beaches Section for the surveys.[50]

Section 35 and Supplement I, Ports and Naval Facilities. This section was of almost universal interest to the armed forces and civil agencies concerned with foreign areas. Inherently a component of logistics, port intelligence had its greatest use in logistics plans and estimates and in all aspects of movement, basing, and supply. A great part of the ONI production effort in connection with the project was devoted to the collection and development of intelligence on seaports and naval facilities. The exhaustive detail found in the Ports and Naval Facilities section of the *National Intelligence Surveys* was in sharp contrast to the more generalized coverage of other topics in the surveys. Because of the comprehensive coverage and the static nature of much of the information, the Ports and Naval Facilities section was of value at all intelligence levels—national, departmental, and operational.[51]

Section 36, Merchant Marine. This section compiled an authoritative list of merchant ships of 1,000 or more gross tons owned by the country covered by the survey. For each ship listed, the following characteristics were given: present and former names, type of ship, gross tonnage, deadweight tonnage, passenger capacity, year built, country in which built, number of crew; types of any special equipment, length, beam, depth (i.e., moulded depth), draft, speed, type of engine, type of fuel, and name of owner.

Section 64E, Shipbuilding. This section was coordinated by the State Department as a part of the economic analysis of the nation under survey. It was a detailed analysis of shipbuilding capabilities and activities.

Section 82, Naval Forces. This section was the contribution by ONI to the chapter of the *National Intelligence Survey* dealing with the entire military establishment of the country under survey. The objective was to present a thorough study covering all phases of the Navy, with an interpretive analysis of those factors that might have an effect on the international military-political scene; in other words, how much assistance it could render as an ally or how much of a threat it would be as an enemy.[52]

It was recognized that the information in the *National Intelligence Survey* on naval forces could not be a final authority for justifying military decisions because new intelligence was constantly being received that would modify to a greater or lesser degree the strategic significance of the *NIS* information. The surveys, however, were intended to serve as points of departure for planning. In addition, full strategic details and estimates in the area covered by the Naval Forces section required higher security classification than was allowed in the surveys.

The initial *NIS* production schedule called for six areas to be completed in the first year, twelve in the second year, and twenty-three in each succeeding year. When only two areas were completed in the first year, the maximum production goal was lowered in 1949 from twenty-three to fifteen per year.

In 1950, the Joint Intelligence Group, speaking for the JCS, prescribed that *NIS* production be stabilized at eight per year. Secretary of Defense Gen. George C. Marshall authorized the Army, Navy, and Air Force to augment their intelligence staffs sufficiently to produce eight *National Intelligence Surveys* per year "on a regular basis undiminished by fluctuations in the world situation."

In 1951, the Director of Naval Intelligence reported to the Chief of Naval Operations that 135 additional personnel would be required in the three Navy components involved in *NIS* production if Secretary Marshall's requirement was to be met. However, in the years 1953 through 1957, the production goal of eight *National Intelligence Surveys* per year was met by a smaller increase in the number of personnel assigned and by judicious use of the available resources.[53]

As of 1956, the *National Intelligence Survey* program had become the largest single production effort by the United States intelligence community. More than 3,200 *NIS* sections had been produced, covering 61 percent of the world. Production efforts had concentrated on JCS-designated high-priority areas; they were 90 percent completed. Of the 3,200 sections, the Navy had produced approximately 450, of which 64 had been completed in the past twelve months. Navy contributions had also been made to 100 sections and subsections being coordinated by other agencies.[54]

National Intelligence Estimates began to be issued in the mid-1950s and primarily served the needs of U.S. national policy planning in the sphere of national security. They represented the coordinated judgments of the Intelligence Advisory Committee (IAC) agencies; the Central Intelligence Agency (CIA); Department of State; the Army, Navy, and Air Force; the Intelligence Staff (J-2) of the Joint Chiefs of Staff; the Atomic Energy Commission; and the Federal Bureau of Investigation.[55]

Special National Intelligence Estimates were specialized estimates that usually involved scientific or economic problems. They were prepared by appropriate permanent subcommittees of the Intelligence Advisory Committee, which included representatives of each cognizant IAC member agency.

Intelligence Production Concepts in the Cold War Era

On 1 June 1949, CNO letter serial 001283P32 enunciated the Navy's "Primary Intelligence Objectives" and included the statement that production of intelligence to satisfy the primary intelligence objectives "should support the assigned missions and tasks of the U.S. Naval Service." In response, ONI Instruction 00382.22/32 gave as ONI's objectives the requirements to determine the following:

1. The intentions, capabilities, and indications of the USSR and its possible allies of making war on the U.S. or friendly nations.
2. The tactical deployment of the armed forces or agents of potential enemies as they affect the missions of the U.S. Naval Service.
3. Present and future capabilities and means in the possession of possible hostile nations to inflict damage upon the naval forces or installations of the U.S.; including most likely time, place, and effectiveness of such damage.
4. The present and future capabilities of possible hostile nations to defend their forces, homelands, and areas of interest against attack by naval forces.
5. The intelligence to support the strategical and tactical plans that would be required by the U.S. Naval Service to wage war against all potential enemies.
6. The climatic, geographic, topographic, and other intelligence needed to support naval operations in possible theaters of action.
7. The present status, trends, and future development in scientific and technical fields as they affect the U.S. Naval Establishment.
8. The activities of foreign, or foreign-sponsored, individuals or agencies in the field of present and future actions in subversion, espionage, and sabotage which may affect the accomplishment of the missions of the U.S. Naval Service.
9. Internal political and economic developments of the USSR and possible allies which may affect the accomplishment of the missions of the U.S. Naval Service.

With the start of the Korean War on 24 June 1950, the volume of information reports increased markedly without an increase in ONI processing personnel. The situation caused considerable delay in evaluating, reproducing, and disseminating intelligence reports. To alleviate the situation, reproduction and dissemination of Intelligence Reports by ONI was carried out without waiting for evaluation, which was disseminated separately at a later date when warranted. A project control system was set up in September 1950 to ensure that the efforts of available processing personnel were applied to tasks consistent with the priorities of customer requirements. A project control element, consisting of two officers and two enlisted personnel, was assigned the mission of coordinating and controlling the production of intelligence, under the immediate direction of the head of the Intelligence Branch (OP-322). The system was analogous to the work-request–job-order system long in use to control and direct work in naval shipyards and repair activities throughout the Navy. It worked very well for ONI.[56]

As of August 1952, missions of OP-322 included informing responsible U.S. officials of the warmaking capabilities of foreign nations, primarily the foreign naval establishments, and providing planners and operators with the best evaluated intelligence for use in strategic planning and for conducting operations.

The priorities, of necessity, fell into two categories: products required by higher authority and assigned a deadline by that authority, and products initiated by OP-322 in the course of carrying out the production responsibilities of the Intelligence Branch. Under the first category were (1) National Intelligence Estimates required by the National Security Council and coordinated by the CIA; (2) Department of Defense estimates and requirements of the Joint Chiefs of Staff coordinated by the Joint Intelligence Committee; (3) NATO (North Atlantic Treaty Organization) estimates and requirements requested by Supreme Allied Commander, Europe and Supreme Allied Commander, Atlantic and coordinated by the NATO standing group; and (4) Naval estimates and requirements requested by the CNO (Director of Strategic Plans) and coordinated by the Navy staff. The requirements levied on the Director of Naval Intelligence were assigned completion dates by higher authority and had equal priority. Normally, there was no discretion granted to the DNI in the assignment of production priorities.[57]

When RAdm. Carl F. Espe was DNI (1952–1956), he enunciated the policy that, where special initial processing techniques, equipment, and training were required, it was better to have activities outside of ONI perform the processing if they already had the capability to do so. An example of such processing was the Naval Photographic Laboratory's initial photo interpretation work. It was ONI's responsibility, Espe believed, to correlate the products as required by the intelligence customers.

A major weakness in Espe's policy was demonstrated when ONI tried to use Bureau of Ships laboratories to perform initial processing for acoustic intelligence (ACINT). Having no particular interest in, or understanding of, the significance of their output for ONI, the laboratories used little or no initiative to exploit the raw data completely or on a timely basis, and they found no justification to expand their processing capacity or to juggle their priorities in order to give preferential treatment to work for others. Espe's policy was probably sound enough, however, when the outside activity or agency was responsible for the collection of the raw material and would, therefore, have had an interest in its initial processing.[58]

Information about the capabilities, vulnerabilities, and probable courses of action of foreign nations was not always considered to be "strategic" intelligence, particularly if it had no specific or immediate military application, or if its military application was only secondary or remote. In such cases, it was called "national" intelligence and was distributed at the highest government levels to those charged with formulating national policy. Strategic intelligence in 1954 was viewed as encompassing knowledge of the capabilities, vulnerabilities, and probable courses of action of foreign governments, as seen from the viewpoint of those charged with planning and executing national security measures in peacetime and military campaigns in wartime.[59]

The ONI Publications Review Committee, established sometime prior to 1956 (referred to in ONI Instruction 5600.6 of 18 April 1956) was designated in January 1960 to carry out the policy of the Secretary of the Navy that nonessential publication and printing be eliminated. The functions of the committee included conducting an annual survey of all existing ONI publications; making recommendations to the Director of Naval Intelligence on their continuance, revision, consolidation, or cancellation; and examining all proposals for new publications and approving or disapproving them based on need, duplication, compliance with policy, and availability of funds.[60]

ONI Series

In the early 1950s, the ONI-32 series of publications was maintained as the U.S. Navy's standard reference on the ships and craft of foreign navies. Classified secret, the series included the following volumes:

32-R *Naval Vessels of the USSR*

32-BC *Naval Vessels of the British Commonwealth*

32-NE *Naval Vessels of Northern Europe*

32-WE *Naval Vessels of Western Europe*

32-MED *Naval Vessels of Mediterranean Countries*

32-FE *Naval Vessels of Far Eastern Countries*

32-LA *Naval Vessels of Latin American Countries*[61]

A new, short-title system for designating ONI publications was established in April 1956 using a designation format employing "ONI" followed by a dash and Arabic numerals according to subject category, followed by another dash and a consecutive number plus a letter for revisions. Thus, ONI-52-11A would have been assigned as the short title for the first revision of the eleventh publication on operational intelligence. The subject categories were as follows:

ONI-10 Global

ONI-11 Western Hemisphere (General)

ONI-12 Eastern Hemisphere (General)

ONI-13 Latin America

ONI-14 Western Europe

ONI-15 Africa-Middle East

ONI-16 Communist China

ONI-17 Far East

ONI-18 USSR

ONI-19 Soviet Satellites, Greece, Turkey

ONI-20 Nuclear Science

ONI-21 Biological Science

ONI-22 Chemical Science

ONI-23 Explosives and Propellants

ONI-24 Technical Science

ONI-25 Naval Engineering

ONI-26 Electronics

ONI-27 Ordnance

ONI-28 Underwater Ordnance

ONI-29 Miscellaneous Technology

ONI-30 Miscellaneous Maritime Publications

ONI-31 Sighting Guides

ONI-32 Foreign Naval Vessels

ONI-33 Foreign Naval Vessels (IBM Listings)

ONI-34 Foreign Naval Vessels (Offset Presentations)

ONI-35 Foreign Naval Vessels (Graphic Presentations)
ONI-36 Sino–Soviet Bloc Merchant Ships (IBM Listing)
ONI-37 Foreign Shipbuilding
ONI-38 Foreign Trade
ONI-39 Estimates
ONI-40 Naval Order of Battle
ONI-41 Amphibious Objective Studies
ONI-42 Special Amphibious Studies
ONI-43/44 Spares
ONI-45 Foreign Port Studies
ONI-46 Miscellaneous Studies—Ports, Naval Facilities, and Shipyards
ONI-47 Collection Requirements
ONI-48 Collection Guides (Ports)
ONI-49 Collection Guides (General)
ONI-50 Collection Aids
ONI-51 Indices and Accession Lists
ONI-52 Operational Intelligence
ONI-53 Evasion and Escape
ONI-54/59 Spares
ONI-60 Telecommunications Censorship
ONI-61 Telecom Censorship Training
ONI-62 Armed Forces Censorship (Navy)
ONI-63 Investigations
ONI-64 Subversion, Espionage, and Counterintelligence
ONI-65 Commerce and Travel
ONI-66/69 Spares
ONI-70 General Intelligence Procedures
ONI-71/79 Spares
ONI-80 Soviet and Satellite Uniforms
ONI-81 East-West Trade, Europe
ONI-82 East-West Trade, China[62]

Merchant Marine Intelligence Production

In late 1951, the Foreign Merchant Marine Intelligence Unit (OP-322F3C) was responsible for plotting and recording foreign merchant-ship movements and seaborne trade. It was headed by Cdr. Joseph A. Meyertholen, who had recently come from duty with Commander in Chief, Eastern Atlantic and Mediterranean, where he had served as a liaison officer to the British Admiralty Naval Intelligence Division Merchant Ship Plot and had been in charge of the "plot."

Information for OP-322F3C came from message traffic and from *Lloyd's Shipping Index* (on a delayed basis), plus naval attaché reports and Navy patrol plane photographs, mainly from the Far East area. Meyertholen felt "lost" without Lloyd's current information, and he convinced Capt. R. N. McFarlane (OP-322) and RAdm. Felix L. Johnson, Director of Naval Intelligence, to contract with Lloyd's for $25,000 a year to airmail three copies of their daily Index plus a typed report on all Soviet and Soviet Satellite ship movements. The report also included movements of all Allied flag ships under charter to the Soviet Union, China, and Korea, and all non-Soviet flag ships declaring for Communist-controlled ports. One copy of Lloyd's Index and the daily typewritten report were passed on to the Operational Section (OP-322Y).

In addition to the contributions made to ONI publications and to *National Intelligence Surveys*, an almost continuous scheduling of high-level briefings developed. Recipients included the Secretary of the Navy every two months, the Secretary of Commerce, the Commandant of the Marine Corps, and the Chief of Naval Operations (Adm. William M. Fechteler and then Adm. Robert B. Carney). On occasion, Cdr. Meyertholen's organization was also consulted for quick-reaction briefings. Adm. Fechteler showed particularly intense interest. At a briefing on 30 December 1952, he quizzed Meyertholen on the feasibility of mining or blockading the Chinese coast, or both. Fortunately, from his coast-watcher experience with Commander Naval Group, China, in 1944–1945, Meyertholen had an intimate knowledge of Chinese ports from Shanghai southward. The next morning, Fechteler called Meyertholen at his home and requested that he come to the admiral's office as soon as possible. DNI Carl Espe also attended. Adm. Fechteler wanted a condensed version of the previous day's briefing to take to the White House that afternoon in preparation for a meeting with British Prime Minister Winston Churchill and Foreign Secretary Anthony Eden to be held by President Harry Truman and Secretary of State Dean Acheson on New Year's Day. Fechteler took Meyertholen to the 31 December briefing for the President and Acheson, in the event that he needed a backup.

A new series of merchant marine reports to the White House, started for President Truman, was continued for President Dwight D. Eisenhower. In a discussion between Truman and Churchill, it had been pointed out that non-Communist ships, predominantly British and Greek, were providing the logistic support for the Communists in the Korean conflict. Without the Allied merchant-ship charter support, the Soviets would have been hard-pressed to supply the Chinese and North Koreans. The Soviets had fewer than 200 freighters of over 5,000 gross registered tons; 50 or so were Lend-Lease Liberty ships. The British did not like the blockade or mining idea, but Truman and Churchill agreed to appoint representatives to study the merchant-ship logistic problem.[63]

Soon thereafter, a meeting was held in London, with the United States represented by two persons from the Central Intelligence Agency, one from the State Department, and Cdr. Meyertholen from ONI. The British had two representatives from Counter-Intelligence Division (CID) and a Cdr. Alywyn from the Admiralty Naval Intelligence Division. According to Meyertholen, the CIA and CID representatives minimized the importance of shipping and over-estimated the tonnage carried by the Trans-Siberian Railroad.

Following the meeting, Adm. Arthur W. Radford requested that Commander in Chief, Pacific be represented at any such subsequent meetings. The second meeting was held six to eight months later in Washington, and LCdr. H. H. Calhoun represented CINCPAC. Calhoun, who relieved Meyertholen as OP-322Y3C in December 1954, also attended the third meeting in London in May and June of 1954.[64]

Project Control

The Project Control System was redefined in ONI Internal Instruction 5201.1C on 23 November 1956 as "the system to coordinate and control the production of intelligence within the Office of Naval Intelligence and to maintain production statistics for use in manpower and budget estimates and justifications." The system was operated by the project control officer (OP-922B2), under the direction of the Assistant Director of Naval Intelligence for Production (OP-922). Production efforts were classified as projects under the cognizance of Project Control whenever one or a combination of the following situations prevailed: (1) a total of four or more work hours was required; (2) more than one branch was involved; (3) expediency dictated designating the effort as a project; or (4) production was associated with CNO special briefings, or lectures and briefings to be delivered outside of ONI, or presentations requested by an outside activity.[65]

Summary of ONI Periodicals, 1915–1967

Periodical and Dates

Semi-Monthly Compilation April 1915–December 1918

ONI Monthly / Quarterly Information Bulletin 15 January 1919–1941 (omitted 1932–1935 for lack of funds)

ONI Weekly 28 January 1942–26 September 1945

International Developments of Naval Interest March 1945–December 1954

ONI Review November 1945–April 1963

Naval Intelligence Quarterly 1948–1959

Daily Information Memorandum 1939–1949 (?)

Daily Summary of World Events April 1949–December 1955

ONI Review Supplement 1954–1957

Weekly Summary of World Events August 1956–November 1961

Scientific and Technical Abstracts and Reports 1953–1967

ONI Operational Notes May–August 1945

Fortnightly Summary of Current National Situations 1 December 1940–15 January 1943

Office of Naval Intelligence Bulletins 1958–?

Early ONI Publications, General Information Series

Issue and Title

No. I, 1883 *Operations upon the Korean Coast, Japanese-Korean Ports, and Siberia*

No. II, 1883 *Report of the Exhibits at the Crystal Palace* [London] *Electrical Exhibition, 1882*

No. III, 1884 *Examples, Conclusions, and Maxims of Modern Naval Tactics*

No. IV, 1885 *Papers on Naval Operations During the Year Ending July, 1885*

No. V, 1886 *Papers on Squadrons of Evolutions and the Recent Development of Naval Materiel*

No. VI, June 1887 *Recent Naval Progress*

No. VII, June 1888 *Naval Reserves, Training and Materiel*

No. VIII, June 1889 *Naval Mobilization and Improvement in Materiel*

No. IX, June 1890 *A Year's Naval Progress*

No. X, July 1891 *The Year's Naval Progress*

No. XI, July 1892 *Notes on the Year's Naval Progress*

No. XII, August 1893 *The International Columbian Naval Rendezvous and Review of 1893, and Naval Maneuvers of 1892*

No. XIII, July 1894 *Notes on the Year's Naval Progress*

No. XIV, July 1895 *Notes on the Year's Naval Progress*

No. XV, July 1896 *Notes on the Year's Naval Progress*

No. XVI, October 1896 *Notes on Naval Progress*

No. XVII, Part I, January 1898 *Notes on Naval Progress*

No. XVII, Part II, April 1898 *Discussion of Questions in Naval Tactics,* by VAdm. S. J. Makaroff, *IRN*

No. XVIII, November 1899 *Notes on Naval Progress*

No. XIX, July 1900 *Notes on Naval Progress*

No. XX, July 1901 *Notes on Naval Progress*

No. XXI, July 1902 *Notes on Naval Progress*

Chapter Notes

1. Capt. Herbert E. Cocke, USN, "History of ONI," MS, Office of Naval Intelligence, 1931, 7–8.

2. *Army-Navy Journal*, 25 Jan 1902, 529.

3. Chief Intelligence Officer, *Annual Report, 1903*, passim; and *SECNAV Annual Reports, 1902*, 95–96; *1903*, 492.

4. Roosevelt's secretary's and Sibsbee's ltrs 14958–2, box 620, Assistant Secretary of the Navy Alpha File, RG 80, NA.

5. See Cocke, "History of ONI," 7, for details on the establishment of the series.

6. Department of the Navy, "Administrative History of the Office of Naval Intelligence in World War II," 10 Jul 1946, unpublished MS, 15–16, hereafter *ONI WWII Admin History*.

7. Ibid., 577.

8. Letter in OA.

9. ONI-8, *Instructions for Intelligence Officers*, 1923, 4.

10. *ONI WWII Admin History*, 575–76.

11. ONI-19, *ONI Intelligence Manual*, 1933, para. 2004, 2005.

12. Ibid., paras. 2006, 2007.

13. Ibid., para. 2009.

14. Ibid., para. 5006.

15. RAdm. Arthur H. McCollum, Oral History, USNI, Annapolis, MD, 1971, 1:141–43.

16. *ONI WWII Admin History,* 586, 597.

17. Ibid., 601.

18. Mrs. R. H. Rice (nee Eunice Willson) ltr to author, 14 Apr 1976.

19. McCollum oral history, USNI, 1:251–52.

20. *ONI WWII Admin History,* 623.

21. "Report of Functions and Workload of the Office of Naval Intelligence, Feb–May 1952," 117.

22. CNO (OP-16F) ltr ser 110P16, 18 Mar 1940, box 6, Naval Info File, Job 3770, OA.

23. *ONI Review*, Sep 1952, 367.

24. Naval Intelligence School, lecture notes, 2 Aug 1946.

25. Director Naval Intelligence (DNI) (SC) ltr A8-2/EF ser 048712, 12 Dec 1940.

26. Booz, Fry, Allen and Hamilton, *Special Survey of ONI*, 27 Aug 1941, 14–15, 57–58. Special Intelligence Section (OP-16-Z) was not discussed in the survey.

27. *ONI WWII Admin History,* 927–32.

28. Ibid., 938–44.

29. ONI-54, Dec 1943, in OA provided the listing.

30. Ibid., 516.

31. Charles A. Rocheleau, taped interview, 1 Oct 1975; and *ONI Roster*, Mar 1943.

32. *ONI Weekly* file, OA.

33. *ONI Operational Notes* in OA contains all four files.

34. *ONI Review*, Nov 1945, note on inside front cover.

35. *Defense Intelligence Digest*, May 1963.

36. *ONI Review*, Nov 1952, 423.

37. ONI Instruction 05602.1 of 29 Nov 1957.

38. *Report of Functions and Workload of ONI*, Feb–May 1952, 117.

39. OP-32C1 ltr, ser 1877P32, 25 Mar 1948. The periodical was revived with the fall 1977 issue (13 December 1977) and is prepared by naval reservists drilling with the Office of Naval Intelligence.

40. Naval Scientific and Technical Intelligence Center (NAVSTIC), *Command History, 1967*, 4.

41. Note on inside cover of Air Intelligence Reports.

42. CNO (OP-32F) ltr, ser 0516P32, 27 Jan 1948.

43. CNO (OP-922F4) ltr, ser 000951P92, 23 Aug 1955.

44. Copies of this publication are retained in OA.

45. ONI Internal Instruction 02110.3 of 28 Jun 1954.

46. ONI Instruction 5213.7A of 10 Nov 1959.

47. *ONI Review*, Aug 1955, 429.

48. *ONI Review*, Jan 1958, 29–30.

49. ONI-70-1, *U.S. Naval Intelligence Manual*, 20 Jun 1956, 28. ONI-70-1 superseded ONI-19B.

50. *ONI Review*, Mar 1952, 121.

51. *ONI Review*, Feb 1952, 74.

52. Ibid., 80.

53. *ONI Review*, Jan 1958, 33.

54. Ibid., Oct 1956, 435.

55. ONI-70-1, 27.

56. Navy Department, "History of Administrative Problems, Korean War," 1955, 2:16–17. The series is in eight looseleaf binders.

57. OP-322 memo, 4 Aug 1952.

58. W.E.W. Howe memo on acoustic intelligence (ACINT), 10 Nov 1958.

59. ONI Instruction 5201.1C of 23 Nov 1956.

60. ONI Instruction 5600.1A of 13 Jan 1960.

61. *ONI Review*, Jul 1956, 310.

62. ONI Instruction 5600.6 of 18 Apr 1956.

63. Cdr. Joseph A. Meyertholen, interviews by author, 24 Mar 1980 and 3 May 1980.

64. Ibid.; and Meyertholen ltr to author, 6 May 1980. OP-322F3C became OP-322Y3c in 1952.

65. ONI Instruction 5201.1C of 23 Nov 1956.

Chapter 11

Technical Intelligence

Beginning of Technical Intelligence in the U.S. Navy

One of the main justifications for establishing the Office of Naval Intelligence was to have an office in the Navy for coordinating and correlating the technical information in foreign books, periodicals, reports, and studies on progress being made in naval science by the maritime nations of the world. Up to that time, such technical information had been collected independently by the various bureaus of the Navy, each according to its own interests and with little or no exchange of collected data.

With the establishment of ONI in 1882 and the placement of naval attachés in appropriate countries, much of the information continued to be collected and processed primarily for use by the technical bureaus. Accordingly, ONI's initial organization, as mentioned in earlier chapters, was functional rather than geographic, with a desk for each of the principal technical bureaus to make sure that information received and studies produced were passed on to the bureaus according to their interests.

Many of the early ONI products pertained to technical and logistic support subjects and were distributed in the ONI General Information Series, which included the highly regarded and much-used annual *Notes on Naval Progress* (see list of early ONI publications in Chapter 10).

In 1890, Secretary of the Navy Benjamin F. Tracy commented in his annual report that, with the increase in construction and the growing work of arming and equipping new ships, the importance of ONI was being felt by every bureau and office of the Navy Department and by those in the service at large.[1]

By 1896, ONI was keeping the Navy Department informed on developments abroad that were likely to affect the construction or equipping of the battleships, cruisers, and torpedo boats that the United States was then building or designing.[2]

It was ONI policy in 1914 to gather technical information with an emphasis on assisting with improvements to the fleet rather than providing intelligence of an operational nature. The office was still organized along functional lines to cover specific subjects such as ships, ordnance, and engineering.[3]

Collection of technical intelligence was carried out quite extensively by the technical bureau of the office of the U.S. Naval Attaché, Paris, during World War I. All technical inventions relating to naval matters that were submitted to the U.S. ambassador or the U.S. naval attaché in France were inspected by Capt. George R. Evans, USN (Ret.), and his technical bureau in Paris. Descriptions of inventions found sufficiently interesting were translated, dossiers were prepared, and the reports were then sent to ONI. Few inventions turned out to be of much use.

Frequently, ONI requested reports on various technical subjects relating to Allied material. The reports were researched extensively, and answers were prepared by a technical bureau, usually in collaboration with technical experts of the Allied government involved.

A technical bureau also made lengthy translations of enemy submarine reports and studies covering activities in the Atlantic Ocean, English Channel, and Mediterranean Sea. The documents came from the French Ministry of Marine. In addition to submarine reports, other documents from the French ministry were screened along with French scientific publications, and the interesting items were forwarded in translated form to ONI in Washington. The technical bureau also arranged for the purchase, or manufacture in the United States, of Allied items desired by the Navy Department.[4]

A problem confronting ONI in the late 1930s was to convince the Navy's technical bureaus that the information collected by ONI sources should be taken seriously. Reports from impeccable sources, and sometimes samples of genuine hardware, would be rejected by the bureaus, based on the assumption that no foreign power could build better than the United States. So, it was assumed the reports must be in error.[5]

Technical Intelligence During World War II

Chief of Naval Operations letter serial 0225716 of 28 September 1940 removed practically all restrictions on the exchange of technical information with the British government. Sending large numbers of scientists and technical engineers to England from the Navy's bureaus and laboratories in conjunction with the order increased the volume of scientific and technical material being received by the British Empire Section of ONI for passing on to the Navy's various bureaus and laboratories. The volume of such material was soon beyond the section's capacity; all it could do was record, duplicate, and disseminate.

The Identification and Characteristics (I&C) Section (OP-16-F-20) of ONI was established on 31 December 1941 as a result of the Director of Naval Intelligence's verbal approval of a memo proposing creation of the organization by Cdr. Charles G. Moore, USN (Ret.), dated 11 December 1941. Capt. William A. Heard, on 8 January 1942, outlined the functions of the new section:

> The Identification and Characteristics Section will . . . collect, evaluate, codify, correlate and disseminate all available information on the characteristics and appearance of all foreign naval and merchant vessels. It will carry as continuing projects the design characteristics of U.S. and foreign vessels and translate them into tables, line drawings and models from which identification studies will be produced for the use of all the United States armed services."[6]

On 9 January 1942, the functions of the I&C Section were expanded to include collaboration on the production of aircraft identification studies. The concept of having a single section responsible for the characteristics and appearance of warships, merchant ships, and aircraft was based on the recognition that one centralized drafting, production, and publication force could better serve the analytical and evaluating specialists in these three hardware fields.

At its inception, I&C consisted of Cdr. Moore who was also head of the Statistical Section (OP-16-F-10), a civilian analyst, and an architect awaiting a naval officer commission.

The requirement to place highly accurate ship identification material in the hands of the operating forces and training activities received first priority. Statistical and photographic information on naval vessels was available to I&C from the various foreign desks and was employed to produce detailed drawings and models for recognition publications and for vulnerability studies.

As related in the previous chapter, the models were built by the David Taylor Model Basin, professional model builder Van Ryper at Martha's Vineyard, Massachusetts, and an expert model maker in I&C. The models were then photographed from all angles by Time, Inc., which was under contract to publish the recognition manuals.

After the recognition documents had been produced, the section was able to put more time and effort into the analysis of the performance and development of statistical characteristics of warships. Many sources had not been previously researched, and the process of searching every prisoner-of-war report, reading every captured document, and rescrutinizing every pertinent photograph culminated in *A Statistical Summary of the Japanese Navy* (ONI-222-J), which was followed by numerous other similar documents.

In early 1943, it was possible to put more effort on the technical aspects of ships' equipment. Files were started on enemy guns, fire-control equipment, radar, and similar developments. There was a tremendous increase in liaison with the Bureau of Ships (BUSHIPS), Bureau of Ordnance (BUORD), and Bureau of Aeronautics (BUAER) and with such special organizations as the Office of Scientific Research and Development and the Naval Research Laboratory (NRL). Joint publications were issued containing information collected by ONI and evaluated by the appropriate bureaus. In the March 1943 reorganization of the Office of Naval Intelligence, OP-16-F-20 became OP-16-P-2, retaining the name, Identification and Characteristics Section.[7]

At about the same time, an Intelligence Analyst Unit was established outside of ONI in the Progress and Planning Section of the Office of the Coordinator of Research and Development. Such a unit had not been included in the original organization of the coordinator's office; it had been assumed that ONI would make preliminary analyses of incoming reports. The magnitude of the task, however, had been very much underestimated. Most reports did not deal primarily with scientific and technical matters, and any clues on weapon developments were obscured by other unrelated matter and were recognizable only by those familiar with research matters. ONI personnel involved in dissemination who were also qualified to screen out information of that kind were limited in number and inadequate to deal with the magnitude of the task.[8]

In early 1944, it had become increasingly apparent that a centralized organization was needed for the collection and dissemination of naval technical intelligence information. There were numerous instances of the failure of such information to reach the technical activity having primary cognizance. Furthermore, with the intensification of the war in the Pacific, the fleet had an increasingly urgent need for timely technical information in a processed format.

Accordingly, RAdm. Roscoe E. Schuirmann, Director of Naval Intelligence, issued a memorandum on 3 October 1944 establishing the Technical Intelligence Center (TIC) within the Publications and Dissemination Branch and designating the center OP-16-PT. The memo stated the duties of TIC to be as follows:

> a. Establish and maintain central technical intelligence files of all information relative to foreign warships, merchant ships, and naval and military equipment for the use of all service activities;
>
> b. Expedite and insure adequate routing and interchange of reports and information on these subjects;
>
> c. Maintain a panel in which representatives of all interested technical bureaus and activities will participate for the purpose of determining requirements of the activities either from incoming material or from the central files; and
>
> d. Collaborate with technical bureaus, through their representatives, in the preparation of technical intelligence articles on foreign naval equipment, on warships and on merchant ships, for dissemination through a common medium.

The I&C Section served as the nucleus around which OP-16-PT was formed. One officer and one yeoman were also added from the Special Activities Branch (OP-16-Z) to control captured enemy equipment.[9]

A Captured Enemy Equipment (CEE) Program, sponsored by the Bureau of Ordnance, assigned field personnel to both Europe and the Pacific, supplied them with cameras and CEE documents, and gave them orders to photograph, properly identify, and serialize every piece of equipment forwarded to the U.S. for exploitation. The field teams had the capability to write preliminary reports on CEE items and disseminate them rapidly to area units when appropriate. For example, in case of booby traps, data were to be disseminated immediately to infantry units after a preliminary checkout by explosives experts.[10]

LCdr. C. H. Watson, USNR, was the first acting head of ONI's Technical Intelligence Center. The center became involved in the filing, translation, distribution, and control of German documents of naval interest that were picked up by the Naval Technical Mission in Europe (NAVTECHMISEU) as elements of Europe were liberated. TIC functioned generally as a library and clearinghouse for the control of the unevaluated documents. A Control Section was established, and later an Estimates Section was formed, but, initially, no formal intelligence studies of the captured documents were undertaken.

The "reading panel" system was adopted to help TIC personnel keep in close touch with representatives of agencies interested in the technical material available at the center. Representatives visited the panel several times a week from the Office of the Chief of Naval Operations, the Office of Research and Inventions, all the Navy technical bureaus, the British Admiralty Delegation, the Joint Electronics Intelligence Agency, the Marine Corps, and the Military Intelligence Service of the War Department representing the Army. Dissemination was also made to the Naval War College, the Navy's General Board, the Ship Characteristics Board, and the Joint Army-Navy Experimental and Testing Board.

On 28 August 1945, an abstract system was inaugurated by TIC to supplement its reading panel. The abstracts consisted of a brief summary of each document, with no evaluation of content. Specialists in the following subjects drafted the abstracts: warships and warship equipment, merchant ships and their equipment, ordnance, electronics, aeronautics and miscellaneous—the last covering synthetics, chemicals, medical intelligence, electrical instruments, etc.[11]

Capt. George R. Phelan, by February 1945, had relieved LCdr. Watson as head of the Technical Intelligence Center. He, in turn, was relieved by Capt. Francis R. DuBorg in December 1945. The designation of the center was changed to OP-23F2 in October 1945 when ONI's designation was changed from OP-16 to OP-23.[12]

Exploitation of German and Japanese Technical Developments

Naval Technical Mission in Europe

The Readiness Division of Commander Naval Forces, Europe (COMNAVEU), a unit that had performed well in the field of technical intelligence in England, prepared extensive plans for the exploitation of the vast sources of German technical information of interest to the Navy. Capt. Henry A. Schade was sent to Europe to investigate the best means of exploiting technical data about the German navy. In May 1944, the War Department had proposed a joint Army-Navy mission, known as the

"Alsos Mission," for technical intelligence work in Europe. (*Alsos*, the Greek word for tree, was a play on words derived from the organization's having been established at the instigation of Army MajGen. Leslie Groves, head of the Manhattan Project.) The mission's primary purpose was to acquire the leading European nuclear scientists and data on the German atomic bomb project; the mission's other scientific data-gathering work was, in effect, a cover for its principal mission.

In August 1944, Capt. Schade was assigned as head of the Navy Section of the Alsos Mission, to report to COMNAVEU and to be under COMNAVEU administrative control. Alsos naval members were to represent COMNAVEU Readiness Division on the continent, and Commander Naval Forces, France provided assistance in personnel, billeting, and office space.[13]

On 4 December 1944, the Secretary of the Navy approved the establishment of the U.S. Naval Technical Mission in Europe. Its mission was to exploit German science and technology for the benefit of the Navy Department's technical bureaus and the Coordinator of Research and Development. The mission's tasks were to coordinate all U.S. Navy activities on the continent of Europe that were exploiting German scientific and technical intelligence and to form a pool of technically qualified personnel under Navy control to operate as field teams, either independently or with Combined Intelligence Objectives Subcommittee teams, Technical Industrial Intelligence Committee teams, Alsos teams, or U.S. Army or British teams exploiting targets of naval interest.

The naval Alsos group that had been established to help in the search for information on, and personnel involved in, Germany's nuclear research served as the nucleus of the personnel pool. The senior Navy representative on the Alsos mission was designated by Commander in Chief, U.S. Fleet to be Chief NAVTECHMISEU. The Navy technical bureaus and the coordinator of research and development provided technical officers, civilian technicians, and the necessary administrative personnel to staff NAVTECHMISEU. An office for the representative of NAVTECHMISEU was established in ONI (OP-16-R) to keep Chief NAVTECHMISEU continuously informed as to the plans and activities of the Technical Intelligence Committee and the technical missions of the War Department.

Chief NAVTECHMISEU was to report directly to COMNAVEU, and to the senior U.S. naval authority in the areas being exploited. He was authorized and directed to:

> (a) travel, and order travel, anywhere in Europe; (b) obtain and expend funds as necessary in procuring technical intelligence; (c) obtain necessary assistance from U.S. naval authorities in Europe; (d) obtain assistance from U.S. Army authorities in Europe, using Alsos Mission channels wherever possible; (e) forward Information Reports (IRs) direct to the Director of Naval Intelligence, with copies to appropriate Navy Department offices and to U.S. activities in Europe, and to communicate directly with the Navy Department regarding the intelligence operations of the missions; (f) ship material to the United States of special interest to the Navy Department; and (g) return to the United States for consultation when necessary.[14]

NAVTECHMISEU was activated on 20 January 1945. Commo. Henry A. Schade was the first chief of the mission and was a direct representative of Commander in Chief, U.S. Fleet, reporting to Commander Naval Forces, Europe, with the designator Commander Task Force (CTF) 128. NAVTECHMISEU absorbed most of the officers from COMNAVEU's Forward Intelligence Unit, Task Group (TG) 125.8, and they became the Intelligence Section of the mission. Civilian technical specialists were provided by Navy contractors. One such civilian was the aviator Charles A. Lindbergh. The administrative headquarters for TG 125.8 was established in Paris, with forward headquarters located variously at Bad Schwabach (mid-April), Heidelberg (late April), Bremen (late May) and Munich (mid-July).[15]

The personnel of the NAVTECHMISEU Intelligence Section (six officers and two enlisted) had been engaged in intelligence collection work on the continent since Normandy D-Day and were the most experienced naval field intelligence officers and men in the European theater. Their language qualifications, previous experience as interrogators of German prisoners of war, and familiarity with U.S. Army field procedures were their principal assets.

At its peak, the Intelligence Section had thirty-eight officers and two enlisted personnel. The additional officers were recruited from CTF 124, the Special Activities Branch, and other naval activities, including the Bureau of Personnel.

Some interpreters were assigned on a semipermanent basis to other NAVTECHMISEU sections, and about half were retained in an interpreter pool. Those assigned to a specific section made trips with officers of that section and later assisted in report writing and translating pertinent German documents.[16]

Various sections of the NAVTECHMISEU found a number of noteworthy German technical developments. The researchers discovered that the Germans had produced hydrogen peroxide, concentrated to 85 percent and solid-free, to support combustion in submarine and torpedo power-plants and in propellants. A captured one-ton-per-day output plant was shipped to the United States. They had also developed fin-stabilized, rocket-assisted projectiles for high-velocity guns, and a number of sophisticated guided missile programs were uncovered. A Mach 4.3 wind tunnel that had been used to conduct initial tests on V-2 rocket models was disassembled and shipped to the Bureau of Ordnance in the United States.[17]

German developments in ship design and engineering were also investigated. High-speed diesel propulsion systems, closed-cycle diesel engines, and a 2,500 horsepower hydrogen-peroxide Walter-cycle turbine capable of propelling a submarine at 26 knots submerged were obtained. Examples of clandestine attack craft and saboteur equipment were also acquired. Nine aircraft were shipped to the United States for exploitation, and, by agreement with the Army Air Force and the British Royal Air Force, the U.S. Navy studied captured German turbojet engines.

German infrared and guided missile electronic systems were investigated through a U.S. joint working group. Other electronics systems recovery and investigative work was performed through the Committee on Captured Enemy Electronic Equipment.[18] NAVTECHMISEU personnel maintained a target information card index file to permit technical officers to brief themselves on information collected previously by other agencies.

The interrogation of German naval personnel was facilitated by Adm. Karl Doenitz's directive that the German navy furnish all information requested after hostilities ceased.[19]

When NAVTECHMISEU was disestablished on 1 November 1945, eleven officers were attached to the Naval Advisor, Office of Military Government, Europe (in Berlin) and given the title U.S. Naval Technical Unit, Europe. The unit took care of any new intelligence objectives and worked to complete joint U.S.-British projects such as torpedo tests and procurement, shipment of heavy armor to the United States for ballistic tests, hydrogen-peroxide supply programs, and obtaining data on the manufacture and tests of German gas turbine engines.[20]

Although NAVTECHMISEU was blocked by the Soviets from collecting information on various German naval installations, such as the torpedo plant in Gdynia and the submarine base at Danzig, it did visit German ships in Russian-occupied Baltic ports and targets in Russian-occupied Berlin and Vienna. To run down leads on German intelligence, investigations were also conducted on a limited scale in Sweden and Switzerland and some visits were made to France, Belgium, Holland, and Norway.

During its eleven-month existence, NAVTECHMISEU faced a number of problems and deficiencies. The initial estimates of personnel requirements were too low; tours of duty for officers and civilian technicians were too brief; investigators were not sufficiently briefed about information that had already been obtained prior to their field projects; inadequate language training had been provided for investigators; interrogation of enemy personnel had not been fully exploited; and interrogation centers were too far from the point of procurement. Furthermore, there was no planning officer to plan and organize priority projects for the most effective exploitation, and difficulties in obtaining U.S. Army clearance for field operations had been experienced.[21]

During its existence, NAVTECHMISEU submitted 240 letter reports and 350 technical reports. In addition, a great amount of material and equipment was sent to the United States for study. A total of 309 officers, 109 civilian technicians, and 340 enlisted men was assigned to NAVTECHMISEU at various times.[22]

Naval Technical Mission to Japan

The U.S. Naval Technical Mission to Japan (NAVTECHJAP) was established on 14 August 1945 by the Chief of Naval Operations, in accordance with the Intelligence Appendix of Operation Blacklist, the operational plan for the occupation of Japan. Capt. Clifton C. Grimes, Fleet Intelligence Officer in Charge of Technical Intelligence for Joint Intelligence Center, Pacific Ocean Areas (JICPOA), was designated chief of mission. The nucleus of personnel came from among those attached to JICPOA who had technical and language qualifications and from technical personnel at other commands. The initial group was designated JICPOA Team No. 29 and entered Sasebo harbor on 23 September on board the attack transport *Shelby* (APA 105) on the date of the initial occupation of Kyushu. Another group, designated JICPOA Team No. 30, joined the Third Amphibious Group in the occupation of certain areas of China. Elements of the intelligence groups of Commander Seventh Fleet joined in Sasebo, and on 28 September all units were consolidated as NAVTECHJAP. The headquarters, initially located at Sasebo, was soon moved to Tokyo to improve coordination with the other occupation activities.

The purpose of the mission was to survey all Japanese scientific and technological developments of interest to the Navy and Marine Corps in Japan, China, and in Korea south of 38° north latitude. The mission's work involved seizure, examination, and study of intelligence material; interrogation of personnel; and preparation of reports.

Before the cessation of hostilities, ONI had prepared a list of all the Japanese technical "targets" it desired, including lists of specific items and information sought by the technical bureaus of the Navy Department. As early as 15 September, copies of "Intelligence Targets Japan" of 4 September 1945, prepared by ONI, were received by NAVTECHJAP, permitting the movement of the mission without much additional planning.

NAVTECHJAP was organized into two departments: Executive (administration, etc.) and Technical. The latter was divided into sections: Ships, Electronics, Ordnance, Medical, Special, and Petroleum. One other section had the job of filing, printing, editing, and distributing intelligence material. The Technical Liaison Section, located at the Intelligence Staff (G-2), Supreme Commander Allied Powers (SCAP), attended policy conferences and other meetings and maintained contact with SCAP headquarters. The Special Intelligence Section exploited any non-technical targets that might be assigned. It also assisted in the completion of the U.S. Strategic Bombing Survey after the departure of the survey's personnel from Japan.

By 1 November 1945, NAVTECHJAP had 295 officers, 125 enlisted personnel, and 10 civilian technicians assigned to it. Among the officers were approximately twenty-three British technical specialists and language officers.

Collection centers were established at Sasebo, Yokosuka, Kure, and Kobe for documents and equipment. Field personnel wrote reports, and the NAVTECHJAP headquarters in Tokyo edited, typed, and/or printed the reports after checking them for completeness, accuracy, and acceptability.[23]

As of June 1946, NAVTECHJAP had finished its work in the field and moved to Pearl Harbor to finish its reports.[24] It was then disestablished on 1 November 1946. During its existence, 350 officers, 260 enlisted, 29 British officers and enlisted, and 16 civilian naval technicians, for a total of 655 personnel, had worked on its projects at one time or another. Approximately 3,500 documents had been seized and shipped to the Washington Document Center and the Navy technical bureaus, and 15,000 pieces of equipment had been shipped to U.S. laboratories for investigation. The largest items were two 18.1-inch guns shipped from Kure, each weighing 180 tons and measuring 75 feet in length.[25]

OVERCAST and PAPERCLIP: German Scientists and the U.S. Navy

As territory was occupied after the European landings, NAVTECHMISEU teams roamed far and wide, sometimes just behind the advancing troops, questioning, searching, and trying to find the answers to Germany's amazing wartime technical progress. One day in April 1945, while one of the teams was searching at Oberammergau in Bavaria, they found a group of German missile designers and their leader, Professor Herbert Wagner. Wagner had been the chief missile design engineer for the Henschel aircraft works and had masterminded the development of the Hs-293, a radio-controlled glide bomb. In the nearby Hartz Mountains, buried blueprints, models, and prototypes were found, enough to fill seven large cases.

By early May 1945, Professor Wagner, his four assistants, and their files were in Washington. Many organizations were interested in exploiting them, including the Navy Bureaus of Aeronautics and Ordnance and the Army Air Force, but none was willing to take custody of the missile team. So they were placed in a Washington hotel, where ONI officers stood watch as Wagner and his men worked to perfect a controlled antiaircraft rocket for use in the continuing war against the Japanese.

The hotel arrangement was too expensive for ONI's staff and funding resources. The Office of Research and Inventions (later known as the Office of Naval Research) and the National Advisory Committee for Aeronautics were asked to help. What was needed was a secluded estate where life would be pleasant but secure. The Guggenheim Foundation was found to have such a place, the Jay Gould medieval castle at Sands Point on Long Island, which became the Special Devices Center of the Office of Naval Research. Initially, its use was kept quite secret; guards were placed at the gate, and no Germans left the grounds except under escort.

In the summer of 1945, the Technical Information Center published *German Technical Aid to Japan* to delineate "those German techniques, devices and weapons, the use of which by the Japanese would have a bearing on the war in the Pacific." The surrender of the German submarine *U-234* to U.S. forces at the time of Germany's collapse contributed significantly to the survey: the submarine had been en route to Japan with a valuable cargo, including complete drawings for the Messerschmitt Me-163 rocket fighter, an entire German electronics library, fire-control equipment, radar, and radio equipment.

Dr. Heinz Schlicke, a German electronics expert, was one of the passengers aboard the *U-234*. He was going to deliver a series of lectures in Japan on German electronic development and had extensive documentary material with him. Arrangements were made for Dr. Schlicke to give the same lectures in the Navy Department between 19 and 31 July 1945.

To ensure a more widespread distribution of the information from Dr. Schlicke's lectures and documents, the Technical Intelligence Center issued 800 copies of *Electronics Research in the German Navy* on 15 September 1945. The book covered a general review of electronic research in the German navy, U-boat camouflage, radar and search receivers on U-boats, infrared in the German navy, communications with submerged U-boats over great distances, direction-finding in U-boat warfare, ships' antennae, prevention of radio transmitted direction-finding (D/F), and some observations on German techniques in the use of centimeter waves and the theory of line transformers.

The Technical Intelligence Center also edited and distributed 12,000 copies of German Admiral Doenitz's essay, "The Conduct of the War at Sea," a review of the German navy's participation in World War II.[26]

The surrender of *U-234* following VE Day brought more scientists to the Special Devices Center. They included Dr. Falck, one of the German navy's top ship designers; the head production engineer for the Messerschmitt works; and several experts on night fighter techniques.

From the information developed by the interrogation of the Wagner Group, Dr. Schlicke, and others, the Navy realized that a most valuable reparation from Germany could be the brains of its scientists. The various technical bureaus of the Navy began to show more interest in acquiring the services of some of the German technical specialists. As the Army and Army Air Force were also interested in procuring German specialists, it was determined that only a joint program under the general administration of the Joint Chiefs of Staff (JCS) could be effective for successful and reasonable exploitation.

In July 1945, the Secretary of State and the Secretary of War agreed to establish Project OVERCAST, a German scientist procurement and exploitation program initially in the rocket and guided-missile research area. The first scientists to arrive under the project came on 8 July 1945 for short-term interrogation and exploitation by the Army, Navy, and Air Force. An additional twenty-two scientists were brought to the United States in September.

On 27 August, the Joint Intelligence Committee recommended that the Joint Intelligence Objective Agency (JIOA) be authorized to set up an interim procedure to coordinate the temporary exploitation of German and Austrian specialists, scientists, and technicians in the United States pending the formation of approved government exploitation policy and procedures. On 12 September, the JCS, with the concurrence of the State-War-Navy Coordinating Committee (SWNCC), approved the interim procedures to be used by the JIC through the JIOA.

The project code name was changed from OVERCAST to PAPERCLIP on 10 November 1945, following the compromise of the former code name. By the end of 1945, 132 scientists had been brought to the United States under Project PAPERCLIP.

The policy and procedures for exploitation of German and Austrian scientists in the United States were submitted by SWNCC to JCS for comment on 26 February 1946; and on 4 March it was approved by SWNCC and sent to the JCS for execution by JIOA.

By April 1946, approximately 155 German scientists and technologists were in the United States under Project PAPERCLIP for exploitation by the military services, all under voluntary contracts not to exceed twelve months. By the end of July, the figure had risen to 190 scientists, and there were over 200 others whose services had been requested by the various technical services of the War and Navy Departments.

A revision of PAPERCLIP was considered urgent. The morale of the first scientists to arrive was low because of the failure to implement an approved policy under which their families could join them in the United States. The 350-scientist ceiling needed to be raised to 1,000 in order to meet the demands of both the technical services and civil research programs. And the top salary of $10.00 per day paid to dependents in German marks had become insufficient inducement for highly qualified German scientists, who were being offered three to five times that amount by the French and Russians in addition to being permitted to take their families with them.

The proposed revision of PAPERCLIP was coordinated in the War Department General Staff and was approved by the Secretary of War on 31 July 1946. The resulting SWNCC policy was recommended to the President by Secretary of State James F. Byrnes in August and was approved by President Truman on 3 September 1946.[27]

In spite of White House approval, there was continuing obstruction to Project PAPERCLIP by elements in the State Department. Samuel Klaus was the State Department representative on the committee set up to formulate a plan to implement the SWNCC policy. Capt. Bosquet N. Wev, head of JIOA, was chairman of the committee. Klaus, whose primary interest was in getting displaced persons from Germany into the United States, looked upon the project as depriving him of immigration quotas. Furthermore, he considered any German scientist who had performed any service in support of the German war effort as having accordingly performed services detrimental to the U.S. war effort and, based on the

War Powers Act, was to be considered prejudicial to the best interests of the United States and thus not eligible for immigration. It apparently made no difference to Klaus that bringing German scientists to the United States had been approved by Secretary of State Byrnes and President Truman.

As a result of Klaus's actions, many of the 1,500 German and Austrian scientists that the U.S. Army and Navy had wanted to bring to the United States were taken over by the Soviet Union for exploitation of their knowledge and experience. The military technical bureaus and laboratories that could have used these highly qualified people kept putting pressure on JIOA for action, and JIOA kept trying to get the word up to the top levels of the State Department to get around the obstructionism there. But each channel seemed to run into someone officially subservient to Klaus.[28]

By 10 January 1947, there were still only 285 Project PAPERCLIP scientists in the United States, and 390 others had been requested for employment. The German scientists working in the United States were either under military employment or working on military projects in industries that had contracts with the War or Navy Departments.[29]

Finally, Capt. Wev, through his own efforts, arranged for a hearing before a Senate appropriations subcommittee in June 1947. As a result of his testimony, appropriations to the State Department were blocked by the Senate until the barrier to Project PAPERCLIP was broken. Adm. Chester Nimitz, then Chief of Naval Operations, heard of Wev's testimony and called him on the carpet. When Nimitz heard the full story, however, he gave Wev a "Well Done."[30]

As of late 1949, there were eighty-two German scientists still employed by the Navy in Project PAPERCLIP. They were located at a number of naval stations on both coasts. BUORD, BUAER, and BUSHIPS each had project scientists at their various laboratories and testing centers. Others were located in numerous industrial cities and were accountable to inspectors of naval material and branch offices of the Office of Naval Research. The Bureau of Medicine had some Germans working on medical research at Bethesda and at the Submarine Base, New London.

Scientists were used in a variety of fields. For example, the Naval Ordnance Laboratory at White Oak, Maryland, employed them in three main areas: (1) designing facilities for research in acoustic, supersonic, and explosive phenomena; (2) conducting basic research in physical optics, fluid mechanics, acoustics, and explosives; and (3) designing specific weapons and weapons countermeasures. Many Project PAPERCLIP personnel working for BUORD were experts in aerodynamics and had had extensive previous experience with the operation of the Kochel supersonic wind tunnel that had been brought to the United States and installed at the laboratory at a cost of $2.5 million.[31]

Establishing a Permanent Technical Intelligence Center

By late 1945, it was becoming obvious that to meet new needs the ONI Technical Intelligence Section needed greater numbers of more diversified technical experts. As a result, additional civilian experts were hired in nuclear energy, aeronautics, naval engineering, and other fields.[32]

One of the first key civilians to be employed was Dr. A. Keith Brewer, a physical chemist who had specialized in isotope operation processes at the Bureau of Standards and who came to ONI in 1946. Brewer was a strong believer in the "scientific method" and pushed for more and better scientifically trained personnel in the intelligence production process. Like all true scientists, he was a skeptic and for a decade urged proof positive in appraising Soviet and Communist Chinese nuclear developments.

Another early civilian employee whose experience had been more in applied sciences was Dr. Jack Alberti, who had served in the Special Activities Branch of ONI in the prisoner-of-war interrogation field throughout World War II. In addition to being a linguist, Alberti was also an archaeologist and a concert pianist. His specialty was captured enemy equipment and scientists.

William E. W. (Bill) Howe, an electronics engineer, came in February 1949 from the Naval Research Laboratory, where he had been involved in testing captured Japanese equipment and in developing a receiver to intercept German missile guidance signals. At ONI, Howe was soon involved in electronic reconnaissance activities.

Mr. R. E. Kinzy, an aeronautical engineer, came from the Glenn L. Martin Company to do some of ONI's earliest Soviet missile threat analysis.[33]

In August 1948, Capt. Francis DuBorg was succeeded by Cdr. Harvey C. Lawder, USNR, as head of the Technical Intelligence Center. In December 1948, the center moved with the rest of ONI from the Main Navy Building to the Pentagon.[34]

Technical Intelligence During the Cold War Era

The Korean War brought technical intelligence to the fore again, with emphasis on potential enemy nuclear capability and on electronics (see also Chapter 15).

One of the particularly outstanding naval intelligence collection achievements of the Korean War was the discovery of several important Soviet technical documents in a cave outside the North Korean city of Wonsan. The documents included material on underwater ordnance, the first instruction book obtained on a Soviet radar, and the first firm technical information on Soviet sonar capability. The Soviet documents were the sources for the first in-depth analysis of Soviet shipborne radar, sonar, and communications and were the basis for a series of ONI reports.[35]

On 18 June 1951, a meeting of representatives of ONI's Special Unit of the Operational Section (OP-322Y1), the ONI Technical Unit (OP-322F2), and the Armed Forces Security Agency was held to discuss the establishment of standardized procedures for processing and exploiting information obtained from naval equipment of Soviet origin. A collection guide for field use was to be formulated by OP-322F2 in coordination with the Collection Section (OP-322H). In the meantime, a team from OP-322F2 was to proceed with the exploitation of foreign equipment already available in the United States.[36]

In mid-1952, the Technical Unit was separated from most of the other intelligence production sections of ONI when it was moved (with the Coast and Landing Beaches Section and the Ports and Naval Facilities Section) to Building 166 at the Naval Gun Factory at the Washington Navy Yard. The move exposed the many disadvantages inherent in separating analysts who need to collaborate in their analytical efforts if they are to produce coordinated analyses that are consistently based on all available pertinent data. The separation also introduced security problems in communications and courier arrangements between the ONI analysts at the Naval Gun Factory and those at the Pentagon.

In drawing a distinction between the intelligence production interests of OP-322F2 and the other units of the Foreign Section (OP-322F) of ONI, the Technical Unit expressed its work in terms of a "research and development stage" versus the production stage. The unit was concerned with the technical aspects of intelligence about weapons, weapon systems, propulsion, and scientific developments. When equipment, processes, and hardware reached the production stage, they became the primary concern of the other units of the Foreign Section.[37]

An accession list of technical intelligence documents that had been received by ONI and were available for loan to technical activities was issued semi-monthly by OP-322F2. The purpose of the list was to help U.S. Navy activities keep abreast of technical developments in other countries.[38]

The Technical Unit made another move in January 1955 when it shifted from the Gun Factory to Building 52 at the Naval Observatory on Massachusetts Avenue in Northwest Washington. One advantage of the move was that the unit was now co-located with the Translations Section, which, in recent years, had increasingly devoted its talents to translating technical documents. Having the technical analysts easily available to the translators increased the efficiency and accuracy of the output of the latter and the value of their products to the former.

In the early and mid-1950s, the Technical Unit became increasingly involved in support of the information collection effort. German scientists and engineers returning to West Germany from the Soviet Union were potential sources of considerable value if properly exploited, and the Technical Unit, with the primary interest in the information thus available, participated directly in some of the interrogations and provided guidance for most of them. Information obtained from the scientists indicated that the Soviets were working on ideas that U.S. scientists had had under way for some time. The information gave the U.S. projects higher priority.

Opportunities for inspection and exploitation of actual Soviet equipment also increased by the mid-1950s, when numerous items were recovered from the Sea of Japan and elsewhere. Examinations of the Soviet hardware made it apparent that the USSR was capable of more rapid technological advances than had been estimated earlier. Many American experts still erroneously believed that the Soviet "peasant economy" could not mass-produce the electronic components needed for the mastery of radar and television technology and, in turn, of missile guidance systems.[39]

In late 1954, an opportunity arose to examine Soviet electronic equipment of more recent vintage than World War II. The Chinese Nationalist navy, in its efforts to blockade the mainland China coast, seized the relatively modern Soviet commercial tanker *Tuapse* and brought it into the Taiwanese port of Kaoshiung. Unfortunately, the U.S. chief delegate to the United Nations, Warren R. Austin, blasted the Chiang government on the floor of the United Nations for what he termed an act of piracy. Austin's remarks complicated and prolonged the dialogue needed for U.S. experts to gain permission to look over the technological plum.

In January 1955, a joint *Tuapse* exploitation team was assembled, made up of Bill Howe of the Technical Section of ONI in charge, two engineers (Lamont Blake and William Main) from the Radar Section of the Naval Research Laboratory, a vacuum tube specialist from the Army's Fort Monmouth Laboratories, and a Russian-born technical translator and another expert from the Air Force's Foreign Technology Division. The group was joined in Japan by two specialists with additional photographic equipment.

The principal items of interest found on board *Tuapse* were the X-band surface search-navigation radar, *Neptun*, which at that time was being installed by the USSR in its larger merchant ships and naval combatants; the NEL-4 echo sounder, the first piece of Soviet operational sonar equipment to become available to U.S. technicians; and assorted communications and navigational devices. The *Neptun* radar was the main justification for the team visit; it was thoroughly scrutinized, as were all the other pieces of equipment. Fortunately, liberal supplies of spares were on hand, particularly for *Neptun*, and thus the *Tuapse* capture was an analysis windfall.

Back in the United States, the spare-part components brought back from the *Tuapse* were further tested and examined, and the resultant reports were disseminated. ONI's report included the results of tests by the Naval Research Laboratory and the Army Signal Corps Laboratory. The radar components were the first components—and remained for many years the only such Soviet components—in U.S. hands. ONI's Bill Howe subsequently was sent on a lengthy tour throughout the United States, briefing various government and technical groups on the results of the team's visit. Having actual hardware, as well as information on technical performance, was extremely useful at the briefings, which, in the Washington area, were arranged by the Bureau of Ships and were presented to the Secretary of Defense and other senior officials.

The important insight gained from the cooperative interagency collection effort was the knowledge that the Soviets had thoroughly mastered the needed technology by 1953–1955 and had high production rates with multiple suppliers of key electronics items. It was also evident that the same technology and many of the components would be utilized in many critical military airborne, shipborne, and missile-related equipments.[40]

In the fall of 1957, Cdr. Felix Caracciolo relieved Cdr. Robert E. Barnhart as the head of the Technical Section, but at approximately the same time, all elements of ONI at the Naval Observatory were combined organizationally and designated OP-922G under the name Basic and Technical Intelligence Branch. Capt. Theodore M. Peterson was placed in charge of the new branch, and Cdr. Caracciolo was assigned as his technical assistant. In the fall of 1958, Caracciolo was promoted to captain and was relieved by Capt. C. A. Stay as technical assistant to Capt. Peterson.[41]

It was under Peterson that steps were initiated to set up an in-house acoustic laboratory. See Chapter 14 for more details. A small electronic intelligence laboratory was also assembled at OP-922G during the late 1950s.[42]

Direct exchange of correspondence between the Basic and Technical Intelligence Branch of ONI and the Scientific and Technical Unit, Frankfurt (via the Commander in Chief, U.S. Naval Forces, Eastern Atlantic and Mediterranean representative in Frankfurt, Capt. Richard H. Tenney), served as a means for the exchange of working-level information between technical analysts of both organizations.[43]

The publication of a monthly journal, *Scientific and Technical Abstracts and Reports* (*STAR*), had been initiated by the Technical Unit in 1953. In early 1960, Harvey Lawder, retired from active duty, was back in a civilian capacity as editor of the *STAR*. In early 1960, Capt. Harry Marvin-Smith relieved Capt. Peterson as head of OP-922G.

NAVSTIC Established: Naval S&T Under the Defense Intellgence Agency

On 30 June 1960, the scientific and technical intelligence elements of OP-922G were set up as the Naval Scientific and Technical Intelligence Center (NAVSTIC) and designated a field command of ONI under the management control of the Chief of Naval Operations. NAVSTIC's staff ceiling was set at sixty-five civilians, ten officers, and eight enlisted personnel. Capt. Marvin-Smith was designated officer in charge and was double-hatted as OP-922B5, Coordinator, Scientific and Technical Matters.[44]

NAVSTIC was under the military command of the Commandant Potomac River Naval Command and the management of the Chief of Naval Operations. Its mission was to produce basic scientific and technical intelligence as directed by the Director of Naval Intelligence for the Chief of Naval Operations.[45]

On 27 April 1964, all organizational activities of the three military services involved in the production of scientific and technical intelligence were placed under the management and direction of the director of the Defense Intelligence Agency (DIA). Administrative control and logistic support for NAVSTIC continued to be provided by the Assistant Chief of Naval Operations (ACNO) for Intelligence, another hat worn by the Director of Naval Intelligence.

The management of scientific and technical matters by the Defense Intelligence Agency, according to NAVSTIC, was considered highly satisfactory and resulted in significant improvement in both the quality and the quantity of intelligence products available to the user.[46]

With the formation of DIA, the administrative workload at NAVSTIC increased sharply, and its direct involvement in collection was reduced; NAVSTIC, however, acquired a stronger voice at the Department of Defense level. As a result of this last development, NAVSTIC was able to stimulate more favorable decisions on its requirements for staff resources in numbers, qualifications, and grade levels.[47]

With the establishment of the Naval Intelligence Command in 1967, military control and logistic support of NAVSTIC was transferred from ACNO (Intelligence) to Commander Naval Intelligence Command. NAVSTIC's relationships with other organizations and activities were not changed, nor was its mission:

> To produce scientific and technical intelligence of naval interest as directed by the Director, Defense Intelligence Agency; to assist the Naval Establishment in analysis of current intelligence; and to maintain close scientific and technical relationship with the research and development efforts of the Department of Defense.

In 1967, NAVSTIC's staff ceiling was 224 civilians, 34 officers, and 33 enlisted personnel, most of whom had the specialized qualifications needed in the production and documentation of scientific and technical intelligence.[48]

Foreign material of naval interest had become increasingly available as of 1968. To achieve a quicker reaction capability in the exploitation of the material, NAVSTIC developed designated points of contact and liaison officers in more than thirty naval research, development, test, and evaluation commands, laboratories, and facilities. OPNAV Notice 5450 of 20 December 1968 changed the title of the head of NAVSTIC from officer in charge to commanding officer.[49]

The NAVSTIC mission statement in 1969 showed a few changes when compared to the 1967 version (highlighted by italics):

> To produce scientific and technical intelligence of naval interest as directed by *CNO* and DIA; *to provide scientific and technical intelligence to the Naval Establishment*; to maintain close scientific and technical relationships with the U.S. *government* research and development *community*.

The former requirement to "assist the Naval Establishment in analysis of current intelligence" had been deleted.

The aid of a group of distinguished scientists was enlisted by NAVSTIC during 1969 to study the problem of how to overcome the lead time gained by the USSR in its development of weapon systems over the U.S. Navy's production of countermeasures. The effort concentrated initially on the field of electro-optics.[50]

In January 1970, NAVSTIC commenced moving most of its staff from the Naval Observatory into the newly completed second floor addition to the Naval Reconnaissance and Technical Support Center (NRTSC) in Suitland, Maryland. As early as the fall of 1967, however, an initial NAVSTIC contingent had been moved to the NRTSC building when ONI reclaimed the responsibility to produce the *Naval Ships Characteristics Handbooks* series from the Defense Intelligence Agency, along with the related, voluminous files; the space available at Building 52 had proved insufficient, and the production group, under Lt. Arthur D. Baker III, USNR, had been transferred to Suitland.[51]

The first links of a new scientific and technical intelligence communications net were established during fiscal year 1972 to provide a secure, all-source teletype circuit, under the control of NAVSTIC, that included the Army's Missile Intelligence Agency and the Air Force's Foreign Technology Division. The Army's Foreign Science Technology Center was also expected to become a subscriber. The net was intended to serve as an informal analyst-to-analyst exchange net and was not used for official command traffic. Through the use of uninhibited dialogue, the net was expected to be a valuable tool to scientific and technical analyses, particularly in the field of missile intelligence.

Soviet naval technology developments of major proportions precipitated the development within ONI in fiscal year 1972 of four new programs designed to ascertain Soviet progress in antisubmarine warfare, certain aspects of communications, antiship missiles, and electromagnetic antenna design.[52]

On 30 June 1972, NAVSTIC and NRTSC were merged; both organizations were already occupying the same building in Suitland, Maryland. The new organization was given the name Naval Intelligence Support Center. The combined organization was intended to

> produce technical intelligence of national and naval interest as directed by the Director, Defense Intelligence Agency and the Director of Naval Intelligence; provide scientific, technical, and imagery intelligence support to the CNO and his staff, to the Naval Material Command and the Systems Commands, to the Navy Laboratories, and to the Fleet; support research, development, test and evaluation of equipment, materials and techniques associated with the production of technical and imagery intelligence; and provide technical guidance for collection activities.

Capt. John P. Prisley, who on 21 April 1972 had relieved Capt. Wallace A. Greene as commanding officer of NAVSTIC, took command of the Naval Intelligence Support Center on 30 June.[53]

Chapter Notes

1. *SECNAV Annual Report, 1890*.

2. *SECNAV Annual Report, 1896*, 48.

3. Department of the Navy, "Administrative History of the Office of Naval Intelligence in World War II," 10 Jul 1946, unpublished MS, 9, hereafter *ONI WWII Admin History*.

4. LCdr. Charles O. Maas, USNRF, "History of Naval Attaché, Paris," 1 Jun 1919.

5. RAdm. Arthur H. McCollum, Oral History, USNI, Annapolis, MD, 1971, 1:146–50, copy in OA.

6. *ONI WWII Admin History*, 928–29.

7. Ibid., 930–33, 940–42.

8. RAdm. Julius A. Furer, *Administration of the Navy Department in World War II* (Washington: Naval History Division, 1959), 791–96.

9. *ONI WWII Admin History*, 948–51.

10. Capt. Chester J. Oleniacz, USNR, notes prepared for the author, 13 Dec 1974. Oleniacz served in NAVTECHMISEU, 1944–1945.

11. *ONI Roster*, Nov 1944; Naval Scientific and Technical Intelligence Center (NAVSTIC), *Command History, 1967*, 9; and *ONI WWII Admin History*, 954–57.

12. *ONI Rosters*, 1945.

13. Commander Naval Forces, Europe (COMNAVEU), "Administrative History, 1940–1946," 342–45, hereafter *COMNAVEU Admin History*.

14. Naval Technical Mission, Europe Historical Section, "Historical Data on NAVTECHMISEU," 1 Nov 1945, 8–10.

15. Ibid., 1–13, 94.

16. Ibid., 36–39.

17. Ibid., 18–21, 34.

18. Ibid., 22–32.

19. Ibid., 36–39.

20. Ibid., 12–13.

21. Ibid., 77–78, 82–83.

22. *COMNAVEU Admin History, 1940–1946*, 343–45.

23. Naval Technical Mission, Japan (NAVTECHJAP), "History of Mission, Report to the Chief of Naval Operations," 1 Nov 1946, 1–8, hereafter *NAVTECHJAP Admin History*.

24. ONI F-Branch contribution to "CNO Report to the Secretary of the Navy," 4 Jun 1946.

25. *NAVTECHJAP Admin History*, 13.

26. *ONI WWII Admin History*, 961–63.

27. *ONI Review*, Feb 1949, 22–26.

28. Capt Bosquet N. Wev, USN, Chairman, Joint Intelligence Objective Agency, testimony transcript, 27 Jan 1947.

29. *ONI Review*, Feb 1949, 27.

30. Capt. Bosquet N. Wev ltr to author, 17 May 1976. Although Wev provided a copy of the testimony transcript, he could not recall the precise name of the subcommittee.

31. *ONI Review*, Dec 1949, 15–16.

32. *NAVSTIC Command History, 1967*, 9.

33. William E.W. Howe memo to author, notes on the development of NAVSTIC, undated [mid-1970s].

34. *ONI Rosters*, 1948.

35. Howe memo on NAVSTIC.

36. OP-322F memo, 18 Jun 1951, file on Conferences etc., box 3, Job 12876, OA.

37. "Review of ONI Functions and Workload, February–May 1952," 128–29. The distinction was not drawn in ONI Instruction 05201.6, "Designation of Material Responsibility within OP-322F" for example, nameplate data analysis was assigned to OP-322F2 by the instruction but was actually performed by OP-322F1.

38. ONI Instruction 5605.4 of 11 Sep 1953.

39. Howe memo on NAVSTIC.

40. W.E.W. Howe memo to author, on the *Tuapse* exploitation, undated.

41. *ONI Rosters*, 1957 and 1958.

42. Howe memo on NAVSTIC.

43. OP-922G ltr, 15 Jul 1958 in H4 REPGER (General) File, 1950–1961, OA.

44. *NAVSTIC Command History, 1967*, 9.

45. SECNAV Notice 5450 of 20 Jun 1960.

46. *NAVSTIC Command History, 1967*, 9.

47. Howe memo on NAVSTIC.

48. *NAVSTIC Command History, 1967*, 9.

49. *DNI Annual Report to CNO*, Jun 1968.

50. DNI memo to CNO, "Naval Intelligence Developments in 1969," undated, in author's files, OA.

51. Naval Reconnaissance Technical Support Center (NRTSC), *Command History, 1970*, 5; information about 1967 arrival from A.D. Baker III.

52. *DNI Report to CNO, FY 1972*, 5-6.

53. Naval Intelligence Support Center (NISC), *Command History, 1972*, 1.

CHAPTER 12

Air Intelligence

Pre–World War II U.S. Navy Air Intelligence

One might consider the Navy's air intelligence as having been first tried out on 3 August 1861 when John La Mountain made his first ascent in a tethered balloon from the Union ship *Fanny* at Hampton Roads, Virginia, to conduct aerial reconnaissance of Confederate batteries on Sewell's Point. The effort apparently proved to have some merit (or, at least its potential was not disproven) for similar aerial reconnaissance efforts were tried several more times during the Civil War.[1]

The airplane's use as a platform for collecting intelligence information needed by the Navy was recognized officially by Secretary of the Navy George von L. Meyer in his annual report for 1912 in which he commented that aircraft could be carried, stowed, and used by all large ships to reconnoiter an enemy's port or to search out the enemy's advanced bases and extend the eyes of the fleet in naval scouting or blockading operations.

The Navy's first use of airplanes for reconnaissance in a combat situation was in April and May 1914, during fleet operations in connection with the occupation of Veracruz, Mexico. Two seaplanes were carried by the pre-dreadnought battleship *Mississippi* and the scout cruiser *Birmingham* to Veracruz and Tampico, respectively. The aircraft's scouting work for the fleet commander in chief assured him of the absence of mines and located underwater obstructions. The aircraft were judged to have been "of inestimable value in scouting for the combined operations of the Army and Navy," much of their operations having taken place over the trenches protecting Veracruz. See Chapter 4 for more details on the collection phase of air intelligence.[2]

At the outbreak of World War I and before the United States became involved, three of the Navy's pioneer aviators were sent as assistant naval attachés to U.S. embassies in Europe to serve as expert observers in aviation matters: Lt. John H. Towers to London; Lt. Victor D. Herbster to Berlin; and 1st Lt. Bernard L. Smith, USMC, to Paris.[3]

The instructions for the intelligence officer at Naval Air Station (NAS), Porto Corsini, Italy, Ens. A. R. Tilburne, USNRF, were typical of the guidance given to intelligence officers at air stations during the latter stages of World War I:

> The Intelligence Officer shall, under the direction of the Commanding Officer, procure and prepare for distribution literature concerning the operation of aircraft. He shall keep posted up to date, a military map and a hydrographic chart of the area coming under the jurisdiction of the station to which he is attached. He shall keep posted up to date on such information as he may be able to obtain of all enemy dispositions within his area. He shall keep on the largest practicable scale map, the position of the Allied, enemy and American aeronautical enterprises. He shall prepare the daily station report and keep on file all statistical data therein required. He shall be responsible for all signal equipment of aircraft and station.[4]

The concept of the status and duties of peacetime intelligence officers at naval air stations under the naval district commandants was explained by Capt. Thomas T. Craven in March 1920, in an article that was repeated in the Office of Naval Intelligence *Monthly Information Bulletin* of 15 September 1920:

> The intelligence officer should be an aid for operations, strategy and confidential information. He should be a line officer, especially trained and, when possible, a pilot or ex-pilot. His duties include keeping track of enemy movements; keeping track of the movements of own forces; keeping custody of confidential books; taking care of communications including radio, telephone, telegraph and pigeons; having cognizance of photography, reports of operations, and meteorology.

Prior to the outbreak of World War II in September 1939, information on foreign aviation reached the Bureau of Aeronautics (BUAER) from many sources, including naval and military air attachés in Europe, foreign representatives of the National Advisory Committee for Aeronautics (NACA), and certain foreign contacts of U.S. aircraft manufacturers. Information collected by naval and military air attachés was received via ONI, and the NACA representative in Paris, LCdr. John J. Ide, USNR, forwarded his technical information to both NACA and ONI. BUAER thus received Ide's reports from both NACA and ONI.

With the collapse of the French armies in June 1940, NACA closed its Paris office. LCdr. Ide was recalled to active duty and assigned to BUAER as head of its Technical Information Section in December 1940. In the following months, he attempted to convince the Chief of BUAER of the need for formally establishing an intelligence section to build up and systematize the work of evaluating, interpreting, and drawing conclusions from air operational and technical information.[5]

To improve the distribution of intelligence to the various Navy technical bureaus, Chief of Naval Operations (CNO) letter serial 981116 of 26 November 1940 directed all bureaus and divisions of the Navy Department to establish organizations to receive and circulate naval intelligence reports to cognizant sections of their organizations. As a result, the chief of BUAER designated a liaison officer to work with ONI.

In the summer of 1941, intelligence reports on radar, fighter direction and antisubmarine warfare, particularly from Cdr. Ralph A. Ofstie and LCdr. John P. W. Vest, Naval Attaché for Air and Assistant Naval Attaché, London, respectively, were not getting through to the proper desks at the Office of the Chief of Naval Operations (OPNAV) and BUAER. The side-tracking of the reports was found to be taking place in the technical bureaus themselves.[6]

Also in the summer of 1941, the air intelligence function in BUAER had grown to where it made up a large part of the workload of the Technical Information Section. On 1 August 1941, LCdr. John Ide proposed that the intelligence and technical functions be separated and that intelligence be given the status of a separate section.[7]

A Chief BUAER letter of 29 September 1941 set up Ide's recommended Aviation Intelligence Branch in the Administration Division of BUAER. The branch consisted of the Air Intelligence Section under LCdr. Ide and the Technical Applications Section under LCdrs. Frank C. Sutton and Steadman Teller. For Ide's section, the principal sources of information were ONI and the Army's Military Intelligence Division (MID), with which close liaison was maintained. The initial functions of the Air Intelligence Section were to collect, index, and distribute information on Allied and enemy aviation within BUAER and to naval air stations. The Technical Applications Section prepared studies for the Chief of BUAER on the development and tactical use of aircraft, radar, night fighters, aircraft carrier complements, etc.[8]

Air Intelligence During World War II

With the U.S. entry into the war, air intelligence was shifted on 26 December 1941 from the Administration Division to the Planning Division of BUAER, thus reflecting an appreciation of the increasing importance of operational intelligence in support of planning.

By early 1942, a large volume of foreign and U.S. information was being received by the Air Intelligence Section from ONI and from British sources. During January 1942, new specialized functions were added, and the Air Intelligence Section was given branch status. As of 24 February 1942, the Aviation Intelligence Branch comprised four sections:

(1) The Foreign Intelligence Section disseminated foreign aviation intelligence to BUAER and to naval aviation shore establishments. It also compiled foreign data and statistics, assisted in the evaluation of all foreign information, and participated in logistic and tactical planning by keeping readily available all information on landing fields and seaplane operating facilities.

(2) The U.S. Information Section collected and collated data on U.S. forces; maintained liaison with the War Plans Division of OPNAV; and prepared periodic reports, records of current operations, daily war diaries, and war maps.

(3) The Tactical Applications Section analyzed all information coming into the branch for implications pertinent to the Navy's current air tactics. It also developed summaries and analyses for the United States from information available to the branch.

(4) The Strategic Information Section interviewed selected officers returning from operating areas and edited and published the information collected.[9]

Four Army-Navy teams were sent out in early 1942 to obtain crashed or captured enemy aircraft and equipment, take pictures, and make special reports to ONI and MID. The teams were assigned to cover India-Burma, China, the Southwest Pacific, and the Pacific Ocean. They furnished commands in those areas with "hot" information of operational value.

Once they had reached combat areas, the air combat intelligence (ACI) officers were sources of valuable information on the characteristics and performance of the various types of Japanese aircraft. The information they obtained was generally acquired through debriefings of combat pilots, examination of captured equipment, and, occasionally, from interrogations of prisoners of war.[10]

The need for an organization to collect, analyze, and distribute intelligence derived from our own air combat experiences also became clearly evident. The information on the Navy's first air combat actions was not adequate or in sufficient detail to permit analysis that would provide guidance for improvement in tactics in future air combat situations. To remedy the situation, it was decided to train specialists in air intelligence in the same manner as had been done by the Royal Air Force, and the Naval Air Combat Intelligence School was set up by BUAER at NAS Quonset Point, Rhode Island, in April 1942. While collecting and reporting air technical intelligence had been recognized as a responsibility of ONI and its attachés since World War I, the collection, analysis, and application of air combat intelligence was of immediate interest to the Navy's aeronautical organization. The Bureau of Aeronautics, therefore, assumed responsibility for meeting the new intelligence requirement and assigned it to its Air Intelligence Branch.[11]

The Aviation Intelligence Branch was renamed the Air Information Branch in December 1942 to avoid confusion with ONI's internal organization of the same name. At the same time, the title air combat intelligence officer was changed to air combat information officer.[12]

In connection with the mutual exchange of intelligence between ONI and BUAER, there was some concern in ONI that air intelligence collection opportunities were being missed during ONI's interrogations of captured German submariners. Many of the Germans were former aviators. An undated memo by LCdr. Henry J. White to Capt. Adolf Von S. Pickhardt of ONI expressed the view that Air Combat Information officers who were technically competent and linguistically qualified should be additionally trained as prisoner-of-war (POW) interrogators. None of ONI's POW interrogators were technically competent in aviation matters. White's memo also expressed the expectation that there would be an increasing number of German submariners who had been pilots and air-crewmen.[13]

Many of the first graduates of the Air Combat Intelligence Officers School were assigned to the South Pacific for the Guadalcanal campaign. Some were put ashore on Guadalcanal, some were sent to Espiritu Santo, and the remainder were assigned to Commander South Pacific Forces at Noumea, New Caledonia. They became involved in many aspects of air intelligence, such as locating downed enemy aircraft in order to salvage equipment and retrieve documents; updating maps and charts of the area and developing new operational maps and charts; developing air-sea rescue procedures; and devising escape and evasion nets for retrieving downed airmen. ACI officers even assisted the torpedo boat squadrons until intelligence specialists could be assigned to those squadrons.[14]

Briefing pilots and air crew before each mission was, of course, a primary duty of the ACI officers. They briefed not only on the target and its defenses, but also on all possible survival, evasion, and escape information that would be of help if the pilots were forced down behind enemy lines. Upon their return from a mission, pilots were interrogated by ACI officers regarding the target, enemy forces encountered, and other details that would be of value to future missions. Men who returned to their units from a successful evasion of capture in enemy-occupied territory were debriefed, and their experiences were tabulated by ACI officers and given speedy dissemination.[15]

By early 1943, the Air Information Branch was processing a tremendous volume of intelligence information. To improve its efficiency and to meet the needs for wider dissemination of aviation intelligence information, the branch was reorganized on 20 January 1943 and expanded to six sections:

(1) The Special Foreign Projects Section (former Foreign Information Section).

(2) The Strategic Air Information Section (former Strategic Information Section).

(3) The Material and Performance Section specialized in technical information on Japanese aircraft and to a lesser extent on German aircraft, and also had general cognizance over all aircraft information.

(4) The General Information Section was organized to assemble information on U.S., Allied, and enemy aviation; supervise and set up machine data card records of air combat reports; and furnish statistical data as requested.

(5) The Dissemination Section collected all excerpts and briefs; evaluated studies and material in the branch for the purpose of editing, duplicating, and publishing material as approved by the head of the branch; and distributed information within BUAER, the Navy Department, aviation shore establishments, and the fleet. It also filled special requests for information and maintained a constant check on the adequacy of the air intelligence distribution system.

(6) The Administration Section, in addition to routine duties of office management, personnel, and

files, supervised the administration of the Air Combat Intelligence Officer program.[16]

As the war progressed, the work of the Air Information Branch of BUAER became less concerned with the technical aspects of air intelligence and more involved with operational intelligence matters. When planning, personnel, and training were consolidated in the new office of the Deputy Chief of Naval Operations (DCNO) for Air, it was decided that the Air Information Branch of BUAER more properly belonged in the new office. On 18 August 1943, it was transferred and reorganized into five branches:

(1) Combat Information Branch (former General Information Section);

(2) Technical Intelligence Branch (former Material and Performance Section), which also took over the responsibility for studying foreign air forces, a task previously performed by the Special Project Section;

(3) Analysis and Statistics Branch, which made statistical studies of U.S. and enemy aircraft, tactics, weapons, loss and damage, and flak analysis and maintained combat statistics;

(4) Photo Interpretation Branch; and

(5) Services Branch.[17]

In September 1943, the Air Information Division under the DCNO (Air) was renamed the Air Technical Analysis Division, a name it retained until it became part of ONI four months later.[18]

On 19 October 1943, the air functions of the Identification and Characteristics Section (OP-16-P-2) of ONI were transferred to the Air Technical Analysis Division of DCNO (Air), and on 13 November the entire master file of eighty-one Japanese, German, Italian, and Russian aircraft drawings and the complete photographic files were assigned to the division.[19]

An example of duties performed by ACI officers were those of Lt. Charles S. Melvin, assigned to Patrol Squadron (VP) 23, which flew PBY-5 Catalina flying boats and was based at Tulagi in 1943. Melvin found that his duties included not only the usual intelligence functions but also service as recognition officer, assistant operations officer, assistant ground training officer, and assistant communications officer. VP-23 was involved in reconnaissance, search and rescue, antisubmarine warfare, coastwatcher supply, and bombing operations. Melvin made up flight schedules, assigned crews and missions, briefed and debriefed crews, wrote up reports, and coded and decoded radio messages. As ACI officer, Melvin was also custodian of classified material, provided charts and maps, and served as material and supply officer for the squadron.[20]

On 24 January 1944, the Air Technical Analysis Division (OP-35) was incorporated as a branch of ONI when the Air Intelligence Group (OP-16-V), was created. It was organized and functioned as follows:

(1) The Evaluation Section (OP-16-VE) performed liaison with BUAER and DCNO (Air), prepared air combat information for air units, and analyzed and summarized air combat action reports.

(2) The Service Section (OP-16-VS) filled needs of air combat intelligence officers in the fleet for intelligence, reproduced and distributed intelligence material prepared by OP-16-V; maintained an intelligence library; collected, analyzed, and distributed information on foreign air facilities; conducted the terrain model workshop in the American Museum of Natural History in New York; and collaborated with the Hydrographic Office in preparation of aviation charts and target data as developed by OP-16-V.

(3) The Analytical and Statistical Section (OP-16-VA) collected and summarized operational data on naval air combat, and studied air technical documents.

(4) The Technical Air Intelligence Section (OP-16-VT) produced data on the performance and characteristics of Allied and enemy aircraft engines and equipment, as well as on design and construction details about enemy aviation material; it also prepared updated drawings, master models, and photography of enemy and Allied aircraft, and it trained, equipped, and supervised technical air intelligence field personnel.

(5) The Photographic Interpretation Section (OP-16-VP) conducted liaison with the Photo Interpretation Center at the Naval Air Station, Anacostia, whose functions included training officers in photogrammetry, map reading, aerial photo interpreting and the making of rubber terrain models (see Chapter 13).[21]

OP-16-VT, in accordance with Secretary of the Navy letter serial 1296916 of 28 June 1944, was moved to NAS Anacostia and became the Technical Air Intelligence Center. Also on 28 June 1944, the Overseas Air Facilities Subsection of OP-16-VS was combined with the Air Movements Branch of the Army Air Force at Gravelly Point in the District of Columbia to form the Air Facilities Branch, AC/AS-2.

On 5 May 1944, management and control of the U.S. Navy Terrain Model Workshop in New York was transferred to the Photo Interpretation Center. The workshop itself remained in New York.

On 30 May 1944, the Flak Intelligence Unit was activated as a subsection of OP-16-VA to study enemy antiaircraft (AA) fire by type, effectiveness, and weapons. A four-week flak analysis course was inaugurated in July 1944 at the Naval Air Combat Intelligence School at Quonset Point, Rhode Island.[22]

Joint Services Air Intelligence

In November 1944, personnel from OP-16-VA who had been working on economic analysis and damage assessment of targets were shifted to the newly established Joint Target Group to provide Navy and Army Air Corps commands with lists of air targets, including detailed target information and the recommended munitions to be employed; standard air objective folders; damage assessments following attacks; and technical studies of effects of different weapons against specific targets. Administrative control of the Navy unit of the Joint Target Group was held by the Director of Naval Intelligence (DNI).

In June 1945, the Flak Intelligence Subsection of OP-16-VA joined the Flak Agency of the Army Air Corps to form the Army-Navy Flak Intelligence Group. Kamikaze tactics, guided missiles, night carrier operations, and other developments expanded the areas of interest and requirements of the group. On 30 October 1945, the Air Intelligence Group was renamed the Air Branch and designated OP-23V (ONI's designator had been changed from OP-16 to OP-23 on 10 October 1945).

To continue joint Army-Navy action in the field of air intelligence, a Joint Army-Navy Air Intelligence Division (JANAID) was approved by Joint Chiefs of Staff directive in JCS 1020/3 of 14 November 1943. JANAID was instructed to prepare continuing estimates of alien air forces and their potentialities, strategic objectives in alien countries, and conditions and installations in alien areas of specialized concern to air operations. JANAID was specifically excluded from the collection of basic data and was intended to replace existing joint Army-Navy air intelligence activities.[23]

Secretary of Defense James V. Forrestal, in a 2 October 1948 memorandum to the Secretary of the Navy and the Secretary of the Air Force, recognized their joint interest in air intelligence production and directed that there should continue to be a joint arrangement for the evaluation and production of air intelligence and that naval personnel should participate in such an arrangement. He further stated, "I wish to emphasize dominant interest does not mean preclusion interest." National Security Council Intelligence Directive No. 3 dated 13 January 1948 stated that "for the purpose of intelligence production," the Department of the Air Force would have "dominant interest" in air intelligence.[24]

As a result of a "Joint Agreement for the Production of Air Intelligence"—dated 29 March 1950 and signed by DNI RAdm. Felix L. Johnson and MajGen. C. P. Cabell, Director of Intelligence, U.S. Air Force—naval personnel assigned to the Directorate of Intelligence, USAF, were completely integrated at all levels within the Air Intelligence Production Divisions (AIPD). Paragraph 3C of the agreement stated:

> It is recognized that new situations may require changes in the organizational structure or the functions of the Air Intelligence Production Divisions. When such changes are of concern to the Navy, the Director of Naval Intelligence will be consulted. Otherwise, such changes are considered to fall in the category of routine Air Force administration, which is a responsibility of the Director of Intelligence, USAF.

One achievement worthy of mention came out of ONI's collaboration with the Air Force in the production of air intelligence during the Korean War. A classical correlation of hundreds of bits of raw intelligence consisting of personalities, places, events, and times led to publication in the *ONI Review*, in 1952, of the first evaluated and collated information on the first of the long series of Soviet antiship missile systems, Komet III. The analysis was followed by appropriate ONI (OP-322V2) recommendations to the Chief of Naval Operations for countermeasures and led to initial funding for a defensive electronic countermeasures systems for ships.[25]

On 29 April 1952, the Director of Intelligence, USAF, with the approval of the Vice Chief of Staff, USAF, unilaterally abolished the Office of the Assistant for Production in violation of the Joint Agreement of 29 March 1950. The Director of Naval Intelligence was not informed until after the fact. Abolition of the office, in effect, disestablished the Air Intelligence Production Divisions. Prior to the forced reorganization, there had been a total of three divisions in the AIPD. The Evaluation Division was headed by a naval officer with an Air Force deputy. The other two divisions, Estimates and Targets, were headed by Air Force officers with Navy and Marine Corps deputies, respectively.

Because of the change in organization, and to ensure a continuation of the desirable and necessary allocation of top billets between the services in the production of air intelligence, the Director of Naval Intelligence wanted a complete and mutually acceptable billet structure approved prior to the signing of a new agreement. However, to assist the Army in obtaining personnel for assignment to air intelligence production and after assurance by the Director of Intelligence, USAF, that a mutually agreeable billet structure would be set up as soon as possible, the Director of Naval Intelligence reluctantly signed the agreement on 16 May 1952.

A major reorganization of the Air Force Directorate of Intelligence was then instituted, with no naval officers as division chiefs and no Army or Navy representatives in a new Policy and Management Group. The group, responsible for "Requirements, Plans and Programs, Development and Management," controlled the production of air target intelligence for use by all three services. Such a unilaterally conceived and controlled intelligence production system was not satisfactory and did not meet the Navy's requirements.

Accordingly, a series of papers were submitted to the Joint Chiefs of Staff by CNO Adm. William M. Fechteler, "in order that an obviously unsatisfactory situation may be corrected, to the end that the production of air intelligence may adequately and satisfactorily serve the best interests of all three Services and the Nation."[26] The Air Force response was that it had no requirement for joint participation in the production of air intelligence.[27]

JCS memorandum 2056/47 of 13 May 1953 provided for the integrated participation by Army and Navy personnel in the Estimates and Targets Directorates of the office of the Assistant Chief of Staff, Intelligence, USAF (AFCIN). An ad hoc committee produced a memorandum of agreement for the implementation of JCS 2056/47, dated 23 October 1953, that was approved by DNI memorandum serial 024676P32 of 18 December 1953.[28]

Following a comprehensive study in 1954 of the mobilization requirements for naval collaboration in joint services air intelligence, a revision of the Navy's representation was recommended:

	New Complement	1954 Allowance
Naval Officers	60	55
Marine Officers	6	6
Civilians	132	132[29]

By authority of the Chief of Staff, USAF; the Chief of Naval Operations; the Chief of Air Staff, Royal Canadian Air Force; and the Chief of Air Staff, Royal Air Force, the United States, Britain, and Canada formed an Air Standardization Coordinating Committee for designating Soviet aircraft and guided missiles to satisfy the requirements of operations, intelligence, communications, and training. The system was to be simple, usable with a limited vocabulary, suitable for voice and radio communications, as descriptive as possible, consistent with security, and adaptable to the inclusion of new Soviet aircraft and guided missiles.

The devised and adopted system employed the initial letter of each selected aircraft nickname to indicate the aircraft's operational role (e.g., "F" for fighter, "B" for bomber); single syllable words were used for nicknames of propeller-driven aircraft and two syllables for jet aircraft. For guided missiles, the initial letters indicated the weapon's operational role (e.g., "SS" for surface-to-surface, "SA" for surface-to-air). The names were chosen by coordinated agreement of the three participating nations.[30]

To implement the Joint Chiefs of Staff Directive requiring collaboration of ONI and the Air Force's Directorate of Intelligence in the production of air intelligence, a separate organization, Naval Collaboration in Air Intelligence (NACAIN), was established by Secretary of the Navy Notice 5450 of 26 August 1957, with an officer in charge under the military command of the commandant of the Potomac River Naval Command. He also reported to the Director of Naval Intelligence for additional duty as the head of the NACAIN Branch of ONI and to the Directorate of Intelligence, USAF, for additional duty as required. His office was under the management control of the Chief of Naval Operations. A joint Navy-Air Force air intelligence production effort had been operating at least since 1948, but the establishment of NACAIN changed the Navy's participating element to a field command, with the officer in charge double-hatted within the Office of the Chief of Naval Operations as OP-922V2.[31]

As of 1962, Radar Target Materials were being produced for use in the preparation for, and the accomplishment of, all-weather missions involving bombing, mining, navigation, and reconnaissance. The materials consisted of graphic, textual, radar-photographic, tabular, and other presentations of radar target intelligence, both from actual and predicted or simulated radar scope photography. Much of the material was produced under the Air Target Materials Program to meet standards and specifications jointly approved by the Navy and Air Force. Some material was produced by the Naval Photographic Interpretation Center (NPIC) to meet unilateral Navy requirements. The Air Branch (OP-922V) of ONI was the CNO–ONI point of contact and coordinator for Radar Target Materials.[32]

Air Intelligence During the Korean War

When the Korean War broke out in June 1950, one immediate requirement was to get qualified air intelligence Naval Reserves back on active duty. No real effort had been made to keep track of Naval Reserve air intelligence officers trained during World War II after they were released from active duty in 1945. Some, however, had reestablished contact when the Air Intelligence Reserve program was activated. By October 1950, a total of fifty-two had been successfully recalled; of these, seventeen were assigned to the Pentagon, mostly in OP-322V2 (the joint ONI-Air Force intelligence group), and thirty-five went to the fleet.[33]

Because there was a lack of qualified air intelligence officers, an effective naval air intelligence organization did not exist in the Japanese-Korean area at the outbreak of the Korean War. By 30 April 1951, conditions had improved to the extent that adequate numbers of air intelligence personnel were in the area, but there was still a shortage of qualified photo interpreters.[34]

Basic information and materials were also inadequate, initially, for the conduct of naval air combat operations. For example, gridded charts suitable for target designation were few to nonexistent. In many instances, different grids were used by the Army, Navy, and Air Force, and place names were spelled differently or varied widely on the charts of the various services. Correlation and verification by geographic coordinates was the only solution. Providing target intelligence to naval shore bombardment forces posed similar problems.[35]

Although U.S. forces had occupied a large part of Korea for several years following World War II and they had been directed to carry out an extensive mapping program during that period of occupation, the quality of Korean maps, as of June 1950, was unsatisfactory.[36] Another initial deficiency was the delay in the receipt of information by naval air units on ground force dispositions, enemy air activity, and ship movements. Escape and evasion information supplied to naval air units was also inadequate. There was a need for aviation officers of sufficient rank and intelligence qualifications to serve on fleet and force staffs to organize and guide the air intelligence effort.[37]

Naval Reserve air intelligence officers with World War II experience who had been recalled to active duty were sent out to join carriers, carrier air groups, and squadrons participating in the Korean War. They received no refresher training before departing from the West Coast, but some were briefed by Commander in Chief, U.S. Pacific Fleet's Intelligence Division at Pearl Harbor and at Commander Naval Forces, Far East in Tokyo before reporting to their assigned commands. Some also spent a day or so at Far East Air Force and Commander in Chief, Far East headquarters to get the picture on Navy air support to, and collaboration with, Air Force and Army operations.[38]

One of the early problems was the need for intelligence support to strategic and tactical interdiction of railroads, highways, tunnels, bridges, etc. RAdm. Ralph Ofstie, Commander Carrier Division (COMCARDIV) Five in *Princeton* (CV 37), had on his staff LCdr. Benjamin H. Fisher, USNR, an air intelligence officer with Carrier Air Group (CAG) 19 (also in *Princeton*) assigned to temporary additional duty as COMCARDIV-5 Task Force Interdiction Officer. There was very little source material from which to obtain the information Fisher needed. Starting from scratch, Fisher arranged for strip photography of the enemy's primary routes. Vertical views were then obtained to pinpoint specific targets, complete photo interpretation was accomplished on primary interdiction targets in the task force's assigned area, and a catalogue of the bridges was prepared. Even before all the preparatory work had been completed, successful strikes were being flown, and the inventory of targets was updated accordingly. In addition, Fisher had to conduct briefings and debriefings on close air support missions, night fighters, hecklers, armed reconnaissance, and photo reconnaissance missions. The daily routine required eighteen to twenty hours every day during the period that the interdiction program was being researched and developed.[39]

Based on experiences in the Korean War, CAG-19 and COMCARDIV-5 intelligence officers in early 1951 developed a list of "Duties of the Air Group Intelligence Officer" to satisfy various briefing and instructional needs. A copy was sent informally to the Air Intelligence School in Washington, D.C., for its possible use. The list of duties included:

1. The Air Group Intelligence Officer is responsible for planning, developing and activating programs relating to Recognition, Air Intelligence, and Evasion and Escape; and for overall supervision and coordination of Air Group personnel assigned duties in these programs.
2. The Air Group Intelligence Officer formulates the daily work schedule of squadron Intelligence Officers and the other personnel mentioned above, after consultation and coordination with the squadron commanders and subject to the approval of the Air Group Commander.
3. Specific duties can be grouped best according to operational situations, as follows:
   a. Period of Organization:
      (1) Ascertain types of aeronautical charts and related material needed in the combat zone and make these needs known to the ship scheduled to carry the Air Group.
      (2) Initiate Recognition Program to include both recognition features and combat characteristics of enemy armored vehicles, submarines, surface vessels and aircraft.
      (3) Develop instruction program on Evasion and Escape Techniques, coordinating this program with the Survival Program developed by the Operations Department.
      (4) Develop displays and prepare lectures on the following subjects:
         (a) Enemy Order of Battle;
         (b) United Nations or United States Order of Battle;

(c) Combat Situation;
(d) Close Air Support;
(e) Interdiction;
(f) Anti-Submarine Warfare;
(g) Enemy Flak;
(h) Enemy Air Force;
(i) Enemy POW Interrogation Techniques;
(j) Area Geography;
(k) Enemy Bases and Air Fields;
(l) Geo-Politics;
(m) Air Sea Rescue;
(n) Other Related Subjects.

(5) Work with the Air Group Commander and the Squadron Commanders in their planning and execution of combat exercises including typical air strike, close air support, amphibious operations, and anti-submarine warfare.

(6) Train squadron Intelligence Officers in briefing and debriefing techniques, using the opportunities presented in the above exercises to provide reality for both pilots and Intelligence Officers.

(7) Work with Air Group Commander and the Squadron Commanders in the development of a program to train pilots in the technique of aerial spotting of the enemy while on area searches and reconnaissance patrols. Here again this program should provide intensive briefing and debriefing practice for the squadron intelligence officers and familiarize pilots with debriefing procedures.

(8) Consult with respective Squadron Commanders regarding preparation of fitness reports for their intelligence personnel, writing up rough drafts for their consideration.

b. Period of Combat Operations:

(1) Distribute aeronautical charts, maps, templates and other materials.

(2) Obtain and distribute any specialized Evasion and Escape gear such as blood chits and area escape and evasion kits not normally available in the U.S.

(3) Emphasize combat related programs, especially Recognition, Air-Sea Rescue, and Evasion and Escape.

(4) Insure proper dissemination of current intelligence information from the ship's intelligence office through the squadron intelligence officers to the respective squadrons, to include the following:

(a) location of enemy forces;
(b) location of friendly forces;
(c) location of friendly airfields;
(d) location of enemy airfields;
(e) activities at these enemy airfields;
(f) location of friendly and enemy surface forces;
(g) location of "bird dogs" and air-sea rescue reference points;
(h) location of advance rescue stations;
(i) location of friendly intelligence parties;
(j) daily operation plans of friendly air;
(k) bombline information;
(l) location of submarine sanctuaries;
(m) location of areas in which air operations are restricted;
(n) location of targets for the day including primary, secondary, weather alternatives, and dump targets;
(o) location of search areas and reconnaissance routes;
(p) enemy air sightings;
(q) enemy's logistics effort including vehicular sightings, train sightings, and troop locations.

(5) Recommend reconnaissance areas and strike targets including weather alternatives and dump targets.

(6) Keep Air Group Commander and his staff informed on the enemy situation.

(7) Serve as liaison officer between the ship's Intelligence Office and the Air Group.

(8) Prepare statistics and reports, including the Aircraft Action Report, the Air Group Action Report, and any others as required.[40]

Photographic intelligence was an essential element of air intelligence in support of carrier air combat operations. The carrier and air group staffs organized their air intelligence, photo intelligence, and photo interpreters into a highly efficient unit to provide the carrier task force with the intelligence it needed to achieve its assigned objectives.

The selection of worthwhile military targets was often a major problem. Some target information was furnished by ground forces, but only sporadically. Several urgent strike requests were received, and the strikes were usually fulfilled with excellent results. In other cases, strike requests were so lacking in target description, or on information on the location of friendly forces relative to the target, that strikes could not be carried out until amplifying information had been requested and received.

Target search was the predominant type of photo reconnaissance coverage required and flown to meet air intelligence needs during the Korean War. Other types of photo coverage were flown to provide material for flak studies, damage assessments, and supply route and airfield activity analyses.[41]

Air Intelligence During the Cold War Era

During the mid-1950s, the aircraft carrier's nuclear strike capability expanded the scope of the air intelligence services and products required by Navy carrier attack squadrons, whose aircraft had both nuclear and conventional weapon-carrying configurations. The gathering of all necessary intelligence materials required to support a squadron's mission was the primary task for the air intelligence officer of any newly organized or reactivated attack squadron. First, a determination was necessary as to what material was available. On the West Coast, Commander Naval Air Forces, Pacific Intelligence Library at North Island provided clues on what should be obtained for shipboard reference. Then came the task of providing pilots with all the graphics and written material necessary for them to approach, identify, and attack successfully each of the many nuclear and conventional targets listed in current contingency plans. Detailed area studies were assembled, with charts and photographs to show navigation aids, alternative courses for low-level approach to the target, and the defenses to be expected—including missiles, which were fairly new at that time.[42]

An Integrated Air Intelligence System (IAIS), comprising an airborne multisensor collection system (using A-5C Vigilante and A3J-3 Skywarrior aircraft) and a shipboard processing installation called an Integrated Air Intelligence Center, was initiated during 1962. Director of Naval Intelligence secret letter 005187P92 of 6 June 1962 assigned to OP-92B4 (Automation Coordination Staff) the task of directing and coordinating the development of a prototype intelligence database for the IAIS. Production of the database was accomplished by the Naval Photographic Interpretation Center, Fleet Intelligence Center, Europe and the Atlantic Intelligence Center in conjunction with the IAIS surface system development at North American Aviation, Inc., and the aircraft carrier *Saratoga* (CVA 60). A master database was maintained at NPIC. In addition to the above, phase I of the program included providing the initial database and programming for the *Independence* (CVA 62) and for the Naval Air Station, Sanford, training installation. OP-922V was responsible for the direction and coordination of the operational production and maintenance of the IAIS database and for its proper distribution. It also maintained a standardized coding and indexing manual for IAIS use. See Chapter 20 for more details.[43]

Chapter Notes

1. Naval Historical Division, *Civil War Chronology, 1861–1865* (Washington: GPO, 1971), 21.

2. *SECNAV Annual Report, 1914*.

3. *ONI Review*, Aug 1956, 346.

4. Instructions in box 395, subj: PA, General Board Subject Files, 1900–1947 (RG 45), NA.

5. Bureau of Aeronautics (BUAER), "Administrative History of the Bureau of Aeronautics in World War II," unpublished MS, 1946, 19:5–6, hereafter *BUAER WWII Admin History*. See U.S. Naval Administrative Histories of World War I, NDL.

6. Office of Naval Intelligence (ONI), "Administrative History of the Office of Naval Intelligence in World War II," 10 Jul 1946, unpublished MS, 1206–7, hereafter *ONI WWII Admin History*. See U.S. Naval Administrative Histories of World War II, NDL

7. *BUAER WWII Admin History*, 6–7.

8. *ONI WWII Admin History*, 1208.

9. *BUAER WWII Admin History*, 8–10.

10. Ibid., 57–58.

11. *ONI Review*, Feb. 1946, 35.

12. *ONI WWII Admin History*, 1215.

13. Memo, file A8-2, box 16, Job 3697, FRC/WNRC.

14. *ONI Review*, Feb 1946, 37.

15. Ibid., Mar 1946.

16. *BUAER WWII Admin History*, 11–12.

17. Ibid., 13–14.

18. *ONI WWII Admin History*, 1210.

19. Ibid., 939.

20. Capt. Charles S. Melvin, interview by author, 11 Jul 1976.

21. *ONI WWII Admin History*, 568, 1218–1321 passim.

22. Ibid., 1221–22.

23. Ibid., 1223–24, 1329–30.

24. CNO (OP-322V) ltr to JCS, ser 00058P32, 5 Feb 1953, 6.

25. C. W. Merrill memo to author, Oct 1974.

26. CNO (OP-322V) ltr, 5 Feb 1953, 2–3, 7.

27. DNI (OP-322V) memo to OP-03, ser 000163P32, 13 Feb 1953.

28. ONI Instruction 5430.1 of 2 Jan 1959.

29. OP-922 memo, 30 Dec 1954, OP-922 file 1954, box 1, Job 12876, OA.

30. OPNAV Instruction 003561.1 of 12 Aug 1954 and 9 Mar 1959.

31. ONI-70-1, *Naval Intelligence Manual*, Change 1, 4 Nov 1957, 23; ONI Instruction 5430.1 of 2 Jan 1959; and *ONI Rosters*, 1957.

32. OPNAV Instruction 03800.10 of 23 Apr 1962.

33. Melvin interview, 11 Jul 1976.

34. Commander in Chief, Pacific Fleet (CINCPACFLT), *Interim Evaluation Report No. 2: Korean War Naval Operations*, 213.

35. CINCPACFLT, *Interim Evaluation Report No. 1: Korean War Naval Operations*, 944.

36. Ibid., 1497.

37. Ibid., 213.

38. LCdr. Benjamin H. Fisher ltr to Capt. Charles Melvin, 9 Jan 1951.

39. Fisher ltr to Melvin, 19 Feb 1951.

40. Fisher ltr to Melvin, 8 May 1951.

41. CINCPACFLT, *Interim Evaluation Report No. 6: Korean War Naval Operations*, 3–28, 3–74, 3–77.

42. Lt. Rudolf A. Fuentes, USNR, interview by author, 11 Jul 1976, 3–4, in author's file, OA. Fuentes was Aviation Intelligence Officer (AIO) for Attack Squadron (VA) 56.

43. ONI Instruction 05430.11 of 3 Jan 1963.

CHAPTER 13

Photographic Intelligence

Early Navy Photo Intelligence

One of the first projects undertaken by Lt. Raymond Perry Rodgers when he took over as Director of Naval Intelligence in 1885 was to write to various men of prominence in science, soliciting their opinions on the value of the camera for surveying and reconnaissance.[1]

A year later, the Chief of the Bureau of Navigation commented, "Much progress has recently been made in instantaneous marine photography, and as the process can now be successfully carried out underway on shipboard, it is my intention to supply our cruising ships with photographic outfits as rapidly as practicable."[2]

Thus, the Navy was off to an early start in the use of the camera in collecting information for intelligence. See Chapter 4 for additional information on photographic collection.

A report by Capt. H. E. Ives, received by the Office of Naval Intelligence in September 1918, described British use of aerial photography during World War I. The British employed hand-held cameras to photograph convoys, suspicious objects, and ships that failed to give proper recognition signals, coastal landmarks useful to aviators, and Allied ships and submarines for identification purposes.[3]

The systematic collection and filing of photographs was begun by ONI in 1936. The War Plan for Photographs gave the Naval Records and Library Branch (OP-16-E) the job of receiving, recording, and distributing all incoming photos to the offices primarily interested. The work was made the primary function of the War Records Section (OP-16-E-2), which at that time also maintained the Naval Historical Photographic Files. Naval Records and Library was headed by Capt. Dudley W. Knox, USN (Ret.), who was also curator for the Navy Department. Prior to 1936, there had been no central photo section.

By OP-16-E-2 serial 11247 of 6 April 1937, the Chief of Naval Operations (CNO) directed the collection of strategic photos and specified the types of photos desired. From 1936 to 1941, however, very little was actively done to acquire photographs of current interest or potential value for operational use.

In 1940, the handling of strategic photography was shifted from the War Records Section to the Strategic Photography Section (OP-16-E-3). On 25 November 1940 the name of OP-16-E-3 was changed to the Graphic Section.[4]

Navy operational exercises expanded in number and scope in 1940 and incorporated the increased use of photographic reconnaissance.[5]

Navy Photo Intelligence During World War II

Washington-Area Organization

The British first recognized the need for, and the military applications of, information extracted from photos taken over enemy-held territory. To learn their techniques in that method of collecting intelligence information, VAdm. Robert L. Ghormley, Special Naval Observer in London, requested in the spring of 1941 that an officer be sent to England from the Bureau of Aeronautics (BUAER). LCdr. Robert S. Quackenbush, Jr. was selected to make the study.

LCdr. Quackenbush arrived in England, saw the scope of the task, and recognized the importance of learning as much as possible about photo interpretation. He requested that the Navy send over additional officers, both Navy and Marine Corps, to increase the number who would be knowledgeable about British photo interpretation methods and procedures. Quackenbush also stressed the need for the establishment of a Navy school in the United States to train officers in the science of photo interpretation. As a result, on 12 September 1941, the

CNO authorized the establishment of a photo interpretation school under the Bureau of Aeronautics, to be located at the Naval Air Station (NAS), Anacostia. LCdr. Quackenbush was made officer in charge and Capt. Charles H. Cox, USMCR, and Capt. Gooderham L. McCormick, USMCR, were appointed executive officer and chief instructor, respectively. The initial class of twenty-eight Navy and Marine officers convened on 5 January 1942.[6]

The first photographic intelligence reports were received from the field by the Army and Navy in the summer of 1942. The Army created the first Photo Procurement Detachment, which was to function in coordination with the British in England. A U.S. Navy representative, briefed by the Navy on its needs, was assigned to the detachment. The arrangement was not satisfactory because all material was forwarded to the Army's Military Intelligence Service before dissemination to the Navy. Ensuring that material selected by the Navy representative in England was received by the Navy in Washington was a continuing task.[7]

In general, handling photo interpretation reports paralleled that for aerial photos. Separate sections were set up in both the Graphic Section of ONI and the Map and Photo Branch of Army Intelligence (G-2) to handle the material. The two sections maintained a constant mutual exchange of systems and techniques.[8]

As more trained photo interpreters became available, photo interpretation units were formed either as part of a photo group or as an intelligence center assigned to an area command. The units varied in size from thirty to one hundred officers and were assigned to air and surface units, amphibious commands, Marine Corps detachments, etc.

At the end of 1942, the Navy's Photographic Interpretation School moved into the newly completed Photo Science Laboratory at NAS Anacostia in the District of Columbia.

On 12 November 1943, the Secretary of the Navy established the Photographic Interpretation Center (PIC) at NAS Anacostia as an activity of the Air Intelligence Group of the Deputy Chief of Naval Operations for Air (OP-35). On 24 January 1944, supervision of the center was transferred to the Division of Naval Intelligence (OP-16-V). The primary functions of PIC were to conduct a school for training photo interpreters, operate a workshop for the manufacture of terrain models in quantity, and maintain a pool of trained photo interpreters to serve the needs of the fleet.

On 27 February 1945, the Secretary of the Navy established the U.S. Naval Photographic Intelligence Center (NPIC) at the Naval Receiving Station, Anacostia, under the management and technical control of the Division of Naval Intelligence.[9]

Pacific-Area Organization

Early in the summer of 1942, the Photo Reconnaissance and Interpretation Section, Intelligence Center (PRISIC) was formed at Pearl Harbor to serve as a pool of photo interpreters to be drawn upon by units of the Pacific Fleet and to perform a more thorough and detailed analysis than did those units concerned with interpretation for immediate operational use. PRISIC additionally became the Photographic Section of the Intelligence Center, Pacific Ocean Areas (ICPOA) in July 1942 and was divided into four sections in April 1944 as part of the Joint Intelligence Center, Pacific Ocean Areas (JICPOA).

The South Pacific Photographic Interpretation Unit was another early unit. Photographic Interpretation Squadron (INTERPRON) One was formed under Commander South Pacific (COMSOPAC) in July 1943 as part of Fleet Air Photo Group One, with headquarters at Guadalcanal until September 1944, at which time it returned to the United States. While at Guadalcanal, INTERPRON-1 furnished photo intelligence for the Solomon Islands campaign and the Peleliu landing. In July 1945, INTERPRON-1 returned to the Pacific, basing at Okinawa.

INTERPRON-2 was formed as part of Photo Group Two under Commander Aviation Forces, Pacific and was based at Eniwetok from April until October 1944, when it was moved to Guam. INTERPRON-2 provided photo intelligence during the westward drive in the central Pacific, including the landings in the Marianas, and the aerial and surface strikes against the Japanese homeland.

Other photographic intelligence units included the Central Interpretation Unit, Southwest Pacific Area, and the Advanced Intelligence Center, North Pacific area. The latter was originally established at Kodiak, Alaska, in October 1942 and was composed of both photo interpreters and air combat intelligence officers; in March 1943 the organization was moved to Adak, where it furnished the intelligence support for the Attu and Kiska invasions and the strikes against the northern Kuriles.[10]

In April 1944, the Joint Chiefs of Staff (JCS) assigned to the Navy Department prime responsibility for the collection of all graphic and photographic material for the Pacific Ocean area. The Navy set up a photographic review panel for its own use as well as for the use of Army G-2, Army Air Force A-2, and other interested activities.[11]

Submarine Photography

Normally, aircraft photos for intelligence purposes were far superior to submarine periscope photos. In many instances, however, the distance between the enemy objective and the closest allied air base was so great that aircraft photographic reconnaissance was impractical or inadequate. Also, the presence of many aircraft performing photographic reconnaissance over an enemy island potentially alerted the enemy as to the probable location of the next landing. Submarines could carry out photographic reconnaissance undetected, and, for that reason or because of the distance involved, they were called upon to substitute for, or to augment, aircraft reconnaissance. Submarines could also check the accuracy and orientation of charts, which was impossible for aircraft to do. For more details on submarine photography, see Chapter 6.[12]

Domestic Photographic Services

The "F" sections of each district intelligence office (DIO) throughout the war forwarded to ONI photos of operational areas, many of which were extremely valuable in supplying information on enemy-held areas. The DIOs were ONI's largest original source of graphic material. Between September 1943 and September 1944, approximately 36,000 pictures from the DIOs were received at ONI's Graphic Section. Duplication was very low, averaging 10 percent, and about one of every thirty pictures submitted eventually found its way into the Graphic Section files. At first, all material was used, but, as the files grew larger, directions were given not to process travel brochures, postcards, and similar material.[13]

The National Geographic Society made available its published and unpublished picture files and its records listing the names of picture contributors. The list was disseminated to the cognizant district intelligence offices, which would contact the individuals for any material of intelligence value that had not been sent to the National Geographic Society.[14]

Pictures were filed according to location. Some descriptive matter and the geographic coordinates were added before reproduction. One copy of the reproduced picture was mounted on a card about 13 inches by 9 inches. Across the top of the card were ten classifications into which each picture could fall: Aerodromes [airfields]; Oil Facilities; Utilities; Docks/Port Facilities; Railroads; Roads/Bridges; Coast/Beach Hydrography; Military and Naval Installations; Lakes, Rivers and Terrain; and Cities and Towns.[15]

Identification and Characteristics Section

A large proportion of the work of the Identification and Characteristics (I&C) Section (OP-16-P-2) of ONI during World War II depended on the interpretation of photos. At first, the pictures were mostly surface photos taken before the war, but, as time went on, more and more information was obtained from high-altitude aerial photos. OP-16-P-2 developed a highly trained group of photo interpretation specialists, carrying the process to a greater degree of competency than the Photo Interpretation School of BUAER could teach in its overall course. By informal agreement with the school, students at the school who were destined to specialize in photos of ships and aircraft spent two weeks of concentrated additional training at the I&C Section before assignment to sea or advanced base duty. During 1944 and 1945, ship-photographic interpreters from the school were assigned to I&C for temporary duty. These men, in general, had had overseas experience and were able to point the work of I&C toward specialized needs in the field and to coordinate the activity of the section with the publishing activities of the Photo Interpretation School.[16]

Post–World War II Organization

Because of the cutback in ONI funding, it appeared probable early in the postwar period that the activities of the Photographic Interpretation Center would have to be severely curtailed, if not eliminated. A SECNAV letter of 16 October 1946 disestablished the Photographic Intelligence Center under ONI and established in its place the Photographic Interpretation Center, retaining the acronym PIC, under the Bureau of Aeronautics. On 13 January 1947, PIC was designated a subordinate unit of the U.S. Naval Photographic Center under the military command and coordination control of the Potomac River Naval Command and under the management control of BUAER.[17]

Photographic Intelligence During the Korean War

In the early phases of the Korean War, photographs of intelligence value were sent back to the Photographic Intelligence Center at Anacostia, the Marine Corps headquarters, and other rear-area processing facilities to the detriment of Naval Forces, Far East (NAVFE) and Pacific Fleet units that had an immediate need for the intelligence information available in the photographs.[18]

During November 1950, a plan was effected within Naval Forces, Far East for more rapid dissemination of photographs within the Navy. After their immediate operational needs were satisfied, all units within NAVFE that were engaged in taking pictures (except public information photography) were to send all negatives and, if practicable, prints to COMNAVFE. Upon receipt of the material, the Intelligence Section of the COMNAVFE staff screened it for photographs of value to other elements of NAVFE, made prints and positive transparencies as required, and sent negatives (and prints, as practicable) to the Commander in Chief, Pacific Fleet. The intelligence section of COMNAVFE filled photo requests from other services within the Far East Command from its own photographic files and made requests to Army General Headquarters and the Far East Air Force (FEAF) to fill the needs of Navy units.

Commander Amphibious Group One had a well-equipped and well-staffed laboratory on shipboard for processing photographs and for reproducing surface photos. In 1950, aerial photography for support of amphibious landings, however, was in need of improvement. The inadequate responses to requests made to FEAF for routine and special photographic reconnaissance demonstrated the need for Navy-oriented photographic intelligence support. For example, in the preparations for the Inchon operations, where the tidal range was almost 30 feet and mud flats extended up to two miles offshore, photographic intelligence was urgently needed, particularly to show the height of seawalls above water level at various tide stages. The commanding officer of the FEAF squadron that was to perform the photographic reconnaissance was briefed on the predicted tidal data and was asked to get pictures of the seawalls at four stages of the tide for several days. The first two sets of photos were obtained, but then the weather interrupted. When the other two reconnaissance missions were finally flown, the same times were used as for the first two sets, thus negating their value for determining the seawall heights.[19]

One of the significant naval aerial photographic reconnaissance developments of 1951 was the increasing reliance of Task Force 77 (fast carrier force) on its own aerial photography for intelligence. By December 1951, it was estimated that TF 77 generated 90 percent of its own intelligence, most of it through photos and pilot debriefings. The unanticipated load on the photographic facilities of the carriers, particularly in the photographic interpretation field, pointed up the serious deficiencies in numbers of trained personnel and in equipment and space. Demands by other commands for photographic intelligence further increased the requirements placed upon the carriers' photographic facilities.

Fast-carrier operations in April and May 1951 increased the use of aerial photography for target planning, strike evaluation, and flak analysis. Commander Task Force 77's photographic reconnaissance plan of 2 May 1951 provided for most important bridge and airfield coverage every fourth day, city coverage once a week, and damage assessment on the day of the attack, whenever possible.

In June and July 1951, the introduction of a K-25 camera package on four AD Skyraider aircraft of each carrier air group resulted in better and more prompt pre- and post-strike damage assessment and permitted jet-powered photo aircraft to concentrate on other missions.

Considerable photography was taken by TF 77 photo aircraft for the use of the surface bombardment forces in counter-battery fire. Mosaics were made and sent to Commander Task Group 95.2 (gunfire support force) to show enemy shore guns and small boat activity.

In August and September, the entire North Korean east coast rail line from Wonsan to the Chinese border was mapped by TF 77. Also in September, pilots were provided with current flak mosaics of key bridges, prior to strikes, in the form of route study booklets to be carried in flight to help them avoid known flak areas.

Complete photographic coverage of Hungnam was made by TF 77 for the surface bombardment of that port on 20–21 November. A gridded mosaic was made, targets were annotated, and all beach defenses were plotted. Oblique pictures were also taken of the waterfront, and, with those two aids, specific targets could be systematically attacked.

The main advantage of TF 77 doing its own photography and photographic interpretation was that the usefulness of the information obtained was ensured by the immediacy with which the prints could be interpreted, the information evaluated, and the indicated desirable strikes launched.[20]

Patrol aircraft photographic equipment consisted of hand-held cameras for use in photographing shipping for identification purposes. As of 31 December 1951, specialized photographic aircraft were not included in the air group complements of the light and escort carriers of Task Element 95.11, whose aircraft complement was too small to accommodate specially configured photo reconnaissance aircraft.[21]

The bottleneck in the production of intelligence from photography in the Korean conflict area in 1951 was caused by a lack of adequate processing facilities and photographic interpretation personnel. The planned use of facilities and personnel at NAS Atsugi was expected to improve the situation.[22]

A shore-based photographic interpretation unit within carrier air distance of the operating area had been recommended for many months to reduce, as much as possible, the work load on carrier photographic laboratories and photographic interpretation personnel. On 1 July 1952, a photographic interpretation unit was established at NAS Atsugi, under the control of Commander Fleet Air Forces, Japan (COMFAIRJAP), to provide photographic interpretation functions as directed by COMNAVFE.

Navy photographic intelligence work in TF 77 in 1952 consisted of three types of operational support:

(1) Target selection: A photo plane covered a reportedly fruitful area. If photographic interpretation of the coverage found profitable targets, a mosaic was prepared and distributed to ships and squadrons. An analysis of target construction, size, nature, and defense was made; the target's military worth was evaluated; and the force required for its destruction was estimated. A conference of intelligence, operations, ordnance, and air group officers decided on the number of planes and the types of weapons to be employed.

(2) Damage assessment: The K-25 or K-17 cameras on the attack aircraft of the strike force provided post-strike coverage, and the ordnance officer would determine the effectiveness of the weapons used.

(3) Program evaluation: Photographic reconnaissance and photographic interpretation played a major role in evaluating the interruption of rail lines, bridge destruction, shift of anti-aircraft defenses, and buildup of rail cars, vehicles and supply revetments.

The K-25 cameras mounted in the AD Skyraider aircraft, which had provided good damage-assessment photographs in 1951, were largely discarded later in the conflict because the short focal length of the camera required the aircraft to fly at low altitudes where flak losses were heavy. Several carriers manufactured K-17, 24-inch focal-length camera capsules that surpassed the K-25 in quality and image size.[23]

In spite of the many tons of ammunition being expended on North Korean targets, very little intelligence was being received about the results. Participation of the naval surface forces in the interdiction program could have been far more effective had photographic intelligence been available to indicate damage being inflicted. Only a limited amount of such support was received from TF 77, which was hard put to provide damage assessment photos for its own interdiction efforts. The surface forces had inadequate facilities and personnel for processing and interpreting the photo coverage that they could obtain. There was an urgent need for such a capability in the ships engaged in shore bombardment. Commands responsible for target selection could not exercise that function satisfactorily without accurate intelligence obtainable only from timely photographic intelligence services.[24]

The special nature of the naval operations being conducted in early 1952 had increased the tasks of the naval photographic intelligence units beyond their normal capabilities. Personnel and facilities for photography, photo production, and photographic interpretation were not adequate, in spite of the excellent intelligence they were producing and disseminating to the fleet.[25]

Corrective action to alleviate the shortage of trained photographic interpreters in fleet units included sending photographic interpreters to the theater of operations ahead of the ships or staffs to which they were assigned. In addition to providing extra personnel for the operating units, the procedure prepared the photographic interpreters to be effective immediately upon the arrival of their own units in the combat area and also permitted the passing on of experiences and lessons learned from one unit to the next.[26]

Photographic intelligence provided a means for verifying agent reports on targets in North Korea and for determining the relative merits of striking those targets, leading to very successful attacks such as those against the oil refinery at Aoji-dong, the iron ore mine works at Musan, the iron foundaries at Chongjin, and various hydro-electric power plants in North Korea. Location of flak traps in the immediate target area led to the assignment of adequate flak suppression missions prior to the arrival of attack aircraft.[27]

The enemy employment of camouflage, using caves, natural cover, and netting, made the photo interpretation of target areas difficult. Accurate attack on such targets was only possible through thorough briefing of the pilots, using annotated photo mosaics. Vehicle track detection proved to be the primary key to spotting some of the more elaborate attempts at concealment.[28]

The elaborate and time-consuming "Touraids," for use in attacks on rail routes, were discontinued. In their stead, flak was plotted on an Army Map Service 1-to-50,000 chart and reproduced photographically to provide each pilot with a copy. Toward the end of January 1953, stereo mosaics were frequently being used to familiarize pilots with flak defenses and target details.[29]

In January 1953, Commander Naval Forces, Far East made provision for daily delivery of TF 77 aerial reconnaissance films to the photographic interpretation unit at Atsugi by means of carrier on-board delivery (COD) plane flights. The work performed by the Atsugi unit was of little direct benefit to the support of carrier operations but did reduce the work load on the carrier photo laboratories. Most of the output of the unit was distributed to the surface units of TF 77 and TF 95, which had formerly depended almost entirely upon the aircraft carriers for photographic intelligence support.[30]

In the procedure developed for furnishing photographic intelligence to the cruiser division commander (CTG 95.2), the carrier, following a photo mission, would issue an early photographic interpretation (PI) message. Then the photos were flown to Atsugi for more detailed study and another message. That took seven to ten days, but it was anticipated that the delay could be reduced to seventy-two hours with experience. The photographs were then annotated and sent to CTG 95.2, where they were usually received about four weeks after the photographic mission. The excessive delay reduced the value and reliability of the photography as attack-planning information because the targets could have been moved or destroyed in the interim.[31]

Surface bombardment units were obviously in need of better damage assessment from photographic intelligence processed from aerial photographic reconnaissance. Evaluation of shore bombardment damage by the individual ships was unsatisfactory, and air gunfire spotters generally could not get close enough to the target to assess damage without receiving concentrated antiaircraft fire. Dust and smoke covering the target area were also a problem. The reports that the ships did make were usually overly optimistic.

To satisfy the need for shore bombardment photographic intelligence, COMNAVFE requested on 24 January 1953 that aerial oblique and vertical photos of target areas be taken regularly at intervals of approximately two weeks. The photographs were to be screened, annotated, and distributed to the fleet units supported by the photographic interpretation unit at Atsugi, which was also to assess the results of naval gunfire, where feasible.[32]

By 1953, the best intelligence reaching the forces afloat was from the photographic efforts of the TF 77 carrier force, which had developed a highly efficient system of photographic intelligence for its own use. Early in 1953, the effort was further expanded by the establishment of Special Intelligence Photo Unit One at NAS Atsugi, using the photographic interpretation unit previously set up there in July 1952 as a nucleus. Photographic intelligence continued to be slow in reaching the surface forces, but what was received was a considerable improvement over the target intelligence provided to ships in the earlier years.[33]

Each aircraft carrier in TF 77 had fairly extensive photographic intelligence capabilities aboard, and each had a complement of three photo-configured jet aircraft, either F2H-2P Banshees or F9F-5P Panthers, and five pilots trained in photographic reconnaissance. A complete processing laboratory operated between fifteen and twenty hours per day developing, printing, and interpreting exposed film.[34]

The greatest limitation on photographic reconnaissance was the weather. Post-strike damage assessment photos were needed for day-to-day planning to determine whether targets should be rescheduled promptly and whether the ordnance used had been suitable. If the target was rescheduled, the undamaged areas of the target were identified for pilot briefings. The photos were also needed to improve the effectiveness of aircraft attack tactics and to maintain pilot morale.[35]

Evolution of the Naval Intelligence Support Center

Creation of NPIC

SECNAV Notice 5450 of 16 June 1953 established the Photographic Intelligence Center as a separate command, the Naval Photographic Intelligence Center. At that time, the center consisted of the following departments: Training, Evaluation, Technical Services, and Production. A Special Projects Department was added in 1956. In December 1957, NPIC moved from Anacostia to new spaces at the Federal Office Complex in Suitland, Maryland. The new building was especially designed for photographic interpretation work with 135,000 square feet of office, laboratory, and vault spaces. During March 1960, the Photo and Graphic Section (OP-922H1/Y3) of ONI moved into the center as co-tenants, bringing with it an additional fifty officers, enlisted men, and civilians.[36]

In 1955, the Photo and Graphic Section (OP-922H1 at this time) was receiving photographs from such sources as Navy and Marine photographic reconnaissance flights, including carrier-based detachments, naval attaché ground and aerial photos, and foreign sources. An example of the latter were the photo collection flights of Chinese Nationalists over the Chinese mainland from Taiwan. Negatives, prints, preliminary photographic interpretation reports, plots, and mission data cards were obtained by the Commander Taiwan Defense Command intelligence officer, Capt. Rudolf J. Fabian. He airmailed them to OP-922H1, and from there they were rushed to the Naval Photographic Intelligence Center for in-depth photographic interpretation used in briefing the Chief of Naval Operations, the Joint Chiefs of Staff, and others. Additionally, photo enclosures to information reports submitted by attachés were selected by OP-922H1 for photographic interpretation at NPIC.

The OP-922H1 photo library, located under the eaves above the Mall Entrance of the Pentagon, was regularly used for research by, or for, the fleet, ONI production analysts, the Navy Hydrographic Office, Army and Air Force intelligence, the CIA, the Coast and Geodetic Survey, NPIC, Marine Corps intelligence, the State Department, and the U. S. Geological Survey.

Relations between OP-922H1 and NPIC were excellent, with visits and phone calls going both ways many times each day at all levels. NPIC had a liaison officer in OP-922H1 who watched for items for NPIC. Although NPIC was under, and funded by, the Bureau of Aeronautics, it worked for the Director of Naval Intelligence through OP-922H1.

In addition to maintaining a photo and graphics library for ONI and other users and providing support to the fleets through photographic research, OP-922H1 was also responsible for handling the Navy's staff work for all Joint Chiefs of Staff papers pertaining to mapping, charting, geodetic data, photography, photographic interpretation, NATO and CENTO standardization papers on those topics, etc. The head of OP-922H1 also had collateral duty as the Navy's representative to the Photo and Survey Section of the Joint Intelligence Group of the JCS.

During 1955–1958, the first effort at reducing the photo and graphic holdings of OP-922H1 was made. The move was necessary to make room for new material and to make the current holdings more easily accessible to researchers. The project for purging the files of obsolete, duplicative, and useless material was given the name HOUSECLEAN. Cdr. Victor A. Moitoret, the head of OP-922H1 at that time, was in charge of the project. His research prior to the start of Project HOUSECLEAN had found great quantities of World War II aerial photographs covering targets that, over ten years later, had no potential value. For Marcus Island, for example, there were 143 different missions on file, each of 100 or more photos, covering 1942 to 1945 and taken from all possible altitudes and at all seasons of the year. Before destroying the aerial photography files, photos of historic value were given to any Navy activity that could use them. An attempt to donate photos to universities was found to be illegal since the pictures were government property, so the material was loaned on a ninety-nine-year basis for use in photogrammetry and photo interpretation courses.

It had been costing the Naval Photographic Intelligence Center hundreds of thousands of dollars each year to process Nationalist Chinese photo material because of the high priority that had been assigned to it, President Dwight Eisenhower being the original customer. Yet by 1954, it had been several years since Eisenhower or Adm. Arthur W. Radford, Chairman of the Joint Chiefs of Staff, had looked at any of the photography. RAdm. Samuel B. Frankel, head of the Joint Intelligence Staff, approved the termination of the project.

Another significant saving was achieved when OP-922H1 succeeded in downgrading from top secret to confidential material from a special Navy-sponsored aerial photo project operated in a Mideast country. The collectors had been automatically classifying any maps or photos top secret, thereupon requiring special couriers for transmission back to the United States and special processing and limited distribution after the material was received. OP-922H1 licked the problem by proposing that the U.S. classification be assigned based on a written description of how it would be protected. The local U.S. naval attaché was able to sell the proposal to the collectors.

A breakthrough in overhead photography was made in the mid-1950s when a camera, mounted in a weather balloon and launched in Europe, drifted with the jet stream, and was recovered in the Pacific. The use of the jet stream to carry fire bombs across the Pacific to the U.S. West Coast was a technique attempted with little success by the Japanese in World War II. Although the balloon collection project gave random coverage, it did provide the first photographs of areas of the USSR and Communist China not previously seen by the Western World.

Material from the balloon program was received in OP-922H1, where it was screened for items of naval interest to pass to NPIC for further processing by its Special Projects Department, set up especially to handle the balloon material. Special channels were established for passing balloon-derived information of operational interest to the fleets, and Cdr. Moitoret (OP-922H1) visited fleet commanders to advise them in person of the prospective receipt of the extra-sensitive material.[37]

In the early 1960s, NPIC's mission was to maintain and operate facilities to prepare photographic and related sensor interpretation reports and studies; to reproduce and store intelligence photographs; to construct terrain models; and to conduct research, development, testing and evaluation (RDT&E) of photographic interpretation techniques, equipment, and materials. NPIC was under the military command of the commandant of the Potomac River Naval Command, and the Chief of the Bureau of Weapons provided management control and technical direction, including staffing authorization and funds. The Chief of Naval Operations assigned the photo interpretation tasks to be performed by NPIC. As of 1 January 1964, the personnel allowance was 57 officers, 332 enlisted personnel, and 216 civilians.[38]

NPIC Becomes NRTSC

On 15 February 1964, certain Navy photographic interpretation and library functions, personnel, and assets were transferred to the Defense Intelligence Agency (DIA), and the responsibility for exercising management control of the remaining Navy photographic intelligence assets was assigned to the Chief of Naval Operations and the Fleet Activities Command. The Chief of the Bureau of Weapons retained funding responsibility, civilian ceiling control, and sponsorship of military personnel. SECNAV Notice 5450 (OP-09B23C serial 508) of 25 February 1964 changed the title of the Naval Photographic Interpretation Center to the Naval Reconnaissance and Technical Support Center (NRTSC). Its stated mission was to maintain and operate facilities to provide photographic and related sensor interpretation support to the CNO; to conduct RDT&E of photographic interpretation equipment, materials, and techniques; to provide intelligence data-processing services in support of the Department of the Navy and the Worldwide Military Command and Control System; and to provide intelligence production support to the operational forces of the Navy in production areas where requirements were in excess of force capabilities. SECNAV Notice 5450 of 29 September 1964 modified the above mission statement to delete the intelligence data-processing service.

On 1 July 1964, NRTSC was transferred to the management control of the Director of Naval Intelligence, but it continued to be responsive to the guidance of the Chief of the Bureau of Weapons in RDT&E matters.

The earlier transfer of some functions to DIA and the expansion of responsibilities in other areas had required a reorientation of NRTSC's departmental structure, including the establishment of a new Data Processing Department in support of the Navy's rapidly expanding Integrated Operational Intelligence System (IOIS).

NRTSC's personnel allowance at the end of 1964, reflecting the transfers to DIA, was reduced to 32 officers, 267 enlisted personnel, and 133 civilians.[39]

On 23 August 1965, RAdm. Rufus L. Taylor, Director of Naval Intelligence, dedicated NRTSC's new Integrated Operational Intelligence Center (IOIC) computer installation. Effective 1 December 1965, the Office of the Chief of Naval Operations "hat" for the commanding officer of NRTSC was changed from OP-922F1 to OP-922RC, and at the same time he absorbed the responsibilities of the former OP-922H1, the ONI Photo and Graphic Section.[40]

NRTSC Mission and Accomplishments, 1966–1972

In 1966, NRTSC assumed full responsibility for IOIC production and maintenance of the IOIC Master Data Base, and the center completed the IOIC *Data Base Manual* for use in standardizing and controlling data within the worldwide IOIC Data Base. Other major NRTSC accomplishments in 1966 included production of a complete set of Airfields and Seaplane Stations of the World aperture cards of the Pacific areas for *Intrepid* (CVS 11); production of thirteen sets of 35mm Foreign Submarine Recognition Training Slides for Commander Submarine Forces, Atlantic; preparation of annotated photographic briefing materials of thirteen specific Mekong Delta areas for OP-090C, Capt. Lothrop, the head of the Mekong Delta Mobile Assault Force; and production of a semi-controlled photo-mosaic of the island of Cyprus at an approximate scale of 1:100,000, using British photography.[41]

Effective 1 July 1967, NRTSC shifted from being a Chief of Naval Operations field activity administered by ONI to a field activity of the Naval Intelligence Command (NAVINTCOM). At the same time, the Merchant Marine Intelligence Unit (less the current movements section) was transferred to NRTSC as a division within the Intelligence Production Department. Intelligence support to targeting, including the Single Integrated Operational Plan (SIOP), was also assigned to NRTSC, being absorbed by the Special Projects Department. Personnel figures changed as follows:

	Officer	Enlisted	Civilian	Total
1 Jan 1967 allowance	32	262	163	457
1 Jan 1967 on board	28	222	147	397
31 Dec 1967 allowance	32	245	212	489
31 Dec 1967 on board	31	206	196	433

The NRTSC mission in 1967 was to provide image interpretation and merchant marine intelligence support to the Secretary of the Navy, the Chief of Naval Operations, and the Defense Intelligence Agency; provide ocean surveillance intelligence information to the Ocean Surveillance Information Center; produce image-derived information in support of the National Tasking Plan for the exploitation of multisensor imagery; provide miniaturized and automated intelligence databases and intelligence production support, including SIOP, to the CNO, COMNAVINTCOM, Navy Planning staffs, and fleet commanders; and conduct research, development, testing, and evaluation of imagery exploitation equipment, materials, and techniques.[42]

The war in Southeast Asia, as well as the 1967 Middle East War, continued to impose intelligence production and support requirements on the operating departments of the Naval Reconnaissance Technical Support Center during 1967. Some of the major accomplishments during 1967 included completion of RDT&E and delivery to the fleet of the AN/AAS-21 infrared (IR) sensor for the RA-5C Vigilante aircraft; delivery of the final two non-IOIC databases to *Shangri-La* (CVA 34) and *Franklin D. Roosevelt* (CV 42) (up to that time, a total of nine IOIC and twelve non-IOIC databases had been delivered to the fleet and were being maintained by NRTSC); continuing support to Commander in

Chief, Pacific by providing photographic identification of all storage areas in and around the city of Haiphong; production of photo interpretation keys for imagery derived from infrared sensors and production of the publications *Friendly Small Boats in Vietnam*, *Waterborne Logistic Craft in North Vietnam*, and *Naval Vessels of the Middle East*; and assumption of the responsibility for maintaining the database for the Worldwide Tactical Target Aperture Card System for the Department of Defense.[43]

Significant accomplishments by NRTSC during 1968 included initiating a Mission Support Mosaic program with DIA and the Office of Naval Oceanography to support operating forces; providing photographic support to the Chief of Naval Operations to assist in search operations for the ill-fated *Scorpion* (SSN 589) (approximately 185,000 photo images were processed and large photo-mosaics of selected debris areas of the ocean floor were constructed to assist the Naval Board of Inquiry and the *Scorpion* Analysis Group); making *Pueblo*-related viewgraphs and briefing charts for the State Department and the Department of Defense; producing, on a crash basis, three-dimensional relief maps of the Khe Sanh and Ah Shau Valley areas for the CIA; completing a comprehensive recognition slide program for fleet units entitled "Ships of the Soviet Navy," with instructor guides and over 300,000 35mm slides produced and disseminated for use in recognition training; preparing and disseminating at the special request of Commander Seventh Fleet a bilingual identification guide that covered the several classes of North Vietnamese trawlers known to be engaged in infiltration operations into South Vietnam; supporting installations of the AN/AAS-21 infrared sensor in RA-5C aircraft, and providing on-site technical assistance and limited training in its use; supporting the at-sea testing of the Type 18 periscope, with NRTSC performing final photographic and video recording evaluations (NRTSC had participated in the development of the Itek & Kollmorgen periscopes from the initial requirements to their technical evaluation); participating in the development of the Tactical Airborne Reconnaissance System (TARS), NRTSC having been tasked to establish the optimum ground resolution, contrast, and other performance characteristics to detect and identify a series of targets; completing testing and evaluation in the development of a micro-linear scaling instrument, used for making high accuracy measurements directly on photographic film to be distributed to shipboard and shore-based units; and completing the development, test, and evaluation of a desk-top, rear-projection viewer capable of projecting all or selected portions of miniaturized-image materials used in the IOIS database.[44]

Significant photographic intelligence activities accomplished by NRTSC in 1969 included providing current intelligence materials covering more than 172 targets in response to high priority requirements from Commander Sixth Fleet (delivery was made within ten days of receipt of the initial request); participating on a special committee under the supervision of RAdm. Hyman G. Rickover and the Naval Ships Systems Command to investigate and develop a detailed analysis of Soviet submarine construction, capabilities, and technologies; supporting a new multisensor airborne ocean surveillance system by providing briefings, analysis of collected imagery, immediate readout reports, and detailed reports; production of photo interpretation keys on soviet intelligence collectors (AGIs); and revising and publishing a fleet support guide, *Range Finding by Superstructure Heights*, for estimating distances to Communist naval ships; and providing continued photographic support in the search for the submarine *Scorpion*, with NRTSC representatives working on board *White Sands* (ARD 20) and the bathyscaphe *Trieste* in July and August. A three-dimensional model of the two major sections of the sunken *Scorpion* was produced. NRTSC was awarded the Navy Unit Commendation in April 1970 by Secretary of the Navy John H. Chafee for its contributions to the *Scorpion* search operations during the period 15 April 1969 to 15 April 1970.[45]

In the latter half of 1970, NRTSC underwent a complete reorganization, changing department titles and billet identifications. The resultant departments were Administration, Exploitation, Production, Photographic, Automated Intelligence, Systems Development, and Evaluation.

Some of the many significant NRTSC accomplishments in 1970 included providing photo interpretation support for the collection of information on the Soviet navy's worldwide exercise Operation OKEAN; producing a recognition handbook, *Communist Naval Ships, Atlantic*, in response to fleet requirements for pictorial identification display material and basic dimensional data; issuing the *Vietnam Infiltration Trawler Identification Guide (INTRIGUE)* in response to a requirment for a bilingual recognition handbook that could be released to the South Vietnamese and that described the various classes of ships involved in covert resupply operations along the Vietnamese coast; performing a photographic survey of the nerve gas disposal ship *LeBaron Briggs*, which had been scuttled in the munitions dumping area off the coast of Florida; and preparing a photo-mosaic of Philippine areas in support of State Department negotiations over water rights.[46]

Significant accomplishments by NRTSC in 1971 included constructing lead-sheet and wood models of Soviet warships and aircraft, including the *Moskva*-class helicopter-carrying cruiser (four copies), the *Kresta* II-class guided-missile cruiser (three copies), the *Krivak*-class guided-missile-frigate (three copies), the KA-25 Hormone helicopter (seven copies), and various submarines (four copies); publishing *North Korean Naval Vessels Key*, a recognition handbook prepared in response to fleet requirements for pictorial identification displays and basic dimensional data on ships of the North Korean Navy; publishing *Middle East and North African Naval Ships*, prepared to provide U.S. naval forces in the Mediterranean with a current recognition aid; providing support for a new worldwide aerial reconnaissance effort; and supporting Navy freedom of navigation deployments to the Black Sea through briefings, provision of photographic equipment and trained personnel, analyses, and preparation of special reports.[47]

During Fiscal Year 1972, the installation of a newly developed digital image processing system was started at NRTSC. The system extracted information from photographs of marginal quality (due to low light or other technical limitations) and rendered the imagery suitable for analysis, a technological breakthrough in photographic intelligence processing.

A forward-looking infrared (FLIR) imaging system designed to collect infrared signature data on ships at sea was developed and tested in 1972. Its primary recording medium was 16mm film, with videotape as a secondary recording method. The system provided a night-time, poor-visibility observation and identification capability. After further refinement, it was scheduled for operational deployment during Fiscal Year 1973.[48]

NRTSC Merges With STIC

On 30 June 1972, the Naval Reconnaissance and Technical Support Center was merged with the Naval Scientific and Technical Intelligence Center (NAVSTIC) to form the Naval Intelligence Support Center (NISC), a field command of the Naval Intelligence Command. Capt. John P. Prisley, the commanding officer of NAVSTIC, became the first commanding officer of NISC.[49]

The mission statement for the newly merged NISC organization was to

> produce technical intelligence of national and naval interest as directed by the Director, Defense Intelligence Agency and the Director of Naval Intelligence; provide scientific, technical, and imagery intelligence support to the CNO and his staff, to the Naval Material Command and the Systems Commands, to the Navy Laboratories, and to the Fleet; support research, development, test and evaluation of equipment, materials and techniques associated with the production of technical and imagery intelligence; and provide technical guidance for intelligence collection activities.[50]

The following served as commanding officers of NRTSSC:

Name	Date
Capt. William G. Matton, Jr.	25 Feb 1964
Capt. Charles D. Payne	17 Dec 1965
Capt. J. W. Taft	30 Jul 1968
Capt. Charles V. Choyce	22 Dec 1970
Capt. John P. Prisley (NRTSC combined with STIC to become NISC)	30 Jun 1972

Chapter Notes

1. *ONI Review*, Apr–May 1945, 175.

2. *SECNAV Annual Report, 1886*.

3. In folder 6147, box 148, RG 36, NA.

4. Department of the Navy, "Administrative History of the Office of Naval Intelligence in World War II," 10 Jan 1946, unpublished MS, 1027–28, hereafter *ONI WWII Admin History*; and Navy Dept. General Orders 96, 179, and 204.

5. Commander in Chief, Pacific Fleet (CINCPACFLT), *Annual Report, FY 1941*.

6. *PI and Radar Target Intelligence Newsletter*, Jan 1962.

7. *ONI WWII Admin History*, 1046.

8. Ibid., 1049.

9. *PI and Radar Target Intelligence Newsletter*, Jan 1962.

10. Bureau of Personnel, *Intelligence for Naval Officers*, (NAVPERS 16047), 1948, 143–44.

11. *ONI WWII Admin History*, 100.

12. *ONI Review*, Nov 1946, 15.

13. *ONI WWII Admin History*, 1036, 1038.

14. Ibid., 1042.

15. Ibid, 1061–62.

16. Ibid, 947.

17. *PI and Radar Target Intelligence Newsletter*, Jan 1962.

18. CINCPACFLT, *Interim Evaluation Report No. 2: Korean War Naval Operations*, 1744.

19. CINCPACFLT, *Interim Evaluation Report No. 1: Korean War Naval Operations*, 1482–83.

20. CINCPACFLT, *Interim Evaluation Report No. 3: Korean War Naval Operations*, 5-47–5-48.

21. Ibid., 19-8–19-9.

22. Ibid., 19-11.

23. CINCPACFLT, *Interim Evaluation Report No. 4: Korean War Naval Operations,* 3-47–3-48.

24. Ibid., 5-24.

25. Commander Seventh Fleet (COM7THFLT), *Report of Operations, 28 Mar 1951–3 Mar 1952*, 37.

26. COM7THFLT, *Report of Operations, 20 May 1952–27 Jul 1953*, 30.

27. Ibid., 87.

28. Ibid., 98.

29. CINCPACFLT, *Interim Evaluation Report No. 5: Korean War Naval Operations*, 3-64.

30. Ibid., 3-66.

31. CINCPACFLT, *Interim Evaluation Report No. 6: Korean War Naval Operations*, 5-127–5-128.

32. CINCPACFLT, *Interim Evaluation Report No. 5*, passim.

33. CINCPACFLT, *Interim Evaluation Report No. 6*, 1-13.

34. Ibid., 3-74.

35. Ibid., 3-77–3-79.

36. *PI and Radar Target Intelligence Newsletter*, Jan 1962.

37. Information for 1955–58 derived from Victor A. Moitoret ltrs to author, 23 Feb 1975 and 7 Jun 1975.

38. Naval Reconnaissance and Technical Support Center (NRTSC), *Command History, 1964*, 1.

39. Ibid., 1–2.

40. *NRTSC Command History, 1965*, 1.

41. *NRTSC Command History, 1966*, 2–3.

42. *NRTSC Command History, 1967*, passim.

43. Ibid., 1, 3.

44. *NRTSC Command History*, 1968, 2–3, 5–6.

45. *NRTSC Command Histories, 1969*, 5–6; *1970*, 3.

46. *NRTSC Command History, 1970*, 1, 8–10.

47. *NRTSC Command History, 1971*, 7, 10.

48. *DNI Report to the CNO, FY 1972*, 6, 23.

49. Naval Intelligence Support Command (NISC), *Command History, 1972*, passim.

50. OPNAV Notice 5450 of 3 Aug 1972.

Chapter 14

Acoustic Intelligence

In 1949, the Technical Unit (OP-322F2) of ONI's Foreign Section began studying acoustics as a part of electronic intelligence (ELINT). The interest in acoustics was stimulated by requirements to provide intelligence support to engineering and power-plant appraisal programs; a ship engineering noise program similar to ELINT, the Sound Surveillance System (SOSUS); and homing torpedo and acoustic fuze research and development programs. It was obvious that the Navy, the only service concerned with acoustic intelligence (ACINT), would have to take on the extensive task of collecting and analyzing the mass of data involved.[1]

Dr. Paul Martin, a physics instructor from Wheaton College, Norton, Massachusetts, joined the OP-322F2 electronics effort in June 1951 and spent a full year plus one summer developing a dedicated sonar analysis effort. After Dr. Martin's return to Wheaton in September 1952, Eldon Bissett was obtained from the Sound Division of the Naval Research Laboratory to become the Technical Unit's principal sonar analyst.[2]

In 1953, an acoustic intelligence program of limited scope was started when Commander Submarine Forces, Pacific collected sonar recordings for analysis by the Naval Electronics Laboratory at San Diego. From that initial effort evolved the Acoustic Intercept Data Analysis (AIDA) Program, which was established in the summer of 1956. The objective of the AIDA Program was to derive information on Soviet ships, particularly submarines, by analyzing sonar recordings of their acoustic signatures. OP-922F2 served as the control point for the Navy to receive the recordings; make preliminary evaluations of the information in the recordings; prepare duplicate recordings to send to Bureau of Ships (BUSHIPS) laboratories (Naval Electronics Laboratory, San Diego, and Data Processing Unit, Brooklyn) for technical analysis; receive laboratory reports and collate the reports with other intelligence; and prepare and disseminate finished intelligence.[3]

The Chief of Naval Operations Acoustic Intercept Committee (AIC), established on 3 July 1956, provided direction for ACINT analysis, and a few people in the ONI Technical Unit screened the recording and performed the daily control-point tasks. Detailed analysis was accomplished primarily by the BUSHIPS laboratories at San Diego and Brooklyn, with minor assistance provided by a few other facilities.[4]

The Acoustic Intercept Committee was composed of two members from the Division of Undersea Warfare (OP-31), two members from ONI, and a non-voting secretary from ONI. AIC was to consider matters relating to the detection and recording of the underwater noise and active sonar signals of foreign ships and equipment, including the analysis, evaluation, and collation of the information obtained, and to disseminate the finished intelligence. AIC was also instructed to make a study of the policies and procedures necessary to achieve optimum effectiveness for the ACINT program.[5]

Underwater acoustic collection programs had been developed and expanded for many years. By the late 1950s, recordings from the programs were being received at an increasingly rapid rate—about 500 tape reels in 1958 and about 700 reels by mid-August 1959. The Basic and Technical Intelligence Branch (OP-922G) of ONI was the designated control point for the acoustic intelligence program, receiving the raw data and recordings, screening and forwarding the recordings of highest priority targets to the BUSHIPS's labs for analysis, receiving the analytical results, and correlating them with the other related available data.[6]

By 1958, the Navy's ACINT collection and processing capabilities became unbalanced. The scale of the processing effort at the two BUSHIPS laboratories was inadequate, and both facilities were remote from intelligence centers. Normal BUSHIPS tasking was sufficient to keep the laboratories fully occupied, and their work for ONI received reluctant support, both from the laboratory authorities and from BUSHIPS. The analysis of ACINT recordings of less than urgent priority was taking up to one year to accomplish.

On 19 November 1958, ONI civilian ELINT analyst William E. W. Howe addressed a letter to Director of Naval Intelligence (DNI) RAdm. Laurence H. Frost, as well as to many senior officers in the Intelligence Division of ONI, explaining the unsatisfactory situation in the Navy's program for the analysis of acoustic recordings. The letter stimulated comments and, in due course, the start of corrective action.

Various inquiries were made and studies developed to find an optimum solution to the ACINT analysis resource problem by setting up an acoustic intelligence laboratory within ONI; improving integration and coordination of the intelligence effort with the Navy's total, in-house, underwater acoustics effort; and placing a suitable contract with an appropriately equipped Washington facility. The belief was that the continued failure of ONI to tackle a uniquely Navy requirement would ultimately lead to other agencies doing the job without being responsible to, or under the direction of, the Navy, which was the primary user.[7]

In 1959, the ACINT requirements that ONI was trying to fulfill included (1) collecting the underwater acoustic signatures of known Soviet submarines in order to help in identifying SOSUS contacts; (2) analyzing the signal characteristics and operational capabilities of Soviet echo-ranging sonars, echo sounders, and underwater communications equipment to assist the naval operating forces in identifying the origin of the signals; (3) collecting acoustic recordings for analysis to determine the technical characteristics of the propulsion systems of Soviet naval ships, so that analysts could evaluate the operational capabilities of those ships and detect new engineering developments; (4) collecting speed versus propeller revolutions-per-minute data; (5) obtaining information on underwater noise characteristics of foreign ships for use by BUSHIPS in designing sonar and sonar countermeasures and for ship-noise studies; (6) collecting the same information for use by the Bureau of Ordnance (BUORD) in designing acoustic mines, torpedoes, and countermeasures; and (7) collecting information on the acoustic noisemakers and torpedo countermeasures used by the Soviets in order for BUORD to design countermeasures.[8]

ONI, however, was unable to fulfill the requirements because of a two-year backlog of unanalyzed data at the two BUSHIPS laboratories. Finally, at a CNO Advisory Board hearing on the budget in early November 1959, support was given to the establishment of a specialized acoustic intelligence analysis facility. Accordingly, Director of Naval Intelligence RAdm. Laurence Frost requested research and development funds from Deputy Chief of Naval Operations for Development to establish the urgently required facility, while efforts continued to insert in future budgets the funds needed to operate it.[9] The Navy had invested heavily on SOSUS; it was illogical to spend an inadequate amount to develop a means to understand the signals being collected.

Consideration was given to various alternative locations for the ACINT analysis facility, and it was finally decided to place it in the southeast end of the upper floor of Building 52 at the U.S. Naval Observatory in Northwest Washington, D.C. The rehabilitation of the space to accommodate the facility and its electronic gear was started in 1960, and the Acoustic Intelligence Analysis Facility became operational on 6 June 1962 as a LOFAR (low frequency acquisition and ranging) spectrum analysis center and as the location for a tape recording playback system. The facility was not computerized, a deficiency that soon made itself felt.[10]

On 1 July 1965, a unit of the Naval Scientific and Technical Intelligence Center (NAVSTIC) (formerly OP-322F2) became the Undersea Warfare Office (STIC-3), and the Acoustic Division of that office, with its ACINT Analysis Facility, was designated STIC-3E. The basic function of STIC-3E was to provide scientific and technical intelligence derived from acoustic recordings. It was also responsible for correlating information derived from the acoustic collection effort with other related information available from technical reports, other intelligence sources, ship-sighting reports, and other sources.[11]

The increased tempo of Soviet submarine operations, coupled with the introduction of a Chinese Communist submarine threat, provided increased targets of opportunity for acoustic recordings. Timely, thorough analysis of the resulting data was required, not only by antisubmarine warfare forces but also by technical activities concerned with ASW devices. It was readily apparent that STIC-3E's acoustic intelligence analysis facility could not keep up with the volume of material requiring processing to meet the intelligence needs of the various operational and technical intelligence customers. Furthermore, STIC's capability had not developed sufficiently to discriminate trends in the radiated noise patterns of Soviet submarines with the degree of accuracy considered necessary to detect and identify them from established statistical norms.

Such a capability was essential to any effort to obtain early intelligence of a Soviet submarine-quieting program.

The David Committee, named for its chairman, E. E. David, Jr., was assembled by Director of ASW Programs (OP-95) VAdm. Charles B. Martell in 1965 to review and recommend solutions to ASW problems. In its final report, submitted on 28 January 1966, the committee summarized major deficiencies in the Navy's ASW intelligence program and specifically highlighted acoustic intelligence as a critical problem deserving an independent research effort to develop advanced, automated processing methods capable of handling the volume of associated raw data. An Acoustic Intelligence Conference attended by fleet representatives was also called in 1965 by VAdm. Martell to define fleet requirements for real-time and long-term acoustic intelligence. The results of the conference further confirmed that STIC did not have the capability to meet the Navy's operational needs.

The David Committee report also recommended that an additional eight to ten forward-area passive sonar installations should be deployed for gathering technical intelligence, operational intelligence, and order-of-battle information on Soviet submarines.

DNI RAdm. Rufus L. Taylor established two technical and management groups within ONI as a result of the David Committee report. One was a working-level group that met weekly and represented all activities concerned with ASW intelligence; it was called the ASW Intelligence Coordinating Committee. The second group, the Undersea Warfare Intelligence Section, was established to implement specific recommendations of the David Committee and to meet the requirements of the Director of the Defense Intelligence Agency to improve collection, collation, and integration of undersea warfare intelligence.[12]

A program to provide NAVSTIC sonar technicians as ship-riders to provide ACINT technical support to designated collection platforms was such a success that a special Navy Enlisted Classification Code ST-0416, acoustic intelligence analysis technician, was established in 1965, and a small cadre of experienced personnel was given the designation.[13]

The acoustic intelligence products generated by STIC-3 within its limited capabilities were dictated by the requirements of the users of the data. Force commanders needed information on the detectability of submarines in order to determine force densities. Strategic planners, who had to develop future Navy force levels and improve antisubmarine warfare capabilities, needed data on submarine noise levels.

In the production of ACINT, most acoustic recording tapes underwent preliminary screening at a sonar information center (SIC) to determine their suitability for further processing by NAVSTIC. The SIC could thus provide early feedback of information and corrective guidance to the collectors. SICs also duplicated selected recordings that might have value for training purposes.[14]

After a four-year development period commencing in 1968, the Acoustic Intelligence Data System was accepted on 10 May 1972 to provide the automated acoustic intelligence processing methods recommended by the David Committee in its January 1966 report. The system provided the capability to analyze the results of the diverse ACINT collection programs in a timely manner. It also provided for automated and interactive data management, multi-user capability, enhanced graphic presentation, and the capability to run large signal-processing and scientific application programs.[15]

In 1966, NAVSTIC began providing Airborne ACINT Field Teams to assist the fleet in all phases of airborne antisubmarine warfare ACINT data collection. As of 1971, the objectives of the field team program were to provide an improved capability in the collection of ACINT by airborne ASW platforms; collection guidance to the operators of airborne ASW systems in their role as collectors of acoustic data on foreign naval forces; liaison on a continuing basis between NAVSTIC and the operating forces; timely feedback to the collectors; and critiques of NAVSTIC analyses of significant detections.[16]

During Fiscal Year 1972, new equipment installed in the NAVSTIC ACINT laboratory (relocated to Suitland, Maryland, in 1968) included a new tape-editing system for the rapid scan of magnetic tape and a fast-time analysis system for rapid processing of acoustic data at several times the previous normal rate.[17]

In February 1972, the Undersea Warfare Department of NAVSTIC developed a new acoustic processing technique for the improvement of SOSUS detection and classification capabilities against Soviet Yankee-class nuclear-powered ballistic missile submarines.[18]

To keep the operational fleets informed of ACINT developments, nineteen *ACINT Newsletters* were published during Fiscal Year 1973. The effort had to be terminated in May 1973 because of lack of support by the Defense Intelligence Agency. Subsequent fleet requests that the newsletter be reinstated were ineffective. The Naval Intelligence Support Center, however, was able to publish a *Surface Ship ACINT Collection Guide* for fleet use that year.[19]

Chapter Notes

1. Naval Scientific and Technical Intelligence Center (NAVSTIC), *Command History, 1967*.

2. William E.W. Howe memo to author, on the development of NAVSTIC, undated [mid-1970s], 3.

3. AIDA Program Brief, 1959, Acoutic Intelligence (ACINT) Historical File, OA.

4. Howe ltr to author, 14 Nov 1958, ACINT Historical File.

5. OPNAV Internal Instruction 003820.1 of 3 Jul 1956.

6. OP-92B3 Study of Jul 1959; and RAdm. Reed, CNO Advisory Board (CAB) Brief, 12 Aug 1959, both in ACINT Historical File,

7. Howe ltr, 14 Nov 1958.

8. OP-92B3 study and Reed brief.

9. Assistant Chief of Naval Operations (Intelligence) ltr, ser 003958P92, 16 Nov 1959.

10. Eldon Bissett, ltr response to author's queries, undated.

11. Bolt, Beranek and Newman, Inc., engineering study, 23 May 1967, ACINT File.

12. OP-951F memo, ser 0064P95, 31 Aug 1966, ACINT File.

13. OPNAV Instruction S1510.9A of 7 Aug 1972.

14. Bolt, Beranek, and Newman study.

15. Naval Intelligence Support Center (NISC), *Command History, 1972*, 9.

16. OPNAV Instruction S3820.37 of 10 Sep 1971.

17. *DNI Report to CNO, FY 1972*, 24–25.

18. *NISC Command History, 1972*, 8.

19. *DNI Report to CNO, FY 1973*, III-4.

CHAPTER 15

Electronic Intelligence

World War II

The need for electronic intelligence (ELINT) began with the introduction of radio communications and electronics into warfare, but the U.S. Navy did not address the need for greater use of these tools until electronic countermeasures (ECM) was considered to be a possible weapon or counterweapon. Radio direction-finding, radio jamming, and radio deception were used in World War I, and the developments in radar by 1935–1938 opened up the non-communication field of electronics. Electronic warfare became a vital element in naval operations during World War II.

For the U.S. Navy, the need for ECM and ELINT was graphically demonstrated in 1943 when it became necessary to determine the characteristics of the guidance systems used on the German Hs-293 radio-controlled glide bombs so that countermeasures could be developed. The Naval Research Laboratory (NRL) developed prototype intercept receivers, signal analysis equipment, and recorders for installation in the destroyer escorts *Frederick C. Davis* (DE 136) and *Herbert C. Jones* (DE 137) for use in determining the radio frequencies and the types of guidance modulation used in the control of the German glide bombs.[1]

The glide bomb intelligence collection effort was successful (although the *Davis* was later lost on 24 April 1945), and, based on the data obtained, NRL developed and procured equipment that, during the Allied Mediterranean and Normandy landings, protected ships by giving repeated false commands to the glide bombs. NRL also designed towed radar decoys (based on ELINT developed on German radar characteristics) to mislead German intelligence and commanders into believing that the major Allied landings were to be at Pas de Calais.[2]

The first efforts in radar countermeasures (RCM) in the Pacific during World War II were almost a complete failure because of the undeveloped state of Japanese radar, the vast areas involved, and the paucity of enemy radar contacts. Furthermore, the U.S. policy of over-classifying RCM publications and techniques succeeded only in hiding many RCM objectives from the fleet.

As of the summer of 1943, only a brief list of Japanese radar frequencies and a few plots of enemy radar locations had been produced. Three badly damaged Japanese radars were captured on Guadalcanal and shipped immediately to NRL for study and to aid in the development of countermeasures devices. In April 1943, an officer had been assigned specifically to direct radar countermeasures activities in the Pacific, first from the Intelligence Center, Pacific Ocean Area and then from Fleet Radio Unit, Pacific.[3]

In November 1943, steps were initiated to make RCM a fleet activity. The steps included the dissemination of RCM intelligence information in a series of *Countermeasures and Detection* (CM and D) *Summaries and Bulletins of Information.*

By January 1945, a Countermeasures Air Analysis Center had been established by Commander in Chief, Pacific on Guam under Commanding General Strategic Air, Pacific Ocean Area. The center's mission was to collect, analyze, and disseminate information on enemy radar locations and electronic equipment. CINCPAC also clarified, simplified, and standardized procedures for reporting on enemy radar characteristics and locations.

The RCM program was well organized by spring. Frequent issues of CM and D summaries kept the fleet informed of enemy radar tactics, new uses for equipment, and the success or failure of RCM operations. Information on enemy radar equipment was also compiled and issued as *Japanese Radar and RCM Equipment*.[4]

It was obvious from World War II experiences that ELINT collection was dependent upon laboratory designed detection devices, special receivers, analyzers, and recorders. In turn, laboratory development of ECM devices was dependent upon ELINT. A mutually beneficial relationship between the Office of Naval Intelligence and NRL was a logical evolution, facilitated by the dedicated personnel involved.

Post–World War II

The rapid demobilization at the end of World War II virtually eliminated, for a while, any thought of further research on ECM. ELINT was similarly reduced to practically a caretaker status. ECM gear and ELINT collection equipment were put on the open surplus market.

When captured enemy equipment became available for study, the pool of research talent still remaining in the Countermeasures Section of the Radio Division of NRL, headed by civilians Edward Speakman and Howard Lorenzen, was particularly interested in German developments in direction finding, intercept receivers, and tape recording. Various members of the Countermeasures Section recognized the vulnerability of radar systems and felt that basic research in signal intercept, analysis, and location (DF) were going to be essential to future ECM and ECCM (electronic counter-countermeasures, or the countering of enemy ECM systems) preparedness.[5]

Electronic intelligence collection "Ferret" flights in the late 1940s showed that a radar net was being built by the Soviet Union to back up the Iron Curtain, and an ECM program was thereupon reestablished. World War II equipment was repurchased from surplus markets, and ELINT was again in demand to support ECM research and development.[6]

U.S. knowledge of Soviet electronics in the late 1940s was negligible. Lists were maintained of known Lend-Lease radar and electronics equipment, as well as of equipment obtained by the USSR through purchase on the open market from the United Kingdom, Germany, Italy, and Japan. Attaché photographs of Soviet electronics installations were of low quality, always taken under anything but ideal conditions. However, a defector from the Soviet Ministry of Shipbuilding, circa 1949, opened up an invaluable window to Soviet developments to be anticipated in the field of naval electronics.[7]

The small group of engineers in the NRL Radio Division were convinced that, as the navies of the world increased their use of electronics in radar, communications, guided missiles, etc., there would be an ever-increasing demand for better ECM to defeat these systems and weapons. Better ELINT was needed again, and one early result of an effort by the group at NRL under Howard Lorenzen was a series of intercept equipment for shipboard and airborne reconnaissance that covered the entire frequency spectrum. Next, the equipment was converted so that they could scan the various frequency bands in seconds, and show a wide-band panoramic display of the spectrum activity to the operator.

Extensive research was done on signal recording. From early wire recordings (which had many inherent problems) to paper-backed magnetic tape, to polyester-backed tape, the NRL group pushed the recording band widths from a few kilohertz, to hundreds of kilohertz, and then to a few megahertz.

The Navy Security Group formed Naval Communications Units in 1948–1949 to participate in electronic reconnaissance missions flown by P2V Neptune and PB4Y Mercator "Ferret" aircraft as the best way to gain a better insight into Soviet electronic developments. The first aircraft were hand-fitted with intercept systems by the Lorenzen group at NRL and provided the first valid assessment of post–World War II electronic progress being made by various nations. For surface reconnaissance, NRL fitted the heavy cruiser *Columbus* (CA 74) with suitable intercept, analysis, and recording equipment and commenced the Special Electronic Search Project.[8]

During World War II, the Germans had developed an extraordinarily effective high-frequency direction finder known as the *Wullenweber*. NRL built a similar array and placed it at the Coast Guard station in Hybla Valley, south of Alexandria, Virginia. The Soviets were also interested in the German development, which was important to antisubmarine warfare in particular and to communication intelligence (COMINT) in general. As the German and Austrian scientists who had worked with the Soviets returned to the West, they were interviewed in detail about the Soviet work on duplicating and improving the *Wullenweber*. As various circuits and instrumentation were described by the returnees, the Naval Research Laboratory would rapidly try out the techniques and thus evaluate their effectiveness. This was another example of the close cooperation between ONI and Navy laboratories, particularly NRL.[9]

A subpanel of the Joint Communication and Electronics Committee was set up in 1948 to aid in the accumulation and appraisal of the information being received by the three services about the Soviet electronic warfare threat. Edward Speakman of the NRL was the first chairman of the subpanel and was succeeded by Howard Lorenzen. The ONI representative initially was LCdr. Irving J. Superfine. Superfine was followed by civilian William E. W. Howe; both were from ONI's Technical Unit (OP-322F2). Other agencies represented were the Air Force's Foreign Technology Division (FTD), then called the Aviation Technology Intelligence Center, and the Army's Signal Corps. Frederick Hitz of the National Security Group served as secretary of the subpanel.

During the period 1948–1952, five annual reports were issued by the subpanel. They contained studies on land-based radar order-of-battle, lists of shipborne electronic suits, and reports on various electronic problems such as IFF (Identification Friend or Foe) equipment.[10]

From the period 1948–1949 on, it was apparent to ONI's electronics personnel that a catalog of radar-emission characteristics, properly indexed by frequency and other parameters, was badly needed by the intelligence community, ELINT operators, and other operational and planning personnel. By the late 1950s the publication had grown into the *Radiation Characteristics of Electronic Equipment* (ONI-26) series, which was issued along with supporting guides to intercept operators. The ONI-26 series was used extensively at intercept facilities from Korea to Germany and by ships at sea, as well as by U.S. allies. A principal problem was compilation of data about U.S. and friendly foreign emitters. The publication predated by years the National Security Agency's *Electronic Parameter List* and ONI's own *Threat Emitter Evaluation Guide*.[11]

In the period 1949–1951, then-Lt. Thomas L. Dwyer was a naval aviator attached to the District Intelligence Office, 17th Naval District (Alaska). Dwyer rounded up some available ELINT collection equipment and proposed flying along the Kamchatka-Bering Sea littoral of the Soviet Union. His reports were forwarded through regular intelligence reporting channels, and feedback from ONI's Scientific and Technical Section was provided in the usual evaluation format. Although the use of only one tuning head was possible per flight, the data was unique for that period of low-signal density. The signals detected were principally metric radars (72 megahertz) and E/F band (3,000 megahertz) coastal surveillance radars, but the effort gave an early indication of the extent of Soviet defensive concerns.

Among the pre-Korean War events in the ELINT collection effort was the loss of a Navy PB4Y-2 Privateer reconnaissance aircraft in 1950 in the Baltic Sea in one of the first of too many U.S. Navy "Ferret" losses in the intelligence war against the USSR. The Navy had Joint Chiefs of Staff authorization for three such Baltic missions. The first mission was successful; an analysis of the data collected showed no particular surprises, except that no air intercept (AI) radars had been detected. The second mission was the one that was lost. The third was apparently not flown, leaving responsibility for subsequent U.S. electronic reconnaissance in the Baltic area to the Air Force. Later reports indicated that the Soviets had salvaged the Privateer's equipment from the waters of the Baltic.[12]

In the late 1940s and early 1950s, prior to the onset of a Soviet naval presence in the Mediterranean, the Navy's surface ELINT collection operations in the Mediterranean were aimed mainly at Albania, Bulgaria, and Yugoslavia and conducted by specially equipped destroyers carrying Naval Communications Units. There were five such units, and although their operations did not provide much data of great strategic interest, they did serve as a valuable training medium for developing naval ELINT talent for future, more productive operations. Two of the Naval Communications Units were, in due course, assigned to shore-based operations. Howard Lorenzen from NRL repeatedly checked out the crews and equipment at the two shore sites in the early 1950s to ensure that their maximum operational effectiveness was maintained.

It was during one of Lorenzen's visits to the intercept sites that he contributed to determining the characteristics of the Soviet "Token" radar. Many skeptics believed that the Token was a dummy installation without radiation. Lorenzen took multiple receivers, and, working with them for a couple of nights, was able to pick up the Token's five high-power, pulse-synchronized beams, which were similar to the U.S. AN/CPS-6 (which Token appeared to mimic), even though the U.S. version had not been lend-leased, sold, or used outside of the United States. The skeptics who didn't believe the Soviets had the capability to match U.S. electronic developments were gradually being shown otherwise.

One of the advantages of the 1949–1950 period was the close contact allowed between the collectors, analysts, and researchers. Many analysts and researchers worked for a while with the collectors, both to learn the conditions and problems of the collector firsthand and to help the collector understand the relative values of the data available and being collected.[13]

Korean War

The Korean War found ELINT on Soviet, Communist Chinese, and North Korean radars fairly well documented. But U.S. combat readiness to exploit it left much to be desired. The preponderance of Communist low-frequency radars, fashioned after U.S. and German World War II equipment, made necessary a frenzied retrieval of World War II-era jammers from surplus dealers.

An indicator of the Chinese intention to intervene in the Korean War was noted by the few ELINT analysts in the armed services in November 1950. An SCR-270 100-megahertz radar of unusual strength had been observed during each Air Force and Navy sortie along the China coast in the late 1940s. Its signal strength made the radar uniquely easy to detect and identify. Around August 1950, the radar ceased to be detected. It was not known whether the SCR-270 had been moved or just wasn't being used until, in October 1950, it appeared at Antung on the Yalu River on the northern border of North Korea. The report on the radar's relocation was widely circulated, and although the move was only one indicator to be considered, it should have generated some follow-up collection effort to determine the reason for the move.

Another Korean War experience presented some lessons for those wishing to learn. One of the cornerstones of Chinese and North Korean air defenses was a 150-megahertz radar known as the Japanese Type 3, a version of the German World War II *Freya*, whose blueprints had been sent to Japan by submarine from Germany. Several Japanese Type 3 sets located along the Yalu River had an odd multiple-pulsing signal pattern. The U.S. Air Force intelligence headquarters and the Air Force FTD put a great deal of analytical effort into trying to detect some novel implication in this anomaly such as the incorporation of a command communication system for nightfighters. During the same period, one of the ONI Technical Section officers, Lt.(jg) R. E. Anderson, USNR, was screening old World War II ECM documents about the operating characteristics of Japanese radars. Anderson noted among the comments on the Type 3 the statement that "double-pulsing is a common fault." Further research into World War II efforts to counter the *Freya* showed the same comments, and the German archives verified the fact. When the Japanese copied the *Freya*, they apparently also copied the anomaly, which didn't affect the radar's performance appreciably. Thus, ONI provided a simple solution to an ELINT problem of operational concern, and an important lesson: Don't destroy historical files.[14]

The Far East Command Joint Electronic Intelligence Center (JEIC) was established on 2 October 1952 to provide a central agency for the coordination, joint processing, analysis, evaluation, and dissemination of electronic (noncommunication) intelligence in the Far East Command. The Navy was a prime mover in initiating action to set up JEIC and provided most of the raw data; the Army provided financial support and at least 50 percent of the evaluators. Far East Air Force intercepts were not made available to JEIC initially, so information about the radar defenses of some areas was incomplete.[15]

Commander in Chief, Pacific Fleet (CINCPACFLT) prescribed the objectives to be fulfilled by the ECM programs during the Korean War to provide a system for effective use of intercept equipment (passive ECM) for early warning, intelligence functions, and radiation-control monitoring. Commander Naval Forces, Far East did much to advance the program by actively supporting and participating in JEIC. Valuable data were accumulated on Communist radar characteristics, locations, and operational procedures. In spite of the limited number of adequately equipped ships and aircraft, it was found that ECM served as the most reliable and earliest means for the detection of submarines.

As often as aircraft carrier operations permitted, passive ECM missions were scheduled to intercept electronic emissions and locate enemy radar stations along the east coast of North Korea. Prospective enemy radar sites in the eastern half of Korea were photographed in an attempt to confirm the locations of the stations.[16]

Cold War Era, 1950–1973

Other joint ELINT collection efforts were made by the Army and Navy in the early 1950s from overseas sites. The first such site was operated by the Army using fixed "inside" intercept equipment provided and tested by the Naval Research Laboratory. Operator training was initially provided by NRL, then by the Army-Navy Electronic Evaluation Group (ANEEG), later retitled the National Technical Processing Center, followed by the National Security Agency (NSA). The operations, which continued unimpeded without security compromise and at a very reasonable cost, contributed tremendously to early U.S. ELINT collection efforts. Early joint Army-Navy efforts during the 1950s also included the use of portable intercept systems supplied by the Army Signal Corps, but this effort was discontinued when a security compromise occurred in the mid-1950s.

ELINT collection equipment was also loaned to friendly navies. The equipment was carried in ships and aircraft, and the return on the modest investment was usually quite high since the collectors normally operated in areas in which U.S. ships could not operate without inducing Eastern European countries to exercise security measures during their presence. The equipment loan program, in addition to increasing ELINT collection, also helped to keep open intelligence exchange channels between ONI and friendly foreign navies.

As mentioned in Chapter 11, interrogation of technically competent Germans who had returned from forced labor in the Soviet Union in the early 1950s was a productive collection method. For ELINT, it proved to be one of the early and particularly valuable sources. The Navy's Scientific and Technical Unit under Commander Naval Forces,

Germany was located at Frankfurt and played a key role in the effort. Some special cases were sent to the United States for exhaustive interrogation by the intelligence community in the Washington area. Defectors also furnished unique electronics information. One former Soviet air force officer with a wealth of knowledge on Soviet electronic warfare got lost in administrative red tape and was "re-found" by John Wallace of the Navy's Frankfurt unit; the Soviet defector's special knowledge enticed key personnel to come from Washington to exploit his memory. Other defector sources justified months of detailed questioning by teams of specialists.

One such case involved a former Soviet naval officer known as Nicholas Shadrin; he had served as the commanding officer of a *Skoryy*-class destroyer and was believed to have had extensive knowledge of shipborne electronics and ordnance. Shadrin eventually became a U.S. citizen and a translator-consultant for the Naval Scientific and Technical Intelligence Center (NAVSTIC) and the Defense Intelligence Agency; he disappeared under mysterious circumstances during a visit to Vienna in the late 1970s.

As a result of the flood of electronic intercept reports accumulating during the Korean War, it was decided to attempt to form a joint analytical-processing-reporting facility for ELINT. The Technical and Operational Intelligence Sections of ONI, plus the Naval Security Group, represented the Navy in the discussions that resulted in setting up the Army-Navy Electronic Evaluation Group at the Navy's Communications Annex on Nebraska Avenue, Washington, D.C. The Air Force, with its own processing center at the Air Technical Intelligence Center at Wright Field (later called the Foreign Technology Division), declined to join.

The failure of the Air Force to be part of ANEEG brought much criticism in the early to mid-1950s from advisory groups such as the Defense Science Board. As a result, the Joint Chiefs of Staff designated a three-service flag officer team to look into the matter. VAdm. Harry D. Felt was the Navy member. The team report, submitted around 1956, brought the Air Force into ANEEG, which thereupon changed its name to the National Technical Processing Center.[17]

In the mid-1950s, a mysterious pair of radomes appeared, one on each side of the foremast on board Soviet *Sverdlov*-class cruisers. Lack of any views of the antenna or transmission lines under the radome made analysis difficult. From fragmentary information, however, it was possible to construct a reasonable determination that the equipment was the first Soviet shipborne radar jammer, and it was given the nickname "Top Hat."[18]

In mid-1957, ONI was partially reorganized in the area of electronic analysis by establishing an ELINT section (OP-922Y4) in the Operational Intelligence Branch. LCdrs Donald S. Lindberg, J. W. Douglas, and F. A. Musial were the first officers assigned to the new section, and William Howe was loaned to it from ONI's Technical Section (OP-922F2). OP-922Y4's responsibilities were mainly concerned with collection equipment, operations, and policy and did not duplicate the ELINT processing functions of the Technical Section. The OP-922Y4 at the same time was gradually developing a small ELINT processing facility. With the only wire transcription capability in the Washington area in the late 1950s, the Technical Section had the unique and unenviable duty of making magnetic tape duplicates for the intelligence community of the Navy's old-fashioned recordings, which were still being made on wire.[19]

Circa 1958, tactical ELINT became useful in the U.S. Navy's efforts to assist the Chinese Nationalist navy in the creation of a cover and deception capability during resupply operations for the island of Quemoy in the face of concentrated heavy gunfire from the Chinese Communist mainland. RAdm. Paul P. Blackburn, Jr., Commander Taiwan Patrol Force (the Navy component of the Taiwan Defense Command), was also in command of Fleet Air Wing One and Commander Task Force 72. RAdm. Blackburn urgently needed to fix the locations of the Communist Chinese search radars involved in directing effective land-based gunfire on Chinese Nationalist resupply forces. As a consequence, Fleet Air Reconnaissance Squadron (VQ) One assets, then under Blackburn's operational control, staged out of Taiwan and produced a current radar order-of-battle for the area. Taiwan Defense Command photo interpreters then used the data in their efforts to pinpoint the actual radar locations. The efforts failed but did establish that the logical sites for coastal surveillance radars (on high elevations) were not occupied. Those factors enabled the CTF 72 electronic warfare officer to develop a number of deception plans to attract gunfire to decoys and away from Chinese Nationalist navy units by indicating the presence of targets at unused beach areas through the use of radar reflectors. In one particular case, only one or two rounds of fire fell in the actual unloading area, and a reported 200 rounds fell in the immediate area of the decoy elements. The efforts contributed to the successful landing of several heavy tanks and other reinforcements for the beleaguered island.

One publication that had its genesis in ELINT and collateral intelligence was the Chief of Naval Operations *Threat Emitter Evaluation Guide*, issued in 1958. The predecessor to the guide had evolved during the days just prior to and following the tense period caused by the Suez crisis. The factor that occasioned the creation of the 1958 document was the need for quick evaluation of tactical electronic intercepts from U.S. Navy Pacific Barrier surface and air units. Following a comparison of the intercepted ELINT parameters with all the emitters expected in the area and a comparison of corresponding radar echoes with flight plans and known surface plot data, the evaluated information was flashed to the North American Air Defense Command (NORAD). The *Threat Emitter Evaluation Guide* presented ELINT, technical intelligence, and order-of-battle data on hundreds of Communist emitters, whereas the original Barrier Forces, Pacific evaluation tool had presented data on only thirteen emitters. The *Threat Emitter Evaluation Guide* became a tactical tool that combined, in a single document, everything known about a threat emitter and its associated weapons and platforms that was needed to evaluate quickly and confidently any potentially hostile intercept in a tactical situation.[20]

Another project of naval origin in the late 1950s, Project PAMOR (Passive Moon Relay), was the brainchild of James Trexler, who worked under Howard Lorenzen at the Naval Research Laboratory. The idea was to use the moon's surface for reflecting communication signals between two points on the earth and also for ELINT collection. The first experiments were conducted at Stump Neck, near Indian Head, Maryland, using modified naval radars as signal sources and a meshed surface in a hollowed-out bowl, dug into the ground as a receiver. A whole series of communications tests proved the idea to be soundly conceived for direct communications between two points on the earth when they were simultaneously in view of the moon.

As a result of the initial PAMOR successes, NRL proposed the construction of a 600-foot trainable dish antenna at Sugar Grove, West Virginia, with a programmed cost of $60 million. ONI strongly supported the proposal because the concept had a demonstrated ability to collect unique intelligence.[21]

As the PAMOR project was briefed throughout the Defense Department and finally presented to Congress, the cost was projected to be $71.1 million plus $7.9 million that had previously been allocated. Congress cut the appropriation request and approved only $60 million for Fiscal Year 1959. When initial planning, however, found numerous cost add-ons for unexpected construction complications for an antenna of such size and for support installations such as housing, computer controls, and security features, the overall funding requirements became prohibitive (an estimated $300 million). A review of the project was called for, and the size of the installation was reduced to a 150-foot dish. Much of the steel already procured for the 600-foot antenna was sold as scrap, and the computer cost was cut from $3.5 million to $1 million. Final drafting of construction plans was begun in 1962, and the 150-foot dish went on the air in the fall of 1968. NRL subsequently used it periodically in its space and other research efforts.

The responsibilities for ELINT within ONI were established by ONI Instruction 03840.1 of 30 June 1958:

a. Operational Intelligence Branch (OP-922Y):

(1) In collaboration with DCNO (Fleet Operations and Readiness), establish and promulgate Navy ELINT policy, doctrine, and objectives.

(2) Establish and coordinate the promulgation of Navy ELINT requirements, including intelligence based primarily on ELINT.

(3) Formulate and publish appropriate recommendations concerning plans and programs for the coordinated development of Navy efforts in the ELINT field.

(4) Prepare and submit recommendations regarding ELINT tasks of the operating forces to the DCNO (Fleet Operations and Readiness) or the appropriate operational commander.

(5) Coordinate within OPNAV [Office of the Chief of Naval Operations] on ELINT matters and effect external liaison for ELINT.

(6) In coordination with OP-05 and OP-03 [ACNO (Naval Aviation) and ACNO (Surface Forces)], provide technical control for air ELINT activities.

(7) Provide coordinated guidance for budgeting and procurement of ELINT equipment for attaché and foreign exchange ELINT programs. Monitor other budget programs as appropriate to insure the timely and integrated development of Naval ELINT capability.

(8) Prepare and disseminate all-source operational intelligence derived from ELINT.

(9) Collate ELINT with special intelligence.

(10) Provide Navy special intelligence liaison to the National Technical Processing Center (NTPC) for ELINT processing.

b. Basic and Technical Intelligence Branch (OP-922G):

(1) Provide technical guidance, technical liaison, and evaluation for attaché and foreign exchange ELINT information.

(2) Provide tasks for attaché and foreign exchange ELINT activities.

(3) Provide Navy Technical Intelligence liaison to the NTPC for the technical analysis and evaluation of ELINT materials. Provide Navy ELINT materials to NTPC from all-sources other than those provided by OP-30G.

c. Collection and Dissemination Branch (OP-922H):

(1) Operate the attaché system in a manner to provide administrative support and maximum utilization for ELINT purposes.

(2) Promulgate standing ELINT Technical Search Requirements.

(3) Receive and distribute ELINT materials within OPNAV.

(4) Disseminate all intelligence derived from ELINT other than that disseminated by OP-922Y.

In 1960, the Secretary of Defense ruled that the services and joint commands should retain their ELINT resources and capabilities. This ruling, in essence, meant that most U.S. ELINT activities would not be turned over to the new National Security Agency. The Navy had strongly supported the joint theater concept and had participated significantly, both as a contributor and as a customer at joint centers. NSA continued to exert pressure within the Department of Defense to gain control of theater command resources and of passive ECM, believing that centralization would improve management of U.S. ELINT resources regardless of the effects the change might have had on operational readiness.[22]

The increased volume of ELINT material collected during 1962 in the Atlantic and Caribbean area, particularly incident to the Cuban missile crisis, required additional analysis capability at the Commander in Chief, Atlantic Fleet (CINCLANTFLT) ELINT Center (CEC). In October, an additional analysis position was acquired, and twenty-four hour operation of the center was commenced.

Operational control of two ground-based ELINT positions was delegated to Commander in Chief, Atlantic (CINCLANT) in June, and the ELINT facilities in *Oxford* (AGTR 1) were similarly delegated in September. At the end of 1962, "hearability" tests were being conducted at Guantanamo Naval Base to determine its suitability as an additional ground-based ELINT collection site.[23]

In 1963, electronic intelligence was defined by ONI as "the technical and intelligence information derived from foreign noncommunications electromagnetic radiations emanating from other than atomic detonations or radioactive sources."[24]

The CEC, staffed by personnel attached to Naval Security Group, Atlantic on the CINCLANTFLT staff, performed preliminary processing of all electronic signals collected by airborne and surface forces, Fleet Air Reconnaissance Squadron units, technical research ships, and other fleet collection assets during the early 1960s. In addition, four cryptologic service shore sites provided ELINT input to the CEC for processing and subsequent introduction of the results into the CINCLANT *Radar-Order-of-Battle* (ROB) through the Atlantic Intelligence Center (AIC). Upon completion of preliminary evaluation, CEC provided feedback to collectors on the results they had obtained and the efforts necessary to correct or improve their collection and reporting. In addition to providing ROB data to AIC, the material collected was also provided to the National Security Agency in IBM-machine records format for use at the national level. As a means of improving fleet proficiency in ELINT collection, CEC personnel were assigned temporary additional duty to numerous fleet units so they could provide technical guidance and on-site training to ECM operators.[25]

Within OPNAV in 1964, responsibilities for electronic intelligence were assigned as follows:

1. To the Director of Naval Intelligence:
   a. Coordinate all ELINT activities within CNO;
   b. Disseminate intelligence derived from ELINT;
   c. Recommend personnel requirements and programs in support of non-cryptologic ELINT activities;
   d. Delineate requirements related to research and development on ELINT equipment for use by non-cryptologic forces;
   e. Monitor all Navy ELINT collection programs to ensure that electronic warfare is supported in an effective and timely manner. Make recommendations on any deficiencies or improvements as concern the equipment or operational aspects of the collection program; and
   f. Provide technical evaluation and support to appropriate operational commanders in all ELINT collection efforts by ground, air, surface or submarine units.
2. To the Deputy Director, Naval Communications for Naval Security Group and the Director, Naval Security Group (NAVSECGRU/DIRNAVSECGRU) (OP-94G):
   a. Program and provide personnel in support of collection activities at naval shore bases and in technical research ships (AGTRs);
   b. Exercise technical control for special surface and submarine ELINT activities, including requirements related to RDT&E [research, development, test and evaluation] for ELINT equipment;
   c. Provide personnel and equipment for specialized ELINT collection requirements and for training of fleet ECM personnel in ELINT collection and analysis;
   d. Provide personnel for joint ELINT processing centers, as required; and
   e. Program and provide personnel in support of specified naval ELINT processing facilities.
3. To the Deputy Chief of Naval Operations (Fleet Operations and Readiness) (OP-03):
   a. Fulfill Navy ELINT requirements through assignment of appropriate operating forces;

b. Monitor all ELINT and special operations insofar as operational feasibility and safety is concerned;
c. Apply ELINT results to Electronic Warfare; and
d. Provide timely feedback to the Director of Naval Intelligence on the results of ELINT support to electronic warfare.

4. To the Deputy Chief of Naval Operations (Air) (OP-05);
   a. In coordination with the Director of Naval Intelligence, establish technical configurations of aircraft for use in ELINT missions and provide such aircraft to the operating forces; and
   b. Provide personnel and programs in support of airborne ELINT collection activities.
5. To the Deputy Chief of Naval Operations (Development) (OP-07):
   a. In collaboration with DNI and DIRNAVSECGRU, determine research and development (R&D) requirements in support of ELINT;
   b. Provide programming and budgeting insofar as ELINT R&D is concerned;
   c. Maintain a Quick Reaction Capability (QRC) program within the ELINT area;
   d. Maintain an effective R&D effort in support of the Navy ELINT Program; and
   e. Provide timely feedback to the Director of Naval Intelligence on R&D in support of the Navy Electronic Warfare and ELINT programs.[26]

Electronic intelligence, in 1964, was being produced by ONI to support the following programs:

1. Navy Electronic Warfare:
   a. In the design doctrine and tactical use of electronic countermeasures, by providing data on unfriendly electronic emitters;
   b. In the development of airborne and missile weapons systems, by providing data on the emission characteristics and vulnerabilities of electronic components of unfriendly anti-air warfare and anti-missile warfare weapons and supporting units;
   c. In the development of shipborne primary weapons systems, by providing data on the emission characteristics and vulnerabilities of electronic components of unfriendly anti-surface and anti-submarine weapons and their controlling components;
   d. For land warfare activities, by providing similar data on the electronic components of land-based weapons;
   e. For Early Warning Systems, by providing data to discriminate between emissions of friendly and unfriendly forces; and
2. National Intelligence Programs:
   a. By providing information of value in the fulfillment of National Intelligence Objectives.[27]

The National ELINT Plan, approved by the Secretary of Defense on 22 September 1964, required, among other things, that each theater commander establish a Joint Reconnaissance Center responsive to the Joint Chiefs of Staff on a 24-hour-a-day basis and a Theater Joint ELINT Center. The plan also increased the number of ELINT missions to be flown by the Strategic Air Command and terminated the missions previously flown by the Marines and the Air Force Tactical Air Command.

The CEC was responsible for tasking the intercept capabilities of VQ-2 Detachment 14 based at Key West, VMCJ-2 at Cherry Point, and USNS *Sgt. Joseph E. Muller* (T-AG 171). In addition, numerous ships and submarines of the Atlantic Fleet served as collection platforms during normal transits along the north and south coasts of Cuba. During 1964, CEC continued to prepare Standardized ELINT Report Format and the Periodic ELINT Activity Report for VQ-2 and VMCJ-2. The processed ELINT information was also used to provide input to the CINCLANT *Caribbean Radar Order-of-Battle* publication produced by the Atlantic Intelligence Center.[28]

In addition to intercepts, ELINT collection included the acquisition and study of equipment. The value of such collection has already been indicated in the account of the exploitation of the Soviet *Neptune* navigational radar in 1956 (Chapter 11). The Arab-Israeli Six Days War in June 1967 afforded another opportunity for ONI personnel to analyze actual equipment. NAVSTIC representatives studied the Soviet SA-2 surface-to-air missile system with its Fan Song radar, the Bar Lock ground-controlled intercept radar, an S-band coastal surveillance set, the AS-1 air-to-surface missile-guidance system, the Styx antiship missile-guidance system, the Atoll and Strela missiles, a video radar data link, and the Fire Can antiaircraft fire-control radar and its associated optics and computer. Some of the captured equipment was checked out and evaluated in the field and some was afforded laboratory analysis, all of which provided invaluable information used in the development of countermeasures that saved lives in Vietnam.

As in the Korean conflict, the Vietnam action accelerated the need for timely tactical ELINT. The Desoto surface ship patrols in the Gulf of Tonkin accentuated the requirement for shipboard passive ECM with good technical backup support from the Pacific Command ELINT Center and the Washington intelligence community. It was an attack on one of the Desoto patrols that launched the air war in Southeast Asia. ELINT intercepts of S-band Skin Head radars on board North Vietnamese motor torpedo boats, along with other information, alerted patrolling U.S. Navy destroyers in August of 1964 that they were targets for an impending attack.

ELINT and ECM made a much greater contribution to survival in the Vietnam War than in Korea because of the greater variety and sophistication in the elctronic elements of the enemy's defenses and weapon systems. One interesting device introduced during the conflict was the anti-radiation missile (ARM). ONI, particularly the Naval Scientific and Technical Intelligence Center, was intimately involved in designing the ARM, particularly the AGM-45 Shrike, from its earliest days to its operational phase. The development of threat-warning devices and jammers was a continuing effort in the ECM/ECCM competition throughout the air war, with ELINT playing a vital part in providing guidance to the weapons and countermeasures developers.[29]

Closer lines of coordination were established within OPNAV and between OPNAV and the systems commands in 1969 to facilitate rapid responses to electronic warfare sensor and sensor-systems requirements. NAVSTIC was rapidly expanding its capabilities to evaluate raw data on foreign navy electronic warfare capabilities. A significant gap had already existed in this field that threatened to widen as systems of increased complexity and sophistication were employed by foreign navies.[30]

During Fiscal Year 1972, NAVSTIC developed and disseminated an intra-Navy working paper to provide up-to-date information on high-threat-associated electronic signals. The Threat Parameter List was limited to high-threat signals, more readily usable by Navy customers and updated monthly.[31]

A new electro-optics (EO) processing facility was established by Commander Naval Intelligence Command during Fiscal Year 1972 as the first such facility in the defense intelligence community. It provided a focal point for all EO intelligence matters, including RDT&E, operational signals analysis, threat assessments, data storage and retrieval, and production and distribution of finished EO intelligence. The new naval intelligence EO capability represented a critical step forward in the development of countermeasures to known foreign EO technology.[32]

An electro-optical detection system was developed by the Naval Ships Systems Command and first deployed in a ship on 10 March 1972 for a six-month evaluation period in the Mediterranean. In May, the first laser signal identified as emanating from a Soviet ship was intercepted.

By the end of Fiscal Year 1972, the first Airborne Dual Detection Indicator laser detectors had been provided to CINCPACFLT by the Naval Intelligence Command. Developed by the Naval Reconnaissance and Technical Support Center in coordination with the Naval Weapons Center, China Lake, California, the detectors were tested and employed successfully in specially equipped antisubmarine patrol aircraft. Three additional sets of the equipment were provided to Commander Fleet Air, Mediterranean, for use in VP (patrol) and VQ (reconnaissance) squadron aircraft in the Mediterranean.[33]

Another ELINT program analysis developed during the early 1970s analyzed Soviet Deceptive Electronic Countermeasure (DECM) signals that had been collected and recorded through modifications to radar receivers on carrier-based aircraft. The Naval Intelligence Support Center, the successor to NAVSTIC, was responsible for the preliminary analysis of DECM signals.[34]

Chapter Notes

1. Howard Lorenzen ltr to author, notes "On ECM at Naval Research Laboratory (NRL)," undated.

2. Ibid.; and Louis Gebhard, *Evolution of Naval Radio Electronics and Contributions of the Naval Research Laboratory* (Washington: Naval Research Laboratory, Jan 1976), 303.

3. Gebhard, 305; Commander in Chief, Pacific Fleet/Commander in Chief, Pacific Ocean Area (CINCPACFLT/CINCPOA), *Command History, 1946*, 346.

4. *CINCPACFLT / CINCPOA Command History, 1946*, 347–48.

5. Howard Lorenzen ltr to author, notes "On DF," undated.

6. Commander Fleet Air Wing, Miramar, "History of Electronic Warfare," undated [ca. 1973], 18.

7. W.E.W. Howe memo to author, on ONI's and the Navy's Participation on Electronic Intelligence (ELINT), 1948–1964, 20 Jun 1973.

8. Lorenzen ltr on ECM; and Howe memo on ELINT.

9. Howe supplemental memo, 1983.

10. Howe memo on ELINT.

11. Howe supplemental memo.

12. Howe memo on ELINT; Howe memo on "Ferret" operations, undated.

13. Howe memo on ELINT, 8–9.

14. Ibid., 4a.

15. CINCPACFLT, *Interim Evaluation Report No. 5: Korean War Naval Operations*, 3–72.

16. CINCPACFLT, *Interim Evaluation Report No. 6: Korean War Naval Operations*, 1–13 and 3-86.

17. Howe memo on ELINT, 4, 6–7, 13.

18. Howe supplementary memo, 1983.

19. Howe memo on ELINT, 5.

20. Clayton W. Merrill memo to author, Oct 1974. Merrill was a former ONI analyst and was responsible for producing the *Threat Emitter Evaluation Guide*.

21. Howe memo on ELINT, 9–10.

22. ONI OPINTEL Briefing Folder, 1 Jun 1963, 4.

23. Commander in Chief, Atlantic (CINCLANT), *Annual Report for 1962*, 51.

24. ONI Instruction 03840.1A of 19 Jul 1963.

25. *CINCLANT Annual Report, 1963*, 56.
26. OPNAV Instruction 003430.2B of 7 Feb 1964.
27. *CINCLANT Annual Report, 1964*, 17.
28. Ibid., 55–56.
29. Howe memo on ELINT, 12–13.
30. *DNI Report to CNO, FY 1963*.
31. *DNI Report to CNO, FY 1972*, 24.
32. Ibid., 9.
33. Ibid., 22–24.
34. *DNI Report to CNO, FY 1973*, II-11–II-12.

CHAPTER 16

COMINCH

Beginning of Organized Operational Intelligence

In January 1941, OP-38W was established in the Office of the Chief of Naval Operations (OPNAV) to furnish the Chief of Naval Operations with a "war room" capable of supplying complete combat or operational intelligence and providing facilities for strategic staff conferences. Two sections were formed, one responsible for U.S. fleet information and one for foreign information. In December 1941, the U.S. section was flooded with demands for special reports on the location of combatant ships, leading to the preparation of a daily ship-location list with notations on ships lost and damaged. The U.S. section also prepared a weekly report of task force organizations.

Adm. Ernest J. King was designated Commander in Chief, United States Fleet (COMINCH) by President Franklin D. Roosevelt on 20 December 1941, and he took command on 30 December. By Executive Order of 12 March 1942, King relieved Adm. Harold R. Stark and took on the added title of Chief of Naval Operations. Operational functions, including operational planning, were soon shifted from OPNAV to the COMINCH staff.[1]

OP-38W became the Operational Information Section (F-35) of the Operations Division of COMINCH headquarters in mid-January 1942. Fleet Intelligence Officer (F-11) Cdr. George C. Dyer was given additional duty in charge of F-35 on 26 March 1942. Inasmuch as F-11 came under the Assistant Chief of Staff for Plans (F-1), the Fleet Intelligence Officer was in the advantageous position of being informed on future planning. Consequently he could guide the activities of the Operational Information Section to meet COMINCH planning requirements for strategic intelligence effectively.[2]

One of the most immediate problems in early 1942 was the German submarine force. Accordingly, a system of tracking enemy submarines, based primarily on communications intelligence (COMINT), was set up in F-35 to facilitate the detouring of merchant shipping around the submarines. Antisubmarine warfare information became the principal business of F-35; other functions included providing official weather forecasts in code and reporting ship movements to postal authorities.

Because a relatively large number of persons were allowed access to the F-35 war room, information plotted there did not include highly classified details on future operations. These details were maintained in the F-11 chartroom for the use of Adm. King and a severely restricted list of officers. The F-35 war room was for the benefit of personnel "entitled to see *something*" but not the full picture on future events.

In November 1942, Capt. Theodore T. Patterson was assigned to the Plans Division as F-15 to act as liaison between COMINCH headquarters and the Office of Public Relations.[3]

Tracking German Submarines

The Atlantic Unit of COMINCH F-11 (later F-21 of the Combat Intelligence Division) for practical purposes was responsible for operational intelligence about all world sea areas except the Pacific Ocean, and its principal function was U-boat tracking. Initially, the section was set up under Lt. George H. Laird, Jr., as a tracking room for U-boats in the western Atlantic, but it soon became apparent that a closer liaison with the British Admiralty was necessary. As an initial step in that direction, the head of the British tracking room, Cdr. Rodger Winn, RNVR, visited COMINCH headquarters in May 1942.

During Cdr. Winn's visit, arrangements were made for a complete correlation and exchange of information between the Royal Navy and U.S. Navy tracking rooms, including a unified system of U-boat designations, a regular exchange of messages between the heads of the two tracking rooms, and the promulgation of daily U-boat estimates. Winn's visit was of inestimable benefit to the U.S. tracking room and resulted in a more accurate and uniform system of estimating German U-boat positions.

In July 1942, LCdr. Kenneth A. Knowles, USN (Ret.), then the head of the U.S. tracking room, visited the Admiralty tracking room in London and the Headquarters, Western Approaches at Liverpool, which controlled antisubmarine operations and trans-Atlantic convoys in the British strategic area of the North Atlantic. Other exchange visits were made by the U.S. and the British assistant tracking officers and the head of the Canadian tracking group, and Winn made another liaison visit to the COMINCH tracking room in the early summer of 1943. The various visits permitted tracking officers to familiarize themselves with the methods used in the other tracking rooms and to maintain the close personal relationships so necessary for ensuring a full and complete exchange of ideas, comments, and information related to U-boat tracking.

With the establishment of the U.S. sea frontier organizations, a U-boat tracking room was set up at each sea frontier headquarters. To facilitate the exchange of U-boat information, a scrambler teleprinter was established between COMINCH (F-21) and the various Atlantic Sea Frontier tracking rooms. The direct system assured more rapid delivery of U-boat contact reports and submarine tracking estimates.[4]

Reorganization and Expansion During 1943

On 8 January 1943, Capt. Homer L. Grosskopf reported as Security Control Officer (F-16) in the Plans Division, and during the same month Cdr. Dyer was relieved by Cdr. Henri H. Smith-Hutton, lately repatriated from Japanese confinement following duty as Naval Attaché, Tokyo, at the start of the war.[5]

With the establishment of the new Combat Intelligence Division (F-2) and the associated reorganization on 1 July 1943, RAdm. Roscoe E. Schuirmann became F-2; Capt. Smith-Hutton became Assistant for Combat Intelligence (F-20); F-35 was disestablished and broken up into the Atlantic Section (F-21), the Pacific Section (F-22), and the Operational Summaries Section (F-23); the Security Control Officer, formerly F-16, became F-24; and the duties of F-15 were transferred to a publicity security officer (F-25).

As a consequence of a memorandum of 12 October 1943 from the Secretary of the Navy to Adm. King, F-2's responsibility for publicity security was cancelled on 15 October. F-25's responsibility for security review of all material to be released for publication, however, continued throughout the war.[6]

There were about fifteen officers and fifteen enlisted men in the F-11 section of the COMINCH staff. Most of their attention was devoted to German submarine activities in the Atlantic. In addition to standing a continuous watch in the antisubmarine plotting room, the officer in charge of the section gave a daily briefing that was attended by Adm. King; his Chief of Staff, VAdm. Richard S. Edwards; and all COMINCH officers concerned with the ASW situation. Frequently, high-ranking Army officers were also present, as were some British officers. The daily briefings usually lasted about fifteen minutes and were held in front of a 40-foot wall chart of the Atlantic that showed convoy positions, known submarine locations, and all actions that had taken place during the previous twenty-four hours.

The information used by the COMINCH intelligence staff was derived from all sources, including the forces at sea. All of the information had to be recorded, estimates prepared, and messages drafted to send the information and estimates to all commands that needed them.[7]

As mentioned earlier, the dual organizational arrangement continued until 1 July 1943, when F-35 was renamed the Combat Intelligence Division, and the Fleet Intelligence Officer became Assistant Chief of Staff for Combat Intelligence (F-2). The new division retained its operational functions and was also responsible for passing current intelligence to naval forces; providing intelligence support to other divisions of COMINCH headquarters; maintaining a strategic plot and chart room for the use of flag and other high-ranking officers (with the plot and chart room to contain all current situation information, including all relevant dispatches and operational plans, past, present, and future); and maintaining a complete library of current intelligence, monographs, air target folders, and other basic intelligence reference material.

As part of the expansion of the COMINCH intelligence organization in July 1943, all radio intelligence functions and personnel were shifted from ONI's Far East Section (OP-16-FE) to COMINCH (F-2), including security control of the radio intelligence material.[8]

RAdm. Roscoe E. Schuirmann, the first F-2 in the new organization, took on the added job of Director of Naval Intelligence on 25 September 1943. Back on 21 June, Schuirmann had proposed the setting up of a Combat Intelligence Division to Adm. King. In his memo of that date, he acknowledged that a line of demarcation between the duties of ONI and those of the Combat Intelligence Division could not be sharply defined but that, generally, the former would supply strategic intelligence and the latter combat (tactical) intelligence. The Combat Intelligence Division, having knowledge of operational plans, would guide ONI as to the direction and priority of its efforts.

As of 20 May 1943, the Tenth Fleet was established under COMINCH command "to exercise unity of control over U.S. anti-submarine operations in that part of the Atlantic under U.S. strategic control."

The Atlantic Section (F-21) of the Combat Intelligence Division, under Cdr. Kenneth Knowles, had the additional duty to serve as operations plot for the Tenth Fleet. It was thus the western Atlantic counterpart of the huge "Western Approaches" operations center at Liverpool. The importance of Knowles's plot to the antisubmarine effort in the western Atlantic cannot be overestimated in its function as a clearing-house for the Navy's shore-based, high-frequency, direction-finding (HF/DF) network, which extended from Jan Mayen Island to the Brazilian bulge. The plot's furnishing of timely, accurate information on the daily U-boat situation in the Atlantic to the Commander in Chief, Atlantic Fleet was vital to the latter's effective operations against the U-boats.[9]

The reorganization of COMINCH headquarters, which became effective on 1 July 1943, gave F-21 the mission "to correlate, evaluate and disseminate naval intelligence of operational or combat significance within the Atlantic Theatre." For that purpose, active contact was maintained with the various divisions of COMINCH headquarters, with other offices of the Navy and War Departments (including the Joint Intelligence Committee), and particularly with ONI.

In fulfillment of its mission, F-21 performed the following functions:

1. U-boat tracking, which comprised the continuous plot of estimated enemy U-boat positions and all related intelligence;

2. Advising the Chief-of-Staff, Tenth Fleet (FX-01) as to the general U-boat situation as it affected the Convoy and Routing Section and antisubmarine measures;

3. Analyzing, evaluating, and disseminating enemy HF/DF fixes;

4. Preparing information on U-boat dispositions and trends for dissemination to operational commands by means of COMINCH Daily and Special Estimates and the COMINCH Biweekly U-boat Trends;

5. Exchanging U-boat and blockade runner intelligence and tracking data with British Admiralty U-boat tracking room;

6. Preparing a monthly article on U-Boat Tactics and Trends for the COMINCH *Anti-Submarine Warfare Bulletin;* and

7. Maintaining general combat intelligence and statistical records on the Atlantic Theatre.[10]

A continuous, four-section watch was maintained in F-21, each section consisting of two officers and one yeoman. In addition to maintaining various logs of U-boat sightings, contacts, and attacks, each watch section was responsible for keeping the current positions of convoys and merchant ships plotted on wall charts; posting HF/DF fixes and ensuring their proper dissemination to operational commands; plotting on wall charts all incidents, sinkings, estimated U-boat positions, and locations of U.S. task groups; maintaining a "hot spot" file of current reports and dispatches of ongoing or recent incidents; and providing information to authorized officers as required.[11]

At the time of the reorganization of the COMINCH staff in July 1943 when F-11 became F-2, the five officers, including LCdr. William J. Sebald, who had been transferred from ONI, became the Pacific Section (F-22). The reorganization improved the processing of information on the Pacific, enhanced its availability to the COMINCH planning and operational sections, and permitted better liaison with the intelligence section of Commander in Chief, Pacific Fleet.

In early 1943, when WAVES (Women Accepted for Volunteer Emergency Service) became available, four of the first ten arriving in Washington were accepted for duty in the COMINCH Intelligence Section to release an equal number of male quartermasters and yeomen for sea duty. This was the first of several groups of WAVES assigned to F-11/F-2, and soon the entire office, which plotted submarine positions, consisted of both commissioned and enlisted WAVES, except for two male officers who were not physically qualified for sea duty.

There was no duplication of effort or rivalry between ONI and the COMINCH Combat Intelligence Division. The latter was a small staff organized for rapid handling of operationally vital intelligence. ONI, on the other hand, being a large organization, couldn't, and was not required to, take rapid action on the vast amount of noncombat-related information it received. The processing of information into area studies, recognition manuals, and technical reports was slow work, and the products did not usually lose their value if not used immediately. ONI's work, however, was important to the antisubmarine effort. For example, its expert interrogation officers prepared detailed reports on tactical and other information obtained from German submarine personnel, and ONI's technical experts provided valuable data on captured submarine equipment. ONI's psychological warfare broadcasts to German submariners were also very effective.

The Combat Intelligence Division was able to get out a warning message to a convoy in the mid-Atlantic within twenty minutes of detecting a German submarine radio broadcast by the Atlantic direction-finder net. A delay of even a few minutes in putting out the warning messages could have meant the loss of additional shipping; the prevention of such losses was the reason that Adm. King had established an intelligence section within his headquarters where he could oversee its work.[12]

Initially, all U-boat tracking was done in F-21's main plotting room. When the German cipher was broken and regular U-boat communication traffic was being received from communications, however, it became necessary for security reasons to limit access to the special intelligence information. A secret room in which to process the special material was established adjoining F-21's plot on 27 December 1942. Eventually, it became standard practice to do most of the U-boat tracking in the secret room (known later as F-211) and then transfer U-boat positions each day to the U-boat pins on the main wall charts in F-21.

The accuracy of U-boat estimates increased with each new intelligence collection capability. However, there were many times when changes in enemy U-boat strategy remained unclear, resulting in different estimates flowing concurrently from the U.S. (F-211) and British tracking rooms. One such period occurred during May 1943: U-boat concentrations had suffered heavily in the North Atlantic that spring, and a major shift in U-boat dispositions was pending. The enemy's traffic was being read only spasmodically, and its disguised grid positions were giving considerable trouble. At that point, the German U-boats' Commander in Chief, Adm. Karl Doenitz, ordered a large wolf pack to a new disguised area. The British tracking room believed the area to be off the Canadian coast at the western focal point of the North Atlantic convoy lanes. The U.S. tracking room believed that the shift was more fundamental, particularly in view of the heavy losses already inflicted on the U-boats in the North Atlantic and the longer periods of daylight in that area, which would aid antisubmarine aircraft operations. Next to the North Atlantic convoy shipping, the most profitable targets for U-boat operations appeared to be the U.S.-Gibraltar convoys carrying much needed troops and supplies to the Mediterranean Theater. On the basis of F-211 estimates, Commander Tenth Fleet ordered an escort carrier group centered on *Bogue* (CVE 9), the first such group to engage in antisubmarine operations under U.S. control, into the estimated area. The group struck the U-boat wolf pack by surprise right in the estimated area. It was the start of a new approach to the antisubmarine war: conducting offensive operations using U.S. CVE groups against U-boat concentrations and refueling areas instead of employing these "jeep carriers" for defensive ocean convoy protection.[13]

Success in the U-boat War and Postwar Dissolution

Another example of antisubmarine warfare cooperation between COMINCH intelligence and ONI took place one afternoon in the spring of 1944. Capt. Daniel V. Gallery, the commander of a hunter-killer escort carrier group in the Atlantic, visited Combat Intelligence in COMINCH and said that his next deployment was to be a sweep into the Azores area. He wanted information that would help him capture a German submarine. Cdr. Smith-Hutton provided him with data on the types of submarines normally operating in that area and assured him that accurate reports of submarine locations in Gallery's area would be sent to him while the hunter-killer group was deployed. Gallery was then taken to the head of ONI's Technical Section, who gave him detailed plans of two types of submarines that might be in the assigned area. Using the plans, Capt. Gallery had two boat crews from destroyer escorts of his group drilled as to the location of sea valves and other fittings in the anticipated types of German submarines, as well as the probable locations of demolition charges. The predeployment intelligence support to Gallery in large measure made possible the capture of *U-505* on 4 June 1944.

In October 1944, RAdm. Schuirmann was relieved as Assistant Chief of Staff for Combat Intelligence (and Director of Naval Intelligence) by RAdm. Leo H. Thebaud, and Capt. Smith-Hutton, Schuirmann's assistant (F-20), was relieved by Capt. William R. Smedberg III. The F-2 Division was by then divided into four sections: the Atlantic Section (F-21) under Cdr. Kenneth Knowles, the Pacific Section (F-22) under Cdr. William Sebald, the Operational Summary Section (F-23) under LCdr. P. N. Culp, and the Security Control Section (F-24) under Capt. Stanley D. Jupp.[14]

On 26 October 1945, the Pacific Strategic Intelligence Division was directed by the CNO to be transferred from the cognizance of the Chief of Naval Communications to the Chief of Naval Intelligence, who would also assume the direction of its work and the administration of its personnel. The title of the section was abolished, and it was made part of the Special Branch (OP-23W) of ONI, where it was headed by Capt. Smedberg. OP-23W remained at the Communication Annex on Nebraska Avenue where it enjoyed the benefit of a close, informal, working-level relationship with the processing personnel of the Communication Supplementary Activity, Washington. See Chapter 18 for additional details.

Chapter Notes

1. FAdm. Ernest J. King biographic summary, OA. The office of the CNO, or OPNAV, was referred to as the Office of Naval Operations prior to World War II; after the outbreak of war, when Adm. King assumed the combined COMINCH–CNO title, the office was referred to as the Office of the Chief of Naval Operations.

2. OP-23Y memo to OP-23B, 29 Apr 1946, George C. Kidd File.

3. Commander in Chief, U.S. Fleet (COMINCH), "Administrative History of COMINCH Headquarters in World War II," 102–3, 144, hereafter *COMINCH WWII Admin History*.

4. COMINCH (F-21) War Report, 15 May 1945, 3–4.

5. *COMINCH WWII Admin History*, 144; and Capt. Henri H. Smith-Hutton, Oral History, USNI, Annapolis, MD, 1979, 2:388, copy in OA.

6. *COMINCH WWII Admin History*, 144–45, 147.

7. Smith-Hutton oral history, USNI, 2:390–92.

8. *ONI WWII Admin History*, 707.

9. Philip K. Lundeburg, "American Antisubmarine Operations in the Atlantic, May 1943–May 1945" (Ph.D. diss., Harvard University, 195), 26–27.

10. COMINCH (F-21) War Report, 5, 5a.

11. Ibid., 11–13.

12. Smith-Hutton oral history, USNI, 2:394–401.

13. COMINCH (F-21) War Report, 20, 24–25.

14. Smith-Hutton oral history, USNI, 2:402.

CHAPTER 17

Files, Records, and Library

Establishing the ONI Filing System

Filing is an essential element in the processing and production of intelligence. It has been said, quite correctly, that an intelligence organization is only as good as its files and the accessibility of those files to the analysts. Additional information on naval intelligence filing activities may be found in Chapters 3, 10, 13, 20, and 21.

When the Office of Naval Intelligence was established, the organization was combined with the Navy Department Library to make the latter's collection of foreign books, periodicals, and technical literature available as ONI's initial files. As mentioned in Chapter 1, Lt. Theodorus B. M. Mason devised a card indexing system for information obtained from the files and other sources, based on the anticipated subjects of concern to ONI's primary customers: the Secretary of the Navy, the bureaus of the Navy, and Congress. Desks were set up by subject according to the interests of each customer, and the researchers at each desk maintained the files on each subject.

Soon after Professor James R. Soley, USN (his formal title), reported from the Naval Academy to head the Navy Department Library in late 1882, he became involved in the collateral duty of assembling the records on the Navy's participation in the Civil War. Consequently, the library took on the new title, Navy Department Library and War Records Office.

The Library's procurement of the "most recent treatises on professional subjects, expensive books of scientific and technical value, and such periodicals, foreign and domestic," was "especially to supply the needs of the Office of Naval Intelligence, which depends mainly upon this library for technical and professional information used in its publications." Separate appropriations for new books for the library during the period fluctuated between $1,000 and $2,000 per year and, of course, were never considered adequate.[1]

Professor Soley served as both Librarian and Superintendent of Navy War Records until he was appointed Assistant Secretary of the Navy in 1890. Prior to this appointment, the Navy Department Library and the Office of Naval Records were reassigned on 21 October 1889 from the Bureau of Navigation to the Office of the Secretary of the Navy (SECNAV). ONI was similarly reassigned in October 1890.

Mason's filing system continued, with minor adaptations to meet new customer requirements, until 1896–1897, when Director of Naval Intelligence LCdr. Richard Wainwright found it necessary to streamline the indexing and files by removing obsolete matter so that the reduced number of assigned personnel could more readily maintain and use the files of current interest.

When ONI was returned to the Bureau of Navigation in 1898 and the library remained under the Secretary of the Navy, the work of both organizations suffered because the former was separated from much of its reference materials and the latter was no longer in a position to accession new materials received by ONI. The separation of ONI from the library continued until 1919.[2]

During 1902, ONI's files were again overhauled and all material not deemed of current value or historical interest was culled. Printed books and pamphlets that were found to be of no value to the office were transferred to the libraries of the Navy and War Departments, the Naval Academy, the Library of Congress, and the Superintendent of Documents.

The register, or general information files, which were the most important files in the office, were divided into a Current Branch and an Obsolete Branch. All registered file numbers bearing a date prior to 1 January 1896 were collected into an obsolete file, and all registered numbers bearing a subsequent date formed the current file. About two-thirds of the material on hand was in the Obsolete Branch. The reference cards were divided at the same date and also separated into current and obsolete sections.[3]

When counterintelligence was added to ONI's responsibilities in 1916, case histories were added to its filing system. With the entry of the United States into World War I and the resultant expansion of ONI, its files also expanded, each section maintaining files of copies of, or extracts from, reports according to the section's designated primary interests. In addition, central files were established in the "D" section, where all reports on naval and military material, personnel, and operations were registered, carded, and filed.

During the first year of World War I, the filing system in use in the general files of ONI was changed to facilitate the handling of a greatly increased daily flow of mail. In the summer of 1918, the mail was reported to have reached a total of 17,000 letters per day. The work force for handling the mail was increased until it reached a peak of 208 personnel.[4]

Establishing the Naval Historical Section

The Naval Historical Section was established in 1918 in the Office of Naval Operations (OPNAV), at the urging of Adm. William S. Sims. A SECNAV Circular Notice dated 18 August 1918 made the Historical Section responsible for collecting historical material. Capt. William W. Kimball was appointed as head of the new section, and in May 1919 he was relieved by Capt. Charles C. Marsh.[5]

On 1 July 1919, a SECNAV order directed that the section of OPNAV and the Library and Office of Naval Records of the Secretary's office be combined within the Naval Intelligence Division of OPNAV. Thus, the Navy Department Library rejoined ONI after being separated for over twenty years, and ONI took on the additional task of being the repository for Navy historical material.[6]

In July 1921, Capt. Dudley W. Knox, who had been assigned to ONI before World War I, relieved Capt. Marsh as officer in charge of ONI's Office of Naval Records and Library. Knox retired on 24 October 1921 but held the post until 1946. The office was divided into three sections: the Library, the Old Records Section, and the World War Section.

In the January 1926 issue of the U.S. Naval Institute *Proceedings*, Capt. Knox, in an article entitled "Our Vanishing History and Traditions," pleaded with former officers, their families, and their descendants to make available any documents of historic value in their "family papers." The article stimulated widespread interest; in due course, it resulted in the establishment of the Naval Historical Foundation, a nonprofit organization that accepts and retains gifts of documents, relics, and other memorabilia for the Office of Naval Records and Library (now the Naval Historical Center).[7]

In December 1928, Knox expressed the mission of the Office of Naval Records and Library as being "to acquire, systematically arrange, and preserve manuscripts, pictoral and technical naval information, and to make such information readily available to the Naval Service and the public."[8]

Another collateral duty of the Office of Naval Records and Library finally received a slight, but encouraging, impetus when a SECNAV order, on 28 April 1930, appointed Knox curator for the Navy Department. As such, he was responsible for the collection and preservation of objects, trophies, and relics of historical and inspirational value to the Navy, except for material permanently assigned to the Naval Academy and other naval stations. He also had cognizance over matters connected with a proposed naval museum.[9]

In 1934, President Franklin D. Roosevelt initiated, and Congress authorized, the publication of documents on the early history of the U.S. Navy. The detailed preparation and editing of the volumes was started in April in the Office of Naval Records and Library under the direction of Loretta J. MacCrindle and the supervision of Capt. Knox. The project continued throughout World War II and continues to this day.[10]

World War II

By 1 July 1941, the Navy Department Library had over 92,600 books in its stacks. Since the declaration of the national emergency, the library had been called upon to perform considerable research work to answer numerous inquiries from new defense agencies, other government libraries, and the Navy Department.[11]

Space problems became so acute in the Main Navy Building in early 1941 that arrangements were made with the National Archives to provide temporary quarters for the Office of Naval Records and Library pending completion of the Navy's Arlington Annex. In March, all personnel, equipment and activities moved to the National Archives building, except for a small reference library and the pictorial section. The move to the Arlington Annex was accomplished in October 1941 when the third wing of the new building was ready. The civilian staff involved in the move totaled fifteen. At the outbreak of direct U.S. involvement in the war, the only officer on duty in the Office of Naval Records besides Capt. Knox was Lt. Sterling T. Dibrell, USNR, head of the Graphic Section.

The Library (OP-16-E1), with a small collection of reference books and the Library Catalogue, remained in two small rooms in the Main Navy Building under Constance D. Lathrop. The main collection of books was located at the Arlington Annex. The Manuscript Section (OP-16-E2), formed by amalgamation of the Old Records Section and the World War [I] Section, was housed entirely at the Arlington Annex, with Alma R. Lawrance in charge. The Graphic Section (OP-16-E3), although originally concerned with photos of historical interest, had "strategic" photographs added to its collection in 1941, and this section remained in the Main Navy Building. The Publications Section (OP-16-E4) under MacCrindle at the Arlington Annex was preparing naval documents on the Barbary Wars for publication.

Following Pearl Harbor, the three immediate objectives of the Office of Naval Records and Library were to procure and commission the most skilled personnel in the field of naval archives and history; survey the entire field of operational records to determine what, in addition to war diaries, would be of most value to the war effort; and arrange and classify documents received from the operating forces so that the pertinent operational data in them would be readily accessible to authorized persons. The above objectives were chiefly the concern of the Manuscript Section.

In spite of some opposition from other parts of the Office of the Chief of Naval Operations, a letter of instruction on the preparation of war diaries was finally drafted and signed by the CNO on 22 February 1942. Even before the directive had been issued, some ships and task forces, particularly in the Pacific, had started keeping war diaries on their own initiative. When the war diaries started to arrive in Washington, the attitude toward them changed, and their high value as a source of current operational information and future guidance was soon recognized.[12]

A microfilm library was set up in April 1942 in the Readiness Section of Commander in Chief, U.S. Fleet (COMINCH) to facilitate dissemination of the information in the war diaries, submarine patrol reports, action reports, and personal narratives to the bureaus and offices of the Navy Department. The microfilm library was turned over to the Office of Naval Records and Library in October 1945, when COMINCH was dissolved.

By October 1942, the Bureau of Naval Personnel needed the space in the Arlington Annex occupied by the Office of Naval Records and Library. An offer of space in the old Smithsonian Building was turned down, even though acceptance would have kept the office together in one location. Instead, the books of the Navy Department Library (and the Naval Historical Foundation) were sent to the Library of Congress Annex in November 1942. All pictures, trophies, and historical relics were inventoried and photographed, packed in 107 large cases, and stored in the Navy's warehouse on South Courthouse Road in Arlington, Virginia. Records in the Manuscript Section antedating 1910 were transferred to the legal custody of the National Archives, and those dated 1910 and later were also stored there, with the requirement that they would remain in the legal custody of the Navy Department. The archivist also provided space for the civil service personnel associated with the above records as well as with the Publications Section and the (noncurrent) manuscript collections. Capt. Knox, his administrative office, and the World War II portion of the Manuscript Section remained at the Arlington Annex.[13]

In 1942, the Counter-Intelligence Section (OP-16-B-7) of ONI was keeping a card file on known or suspected Communists and "fellow travelers." Also in 1942, the Case History Section (OP-16-A-7) established its Communist Geographic Files. Beginning in early 1943, a consolidation of the B-7 files in A-7 was begun and continued into 1945. At first the analysts in B-7 were somewhat opposed to the consolidation because of the need to have the files readily at hand and also because A-7 files were arranged by subject and not by the categories required by B-7. Numerous conferences resolved most of the problems, although B-7 continued to maintain category files relating to currently active subjects.

In addition to the Communist Geographic Files established by the Case History Section in 1942, Communist Category Files were created in April 1943. During the period 7 February to 13 May 1944, OP-16-A-7's Communist Category Files were reviewed by representatives of the Counter-Intelligence Section, and a revised Category Index was adopted on 14 March 1944. On 6 October 1944, the OP-16-B-1 Communist Activity Files were transferred to OP-16-A-7, and a Communist Party "Variadex" (Topical) File was begun. On 28 July 1945, the Communist Category and Communist Activity Files were combined under OP-16-A-7.

Following a further revision to the organization of the files maintained by OP-16-A-7, the Case History Section (which became OP-32C242 postwar), the files were frozen on 3 June 1948. On that date, the Communist files were reorganized under four headings: 100 Series, Communist Activities in the United States; 200 Series, Foreign Activities in the United States; 300 Series, Russian Communist International Activities; and 400 Series, Communist Party by Country. On 7 March 1949, the 300 Series files were transferred to the new Intelligence Files Index (IFI) system (see below) under the geographic heading "USSR;" and the 400 series files were transferred to the individual country files under the IFI system.[14]

By the summer of 1943, the Naval Records Section of ONI was filing the bulk of all operational reports received from the fleet. The reports included action reports, war diaries, patrol reports from submarines, and a considerable amount of material from the various offices of ONI, OPNAV, COMINCH, and SECNAV. The ONI Library was also maintaining books of current interest, copies of unrestricted ONI publications, the confidential *ONI Weekly*, and combat narratives.[15]

In the March 1943 reorganization of ONI, the Graphic Section was transferred to the Publications and Distribution Branch where it could more directly provide graphic support to the production of all ONI publications. Its new designation was OP-16-P-5.

A SECNAV letter of 12 July 1944 designated Adm. Edward C. Kalbfus, USN (Ret.), as Director of Naval History. On 2 August, Kalbfus, in a letter to all bureaus and offices of the Navy Department, designated retired Capt. Dudley Knox as Deputy Director of Naval History. This was in addition to Knox's duties as officer in charge of ONI's Office of Naval Records and Library.[16]

Acquisition of German Naval Records

In May 1945, at about the time hostilities in Europe ceased, a joint British-American naval intelligence team found the German naval archives at Tambach Castle, not far from Nuremberg. The material included the entire basic archives of the German admiralty for World Wars I and II and for other periods back to about 1850. Adm. Harold R. Stark, Commander Naval Forces, Europe (COMNAVEU) notified Washington of the valuable find. When Capt. Knox saw Stark's message on 7 May, he set the wheels in motion to begin the massive program of microfilming the German records for the Office of Naval Records and Library in Washington.

On 8 May, in a memo to the Director of Naval Intelligence, Knox wrote that "such records are obviously of outstanding importance as supplements to our own records, to serve the future purpose of historical research, study and composition, as well as utility in future education and planning." RAdm. Leo H. Thebaud, the director, agreed with Knox and gave him a green light on the program. On the same day, Knox wrote Cdr. Tracy B. Kittredge on Adm. Stark's staff to emphasize the tremendous importance of the German historical archives and to propose microfilming them.

Some of the records had already been moved to London, but many still remained at Tambach Castle, where they were exposed to looting and uncontrolled dispersal. In June, two naval Lieutenants, one British and one American (Lt. H. P. Earle, USNR) were sent from London to inspect the castle. They found the 15 to 20 tons of remaining archives scattered in various private apartments in the castle and surrounding farm houses. The officers centralized the material in two locked rooms and arranged for its safe shipment to London.

Knox next defined the general policy on the selection of documents for microfilming. On 13 June, he wrote Kittredge:

> In general, I would want all operational material in all dates. I suggest the selection should be by classes, rather than individual documents, and that everything in the selected classes should be filmed. Trying to pick and choose within an operational group is too time-consuming and too subject to error of judgement. What may seem of no consequence today may be of great interest tomorrow.

Knox also noted that some large groups of records, such as muster rolls and internal administrative files, should not be filmed. This decision to be mildly selective and not microfilm everything was fortunate. If all documents had been included, the program might have been stopped, because additional time, money, and people would have been needed when military funds were being drastically cut back.

Obtaining staff and funding was the next problem faced by Knox. On 19 June, Kittredge estimated that it would take ten people operating five microfilm machines twenty-one months to complete the filming outlined by Knox. Capt. Roland Krause was chosen by the Director of Naval Intelligence to oversee both the historical and intelligence phases of the program. A temporary staff was assembled in London, initially under Lt. K. M. Davee, USNR, from the Navy's Office of Records Administration; filming began in August 1945. Shortly thereafter, the staff was placed under the direction of LCdr. Samuel R. Sanders, USNR, who had had previous experience in microfilming selected German technical documents for ONI.

After the Tambach Archive material had been registered by the British Admiralty, U.S. translators listed and described it on accession lists. Then other members of Sanders's staff prepared a subject card index as the documents were filmed. On 20 September, the first shipment of thirty-three reels, with associated accession lists and subject index cards, was sent from London to the Office of Naval Records and Library in ONI.

In October 1945, Capt. Krause arrived in London on temporary duty to establish a permanent organization. When he left in December, it consisted of seven translators, ten enlisted photographers operating five cameras on two shifts, four typists, and five seamen assisting the photographers. Krause also made the final review and selection of the post-1930 files for microfilming, earmarking 80 percent for copying. He also found that the pre-1930 Tambach files were at Bletchley Park, outside London, where they were relatively inaccessible to the American microfilming unit and probably wouldn't be registered by the British because of a personnel shortage. Krause reported the situation to Knox in a letter of 13 November. Knox rejected one proposal to bring the pre-1930 records to Washington and suggested they be taken to the U.S. Navy Headquarters at 20 Grosvenor Square, London, where Sanders's staff could register and film them. That was done, and by March 1946, Sanders reported that the processing had been started.

By late June 1946, the cataloging was far enough along to enable Capt. Krause to visit London again to make the final selections for microfilming. Over the next year, the project progressed rapidly. At one point, nine machines were being operated on two shifts and were producing an average of eighteen microfilm reels per day. The filming was finished in July 1947, the project having produced 3,905 reels of microfilm.[17]

In August 1946, Krause drafted a letter for signature by the Director of Naval Intelligence to the Deputy Chief of Naval Operations for Personnel requesting as an extra allowance for the COMNAVEU staff one commander, one lieutenant, and nineteen enlisted men to permit continuation of the project to microfilm the German naval archives then held in London. It was pointed out that the documents had been found to be a rich source of intelligence information. The microfilming team then in London was composed of Cdr. Sanders as Officer in Charge; Lt. V. A. Tetrault, USNR, assistant officer in charge; one chief yeoman in charge of the office; ten photographer's mates or seamen in the Microfilming Unit; five photographer's mates or seamen in the Developing Unit; and three seamen serving as assistants for breaking, assembling, and moving the documents.[18]

A translation and study program that paralleled the microfilming project consisted of two groups. One in London was composed of eight civilian translators who worked on diaries of senior submarine commanders and other documents of special interest such as the Fuehrer Directives and the Fuehrer Conferences on Naval Matters. The other at Minden, Germany, under LCdr. R. H. Rathman, comprised a large number of German nationals who worked on the *Seekreigleitung* (war operations staff of the German Admiralty) diaries, probably the most important single source of information on German naval operations in World War II. The London group was paid with funds from ONI, and the group in Germany was included under the costs of the German occupation. The program, up to July 1949, yielded 30,000 pages of English-language text.[19]

Post–World War II

The Navy Department Administrative Office, in a letter to the Chief of Naval Intelligence on 9 May 1946, mentioned President Harry Truman's recent directives to reduce space occupied by non-current files and the Secretary of the Navy's requirement to reduce by 50 percent the space occupied by files by June 1946. The letter also stated that

> at present, approximately 250 file cabinets in the ONI, Room 2600, L Building, contained Attachés' reports covering the period approximately from 1898 to the present. . . . A records disposal schedule now in the process of becoming law, provides for the forwarding of the Attachés' reports six years old to the Navy Department Records Center and thence to the National Archives for permanent retention.

Capt. John B. Heffernan, having relieved Kalbfus as Director of Naval History, also relieved Knox in June 1946 as officer in charge of the Office of Naval Records and Library. Soon thereafter, the recommendation that the two activities be consolidated was approved, and on 2 August, they became the Naval History Division (OP-29) under the Deputy Chief of Naval Operations for Administration. Thus ended the long and close association between ONI and the Office of Naval Records and Library.[20]

On 1 July 1948, the Intelligence Files Index system was adopted to replace the Monograph Index Guide for processing and maintaining foreign intelligence files. The IFI was a list of numbers for all subjects of naval intelligence interest, and all collectors were to assign IFI numbers to their reports accordingly.

A Director of Naval Intelligence memorandum of 12 April 1949 implemented the ONI Central Filing System directing that

> it shall be the responsibility of cognizant Section and Unit Heads, and principal analysts and their associates, by periodic visits to the Intelligence File Unit (OP-323M5), to ensure that the Intelligence Files Index material is properly maintained and filed; to examine such material for gaps and omissions; and to initiate appropriate want lists to remedy such gaps.

The opinion expressed by intelligence processing personnel in OP-322F1 about the Central Filing System in use in ONI in 1952 was that, insofar as the foreign intelligence files in OP-323M5 were concerned, they existed only for the use of OP-322F1. Since the desks of OP-322F1 had their own reference files, the central file for foreign intelligence material was an unnecessary and undesirable complication of their tasks.

There was considerable variation in the form and content of the analysts' files. Most of them used the IFI system; one analyst maintained current files according to titles he devised himself; another had most of his material filed according to the previously used Monograph Index Guide. All analysts felt that they had to keep complete files of their own but seemed to be complying with what they considered to be the letter of the law in regard to OP-323M5. Suggestions were made that Central Files should contain only "Case History," Domestic Intelligence Material, and those reports not readily assignable to the cognizance of an individual desk, and that the desks should keep all raw source material over which they had primary cognizance.

The Board for Review of ONI Functions and Workload believed that it was evident that the Intelligence File Unit was not being used effectively by OP-322F1. The purpose of filing foreign intelligence material in the central files was to make the material available to other components of ONI and other agencies without the necessity of calling upon the individual desks for assistance. Its continued use would also assure a uniform filing system.[21]

In October 1963, ONI's Foreign Intelligence Library, consisting of original Navy Information Reports, enclosures to information reports, and other agency intellience publications (except for Central Intelligence Agency and Defense Intelligence Agency (DIA) finished intelligence publications) was transferred to DIA. Personnel from OP-923M5 who had performed library and loan functions were similarly transferred on 5 November. All intelligence material thus transferred, including documents retired by OP-923M5 to the Federal Records Center, was thereafter to be maintained and serviced by DIA. Needless to say, analytical organizations remaining within ONI after the creation of DIA continued to maintain their own files, and the conflict between centralized filing systems and the analysts' personal filing systems continued. The gradual introduction of automated data retrieval systems, in time, has brought about improved centralization of storage and access.[22]

Chapter Notes

1. *SECNAV Annual Report*, 1898.

2. Department of the Navy, "Administrative History of the Office of Naval Intelligence in World War II," 10 Jul 1946, unpublished MS, 1152, hereafter *ONI WWII Admin History*.

3. *SECNAV Annual Report, 1902*, 95.

4. Capt. Herbert E. Cocke, USN, "History of ONI," MS, Office of Naval Intelligence, 1931," 29.

5. *ONI WWII Admin History*, 1145.

6. Ibid., 1147.

7. Ibid., 1165.

8. Ibid., 1154.

9. Ibid., 1160.

10. "History of the Office of Naval Records and Library," 1946, Appendix B, 6, MS history in Navy Department Library.

11. *CNO Annual Report, FY 1941*, 12–13.

12. *ONI WWII Admin History*, 1167–71.

13. Ibid., 1181–82.

14. LCdr. F.C. Caskey, USNR, material in Naval Investigative Service (NIS) files, OA. Caskey was head of OP-16-B-7 at the end of World War II.

15. Military Intelligence Section, Office of Naval Intelligence (MIS–ONI), Joint Editorial Board memo, 27 Aug 1943, encl. 1, box 8, Job 5938, FRC/WNRC.

16. Loretta L. MacCrindle, "History of Operational Archives," Apr 1958, 3 vols., 1:7, 9. MacCrindle was a former head of OA.

17. All information in this section is from RAdm. Ernest M. Eller's paper on the Tambach Archives in OA; hereafter Eller MS.

18. OP-32C memo, ser 01077P32, 26 Aug 1946, Accession 3770, box 1, ONI Day File, OA.

19. Eller MS.

20. MacCrindle, "History of OA," 1:23.

21. Review of ONI Functions and Workload, Feb-May 1952, 119–21, OA.

22. ONI Notice 5070 of 29 Nov 1963.

CHAPTER 18

Operational Intelligence

Operational intelligence (OPINTEL) for the Navy is definable as that intelligence needed by naval commanders for planning and conducting operations, including battle. Although this definition is properly all-inclusive, in practice the emphasis is on the "now" situation—information that may have been needed *yesterday* for today's command decisions.

Because OPINTEL requires rapid communications between the collector and the user of the intelligence, and the processing phase must be carried out with minimum delay, it is a relatively new intelligence concept that came into its own, subsequent to the development of radio.

Sampson and Dewey would not have groped around for the Spanish naval forces if they had been supported by operational intelligence; without radio communications, however, such support was not possible. Intelligence of operational value in those days was gathered mainly by the operating forces themselves, using converted merchant ships of high speed and endurance as scouts, but their sightings lost much of their potential value because of the delay in getting the information to the operational commander who could use it in planning and conducting his counteraction.

Operational Intelligence Before World War II

In the period between the Spanish-American War and World War I, the Navy's scouting forces were recognized as serving an intelligence-gathering function for the fleet commander. In 1915, Secretary of the Navy Josephus Daniels was advised by the General Board that, ideally, the fleet had to meet and defeat the enemy before he reached the neighborhood of friendly coasts. But to do that, the fleet had to have an adequate information service to provide early and continuous intelligence on the enemy's movements.[1]

When the United States entered World War I, the Allies already had operational intelligence systems functioning in support of their convoy routing and antisubmarine operations. Allied operational intelligence was made available to the United States, making it unnecessary for the Office of Naval Intelligence to become extensively involved in processing operational intelligence.

At the end of World War I, RAdm. William S. Sims—who had been Commander U.S. Naval Forces in European Waters as well as Naval Attaché, London, and had had direct access to the Royal Navy's operational intelligence—recommended that to meet its two-fold purpose of serving the Navy Department and all the individual naval forces in all areas, the U.S. Navy's intelligence service be divided into groups based on the disposition of forces: "Each group should be under the immediate command of the senior commanding officer of the forces in the area and should have an intelligence officer with an adequate intelligence staff at his command headquarters, whether afloat or ashore."[2]

Also following World War I, Assistant Secretary of the Navy Franklin D. Roosevelt circulated a letter dated 24 March 1919 that touched on operational intelligence in naval districts defining the first duty of an intelligence officer in time of war as being "the collection and compilation of prompt, reliable, and accurate information concerning the approach, arrival, movements, and position of enemy naval forces . . . [and] the prompt dissemination of the above information" to the commandant of the relevant naval district, the Navy Department, and the fleet operating in the waters adjacent to the district.

Lessons learned in World War I about the need for close cooperation and collaboration between operations and intelligence had been long forgotten by World War II. The British Navy had its operational intelligence organization functioning when the U.S. Navy set up the Neutrality Patrol in the fall of 1939, and it was not long thereafter that intelligence information of an operational nature was drifting in to ONI from the increasing number of U.S. naval observers and liaison officers assigned to British naval activities.

ONI, however, was preparing for World War II following the basic concept that it had two functions: (1) gathering primarily strategic information about foreign countries, and (2) protecting naval installations against espionage and sabotage by foreign agents. How the various parts of the Navy used the information about foreign countries was their prerogative and was not considered ONI's responsibility.[3]

Consequently, when the United States entered World War II, ONI was unprepared to provide timely tactical intelligence support to operational commands. Commander in Chief, U.S. Fleet (COMINCH) set up its own operational intelligence organization, and ONI didn't really become involved in—or assume cognizance over—operational intelligence until COMINCH was dissolved after World War II. A few faltering steps were taken during World War II to try to stimulate ONI's interest and action in operational intelligence, but with only temporary or partial success.

ONI and OPINTEL During World War II

The need for intelligence to support operations was obvious, but the need for intelligence on friendly operations was not as well appreciated by operations personnel. Frequently, enemy actions are reactions to friendly actions. As such, they have a far different meaning or significance than if the enemy's actions are spontaneous and based on its own initiative. Thus, operational intelligence cannot be complete until it includes an interpretation of any pertinent input from its own forces' operations staff.

The term "combat intelligence" was originally borrowed from the U.S. Army and defined for naval use by COMINCH as "information about enemy forces, their strength, disposition and probable movements." It was soon found that the definition unduly limited the scope of the intelligence output desired, and the term operational intelligence was adopted as more fully identifying the functions that naval intelligence should play in naval warfare. Combat intelligence, by 1945, was considered merely a phase of operational intelligence and was defined as that intelligence needed by commanders of forces before, during, and immediately after battle. When strategic intelligence is used in conducting operations against an enemy, it becomes operational intelligence. On the other hand, much information obtainable during combat operations is of future strategic intelligence value.[4]

An officer with broad antisubmarine warfare (ASW) experience, in response to a request to name the three most important factors contributing to World War II ASW, emphasized the importance of operational intelligence:

> Harking back to the stated mission for ASW, no matter what line of advance is taken, we always get back not to the "hunt them down and kill them" statement which inspires the fire breathers, but to the basic fact that however the skin is taken off this particular cat, the basic accomplishment must be to "deprive the enemy of effective use of his submarines." Even with discovery ranges of, say, 40,000 yards, we would still, on an open ocean basis, need information as to where to put our searching units, [and] we would still need information as to how to route our most vital shipping. I do not believe that there is any way around the fact that the single most important point which must be covered is the maintenance of a high degree of effective operational intelligence for use in combat. Combat intelligence multiplies our effective forces by factors which are impossible to achieve by simply building more units and training more men. Examples which are known to me are the German evaluation of the number of active hunter-killer groups we were operating in the Atlantic during the war. They estimated 200 operating groups at a time when, in actual fact, we had six operating groups plus a high degree of operational intelligence. . . . Another example is the effect of the performance of USS *England* (DE-635) when she accounted for six Japanese submarines in nine days as a direct result of good operational intelligence. Her performance led the Japanese to believe that a whole fleet had come through the area. Other examples . . . are the performance of our own submarines in hunting down Japanese submarines. What would you estimate as a multiplication factor given you by intelligence, as compared to attempts to perform the same feats without that intelligence?
>
> The Battle of the Atlantic was, in a large measure, a battle of wits in which intelligence played the major role. Unfortunately, this fact is fully understood only by a relatively small group of officers because of the highly classified nature of the subject.[5]

Coastal information sections were established in naval district intelligence offices to perform operational intelligence functions for locally based naval coastal defense forces. By April 1941, it was found necessary to activate ONI's Coastal Information Section (OP-16-B-8) to help support the nascent operational intelligence activities in the naval districts. The new ONI section was placed in the Domestic Intelligence Branch because all other contacts with district intelligence offices were handled from there.

The officer in charge of OP-16-B-8, LCdr. Charles F. Baldwin, USNR, began by making a study of British navy operational intelligence procedures. Baldwin's studies convinced him that intelligence support to operations would not be possible without the closest cooperation between his organization and the operational organizations within the Navy. In June 1941, he recommended steps be taken to coordinate certain operational and intelligence activities to assure the timely exchange of information, as the British navy was already doing.[6]

In May and July 1941, orders were sent from ONI to the naval districts that the coastal information sections of the district intelligence offices should be placed in an advanced state of readiness. The orders caused some confusion, because few people in the naval districts had had any thoughts on what the coastal information sections were supposed to do. On the day after Pearl Harbor, another directive was issued defining in more detail the scope of coastal information and prescribing operating procedures. The main problem in the naval districts was getting operations organizations to accept officers from the B-8 (coastal information) sections of the district intelligence offices as OPINTEL officers or to make use of B-8 facilities and information. On 14 April 1942, Commander Eastern Sea Frontier issued a directive excluding coastal information officers from operational intelligence duties. This was obviously contrary to what the Director of Naval Intelligence (DNI) had been directing. To correct the conflict, the Chief of Naval Operations (CNO) issued a directive on 29 May 1942 that placed coastal information officers as operational intelligence officers for each Inshore Patrol Section Base or other surface operations center. Passive resistance continued, and finally, on 13 November 1942, the Vice Chief of Naval Operations wrote personal letters to all district commandants and sea frontier commanders referring specifically to the potentially hazardous results of inadequate cooperation between operations and intelligence.[7]

The work of an intelligence plotting room is part of the operational intelligence function, and it existed in a rudimentary sense prior to U.S. entry into World War II. The Situation Room in ONI was not established, however, until 12 January 1942. Prior to that time, some of the activities later performed by the Situation Room were carried out by various units of the F (Foreign Intelligence) Branch of ONI. One of these, OP-16-F-a, was formally established on 6 August 1941, although it had already been operating for some time. OP-16-F-a's job was to prepare a daily Information Memorandum on the war situation based on a digest of Navy, Military Intelligence Division, and State Department dispatches, press reports, and other material. The memoranda, plus similar digests by several of the F Branch theater (geographic) sections, were delivered to DNI RAdm. Alan G. Kirk each morning for his use in making a daily situation report to the Secretary of the Navy.

On 11 January 1942, a DNI directive abolished OP-16-F-a and established the C Branch (Fleet Intelligence), effective 12 January 1942. C Branch (OP-16-C) included C-1, the Intelligence Center, which was to "process, evaluate, plot and disseminate current information from all sources." It was contemplated that C-1 would constitute a complete operational intelligence center for the use of the then still-separate COMINCH and CNO organizations. C-1 continued to produce much of the *Daily Summary*, and its situation room was used by ONI as a display room for current combat intelligence. C-2, the Information Center, produced the *ONI Weekly* and other publications.[8]

When COMINCH Ernest J. King was additionally designated Chief of Naval Operations in March 1942, it was decided that the C-1 Center should include a coastal information plot that would be OP-16-B-8's responsibility to maintain. OP-16-B-8 considered it essential that the OP-16-C plot be near COMINCH Operational Information Section which later became the Operational Intelligence Section. OP-16-C, however, remained physically and organizationally separated from COMINCH, and the Operational Information Section of COMINCH became the Combat Intelligence Division of COMINCH (see Chapter 16).[9]

In June 1942, LCdr. Baldwin recommended that selected officers be trained for operational intelligence duties, and, as a result, he was directed in December 1942 to develop an advanced OPINTEL training program.

The need for operational intelligence on the part of sea frontier, fleet, and advanced base commands grew in direct proportion to their tempo of operations, and ONI was urged by intelligence officers assigned to those commands to establish a true OPINTEL organization within itself to help fulfill the need.

In August 1942, the ONI F Branch theater sections took over exclusive preparation of all parts of the ONI *Daily Summary*, with the exception of the merchant shipping situation report, which continued to be reported by C-1.[10]

In the spring of 1943, the task of making the daily situation report had been delegated to the head of the F Branch. A new situation room was constructed, and, on 20 March 1943, C Branch was abolished. C-1 was redesignated OP-16-FP, the Foreign Plot Section. On 25 March, the new situation room was put into use for the Secretary of the Navy's morning conferences, and it continued to be used as such for the rest of the war.

The principal attendees at the morning conferences were the Secretary of the Navy, the Assistant Secretary, the Under Secretary, the Assistant Secretary of the Navy for Air, the Commander in Chief, U.S. Fleet, the Deputy COMINCH, the COMINCH Chief of Staff, the Vice Chief of Naval Operations, the Commandant of the Marine Corps, the Commandant of the Coast Guard, various Navy bureau chiefs, and the division directors in the office of the CNO. It was a matter of policy that the FP Section would not handle or plot any material concerning the strength or disposition of Allied forces, nor would it address some categories of "specially reported material."[11]

The Advanced Naval Intelligence School in New York City was established in January 1943 and started training officers in operational intelligence in February.

In March 1943 LCdr. Baldwin recommended the establishment of an OPINTEL unit in ONI and the shift to the new unit of the Coastal Information Section from the Counter Intelligence Branch. Baldwin's recommendations were approved by DNI RAdm. Harold C. Train, and the Operational Intelligence Section (OP-16-FO) was established as part of an ONI reorganization on 20 April 1943. OP-16-FO was divided into three subsections: Pacific, Euro-African, and American.

The mission of the first Operational Intelligence Section of the Office of Naval Intelligence was stated in DNI letter serial 01020916 of 21 April 1943:

> (a) To insure that information acquired through the facilities of the Naval Intelligence Service, which is of value to naval operating forces, is properly processed and promptly made available to such forces;
>
> (b) To insure that the personnel and facilities, established by the Office of Naval Intelligence for the performance of operational intelligence activities, function efficiently.

OP-16-FO operated for four and one-half months, during which time it prepared an OPINTEL manual and periodic bulletins and supplied valuable data to intelligence officers in combat areas, assigned 180 graduates from the Advanced Naval Intelligence School to combat theaters, and helped the school develop an effective OPINTEL course. OP-16-FO also obtained COMINCH approval to establish a naval intelligence mission in North Africa to provide intelligence support to naval forces in northwest African waters, assign an intelligence officer to each motor torpedo boat squadron, and publicize ONI's program to train and furnish officers for intelligence duties afloat to all fleet commanders.

From its inception to its establishment as a separate organization, operational intelligence faced strong opposition. The main objections were that such an organization was not in accord with the existing war plans and that its functions crossed those of other sections and usurped their prerogatives. The establishment of an OPINTEL organization had also disrupted the naval district intelligence organizations.

In August 1943, Deputy DNI Capt. Ellis M. Zacharias, who had supported the establishment of an operational intelligence section, was relieved by Capt. Adolph von S. Pickhardt. At the time, Cdr. Baldwin was on an inspection trip to Great Britain and the Mediterranean. While Baldwin was away, his opponents convinced RAdm. Train (DNI at the time) that setting up OP-16-FO in April had been a mistake. When Baldwin returned, he found that his office had been abolished and its functions delegated to other sections.

By DNI (OP-16-X-1) serial 01924316 and Assistant Director, Intelligence Group (OP-16-1-F) Memorandum No. 7, both of 9 September 1943, the Operational Intelligence Section was disestablished, and responsibility for all intelligence within the North American area was transferred to the North American Theater Section (OP-16-FN). Operational intelligence activities were to be administered thereafter by the head of the Intelligence Theater Section in which such activities were being conducted. Thus, procurement of operational intelligence personnel for the forces afloat was made the responsibility of the ONI Services Branch (OP-16-A).

The effort by the opponents of OPINTEL to disestablish the Advanced Naval Intelligence School almost succeeded, except that Adm. King had informed fleet commanders about the school and had requested advice on the number of operational intelligence officers they would need. Their responses kept the school (and operational intelligence) alive.[12]

The school eventually graduated 1,300 officers qualified for assignment to operational intelligence billets. Approximately 750 were assigned to billets outside the United States, others were sent to sea frontier staffs, and some made up a pool from which emergency and future requirements could be met. The lack of established doctrine gave the graduates assigned to ships and afloat staffs an opportunity to use their initiative to make operational intelligence of benefit to the commands to which they were assigned, and they did. The amphibious forces in both the European and Pacific theaters relied heavily on the intelligence sections of their staffs to perform research during planning stages and evaluation during operations. OPINTEL officers assigned to battleships, cruisers, and other individual commands afloat and ashore proved of value for planning and operations.[13]

The next active proponent for operational intelligence was LCdr. S.A.D. Hunter, USNR, who returned to ONI in January 1944 from an extended tour of duty as an intelligence officer with the fleet in the Mediterranean. From his experiences, he advised Deputy DNI Pickhardt of the "great and increasing need for intelligence officers in connection with tactical operations" and the need for an adequate sustaining program for them in ONI. Hunter was then directed by the prospective Deputy DNI, Capt. William A. Heard, to draw up specific proposals for such a program.

On 14 February 1944, LCdr. Hunter submitted his written proposals. With Capt. Heard, he had several sessions with RAdm. Roscoe E. Schuirmann, who had relieved RAdm. Train as Director of Naval Intelligence in September 1943. The same opponents who had frustrated Cdr. Baldwin were still present, and Schuirmann was reluctant to override the majority of his captains in favor of a lieutenant commander. Consequently, the proposals were shelved for the duration of Schuirmann's directorship.

After RAdm. Leo H. Thebaud relieved Schuirmann in October 1944, Hunter reintroduced the subject. Thebaud made a thorough investigation of the ONI organization and the requirement for operational intelligence, and he came to the conclusion that ONI was deficient in its capacity to fulfill fleet intelligence requirements.

To correct the situation, Thebaud designated Cdr. Frank P. Morton, USNR, who had just returned from duty as an amphibious intelligence officer in the Mediterranean, to be the head of an operational intelligence organization in ONI and directed him to draw up appropriate plans. After conferring with various officers, including LCdr. Hunter and several air combat intelligence officers, Cdr. Morton presented a memorandum to the Deputy Director on "Establishment of Operational Intelligence Sustaining Program" dated 16 November 1944. The program, with some slight modifications, was approved, and on 7 December 1944 an Operational Intelligence Section was again established, this time in the Administrative Branch, where it was designated OP-16-A-6. Its mission was stated to be "to support the operational intelligence personnel afloat, abroad, and in training at the Advanced Naval Intelligence School" (but not in the naval districts).[14]

Cdr. Morton and Capt. Herman E. Keisker, USNR, head of ONI's counterintelligence effort and the leader of those opposed to OPINTEL, continued to push their different views, particularly as they related to the operational intelligence organizations in the naval districts. Finally, on 8 February 1945, the director resolved the matter by establishing the Operational Intelligence Branch (OP-16-O) and giving it cognizance over OPINTEL units under the sea frontier commanders. A follow-up memorandum of 26 February 1945 excluded from the jurisdiction of the new branch those personnel performing operational intelligence work purely for the naval districts.

Another function was added to OP-16-O's duties on 25 May 1945 when the OPINTEL Branch was directed to support, and assume cognizance over, personnel assigned to military government duties in occupied territories. OP-16-O continued to function effectively through the remainder of the war as an administrative home base for personnel assigned to operational intelligence billets with naval operating commands.[15]

ONI Becomes Formally Involved in OPINTEL

Following the cessation of hostilities in the war with Japan, COMINCH headquarters was disestablished effective 10 October 1945. The elements of the COMINCH staff that were continued became the Operations Division (OP-03) of OPNAV. Combat intelligence, one of the elements continued, was renamed the Operational Information Section (OP-32). The section was initially organized as follows:

Head of Section (OP-32), Capt. William R. Smedberg III
Dissemination (OP-32D), Cdr. W. R. Brandt
Pacific Subsection (OP-32P), Cdr. William J. Sebald
Atlantic Subsection (OP-32L), Cdr. Kenneth A. Knowles
Chart Room (OP-32C), Cdr. F. M. Curran

As of 30 October, Capt. Smedberg was given the additional designation of OP-23W, Special Branch of ONI, preparatory to the merging of OP-32 with ONI. Also on 30 October, when ONI's designation was shifted from OP-16 to OP-23, the Operational Intelligence Branch (OP-16-O) became the Operational Branch (OP-23Y).[16]

On 15 February 1946, the transfer to ONI of OP-32, the former Combat Intelligence Section of COMINCH, took place, and a major change was made in the organization and functions of OP-23Y. Some of the functions of the OP-16-0 were retained in the new OP-23Y and some were transferred to the Training Section (OP-23C3). OP-23W was abolished, and its functions were shifted to the new Operational Branch (OP-23Y). The former COMINCH Combat Intelligence Section was designated OP-23Y2, and its head, Capt. Smedberg, was placed in charge of the new OP-23Y. The former Pacific Strategic Intelligence Section of COMINCH, temporarily part of OP-23W, was named Special Section and designated OP-23Y1.

The mission of OP-23Y was to (1) disseminate all necessary intelligence to the Operations Division of OPNAV and to the commanders of operating forces; (2) coordinate intelligence activities of the operating forces with ONI; (3) control covert intelligence activities; and (4) control material above the classification of secret.

The mission of the Special Section comprised (3) and (4), above, plus maintaining liaison for ONI in covert intelligence matters and maintaining "monographs" (information folders) on foreign intelligence services. The mission of the Operational Intelligence Section was (1), above, plus maintaining information on the strength, disposition, and movements of foreign naval and air fleets and assisting ONI in collecting information through the intelligence officers attached to the operating forces.[17]

In July 1946, the Fleet Support Section (OP-23Y3), headed by Cdr. David J. MacDonald, was established in ONI's Operational Branch to produce intelligence manuals based on World War II experience. An *Operational Intelligence Manual* (ONI-Y-1) was one of the first products. Knowledgeable Naval Reserves on their two-week active-duty training were used to assist in preparing the manuals. Other manuals covered subjects such as interrogation and technical intelligence. ONI-Y-1 was edited by Capt. Carl F. Espe (OP-23Y), and he delivered the finished product to the Director of Naval Intelligence, who hand-carried it to the Chief of Naval Operations. (The date of distribution has not been determined; Espe was detached in May 1948.)[18]

The initial edition of ONI-Y-1 had many obvious shortcomings, but it was a start, and it stimulated comments. This prototype manual didn't cover some aspects of operational intelligence, and it over-emphasized others. There were technical points not in keeping with accepted doctrine or good practice. Under a letter of promulgation dated 1 January 1951, a revised manual was issued for the guidance and information of officers concerned with the intelligence aspect of planning for and executing naval operations. The revision was given the same designation as the original, ONI-Y-1.[19]

When ONI was transferred from the Administrative Division of OPNAV to the Operations Division on 1 August 1946, the Operations Chartroom (OP-31C) became part of ONI and was given the designation OP-32-Y-23 in the Operational Intelligence Branch.[20]

A statement of concept prepared in 1948 described operational intelligence essentially as follows:

> There is no sharp line of demarcation between operational and strategic intelligence; one flows into the other. However, there are certain characteristics which definitely distinguish operational intelligence:
>
> It is directly concerned with the operating forces;
>
> It is intended for use by the operating forces in the near or immediate future and is shaped for this use by every echelon of command;
>
> It is the practical application of intelligence from all available sources to solve a specific operational problem; and
>
> It requires precise and detailed information about physical conditions and situations within restricted areas, primarily those areas within the assigned mission.[21]

On 25 October 1948, DNI RAdm. Thomas B. Inglis and RAdm. E. W. Longley-Cook, the British Director of Naval Intelligence, reached an agreement, thereafter called the Inglis/Longley-Cook Agreement, whereby an active operational intelligence liaison would be maintained between Commander in Chief, Eastern Atlantic and Mediterranean and the Naval Intelligence Division (NID) of the British Admiralty. The liaison channel was to be in addition to the normal Naval Attaché, London, collection and exchange channel between ONI and NID. The need for a combined operational intelligence plot in wartime in London was contemplated, with the U.S. Navy to provide three lieutenant commanders in peacetime who would be assigned to NID as a nucleus for a wartime organization.[22]

As CNO (1949–1951), Adm. Forrest P. Sherman required a daily, private briefing by ONI's Operational Intelligence Branch (OP-322Y). Sometimes the briefer was supplied by the Special Intelligence Section (OP-322Y1) then located at Arlington Hall, Virginia, but usually the briefer was from Fleet Intelligence Unit (Y2) at the Pentagon. Adm. Sherman encouraged the briefers to comment on the degree of reliability of their information. At one of the briefings, at which then-Cdr. Rufus L. Taylor was the briefer, Sherman reportedly said:

> I want to know what you really know when you have solid evidence to back it up, and I'm going to hold you responsible for that. Then I want to know what you suspect and what you think is probable and so on and why, and I won't hold you responsible for that because that's in the field of opinion. And then I want to know anything you have in the way of hunches or guesses and, if you've got anything to base them on, what it is that it is based on. Don't worry about that either. I can put all those things together, and you'll be held responsible only for that information on which you say you

have solid evidence and you think you can produce logical proof.[23]

Later in his tour as CNO, Adm. Sherman became so interested in analysis techniques that he had one of the officers from Y1, LCdr. J. W. Logan, who was responsible for intelligence on Soviet air movements, come to his office once every two weeks so that he could go over, card by card, the files that Logan kept. As a result, Sherman became a fairly proficient analyst himself. He took a very keen, direct, and intimate interest in the intelligence process and the information produced. In the opinion of VAdm. Taylor, a later Director of Naval Intelligence, who served in a number of operational intelligence billets during the period, once senior officers realized the number and nature of intelligence sources and the amount of material from those sources that had to be processed, those who took a real interest in intelligence recognized that the intelligence staff had to be kept informed on any plans and operations being considered if the staff was to be responsive to command requirements.[24]

With the outbreak of the Korean War, Commander in Chief, Far East was authorized by the Joint Chiefs of Staff to disseminate operational intelligence to British, Canadian, Australian, and New Zealand forces under his command. The attention of Commander Naval Forces, Far East was directed to the requirement for passing relevant, appropriately disguised signal intelligence (SIGINT) to all units under his command.

Relative to the OPINTEL exchange arrangements prescribed in the Inglis/Longley-Cook Agreement of 25 October 1948, DNI Radm. Felix L. Johnson believed that, with the establishment of the Department of Defense (DOD) and the execution of the U.S./U.K. "Burns/Templer" Agreement at the DOD level, the latter agreement should be the basis for implementing intelligence support measures to the forces involved in the Korean War. The National Security Act of 1947 had left the service secretaries with all powers and duties relating to their departments that had not been specifically conferred upon the Secretary of Defense. The National Security Act Amendments of 1949 stated that "the Departments of Army, Navy, and the Air Force shall be administered by their respective Secretaries under the direction, authority, and control of the Secretary of Defense." The Inglis-Longley-Cook Agreement pertained to the exchange of naval intelligence between the United States and the United Kingdom upon commencement of hostilities or the declaration of a national emergency. The Burns-Templer Agreement covered the exchange of *all* elements of intelligence under *all* conditions and hence included the conditions for implementing the provisions of the Inglis-Longley-Cook Agreement.[25]

In a study of ONI's functions and work load made by a panel of ONI senior officers in 1952, several comments on the Operational Section (OP-322Y) were made that are pertinent to an understanding of its status at that time:

> Functions peculiar to OP-322Y, such as the operation of the War Room, briefing activities, participating in covert, clandestine, and related sensitive operational activities, are performed by the Army G-2 in their Collection and Dissemination Division; OP-322Y makes every effort to satisfy the needs of the operating forces. The rest of ONI does not operate on the premise of responsibility toward any particular segment of the Naval Establishment.
>
> In practice, the mission of OP-322Y is broad and not well understood . . . its actual existence as a separate entity has a historical basis, stemming from the inability of ONI to "serve the fleet" expeditiously during World War II.
>
> A feature which has assured its separate existence as an agency within ONI is its exclusive control over sensitive material.[26]

The intelligence information employed in support of the planning and execution of day-to-day carrier operations during the Korean War was obtained primarily from carrier photo reconnaissance (see Chapter 13). For the surface bombardment units, the intelligence sources included photo intelligence and agent reports (see Chapter 8). As reported by various units involved, the chief deficiencies in gunfire support intelligence included target lists, which for the most part were merely listings of coordinates and were too extensive because targets were added but never deleted. In addition, no record was maintained on how often which targets had been fired upon or on the success achieved; post-firing photography to identify permanently damaged targets was neither adequate nor timely; the relative importance of targets was not indicated; and no coordination of target intelligence was performed by the various commands.[27]

DNI RAdm. Carl Espe expressed his view on the importance of close coordination of intelligence and operational planning in a May 1953 letter: "The control and evaluation of Operational Intelligence should not be in the hands of an agency which has no responsibility for the success or failure of operations based thereon."[28]

Espe's view of the proper way to handle operational intelligence within the Supreme Allied Command, Atlantic (SACLANT) was as follows:

> OPINTEL Centers ("OICs") belonging to SACLANT, CINCWESTLANT, [Commander in Chief,

> Western Atlantic Area], CINCEASTLANT/CINC-AIREASTLANT [Commander in Chief, Eastern Atlantic Area/Commander in Chief, Air Forces, Eastern Atlantic Area], and possibly other major subordinate commanders, [should] collect, exchange, evaluate, and disseminate intelligence in their spheres of operational responsibility. Backup and assistance would be given OICs from national agencies, such as Admiralty and ONI. Control of intelligence would follow the chain of command. Exchange of intelligence between SACLANT, SACEUR [Supreme Allied Commander, Europe], and CINC-CHAN [Commander in Chief English Channel Forces] and their subordinate commands would be as mutually agreed. Exchange of intelligence between non-NATO area commanders would be via appropriate national channels.[29]

In 1954, operational intelligence was considered to consist of processed information that was applied by naval commanders to the planning and execution of specific operations. OPINTEL thus included the information concerning the hour-by-hour conduct of an operation that was commonly referred to as combat intelligence. Supervision of the planned action required the best possible operational intelligence if the commander's decisions were to be based on fact rather than conjecture. The relationship between strategic and operational intelligence was one of mutual assistance. In one direction, the information gained in the course of ferreting out strategic intelligence might uncover trends that demanded military action or that would influence the conduct of a contemplated operation. In the other direction, operational units were frequently in a position to collect information of strategic or national value that would otherwise be unavailable.[30]

The distinction between strategic and operational intelligence frequently was (as it had been and continues to be) difficult to discern. Often the information used as strategic intelligence by a planner was also used by an operational commander in the execution of his mission; in the latter case, the information became operational intelligence. Thus, the name applied to intelligence depended on the user or the purpose for which it was being used.

One of the DNI's greatest problems in 1954 was to ascertain the intelligence requirements of the operating forces. Within ONI, OP-322Y acted as a central point of contact for the intelligence staffs of the operating forces and had responsibility to determine the intelligence needs of the fleets and, as possible, to assure their satisfaction. To accomplish this mandate, OP-322Y relied primarily on the "Report on Organization and Operations of Staff Intelligence Activities" (OPNAV 5440-2), which all fleet intelligence activities submitted quarterly to the Chief of Naval Operations (specifically to the Director of Naval Intelligence). In addition to providing other information, the quarterly report outlined the support required from ONI.

To gain a better understanding of the fleets' problems and of the intelligence support which they required, DNI Carl Espe during 1953–1954 sent teams of officers to visit the intelligence sections and units of the operating forces in the Atlantic and Pacific. Fleet requirements for intelligence were also ascertained by studying the fleet operation plans then in effect. ONI reviewed the intelligence annexes received from the operating forces, particularly the essential elements of information (EEI) and their collection plans. Any information held by ONI that would fully or partially fulfill any EEI was forwarded to the command concerned.[31]

During the mid-1950s, the district intelligence officer in each naval district that was also a part of a sea frontier was required to develop and maintain an operational intelligence unit ready to provide the sea frontier commander, or appropriate subordinate commanders, with operational intelligence obtained within the district and adjacent sea areas or from naval intelligence sources. The unit was kept in an inactive status in peacetime but was to be ready for prompt activation, on instructions from the Director of Naval Intelligence in time of emergency or war.[32]

Special Intelligence Section (Y1) Becomes NFOIO

By SECNAV Notice 5450 of 26 August 1957, the former Special Intelligence Section (Y1) was established as the Navy Field Operational Intelligence Office (NFOIO), a field activity "to exploit all sources of special intelligence for the purpose of producing operational intelligence for timely dissemination to commanders of operating forces of the Navy and other designated recipients." NFOIO's mission, tasks, and reporting channels were unchanged from those of Y1.

The ONI Current Intelligence Section (Y3) was redesignated OP-922B4 as a staff organization of the Assistant Director of Naval Intelligence, Intelligence Production Division, effective 21 November 1957. Its mission and tasks remained unchanged: to maintain a twenty-four-hour watch in Intelligence Plot for support of the OPNAV duty officer and to give the daily intelligence briefing to the Chief of Naval Operations and other appropriately cleared officers. The other sections of the Operational Intelligence Branch were Operational Intelligence Evaluation and Dissemination (Y2) and Electronic Intelligence (Y4).

In late 1957, NFOIO was moved from Arlington Hall Station in Northern Virginia to Fort Meade, Maryland. The benefits of being co-located with the National Security Agency (NSA) were many: NFOIO analysts were able to provide NSA analysts with valuable general service information of assistance to them in their work; NSA analysts were able to give NFOIO analysts bits of information that had not been published and their own interpretations of unique data; there were opportunities for the mutual exchange of ideas through informal discussions; NFOIO analysts were able to use the extensive files of NSA's Office of Central Reference in the course of their research; all NFOIO reports and studies, including art work, were reproduced by the NSA Document Services Group on an informal basis (since the appropriate ONI facilities in the Pentagon could not handle special intelligence materials); and the Fort Meade location facilitated research by NFOIO analysts, who were relatively free from Pentagon briefing requirements.[33]

At an Operational Intelligence Conference of representatives of ONI and the British Naval Intelligence Division, held from 10 to 14 November 1958, it became obvious that the flow of operational intelligence from NID to ONI had practically ceased. One of the reasons for the deterioration of the Inglis/Longley-Cook Agreement of 1948 was the difficulty that CINCNELM was having in keeping the three U.S. Navy lieutenant commander billets at NID filled. In 1958, for example, there were only two officers (one lieutenant and one lieutenant junior grade) assigned by CINCNELM to conduct liaison with NID's merchant shipping plot, and no U.S. Navy officers were assigned to the NID navy ship plot. Another problem was that the official channel for much intelligence passed to ONI went from the British Joint Intelligence Bureau to the Central Intelligence Agency to ONI, a slow process that did not include NID's evaluations, and when reports included enclosures, such as photographs, they were seldom still attached by the time they reached ONI.[34]

The name of the Operational Intelligence Branch (OP-922Y) is believed to have been changed to the Composite Support Branch in 1962. Its mission and functions were not changed despite the loss of many key personnel to the new Defense Intelligence Agency. The organization's principal activities continued to be related to communications intelligence, electronic intelligence, close support to fleet operations, and certain special projects involved with the planning, staffing, and monitoring of intelligence collection by the operating forces.[35]

In 1969, the missions of Navy Field Operational Intelligence Office were stated as follows:

> To exploit processed signal intelligence in the production of finished operational intelligence tailored to the requirements of the Defense Intelligence Agency, the military departments and the unified, specified, and fleet commands;
>
> To provide for the timely dissemination of such intelligence so as to permit its effective use by authorized recipients;
>
> To assist the Defense Intelligence Agency and the Navy Department staff elements in planning and policy matters; and
>
> To provide an interface between the National Security Agency and naval intelligence.[36]

On 18 January 1971, as part of a Naval Intelligence Command/NFOIO realignment of resources, the Ocean Surveillance Branch of NFOIO was relocated from Fort Meade to the Naval Reconnaissance Technical Support Center/Naval Scientific and Technical Intelligence Center building at Suitland, Maryland, to become the nucleus of the Ocean Surveillance Information Division of the Naval Ocean Surveillance Intelligence Center (NOSIC). The division became operational on 1 February 1971.

NOSIC's Intelligence Analysis Division remained at Fort Meade and began to put greater emphasis on undersea warfare intelligence production. Information processing, filing, and retrieval systems were installed to enhance data recall capabilities.[37]

Table 18.1.
Heads, Special Intelligence Section, ONI

Starting Date	Designator	Name	In Charge
Aug 46	OP-32Y1	Special Section	Capt. Laurence H. Frost
May 48			Cdr. Rufus L. Taylor
Oct 48	OP-322Y1	Special INTEL Unit	
Nov 51			Cdr. Peter Belin
1953			Cdr. John L. Holmes
Sep 55	OP-922Y1		
1955			Cdr. Thomas C. Jones
Jun 57			Cdr./Capt. Donald M. Showers
Aug 57	OP-922Y1	NFOIO	
Jul 60			Cdr./Capt. John Q. Edwards
Jul 64			Capt. Arthur S. Osgood
Jul 67			Capt. William R. Quisenberry
May 69			Capt. Emory R. Sourbeer
Aug 70			Capt. Donald P. Harvey
Nov 71			Capt. Richard W. Bates
Jun 74			Capt. John L. Butts
May 76			Cdr. Chauncey F. Hoffman
Sep 76			Capt. George B. Pressly

Chapter Notes

1. *SECNAV Annual Report, 1915*, 75–76.

2. Department of the Navy, "Administrative History of the Office of Naval Intelligence in World War II," 10 Jul 1946, unpublished MS, 14, hereafter *ONI WWII Admin History*.

3. Ibid., 824–26.

4. *ONI Review*, Feb 1948, 46.

5. VAdm. Francis F. Low, *Study of Undersea Warfare* (The Low Report), 22 Apr 1950, Top Secret Annex, 8–10, Post 1 Jan 1946 Command File, OA.

6. *ONI WWII Admin History*, 825, 487.

7. Ibid., 490–94.

8. Ibid., 770–71.

9. Ibid., 488.

10. Ibid., 825–27.

11. Ibid., 772–73, 788–97.

12. Ibid., 827, 830–32.

13. *ONI Review*, Nov 1945, 39.

14. *ONI WWII Admin History*, 834–38.

15. Ibid., 843.

16. OP-23C ser 573P23, 30 Oct 1945; and ONI and OPNAV rosters.

17. *ONI WWII Admin History*, 846–48.

18. VAdm. Carl F. Espe ltr to author, 25 Sep 1974.

19. ONI-Y-1, *Operational Intelligence (OPINTEL) Manual*, 1 Jan 1951.

20. DNI memo, ser 7073P23, 31 Jul 1946, Accession 3770, box 1, ONI Day File, OA.

21. *ONI Review*, Feb 1948, 46.

22. OP-922Y1 memo to OP-92, 20 Nov 1958.

23. VAdm. Rufus L. Taylor, interview by author, 19 Sep 1975, tape recording in author's files, OA.

24. Ibid.

25. DNI ltr to CSO, British Naval Staff, Washington, ser 000334P32, 10 Jul 1950; and DNI ltr to CSO, British Naval Staff, Washington, ser 000464P32, 7 Aug 1950.

26. ONI, *Report of Functions and Workload*, 1952, 190–94.

27. Commander in Chief, Pacific Fleet (CINCPACFLT), *Interim Report No. 5: Korean War Naval Operations*, 5-130; *Interim Report No. 6*, 5-164–5-166.

28. DNI ltr to U.S. Liaison Officer, Supreme Allied Commander, Atlantic (SACLANT), ser 000404P32, 8 May 1953.

29. Ibid.

30. Naval Warfare Publication (NWP) No. 10, 1954, para 651.

31. *ONI Review*, May 1954, 179.

32. ONI-70--1, *ONI Manual*, 20 Jun 1956, 67.

33. ONI OPINTEL Branch, Office Methods Survey, 21 Jan 1958.

34. OP-922Y1 memo to OP-92, ser 00479-Y-1, 20 Nov 1958.

35. OPINTEL Briefing Folder, 1 Jun 1963.

36. Naval Intelligence Command, *Command History, 1969*.

37. Navy Field Operational Intelligence Office, *Command History, 1971*, 1–2.

CHAPTER 19

Unified and Joint Intelligence 1939–1971

Joint Intelligence in Washington During World War II

On 8 September 1939, Director of Naval Intelligence (DNI) RAdm. Walter S. Anderson drafted a memorandum to President Franklin D. Roosevelt based on information received from England and proposing the establishment of a National Defense Committee. The memo drew attention to the British Committee of Imperial Defense, which was headed by the prime minister and composed of representatives of the British Army, Royal Navy, Royal Air Force, Foreign Office, Treasury, Board of Trade, and other departments as appropriate for the development of British national defense plans. In the memo, RAdm. Anderson proposed that the President, as Commander in Chief, should have a similar committee made up of the Secretaries of State, War, Navy, and Treasury, the Chief of Staff of the Army, the Chief of Naval Operations (CNO), and the heads of other executive departments, as required, to unify and ensure completeness of national defense plans, not only for the armed services but for all phases of national life.[1]

On 14 July 1941, the Army Assistant Chief of Staff for Intelligence (AC/S, G-2) and DNI RAdm. Alan G. Kirk recommended the establishment of a Joint Intelligence Committee (JIC) to serve the military services' Joint Board. The proposal was considered by the Joint Planning Committee and resulted in a report (Joint Board No. 329 of 10 September 1941) recommending that JIC should be co-equal with the Joint Planning Committee and have the primary task of preparing daily summaries of military and related intelligence for the use of the President, the Secretaries of War and Navy, and certain other high officials. In addition, JIC was to prepare special information and intelligence studies as might be required. The Joint Intelligence Committee was established shortly after the report was issued and comprised four representatives from the Army's Military Intelligence Division (MID) and three from the Office of Naval Intelligence; the senior representative acted as chairman. Five members served full time, one of whom as secretary. The offices of the committee were established in the Main Navy Building, adjacent to the offices of ONI.[2]

As the United States found itself suddenly projected into a global war, immense gaps in the knowledge available on foreign countries became readily apparent. The word "intelligence" took on a fashionable connotation. Each new wartime agency, as well as many of the older departments, blossomed out with an intelligence staff of its own, each producing a mass of largely uncoordinated information. The resultant competition for funds and specialized personnel was a monumental example of waste. The War and Navy Departments developed full political and economic intelligence staffs, as did the Research and Analysis Division of the Office of Strategic Services (OSS). The Board of Economic Warfare and its successor, the Foreign Economic Administration, also delved deeply into fields of economic intelligence.

When officials, for example, requested a report on the steel industry in Japan or the economic conditions in the Netherlands East Indies, they had the reports of the Board of Economic Warfare, the Army's Military Intelligence Service (G-2), ONI, and OSS from which to choose. Because the agencies had competed to secure the best personnel, each felt that its particular report was the best available and that the others could be disregarded.

Although there had been much informal contact between ONI and MID, the first official relations were established on 19 April 1942 when the Director of Naval Intelligence named RAdm. Neil B. Nichols, USN (Ret.), as the ONI liaison officer with the Army Assistant Chief of Staff for Intelligence.

His office was set up in the Munitions Building, where MID was then located. Ill health limited Adm. Nichols's service in the liaison capacity to only one month. On 1 July, Capt. Robert Henderson, USN (Ret.), then on duty in the Industrial Incentive Section of the Navy's office of Public Relations, was transferred to the vacant post. Some months after Henderson's appointment, the title was changed from ONI Liaison Officer to Representative of ONI with the AC/S, G-2.

The liaison office moved to the Pentagon on 28 September 1942. It acted as an information bureau, maintaining a file of naval publications and answering questions received from the Army on naval activities. In November, Ens. Frederick Holdsworth, Jr., was added to the office. In addition to his other duties, Holdsworth was charged with handling secret dispatches routed from the Army's Military Intelligence Services to the Navy. One of the most important functions of the office was assisting officers in ONI or MID to reach the appropriate persons in the corresponding sections of each agency and perfecting the cooperation and collaboration between the two organizations. Capt. Henderson also arranged for Ens. J. W. Woodburn, USNR, of ONI's Intelligence Plot, to make a daily submarine report at the morning situation presentation in the office of AC/S, G-2, and for Lt.(jg) R. T. Bates, USNR, from ONI, to serve on permanent duty with the Army's Order-of-Battle Section to represent the naval part of the activity.[3]

Adm. Ernest J. King, Commander in Chief, U.S. Fleet (COMINCH), in a memorandum to Gen. George C. Marshall, proposed a survey on the ways and means of merging intelligence activities so that duplications could be eliminated and headway might be made toward a unified intelligence agency. King stated further: "It would be well for agreement to be reached whereby ONI and MIS each undertake certain functions on behalf of both activities. I would expect this survey to lead in the direction of a unified intelligence agency which could be called [the] Joint Intelligence Agency." On 25 November 1942, Marshall agreed.[4]

Committees were appointed by ONI and Army Intelligence, and, as a result of several meetings, the Army AC/S, G-2, Gen. George V. Strong, and DNI RAdm. Harold C. Train submitted a joint memorandum to Marshall and King on 6 December 1942 incorporating preliminary recommendations for a joint agency. They suggested that such an agency should comprise all intelligence activities of the Army and Navy and of the other intelligence agencies at that time under the Joint Chiefs of Staff (JCS). The proposed Joint Intelligence Agency would include OSS, with the exception of the secret intelligence activities it needed for the discharge of its special operations. The memorandum also recommended that the merged organizations be housed under one roof.[5]

On 15 March 1943, the Army Assistant Chief of Staff for Intelligence and the Director of Naval Intelligence presented a memorandum to the Joint Chiefs outlining what had already been accomplished toward merging the two intelligence services: (1) close cooperation between geographical sections of ONI and MIS through personal contacts; (2) consolidation of mapping and photographic activities; (3) interchange of information on the production of monographs; (4) plans for issuing a "Joint Army-Navy Daily Intelligence Report"; and (5) a permanent interchange of officers between the counterintelligence groups of the two services. The memorandum, JCS 163, also enclosed the draft of a directive for establishing a Joint Intelligence Agency. The directive proposed the merging and placing under the control of the Joint Intelligence Agency prisoner-of-war interrogation, military and naval attachés and observers and joint intelligence collection agencies, mapping and photographic activities, liaison with other government agencies, preparation and dissemination of publications, and preparation of monographs and surveys.[6]

On 23 March 1943, JCS 163 and its associated papers were referred to the Deputy Chiefs of Staff for study and appropriate recommendations to the Joint Chiefs. Meanwhile, the exigencies of war demanded immediate practical measures toward cooperation. Consequently, the spring of 1943 saw the initiation of several joint Army-Navy enterprises, facilitated by the March reorganization of ONI, which brought ONI more closely parallel in organization to the Military Intelligence Service.[7]

In the reorganization, the Planning Branch was abolished and the Planning Group was established that was composed of the Deputy Director of Naval Intelligence, the three assistant DNIs, and such others as might be appointed. Discussions concerning questions of the proposed Army-Navy intelligence integration and merger were entered into by the Planning Group almost from its first session. In an ONI Planning Group (OP-16-X) confidential memorandum of 16 June, it was agreed that the Army, Air, and Navy intelligence agencies should have coordinating subcommittees under the Joint Intelligence Committee to prevent duplication of effort and to effect the integration of activities.[8]

In addition to the general ONI-MID liaison carried on by Capt. Henderson, a special liaison developed between the Report Center (later known as the Reading Panel) of the Dissemination Unit of MID and the Foreign Intelligence Branch of ONI. The first officer assigned in 1943 to the liaison duty was Lt.(jg) W. T. Lowry, USNR. His duties included attending the Army's daily intelligence panel to review reports and information in order that a proper selection of items of interest to the Navy could be made. Lowry was to determine not only what reports were of interest but also the number of copies needed for proper dissemination within the Navy Department.[9]

On 30 March 1943, the Army and Army Air Force, disregarding all the prior progress toward collaboration, proposed setting up (1) an Army-Navy American Intelligence Service, headquartered at Miami Beach, to be under the operational control of the War Department; (2) an Army-Navy Far Eastern Intelligence Service under the operational control of the Navy Department; and (3) an Army-Navy Atlantic and Middle Eastern Intelligence Service under the War Department.

RAdm. Train did not agree with the Army-Air Force proposals, particularly the set-up in Miami. Neither did he concur with placing intelligence services, charged with specific parts of the world, under either the Army or Navy because to do so would risk excluding the other service from proper participation. Train recommended that decisions on the proposals be deferred pending reports of surveys being conducted by management consultants Rawleigh Warner and Associates, and the Booz-Fry-Allen & Hamilton organization.

Adm. King had requested the Rawleigh Warner and Associates survey to analyze the functions of the Navy Intelligence organization and make recommendations. Their study, entitled *Summary Report of Intelligence Functions*, was submitted to King on 29 April 1943. It recommended (1) creating a Combat Intelligence Branch on the staff of COMINCH; (2) assigning all investigative work to the FBI, except those investigations of service personnel in which the services of naval officers were necessary; (3) combining the foreign intelligence functions (with the exception of certain functions) of ONI and MID, including files and personnel involved in monographing and strategic survey processes, with the Research and Analysis Branch of OSS; and (4) creating a new JCS Joint Intelligence Committee, responsible directly to the Joint Chiefs and working with the Joint Staff Planners.[10]

RAdm. Train did not concur with the Warner proposals either, since they would, in effect, abolish ONI. Such a radical change during a war, he felt, would have "a seriously disruptive effect upon the war effort," except for the first recommendation and part of the second.[11]

Despite its usefulness, the office of the ONI Representative with the Army Assistant Chief of Staff for Intelligence was discontinued on 10 February 1944 because of a staff shortage. In evaluating the work of the ONI representative to Army G-2, it should be pointed out that the Army used the services offered more frequently than did the Navy. One important accomplishment of the office was bringing together Army and Navy intelligence officers. Up to the time of the office's creation, ONI officers had not formally met with their opposite numbers in Army Intelligence. Another significant achievement was the aid consistently given to the movement to amalgamate the naval and military intelligence services. Perhaps the reason the Navy did not use the office as much as did the Army was its location in MID, which was more conveniently accessible to the Army.

After a few months, the Army, sensing the need for a continued relationship between the two intelligence divisions, arranged to provide a liaison officer who would spend a major part of his time at ONI. In June 1944, a LtCol. Cranwell, USA, was assigned to the duty and continued in the billet for the remainder of the war. In addition to his specific assignment to ONI, Cranwell acted as general liaison in all Army and Navy matters. His principal duties were to procure from the Navy Department information needed by the Army that would not come through ordinary channels, to expedite important requests, and to straighten out occasional differences. Cranwell also rendered important services in the establishment of ONI's Technical Intelligence Center (OP-16-PT).[12]

Joint Intelligence Outside the Washington Arena, 1942–1945

Joint Intelligence Collection Agency System

The joint intelligence collection agencies in World War II and their central and controlling organization, the Joint Intelligence Agency Reception Center, constituted an almost worldwide organization for coordinating the collection of intelligence materials by the U.S. intelligence agencies with a central clearinghouse for appropriate distribution in Washington.

Indirectly, the organization resulted from the movement in the latter part of 1942 toward greater coordination and eventual integration of ONI and the Army's military intelligence services. In his letter to the Vice Chief of Naval Operations (VCNO) on 13 December 1942 on the subject of intelligence for amphibious operations, Adm. H. Kent Hewitt, Commander Amphibious Forces, U.S. Atlantic Fleet, gave considerable impetus to the concept of developing joint intelligence collection agencies. Hewitt pointed out some of the inadequacies of intelligence for the North African operations and recommended that

"Naval Intelligence Officers, well qualified in the Italian language and instructed in classes of information important for Amphibious Operations, be sent . . . to North Africa." Adm. Royal E. Ingersoll, Commander U.S. Naval Forces, Atlantic, observed in his forwarding endorsement that "the Office of Naval Intelligence should be the clearing house for necessary information obtained from all other agencies, required by any task force of the fleet."

As a result of Hewitt's letter, a VCNO letter (OP-16-B-8 serial 02762316 of 21 December 1942) to COMINCH Adm. King proposed establishing an advanced intelligence center for the Northwest African Sea Frontier. In his reply of 1 January 1943, King approved the proposal but added, "If agreeable to the Chief of Staff, U.S. Army, the center should be a joint activity with appropriate Army and Navy representation."

On 26 January 1943, General Dwight D. Eisenhower, Commander in Chief, Allied Expeditionary Force, approved the establishment of the Joint Intelligence Collection Agency, North Africa (JICA/NA). The mission of the Naval Section of JICA/NA was to perform the intelligence activities required in connection with operations of U.S. naval forces in the North African theater of operations, to obtain information required by the Navy Department for planning, and to obtain counterintelligence information.

The intelligence collection tasks prescribed for the Naval Section of the JICA/NA by VCNO letter serial 098716 of 19 February 1943 were to collect and disseminate

a. Economic, political, geographic, ethnologic, social, and military information;

b. Information required by Naval Task Force and Task Group Commanders and by the Navy Department concerning enemy and Allied ship and plane movements;

c. Counterintelligence information concerning the enemy;

d. Information obtained by interrogation of prisoners of war;

e. Air combat information, including air reconnaissance;

f. Enemy material and equipment, including ship and plane identification data;

g. Merchant shipping information, including port security; neutral ship movement; interrogation of ship masters, crews and passengers; crew and passenger control; and ship routing data;

h. Information obtained through radio intercept agencies; and

i. Information from all adjacent Naval and Military Attachés and Observers, and from Intelligence Units of other U.S. and Allied Agencies.

The Army consistently took a narrower view of JICA/NA functions. The Navy directives were viewed as infringing on the activities of the Army's theater G-2.

JICA/NA was established at the Allied Forces Headquarters in Algiers in February 1943. On 23 April 1943, Gen. Marshall directed the Commanding General, U.S. Army Forces in the Middle East to establish a JICA within that command, with headquarters at Cairo. Intelligence teams were to be located at key points throughout the Middle East, as conditions demanded. It was specifically provided that the Office of Strategic Services was to be represented in the Joint Intelligence Collection Agency/Middle East (JICA/ME) and that its intelligence-gathering activities were to be coordinated with those of JICA/ME to eliminate duplication.

On 4 May, Adm. King approved the Navy's participation in JICA/ME. On 9 June, the Vice Chief of Naval Operations directed all naval attachés, naval observers, and naval liaison officers in the Middle East to forward all intelligence reports to JICA/ME for evaluation and transmittal to Washington.[13]

Subsidiary offices of JICA/NA were subsequently opened at Oran, Casablanca, Port Lyautey, and Tunis. Personnel for the naval sections were supplied by ONI. The agency was particularly active in securing the information used for planning the invasions of Sicily and Italy.[14]

On 3 May 1943, Gen. Joseph W. Stilwell, in Washington to support China's military needs, concurred in the proposal to establish a JICA in the China-Burma-India theater. Gen. Marshall and Adm. King also concurred, and JICA/CBI was established at Delhi by JCS directive (JCS 441) issued 5 August 1943. The same directive established standard procedures for theater commanders concerning JICAs and approved tables of personnel for contemplated JICAs in the South Pacific, Southwest Pacific, and Pacific areas.[15]

On 30 May 1943, Gen. Eisenhower's headquarters issued General Order No. 37, which defined the mission and duties of JICA/NA. It added a positive prohibition: "The JICA will not collect combat intelligence from units in the field, nor will it be charged with counterintelligence activities." Eisenhower's order prohibited the Navy team from complying with the Navy directive, particularly in connection with port security and counterintelligence at ports under U.S. control. Because of the conflict, Commander U.S. Naval Forces, Northwest African Waters (COMNAVNAW) requested clarification in his letter of 18 June 1943 to Eisenhower. Army Forces headquarters memo of 24 June to COMNAVNAW reaffirmed General Order No. 37. COMNAVNAW (Hewitt) felt it was important that both combat and counterintelligence activities be continued by the specially trained personnel of the Navy Section of JICA/NA. Upon Hewitt's recommendation, many of the personnel of the Naval Section were removed from the JICA organization and used to establish a Naval Intelligence Unit directly under him.[16]

JCS 441 of 5 August, which established JICA/CBI, also resolved the same conflict in the China-Burma-India area by stating for JICA/NA: "Nothing [in this directive] shall preclude the Navy Section of JICA/NA from performing such intelligence or counterintelligence activities as are required by the Navy Department and which cannot be performed by the intelligence organization of a Naval Command within the Theatre."[17]

Following the separation of the major part of the Naval Section of JICA/NA and establishment of the Naval Intelligence Unit under COMNAVNAW, a skeleton staff of two officers was left as the Naval Section. JICA/NA later became known as JICA/AFHQ (Joint Intelligence Collection Agency/Allied Forces Headquarters).[18]

Because of the frictions between JICA and Army G-2 personnel in the various operational theaters, there was a strong desire, particularly in the Army, for the abolition of the JICA concept. As a result, Gen. Strong (G-2), Gen. Bissel (Air Intelligence [A-2]), DNI RAdm. Roscoe E. Schuirmann, and Whitney Shepardson (Chief of the Special Intelligence Branch, OSS) met in Washington to discuss the issue. They recommended a three-month trial from 1 November 1943 to 1 February 1944; during this period several testimonials were received affirming the value of the JICA organization, and it was decided to continue the JICA system.[19]

In the summer of 1944, after the capture of Rome and the transfer of Allied Forces headquarters to Italy, JICA/NA (AFHQ) was moved to Naples. At the same time, JICA/ME was given the added responsibility for North Africa, and the branch office of JICA/NA at Algiers was placed under it.[20]

Effective 24 October 1944, the China-Burma-India theater was divided into two theaters consisting of the India-Burma (IB) theater, with headquarters at New Delhi, and the China theater, headquartered at Chungking. There were conflicting opinions on the effect of the change on JICA/CBI. The chairman of the JICA/CBI wanted to retain his organization and serve both theaters. The Army Military Intelligence Service thought the reasons for splitting the theater justified splitting JICA. The China theater commanding general wanted a JICA/China, and the commanding general in the India-Burma theater wanted JICA/CBI to remain as it was. On 3 January 1945, the Director of Naval Intelligence and the Army Assistant Chief of Staff for Intelligence forwarded a study to the Joint Chiefs recommending the division of JICA/CBI to conform to the new theater boundaries. The Joint Deputy Chiefs of Staff on 7 April 1945 approved the request of Commanding General U.S. Forces, China Theater for a separate JICA for the China theater, and on 27 April, Rear Echelon Headquarters, U.S. Force, China Theater, established a separate JICA/China by its General Order No. 57.

In the spring of 1945, the JICA/AFHQ that had transferred to Naples, together with the majority of the personnel of COMNAVNAW's Naval Intelligence Unit (which had moved to Naples in the spring of 1944), became JICA/MED. Headquarters remained at Naples.[21]

On 27 August 1945, JCS Directive 441/4 delegated to the Army Assistant Chief of Staff for Intelligence and to the Director of Naval Intelligence the authority to "make such disposition of the JICA organization in the Mediterranean, Africa-Middle East, and India-Burma Theatres as they jointly consider to be in the best interests of maintaining U.S. intelligence in those areas." A subsequent memorandum of agreement of 15 September 1945 by the Army Assistant Chief of Staff for Intelligence, the Director of Naval Intelligence, and the Deputy Director of OSS, abolished JICAs MED, ME, and IB as of 1 October. JICA/China was temporarily continued to 30 November 1945.

The main weakness of the Joint Intelligence Collection Agency organization was the lack of a central agency in Washington with adequate authority to give positive direction to activities in the field.[22]

Joint Intelligence Center, Pacific Ocean Areas

On 24 March 1942, the Commandant of the Marine Corps, in a letter to COMINCH Adm. King, had suggested the establishment of a joint intelligence center at Pearl Harbor and advanced joint intelligence centers at four other locations in the Pacific (Dutch Harbor, Pago Pago, Auckland, and Brisbane). The Chief of Naval Operations endorsed the letter favorably on 31 March and directed that a plan be submitted. The Commandant drew up a plan on 11 April and submitted it to the CNO and Commander in Chief, Pacific (CINCPAC). The plan was very similar to what finally evolved as the Joint Intelligence Center, Pacific Ocean Areas (JICPOA).

In connection with the Commandant of the Marine Corps' recommended plan for joint intelligence centers in the Pacific, the Chief of the Bureau of Aeronautics (BUAER) on 1 May 1942 proposed to COMINCH that aviation intelligence units composed of air combat intelligence officers be included in the organizations. A photographic interpretation unit was also recommended to be included at each center.

On 14 May 1942, Adm. King approved the general plan for aviation intelligence units at the joint intelligence centers as proposed by the Chief of BUAER and directed that BUAER train personnel for twelve aviation intelligence units.[23]

On 28 May 1942, CINCPAC Adm. Chester W. Nimitz, in a letter to Adm. King, approved the suggestion for a joint intelligence center, but recommended that the establishment of advanced intelligence centers be delayed until the main center at Pearl Harbor was in operation.[24]

On 26 June 1942 the Vice Chief of Naval Operations responded to the 28 May CINCPAC letter that in "consideration of the difficulties inherent in initiating directly a joint project as such, it appears preferable to constitute this Intelligence Center as primarily a naval center." It was understood, however, that CINCPAC could arrange for inclusion in the center such Army participation as appeared desirable.

On 24 June 1942, CINCPAC had directed the Commandant of the 14th Naval District (COM-14ND) in the Hawaiian Islands to set up an intelligence center. On 19 July, COM14ND advised CINCPAC that the Intelligence Center, Pacific Ocean Areas (ICPOA) had been established and was functioning. The first officer in charge was Cdr. Joseph J. Rochefort (of Battle of Midway code-breaking fame), and the nucleus of the new center was the Combat Intelligence Unit (which included the Radio Intelligence Section). In addition, the center had four ensigns as plotting officers, and thirty-one officers and ninety-one enlisted men were assigned to its photographic section, which was known as the Photo Reconnaissance and Interpretation Section, Intelligence Center (PRISIC).

The first location for ICPOA was with the Combat Intelligence Unit in the basement of the COM14ND Administration Building at the Navy Yard, Pearl Harbor. PRISIC, however, was headquartered on Ford Island and included in its organization photo interpretation officers assigned on temporary additional duty orders to aircraft carriers.

On 24 July 1942, in response to a VCNO personnel requirement estimate that 81 officers and 121 enlisted men would be needed for the center, CINCPAC wrote that the proposed staffing seemed "excessive." Faced with inadequate housing and office space, Nimitz wished to keep personnel levels at an "absolute workable minimum." As a result, only 17 officers and 59 enlisted men were ordered to ICPOA from Washington between 20 and 29 July 1942.

In September 1942, Capt. Roscoe H. Hillenkoetter relieved Cdr. Rochefort as officer in charge of the Intelligence Center, Pacific Ocean Area. On 25 September, part of ICPOA moved from its crowded basement spaces in the Naval District Administration Building to the new Navy Yard Supply Building 167. Sections that were moved to the new quarters included Administration, Air Combat Intelligence, Army Liaison, and Marine Liaison. The Combat Intelligence Unit, with its radio intelligence functions, remained in the administration building and made its reports directly to CINCPAC (usually to Adm. Nimitz's intelligence aide, Cdr. Edwin T. Layton, or his assistant). PRISIC remained on Ford Island until 15 October, when it moved to the Kodak Hawaii facilities in Honolulu.

On 15 October, CINCPAC decided that a plotting room at ICPOA would duplicate work at CINCPAC's plot and ordered that plans for such a section at ICPOA be abandoned. Of the twenty-one officers standing by to staff the proposed ICPOA plot, seventeen then requested and received transfers to other activities.[25]

In the fall of 1942, very little information was being received by ICPOA other than the highly classified information from the Radio Intelligence Section. There were few captured documents, few prisoners to be interrogated, and few aerial photographs of enemy-held territory. The Bishop Museum and the University of Hawaii Library, both in Honolulu, were the main sources of background information about Japanese-held islands.[26]

ICPOA gradually became the despository for all strategic intelligence about the islands of the Central Pacific received by CINCPAC, prompting the establishment of the Objective Data Section of ICPOA in October 1942, with Lt. George Leonard in charge.[27]

During the Battle of Midway, CINCPAC War Planners, much to their embarrassment, sent more B-17 bombers to Midway than the island could accommodate. The War Plans Division insisted that Intelligence should keep War Plans informed on facilities at U.S. bases as well as those at enemy bases. Actually, Operations Division had better and more easily accessible sources on such information than did Naval Intelligence. When the Army transport *President Coolidge* hit a mine and sank in a U.S. defensive minefield at Espiritu Santo, the argument ended. The Objective Data Section of ICPOA started compiling the necessary information, and Lt. John P. Lee and two yeomen were assigned to do the work.[28]

In November 1942, the Objective Data Section of ICPOA began to issue publications containing information on Allied bases and also published *Secret Sailing Directions for United Nations Bases, Central and South Pacific*. At the same time, the nucleus for the Drafting and Production Sections of the future JICPOA was activated in the Objective Data Section. Late in 1942, the services of the 64th Army Engineer Topographic Company were made available to ICPOA for the production of maps. The Army unit worked closely with PRISIC under the direction of the officer in charge of ICPOA and the Army liaison officer.

Prior to the end of 1942, the Marine liaison officer at ICPOA became responsible for the study of captured enemy ground equipment. With the receipt of that new responsibility, his section was retitled the Enemy Land Section.

Early in December 1942, the Air Combat Intelligence Section was taken over by Commander Naval Air Forces, Pacific (COMAIRPAC) and was moved to Ford Island. One air combat intelligence officer, Lt. Richard W. Emory, remained at ICPOA to keep records on Japanese air order-of-battle.

At the start of 1943, ICPOA consisted of the Radio Intelligence Section, the Combat Intelligence Unit, the Objectives Data Section, PRISIC, the Enemy Land Section, the Army liaison officer, and the Administration Section. With the exception of the Radio Intelligence Section and PRISIC, ICPOA was seriously understaffed. No officers had reported since the twenty-one who had arrived in September 1942, and only four remained. To continue operations, the center borrowed four officers from COMAIRPAC and five from PRISIC.

In February 1943, the first contingent of twenty graduates from the Navy's Japanese Language School at Boulder, Colorado, arrived at ICPOA, and the nucleus of the Translation Section was formed with Lt. Forrest R. Biard, a Fleet Radio Unit, Pacific (FRUPAC) language officer, in charge. Previously, when the first thirty prisoners of war had arrived in June 1942 from the Battle of Midway, Japanese interrogators had to be borrowed from other activities. The new Japanese language officers were permanently assigned to ICPOA and began one of the most important intelligence collection tasks performed during the war by the center—the processing of captured Japanese documents.

In April 1943, ICPOA made its second move, to the new FRUPAC building on the edge of Makalapa Crater. In May, the Objective Data Section was split up into the Enemy Bases Section and the Allied Bases Section. In July, a request was made to the Vice Chief of Naval Operations for twenty-three additional officers, and they arrived in the fall of 1943. In addition, twenty-four of forty-four photo interpretation officers reporting to PRISIC were assigned to the Objective Data Section.[29]

As long as there were no large-scale offensive operations planned for the Central Pacific, ICPOA had had few assigned responsibilities. When VAdm. Raymond A. Spruance was detached as chief of staff to Adm. Nimitz on 5 August 1943 to assume command of CENPAC (Central Pacific) Forces and RAdm. Richmond K. Turner on 20 August was ordered to command the Fifth Amphibious Force and plan and conduct landings in Micronesia, great changes were introduced at ICPOA.

The Enemy Bases Section had the task of assembling and preparing information bulletins to assist CINCPAC War Plans in selecting objectives for the first offensive. The section was short-handed for the task, but fortunately, as mentioned above, PRISIC had a temporary surplus of officers. The Photo Interpretation School at Anacostia had foreseen the future need, and graduates had arrived at PRISIC before there were many photographs to interpret. Some of the temporarily surplus officers were assigned to the Enemy Bases Section.

Each analyst in the Enemy Bases Section was made responsible for a specific small area for which he would assemble all available information and produce preliminary bulletins. Students for the Photo Interpretation School had been selected from among geologists, foresters, architects, and other professions familiar with the objects expected to be identified in aerial photographs. They were intelligent, well-educated, and adaptable young men. As the interest of the War Plans organization narrowed to specific areas, the appropriate analysts and their files were moved temporarily to CINCPAC headquarters to work directly with the planners.[30]

In September 1943, ICPOA was designated a joint Army-Navy-Marine organization by a CINCPAC directive (serial 001134 of 7 September) and was given the name Joint Intelligence Center, Pacific Ocean Areas. It was placed under the direction of the Assistant Chief of Staff for Intelligence (J-2), Joint Staff, CINCPAC, and CINCPOA (Commander in Chief, Pacific Ocean Areas). Its mission was defined as "the collection, collation, evaluation and dissemination of strategic and tactical intelligence for the CINCPOA and as directed by him." The Radio Intelligence Section was shifted to the Commander in Chief, Pacific Fleet (CINCPACFLT) the day before the establishment of JICPOA, but the Combat Intelligence Unit was included in the transfer and became the Estimate Section of JICPOA. Col. (later BGen.) J. J. Twitty of the Army was assigned as officer in charge, and Cdr. Wilfred J. (Jasper) Holmes, who headed the Estimate Section, was later designated his deputy.[31]

JICPOA was unique among field intelligence organizations. It was staffed by representatives of the Army, Navy, Marine Corps, and Coast Guard. Its strategic studies and estimates were outstanding, and its field operations were incredibly effective.

For example, when the Marines went ashore at Saipan, they landed at Charon Kanoa, the headquarters of the Japanese army on Saipan. The intelligence team from JICPOA went ashore with the first wave of Marines and moved into the schoolhouse that had been the Japanese general's command post. The Japanese had obviously left the post in a hurry and failed to destroy all their documents. The JICPOA team spent all night scanning and translating important documents disclosing where the enemy artillery batteries were sited, where their tanks were dispersed, and what their plans were for counterattack. The information was passed on to the Marines and the bombardment force, and the Japanese tanks were destroyed and their counterattack cut to pieces before it could get started.[32]

For each of the island invasions in the Pacific, instructions were issued to all forces on how to handle captured documents and personnel. Units from JICPOA were assigned for each amphibious assault to examine prisoners and documents for intelligence of immediate tactical value. Instructions stressed that captured documents were often of vital importance, particularly when showing locations of troop concentrations, artillery, or defenses and that documents were not to be pocketed as souvenirs but turned in for examination by intelligence personnel. At Attu, in May 1943, an Army unit found some documents, but they put them in their pockets instead of turning them in, as required by instructions. As the unit advanced, they were taken under fire, and one man lost a leg. Medics took him to the dressing station, where a Japanese document was found in his pocket. Examined by intelligence, the document disclosed the location of the Japanese artillery and mortars that had fired on the Army unit.[33]

On 1 September 1943, the Joint Chiefs of Staff directed CINCPAC to retake the Gilbert Islands and Nauru. The terrain model makers at PRISIC made models of Betio on Tarawa atoll and of the island of Nauru. There were plenty of photographs of Tarawa, both vertical and oblique, and the model was accurate. There were few photos of Nauru available, however, and the first model, based on considerable guesswork, was inaccurate according to a former resident of the island, who had been the engineer of the phosphate works on that island and knew it well. A new model was made, based on his knowledge plus a large-scale contour map that he had completed just before the Japanese landed. The planners had never been too happy with the selection of Nauru as an amphibious objective. When they and Adm. Nimitz saw the new model and the difficult terrain that it revealed, the planners recommended that Makin be substituted for Nauru. On 27 September, the Joint Chiefs of Staff approved the change, thus doubtlessly saving many lives.

Army, Navy, and Marine Corps shore-based planes made photo reconnaissance flights of the Gilberts, flying from the islands of Canton and Funafuti. On 19 September, a carrier task force raided the Gilberts and at the same time took low oblique photos of Tarawa. The submarine *Nautilus* (SS 168) took 2,000 photos of beaches at Tarawa, Makin, and Abemama during an 18-day surveillance mission, returning to Pearl Harbor in late October. PRISIC worked day and night to process the many photographs and produce the photo mosaics and maps needed for the landing operations.[34]

Two months after JICPOA was formed, the United States invaded the Gilbert Islands. Intelligence support to all subsequent amphibious operations in the Central Pacific and the Philippines was supplied, or contributed to, by JICPOA. Also in the Gilbert Islands landing, the first JICPOA team accompanied the invasion forces to assist the intelligence staff of the senior ground commander and to collect and study captured enemy equipment and documents. Similar teams participated in each succeeding invasion, and the Assistant Chief of Staff for Intelligence was made responsible for all captured enemy equipment in the Pacific Ocean areas.[35]

Photographs, sketches, and descriptions of Japanese installations made by the JICPOA teams that accompanied the landings on the Gilbert Islands, when compared with the interpretations of aerial photographs of the same islands made prior to the landings, greatly improved the accuracy of identification of similar installations subsequently found by photo interpreters in photographs of the Marshall Islands.[36]

The Joint Intelligence Center, Pacific Ocean Areas was a unique organization; it was the only U.S. agency in which Military and Naval Intelligence were formed into a single comprehensive organization servicing all the intelligence needs of ground, air, and naval forces of a theater command. All enemy source material for intelligence, including documents and equipment, was assembled at JICPOA, where the material was evaluated. Initially, the intelligence products of JICPOA received no CINCPAC-CINCPOA authentication. After July 1944, however, documents were prepared under the imprint of CINCPAC-CINCPOA, and the title JICPOA was used only for administrative purposes.[37]

When Adm. William F. Halsey moved from Noumea in the late spring of 1944, a number of his South Pacific Command (SOPAC) staff intelligence officers were ordered to duty at JICPOA. Among them were LCdr. Logan Jenkins and Lt. John Goodbody, who had been involved in the production of a series of publications for Commander South Pacific on "Know Your Enemy." Their arrival at JICPOA made possible the activation of the Bulletin Section to produce the *CINCPAC-CINCPOA Weekly Intelligence Bulletin* for the mass distribution of intelligence material. The publication took the place of the *Pacific Fleet Intelligence Bulletin*, which had been published spasmodically, first by the Fleet Intelligence Office and then by JICPOA, but which had been discontinued in November 1943 due to a staff shortage. Volume I, Number 1, of the *CINCPAC-CINCPOA Weekly Intelligence Bulletin,* in 2,000 copies, was issued on 14 July 1944. Demand almost immediately jumped the circulation to 6,000 copies, and, by the end of the war, the number of copies of each issue was 14,000.[38]

For each prospective island targeted for amphibious assault, JICPOA prepared an information bulletin. The bulletin for Palau incorporated the information gained from submarine and aerial photographic reconnaissance conducted in July and August 1944. The Palau landings began on 15 September at Peleliu. Information on the beaches was good, but behind the beaches, hidden under tropical foliage, were jagged limestone ridges honeycombed with caves, features that had not been spotted by photo interpreters prior to the landing. The resultant cost in lives in the capture of Peleliu was excessive in large part because of the lack of intelligence on the island's terrain.

By collating information obtained from numerous sources, including U.S. Navy action reports, excellent intelligence on Japanese naval losses was maintained by JICPOA. The loss of each Japanese naval vessel larger than an escort destroyer was known, both as to where and when it had taken place. One morning, during the period immediately following the Peleliu landing when Halsey and the Third Fleet were conducting strikes on Luzon, a message was received from Chungking. It relayed a report by a Chinese observer that a powerful Japanese surface force had departed Amoy for Luzon. Fortunately, the names of the ships in this force were provided by the observer. The lack of prior information on the existence of such a force made the report immediately suspect. By checking the names of the ships involved, JICPOA found that they were all ships that had been previously sunk but were losses the Japanese apparently believed were still secret. For example, one of the ships in the force was alleged to be the battleship *Mutsu*, which was known to have been sunk by an internal explosion in the Inland Sea. The bogus report, obviously planted, was intended to divert Halsey from his Luzon strikes. But Capt. M. C. (Mike) Cheek and Lt. Harris Cox, Halsey's intelligence officers in the battleship *New Jersey* (BB 62), had good current information on Japanese naval order-of-battle and quickly spotted the discrepancies. Capt. Edwin Layton, CINCPACFLT Intelligence Officer, also sent a message to Halsey discrediting the report. The effort at deception didn't work because the Japanese underestimated the capabilities of U.S. intelligence.[39]

The increase in the numbers of prisoners of war and captured documents required setting up a Translation Section and an Interrogation Section at JICPOA. A Target Analysis Section was also organized from elements of the Enemy Bases Section in September 1944.[40]

The Translation Section of JICPOA was organized into subsections, eventually numbering fifteen. Each subsection worked on captured documents relating to one particular subject, permitting individual language officers to become especially proficient in those aspects of the Japanese language dealing with their designated subject. The subject-oriented organization also encouraged many of the language officers to take courses in their particular subjects and to visit the Pearl Harbor Naval Shipyard to view related equipment.

The translators were inundated with documents captured on Kwajalein, and one shipment from Saipan contained fifty tons of Japanese documents. Items of low current intelligence value were shipped back to Washington. The JICPOA Translation Section concentrated on documents of direct importance to operations in the Pacific. Notices to mariners were of particular interest for their information on Japanese defensive minefields.[41]

The rapid growth of JICPOA generated a space problem, and in early 1944, plans were made for a new JICPOA building of approximately 40,000 square feet to be located just east of the FRUPAC building. On 16 May 1944, the new JICPOA building was completed, and all sections of JICPOA, except the Estimate Section, moved into the new quarters. The Estimate Section remained on the ground floor of the FRUPAC building. PRISIC also shifted into the JICPOA building from Kodak Hawaii in Honolulu, thus making possible the much-needed integration of their work with that of the other JICPOA sections. With the move, PRISIC was broken up into the Photo Interpretation Section, the Photographic Laboratory, the Model Shop, and the Distribution Section. For the first time, all JICPOA sections involved in research, analysis, production, and dissemination were housed under one roof.

In June 1944, a Propaganda Section was organized to plan and execute strategic and tactical psychological warfare in the Pacific Ocean area.

The first leaflets prepared by the Propaganda Section were used in the Marianas campaign. Also in June 1944, a Translation Section Annex was established in Honolulu, where Nisei translators could be employed on the routine analysis of captured notebooks and diaries.

In July 1944, personnel of the Escape and Evasion Section of the Army's Military Intelligence Service in Washington were transferred to JICPOA and set up as the MIS-X Section. In August, the Cartographic Section was officially organized. In September, distribution units for JICPOA material were established at Eniwetok and Guam. The first joint Army-Navy Flak Intelligence Section in the U.S. armed services was organized at JICPOA in November 1944. In December 1944, the responsibility for technical air intelligence was transferred to JICPOA from COMAIRPAC and was taken over by the Air Section. The move made JICPOA the central agency in the Pacific Ocean area for the collection and preliminary study of all captured aircraft equipment. By the end of 1944, JICPOA had grown to 500 officers and 800 enlisted men.[42]

The many tasks and functions assigned to JICPOA can best be appreciated by summarizing its 1945 sectional organization. The first group of sections dealt chiefly with static information on objectives and enemy bases. They included the Geographic, Photo Interpretation, Reference, Terrain Model, Target Analysis, Medical, Hydrographic, and Cartographic Sections.

The second group handled the constantly changing information on enemy ground, air, and naval forces. They were designated the Estimate, Enemy Air, Enemy Shipping, Enemy Land, and Flak Intelligence Sections. Of these, Estimate was the most important because it had access to top secret material. The principal duty of this section throughout the war was the preparation, for wide distribution, of weekly and monthly estimates of strength and location of enemy units of all services. The Estimate Section also prepared special estimates before each major operation.

Two additional sections, somewhat aloof from the others, were Psychological Warfare and Escape and Evasion; both were more interested in contact with the enemy than in information about the enemy. The former section was not actively established until June 1944 when, in cooperation with the Office of War Information, it embarked on propaganda and leaflet campaigns aimed at the Japanese homeland and at enemy troops and civilians on both bypassed islands and islands in the process of being captured. On 8 November 1944, the psychological warfare system became a separate Pacific theater agency under the officer in charge of JICPOA.

The remaining group of sections in JICPOA were those engaged in the publication of intelligence material. They were designated the Bulletin, Translation, Interrogation, Operational Intelligence, Production, and Administration Sections. The Translation and Interrogation Sections were staffed by specially trained language officers and were primarily concerned with the exploitation of captured documents for other sections.[43]

In early January 1945, when Cdr. Jasper Holmes was promoted to captain and given additional duties on the CINCPAC staff, he turned over most of his responsibilities in the Estimate Section of JICPOA to Lt. Donald M. Showers, USNR, who had been in the section since mid-February 1942. When Capt. Layton, the fleet intelligence officer, moved with Adm. Nimitz to Guam, he took Showers with him to set up a Fleet Combat Intelligence Center. Lts. Paul Yardley and Alex Johnson took over Showers's job in the Estimate Section.

Commander Submarine Forces, Pacific (COMSUBPAC), RAdm. Charles A. Lockwood, moved his operational headquarters from Pearl Harbor to the submarine tender *Holland* (AS 3) in Apra Harbor, Guam, on 24 January 1945, taking Capt. Richard G. Voge, his operations officer, with him. The move ended the daily conference that Voge had held at the JICPOA Estimate Section since early in the war. To replace Voge's daily conference, a special code was issued, held only by the Estimate Section and the Submarine Force Operations office at Guam, to be used for the radio communication exchange of intelligence between COMSUBPAC and the Estimate Section. Later, Voge arranged to have the code issued also to Commander Submarine Forces, Southwest Pacific at Fremantle to permit the submarines operating from that area to benefit from, and participate in, the exchanges.[44]

When the naval war in Europe drew to a close, the British sent a carrier task force to join the operations against the Japanese. The British force arrived in time to participate in the Okinawa campaign. As it had done for U.S. naval forces moved from the Atlantic to the Pacific, JICPOA had to provide the British with a complete new intelligence library so that they could receive the same intelligence support as the American forces. The British sent a lieutenant commander to JICPOA at Makalapa to help select material and arrange for its shipment to the staff and ships of the Royal Navy task force.[45]

In January 1945, the Enemy Bases Section was divided into the Geographic Section, the Reference Section, and the Production Section. At the same time, the Photo Interpretation Section was merged with the new Geographic Section, and photo interpretation officers were assigned to work with the intelligence officers on the various area desks. Also in January, an Enemy Shipping Section was established.

On 28 January 1945, the first personnel for the Advance Intelligence Center (AIC), established at the CINCPAC Advance Headquarters at Guam, were sent from JICPOA. More followed in February, bringing the total staff for the Advance Intelligence Center to sixty officers and fifty enlisted men at the time it became operational on 1 March 1945. Cdr. Richard O. Greene, the fleet photographic officer on the CINCPAC staff and executive officer of JICPOA, became the officer in charge of the Advance Intelligence Center. Initially, it was intended that most of the JICPOA functions of supplying immediate operational intelligence would ultimately be transferred to AIC, but, due to the crowded conditions on Guam and the difficulties involved in moving equipment, the objective was never achieved.[46]

When the CINCPAC Advance Headquarters was established at Guam, an Operational Intelligence Section was added to the Advance Echelon of JICPOA. The Operational Intelligence Section was responsible for procurement and distribution of information obtained by visual and photo reconnaissance; preparation of reports of combat operations; preparation of target and objective data for air bombardment, air support, and amphibious operations; coordination of intelligence reproduction facilities in the Forward Area; and maintenance of liaison between the CINCPAC staff, JICPOA, and the operating headquarters in the Forward Area of the fleet, air, and ground forces. The section was staffed by four or five junior officers, all previously trained at the Air Combat Intelligence School at Quonset Point, Rhode Island. In effect, the Operational Intelligence Section at Guam functioned as a section of the Advance Intelligence Center, serving principally to provide intelligence material for the deputy chief of staff and to coordinate pertinent activities in the Forward Area.[47]

In March 1945, the Allied Bases Section of JICPOA was designated the Hydrographic Section, and in April, personnel and equipment from the Terrain Model Unit arrived from Washington, combining their activities with those of the smaller model shop at JICPOA.

In the spring of 1945, a system was initiated by which officers from the Advanced Naval Intelligence School at New York were ordered to JICPOA for further training and assignment. At first, the Personnel Section of JICPOA took over the training task, but, in July, it was made the responsibility of the Operational Intelligence Section.[48]

When peace came in August 1945, JICPOA at Makalapa was staffed as follows:

	Navy	Army	Marines	WAVES	Total
Officers	409	73	51	11	544
Enlisted	931	182	49	61	1,223
Totals	1,340	255	100	72	1,767

JICPOA's production of studies, maps, and charts averaged two million printed sheets per week and the photographic laboratory was producing nearly two million photographic prints per quarter.[49]

A Central Intelligence Agency

The proposal to establish a central intelligence agency to provide a unified intelligence service for all intelligence agencies of the government was first discussed by the Navy Planning Group (OP-16-X) in November 1944. At that time, the group received from the Joint Intelligence Agency Committee a report opposing the postwar establishment of a central intelligence agency. Shortly thereafter, however, proposals were received from the Foreign Economic Administration and the Office of Strategic Services that looked toward the creation of such a joint agency after the war.

Coincident with its study of Army-Navy integration, the ONI Planning Branch carried forward a study on the possible establishment of a national intelligence agency. In December 1944, the Planning Branch, which had been reestablished on 14 December, prepared a memorandum for the Director of Naval Intelligence that analyzed both the OSS proposal and the joint Army-Navy proposal (JIS 89). It suggested that the latter was the more acceptable. Approval by the JCS of the idea of a national intelligence agency intensified the work on the project in ONI. In November 1945, RAdm. Sidney W. Souers, USNR, who had been head of the Planning Branch, was named Deputy Chief in ONI with special duties in connection with joint Army-Navy Intelligence and the proposed national intelligence agency. Souers became the first Director of Central Intelligence (DCI), serving from January to July 1946.[50]

President Harry S. Truman, by his letter of 22 January 1946 to the Secretaries of War, Navy, and State, directed the immediate establishment of a National Intelligence Authority (NIA) to be composed of the addressees plus a representative of the President. The original members were Robert P. Patterson, James V. Forrestal, James F. Byrnes, and FAdm. William D. Leahy, the last as Truman's representative. DCI RAdm. Souers was a non-voting member. The presidential letter also directed that an intelligence advisory board be established to support the NIA.

On 8 February 1946, NIA Directive No. 1 was issued to promulgate the policies and procedures governing the Central Intelligence Group (CIG) and to establish the Intelligence Advisory Board. The board was to be composed of the special assistant to the Secretary of State, in charge of Research and Intelligence; the Assistant Chief of Staff, G-2, War Department General Staff; the Director of Naval Intelligence; and the Assistant Chief of Air Staff for Intelligence, as permanent members. In addition, the DCI could invite the head of any other intelligence agency having functions related to national security to sit as a member on matters within the province of his agency.

Also on 8 February 1946, NIA Directive No. 2 was issued to establish the organization and functions of the Central Intelligence Group.

Joint Communications Intelligence

The Army-Navy Communication Intelligence Board (ANCIB) was established by a joint memorandum signed by Gen. Marshall and Adm. King on 10 March 1945. It was to function outside the framework of the Joint Chiefs of Staff and to report directly to the Chief of Staff, Army and the Commander in Chief, U.S. Fleet. The mission of ANCIB was to coordinate the plans and operations of communications intelligence (COMINT) organizations of the Army and Navy relating to collection, research, production, compilation, dissemination, and security of COMINT matters. On 11 December 1945, the State Department was invited to participate in ANCIB, and it accepted on 20 December 1945. The name of the board was accordingly changed to the State-Army-Navy Communication Intelligence Board (STANCIB).

On 4 June 1946, the FBI was invited to participate in STANCIB and accepted on 11 June 1946. As a result, the name of the board was again changed, this time to U.S. Communication Intelligence Board (USCIB). Next, CIG was invited on 3 July 1946 to join USCIB and accepted on 10 July 1946. No change in the name of the board was required but, in due course CIG was changed to Central Intelligence Agency (CIA). On 7 May 1947, in anticipation of the establishment of a separate U.S. Air Force, the Army Air Force was invited to membership on the board and accepted on 29 May 1947.

On 25 September 1947, the FBI withdrew from USCIB membership on the grounds that matters discussed by the board were primarily of interest to the armed services and the FBI did not consider its time well spent in its participation in the board's decisions.

The USCIB was formally established by National Security Council Intelligence Directive (NSCID) No. 9 on 1 July 1948 "to effect the authoritative coordination of the COMINT activities of the government." Membership was "not to exceed two members from each of the following departments or agencies: State, Army, Navy, Air Force, and CIA." USCIB decisions were to be reached by unanimous approval. Where unanimity could not be reached, the matter was presented to the National Security Council (NSC), except that, when lack of unanimity was between military members, the problem was to be presented to the Secretary of Defense before being submitted to the NSC.[51]

The Armed Forces Security Agency (AFSA) was established by the Secretary of Defense on 20 May 1949, and on 1 October 1949, the Director of AFSA assumed control under the Joint Chiefs of Staff. RAdm. Earl E. Stone, the first director, served from July 1949 to July 1951. Each service had one deputy director, and RAdm. Joseph N. Wenger was the Navy's first deputy, serving from July 1949 to July 1951.

The major provisions of the Secretary of Defense directive of 20 May 1949 establishing the Armed Forces Security Agency were:

> (a) The conduct of COMINT and COMSEC [Communications Security] activities (cryptologic activities) of the National Military Establishment is consolidated under one authority, except that the Departments of the Army, Navy and Air Force continue to be responsible for all such cryptologic activities as are required by intra-service or joint needs and are determined by the JCS not to be the sole responsibility of AFSA.
>
> (b) Subject to the authority and direction of the JCS, the Director of AFSA will formulate and, after approval, implement plans, policies, and doctrines relating to cryptologic matters for the Armed Forces and perform such other functions as the JCS may direct.
>
> (c) An Armed Forces Communications Intelligence Advisory Council (AFCIAC) is established within AFSA, with the major responsibility of recommending to the JCS policies, operating plans, and doctrines for the cryptologic activities of the National Military Establishment.
>
> (d) The Departments of Army, Navy and Air Force are charged with taking necessary action to facilitate the efficient and economical operation of AFSA, such action to include assignment of personnel and furnishing of facilities, equipment, and fiscal and logistic support.

The creation of the AFSA altered the previous relationship between the service intelligence staffs and their respective COMINT organizations. The AFSA was not responsible directly to the service intelligence staffs for collecting and processing foreign communications for intelligence purposes. Consequently, it was necessary to establish COMINT exploitation groups from the armed services, CIA, and Department of State "under the same roof" with AFSA, but the groups remained under the operational control of their respective intelligence staffs and under the administrative control of the Director of AFSA.[52]

AFSA was superseded by the National Security Agency (NSA), in accordance with a revision of NSCID No. 9 dated 28 October 1952.

The Korean War and Joint Intelligence in the Pacific Area

Joint Operations Center (JOC), Korea existed in the advanced headquarters of the Fifth Air Force when the Korean War broke out in 1950. JOC was primarily intended for coordination of air-ground action. Naval participation in JOC was initially weak because the senior naval representative was a relatively junior commander on temporary additional duty from Task Force 77 (the fast carrier force) for a period of sixty to ninety days. On 23 September 1951, a Navy captain with permanent orders assumed the position as a representative of Commander Seventh Fleet. Naval participation improved immediately.[53]

The primary function of the Navy Liaison Group at the Joint Operations Center was to act as an intelligence staff. It distributed naval intelligence in the center, made requests to the center for specific intelligence needed by the naval operating forces, and screened intelligence available at the center for items of naval interest. Naval requests for photographic reconnaissance coverage by the Fifth Air Force were handled through the JOC Naval Liaison Group. In addition, the Liaison Group functioned as an operations staff by screening JOC requests for Navy operations.

In December 1951, the Navy Liaison Group at JOC consisted of one aviator captain with experience in all phases of naval air operations; one commander, two lieutenant commanders, and one aviation lieutenant, all experienced in carrier operations; one lieutenant (air intelligence officer); one Royal Navy lieutenant commander; and two yeomen.[54]

Capt. Edwin Layton was Assistant Chief of Staff for Intelligence on the CINCPAC/CINCPACFLT staff from January 1951 to June 1953. Adm. Arthur W. Radford was CINCPAC/CINCPACFLT.

As the result of a tripartite conference in Singapore in May 1951 attended by representatives of CINCPAC, the British commander in chief, Far Eastern Station, and the French commander in chief in the Far East, an agreement was reached to hold an intelligence conference by representatives of the three commands every six months. The purpose of the conferences was to facilitate the exchange of intelligence information on the Viet Minh operations in French Indochina, hunter-killer (HUK) guerrilla warfare in the Philippines, and the guerrilla-bandit activities against the British in Malaya. All of the Communist expansion efforts were going on concurrently with the Korean fighting and supported intelligence reports that the Communist aim was to gain control of French Indochina and then all of Southeast Asia.

The Chief of Naval Operations directed Adm. Radford to furnish a U.S. representative to the tripartite intelligence conferences, and Radford named Capt. Layton as senior U.S. representative. Layton, in turn, made arrangements for the U.S. military, air, and naval attachés assigned to Thailand, Singapore, Indochina, Hong Kong, and Burma to be invited. Representatives from Army and Air Force Intelligence, ONI, and the Joint Intelligence Group of JCS were also assigned to the U.S. Delegation, and CINCPAC invited representatives from CINC Far East (Gen. Douglas MacArthur's command).

The First Tripartite Intelligence Conference was held 9–10 November 1951 in Saigon. Organization and procedures were established for carrying out the directives of the various nations represented. As far as intelligence in the Pacific was concerned, Capt. Layton was authorized to speak for CINCPAC, and as such, for the United States.

The Second Tripartite Intelligence Conference was held in Singapore, and the British recommended that the Australians and New Zealanders attend officially as observers. The presentation procedure followed at this and subsequent conferences was for the French to brief on happenings in French Indochina, the British on Malaya, and the United States on the Philippines. A subcommittee would then summarize "the high points" of the intelligence that had been developed. Aside from the formal record and exchange of information, the more important benefits gained from the Tripartite Intelligence Conferences were the personal contacts made between the various delegations.[55]

The Third Tripartite Intelligence Conference was held in Saigon in June 1952. Thereafter, Australia and New Zealand became full members. The Fourth Tripartite Intelligence Conference was held in Singapore on 10–12 December 1952.[56]

The Fifth Tripartite Intelligence Conference was held at Pearl Harbor on 10–12 June 1953. In accordance with a recommendation of the conference, future intelligence conferences on Southeast Asia would be named "Quinpart Intelligence Meetings," with the short title of QUINTEL.[57]

At the time of the fifth conference, the Southeast Asia Treaty Organization (SEATO) was in the process of being set up. The Chief of Naval Operations advised that since there had been mutual agreements by the British, French, Australian, New Zealand and U.S. governments to participate in SEATO, a motion should be made at the fifth meeting that any further intelligence conferences or exchanges of intelligence on Southeast Asia should be done under the auspices of SEATO. The motion was adopted, thus making the fifth meeting, held at Pearl Harbor, the last Tripartite Intelligence Conference for Southeast Asia.[58]

The U.S. Taiwan Defense Command (TDC) was established in 1955, with Commander Seventh Fleet assuming that title as a "second hat." Initially, the command was called the Formosa Liaison Center. Capt. Rudolf J. Fabian arrived in April from the Naval War College to head TDC, after visiting Commander in Chief, Pacific at Pearl Harbor to receive guidance and arrange logistic support. Fabian reported to Commander Seventh Fleet, and with the latter's chief of staff, searched Taipei for a building in which to set up shop. Supplies and equipment had already started to arrive, so Fabian took over a building recently vacated by the Naval Auxiliary Communications Unit, even though the structure was in miserable condition.

To expedite readiness for his staff when they arrived, Capt. Fabian personally assembled desks, cabinets, etc. Fabian also relieved a Col. Miller, USMC, as the intelligence officer on the staff of the Military Assistance Advisory Group on Taiwan. In that position Fabian acquired a staff consisting of one Army lieutenant colonel and one Chinese secretary.

Finally, the Taiwan Defense Command staff started to arrive, but they were able to move into the Nationalist Chinese naval headquarters, where Col. Miller's office had been. Then it was decided that TDC should be a full-fledged unified command and not just an intelligence collection activity.[59]

Coordinating Intelligence, 1958–1971

Commencing with the 14 January 1958 issue of the *Central Intelligence Bulletin* (CIB) published by the CIA, the bulletin was modified to the extent that its contents represented material coordinated by all agencies of the intelligence community. Also, its distribution was expanded to include the highest levels of the U.S. Government. ONI was thus afforded the opportunity not only to participate directly in the production of current intelligence for high-level readers, but also to present intelligence on subjects in which the Navy had a primary interest. The ONI Current Intelligence Coordinator (OP-922B4) reviewed current intelligence for the purpose of selecting appropriate material for the CIB and served as the coordinator for all Navy-sponsored CIB articles.[60]

In September 1958, at the recommendation of the President's Board of Consultants on Foreign Intelligence Activities, all National Security Council Intelligence Directives were redrafted, and the coordinating role of the CIA was increased. The United States Intelligence Board (USIB) was formed by combining the former Intelligence Advisory Committee (IAC) and the U.S. Communications Intelligence Board for the purpose of consolidating all intelligence voices under the Director of Central Intelligence.[61]

The Defense Intelligence Agency (DIA) was established in 1961 in an effort to create a mechanism to solve the problems presented by the disparate intelligence estimates being produced and the duplicate intelligence efforts by the military departments (Army, Navy and Air Force). DIA was assigned responsibility for:

1. Organizing, directing, managing, and controlling of all DOD intelligence resources assigned to or included within DIA.

2. Reviewing and coordinating those intelligence functions retained by or assigned to the military departments.

3. Supervising the execution of all approved plans, programs, policies, and procedures for intelligence functions not assigned to DIA.

4. Obtaining the maximum economy and efficiency in allocation and management of Defense Intelligence resources.

5. Responding directly to priority requests levied upon DIA by USIB.

6. Satisfying the intelligence requirements of the major components of the Department of Defense.

Concurrent with the establishment of DIA, the Directorate of Intelligence (J-2) of the Joint Staff was disestablished and its functions assigned to the director of the Defense Intelligence Agency. The established reporting line for DIA was through the Joint Chiefs of Staff to the Secretary of Defense.[62]

The military departments were charged with organizing, training, and equipping intelligence forces for assignment to combatant commands and with conducting those intelligence functions that peculiarly related to departmental missions, including the development and support of intelligence systems organic to combatant forces. In addition, each department retained responsibility to manage and operate certain types of intelligence activities, including counterintelligence and investigative services; scientific and technical intelligence; mapping, charting, and geodesy; and their respective cryptologic agencies.

In March 1962, the Joint Chiefs of Staff issued a memorandum to the Defense Intelligence Agency on "Actions to Strengthen the Intelligence Capabilities of the Unified and Specified Commands," and a memorandum to the commander of each unified and specified (U&S) command on "Authority to Strengthen Intelligence Capabilities of Commanders of Unified and Specified Commands." The memoranda directed that intelligence staffs and activities be established, including an intelligence requirements-collection office at each U&S command headquarters to perform functions compatible with the requirements and collection functions of DIA. DIA was directed to issue guidance as to policies, procedures, format, and priorities of intelligence requirements to achieve standardization of requirements processing throughout the Department of Defense. The requirements flow followed command channels to DIA for validation and collection-levying. A JCS message of March 1962 to the U&S commands further expanded on the memoranda. The original intent of the actions was to take the staffs and activities from the component commanders and move them one echelon higher, but this did not happen, since the military departments were still authorized direct access to the component commands.

The JCS memoranda also directed the U.S. commands to establish and operate activities to perform intelligence functions of common interest. The order was amplified in a JCS memorandum of July 1962 that required the establishment of a current intelligence/indications function; intelligence production, including the estimates function; and the target intelligence function.

Throughout the 1960s, the various intelligence elements of the military departments complained about the time taken for an intelligence "requirement" (request for information) to be validated by DIA. To avoid duplication of a requirement, DIA had to research to ensure that the desired information was not within the files of DIA or in the files of other agencies in the intelligence community prior to levying a specific collection requirement. If a requirement had to be levied on a national agency, it took up to six months before DIA was notified that a requirement had been accepted for collection by the agency.[63]

Until March 1964, the members of the U.S. Intelligence Board were the heads of the intelligence organizations of the following departments or agencies: CIA, State Department, DIA, Army, Navy, Air Force, NSA, FBI, and the Atomic Energy Commission. In March 1964, by agreement between the Secretary of Defense and the Director of Central Intelligence, the Army, Navy, and Air Force ceased to be members of USIB and were thereafter designated observers. However, the DCI stated explicitly in writing that he desired the armed services to continue to participate in USIB affairs and to continue their representation on USIB subcommittees. Also, the National Security Council amended its principal intelligence directive to ensure that when a service intelligence chief had a dissenting opinion on a national intelligence paper, that opinion would be carried in the paper.[64]

In 1966, DIA issued the Defense Intelligence Plan (DIP), which had been extensively coordinated with the military departments and the unified and specified commands. DIP delineated the intelligence responsibilities and relationships of Department of Defense components and instituted a DOD-wide system for review and analysis of intelligence operations to facilitate mutual support and eliminate wasteful duplication. One of the principles that the plan spelled out was that "intelligence produced at higher echelons must be supplemented by local production of lower echelons in order to satisfy particular command requirments." The plan further acknowledged that "some degree of parallel and overlapping effort is normal and necessary." Thus the primary role of DIA shifted from the production of *all* defense intelligence to the production of some strategic intelligence, or to that intelligence used at the JCS/Secretary of Defense/national level.[65]

A Secretary of Defense memo of 1 August 1969 assigned additional responsibilities for intelligence to Assistant Secretary of Defense for Administration Robert Froehlke in order to improve the effectiveness and efficiency of defense intelligence. Froehlke's responsibilities included the establishment of an intelligence resource review and decision-making process; improvement of intelligence communications among Department of Defense agencies and between the Department of Defense and other agencies; evaluation of intelligence organization relationships, roles, and missions; and review of security policies to eliminate unnecessary classification and compartmentation.

The Office of Assistant Secretary of Defense for Intelligence was established on 1 November 1971 for the purpose of bringing together the management authority of the two major Department of Defense intelligence agencies, the Defense Intelligence Agency and the National Security Agency. DIA and NSA managed programs involving about 85 percent of all intelligence resources. In addition, a presidential memorandum of 5 November 1971 broadened the review and coordination authority of the Director of Central Intelligence by giving him the responsibility for "planning, reviewing and evaluating all intelligence activities and the allocation of all intelligence resources."[66]

Chapter Notes

1. VAdm. Walter S. Anderson ltr to author, 5 Sep 1974.

2. Department of the Navy, "Administrative History of the Office of Naval Intelligence in World War II," 10 Jul 1946, unpublished MS, 80, hereafter *ONI WWII Admin History*.

3. *ONI Review*, Dec 1947, 19.

4. *ONI WWII Admin History*, 88.

5. JCS 163 memo, cited in *ONI WWII Admin History*, 89.

6. JCS memo JCS 163/1, 16 Mar 1943, encl. A and B, cited in *ONI WWII Admin History*, 90.

7. Ibid., 1426–27.

8. Ibid., 1396, 1419.

9. Ibid., 77.

10. Ibid., 91–93.

11. RAdm. Train memo to Vice Chief of Naval Operations, 15 May 1943, cited in *ONI WWII Admin History,* 93–94.

12. Ibid., 74.

13. Ibid., 1470–77, 1479.

14. Ibid., 659.

15. Ibid., 1477–79, 1487.

16. Ibid., 1480.

17. Ibid., 1482–83.

18. Ibid., 1489.

19. Ibid., 1485–86.

20. Ibid., 667, 1490–92.

21. Ibid., 1490–92.

22. Ibid., 1508–10.

23. Bureau of Aeronautics, "Administrative History of the Bureau of Aeronautics in World War II," 1946, unpublished MS, 19:23–25, hereafter *BUAER WWII Admin History*.

24. Capt. Wilfred J. Holmes, "History of JICPOA," MS history, 1946, 3–4, OA. The manuscript became the basis for Holmes's later book, *Double-Edged Secrets* (Annapolis: Naval Institute Press, 1979).

25. Holmes, "History of JICPOA, 4–5.

26. Ibid., 177,

27. Ibid., 5.

28. Ibid., 202–3.

29. Ibid.

30. Ibid., 202, 214–15.

31. Ibid., 6.

32. RAdm. Edwin T. Layton, Oral History, USNI, Annapolis, MD, 1970, 129–30, copy in OA.

33. Ibid., 162–63.

34. Holmes, "History of JICPOA," 215, 217–20.

35. Ibid., 6.

36. Ibid., 230.

37. Ltr of transmittal for ibid., 3.

38. Ibid., 7, 281–82.

39. Ibid., 289–92.

40. Ibid., 6.

41. Ibid., 257–58.

42. Ibid., 6–7.

43. Commander in Chief, Pacific Fleet/Commander in Chief, Pacific Ocean Area (CINCPACFLT/CINCPOA), *Command History, 1945*, 324–26.

44. Holmes, "History of JICPOA," 302–4.

45. Ibid., 314.

46. Ibid., 7.

47. *CINCPACFLT / JICPOA Command History, 1945*, 328–29.

48. Holmes, "History of JICPOA," 7.

49. Ibid., 8, 61.

50. *ONI WWII Admin History,* 1421.

51. OP-322Y1 ltr, 29 Aug 1949, a brief on the U.S. Communications Intelligence Board (USCIB) for the prospective head of OP-32.

52. ONI-322Y1 memo to Capt. J. M. Ocker, 16 Aug 1949.

53. CINCPACFLT, *Interim Evaluation Report No. 3: Korean War Naval Operations*, 5-60, 5-88.

54. Ibid., 5-90.

55. Layton oral history, USNI, 175–77.

56. CNO (OP-322) endorsement to CINCPAC ltr ser 003, 9 Jan 1953.

57. CNO (OP-322) ltrs ser 003511P32, 30 Jul 1953; and ser 003518P32, 1 Aug 1953.

58. Layton oral history, USNI, 177–78.

59. Capt. Rudolf J. Fabian, tape recorded response to author's queries, 21 Jan 1976, author's files, OA.

60. ONI Instruction 03820.70 of 3 Feb 1958.

61. Notes in the History File of Dr. George E. Kidd (OP-922X), OA. Dr. Kidd was Special Executive Assistant to the Assistant Director of Naval Intelligence for Foreign Intelligence.

62. President's Blue Ribbon Defense Panel (Foster Committee) Report, 1 Jul 1970, 44–45.

63. Ibid., Classified Annex, 31–33, 37.

64. ONI briefing for Naval Attachés, 1970. The briefing was maintained for prospective naval attachés and assistant naval attachés to read during their indoctrination at ONI in the early 1970s.

65. Foster Committee report, Classified Annex, 38.

66. *DNI Annual Report to CNO, FY 1972*, 2.

CHAPTER 20

Automation

Several factors generated the need for automation in processing intelligence following World War II: A large mass of data was being collected, particularly by sensor devices, and computer analysis was required to put the data in a format that would permit its correlation with information collected by other sources; an even greater mass of data required manageable storage that could only be achieved by automated data processing (ADP) methods if the data's future value in determining the significance of new information was to be exploited on a timely basis. In addition, more sophisticated weapons systems being developed by the Soviets and the United States introduced the need for new, equally sophisticated intelligence collection techniques with associated processing and dissemination methods to give real-time intelligence support to operating forces.

The Office of Naval Intelligence's first venture into automation as an aid to analysis took place in 1954 when electronic accounting machinery (EAM) was first used to compile data on merchant ship characteristics.[1]

The Technological Capabilities Panel in the Killian Report of 14 February 1955 stated:

> The growing principles and technology of information retrieval will advance and may revolutionize the handling of large masses of intelligence data. Their applications should be sought out more actively.

ONI's comment on the recommendation (No. 2C9) was that

> this is a valid and desirable recommendation. The development of mechanical equipment to assist in the performance of intelligence tasks would be highly desirable. The growing problems of receiving, storing, and processing intelligence information are straining present systems to a breaking point. More personnel, time, and space are required to convert the vast amount of information received into intelligence and to file it for future use.[2]

An informal ONI Committee on Mechanization was formed in January 1959 and consisted of three representatives each from the Intelligence Production Division (OP-922) and the Administration Division (OP-923) and one representative from the Security Division (OP-921). The David Taylor Model Basin, under the Bureau of Ships, began designing a merchant ship activity file and a general-purpose information storage-and-retrieval system for ONI. The file system later became Project SEXTANT. The Office of Naval Research conducted a survey of available data-mechanization systems for use in Naval Intelligence processing and submitted its report to ONI in December 1959. ONI mechanization policy was promulgated by ONI Instruction 010462.1 of 16 December 1959.[3]

The United States Intelligence Board (USIB) Committee on Documentation (CODIB) was established in 1959 to review and ensure the compatibility of the various intelligence mechanization programs within the intelligence community. ONI's representative on the committee was the chairman of ONI's mechanization committee. In February 1960, CODIB published the jointly approved Intelligence Subject Code (ISC), which was designed for intelligence community-wide cataloging of intelligence reports and documents.[4]

Exploring the possibility of mechanizing its name files was one of ONI's early ventures into automation. In early 1960, $25,000, originally intended for the purchase of file cabinets, was set aside by OP-921 and used to contract with the Astro Electronics Division of Radio Corporation of America (RCA) on 1 June 1960 to study ONI's name file problems. The report of the study, dated 31 October 1960, included an interim plan and a final recommended system, the latter dependent on the availability of adequate space for the necessary computers and automated data processing equipment to be used in the storage and rapid retrieval of name file data.

ONI's General Services Branch (OP-923M) started the Naval Intelligence Project STORE in January 1961 as an experiment to simplify the processing of the some 100 documents received each day by ONI. The Committee on Mechanization conducted studies of many different systems in use or being tried by other agencies. The most practical system found was the Lodestar 15-mm microfiche cartridge system used by the Social Security Office in Baltimore.

The Project STORE system operated essentially as follows:

— Incoming documents were sequentially numbered on each page with reel and frame numbers in the order in which the documents were received from the mail room. The documents were then microfilmed, with an average of 2,300 frames per 100-foot reel.

— Code sheets were attached to each document showing the reel and frame where a microfilmed document could be found, plus originator, date and serial, classification, etc.

— The document was coded for every subject and geographic area mentioned. About 10,000 different general subject codes were in use, plus approximately 100,000 numeric breakdowns of general subjects.

— The code sheet information was put on punch cards, with an average of twelve cards per document. The cards were then filed by subject, geographic area, originator, latitude and longitude, and vessel (the last for foreign merchant ships only).

The two storage methods used in the STORE system were index cards and microfilm. The cost of the project in 1961 was $20,000; in 1962 the cost was $30,000 for such expenditures as cameras; ready printers and upkeep; other equipment rental, such as punch card equipment, verifiers, sorters, and a collator; and an interpreter. In 1961, an average of five persons was involved in the operation of the STORE system. In 1962 the average was seven.[5]

In early 1961, ONI and various fleet commands were engaged in a number of projects designed to apply electronic data processing to naval intelligence problems. The efforts required close coordination on a worldwide basis in order to provide maximum return for the funds expended, avoid duplication of research and development, and ensure system compatibility. To accomplish the coordination, the following changes in ONI's mechanization effort were instituted, effective 20 April 1961:

1. Capt. Harold G. Bowen, Jr. (OP-92C) was assigned responsibility for all ONI mechanization efforts and, insofar as they pertained to mechanization, for all of ONI's relations within the Navy Department, with naval commands, and with the other armed services and government agencies.

2. Capt. D. C. Higgins (OP-923M) was assigned additional duty to OP-92C as administrative assistant for mechanization. His duties included organizing a mechanization staff to be directed by OP-92C; establishing military and civilian billets for mechanization throughout ONI; coordinating with the Training Branch (OP-923T) to establish a mechanization training program for ONI; coordinating with the Comptroller Branch (OP-923R) to develop and monitor a mechanization budget for ONI; procuring all equipment and supplies required by ONI for mechanization; coordinating the establishment of work schedules and priorities for all ONI machine processing projects once they became operational; and ensuring adequate security protection of intelligence material involved in mechanization projects, including material made available to private contractors.

3. Dr. Maurice H. Hellner was assigned as technical assistant for mechanization, reporting directly to OP-92C. He was responsible for determining ONI's data mechanization requirements; developing a worldwide naval intelligence data processing system, especially in support of the Operations Control (OPCON) Center and Naval Tactical Data System (NTDS) programs; coordinating and monitoring mechanization projects in ONI up to the point where they became operational; monitoring intelligence mechanization projects in naval commands; coordinating within the Office of the Chief of Naval Operations (OPNAV) and the appropriate bureaus the drafting of a technical development plan for naval intelligence mechanization; monitoring ONI mechanization contracts; and coordinating with other armed services and federal agencies involved in the field of intelligence data processing.[6]

The Foreign Merchant Marine Section (OP-922N2) was directed by the Director of Naval Intelligence to establish a Merchant Marine Data Processing Unit (OP-922N2M) commencing 1 May 1961 to receive, process, and store, by means of electronic data processing techniques, intelligence concerning world merchant marine movements, characteristics, history, and special operations. OP-922N2M was also to provide other support that might be required for the Naval Information Center (NAVIC) in accordance with Chief of Naval Operations (OP-335D2) letter serial 03518P33 of 13 April 1961.[7]

Effective 18 September 1961, an ONI Automation Coordination Staff was formalized from the existing Committee on Mechanization. Designated OP-92CB, it came directly under OP-92C, RAdm. Harold Bowen. OP-92CB was charged with the overall planning, development of concepts, scheduling, and coordination of the naval intelligence data-processing effort worldwide. Bowen retained his responsibility for overall supervision and direction of all automatic data processing efforts in ONI.[8]

The Fleet Intelligence Center, Europe (FICEUR), at Port Lyautey, began using an IBM 407 and other punch card equipment in 1961 and planning for a 1401 computer installation for storage and retrieval of formatted and unformatted raw data and finished intelligence.

On 1 March 1961 HRB Singer started the design of Project HORIZON for ONI. Project HORIZON was an automated data processing system intended to accumulate information on all world navies and their activities of interest to the U.S. Navy.

The Foreign Intelligence Division (FID) Automation Group, established by FID Instruction 05400.1 of 3 March 1961, was responsible for assisting the special assistant for mechanization (OP-922X2) to coordinate all automation activity of the Foreign Intelligence Division within the overall program as set forth by the ONI Mechanization Committee. The Automation Group was made up of representatives from each OP-922 branch and from the Naval Scientific and Technical Intelligence Center (NAVSTIC).[9]

In an effort to develop an efficient program for introducing ADP into naval intelligence, a 90-day worldwide survey was undertaken by Dr. Hellner and Cdr. David F. Seaman commencing in April 1961. The survey report, distributed on 22 August, served as a guide for future ADP development in naval intelligence. According to the report, the major factor making automation in intelligence processing necessary was the drastic reduction in the time available for processing in order to meet command requirements. This reduction in available time was caused by the introduction of new technologies into naval operations and the enormous increase in the volume of data being collected by new, sophisticated collection devices and techniques.

Hellner and Seaman's survey report anticipated that naval intelligence would be involved in five types of computer installations: (1) major fixed installations, such as NAVIC and the three fixed Operational Control Centers; (2) mobile and transportable computer installations, such as the Mobile OPCON Centers and the new NTDS that was beginning to be used on command ships; (3) Fleet Intelligence Center installations; (4) ONI installations; and (5) specialized installations at certain collection sites, such as for patrol aircraft (VQ) squadron bases. A major conclusion from the survey was that few things could be more disastrous for naval intelligence than to introduce ADP equipment at various places around the world without any overall concept of how they should be tied together. ONI had to take the lead in developing that concept.[10]

When the ONI Security Division moved into the Fairmont Building in Arlington, Virginia, in August 1963, it carried out some of the major recommendations made by RCA, including the conversion of file cabinet storage to open-shelf filing. These changes, plus the additional space and personnel, helped reduce the huge backlog in filing that amounted to 196,000 documents as of 21 January 1963. The volume of material already stored in the Security Records Branch also presented a special obstacle to microfilm operations. As of January 1963, there were approximately 5 million subjects and approximately 20 million pages, with a growth rate of 700,000 pages per year.[11]

A requirement for an afloat capability to collect various items of tactical information and to process, analyze, and correlate the information for near real-time use by the operating forces was recognized in the late 1950s and resulted in the development of the Integrated Operational Intelligence System (IOIS). Initially called the Integrated Air Intelligence System, the IOIS included the RA-5C Vigilante aircraft as a multisensor collection vehicle—electronic intelligence (ELINT), photographic, infrared and side-looking radar—and the Integrated Operational Intelligence Center (IOIC), located on board the aircraft carrier from which the RA-5C operated and capable of processing the data collected by the aircraft. The IOIS became operational in November 1962 in *Saratoga* (CVA 60), although the RA-5C did not become operational in the fleet until late 1963 (see also Chapter 4).

Within a year, the potential of onboard automated intelligence processing was realized, and IOIC, with its rapid storage and retrieval of intelligence, was seen as desirable in other ships to support staffs afloat. The initial IOICs were operated by seven officers and twenty-four enlisted personnel from the aircraft carrier ship's company, plus three air intelligence officers and twelve enlisted men who reported aboard with the carrier air group. IOIC's intelligence database was supplied by ONI, the Naval Photographic Interpretation Center, FICs, the Defense Intelligence Agency, and the Central Intelligence Agency. Most of the database was supplied in automated data processing format (punch cards), 16- and 35-mm microfilm aperture cards, miniature transparencies, and magnetic tape. Additional storage forms included 70-mm and 5-inch roll photography. The computer for IOIC was the AN/USQ-20, the same equipment used in the Naval Tactical Data System. There were two basic databases, one for ships operating in the Atlantic and one for those in the Pacific.

In 1964, a specific operational requirement was approved that called for a naval intelligence processing system (NIPS) to put automated intelligence systems in additional ships and to link the ships to the FICs, which provided an automated intelligence database to the shipboard centers.[12]

By April 1963, the Automation Coordination Staff (OP-92B4), from its experiences to date, could comment on ONI's automated data processing system that no civilian personnel had been replaced or reassigned in or from ONI as a direct result of their functions being taken over by ADP equipment. In addition, insofar as naval intelligence was concerned, the use of ADP technology had increased the scope, depth, and timeliness of data processing and permitted a greatly expanded intelligence support to Navy operational forces. The total volume of work to be accomplished by human resources had increased and could logically be expected to continue to increase as new and more sophisticated systems became operational.

OP-92B4 also reported that the entire ADP effort was aimed at automating intelligence source data as well as current operational intelligence. The very nature of intelligence made all collected information "source data" for future analysis. In addition to automating source data and current intelligence, each naval intelligence automation project provided for the efficient, expeditious exchange of machine-structured data with other intelligence activities and the automated command system used by the operational forces. The automation effort was being continuously directed toward improving the quality of intelligence support to the operating forces and toward improving the capability of the Director of Naval Intelligence to manage worldwide naval intelligence.[13]

As of February 1963, a naval intelligence data processing system consisted of:

1. The intelligence subsystem of NAVIC, located at the Pentagon to serve the Chief of Naval Operations and employing IBM-7090 and 1401 computers and appropriate peripheral equipment.

2. The intelligence subsystems of the fixed operational control centers at Commander in Chief, Pacific (CINCPAC) at Kunia; Commander in Chief, Atlantic (CINCLANT) at Norfolk; and Commander Naval Defense Force, Eastern Pacific at San Francisco.

3. The computer installations at electronic countermeasures aviation squadron (ECMRON) bases for VQ-1 at Atsugi, Japan, and for VQ-2 at Rota, Spain, using IBM 650 computers to process results of ELINT reconnaissance flights.

4. Fleet intelligence centers, which had extensive punch card installations (FICEUR also had a 1410 computer at Port Lyautey).

5. The intelligence formatted for input into the naval tactical data system (NTDS).

6. The intelligence support to IOIS.

7. Project STORE, the central files of ONI.

Examples of the naval intelligence subjects being automated as of February 1963 included Soviet merchant, fishing, and scientific ship activity; Soviet satellite-country and Chinese Communist merchant ship activity; Communist-chartered western merchant ship activity; all other foreign merchant ship activity; Soviet submarine activity; Soviet surface ship naval activity; Soviet satellite country and Chinese Communist naval activity; all other foreign naval activity; coasts and landing beaches; foreign ports and harbors; and ELINT.[14]

The Fleet Intelligence Center, Pacific (FICPAC) received its IBM 1410 installation in February 1964, and the Atlantic Intelligence Center received its IBM 1410 in April 1964. In June, the intelligence subsystem of the Fleet Operational Control Center, Pacific (FOCCPAC) received its computer equipment, a CDC-160A with tape and disc recording capabilities. The intelligence subsystem for the Atlantic (FOCCLANT) received its ADP equipment (similar to that in the Pacific) in 1965.

Secretary of the Navy Notice 5450 of 9 March 1964 established the Naval Intelligence Processing System Support Activity (NIPSSA) as an ONI field activity, with its commanding officer "double-hatted" as OP-92B4. NIPSSA's mission was expanded to include ADP operations in addition to research and development. NIPSSA had sixty-nine authorized civilian billets, and planning was started for an ONI ADP center following authorization by the Defense Intelligence Agency on 8 April.

In October 1964, the ONI ADP center at Arlington Hall Station received its computers and began operations. The Defense Intelligence Agency adopted the FICEUR-developed formatted file system as standard for use in all 1410 intelligence computers.

The personnel authorization of NIPSSA was increased in 1965 to twelve officers, eighty-nine civilians, and twenty-seven enlisted personnel. The overly ambitious Project HORIZON, started in 1961, was terminated in April 1965 as an unsuccessful research and development effort.

In December 1964, the Director of Naval Intelligence approved an OP-92B4 staff study through the Defense Intelligence Agency on an ocean surveillance intelligence information processing system to be developed and operated by ONI in support of the fleet commanders and the national command authorities. As a result of the study, ONI's Ocean Surveillance Intelligence System (OSIS) was developed.

Also in December 1965, Project COMPASS, a processing system for foreign naval activity and contact reports that was fully compatible with Project SEXTANT became operational.[15]

In January 1966, Project SEXTANT was expanded to include processing reports on the activities of merchant ships of all nations. A secure, high-speed data link became operational in the spring of 1966 between the ONI ADP center at Arlington Hall Station and the Chief of Naval Operations intelligence plot at the Pentagon. In June 1966, a chart-plotting printer capability was established, enabling the presentation of ship activity information of any category, in graphic form, on a worldwide basis. In August, information from operational sensor systems was added as an input to OSIS.

In January 1967, the submarine contact file became an input to OSIS. Also in January, OSIS started receiving data on a daily basis via the Fleet Weather Central circuit. In March 1967, the Movement Report Control Center started sending daily position reports for input to OSIS, and in April the Fleet Operational Control Centers for the Atlantic and the Pacific started sending merchant ship reports to OSIS. In July 1967, a high-speed, special intelligence UNIVAC 1004 data link was established to permit OSIS to fulfill the data requirements of the CINCLANT and CINCLANTFLT organization.[16]

The success of the Integrated Operational Intelligence Centers on board aircraft carriers provided the impetus in Fiscal Year 1968 to mechanize the intelligence centers in certain other types of ships. IOICs were operational in CVAs 59 through 66, and an IOIC was installed in CVA 67, which was due to become operational in Fiscal Year 1969. Automated intelligence centers were being installed in amphibious command ships *Blue Ridge* (LCC 19) and *Mount Whitney* (LCC 20), under construction and due to become operational in Fiscal Years 1970 and 1971, respectively. The NIPS program change request proposed the extension of the tactical intelligence system to intelligence centers in other ships.

The master automated database for NIPS was maintained by ONI's Naval Reconnaissance and Technical Support Center (NRTSC). The time required to prepare and disseminate updated material to NIPS databases had been reduced from five weeks to two weeks by 1968.

An automated wall display was installed in ONI's intelligence plot and became operational in October 1968. It was intended to be capable of a nearly instantaneous display of ONI computer-derived ship track and movement information in variable colors for all ships and oceans and from all sources.

High-speed data link terminals for exchanging communications intelligence (COMINT) between computers were installed at CINCLANTFLT, the Navy Field Operational Intelligence Office (NFOIO) at Fort Meade, Maryland, and ONI. Plans called for the network to be expanded to include CINCPAC (with the terminal at FICPAC) in Fiscal Year 1969 and Commander in Chief, U.S. Naval Forces, Europe and Commander Western Sea Frontier in Fiscal Year 1970.[17]

The data link installation between the ADP Center and NFOIO Fort Meade was completed and became operational in August 1968.[18]

A revised statement of functions for NIPSSA was issued in August 1968:

> 1. Operate the Naval Intelligence Command ADP Center;
>
> 2. Maintain liaison with other services, agencies, departments and technical bureaus in the field of Automatic Data Processing of intelligence to ensure awareness of the state-of-the-art, to assist in optimizing compatibility and exchange of intelligence ADP programs and concepts, and to further the standardization and avoidance of duplication;
>
> 3. Assist in reviewing and coordinating Naval Intelligence Command requirements for the automated handling of intelligence data;
>
> 4. Provide technical assistance in the management of automated naval intelligence processing projects;
>
> 5. Establish and monitor specialized ADP training programs, both formal and on-the-job, for selected personnel in the Naval Intelligence Command;
>
> 6. Participate in the negotiation of Naval Intelligence ADP contracts with private companies, and monitor such contracts until completion;
>
> 7. Participate in the formulation of security policies for the automated processing and handling of Naval Intelligence data;
>
> 8. Accomplish the computer programming necessary to provide for the implementation of assigned Automated Naval Intelligence projects;
>
> 9. Manage NIPSSA's in-house IDHS [Intelligence Data Handling System] funds; and
>
> 10. Maintain the primary automated Ocean Surveillance information processing system for Naval Intelligence, assisting, developing, and expanding it to meet the needs of all users.[19]

In 1968, NRTSC assisted the Integrated Sealift Study Group in establishing the Joint DOD/DOT/MARAD (Department of Defense/Department of Transportation/Maritime Administration) Shipping Information System, providing computer tapes of its merchant ship database and printouts of changes to the database.

A faster, more flexible computer program package called the Command Ship Data System (CSDS) was introduced in four West Coast aircraft carrier IOICs during 1968. Shipboard training, checkout, and maintenance of the new programs were conducted to ensure responsiveness to the tactical environment of Southeast Asian operations.

Six new non-IOIC and one IOIC databases were prepared and delivered by NRTSC in 1968. Maintenance was continued for seventeen non-IOIC and eleven IOIC database holders.[20]

During 1969, NRTSC installed and indoctrinated additional users of the NIPS database, including CINCPACFLT, Commander Cruiser-Destroyer Force, Pacific and its associated flotillas, and Commander Amphibious Forces, Atlantic and its associated amphibious groups. NRTSC also installed the Command Ship Data System software for IOICs and indoctrinated Sixth Fleet and Second Fleet aircraft carriers in its use.[21]

The Master Control Subsystem (MCS) for CSDS was modified in 1970 at NRTSC to enable the CSDS to operate from a single-code tape on both the AN/USQ-20A and AN/USQ-20B computers. An additional modification provided a means to pack more data into a physical tape record, thus increasing the storage capacity of a magnetic tape. In conjunction with the MCS modification, five Tactical Installation File (TIF) CSDS software packages were delivered and installed in four aircraft carriers and at the Naval Intelligence Processing Systems Training Facility in Albany, Georgia. Each delivery and installation included program decks, documentation, and on-site instruction and training of IOIC personnel. Two new documentation and procedures manuals related to the modification were completed and delivered to the users.

A NIPS Phase II prototype database was forwarded to North American Rockwell at Columbus, Ohio, to serve as a vehicle for testing software capabilities. Test results led to software designs that enabled NRTSC to generate a CSDS master file and to transmit a processable update usable with Phase II software. The first completed Phase II File was delivered to the guided missile *Providence* (CLG 6) in December 1970 for test and evaluation. The delivery to the cruiser included initial installation of a Phase II technical file consisting of 2,927 aperture cards that conveyed characteristics, performance, and recognition data on 1,742 militarily significant objects.

Appendix A to the *IOIS Data Base Manual* was completely revised and distributed to all users during 1970. The appendix provided recipients of the NIPS database with a comprehensive compilation of key-words, intelligence subject codes, and numeric subject modifiers to facilitate search and retrieval.

Sixteen database users were added to the distribution list during 1970, making a total of ninety users. Approximately 15,000 aperture cards and microfiche were added to the NIPS worldwide Master File, for a total of 86,735 units.

The Naval Intelligence Command ADP Center was moved from Arlington Hall Station to the renovated basement of the NRTSC building at Suitland, Maryland, in June 1970. In July, the Naval Ocean Surveillance Information Center was established in a newly completed Butler hut prefabricated building in the east courtyard of the NRTSC building. By December 1970, all computer hardware (forty-three units) had been moved into a new site at the NRTSC building, providing approximately 50 percent more usable floor space and thus more efficient placement of equipment.[22]

In October 1970, Fleet Intelligence Center, Europe began processing the NIPS I database, using programs written for both the IBM 1410 and IBM 360 systems (the IBM 360/40 digital computer had been received in 1969). FICEUR's input to the database in 1970 consisted of order-of-battle files and a target intelligence file. The database, produced in card form, was extremely bulky and required time-consuming computer processing. The system was improved in January 1971 with the advent of the NIPS Phase II database, which had the capability to provide a tape output in place of punch cards.[23]

Work was started on defining the NIPS Phase II database in 1971 under the CSDS, which was expected to permit all non-NIPS Change 12 IOICs to process the NIPS Phase II database approximately five to six years ahead of the previous schedule.[24]

As of 1972, NIPS was described as "a program to provide semiautomated intelligence centers to major ships and major commands afloat and to make the databases, produced and maintained by the Fleet Intelligence Centers, accessible to the afloat Intelligence Centers." The intelligence centers were designed to receive and reduce all data from tactical support sensor systems to an intelligible format, to collate and correlate the information with previously received intelligence, and disseminate the finished intelligence as needed. Data link and teletype systems, similar to Naval Technical Data System links, provided near real-time dissemination to commands in the local area that had the capability to receive it.

Ships equipped with NIPS as of April 1973 were the aircraft carriers *Forrestal* (CVA 59), *Saratoga* (CVA 60), *Ranger* (CVA 61), *Independence* (CVA 62), *Kitty Hawk* (CVA 63), *Constellation* (CVA 64), *Enterprise* (CVAN 65), *America* (CVA 66), and *John F. Kennedy* (CVA 67), and the amphibious command ships *Blue Ridge* (LCC 19) and *Mount Whitney* (LCC 20).

NIPS support was enhanced by the addition of data entry/display consoles in operational centers within NIPS-configured ships. Remote stations in the Combat Information Center, the Flag Command Center, and other locations thus became capable of obtaining direct retrieval of intelligence from the NIPS database. Changes made to the NIPS on board *Independence* allowed the carrier to support amphibious and land operations for the first time. Another innovation during Fiscal Year 1973 made it possible for all NIPS-equipped ships in the Mediterranean to have their databases updated by message. Formerly, updates were delayed until they were brought to the ships by courier. In the Pacific, all NIPS-equipped carriers were given a new "SEAWING" ELINT readout capability, which allowed for processing ELINT tapes on board rather than only at shore installations.[25]

Because of the growing interrelationship between communications and data processing during Fiscal Year 1973, NIPSSA was assigned the additional responsibility for telecommunication planning and management for intelligence. The Navy Automatic Relay Controller system was placed in operation to provide automatic handling and distribution of incoming and outgoing messages from the AUTODIN/Defense Special Security Communications System.

In September 1972, a Department of Defense standard for microfiche was established. At NFOIO, refinement of the microfiche database to meet the changes resulted in a capability to generate a computer index of all NFOIO hard copy publications from 1968 onward. The centralized storage and retrieval of microfiche greatly increased the effectiveness of analytical work. Commander Naval Intelligence Command also established a portable microfiche intelligence support package that gave near-instantaneous deployment readiness to any area of the world.[26]

The following commanding officers of NIPSSA served from 1964 to 1975:

Name	**Starting Date**
Capt. David F. Seaman*	March 1964
Capt. Alfred R. Olsen, Jr.	September 1964
Capt. Wendell J. Furnas	September 1965
Capt. Sumner Shapiro	October 1969
Capt. Leonard E. Tillerson	July 1972
Capt. Fred A. Hull	August 1975

*Was senior officer of OP-92B4 and participated in the mechanization study by ONI that led to the establishment of NIPSSA.

Chapter Notes

1. Naval Intelligence Processing Systems Support Activity (NIPSSA), "Chronology of Naval Intelligence Automation," hereafter *NIPSSA Chronology*.
2. OP-992N memo, ser 000491P92, 20 Apr 1955.
3. *NIPSSA Chronology*.
4. OP-923M, "Review of ONI's Efforts in Intelligence Mechanization," 25 Jan 1961.
5. Naval Intelligence School presentation, Feb 1963.
6. ONI Notice 10462 of 21 Apr 1961.
7. ONI Notice 05230 of Apr 1961.
8. ONI Notice 010462 of 15 Sep 1961.
9. *NIPSSA Chronology*.
10. OP-92C, "Worldwide Survey Report," 22 Aug 1961.
11. Naval Intelligence School presentation, Feb 1963.
12. Lecture by Cdr. Frank Hofer, ca. 1967; and ONI lecture on ADP at the Naval Intelligence School, 6 Feb 1963.
13. OP-92B4 ltr ser 127–63, 26 Apr 1963, "Info on ADP for Congressional Hearings."
14. ONI ADP presentation at the Naval Intelligence School, 4 Feb 1963.
15. *NIPSSA Chronology*.
16. *NIPSSA Command History, 1967*.
17. DNI Report to CNO, "Status of Major ONI Programs," 20 Jun 1968.
18. *NIPSSA Command History, 1968*.
19. OPNAV Instruction 05450.118A of 19 Aug 1968.
20. Naval Reconnaissance and Technical Support Center (NRTSC), *Command History, 1968*, 4–5.
21. *NRTSC Command History, 1969*, 8.
22. *NRTSC Command Histories, 1970*, 11–12; and *1970*.
23. Commanding Officer, Fleet Intelligence Center, Europe ltr, 4 Dec 1973.
24. *NRTSC Command History, 1971*, 14–15.
25. *DNI Report to CNO, FY 1973*, III-7.
26. Ibid., II-18, II-19.

Chapter 21

Counterintelligence

Counterintelligence is that aspect of intelligence activity devoted to destroying the effectiveness of inimical foreign intelligence activities and protecting information against espionage, personnel against subversion, and installations and material against sabotage. Counterintelligence activity involves investigations and other measures to collect, process, and disseminate related information.[1]

This part of the history of U.S. naval intelligence includes items on the investigative activities of ONI and the Naval Investigative Service (NIS); security activities, plant protection, censorship, foreign disclosure; and the various organizations involved in these activities either directly or in supporting roles. Other chapters cover the topics of plant protection, censorship, and foreign disclosure in greater detail.

As with other elements of naval intelligence, there is much cross-fertilization and mutual dependence between the various parts. But, as a basic rule (with many exceptions), intelligence is concerned mainly with foreign countries and their activities in foreign areas. Counterintelligence is concerned mainly with foreign countries and their intelligence collection, sabotage, and subversive activities particularly in the United States and at ports and U.S. Navy facilities overseas or in their vicinity.

Initially, most of the Navy's counterintelligence work was carried out domestically; as a result, the term "domestic intelligence" is sometimes used interchangeably with "district intelligence" and "counterintelligence."

Origins of U.S. Navy Counterintelligence in World War I

ONI didn't get involved in counterintelligence until World War I. Even as late as 1913, when plans were stolen from the battleship *Pennsylvania* (BB 38), the Navy called on the Burns Detective Agency to investigate.[2]

On 31 March 1916, in testimony before the House Committee on Naval Affairs, Sixty-fourth Congress, First Session, Secretary of the Navy Josephus Daniels requested $50,000 for the collection of information at home by the naval districts. The naval appropriation bill of 26 August 1916 made available $30,000 for this purpose, and the appropriations were increased in both 1917 and 1918.

Assistant Secretary of the Navy Franklin D. Roosevelt was the true instigator of the move to have ONI engage in domestic investigations.[3] Consequently, ONI was the first of the federal intelligence agencies during World War I to have undercover agents.

On 27 July, in compliance with a CNO directive of 18 April 1916, the Director of Naval Intelligence (DNI) submitted "confidential detailed plans for the establishment of the information service and the collection of information for the use of the officer in charge of Naval Districts." On 22 September, the plans were referred by Acting Secretary of the Navy Franklin Roosevelt to the General Board for comment and recommendation. On 5 October, Adm. George Dewey, president of the General Board, favorably endorsed the plans, and Secretary of the Navy Daniels approved them on 6 October (see Chapter 22).[4]

Thus was inaugurated the Naval District Information Service and the establishment of an aid for information in each naval district (the term "aid" was originally used in official correspondence, but the more conventional spelling was in general use by the end of the war). This was the start of the naval district intelligence offices, which became major elements of the Naval Investigative Service fifty years later. The aids acted as direct representatives of the district commandants. Each aid was responsible for supervising intelligence work in his district in conjunction with, and under the guidance of, ONI. He gathered information about shipping as well as information needed to protect shipping against hostile acts by agents or sympathizers of the Central Powers. The aid also arranged for the procurement and placement of coast observers and for their reporting suspicious ship or coastal activities.[5]

Other duties for the aids included detecting and acting against espionage and sabotage along the waterfronts, in navy yards, and in factories and other work areas associated with the navy yards; investigating Navy personnel within the naval district; detecting illegal radio stations; placing guards on each ship entering U.S. ports and while in U.S. ports; checking and inspecting cargoes and manifests; and searching for and locating enemy goods in storage. Much of the ship inspection work was eventually taken over by representatives of the Customs Division of the Treasury Department, but the Navy continued to work in collaboration with Customs.[6]

In a major reorganization plan, developed in ONI by Maj. John H. Russell, USMC, and Cdr. Dudley W. Knox and approved by the Secretary of the Navy on 1 October 1916, ONI was split into four divisions. Division A, Organization & Control of Agencies for the Collection of Information, had suborganizations concerned with counterespionage and secret service activity within the United States. Great emphasis was put on domestic intelligence.[7]

In early 1916, Cdr. Edward McCauley, Jr., Assistant Director of Naval Intelligence, asked Spencer Eddy in New York City if he would perform undercover work of assistance to the ONI. Eddy agreed and established an office at 2 Wall Street at his own expense. Eddy found that the workload was more than he could handle by himself. So, with McCauley's permission, he solicited the help of A. Duer Irving and John C. King in the early fall. All requests for investigations for information went directly from Cdr. McCauley to Eddy; the reports of the small organization's investigations were sent directly to McCauley.

In December 1916, those assigned to Eddy's office were officially designated voluntary agents of ONI but continued to work undercover. In mid-December, William C. Van Antwerp was added as a voluntary agent.

On 6 January 1917, Cdr. McCauley called all voluntary agents of the office to Washington, recruited them into the United States Naval Reserve Force (USNRF), and gave them the temporary rank of lieutenant (jg). Spencer Eddy was in Florida at the time and McCauley requested he come to Washington. Upon arrival in Washington, he was also enrolled and given the rank of lieutenant commander, effective 6 March 1917.

Thus, the nucleus for the New York branch office of ONI was established at 2 Wall Street under LCdr. Eddy and with Lts. (jg) Van Antwerp, Irving, and Albert R. Fish, and voluntary agent John King. The men were also designated as special agents of ONI.

On 6 April 1917, with the outbreak of war, all of the above personnel were called to active duty, with Eddy in charge. The office also had one stenographer, Frances E. Reid.[8] The New York office was used as a model for other branch offices that were set up later in Philadelphia, Boston, Baltimore, Chicago, San Francisco, and Pittsburgh. All came directly under ONI and took on operations that could not be turned over to the naval districts, such as the surveillance and guarding of plants handling Navy contracts, investigations of sabotage cases, shipping security, censorship, location of illicit radio transmitters, and investigations of naval civilian and service personnel. Over 5,000 manufacturing plants were (at least theoretically) under Navy protection, and many aliens and active enemy agents were removed from the plants before they were able to fulfill their missions. The branch officers were also responsible for directing and supporting many of the secret agents who were operating in the United States under ONI supervision.[9]

The responsibilities of the branch offices and aids for information overlapped in many respects, and there were occasional conflicts and misunderstandings between the two organizations. But the overlapping of an overt by a covert organization also had many unique advantages and gave a desirable flexibility to the methods of surveillance or investigation and the channels available for the prosecution of cases, which, in turn, made for more effective and rapid solutions.

The General Board drew up a basic plan, identified as Serial 666, General Board No. 425, of 4 February 1917, "Steps to be taken to meet a possible condition of war with the Central European Powers." Among its provisions for readiness for war, the following applied to U.S. naval intelligence and investigative capabilities:

> Organize a comprehensive system of intelligence service covering the whole theatre of war in accordance with the plans of ONI. Place under surveillance all citizens of the Central Powers in the Navy, or in Government employ in naval establishments, and remove them from positions in which they may do possible harm.

On 8 January 1918, Chief of Naval Operations RAdm. William S. Benson directed all ships and stations to appoint an officer (lieutenant or above) to serve as confidential intelligence officer (CIO) as a collateral duty. The CIO's identity was to be known only to the commanding officer and the executive officer. The CIO was to make confidential reports to

ONI, noting particularly "the officers and enlisted men whose records, nationality, friendships, associations, or habits, would tend to the probability of their being direct or indirect agents of any enemy government or of any enemy subject or sympathizer." The CIO was also required to investigate any related suspicious cases that came to his or the commanding officer's notice.[10]

Between March and October 1918, ONI issued a weekly *Confidential Bulletin* (Nos. 1–28) containing information on "suspicious individuals and firms" as well as more general information on subversion.[11]

On 14 August 1918, a memo from Secretary of the Navy Daniels cautioned aids for information about "isolated cases of ill-advised zeal" in dealing with certain labor leaders, such as subjecting agitators and union representatives to severe cross-examination or intimidation. He ordered aids "to investigate and report promptly upon labor troubles affecting work for the Navy, particularly those inspired by enemy influence" and not to take sides.[12]

During World War I, eighteen German agents who had been in constant communication with Germany were uncovered in the United States. All were arrested and their papers confiscated, leading to the discovery that wireless radio devices were being manufactured in New York for the German government. Many of the people connected with the illegal wireless device fled, but others were arrested. Again, their papers revealed codes and other secret means of communicating with Germany.[13]

Before the United States entered the war, a careful watch had been maintained on possible German secret service representatives in the Navy. For example, one of the battleships in the Atlantic Fleet reported that a chief petty officer (CPO) on board was suspected of being a German agent. He had served for years and had had an excellent record. He spoke German fluently and when on liberty associated closely with Germans. His duties on board ship gave him access to technical equipment and information. An ONI agent was enlisted as a yeoman and ordered to the ship in the normal way. The agent gradually made friends with the suspect and eventually was invited to accompany him on liberty. The agent yeoman found insufficient evidence to convict the CPO of espionage but was convinced that he was a German agent. Consequently, the CPO was transferred to duty in the interior of the United States where he would not have access to sensitive information. A few days after the United States entered the war, a coded telegram from Holland, addressed to the CPO and confirming his foreign agent activities, was intercepted and forwarded to ONI.[14]

The importance of security control and the supervision of merchant ships, their officers, crews, and passengers was apparent to ONI early in the war. Until July 1917, this control was exercised sporadically and only by the aids for information in the naval districts. Then attention was attracted to the frequent naval attaché reports concerning smuggling, letter carrying, and enemy agents traveling as passengers or in the crews of merchant ships. At that time, the Navy had no legal authority in such matters. The Treasury Department did have such authority, but didn't realize its responsibility or the importance of counterespionage work. Consequently, ONI, recognizing the need, went ahead with the development of an organization to coordinate and support the necessary counterespionage effort (see Chapter 24).[15]

In a report made during the war, DNI RAdm Roger Welles, Jr., discussed the work of ONI and detailed the conditions under which the work was performed, as he viewed them. The report was placed in the record during the testimony of Secretary Daniels on 20 May 1920 in connection with the Senate's investigation of the Navy Department's conduct of the war. The following are some of the portions of that report which relate to counterintelligence:

> It was well known in this country that the Germans had established a wonderful spy system through which Berlin was being informed of the activities in every branch of industry in the country. It is probable that there was not a manufacturing establishment here that did not have at least one paid agent of the German government who kept that government informed of everything that was going on. There is no doubt that, even in the departments in Washington, German agents were at work at all times. It was supposed that bases of some sort for the supplying of gasoline [*sic*] and supplies to German submarines were being secretly established in different points along the coast of Mexico, Central and South America. Before the United States entered the war, Germans were allowed to enter this country freely.
>
> The day the United States declared war, ONI's activities were increased tremendously, for it became the duty of this office not only to continue its peace-time activities, but to form an investigation section to seek out the Germans who had been active in propaganda in favor of Germany, who were attempting to prevent by sabotage, by explosions, or by fomenting strikes, and by many other means the manufacture of munitions of war; who were making bombs for the purpose of blowing up our ships and factories; and in general to prevent the activities of Germans and German sympathizers from continuing their nefarious pursuits.

> This meant expanding tremendously the office in Washington, reorganizing its personnel, and extending its activities to every country of the globe, as well as covering every state in the Union.[13]

All U.S. naval attachés were involved in counterintelligence in varying degrees during World War I. The office in Paris was probably more broadly involved than any other because of ONI's activities in regard to the security of shipping. Most shipping was between France and the United States.

The office in Paris built up a large filing card system on suspects. It controlled the travel of all persons requiring American visas, controlled the issuance of all American passports applied for in France, reported suspects and enemy agents to ONI and other Allied organizations, and controlled crew lists. A black-list of commercial firms was also maintained.

The Inter-Allied Bureau for uniting the Allied counterespionage system was formed in September 1917. No U.S. Navy representative was allowed to attend, and mutual distrust within the bureau generally caused it to fail in practice. The board did serve one useful purpose for the naval attaché office in Paris by publishing the Inter-Allied List of Suspects, which helped the attaché develop his card system and perform his other counterintelligence functions, including investigations of all persons applying for overseas civilian employment with the Navy.

Duplicate reports on suspect persons living or traveling in other Allied nations were constantly being sent back and forth between U.S. naval attachés in Madrid, Berne, London, Rome and Paris. The cooperation between these offices was cordial and effective.

Counterintelligence activities, separate from Allied investigations, were conducted by a bureau set up in the naval attaché's office at Paris under William Chandler, who employed agents to make investigations of suspect individuals. Chandler's agents never knew of his connection with the naval attaché.

In matters relating to suspects, it was the policy of the Naval Attaché, Paris, to make a preliminary investigation, even when the subjects appeared not to be of concern to the Navy. If it was not of Navy concern but worthy of further investigation, and if the case was serious, the facts were sent to the U.S. government agency concerned, usually the U.S. Army. In that manner, people were investigated, detained, watched, and even deported from France.

Closely allied with counterespionage work was the investigation of U.S. Navy deserters and also, on occasion, German deserters found mostly in Switzerland by Naval Attaché, Paris, agents operating there. Information from the German deserters tended to be unique and of special value from a technical aspect.

Carriers of questionable letters found on merchant vessels were also investigated. Also, close control was made of passport issuance, especially when a passport had supposedly been lost or stolen. Several cases involving passports being purchased by German agents were discovered.[17]

The types of investigations conducted by ONI and its various field activities during World War I were listed as follows:

A. Naval Personnel
   1. Deserters, stragglers, imposters
   2. Suspicious persons attempting to enlist
   3. Collusion between contractors and Navy personnel

B. Navy Yard Employees
   1. Navy Yard suspects
   2. Pro-German activity at the Navy Yards
   3. Thefts
   4. Cases referred by the Commandant
   5. Alien or enemy agitation

C. Miscellaneous
   1. Cases referred by Mail and Cable Censorship
   2. Suspicious individuals reported in the vicinity of Navy piers, wharves, and docks
   3. Applicants for marine pilots licenses
   4. Cases involving radio apparatus
   5. Suspicious fires in Navy areas
   6. Protection of shipyards doing Navy work
   7. Protection of manufacturing plants with Navy contracts
   8. Enemy agents and sympathizers and civilians concerning activities inimical to the interests of the Navy[18]

During the first six months of the war, the Navy rounded up some 600 "spies" in the Great Lakes area alone. It is probable that nearly all of the so-called spies were merely aliens of enemy country origin working in plants having Navy contracts and therefore considered vulnerable to enemy agent recruitment.[19]

At peak load, ONI was processing 1,000 names a day in its security checks, and its suspect list eventually reached a total of 105,000 names.[20]

On 19 November 1918, all branch intelligence offices were instructed by DNI RAdm. Welles to close their pending business by 1 December, if practicable, and to recommend which portions of their files should be turned over to an aid for information and which should be forwarded to ONI.[21]

During World War I, ONI had prepared identification cards for issue to operatives as required to assist them in their work. Cards were restricted principally to agents sent out directly from ONI, but some were issued for special cases at the request of district intelligence officers (DIO). DIO was the new designation for the former aids for information, effective 24 March 1919.[22]

Counterintelligence and Investigation Between the World Wars

In 1920, a morals scandal came to light in Newport, Rhode Island. Secretary of the Navy Daniels ordered an investigation, and Assistant Secretary Franklin Roosevelt reported that the matter had been assigned to ONI and then to his office. DNI RAdm. Albert P. Niblack had not wanted to handle a situation that was not his idea of intelligence. As he later told a congressional committee:

> One of the greatest things I have had to contend with has been to get ONI away from some wartime activities which grew with and had to do with enemy agents. I had absolutely nothing to do with the investigation, and I had refused to touch it. I have positive assurance from the Secretary of the Navy that ONI will not be required to do anything of that kind except in great emergency. In the main, my endeavor has been to get back to the old fashioned system with a naval attaché who is a member of the diplomatic corps and who conforms to all the conventionalities.[23]

In the general field of security, ONI had responsibility for the security of naval information during the 1920s and 1930s. However, one of the chief areas of danger, lax control of communications, was the responsibility of the Assistant Director of Communication Security (OP-20-G), whose activities involved checking on violations of regulations in the coding of messages. The only cases referred to ONI were those in which there was a question of classification. The procedure served no useful purpose, since ONI had no authority to determine classification, which was the responsibility of the originator of the document.

Director of Naval Communications (OP-20), also concerned with security violations in the handling of registered publications, initiated a survey of security conditions in the offices of U.S. naval attachés. The survey established that, quite apart from the possibility of attaché office safes being burglarized, all commercial communication companies retained copies of dispatches sent and received by U.S. attachés. It was thought that at several capitals, including London, Paris, and Tokyo, all such messages were being routed through local government offices.

In May 1920, Secretary Daniels described ONI's security and counterespionage functions as "war activities and not previously recognized as legitimate functions of that office." At the same time he assured congressional investigators that "the naval appropriation bill for next year restricts the activities of the ONI in the matter of collecting information at home and places the office on its original footing prior to the War." There seems little doubt that the sudden demise of the Counterintelligence Branch (B-Branch) was in part the result of the understandable hostility of Congress toward a naval secret service.

It was not possible, however, to bring to an end all the activities previously performed under the jurisdiction of B-Branch; it was still necessary to provide for the security of the naval establishment, including investigation of suspected violations of security regulations. The Navy as a whole, however, was little interested in security, and the investigations requested were both infrequent and trivial. Moreover, the surviving organization was so small that the response to inquiries undertaken could not always be satisfactorily completed. There is little evidence that ONI itself originated any measures for security during the post–World War I period.

For a satisfactory solution, every security problem that came to ONI for attention required a trained investigative agency that ONI didn't have. Theoretically, ONI had at its disposal the various district intelligence officers and their organizations, as well as the inspectors of naval material. It could also call on the investigative agencies of other government departments for assistance. Liaison with other departments was indirect and usually faulty, and the district organizations were inadequate. Not every district had an intelligence officer, and in those that did, the officers were assigned other duties and were often not qualified or trained for investigative work. DIOs were supposed to make use of reserve officers for investigations, but the arrangement, which was intended to provide useful experience for the Reserves, was generally unsatisfactory in its results.

The most useful agents available to ONI within the Naval Establishment were the inspectors of naval material, who sometimes reported directly to ONI and at other times to their respective technical bureaus.

The disproportion between ONI's responsibilities and resources was compounded by ONI often being called upon to perform investigations that had no connection with security.

When a matter requiring investigation was brought to the attention of ONI, the normal procedure was to refer it for investigation to the commandant of the naval district in which the incident occurred. The commandant was to use whatever resources he had available and as he saw fit. There apparently was no follow-up on the referrals. Those conditions continued throughout the post–World War I years until 1935.

All communications between ONI and other departments had to be transmitted between the Secretary of the Navy and the head of the other department involved. Furthermore, letters to the Attorney General, which included all those to the FBI, had to be routed via the Navy's Judge Advocate General.[24]

In ONI in 1921, there was a domestic section for "counterespionage" under Cdr. Royal E. Ingersoll. Ingersoll also had the Japanese Espionage Desk, whose principal activity was to follow visiting Japanese. The Russians were considered to be no particular problem at that time. Although ONI was aware of the possibility of Soviet efforts to subvert Navy crews, the problem had not reached the serious proportions it did later.[25]

ONI was keeping track of Japanese activities in South America and the Panama Canal Zone in the early 1930s. Much of the information on Japanese activities in the Canal Zone was collected by the U.S. naval attaché in Buenos Aires. Close cooperation in regard to Japanese activities was also maintained with the FBI. J. Edgar Hoover and Ingersoll frequently exchanged visits and information.[26]

The Commander in Chief, U.S. Fleet, in his annual report for Fiscal Year 1933, noted the problem of Communist literature being distributed to fleet personnel at U.S. West Coast ports. To locate and collect Communist handbills left on mess tables and lockers in the crew's quarters, it became standard practice to search ships following visiting periods for the general public.[27]

To provide detailed information on espionage techniques and how to counter them, ONI, on 18 February 1935, published ONI-22, *Notes on Espionage, Counter-Espionage and Passport Control*. It was a secret, registered publication issued under the signature of RAdm. Joseph K. Taussig, Assistant Chief of Naval Operations. ONI illustrated the need for counterespionage investigations with the example of an investigation being conducted during February 1935 of two crewmen on board one of the newest U.S. Navy cruisers. The crewmen were accused of selling information about the ship to Japanese agents for $500 during the cruiser's shakedown cruise.[28]

Preparing for Wartime Counterintelligence

As early as 1936, ONI began assigning a few officers for training in investigative work. Arrangements were made for naval officers to attend the FBI school, but the number trained was pathetically small: four each in 1936, 1937, and 1938, and two in 1939. In June 1939, the need for expanded investigative resources was suddenly recognized and led to the formulation of a plan calling for four types of investigative personnel: special agents, agents, investigators, and special employees. The first three categories were to be filled by category I-V(S) Naval Reserve officers who had had the requisite training for investigative work; the fourth category was to be filled by civilian experts (toxicologists, chemists, etc.) needed in connection with special types of investigations. (Women could be employed in this category, but their number was to be kept to a minimum.)

In 1939, it was estimated that 209 persons (179 commissioned and 30 warrant officers) would be required to staff the Naval Intelligence Investigative Service (NIIS) upon mobilization. NIIS was to be a completely separate agency subordinate to the Director of Naval Intelligence. In September 1939, however, the idea of a separate investigative service was discarded, and district commandants were instructed that personnel who had already been assigned to NIIS for mobilization purposes were to be absorbed by the Investigative Section (B-3) of the district intelligence office or Section B-3 of ONI.

In October 1940, after considerable discussion, it was decided that there would be only two classes of operatives—agents and special agents—and that they could be either officers or civilians. Civilian agents were to receive an annual salary of $1,500 to $3,600, and civilian special agents were to be paid $1,800 to $4,500. The low pay scale soon proved to be inadequate, and all five civilian agents hired between 20 June and 4 September 1940 terminated their contracts for "more remunerative positions." Accordingly, it was decided to revert to the plan of using Naval Reserve officers. In December 1940, six I-V(S) reservists who had had investigative experience in civilian life were ordered to active duty.[29]

Counterintelligence in the late 1930s was deemed by ONI to require close and cordial relations with the various "patriotic" societies in their efforts at combating persons whom they believed to be conducting subversive, pacifistic, and defeatist activities. Radical elements were increasing their efforts to subvert naval personnel, and foreign espionage continued to increase.

German espionage activities against the United States prior to Pearl Harbor consisted primarily of building up the German intelligence data base to replace the out-of-date files still retained from World War I. Much of the Nazi collection effort was performed by the German foreign trade offices. German intelligence also interviewed returning German businessmen and merchant marine sources. There is no indication that Germany had much success in establishing espionage networks in the United States, but extensive networks *were* established in Mexico, Brazil, Argentina, and Chile to relay information on Allied shipping, and the work resulted in some ship losses. Most of the prewar activities were carried on by *Abwehr* agents.[30]

The Counterintelligence Branch (OP-16-B) in 1939 was organized into the following sections: Naval Censorship (B-2); Investigations (B-3); Security of Naval Information (B-4); Commerce and Travel (B-5); Sabotage, Espionage, and Counterespionage (B-7); and Coastal Information (B-8). The specific tasks of the B-Branch were to determine (1) enemy plans and organizations for espionage and sabotage; (2) the kind of information and intelligence the enemy was getting; (3) the kind of information and intelligence the enemy needed and especially wanted; (4) The connections or channels between the legitimate and proper sources or custodians of information and intelligence and the enemy's intelligence organizations; (5) the methods used to transmit such information and intelligence to the effective enemy destination; (6) the personnel, organization, and methods used by, or available to, the enemy for sabotage directed against the U.S. Navy, including propaganda; (7) the plans and methods for denying information and intelligence about U.S. naval war operations to the enemy and for preventing interference with those operations by the enemy; and (8) the dissemination of intelligence on (1) through (7) above to the proper action agency or agencies, with recommendations for countermeasures.[31]

The first official action taken to resolve overlapping functions and conflicting jurisdiction over national counterespionage activities was a confidential memorandum from President Franklin Roosevelt, dated 26 June 1939 and addressed to the Secretaries of State, Treasury, War, Navy, and Commerce, the Attorney General, and the Postmaster General. It declared that "the investigation of all espionage, counterespionage and sabotage matters is to be controlled and handled by the FBI of the Justice Department, MID [Military Intelligence Division] of the War Department, and ONI of the Navy Department. The Directors of these three agencies are to function as a committee to coordinate their activities."[32]

On 6 September 1939, the President issued a formal statement that instructed the FBI to

> take charge of investigative work in matters relating to espionage, sabotage and violations of the neutrality regulations. This task must be conducted in a comprehensive and effective manner on a national basis, and all information must be carefully sifted out and correlated in order to avoid confusion and irresponsibility. To this end, I request all police officers, sheriffs, and all other law enforcement officers in the U.S. promptly to turn over to the nearest representative of the FBI any information obtained by them relating to espionage, counterespionage, sabotage, subversive activities and violations of the neutrality laws.[33]

Counterintelligence During World War II

During the 1939–1942 period, the work of ONI's Latin American Desk (OP-16-FL) focused on counterintelligence and its related activities. Not only were there believed to be approximately 2.5 million Axis-origin aliens residing throughout Latin America, but their well-entrenched influence upon the political, social, economic, and military institutions of the Latin American republics created the single largest obstacle to effective cooperation by those republics with the United States.[34]

The first formal agreement delimiting the responsibilities for investigation of all espionage, counterespionage, sabotage, and subversive activities was titled "Proposal for Coordination of FBI, ONI and MID." It was dated 5 June 1940 and was signed by J. Edgar Hoover, RAdm. Walter S. Anderson, and BGen. Sherman Miles, USA, as the heads of the three agencies involved. The FBI assumed responsibility for all investigations of cases involving civilians in the United States and in U.S. territories except the Panama Canal Zone, Guam, Samoa, and the Philippine Islands. The FBI was also responsible for cases "directed from foreign countries on those occasions and in those situations in which the State, War or Navy Departments specifically request investigation of a designated group or set of circumstances." ONI assumed responsibility for investigation and disposal of all cases in the naval establishment, including civilians under naval employment or control and all civilians in Guam and American Samoa. The Army's MID assumed responsibility for investigation and disposal of all cases in the military establishment, including civilians employed on military reservations or under military control, and for cases involving civilians in the Canal Zone, the Republic of Panama, and the Philippine Islands. The joint FBI/ONI/MID agreement declared that "responsibility assumed by one organization in a given field carries with it the obligation to provide a pool of all information in that field, but it does not imply the responsible agency is interested in or will work alone in that field. Close cooperation between the three agencies in all fields is a mutually recognized necessity."[35]

On 8 January 1941, a CNO letter to all district commandants forwarded a 12 December 1940 supplement to the Delimitation Agreement of 5 June 1940. The supplement instructed the field services of the three intelligence agencies to "maintain close personal liaison between those offices and their representatives," to include "a meeting of representatives of the three agencies, preferably the O-in-Cs [officers in charge], at least once a week, for the purpose of discussing pending and contemplated investigative activities and any other subjects necessary to insure that there is proper coordination of their investigative work." The personal liaison at all times was to "insure that there is no duplication of effort in any field and that a proper coverage of the whole investigative field is maintained. Particular attention should be paid to avoiding any duplication in connection with the use of informers."[36]

At a meeting in the office of the Secretary of the Navy on 19 May 1942, RAdm. Stanford C. Hooper (OP-14) discussed the danger of Communist party "cells" in the transportation and communications industries and in the armed services. He also pointed out that it was time to prevent formation of such cells and to eliminate those already formed. The temporary military alliance with the USSR was not justification for condoning the establishment of such cells in the United States. A change in the international political situation might occur at any time without advance notice, at which time it would be too late to abolish the Communist cells. A decision was needed as to whether or not the Departmental Qualification Board for Commercial Radio Communications Personnel should continue to disapprove employment of Communist radio operators and whether Communists should be accepted as members of the Defense Communications Board Committees. Secretary Knox reportedly replied that he held no brief for the activities of the Communist party, but President Roosevelt had stated that, considering the United States and Russia were allies at that time and the U.S. Communist party's efforts were now bent toward winning the war, the United States was bound not to oppose the Communist party activities and, specifically, not to disapprove the employment of any radio operator for the sole reason that he was a member of the party, or that he was active in party affairs. The Secretary further stated that this was an order and must be obeyed without mental reservations.

RAdm. Adolphus Staton from the office of the Under Secretary of the Navy then said that, in view of the perceived change in policy, the instructions to the Departmental Qualification Board should be modified. At the time, the instructions stated, "The Board will determine whether such service by the person concerned would be detrimental to the national defense and national safety." Staton said that the board members could not bring themselves to believe that the employment of militant Communist party members as radio operators would not be detrimental to national safety. Secretary Knox agreed to take up the change to the instructions with the Judge Advocate General.

Director of Naval Intelligence RAdm. Theodore S. Wilkinson asked the Secretary if membership in the Communist party constituted a general "whitewash" for all sorts of illegal and other subversive activities. The Secretary answered that it did not.

Knox repeated the order from the President and reiterated that it must be obeyed by all officers without mental reservation. RAdm. Hooper replied that, in the Navy, an order from a superior officer was always obeyed without mental reservation. However, upon receipt of the order (which he considered ill-advised) he had felt it his duty to say so.[37]

In 1942, the policy on the use of confidential informants was expressed in part as follows:

> As a general rule, before using informers, their loyalty and general reliability should be ascertained by an appropriate investigation. When such basic qualifications cannot be checked, information obtained from them should be accepted with reserve. Caution must be exercised regarding their motives, to assure that they don't spring from a grudge or a desire to inflict damage on a competitor.
>
> Informers may be volunteers or work for pay. No person in the Naval Service shall ever be given extra pay as an Informer but may be reimbursed for actual, necessary and extraordinary expenses incurred in obtaining or transmitting information. As a general rule, information should be paid for only on a C.O.D. basis, after verification. Payment of a regular salary to an Informer is a waste of funds.
>
> A signed receipt should be obtained and filed for each payment to an Informant. An alias or other designator may be used, but fingerprints should be obtained from these receipts even when the Informant's real name is not known.[38]

On 9 February 1942, the original Delimitation Agreement was revised. The spheres of responsibility for the FBI, MID, and ONI remained essentially the same, except that ONI now had added responsibility for cases involving civilians on Palmyra, Johnston, Wake, and Midway Islands. MID and the FBI were also responsible for additional territories, including Alaska. New paragraphs covered conditions for operations under a "Period of Martial Law," and "Periods of Predominant Military Interest Not Involving Martial Law," and "Periods of Normal Conditions." The agreement remained in effect without further revision throughout World War II.[39]

Some friction developed between district intelligence officers and the representatives of the other wartime intelligence agencies, especially with FBI field agents. On 9 December 1942, the heads of MID, ONI, and the FBI issued a joint letter to the field offices of the three agencies, calling attention to the Delimitation Agreement as an instrument intended to benefit each of the subscribing agencies and devised to eliminate friction. Representatives were urged to apply its terms in a sensible manner, and they were warned that any attitude other than a cooperative one would not be tolerated. Field relations between FBI and ONI subsequently improved.

The first official liaison between the respective counterintelligence sections of ONI and MID was established early in December 1942. Three Army officers were detailed from MID for duty in the following sections of ONI's Domestic Intelligence Branch: Investigations (B-3), Commerce and Travel (B-5), and Sabotage, Espionage, and Counterintelligence (B-7). Also, three naval officers from those sections were detailed to the Army counterparts of those sections. Although the B-7 liaison was discontinued in early 1944 and the one with B-5 in March 1944, liaison with B-3 existed until the end of the war, with a naval representative on duty in MID for the duration.[40]

In 1943, to simplify classifying and identifying investigative cases by types, the following designators were prescribed:

I. Personnel Investigations
  (a) Service Personnel
  (b) Civilian Personnel
  (c) Applicants (service and civilian)
  (d) Private contractors' employees working in Naval Establishments
II. Sabotage Investigations
III. Espionage Investigations
IV. War Fraud Investigations
V. Investigation of Naval Contractors
VI. Miscellaneous Investigations[41]

The mission of ONI's Case History Section (OP-16-A-7) was to establish and maintain a central file serving all branches and sections of ONI. OP-16-A-7's files contained information on naval and civilian personnel, the heads and executives of business organizations, leaders of various groups (political, possibly subversive, etc.), foreigners of naval interest, and others. The files were used primarily for counterintelligence purposes. The Russell "Soundex" system of indexing was used, as it was adaptable to Japanese names as well as to variations in the spelling of all names. The types of information maintained included case histories that contained a minimum of four evaluated items of significant information on an individual or organization; a visible index of names on which fewer than four items of documented information existed; category files covering various subjects for convenience; and files in a transferred status. In addition to ONI, some twenty to twenty-five outside agencies used the files each month during World War II. During a typical twelve-week period, over 120,000 name checks were made, with an average search time per name of less than six minutes.[42]

Post–World War II Period: Counterintelligence Retained

At the end of World War II, in order to avoid repeating the negative and confused situation that followed World War I relative to ONI's investigative responsibilities, Secretary of the Navy James V. Forrestal issued a letter dated 1 November 1945 to all ships and stations:

> Naval Intelligence personnel are currently authorized to conduct investigations of naval personnel and civilians under naval control in cases of actual or potential espionage, sabotage, or subversive activities, and in those cases which relate to the security of classified naval information.
>
> The investigative jurisdiction of the naval intelligence organization is hereby broadened to permit the use of naval intelligence personnel and facilities to investigate:
>
> (1) Naval personnel
>
> (2) Civilians under purely naval administrative control
>
> (3) Matters under purely naval administrative control in cases not specifically and exclusively within the investigative jurisdiction of other Government Departments or Agencies and subject to the limitations set forth [herein].
>
> Authority to administer, operate, and maintain an investigative service for the Navy to accomplish the purpose outlined in this letter is hereby assigned to the Chief of Naval Intelligence acting under the Chief of Naval Operations.
>
> Investigations . . . shall not be undertaken except on specific request to the Chief of Naval Operations or a District Commandant by competent naval authority.[43]

Experience had shown that the Navy might have an interest in many organizations, groups, trends, and situations that, when they first attracted attention, did not have any discernible immediate naval interest. Thus, the policy in the immediate postwar period was that information for reference and background be compiled by the Sabotage, Espionage, and Countersubversion (SEC) Section (OP-23D4), particularly information about organizations that solicited naval personnel for membership. OP-23D4 was soon renamed the Counterintelligence Section, but it continued to be referred to as SEC on ONI rosters.

Naval authorities were to be advised of threats or dangers to the Naval Establishment by three methods: official communications by dispatch, letter, or memorandum to the appropriate naval command to give warning of any specific immediate danger or threat; periodic studies summarizing subversive trends, to be disseminated usually to the district intelligence officers and, in any case, to commands concerned within the naval service; and special topical studies issued from time to time on special problems. The special studies might be disseminated outside the naval service, depending upon the nature of the study.[44]

With the broadening of ONI's investigative jurisdiction in 1945, the demands made upon the naval intelligence investigative organization steadily increased. This led to the issuance of another letter signed by Forrestal in March 1947, which stated:

> At present there are two classes of cases which are of particular urgency. They are: (1) subversive investigations, and (2) investigations to determine the loyalty of Naval employees and applicants. Other investigations, not of a direct intelligence interest, must be subordinated under present working conditions to permit concentration on those types which directly affect the security of the Naval Establishment.
>
> The need for adequate security of the Naval Establishment is paramount; consequently, in the best interests of the service, requests to Naval Intelligence for investigations of a direct non-intelligence nature must be cleared with the Commandant of the Naval District concerned to assure that the intelligence organization of that district can assume such investigations without jeopardizing the completion of its other work.[45]

A Special Observer–Merchant Marine (SOMM) Plan was issued by a Chief of Naval Operations letter in May 1947 to place informants on U.S. registry merchant ships on foreign runs for the purpose of identifying crew members suspected of subversive activities. The plan had been coordinated with, and formally approved by, the FBI. Implementation was primarily through the district intelligence offices, but naval attachés could become involved.[46]

Executive Order 9835, issued in 1947, established a loyalty program within the federal government. It provided that the FBI check its records for each incumbent employee. If derogatory information from the standpoint of loyalty was uncovered, an investigation would be made.[47] At a Cabinet meeting on 30 October, President Harry S. Truman emphasized that Executive Order 9835 did not mean that a full investigation should be made of every U.S. Government employee. He also pointed out that any department, within the limitations of its organization and funds, could make whatever investigation of employees it considered necessary, but that once evidence of disloyalty was uncovered, the investigation was to be placed in the hands of the FBI, the only agency empowered to conduct loyalty investigations. No funds had been appropriated for any department or agency other than the FBI to conduct loyalty investigations.

As of February 1948, ONI had a backlog of 15,000 investigations pending. This backlog was increasing at the rate of 850 investigations per month. ONI's investigative jurisdiction included all personnel, civilian and uniformed, of the Naval Establishment for any purpose connected with security or the detection of crime. It did not include the President's loyalty program. When an ONI investigation of a Civil Service employee uncovered a suspicion of disloyalty, the FBI was notified and took over the case in accordance with the President's directive.

Several factors stimulated the increasing demand for investigations. Paramount was the atmosphere of suspicion in which the Soviet Union forced the free world to live. Commanding officers consequently demanded investigations of more people than they had in the past. The nation and the Navy became very security conscious. Also, because the weapons being developed for and by the Navy were becoming more complex, the employment of individuals with unique scientific abilities but obscure backgrounds was required. U.S. Navy research authorities were impatient with the delays in research caused by the lengthy investigative process and the excessive backlog.[48]

Beginning in July 1948, ONI used the polygraph as an investigative technique. From 1948 to 1951, ONI polygraph operators were trained at the Leonarde Keeler School in Chicago, but from July 1951, ONI conducted its own training course. Individuals selected by ONI for polygraph training were authorized by the Director of Naval Intelligence to conduct polygraph examinations (but not to provide training in polygraph operation) when the technique was applicable to investigations conducted by ONI. Operators who received such authorization might be required to requalify periodically at ONI to retain their authorizations, which could be withdrawn by the director at any time.[49]

The categories of investigations specified in 1943 were modified on 5 April 1950 to give a more accurate reflection of the scope of the investigative work being conducted by Naval Intelligence. The category designations were:

Class 1. General Background Investigations
- 1(a) Service Personnel
- 1(b) Civilian Personnel
- 1(c) Applicants
- 1(d) Employees of Contractors with Classified Naval Contracts
- 1(e) Private Contractors

Class 2. Sabotage

Class 3. Espionage and Subversive Activities

Class 4. Fraud, and Conspiracy to Commit

Class 5. Security Investigations
- 5(a) Compromise, Leakage or Unauthorized Disclosure of Classified Information
- 5(b) Loss of Classified Matter
- 5(c) Surveys (of space, equipment, security procedures, etc.)

Class 6. Criminal Investigations
- 6(a) Arson
- 6(b) Assault
- 6(c) Black Market Activities
- 6(d) Forgery
- 6(e) Homicide
- 6(f) Impersonation
- 6(g) Mutiny
- 6(h) Narcotics, Customs, Postal, Currency Violations
- 6(i) Perjury
- 6(j) Sex Offenses
- 6(k) Theft and Embezzlement
- 6(l) Other Offenses

Class 7. Special Investigations
- 7(a) Reciprocal
- 7(b) Special Inquiries.[50]

Korean War

With the outbreak of war in Korea in June 1950, the Navy had to make a sudden shift from reductions in personnel to rapid expansion, putting a severe strain on the Navy's personnel security program. Security clearances for the majority of inductees being placed in sensitive billets had to be based on file checks instead of background investigations. The reduced effort represented a compromise of policy.[51]

The counterintelligence and investigative efforts of the U.S. Navy in the Pacific were a function of command. At such activities as Naval Forces, Japan, and Naval Forces, Philippines, the Assistant Chief of Staff for Intelligence developed an investigative capability within his organization. At the start of the Korean War, Commander Naval Forces, Far East (COMNAVFE) at Yokosuka established unit intelligence offices at Naval Air Station (NAS) Atsugi, NAS Iwakuni and Naval Base, Sasebo. Shortly thereafter, Special Agent Douglas T. Wada of the District Intelligence Office, 14th Naval District (DIO-14ND), who held a commission as a lieutenant commander in the Naval Reserve, was ordered to temporary active duty at COMNAVFE and thus became the first professional investigative officer to be assigned by the Navy for duty on foreign soil.

Soon after the Entry Clearance Program was established for the Trust Territory of the Pacific Islands in 1951, former Special Agent Ray Kinnery was transferred from DIO-4ND to the staff of Commander Naval Forces, Marianas, at Guam and became the first naval intelligence special agent assigned in the Pacific on a permanent basis. As requests for investigative assistance in the criminal field increased and investigation requirements for personnel security became more numerous, additional special agents were assigned to Far East naval activities. It became clear that offices similar to the traditional DIO would be needed in all overseas areas where Navy and Marine Corps personnel were assigned in appreciable numbers.[52]

Until 1951, ONI was responsible for performing naval investigative duties related to "Seat of Government" cases, and the Intelligence Office of the Potomac River Naval Command (PRNC) was responsible for investigations for PRNC field activities. In 1951, the Chief of Naval Operations directed that as much investigative work as possible be performed by naval intelligence field activities. Accordingly, ONI transferred the responsibility for conducting the investigation of cases originating at the Seat of Government to PRNC. These included cases such as those for bureaus, boards, and offices of the Navy Department, but not those for the Office of the Chief of Naval Operations (OPNAV) and fleet activities. The transfer was to be completed by the end of 1951 through an orderly shift of files and personnel from ONI to PRNC.[53]

Investigation and Counterintelligence in the Cold War Era

In 1953, consideration was being given to detaching the Investigative Section of the Security

Branch from ONI to make it a separate organization to handle all investigations. RAdm. Carl F. Espe, Director of Naval Intelligence at the time, expressed his views on the subject:

> Although it would be possible to place the investigative function of the Security Branch under the Inspector General in a manner similar to that in the Air Force, this would be inadvisable because of the close inter-relation between both positive and counterintelligence and the investigative effort. This relationship is on a continuing day-to-day basis, and if separated could not possibly be as effective. Additionally, such separation would complicate the administration and management of the DIOs, which are concerned with intelligence and counterintelligence as well as security.[54]

The proposed detachment received no further consideration at that time.

The Interdepartmental Committee on Internal Security (ICIS) was responsible, by charter from the National Security Council, for coordinating all phases of internal security except those functions that were assigned to the Interdepartmental Intelligence Conference (IIC) that related to the investigation of domestic espionage, counterespionage, sabotage, and subversion. One of the ICIS functions was to make recommendations on legislation, executive orders, and regulations related to internal security. On several past occasions the internal security implications of various national policies, treaties, proposed legislation, etc. had not been sufficiently considered in the executive branch, resulting in subsequent difficulties in the handling of such matters.

The Navy member of IIC was the Director of Naval Intelligence, and the head of the Security Branch of ONI (OP-321) was the Navy member on the working committee of IIC. There was no Navy member on ICIS, but OP-321 was designated as the Navy liaison officer to the Department of Defense representative on ICIS.[55]

Articles containing information considered to be of a classified nature were being published in the U.S. press, magazines, and other news media with alarming regularity in the early 1950s. It was apparent that such classified information was often being published without prior clearance by appropriate military public information agencies. Quite possibly it was being leaked to reporters from some high-level person. In order that the Director of Naval Intelligence could evaluate possible violations that were of concern to the Navy and advise higher authority about the perceived violations, a consolidated file on such matters was maintained by ONI's Security Control Section (OP-321K). No active investigation of leaks, however, was to be undertaken without specific authority of the DNI.[56]

SECNAV Instruction 5430.13 of 18 July 1953 restated in broad general terms the Secretary's letter of 1 November 1945 that had permitted the use of the investigative facilities of naval intelligence in any matter in which there was a naval interest and not within the exclusive investigative jurisdiction of another government department or agency.

The 1945 letter had lacked a statement of specific policy concerning the types of investigations for which it would be desirable to utilize naval intelligence facilities. Such a statement was necessary by 1954 in order to confine the investigative efforts of naval intelligence to matters requiring the application of professional investigative techniques and to eliminate those investigations susceptible to administrative solution within a command. Accordingly, SECNAV Instruction 5430.13A of 10 August 1954 directed that naval intelligence investigative facilities be used in, and limited to, the following cases: those involving actual or potential espionage, sabotage, or subversive activities, as prescribed by the Delimitation Agreement of 23 February 1954; matters pertaining to fraud against the government; major violations of the Uniform Code of Military Justice; personnel background investigations; security investigations; and special investigations comprising those cases containing unusual circumstances or aspects of sensitivity that might require specialized techniques and a high degree of discretion.[57]

The mission of the Investigation Branch by 1954 was to provide a centrally directed and controlled investigation service for the Naval Establishment. Its operations were entirely overt and were carried out by accredited agents. The headquarters organization of the Investigative Branch monitored the activities of the field units located in each naval district and at COMNAVFE at Yokosuka; Commander Naval Forces, Philippines (COMNAVPHIL) at Sangley Point; Commander Naval Forces, Marianas on Guam; and Commander Naval Forces, Eastern Atlantic and Mediterranean in London.[58]

The greatest single deficiency in the counterintelligence field in 1954 was the lack of worldwide assets to protect the Navy from sabotage, espionage, and subversion. Personnel assigned to duty in counterintelligence were restricted, for the most part, to investigations of physical security and personnel security. The Army and Air Force had previously assisted the Navy to the extent practicable but were at that time obliged to curtail their help because of their own redeployed and reduced resources.

In many areas, the Navy's $13 billion-a-year overall budget investment was almost wholly dependent overseas upon the CIA for counterintelligence protection. This was not a satisfactory situation. In Korea, where the CIA had overriding long-range commitments, the agency was not able to provide the military services with the tactical and strategic counterintelligence assistance required by the operational commanders.[59]

In 1955, "wiretapping" was the term commonly used to designate the procedure of employing electronic, mechanical, or any other means for the surreptitious interception of telephone traffic or traffic carried over common carrier radio facilities. A clear distinction existed between wiretapping and the installation of a microphone unconnected in any manner to the telephone system. Evidence obtained through the latter technique was normally admissable in court if no offense (such as trespassing) had been committed while the microphone was being emplaced.

ONI policy permitted the district intelligence officers to exercise their own discretion in determining whether to use microphone installations. Referral to the Director of Naval Intelligence was necessary only in cases of peculiar sensitivity or importance. The governing criteria were as follows: orthodox investigative techniques that were clearly not sufficient, and information being sought that was necessary to the resolution of a case; a case of sufficient importance to justify the use of the technique, and a gain expected to outweigh any risk involved; and the availability of personnel thoroughly trained and experienced for use in the installation of the equipment and the supervision of the operation.

When a DIO believed that an investigation could be resolved only through wiretapping, the matter was referred to the Director of Naval Intelligence. Normally, approval to proceed with the operation would not be given except in cases of importance relating to national security.

All correspondence relating to microphone installations and wire tapping was to be classified secret. These installations were not to be referred to as such in reports of investigations. Rather, the sources of information obtained from clandestine intercept methods were to be referred to as a "confidential informant of known reliability."[60]

In 1956 the Interdepartmental Intelligence Conference coordinated national counterintelligence. Its members were Director of the FBI, Assistant Chief of Staff for Intelligence (U.S. Army), Director of Naval Intelligence, and Director of Special Investigations (U.S. Air Force). The IIC reported directly to the National Security Council and was charged with coordinating the investigation of all domestic espionage, counterespionage, sabotage, and subversion, and other related intelligence matters affecting internal security.[61]

Under the general category of sabotage, ONI completed thirty-one investigation cases in Fiscal Year 1956, sixty in Fiscal Year 1957, and fifty-seven in Fiscal Year 1958. In none of the cases was any underlying inimical foreign influence discovered. Motivation for the acts stemmed from a variety of causes, including mental derangement, intoxication, disgruntlement, and a desire to delay the sailing of a ship. Regardless of the reasons, the net effect of the sabotage caused serious damage to naval vessels and property, impeded scheduled operations, and, in some instances, endangered the lives of others.[62]

In 1957, a survey was made by ONI to determine the status of the Special Observer–Merchant Marine Plan that had originated in 1947 for the purpose of collecting counterintelligence information in the maritime industry. The survey indicated that in most of the naval district intelligence offices a formal SOMM Plan collection program was not being implemented. Among the reasons for this situation were inadequate personnel resources, lack of response on the part of designated observers, cumbersome administrative procedures, and problems connected with maintaining repeated contacts with highly mobile sources. The SOMM Plan was formally canceled in April 1960.[63]

In other U.S. counterespionage efforts, the Technical Surveillance Countermeasures Committee (TSCC), established by the National Security Council pursuant to the National Security Act, conducted a continuing study and review of those domestic and foreign threats to the security of classified matter posed by clandestine technical surveillance devices installed in quarters or facilities used by U.S. Government personnel.

TSCC was composed of representatives of the Secretary of State, Secretary of Defense, Director of Central Intelligence, Director of the FBI, Army Assistant Chief of Staff for Intelligence, Director of Naval Intelligence, Director of Air Force Special Investigations, Marine Corps Assistant Chief of Staff (G-2), and the Director of the National Security Agency. Through the operations of TSCC, policies designed to counter the threat from clandestine technical surveillance devices were established and coordinated to ensure a continuing exchange of information and to ensure that uniform procedures would be followed upon discovery of such devices.

In 1957, the responsibilities of ONI in the collection, production, and dissemination of intelligence on espionage, sabotage, and subversion were viewed as separate and distinct from the direct action responsibility of Navy commands to frustrate threats to the security of the Department of the Navy. In the latter regard, the mission of the Director of Naval Intelligence extended only to the development and promulgation of policies for the protection of classified matter.

The responsibilities of naval intelligence with respect to information of counterintelligence value were considered fulfilled when the information had been collected, evaluated, and disseminated to the authorities responsible for taking action. ONI's responsibilities for security policy development did not extend to administering the security functions within the various commands.[64]

At the end of 1957, ONI was assigned the responsibility for conducting background investigations of military personnel whose assignments placed them in the immediate vicinity of the President and First Family. These investigations had been conducted by the U.S. Secret Service, but its limited investigative resources had resulted in unacceptable delays. The naval aide to the President, with the concurrence of the special agent in charge of the U.S. Secret Service, requested that the Chief of Naval Operations have the background investigation cases conducted by ONI. The CNO concurred and directed that such cases be given priority handling to ensure completion within sixty days. Camp David provided a big caseload because it was commanded by a Navy officer under the military command of the President's naval aide. Approximately 160 Marines and 15 Navy personnel were assigned there each year.[65]

Since 21 August 1946, a counterintelligence summary report, *Subversive Trends of Current Interest*, had been prepared on a weekly or monthly basis by the district intelligence officers to keep their respective commandants and the DNI informed of subversive activities discovered in their investigations. To develop and promulgate uniform guidance for the preparation and distribution of the *Trends* reports, ONI Instruction 03850.1 was issued on 9 September 1959. Matters to be covered in the *Trends* reports included shifts in Communist party tactics, policies and plans, Communist party position or utterances on the developing international situation; agitation and propaganda from all groups whose objectives were inimical to the national defense; subversive elements in strike situations (legitimate labor activities were not to be reported); attempts to capitalize on racial or religious problems; the growth and direction of "hate" and "front" groups; and new organizations that had subversive overtones and a membership containing individuals of counterintelligence interest.

A manual, *Administration of Naval Intelligence Agents*, was issued on 27 January 1960 as ONI Instruction 3820.71, superseding ONI-70-2. It established policies and procedures for the employment and administration of agents.

In 1960, ONI reiterated its requirement to collect and disseminate timely counterintelligence concerning the U.S. merchant marine. ONI Instruction 003821.1 of 21 April 1960, although canceling the SOMM Plan, encouraged the development and cultivation of the maximum practicable number of contacts within the maritime industry who could be knowledgeable on matters affecting shipping security. The objective was to identify those individuals who could provide information, particularly in wartime, on incidents of subversive activity within the maritime industry and to coordinate the information, as appropriate, with the FBI.

In the late 1950s, a group known as the Polaris Action Committee was established to subvert U.S. sailors and junior officers, particularly those in the crews of Polaris submarines, and to interfere with the launching and operations of nuclear submarines generally. CNO Adm. Arleigh A. Burke was especially concerned about committee activities in New London, where members were protesting, distributing anti-Polaris literature, and using boats or swimmers to obstruct submarines getting underway from the Naval Submarine Base. Martin Randisi, special agent at DIO-3ND, was called to Washington and instructed by DNI RAdm. Vernon L. Lowrance to develop counterintelligence defensive briefings for submarine crews at New London. After the Pentagon accepted Randisi's presentation, he went to New London every time a new crew was assembled and briefed both the blue and gold crews, the commanding officers, the officers, and the enlisted, separately. Using classified information on Soviet espionage interests, intentions, and methods of espionage and subversion, Randisi's program was well received and effective in forewarning the submariners of what abuse, harassment, threats, and solicitation they should anticipate, and what to do about them. As a consequence of the briefings, the antics of crowds against the crew of one of the first Polaris submarines to arrive at Holy Loch in Scotland were assessed as being less effective than they might have been.[66]

Prior to 1961, the Navy's investigative and counterintelligence effort was fulfilled within the naval districts in the continental United States, Puerto Rico, Alaska, and Panama by district intelligence offices. The offices were under officers in charge. Outside the districts, the effort was the responsibility of the Assistant Chief of Staff for Intelligence of the various fleet commanders.

In 1961, a break was made from the traditional practice. At the instigation of the Commander in

Chief, Pacific Fleet (CINCPACFLT), the counterintelligence and investigative resources throughout the Pacific Fleet area were separated from the various N-2 staff (Intelligence) elements (e.g., COMNAVPHIL, COMNAVFORJAP) and established as separate naval activities under officers in charge, each responsive directly to the local theater or force commander. A coordination and support group, Headquarters Naval Investigative Support Group (NISG), was established in Washington, D.C. Responsible militarily to the Chief of Naval Operations, the headquarters received its support and guidance through the Director of Naval Intelligence (specifically, the Assistant DNI for Counterintelligence and the Assistant DNI for Administration, the latter providing logistic, fiscal, and personnel support). The investigative and counterintelligence effort in the Commander in Chief, U.S. Naval Forces, Europe (CINCUSNAVEUR) area continued under the Assistant Chief of Staff for Intelligence of that command.[67]

SECNAV Notice 5450 of 28 March 1961 established the "activities of the U.S. Naval Investigative Support Organization." It included Headquarters NISG; NISG Pacific; NISA (Activity) Marianas, Philippines, Taipei (Republic of China), and Japan; and NISU (Unit) Atsugi, Sasebo, Yokohama, Iwakuni (Japan), Naha (Okinawa), and Subic Bay.

SECNAV Notice 5450 of 26 May 1961 redesignated the U.S. Naval Investigative Support Group to Headquarters NISG, 8th and South Courthouse Roads, Arlington, VA. Its mission was to supervise and coordinate criminal, counterintelligence, and special investigative services for all overseas naval activities; and to collect, analyze and disseminate information of counterintelligence significance.

DNI Lowrance had requested that the Secretary of the Navy create NISG. Cdr. Paul Mulvihill, USNR, did much of the initial spadework and was the first officer in charge of NISG Pacific and NISA Japan, based at Yokosuka. Capt. Francis (Frank) Klaveness was the first Director, Headquarters NISG in Arlington. On 10 October 1961, NISA Iceland was established, and NISU Saigon was activated in late 1962. An office was also set up in Guantanamo, Cuba, during that period.[68]

OPNAV Instruction 05450.96 of 8 December 1961 provided a statement of the mission and functions of the Headquarters Naval Investigative Support Group, and OPNAV Instruction 05450.97 of the same date provided a similar statement for the group. OPNAV Instruction 04000.59, also of 8 December 1961, prescribed the logistic support for the NISG.[69]

On a day-to-day basis, many formal and informal requests were being received at all ONI echelons for characterizations of organizations, evaluations of movements or trends at either end of the political spectrum, reviews of books, information about authors, and assistance in preparing and presenting programs on current Communist tactics. To emphasize that the mission of ONI included providing consultative assistance to Navy commands in evaluating counterintelligence information, ONI Notice 03850 was issued on 20 April 1962. It prescribed that such requests be answered by a record check, analysis, or inquiry conducted by counterintelligence specialists. The notice emphasized that decisions made by Navy commands on such matters might be subject to judicial review, congressional inquiry, discussions by the news media, etc., and therefore that "Naval Intelligence has an obligation to ensure that it gives command authority the best possible service with regard to factual data and the best possible advice regarding potential or actual inimical forces."

Organization and Reorganization in the 1960s

The Naval Investigative Support Unit established in Saigon in November 1962 was a subordinate element of Naval Investigative Support Activity, Philippines. Lt.(jg) Robert Siler, USNR, the first officer in charge of the unit in Saigon, was assisted by four enlisted investigators, a chief petty officer, and two yeomen for clerical and administrative support. Saigon NISU was responsible for meeting the investigative needs of the U.S. Naval Headquarters Support Activity, Saigon, and the U.S. Naval Advisory Group, Vietnam. The first special agent assigned to Vietnam, Robert M. Kain, arrived on 11 August 1963 and assumed duties as Senior Agent, Saigon.[70]

During August 1962, Headquarters Naval Investigative Support Group moved from Arlington to Curie Hall, near the Lincoln Memorial in Washington. Capt. Robert P. Jackson, Jr., became the second director of Headquarters NISG during the summer of 1962. Cdr. David C. Reid, USNR, relieved Cdr. Mulvihill as Officer in Charge, NISG Pacific in Japan on 22 August 1962.

From its creation, Headquarters NISG exercised for the Chief of Naval Operations (OP-921) the management control of the entire NISG overseas organization. This administrative structure caused no difficulty as long as OP-921 had the "second hat" as Director, Headquarters NISG. On 27 July 1963, Capt. J. O. Johnson, USNR, relieved Capt. Jackson as Director, Headquarters NISG, while Jackson remained as OP-921. The headquarters for NISG moved again in August 1963 to the Fairmont Building in Arlington.[71]

The Naval Investigative Support Group was renamed the Naval Counterintelligence Support Center (NAVCINTSUPPCEN) by SECNAV Notice 5450 of 31 October 1963. The name change was made because counterintelligence was considered a more descriptive title for the duties being performed.

The October SECNAV notice provided that the management control of the overseas NISG organization be exercised by the Chief of Naval Operations. Thus Director NAVCINTSUPPCEN was relieved of his management control responsibilities toward the overseas organization; NAVCINTSUPPCEN provided centralized counterintelligence services (as in the district intelligence officer system), as directed by OP-921. For the first time, OP-921 and his staff were free of routine and detailed tasks and thus were able to concentrate on overall management and directive responsibilities related to the whole investigative field organization; it included district intelligence offices and their subunits, intelligence security units (under the Intelligence Officer CINCUSNAVEUR), NAVCINTSUPPCEN elements overseas, and NAVCINTSUPPCEN itself.

Although enlisted investigators from the NISG/NCSG (Naval Counterintelligence Support Group) Pacific had made a considerable contribution to the counterintelligence and investigative needs of the fleet, an assessment of their cost, training time, short availability, and the intricacies of their "cover" led to the conclusion that a better return could be obtained through the use of professional civilian agents. In 1964, the Director of Naval Intelligence indicated his intention to replace the fifty-three enlisted investigators, as they were phased out, with thirty trained civilian agents.[72]

In order to provide competent authority with all data available on personnel of counterintelligence interest connected in any way with the Naval Establishment, indices were maintained by ONI (and by district intelligence offices to meet their local requirements) in the following categories:

1) Active/inactive/retired Navy and Marine Corps personnel;

2) U.S. merchant marine personnel;

3) Industrial personnel employed by facilities assigned to the Navy for industrial security cognizance;

4) American Red Cross personnel stationed at Navy and Marine Corps activities;

5) Civil Service personnel employed by the Navy and Marine Corps; and

6) Potential hostage (blackmail) situations (i.e., military and civilian personnel within the naval establishment having relatives in Communist-dominated countries).[73]

A study, conducted to determine the advisability of relocating the Naval Counterintelligence Support Group, Pacific from Japan to the Hawaiian area, concluded that close supervision of the counterintelligence support units in the Far East was not necessary. With increased involvement by the Defense Intelligence Agency and Commander in Chief, Pacific in counterintelligence matters, it would be advantageous to have the group headquarters personnel physically located with CINCPACFLT at Pearl Harbor.[74]

The relocation took place in July 1965 upon the relief of the incumbent officer in charge, Cdr. Reid, by Capt. Thomas L. Stevens. The latter was also assigned additional duty on the staff of CINCPACFLT as fleet counterintelligence officer. Special Agent Clyde J. Roach preceded NCSG Pacific to Pearl Harbor to arrange the procurement of space and equipment and was assigned during the interim period to CINCPACFLT for duty. Roach is believed to have been the first Navy civilian special agent assigned to the CINCPACFLT staff.[75]

The assassination of President John F. Kennedy in November 1963 and the publication of the Warren Commission's report generated a continuing interest in the prompt, accurate, and complete reporting of information on any and all Americans who had defected or had attempted to do so. ONI was under a standing requirement to report to the Department of Defense and, as appropriate, to the FBI and other government agencies and departments pertinent information on the defection of any person subject to Navy or Marine Corps jurisdiction. District and area intelligence officers were directed to assign a high priority to the handling of any cases under their jurisdiction that involved defection. Defection was defined as an attempt, successful or not, or a credible and clear demonstration of intent, by an individual to seek asylum in any foreign country.[76]

Naval Investigative Service Established

In May 1965, the Secretary of Defense, following an in-depth survey of Army, Navy, and Air Force investigative and security organizations, directed a reorientation of the Navy's approach to its counterintelligence requirements. Specifically, he directed that the commander of the Navy's investigative organization be the "Commander in fact as well as in name." Pursuant to the directive, the Secretary of the Navy, on 4 February 1966, issued his Notice 5450 creating the Naval Investigative Service (NIS), changing district intelligence offices and naval investigative support activities to naval investigative service offices, and naval investigative support units to naval investigative service resident agencies (NISRA), and concurrently disestablishing all DIOs, investigative support groups, etc. The notice also provided the basis for the transfer of the counterintelligence and investigative staffs of the fleet commanders to the Naval Investigative Service. The Director NIS was given direct command and control (including personnel and other resources) over the entire Navy investigating effort. The first director was Capt. J. O. Johnson, USNR. The new organization devoted its first year to establishing command lines and to rearranging fiscal and other organizational relationships.

In July 1966, Capt. Edward G. Rifenburgh relieved Capt. Johnson as Director Naval Investigative Service, and in July 1967, when the Naval Intelligence Command (NAVINTCOM) was created, NIS became a field element of that command, and the Director NIS was double-hatted as OP-92C (Counterintelligence Plans and Policy).

The mission for the Director NIS, under Commander NAVINTCOM, was expressed as follows: "to maintain, command, and operate a worldwide organization to fulfill the investigative and counterintelligence responsibilities of the Department of the Navy (except those combat-related counterintelligence matters within the functional responsibilities of the Marine Corps)." The personnel allowance for NIS in 1967 included 115 officers, 265 enlisted men, 684 Civil Service civilians, and 905 contract agents.

Several new resident agencies were created in close proximity to important naval commands. Agents were also assigned to San Miguel in the Philippines; Nea Makri, in Greece; Stuttgart, Germany; and Hong Kong. An agent was assigned to *America* (CV 66) for a six-month Mediterranean cruise; he was attached to Commander Sixth Fleet's staff to be on the spot for any investigative requirements. The afloat agent concept was so successful that Commander Sixth Fleet requested that an agent be assigned to each aircraft carrier in commission.[77]

Vietnam War Era

The Vietnam War created special intelligence requirements. In June 1965, another counterintelligence support office had been opened in Vietnam, at Danang. It was also a subordinate element of the Naval Investigative Support Activity, Philippines. When NIS was established in February 1966, the former naval investigative support units at Saigon and Danang became resident agents of the Naval Investigative Service Office (NISO), Philippines, with Lt. Joseph R. Dothard as the Philippines Representative, Vietnam.[78]

On 15 April 1966, U.S. Naval Forces, Vietnam was established, and on 1 June, NISO Vietnam was also set up, taking over the personnel and resources of the Saigon and Danang offices, with a commanding officer reporting directly to NISO Pacific Fleet. On 18 July 1966, LCdr. William H. J. Manthorpe, Jr., reported as the first commanding officer of NISO Vietnam. On 20 July, special agent K. W. Nickel was designated supervising agent, and special agents Charles A. Baldwin and Frank J. Elmonds were assigned tentatively as senior resident agents, Danang and Saigon, respectively. A year later, LCdr. William F. Brubaker, USNR, relieved LCdr. Manthorpe as Commanding Officer NISO Vietnam.

In response to an earlier request by the U.S. Naval Support Activity, Danang, there was an increase, by 1967, of seven special agents in the Danang office, which permitted setting up one-agent satellite offices at Camranh Bay, Vung Tau, and in the Mekong Delta area. The satellite offices provided expanded investigative support to the widely deployed naval forces within the Republic of Vietnam.[79]

OPNAV Instruction 5450.96B of 2 March 1966 stated the mission for the Naval Investigative Service Headquarters as, under the command, direction, and control of the Director NIS, to provide the necessary headquarters staff and operational support to assist in the execution of the mission of NIS.

OPNAV Instruction 5450.97A of 28 March 1966 then provided a statement of mission and functions for the Naval Investigative Service offices, the successors to the DIOs and naval counterintelligence support activities disestablished by the SECNAV notice of 4 February 1966. Included on a temporary basis (until officially relieved of the responsibility) were the following noninvestigative functions that had previously been performed by the disestablished offices and activities:

> 1. Collect, evaluate, and disseminate information and intelligence (other than counterintelligence) of value to the Operating Forces of the Navy, the Shore Establishment, and Marine Corps activities within the assigned geographical area;
>
> 2. Provide intelligence (other than counterintelligence) required by the Director of Naval Intelligence;
>
> 3. Provide the Sea Frontier Commander and subordinate commanders with operational intelligence as required;
>
> 4. Exercise technical guidance and direction in intelligence matters (other than counterintelligence) within assigned geographical area;

5. Perform assigned tasks relating to the disclosure of classified information;

6. Plan for and, when directed, initially operate telecommunications censorship within assigned geographical area until relieved by an appropriate telecommunications censor; and

7. Direct, supervise, and participate in planning for the Naval Reserve Intelligence and Censorship programs within the assigned geographical area.

The majority of the temporary noninvestigative functions previously retained by NIS offices were assigned to the district staff intelligence officers by OPNAV Notice 5450 of 17 March 1967, with the proviso that NIS offices would provide certain essential help on an interim basis. Those residual intelligence functions not transferred had been, or were in the process of being, assigned to other elements of naval intelligence.[80]

To ensure the coordination and integration of counterintelligence, investigative, and security policy and program matters within the Department of the Navy, and to establish a focal point for such matters, both for intra- and extra-Navy relationships, SECNAV Instruction 5500.28 was issued on 5 June 1967. It first quoted paragraph 6.c. of General Order 511: "the Chief of Naval Operations, under the Secretary... shall, except for those areas wherein such responsibility rests with the Commandant of the Marine Corps, exercise overall authority throughout the Department of the Navy in matters essential to naval military administration, such as security [and] intelligence." The instruction then pointed out that the Chief of Naval Operations had assigned staff responsibility for counterintelligence (an aspect of intelligence) to the Director of Naval Intelligence, who was responsible for the development of various counterintelligence and security plans, policies, programs, and operations, and for relationships with other counterintelligence and investigative agencies. Accordingly, it was directed that

(1) All security policy proposals, recommended program changes, etc. would be referred to the Director of Naval Intelligence, who would coordinate the effort to develop a consistent and unified Navy position;

(2) The Director of Naval Intelligence would serve as the Navy point of contact with non-Navy agencies, and, therefore, would be the channel for the receipt and development of Navy comment on security policy and program proposals from outside agencies;

(3) The Director of Naval Intelligence would monitor the operation of all intelligence related programs and bring to the attention of the program sponsor any deficiencies he perceived, together with recommendations for changes; and

(4) The Director of Naval Intelligence would continue his current practice of providing appropriate information on threats to Navy security ascertained through Navy investigative and counterintelligence operations or otherwise.

It was known in 1967 that thousands of pieces of Communist propaganda were being disseminated annually through the U.S. postal system, and that a considerable amount of the subversive literature was addressed directly to naval personnel both within the United States and overseas. Commands that received such material were instructed to forward it to the nearest representative of the Naval Investigative Service, along with a report of the circumstances involved.[81]

A growing number of vendors with international criminal records had established laundry, tailor, restaurant, and other businesses near U.S. Navy installations and at ports frequented by U.S. Navy ships throughout the world. Their unethical methods of doing business and their known association with narcotics traffickers, Communists, or other subversive elements made their presence on or near Navy facilities and ports a security concern.

To assist commanding officers in their decisions on which vendors to authorize or deny access to their commands, ONI compiled a list of undesirable vendors and issued it as an enclosure to OPNAV Instruction 05510.96A of 29 March 1967. Most of the vendors were located in Hong Kong, Yokosuka, and Naples, but some had branch facilities in Norfolk and New York. Commanding officers were requested to forward to the nearest representative of ONI any new or additional derogatory information concerning vendors that should be added to the list.[82]

DOD Directive 5200.24 of 17 August 1967 set forth new policies and restrictions governing telephone interception and eavesdropping by DOD personnel. SECNAV Instruction 5520.2 of 25 November 1967 instructed the Under Secretary of the Navy to exercise overall supervision for the Secretary of the Navy with respect to telephone interception and eavesdropping matters and prescribed the responsibilities and actions required within the Naval Establishment to implement the DOD directive.

The new policy terminated interception or eavesdropping within the Department of Defense, except in special circumstances and subject to certain procedural requirements. The exceptions related to investigative and counterintelligence activities within the mission of the Director Naval Investigative Service, who was required to supervise and coordinate all operations involving wiretapping or eavesdropping within the Department of the Navy and could authorize such activities in those cases not requiring approval outside the Department of the Navy. He could not further delegate the duties and responsibilities.[83]

The Naval Investigative Service pioneered the application of automated data processing (ADP) techniques in the U.S. investigative and counterintelligence community, and the results of its efforts served as the model for ADP developments in related agencies. A high-speed communications system was developed and put into operation within the continental United States, and the necessary approval to expand it into a worldwide system by early Fiscal Year 1970 was obtained from the Assistant Chief of Naval Operations for Communications (OP-94), the Defense Communications Agency, and the Joint Chiefs of Staff.

Placing more agents at major shore installations and on board major fleet units in the Mediterranean, and authorizing them to accept investigative requests from, and report directly to, the commands served, significantly decreased the time needed for investigations. It also enhanced responsiveness to command requirements and decreased the incidence of security and disciplinary problems through the mere presence of agents. The agents briefed ships' crews and officers on danger spots in overseas ports, warned them of the hazards from the use of drugs and narcotics, and provided related crime prevention guidance.[84]

A NISRA was established in Sydney, Australia, and became operational on 20 February 1968 to support the rest and recreation program there for ships from the Vietnam operating area. Like NISRA Hong Kong, it was subordinate to NISO Philippines.

On 23 January 1968, when *Pueblo* (AGER 2) was captured by the North Koreans, Capt. C. Dale Everhart, commanding officer of Naval Investigative Service, Pacific (NISPAC), was designated by CINCPACFLT as the action officer for the preparation of plans for the intelligence debriefing of the *Pueblo* crew upon its release and return. The plan, activated on 23 December 1968, was completed by mid-January 1969 in what was judged to be a highly successful manner. Special Agent Clyde Roach received the Meritorious Civilian Service Award for his part in the preparation of the detailed planning, and NISPAC received the Meritorious Unit Commendation from the Chief of Naval Operations for its execution of the plan.[85]

On 25 January 1968, the Department of Defense reissued its Directive C-3115.1 of 14 January 1959 concerning responsibilities relating to internal security functions assigned to the Department of Defense by interdepartmental organizations. The organizations were:

(1) The Interdepartmental Intelligence Conference, consisting of the Director of the FBI as chairman and representatives of the Secretaries of Army, Navy, and Air Force. It was responsible for coordinating the investigation of all domestic espionage, counterespionage, sabotage, subversion and related matters affecting U.S. internal security. IIC functioned through a working group composed of alternates to the principal members and an executive secretary. It had no permanent subcommittee structure but established ad hoc committees to work on specific problems. The Assistant Chief of Naval Operations for Intelligence (Director of Naval Intelligence) was designated as the Secretary of the Navy's representative on the IIC working group.

(2) The Interdepartmental Committee on Internal Security with membership consisting of a representative of the U.S. Attorney General as chairman and representatives of the Secretaries of State, Defense, and Treasury. ICIS was responsible for all phases of U.S. internal security except the investigation of domestic espionage, counterespionage, sabotage, and subversion. It had five permanent subcommittees and could establish ad hoc committees as appropriate and participate with IIC in joint committee activities of mutual concern. The permanent ICIS subcommittees were Unconventional Attack, Entry and Exit Problems, Foreign Diplomatic and Official Personnel, Protection of Classified Government Data, and Industrial Security. The Assistant Chief of Naval Operations for Intelligence was responsible for designating the Department of Defense member for the Foreign Diplomatic and Official Personnel Subcommittee, and for designating the point of contact within the Navy for coordination of ICIS matters.[86]

Regular rotation of the commanding officers of the Naval Investigative Service Office, Vietnam saw LCdr. William. A. Armbruster relieving LCdr. Brubaker on 5 July 1968, and LCdr. Thomas A. Brooks taking over on 20 June 1969.

Investigative support to U.S. Navy elements in Bangkok was accomplished by temporary additional duty assignment of special agents from NISO Vietnam as required. Satellite offices were opened at Chu Lai on 1 May 1969 and Quang Tri on 15 May 1969; both were closed on 31 December 1969.[87]

On 1 July 1969, all special agents employed under contract by the Naval Investigative Service were converted to Civil Service employees. Prior to that time, special agents had been paid from special funds that permitted complete freedom and flexibility in the hiring, firing, and use of special agents. The retirement and other benefits provided had been comparable to those of the Civil Service system, but the Civil Service administrative and other inflexible procedures, with their alleged security weaknesses, were avoided. After considerable opposition from ONI, the shift was made when it was ruled that the payment of salaries of special agents was not a legitimate charge against the funding category being employed. For the first time, pay was authorized for overtime work by Navy special agents.

By 1970, the Naval Investigative Service consisted of its headquarters and a number of field command components known as Naval Investigative Service Offices. Except in the Pacific area, commanding officers of NISOs reported directly to the Director NIS. In the Pacific area, the commanding officer of NISPAC served as an intermediate command echelon. In addition to NISOs, there were numerous operational units of one to twenty personnel known as resident agents.

NISOs were located at each naval district and at major fleet and force commands. Their mission, under the command, direction, and control of the Director NIS, was to fulfill Department of Navy investigative and counterintelligence requirements in their assigned geographic areas, and to provide direct investigative and counterintelligence support to naval commands in their areas. In the Pacific, the commanding officer of NISPAC coordinated the activities of NISOs in the CINCPACFLT area (Hawaii, Japan, the Marianas, the Philippines, and Vietnam).

NIS programs, as of April 1970, consisted of the following:

(1) Investigations:

(a) Internal Security (instances of actual or suspected espionage, sabotage, subversion, compromise of classified information, defection of personnel, and duress against Navy personnel);

(b) Criminal (violations of the Uniform Code of Military Justice by Navy and Marine personnel);

(c) Personnel Security (limited or full background investigations);

(d) Security Services (special inquiries, security surveys, technical examinations, crime prevention lectures, and protective services);

(2) Technical Surveillance Countermeasures (detection and neutralizing of hostile intelligence use of sophisticated listening, recording, and transmitting devices);

(3) Counterintelligence (collection and analysis of information on foreign espionage, sabotage, and subversion against the United States and its allies);

(4) Intelligence Collection; and

(5) Security Policy Support (development and promulgation of security policy, assistance to the Navy Inspector General in security inspections, monitoring of security compromises, security education, and classification management).

Occasionally, the Naval Investigative Service was called upon to perform tasks outside, or peripheral to, its assigned mission and normal programs. Such tasks included requests from other agencies of the federal or local governments, such as assistance in providing security to VIPs from other countries at the request of the State Department; protection of the President or Vice President at the request of the Secret Service; drug abuse education/lectures at the request of the Department of Defense Drug Abuse Committee (a NIS representative sat on the committee); and investigations for non-Navy federal and non-federal agencies, often on a reciprocal or goodwill basis, or as a logical assignment (such as performing the background checks on, or investigations of, naval personnel assigned to the White House, Joint Staff, Department of Defense, or unified and specified commands).[88]

On 1 May 1972, the Naval Investigative Service Resident Agency, Saigon was reestablished as a component of NISO Philippines, and LCdr. William D. Derryberry became the NISO representative in Saigon. The disestablishment of NISO Vietnam was part of the overall reduction of U.S. forces in Vietnam.

Commanding officers of NISO, Vietnam, during 1971–1972 were Cdr. Donn T. Burrows (28 May 1970–17 May 1971), Cdr. Nelson E. Moore (17 May 1971–30 Apr 1972), and LCdr. Derryberry (30 Apr 1972–1 May 1972).

Investigation in the 1970s: Defense Investigative Service Established

During Fiscal Year 1972, the Naval Investigative Service continued an ongoing counterintelligence effort in Iceland. Operations were conducted to determine Soviet monitoring of U.S., NATO, and Icelandic communications, as well as to determine any possible Soviet involvement in the construction of an unusual "summer house" in the north of Iceland. Also, photographic coverage of the Soviet Embassy building in Iceland revealed that modifications had been made to allow a portion of the roof to slide open, apparently to permit observation of satellites.

In response to tasking by Assistant Secretary of Defense for International Security Affairs, a Navy task force headed by NIS representatives assessed the potential vulnerability of the cities of San Diego and Sacramento to hostile shipborne intelligence operations. Naval Security Group, Air Force security services, communications security and electronics security assets were used to determine the ambient electromagnetic environment. The assessment conclusively demonstrated that both ports were vulnerable to intelligence collection by Soviet merchant ships, San Diego much more so than Sacramento.

During Fiscal Year 1972, NIS submitted various data, studies, and analyses to the House Committee on Internal Security in connection with committee hearings on subversive influences affecting the military services. Inputs included information on the activities of individuals and organizations engaged in efforts to halt the deployment of U.S. Navy ships and personnel to Southeast Asia; underground newspapers and their influence on Navy and Marine Corps personnel; and an evaluation of the impact of the *Beheiren* (a Japanese organization) on U.S. military personnel in Japan.

In January 1972, NIS prepared an assessment of the civil disturbance threat for Puerto Rico and the Virgin Islands in response to a request by Commander in Chief, Atlantic Fleet for use in a revision of his contingency plans.

On 5 November 1971, the President had directed the consolidation of all Department of Defense activities involved in personnel security investigations into a single Defense Investigative Service (DIS). The new organization, established on 1 January 1972, commenced operations on 1 April 1972. On that date, control of all personnel security investigations, the Department of Defense National Agency Check Center, and the Defense Central Index of Investigations were assigned to the Defense Investigative Service. As part of the creation of DIS, a total of 605 personnel spaces were identified for transfer from the naval investigation services. Responsibility for counterintelligence support and for criminal investigations, however, remained with the military services, subject to further study by the Secretary of Defense.[89]

NIS investigative and counterintelligence support to major fleet elements was expanded during Fiscal Year 1973 to include assigning special agents to all sixteen aircraft carriers in commission. The Agent Afloat Program, which began in 1967 with the assignment of a civilian special agent to each carrier deployed to the Mediterranean, had next been expanded during the Vietnam conflict to include carriers on Yankee Station. Finally, in 1972-1973, special agents were placed on board each carrier, whether deployed, operating in home waters, or in port. Only fully qualified and experienced agents were assigned to provide the fleet with the capability for investigating major crimes and for countering the activities of hostile intelligence and dissident groups. The special agent also provided orientation lectures to the officers and crew of a ship on such topics as narcotics, physical security, and counterintelligence.[90]

During Fiscal Year 1973, NIS conducted eleven port security vulnerability surveys for the Secretary of Defense. As a result of the surveys, Norfolk and the Hampton Roads area in Virginia were closed to Warsaw Pact and Communist Chinese shipping, and a decision was being formulated at the National Security Council level on whether to close Charleston, South Carolina. Tentative analyses of seven other ports surveyed suggested that they also should be closed. Curiously, Jacksonville, Florida, was not found to be sufficiently vulnerable to hostile shipborne intelligence collection operations to justify permanent closure, despite its proximity to the major Navy facilities at nearby Mayport.[91]

During Fiscal Year 1972, Commander in Chief, Atlantic Fleet had requested that a counterintelligence supplement be prepared for inclusion in his foreign port directories. In response to the request, NIS compiled thirty-eight classified counterintelligence supplements for port directories worldwide and forwarded them to the appropriate fleets. Increased collection tasking was levied on NIS field elements and Defense attaché offices to fulfill requirements for additional information for the counterintelligence supplements.[92]

Also during Fiscal Year 1973, NIS published a number of studies on a campaign by an antiwar group involved in impeding the deployment of U.S. Navy ships to the Western Pacific and the continuing efforts of antimilitary activists to foster dissent and disruption within the Navy and Marine Corps, particularly in Southeast Asia. A quarterly publication based on NIS investigations entitled *Damage Incidents Affecting the Department of the Navy* was initiated. It set forth significant statistics on the commands affected, the nature of the damage, and the results of the NIS investigations.[93]

After the Defense Investigative Service became operational, NIS transferred its personnel security investigation function, including 580 civilians and 25 military personnel, to the new agency in October 1972. As a result, NIS was reorganized to streamline its operations, five field offices were closed, and the number of resident agents was reduced from 240 to 127. While the investigative workload during Fiscal Year 1973 had increased 32 percent over the previous year, and the requirements of new programs (port vulnerability, special agent afloat) had increased, the personnel and fiscal resources of NIS decreased significantly. Savings from the consolidation of offices and other organizational changes had been exhausted.[94]

The following officers served as directors of naval counterintelligence offices between 1962 and 1971:

Director	Dates of Service
Headquarters Naval Investigative Support Group	
Capt. Frank A. Klaveness	May 1961–Aug 1962
Capt. Robert P. Jackson, Jr., USNR	Aug 1962–Jul 1963
Capt. J. O. Johnson, USNR	Jul 1963–Oct 1963
Naval Counterintelligence Support Center	
Capt. J. O. Johnson, USNR	Oct 1963–Feb 1966
Naval Investigative Service	
Capt. J. O. Johnson, USNR	Feb 1966–Jul 1966
Capt. E. G. Rifenburgh	Jul 1966–Jul 1970
Capt. John Q. Edwards	Jul 1970–Jun 1973
Capt. Barney Martin	Jun 1973–Apr 1976

Chapter Notes

1. OPNAV Instruction 03850.1A of 30 Jan 1968.

2. *New York Times*, 13 May 1913, 1.

3. Capt. Edward McCauley, interview, 19 Nov 1943, author's files.

4. CNO ltr, ser 24514–39, 14 Oct 1916, "Naval District Information Service."

5. Capt. Herbert E. Cocke, USN, "History of ONI," MS, Office of Naval Intelligence, 1931, 33–34.

6. Capt. Edward McCauley, "Naval Intelligence Before and During the War," 1920, typescript in File ZU, box 842, Naval Records Collection (RG 45), NA.

7. Department of the Navy, "Administrative History of the Office of Naval Intelligence in World War II," 10 Jul 1946, unpublished MS, 11, hereafter *ONI WWII Admin History*.

8. Cdr. Spencer Eddy, *History and Anecdotes of the New York Branch of ONI, 1916–1918*, privately reproduced manuscript, 1919, passim.

9. *ONI WWII Admin History*, 153.

10. Folder marked "Enemy Suspects in Russia," box 686, RG 45, NA.

11. File WX-1, boxes 729–31, RG 45, NA.

12. File 161, Records of the Office of the Chief of Naval Operations (RG 38), NA.

13 McCauley, "Autobiography of Capt. Edward McCauley," MS extracts in McCauley ltr to C. R. Wilson in ONI, 7 Dec 1943.

14. Ibid.

15. RAdm. Albert P. Niblack, USN, *The History and Aims of the Office of Naval Intelligence* (Washington: GPO, 1920), 24–25.

16. Testimony of Josephus Daniels, 20 May 1920, Senate, *Investigation of the Conduct of the Navy in World War I: Hearings Before the Subcommittee of the Committee on Naval Affairs*, 66th Cong., 2d sess., 1921, vol. 2, pt. 2, 2710.

17. LCdr. C. O. Mass, USNRF, "History of NA Paris," MS, 1 Jun 1919, passim.

18. Testimony of Josephus Daniels, 2711.

19. *The New York Times*, 11 Nov 1917, 5.

20. Niblack, *History and Aims of ONI*, 22.

21. National Archives, Inventory of Record Group 38, 1966, 79.

22. ONI-22, *Notes on Espionage, Counter-Espionage and Passport Control*, 1935, 5–9.

23. Testimony of RAdm. Albert P. Niblack, 14 Feb 1920, House, *Estimates Submitted by the Secretary of the Navy: Hearings Before the Committee on Naval Affairs of the House of Representatives*, 66th Cong., 1st sess., 1920, 833.

24. *ONI WWII Admin History*, 150–52, 154, 155, 177, 179–89.

25. Adm. Royal E. Ingersoll, Oral History, Columbia University, New York, NY, 1965, 46.

26. Ibid., 104–5.

27. *ONI WWII Admin History*, 55.

28. ONI-22, *Notes on Espionage, Counter-Espionage and Passport Control*, 1935, 5-2.

29. *ONI WWII Admin History,* 179, 264–66.

30. *ONI Review*, Jan 1946, 33–38.

31. *ONI WWII Admin History*, 55, 446.

32. Ibid., 67–68.

33. Ibid., 68.

34. OP-16-FL memo, 20 Nov 1943, "Naval History of WWII," 5.

35. *ONI WWII Admin History*, 68–69.

36. OP-16-B ltr, ser 04816, cited in OP-16-B-3 ltr, ser 01640116, 14 Aug 1945, Naval Investigative Service (NIS) History Files, OA.

37. Memo of Information of 19 May 1942 for OP-14 Files, Naval Security Group History Papers, encl. 4, OA.

38. OP-16-B-3 ltr, ser 01640116, 14 Aug 1945.

39. *ONI WWII Admin History*, 70.

40. Ibid., 71–72.

41. DNI ltr, ser 02997216, 22 Nov 1943, NIS History Files, OA

42. *ONI WWII Admin History*, 139–41.

43. SECNAV Ltr of Instruction, ser 0223P23, 1 Nov 1945, NIS History Files.

44. ONI-D-4-T, Naval Intelligence (NAVINTEL) Training Directive, 1946, NIS History Files.

45. SECNAV ltr (OP-32-D) ser 02416P32, 31 Mar 1947, NIS History Files, OA.

46. ONI Instruction 003821.1 of 21 Apr 1960. The letter was CNO ser 003749P32, 8 May 1947.

47. Assistant to the President memo to Secretary of Defense, 3 Nov 1947.

48. DNI presentation of 17 Feb 1948 for Congressional Committee, in OP-32 Top Secret file for 1948–1949, OA.

49. ONI Instruction 5520.72 of 2 Oct 1961.

50. DNI ltr, ser 06173P32, 5 Apr 1950, NIS History Files.

51. Commander in Chief, Pacific, *History of Administrative Problems, Korean War*, 2:15–16.

52. Naval Investigative Service, Pacific (NISPAC) Instruction 5400.1D, I-IV-1.

53. ONI Instruction 05520.3 of 8 Aug 1951.

54. DNI memo, ser 025308P32, 19 Nov 1953, NIS History Files, OA.

55. SECNAV Instruction 05500.5 of 29 Sep 1953.

56. ONI Internal Instruction 05010.1 of 28 Dec 1953.

57. SECNAV Instruction 5430.13A of 10 Aug 1954.

58. Briefing on NIS, Aug 1954, NIS History Files, OA.

59. Briefing to the Clark Subcommittee of the Hoover Commission, ONI, 22 Nov 1954, NIS History Files, OA.

60. ONI Instruction 005520.41 of 14 Feb 1955.

61. ONI-70-1, *U.S. Naval Intelligence Manual*, 20 Jun 1956, 28.

62. ONI Instruction 05520.63 of 17 Nov 1958.

63. ONI Instruction 003821.1 of 21 Apr 1960.

64. ONI-70-1, Change 1, 4 Nov 1957, 85.

65. ONI Instruction 05520.57 of 2 Dec 1957.

66. Capt. Martin Randisi, USNR, transcript of interview made for author, 23 Jul 1975, 27–29.

67. *NIS Command History, 1967*, 1–3.

68. Navy Counterintelligence Support Center Instruction P5450.1 of 20 Jan 1964.

69. ONI Notice 5200 of 3 Apr 1963.

70. LCdr. William D. Derryberry, briefing folder on Vietnam, 1 May 1972.

71. Navy Counterintelligence Support Center Instruction P5450.1 of 20 Jan 1964.

72. Commander in Chief, Pacific Fleet (CINCPACFLT), *Annual Report, 1 Oct 1963–26 Jun 1964*, 53.

73. ONI Instruction 05521.14A of 27 May 1964.

74. *CINCPACFLT Annual Report, 26 Jun 1964–30 Mar 1965*, 55.

75. NISPAC Instruction 5400.1D of 1 Apr 1975, I-IV-3.

76. ONI Instruction 3850.2 of 9 Feb 1965.

77. *NIS Command History, 1967*, 2–6.

78. Derryberry briefing folder.

79. Ibid.

80. OPNAV Instruction 5450.97A of 28 Mar 1966, and Change 1 of 2 May 1967.

81. OPNAV Instruction 5700.8A of 27 Mar 1967.

82. OPNAV Instruction 05510.96A of 29 Mar 1967.

83. OPNAV Instruction 5520.2 of 25 Nov 1967.

84. DNI Report to CNO on "Status of Major ONI Programs," 20 Jun 1968.

85. NISPAC Instruction 5400.1D of 1 Apr 1975, I-IV-5.

86. OPNAV Instruction 5500.58 of 8 Mar 1968.

87. Derryberry, briefing folder on Vietnam.

88. NIS, "Shore Establishment Review," 15 Apr 1970, 2, 7, 9, 14–16, 21–43.

89. *DNI Report to CNO, FY 1972*, 13, 14, 59.

90. *DNI Report to CNO, FY 1977*, III-8.

91. Ibid., II-13.

92. Ibid., III-5.

93. Ibid., III-8.

94. Ibid., II-14–II-15.

CHAPTER 22

Naval District Intelligence Activities

The information in this chapter on the intelligence activities in the naval districts is divided into a general section followed by sections on individual districts.

The organizational structures of the district intelligence offices (DIO) were essentially the same. The organization, therefore, is described completely only for the District Intelligence Office, 1st Naval District (DIO-1ND) and is not repeated for the other districts, except where some significant difference has been found.

The surviving records of the district intelligence offices of the 5th Naval District, Norfolk, Virginia; 13th Naval District, Seattle, Washington; and 17th Naval District, Adak, Alaska, were not researched for this chapter. Only very scant documentation on the activities of the District Intelligence Office, 16th Naval District, Cavite, Philippines was located. Chapters 21, 24, 26, and 29 contain additional information on activities in the naval districts.

The district intelligence offices were replaced by Naval Investigative Service (NIS) offices and district staff intelligence officers (DSIO) in March 1966.

Origins of the District Intelligence Office System

The district intelligence office system was established in 1916 to cope with what was predicted to be a rising volume of counterespionage requirements. Counterespionage was a new field for the Office of Naval Intelligence, and the instructions that set up the DIO system are the best single source of descriptions for the duties and responsibilities assigned to the newly established district intelligence officers, who were initially referred to as aides for information. The "Instructions for Information Service" were forwarded to naval district commandants by Chief of Naval Operations (CNO) RAdm. William S. Benson on 14 October 1916, to become effective upon receipt:

> 1. The Aid [*sic*] for Information shall, under the immediate direction and control of the Commandant of the Naval District, be charged with the active administration and supervision of the Naval Information Service within the limits of the Naval District to which assigned.
>
> 2. He shall, if practicable, be ordered to temporary duty in the Office of Naval Intelligence preliminary to assuming his duties as Aid for Information in a Naval Defense District.
>
> 3. He shall be charged with the acquisition, compilation, and dissemination of information as specified, observing the instructions issued to him by or through the Commandant of [the] Naval District.
>
> 4. He shall represent the Commandant of the Naval District in matters connected with such information.
>
> 5. In preparation for war the Aid for Information will undertake the following:
>
> a. Prepare and keep posted to date a secret war portfolio, containing all papers and data relating to the war information service, and provide a secure place for filing the same;
>
> b. Seek the general cooperation of the Aid for Information for Communications in the work of organization and preparation, in order to secure prompt and efficient communication with the sources of information, with the Navy Department and with other points as may be required;
>
> c. Familiarize himself generally with the Naval District and its sections and acquire all necessary knowledge in connection with it;
>
> d. Make recommendations for the improvement of the plans for the information service in naval districts;
>
> e. Familiarize himself with the written instructions for the establishment and organization of the war information service in his Naval District, the agencies, sources, and means by

which the information is to be collected, and the facilities available for its transmission and dissemination;

f. As opportunity offers, visit or interview confidentially each of the heads of local branches of departments of the federal government, the officials of the municipal government, heads of steamship and commercial companies, and other individuals, whose cooperation and assistance are contemplated in the war information service;

g. Arrange for the cooperation of the above and ascertain the extent to which they will cooperate and the manner and details of such cooperation;

h. Confer with and arrange for the cooperation of the local military authorities in matters pertaining to information (an officer of the Army is to be detailed on the staff of the Commandant for this purpose);

i. Make written plans for the utilization of all sources of information as specified;

j. Make tentative written plans in cooperation with those directly concerned, for the use of codes, the routing of messages, visits to incoming merchant vessels, instruction of outgoing merchant vessels, the safeguarding of confidential information, the spreading of false information, etc;

k. Make tentative written plans for the supply, control, and expenditure of funds to accomplish the administration, organization, and development of the war information service in his district.

l. Prepare and keep up to date lists of the officials, organizations, firms, individuals, etc., included in the plan for war information service in his district;

m. Draw up written plans for expanding the office force, and if necessary the office quarters on the eve of war, including the installation of telephones, etc., and arrange for a continuous day and night service in the office;

n. As opportunity offers, inform confidentially those concerned of the parts of these plans which concern them, and of the specific duties which will be required or expected of them;

o. Actually write all telegrams, letters, instructions, etc., the need for which can be foreseen, leaving only the date and signature blank, including those for the Commandant to sign; and

p. Take, or send by the hands of an officer, duplicate of war portfolio to the Office of Naval Intelligence, when suitable opportunity presents itself. Take every precaution to prevent the portfolio from being seen by unauthorized persons. The above work will be considered the primary work of the Aid for Information during peace.

WAR INFORMATION SERVICE

6. The mission of the War Information Service in a naval defense district is as follows:

a. The collection and compilation of prompt, reliable and accurate information concerning the following:

(1) Approach, arrival, movements, and position of enemy naval forces;

(2) Approach, arrival, loading, and departure of neutral shipping, whose cargoes may contain contraband of war or articles belonging to the enemy or destined directly to the enemy, his citizens or subjects;

(3) The approach, arrival, and departure of all U.S. merchant vessels;

(4) The identity, nationality, and activities of officers, crews, and passengers of merchant vessels, whether neutral or U.S., arriving in any ports within the limits of the Naval District;

(5) The presence, identity, and activities of enemy agents, citizens, or subjects;

(6) The conduct, progress, and events of the war, information bearing on the enemy, his government, policy, forces and their composition, his subjects or citizens, commerce, finances, and general activities;

b. The prompt dissemination of the above information to the proper authorities, as follows:

(1) The Commandant of the Naval District and through him to:

(2) The Navy Department;

(3) The Fleet, operating in the waters adjacent to the Naval District;

(4) The Commandants of other Naval Districts, of such as may concern their Districts;

(5) The Commanding Officer of the Army within the Naval Districts, of such as applies to the Coast Defense or of concern or interest to the Army; and

(6) The local heads of the Civil Departments of the Government, of such as is of concern or interest to those Departments.

SOURCES OF INFORMATION

7. The personnel available in connection with the mission are as follows:

a. All persons under the direct control of the District Commandant;

b. All persons under the control of other departments of the Government;

c. All persons in the employ of state, city, county, or township; and

d. All civilians.

8. Observers Afloat:

a. The Fleet;

b. The Naval Patrol*;

c. Naval Militia and auxiliary organizations (power-boat squadrons, etc.)*;

d. The Coast and Geodetic Survey (Dept. of Commerce)*;

e. The Bureau of Fisheries (Dept. of Commerce)*;

f. The Bureau of Lighthouses (Dept. of Commerce)*;

g. Coast Guard (Dept. of the Treasury)*;

h. Customs Service (Dept. of the Treasury);

i. Public Health Service (Dept. of the Treasury);

j. Secret Service (Dept. of the Treasury);

k. Bureau of Investigation (Dept. of Justice);

l. Bureau of Immigration (Dept. of Labor);

m. Merchant Marine, U. S. and neutral where possible; and

n. Officers and men of private vessels.

9. Observers ashore:

a. United States Navy (including aircraft);

b. United States Army (including aircraft);

c. Coast Guard-Life Saving Service (Dept. of the Treasury*);

d. Bureau of Lighthouses (Dept. of Commerce*);

e. Weather Bureau (Dept. of Agriculture);

f. Bureau of Investigation (Dept. of Justice);

g. Customs Service (Dept. of the Treasury);

h. Public Health Service (Dept. of the Treasury);

i. Secret Service (Dept. of the Treasury);

j. Bureau of Immigration (Dept. of Labor);

k. Postmasters and Inspectors (Post Office Dept.);

l. Shipping Agents;

m. Steamship agents and representatives of steamship lines;

n. Civilians, importers, bankers, financial and commercial men, etc.;

o. Local police and detective branches;

p. Private secret service agencies;

q. Representatives of the press;

10. Observers abroad.

These sources are covered by Office of Naval Intelligence.

*The Marine personnel and ships of the Department thus marked to be under the control of the Navy after mobilization;

SECRET SERVICE SUPERINTENDENT

11. There shall be detailed as Assistant to the Aid for Information an active, trustworthy and experienced secret service operative of a Government Department. Upon recommendation of the Aid for Information, this detail will be arranged for by the Office of Naval Intelligence.

12. He shall have charge, under the Aid for Information, of the Naval Secret Service within the Naval District and the secret service agents detailed to or acting for the Navy.

13. He shall assist and advise the Aid for Information in matters pertaining to secret service within the Naval District.

COOPERATION WITH THE AID FOR COMMUNICATIONS

14. A close and efficient cooperation is necessary between the Aid for Information and the Aid for Communications. The prompt delivery of communications to the District Headquarters from the different sources of information and the prompt forwarding of information to its proper destinations are essential. Arrangements should be made to insure the confidential character of information where secrecy is necessary or desirable and for direct communications where desirable.

COOPERATION OF THE AID FOR INFORMATION WITH REPRESENTATIVES OF OTHER DEPARTMENTS OF THE GOVERNMENT

15. In general, the civil departments of the federal government will assist the Aid for Information in obtaining information, as follows:

a. Department of Justice: Activities of enemy secret agents within the District. Observations of enemy's citizens or subjects, or enemy sympathizers within the District;

b. Post Office Department: Control and supervision of the mails, and resulting information; foreign money orders, etc. Names and location of enemy's citizens or subjects permanently or temporarily in the District.

c. Department of the Treasury: Division of Customs: Clearance, arrivals and departures of vessels; supervision of officers, crews, and passengers; Division of Secret Service: Activities of enemy agents within the District, particularly maritime; supervision of officers, crews, and passengers of merchant vessels; and

d. Department of Labor: Bureau of Immigration: Supervision of passengers of merchant vessels.

COOPERATION OF THE AID FOR INFORMATION WITH REPRESENTATIVES OF LOCAL ORGANIZATIONS

16. Municipal authorities and other local civil organizations, companies, etc., will assist the Aid for Information in obtaining information, as follows:

a. Police Departments: Activities of enemy agents within the District; river and harbor patrol. Location and observation of enemy citizens or subjects permanently or temporarily in the District;

b. Arrivals and departures of vessels; supervision over passengers; information on officers, crews, clerks, and attachés. Information from abroad; and

c. Commercial Companies: Information received from representatives and others in the course of business, letters, conversations, etc., at home and abroad; information from other sources, correspondence, clerks, etc.; information regarding drafts or money orders.[1]

A report made during World War I by Director of Naval Intelligence (DNI) RAdm. Roger Welles, Jr., detailed the wartime work of the district organizations:

a. Navy Personnel:

(1) Apprehension of deserters and stragglers; investigations and surveillance of enlisted men reported to the commanding officers of all U.S. ships; reported imposters appearing in the uniform of the Navy.

(2) Suspects attempting to enlist in the U.S. Navy or U.S. Naval Reserve Force.

(3) Collusion between firms holding Navy contracts and [Navy] enlisted men.

(4) Cooperation with other naval districts in the investigation of cases reported by them which fall within the field covered by the Aid for Information.

b. Navy Yard Employees:

(1) Investigation and surveillance of Navy Yard suspects.

(2) Investigation of reported pro-Germanism of Navy Yard employees.

(3) Thefts from the Navy Yard.

(4) Cases referred by Commandant relating to the Naval establishment.

(5) Investigations of labor agitation connected with the Navy Yard.

c. Miscellaneous Investigations:

(1) All cases referred by the Mail Censorship Bureau.

(2) Investigation of suspicious individuals reported in the vicinity of Navy piers, wharves, docks, warehouses, etc.

(3) Investigations of applicants for pilot licenses.

(4) Investigations of cases involving radio apparatus.

(5) Investigations of suspicious fires on piers, docks, and wharves under the Navy Department.

(6) Protection of shipyards within the naval district doing Navy work and of naval vessels building or repairing within those shipyards.

(7) Protection of the operation, product and personnel of plants manufacturing munitions or other material for the Navy.

(8) Investigation of enemy agents and sympathizers, and civilians, concerning any activities inimicable to the interests of the Navy.

(9) Investigation of addresses of such cables as may be referred to the Aid for Information by the Cable Censor.[2]

DIO Organization Between the World Wars

On 24 March 1919, Acting Secretary of the Navy Franklin D. Roosevelt directed that the Aid for Information was to be called the District Intelligence Officer (DIO). The directive also set forth the DIO's tasks.[3]

The duties and responsibilities of the district intelligence offices during peacetime were set forth in the *District Manual* and in ONI-19: maintainance of press relations for district headquarters; liaison with the investigating units of federal, state, and city agencies within the naval district; liaison with public and private research agencies and with business interests having information in intelligence fields; liaison with ONI and the intelligence services of the other naval districts, and with forces afloat within the district; counterespionage, security, and investigations; collection, evaluation, and recording of information regarding persons or organizations of value (or opposed) to the Navy; preparation and maintenance of intelligence plans for war; and administrative supervision over the recruiting, training, and activities of the appropriate personnel of the Naval Reserve within the district. Naval intelligence reserve officers were designated I-V(S), meaning Intelligence Volunteer (Specialized); the designator I-V(S) was replaced by S(I)—Special Duty (Intelligence)—in September 1944.

Contacts between the district intelligence service and ONI were almost entirely confined to matters relating to investigations; visits of foreigners; routing and dispatch of correspondence within the naval intelligence service; procurement of funds, special equipment, and civilian assistants for district intelligence officers; and matters connected with the enrollment of I-V(S) Naval Reserve officers. There was no active unit in ONI charged with general administration and coordination of the DIO activities.[4]

As of September 1937, special agents had been employed in the various DIOs as follows: one each in the 1st, 4th, 9th, and 14th Naval Districts; two each in the 3rd and 12th; and three each in the 11th and 13th, for a total of fourteen.[5]

In 1938, the District Intelligence Services consisted of an officer and a clerk in the 1st, 4th, 5th, 9th, 12th, 13th, and 14th Naval Districts; two officers and two clerks in the 3rd and 11th Districts; one officer and one clerk in the 6th District to handle the Naval Intelligence activities of the 6th, 7th, and 8th Naval Districts; and one officer with additional duty as the DIO in the 15th and 16th Districts and at the Navy Yard, Washington, D.C. In addition, a chief yeoman was assigned to the 14th District, two inspectors were assigned to the 3rd District, and civilian assistants were available in various districts. Officers in charge of branch hydrographic offices and of recruiting divisions and offices were ordered as an additional duty to assist the intelligence service of the districts in which they were located.[6]

World War II

Censorship was not intended to be a wartime function of the district intelligence offices, but they were responsible for staffing, selecting station locations, enlisting and training personnel, and providing logistics. Many problems ensued, and it wasn't until March 1942 that all stations were fully staffed and the censorship functions were turned over to a director of censorship who was not associated with the district intelligence office.[7]

In 1939, the DIO-1ND had no counterintelligence section. Such work consisted almost exclusively of adding material to the files; no one evaluated information in advance. Evaluation had to be done as the demand arose. Most information was inadequate, and very little of it could be checked for accuracy.[8]

Close liaison with the FBI and the Army Corps Headquarters, which had been directed by Director of Naval Intelligence (DNI) RAdm. Walter S. Anderson in December 1939, was maintained by occasional, unscheduled meetings.[9]

On 4 October 1940, RAdm. Anderson sent a teletype message to all district intelligence officers directing them to take immediate steps to locate and recommend agents and intelligence reserve officers for active duty as needed to establish intelligence units in the Navy yards and the principal naval activities. This message was followed by a CNO (OP-16-B) letter of 8 October 1940 pointing out "the gravity of the present situation" and the need to place "the Naval Intelligence Service in an advanced state of readiness." The letter required that district plans, estimates of requirements, etc., for naval intelligence be completed at an early date and that district intelligence personnel be augmented as specified in the 4 October message. On 11 October, the Director of Naval Intelligence requested that the names of I-V(S)-designated Naval Reserve officers qualified to make industrial facility security surveys be forwarded to ONI not later than 16 October. A related Secretary of the Navy message of 22 October was sent to the commandants of all the naval districts instructing them to "complete plant protection surveys" of all naval shore establishments in their districts "at the earliest moment."

On 23 October, the Director of Naval Intelligence teletyped to all continental DIOs a request to submit by airmail a summation of all information in their files on Japanese, German, and Italian spies and saboteurs, actual or potential, and any other individuals whose activities were of an undercover nature believed to be inimical to the national defense of the United States.

All of these directives placed major work loads on the DIOs and required immediate expansion of the district intelligence organizations, not only to handle the projects but also to expedite the personnel investigations necessitated by the expedited augmentations.[10]

The Delimitation Agreement between the FBI and the military intelligence services, discussed in Chapter 21, made reference to four categories of investigations: espionage, counterespionage, sabotage, and subversion. Actually, the DIO organization was expected by ONI and other naval activities to conduct any investigation requested.

On 1 November 1940, the Director of Naval Intelligence sent a teletype to the DIOs: "You are not restricted to any particular field of investigative effort by the delimitation agreement with the FBI." DNI RAdm. Harold C. Train further elaborated on that point in a letter (OP-16-B serial 01640316 of 21 August 1942) to all DIOs:

> It will be noted that no attempt was made to delimit investigative responsibility in cases falling outside the four categories. Certain of these cases, involving violations of federal statutes, fall definitely within the investigative jurisdiction of the FBI, such as kidnapping and bank robbery; the Post Office Department, such as the use of the mail to defraud; the Treasury Department, such as narcotics and customs violations; and the Secret Service, such as threats to the President. Except for such cases, however, investigations predicated on purely naval interest, . . . or any other cases outside the four categories specifically covered by the Delimitation Agreement, may be conducted by Naval Intelligence.[11]

In January 1941, the domestic intelligence field offices included the district intelligence offices of the fifteen naval districts plus offices of the Potomac River Command and at Guam, American Samoa, and Naval Station, Guantanamo, Cuba. The Domestic Intelligence Branch of ONI served as the national headquarters for the field offices. Although the commandants of the naval districts provided the district intelligence organizations and normally operated them, the organization as a whole was an activity under the jurisdiction of the Chief of Naval Operations.[12]

Zone intelligence offices (ZIO) and intelligence units established at Navy yards, bases, stations, and hospitals were considered field units of the district intelligence offices. The relationship between the district intelligence officer and the field units in his district was the same as the relationship between the Director of Naval Intelligence and the district organizations.[13]

On 19 May 1941, in anticipation that general mobilization, when called, would stimulate acts of sabotage, Secretary of the Navy Frank Knox ordered that the expansion of naval intelligence services continue in a progressive and orderly manner until, with the exception of the Censorship Branch, a state of readiness equal to that of mobilization had been reached.

On 18 July 1941, the Bureau of Navigation ordered all commandants of naval districts to nominate for active duty all I-V(S)-designated Naval Reserve officers other than those intended for service in the Censorship Branch and those living abroad, and to effect the minimum procurement of I-V(S) officers.[14]

During 1941, three functions were removed from the DIO list of responsibilities. In May, the public relations functions in the naval districts were set up as separate offices, and in December, plant protection was assigned to the district security officer. Censorship was activated on 8 December and was readied for transfer to the Office of Censorship when it was established by Executive Order on 19 December. In spite of the losses of functions, however, the DIO work load increased unabated.[15]

On 21 May 1941, the CNO made naval district commandants responsible for plant surveys and inspections of naval establishments within their districts and of those private commercial plants designated by ONI that were engaged in the production and/or repair of naval material. The DIOs were also made responsible for plant inspections, in conjunction with the Army, of private commercial plants having both Army and Navy contracts, after the plants had been surveyed by the FBI.[16]

On 7 December 1941, a Secretary of the Navy message to all naval district commandants directed all industrial establishments working on naval contracts to double their guards and to take all necessary precautions against sabotage. It was followed on 9 December by a CNO (OP-16-B-6) letter setting forth minimum general security measures to be taken for the protection of private plants of importance to naval procurement. Information on the enhanced security measures was passed to all private plants holding Navy contracts.

When the plant protection function in the districts was passed to the district security officers, the responsibility for the screening of aliens in naval and private plants was retained by the counterintelligence (B-7) sections of the district intelligence offices.[17]

On 2 July 1943, ONI queried all DIOs as to what counterintelligence studies they were working on and listed subjects for future studies that would be of interest. Some examples of the subjects suggested for study were the use of enemy-inspired "front" organizations for peace; the development by the Japanese and Communists of tensions over racial discrimination; possible activities related to the Middle East by Americans, Moslems, Arabs, and Zionists; activities of the Irish in the United States, with particular reference to Irish Republican Army collaboration with the Germans; the use of neutral seamen as enemy couriers; the efforts of the Communists to infiltrate the Navy, either as service personnel or as employees of naval shore facilities; and the smuggling of war materials of concentrated value, such as industrial diamonds and platinum, for the benefit of Germany, Italy, and Japan.[18]

As indicated by the above, the interpretation of "naval interest" was almost all-inclusive in the counterintelligence field. To correct the situation, DNI RAdm. Roscoe E. Schuirmann advised all DIOs on 25 February 1944 that the "expansion of other agencies, enabling them now to assume and discharge their individual responsibilities, coupled with the manpower shortage, makes it mandatory that the restriction to direct Naval interest in CI [counterintelligence] activities be rigidly applied."[19]

The Transition to Peacetime and the Cold War: Administration of the Districts

During the war, the intelligence organizations in the various districts, sea frontiers, and sea frontier groups functioned with only a slight attempt toward standardization, thus bringing about the birth of local terms and local definitions of the intelligence mission and intelligence organizations. To correct the lack of standardization in concept, definitions, and vocabulary, and to implement and support ONI policy, a program of standardization was implemented in 1946. All local operating manuals were recalled, and ONI-19(A), the *Naval Intelligence Manual*, was published in May the following year.[20] In November 1946, all district intelligence offices were requested to submit monthly summaries of their activities to ONI.[21]

The revised *Naval Intelligence Manual* described the mission of the district intelligence officer as follows:

> a. To administer, operate, and maintain the District Intelligence Organization as an integral part of Naval Intelligence;
>
> b. To supply the Commandant and the subordinate commands of the District with the intelligence required to perform their duties, with special reference to situations and trends threatening the security of the Naval Establishment;
>
> c. To supply the Sea Frontier Commander or Sea Frontier Task Group Commander with operational intelligence, as required;
>
> d. To provide intelligence needed by ONI; and
>
> e. To assist in executing the mission and objectives of the Chief of Naval Intelligence.[22]

By the mid-1950s, in accordance with General Order No. 19, the commandant of each naval district was required to maintain within his district an efficient intelligence service, including such counterintelligence functions as might affect the security of naval activities within the district and action on any operational intelligence matters required by the commander of the sea frontier in which the district was located. The directive also required that close relations be maintained with intelligence officers in the naval forces afloat. [23]

In fulfilling the above requirements, the primary functions of the district intelligence offices were the conduct of counterintelligence and the implementation of security policies. The district intelligence officers served on the staff of the naval district commandant, and in some designated districts they had additional duty on the staff of the local sea frontier commander. All district intelligence offices, however, remained under the management control of the Director of Naval Intelligence.[24]

The sea frontier commands used the facilities of the district intelligence organizations to meet their intelligence requirements. When it was necessary to coordinate the operational intelligence activities of two or more districts, the sea frontier commander had an officer on his staff designated as sea frontier intelligence officer. Normally, if a district commandant was also the sea frontier commander, the DIO of that district was designated as the sea frontier intelligence officer.[25]

Inspections of district intelligence offices by ONI representatives in 1955 and 1956 disclosed variations in organization and differences in the allocation and utilization of personnel, both military and civilian. As a result, a basic standard organization was prescribed for adoption by district intelligence offices in the continental United States by 6 November 1956. Guidelines for the use of assigned personnel included the following provisions: the military administration of the Investigations Branch, Intelligence Branch, and the Administrative Staff was to be exercised by the assistant district intelligence officer (ADIO); junior officers were not normally to be assigned to supervisory positions but were to be rotated as practicable through the various sections so as to acquire a better knowledge of the operations of the DIO; the function of the security analyst was to advise and assist the DIO in every area of DIO operations exclusive of those of a purely military character (as the technical advisor and assistant to the DIO and ADIO, the security analyst was to provide continuity for attaining the objectives and fulfilling the programs of Naval Intelligence and the district intelligence office); the supervising agent was to supervise and direct all professional and technical phases of the investigative operations of the office; the intelligence analyst (when assigned) was to supervise and coordinate, under the military administration of the ADIO, all district counterintelligence and intelligence collection activities and programs (other than investigations) and, subject to the overall guidance of the security analyst, to provide continuity for operational intelligence and censorship planning; and the ADIO, in addition to his primary duties in the overall operation of the district intelligence office, was to provide direct supervision of all administrative functions.[26]

Commanders of Marine Corps activities physically located within naval districts or river naval commands were required to have their counterintelligence personnel report in person to the appropriate DIO or intelligence officer of a river naval command, upon assignment to counterintelligence duties, for instructions concerning the extent of their participation in investigations for which ONI was responsible and for technical control and guidance in other counterintelligence matters. A close personal relationship was to be maintained on a continuing basis.[27]

In a follow-up to the above, the Director of Naval Intelligence noted that Marine Corps commands were a significant source of requests requiring investigative action at Marine Corps activities or in adjoining areas and that controlled use of Marine Corps counterintelligence personnel in such cases might result in better service to Marine Corps commands as well as to other commands. Furthermore, the availability of qualified Marine personnel in locations where there was no ONI agent (as on Okinawa) would enhance the resources of the naval investigative program. Intelligence officers were directed to make local arrangements, where possible, to use Marine Corps investigative personnel on a "when available" basis, not to interfere with their regularly assigned duties.[28]

As of July 1958, the allowance of civilian contract agents was as follows: 21 in DIO-1, 65 in DIO-3, 37 in DIO-4, 32 in DIO-5, 38 in DIO-6, 22 in DIO-8, 57 in DIO-9, 3 in DIO-10, 31 in DIO-11, 20 in DIO-12, 14 in DIO-13, 8 in DIO-14, 2 in DIO-15, 3 in DIO-17, 45 in the Potomac River Naval Command, 7 in Japan, 5 in the European area, 4 in Guam, 4 in the Philippines, and 19 at ONI Headquarters, for a total of 437.[29]

SECNAV Notice 5450 of 22 March 1960 established the district intelligence offices in the continental naval districts and the intelligence office of the Potomac River Naval Command as separate activities, each under an officer in charge. The change in status was the culmination of several years of effort to gain formal recognition of the situation that had in fact existed for many years: the DIOs were, for all intents and purposes, separate and distinct naval activities and field appendages of ONI. The change in status was intended in no way to reduce the services previously provided to the naval district commandants and other naval activities by the DIOs. The officers in charge of the DIOs were issued orders assigning them additional duty as assistant to the commandant for intelligence.[30]

On 3 August 1960, the extra-continental DIOs, those of the 10th, 14th, 15th, and 17th Naval Districts, were also made separate actitivies, again with each under an officer in charge. The district intelligence offices were placed under the military command of their respective commandants and under the management control of the Director of Naval Intelligence.[31]

On 24 September 1962, as a result of additional funds received to assist in eliminating an investigative caseload backlog, new contract agent allowances were established as follows: 28 in DIO-1, 63 in DIO-3, 54 in DIO-4, 48 in DIO-5, 53 in DIO-6, 36 in DIO-8, 71 in DIO-9, 47 in DIO-11, 30 in DIO-12, 19 in DIO-13, 16 in DIO-14, and 58 in the Potomac River Naval Command, for a total of 523.[32]

Naval Investigative Service and the Naval Districts

When the Naval Investigative Service was established in March 1966, the investigative and counterintelligence duties formerly performed by a district intelligence office were assigned to a Naval Investigative Service Office (NISO). The noninvestigative functions were temporarily retained by the NISOs until the district staff intelligence officers (DSIO) were established by OPNAV Notice 5450 of 17 March 1967.

The DSIOs were not provided with collection resources; neither were they to be formally charged with collection requirements. The Navy's positive intelligence collection program in the naval districts was to be performed by local assets of the Naval Field Operational Support Group.[33]

Naval District Intelligence Offices in World Wars I and II

In the following section, the activities of the district intelligence officers are discussed in numerical order. As stated in the beginning of this chapter, the discussion for the 1st Naval District is presented in greater detail here as being typical of the experiences of the other DIO organizations.

1st Naval District, Boston

The district intelligence officer in the 1st Naval District in September 1939 was Cdr. Robert P. Hinrichs. His staff consisted of one full-time investigator and one secretary, Mildred Burke. Inactive Naval Reserve officers also helped occasionally when their civilian occupations would permit.[34]

The responsibilities of the district intelligence officer of the 1st Naval District at that time were to make plans for censorship in wartime; make recommendations for maintaining the security of plants in the naval establishments as the result of surveys and inspections; facilitate the flow of permitted information concerning the Navy to the press, but erect safeguards against the release of information that would be detrimental to the Navy and to the public interest; collect information of value to the Navy from all available sources (including merchant mariners, government agencies, and individuals); investigate espionage agents, saboteurs, and, as requested, Navy Department employees handling classified material; aid federal agencies in preventing the infiltration into the United States of spies and saboteurs from abroad; and set up and maintain counterespionage measures.

On 8 September 1939, CNO (OP-16-X) distributed basic orders for the District Intelligence Organization for the period of neutrality. The Director of Naval Intelligence directed the DIO-1ND to draw up a specific organizational plan for CNO approval, to proceed promptly with the enrollment of proper personnel to complete allowed mobilization assignments in the 1st Naval District, and to instruct and train the enrolled officers for their assigned duties. DIO-1ND was also expected to enroll Naval Reserve officers for duty with ONI and for service at stations outside the continental United States. The mobilization quota for DIO-1ND was 116 commissioned officers and 38 warrant officers. The guides for planning and training were manuals and other publications, such as ONI-16, *Instructions and Orders for Port Guards and Naval Ship Inspectors*; ONI-19, *ONI Intelligence Manual*; ONI-22, *Notes on Espionage, Counterespionage and Passport Control*; ONI-35, *Cable and Radio-Censorship*; and various security letters.

To carry out the recruiting program, the district intelligence officer for the 1st Naval District needed help, and Cdr. Hinrichs requested orders to active duty for LCdrs. Herbert A. Ellis, USN (Ret.) and Earl M. Major, USNR, who reported on 9 and 13 October 1939, respectively. An organization plan was prepared and forwarded to ONI for approval on 31 October 1939.

Hinrichs rarely knew how many officers he would be expected to enroll because the DIO didn't know how many of those called would pass the physical exam or how many of those enrolled would be detailed to ONI or to posts outside the United States. Of the thirty-six naval intelligence reserve officers available in the 1st Naval District in June 1939, only twelve ultimately reported for active duty in DIO-1ND.

By September 1940, only four officers and two civilian agents had arrived at DIO-1ND for active duty. Two civilian secretaries provided clerical support. The inadequate personnel situation made the DIO's task extremely difficult. Not until the summer of 1940 did funds become available for paying Naval Reserve officers for active training duty. With this financial inducement, those reserve officers available and willing were given two weeks' active duty, and they proved very helpful.[35]

Various directives in October 1940 from the Secretary of the Navy, the Chief of Naval Operations, and the Director of Naval Intelligence increased the urgency and volume of the work of the DIOs and required an immediate expansion of the District Intelligence Organization. Orders to active duty were requested for officers and enlisted men. Between 1 November and 31 December 1940, DIO-1ND received fifteen officers. Also, as a result of a Bureau of Navigation authorization, fourteen enlisted yeomen were on duty at DIO-1ND by 3 December 1940. On 26 December, the DIO headquarters was moved from the Boston Navy Yard to Room 1217 at 150 Causeway Street, Boston.[36]

A CNO (OP-16-B) letter of 8 January 1941 to all naval district commandants reiterated the requirement for meetings at least once a week of the field representatives from the FBI, ONI, and the Army Military Intelligence Division (MID). The first weekly meeting was held in the 1st Naval District in January 1941. The DIO received excellent cooperation from the local representatives from the other two agencies. The closest contact, of course, was between the headquarters of the three agencies in Boston. Additionally, Zone V headquarters and intelligence officers at the Naval Air Station, Quonset Point, and Naval Torpedo Station, Newport, Rhode Island, maintained close liaison with the FBI in Providence; officers operating in Vermont were frequently in contact with the FBI in Albany, New York.[37]

On 31 March. 1941, DIO-1ND found it necessary to establish a continuous twenty-four-hour watch at his headquarters.[38]

The Bureau of Navigation order of 18 July 1941 directed DIOs to effect minimum procurement of I-V(S) Naval Reserve officers, which for DIO-1ND was sixty-eight. On 8 October, the CNO gave approval for a DIO-1ND personnel allowance of 126 commissioned officers, 34 warrant officers, 110 enlisted personnel, and 74 civilians. When the public relations function was separated from the DIO effective 26 May 1941, three officers were detached to set up the public relations office of the 1st Naval District.[39]

Based on authorization from the Chief of Naval Operations, the commandant had established a district security office on 24 October 1941. It was not until 19 December, however, that the new office was sufficiently organized to receive the files and personnel from the Plant Protection (B-6) Section of DIO-1ND. By that time all naval facilities, except hospitals, had been surveyed and inspected, and sixty-six industrial plants had been surveyed either by the DIO or the FBI and later inspected by B-6.[40]

Censorship was not a wartime DIO function, but it was a DIO peacetime function to prepare naval censorship stations for staffing, select locations for the stations, enroll and enlist the necessary personnel, send them to special classes for censorship instruction, and arrange for the required equipment and space. Between 1 September and 7 December 1941, DIO-1ND had selected three stations: Tropical Radio Company at Hingham, Massachusetts; Radio-Marine Corporation at Chatham, Massachusetts; and Mackay Radio at Thomaston, Maine.

On 8 December, the Bureau of Navigation authorized all district commandants to call to active duty all naval personnel required for censorship. When the three DIO-1ND stations were staffed on 7 and 8 December, only two of the nine officers had been trained in censorship duties. All others who had been recruited and trained in the 1st Naval District had been taken over by ONI for censorship duties elsewhere.

Executive Order 8985 established the Office of Censorship on 19 December 1941, but it was not until March 1942 that the director of censorship was able to take over full responsibility. Effective 1 March 1942, all officers then assigned to censorship duties reported to Commander 1st Naval District, and by letter to Cable Censor, New York, for censorship duties. The personnel involved included seven I-V(S) and two deck volunteer general (D-V(G)) designated officers, four yeomen, and two civilian agents whose applications for commissions in the Navy were pending. Two censors at Naval Operating Base, Argentia, continued to be carried on the DIO-1ND rolls for many months thereafter.

In spite of the losses of functions discussed above, the demands made upon the DIO-1ND increased rapidly. The office was swamped with information that required evaluation and dissemination. Vessels in every important port in the district had to be boarded by intelligence officers. All travelers returning from European war zones had to be interviewed by a joint panel of ONI, MID, and FBI representatives. Innumerable large and small craft had to be carefully cataloged in order to keep all information up to date and readily accessible for use by other naval activities, when and as requested. Evaluated information pertinent to the security of the waterfront and the extensive New England coastline had to be kept at hand to aid the Coast Guard and other agencies in evaluating suspicious incidents off the coast from New Brunswick to the Connecticut-Rhode Island border. An extensive system of observers and informants offshore, inshore, and on shore had to be built up, organized, and given direction.[41]

DIO-1ND maintained an operational intelligence watch to keep a plot of coastal and shipping activity in North Atlantic waters. Contacts for the exchange of information were made with the 3rd Naval District (also Commander Eastern Sea Frontier), the Coast Guard, and the Army. Other sources of information included debriefs of shipboard and coastwatcher contacts.

Information from the 1st Naval District Plot was used to provide local commands with data for conducting aerial reconnaissance of unidentified contacts. Prosecution of an enemy contact could not be carried out without first receiving authority from the 3rd Naval District, the coordinating command for the whole U.S. East Coast, unless, of course, the contact presented an immediate threat to shipping.[42]

As a result of encouragement from ONI to collect and report on ninety-six organizations listed as being of particular counterintelligence interest as of October 1942, the Counterintelligence (B-7) Section of DIO-1ND in 1943 produced a large number of studies whose only naval interest was that they afforded digested information for use in judging the associations of suspected individuals. Examples of the studies are "Arabs in the 1st Naval Districts," "Fascist Organizations in 1ND," "Anti-Semitism in 1ND," "Negro Civil Disobedience Movement," "Communist Activities in Newport, R.I.," and "Albanian Activities in 1ND."[43]

The start of a reduction in district intelligence office personnel in the summer of 1943 dictated a narrowing of the counterintelligence interpretation of "naval interest." The processing of available information about national and ethnic groups and about "front" organizations was discontinued. Attention was focused on evaluating information about individuals. Interest was limited to German, Italian, and other suspects in the 1st Naval District naval establishments. Because there were so few Japanese in the district, it was possible to keep track of all of them.[44]

The "B" Branch of DIO-1ND included all of the Intelligence Sections of the office, as distinguished from the "A" or Administrative sections. "B" Branch, at various times during World War II, comprised 10 sections: B-1, Dissemination (routing officer only); B-2, Censorship (never officially activated); B-3, Investigations (combined with B-7 in May 1944); B-4, Security; B-5, Commerce and Travel; B-6, Plant Inspection (absorbed by the district security officer in December 1941); B-7, Counterintelligence; B-8, Coastal Information (later O Branch); B-9, Training; and F, Strategic Information.[45]

No case of enemy-inspired sabotage was uncovered by investigation in the 1st Naval District during World War II. The few cases of actual sabotage that were uncovered by DIO-1ND investigators were committed by individuals motivated generally by malice against supervisors, "feeble-mindedness," or other petty and purely personal reasons. Although a number of investigations did not prove sabotage had been committed, they did expose faulty industrial practices and carelessness, resulting in the correction of conditions that endangered the overall security of naval facilities and equipment. For example, the investigation in March 1943 of a flash fire on board the aircraft carrier *Lexington* (CV 16) at the drydock in south Boston exposed a hitherto unrecognized explosive hazard in the use of sprayed molten metal to coat the interiors of gasoline storage tanks. One investigation of actual sabotage established that enlisted men on board the British ship HMS *Seychelles* had sabotaged the engines during sea trials on 24 September 1944 in order to delay the vessel's departure from Boston.[46]

Surveys were conducted in 1943 to determine the points along the 1st Naval District's coastline where it might be possible for enemy agents to land. The results of the surveys were of considerable value to the district intelligence office, the FBI, and other agencies in assigning patrols and informants to cover the vulnerable areas. One of the areas, Hancock Point in Frenchman's Bay near Winter Harbor, Maine, was actually used in the successful landing of the German agents Colepaugh and Gimpel from the German submarine *U-1230* on the night of 29 November 1944. Because of the excellent surveys and the precautionary actions, DIO sources informed the FBI shortly after their landing, and Colepaugh and Gimpel were apprehended in New York a week or so later.[47]

The investigative work load, expressed in the number of cases closed each month by the B-3 Section of DIO-1ND, rose from a low of 80 in June 1941 (the first month of keeping such statistics) to a peak of 920 in May 1943. It then gradually declined to 197 in August 1945. From June 1941 to August 1945, a total of 30,519 cases were closed. In addition, B-3 completed between 1,000 and 8,000 name checks each month during the same period.[48]

The merging of B-3 and B-7 Sections in May 1944 resulted in greater efficiency with fewer personnel and achieved a more desirable union of the investigative and counterintelligence functions in DIO-1ND.[49]

DIO-1ND had five zones plus unit offices at principal shore establishments in the district:

Zone		Headqtrs.	Other Offices
I	Maine	Portland	Bath, Eastport, Ellsworth
II	New Hampshire	Manchester	(None)
III	Vermont*	(None)	(None)
IV	Massachusetts, less southern counties	Boston	Springfield, Northampton†
V	Rhode Island plus rest of Massachusetts	Providence	Hyannis, Martha's Vineyard, Nantucket, New Bedford, Plymouth, Provincetown

* Vermont was covered by Zones II and IV or by officers resident in that state.

† The Springfield office was moved to Northampton on 1 July 1943.

Unit Intelligence Offices
1st Naval District

Activity	Date
Drydock, South Boston	5 May 1941
Naval Operating Base, Newport, R.I.	7 May 1941
Naval Operating Base, Argentia, Newfoundland	14 Jun 1941*
Naval Air Station, Quonset, R.I.	23 Jun 1941
Naval Air Station, Squantum, Mass.	3 Oct 1941
Naval Ammunition Depot, Hingham, Mass.	5 Nov 1941
Naval Operating Base, Argentia, Newfoundland	8 Dec 1941
Advance Base Depot, Davisville, R.I.	11 Jun 1942
Section Base, Portland, Me.	17 Jul 1942
Section Base, Woods Hole, Mass.	17 Jul 1942
Section Base, Boston, Mass.	18 Jul 1942
Naval Fuel Depot, Melville, R.I.	1 Oct 1942
Naval Air Station, Weymouth, Mass.	1 Oct 1942
Section Base, Bar Harbor, Me.	1 Oct 1942
Section Base, Newport, R.I.	1 Oct 1942
Section Base, Rockland, Me.	5 Oct 1942
Naval Air Station, Brunswick, Me.	8 Apr 1943[50]

* Censorship only.

District Intelligence Officers
1st Naval District
1939–1945

Cdr./Capt. R. P. Hinrichs	Aug 1939–13 Nov 1940
Cdr./Capt. E. M. Major	14 Nov 1940–17 Feb 1943
Capt. E. F. Jewell	17 Feb 1943–27 Mar 1944
Cdr. T. H. Mitchell	27 Mar 1944–6 Aug 1944
Capt. E. F. Jewell	6 Aug 1944–14 Nov 1945
Cdr. T. H. Mitchell	14 Nov 1945–[51]

2nd Naval District, Newport

The 2nd Naval District was headquartered in Newport, Rhode Island during World War I. The aide for information was LCdr. S. Davis, USNRF. Other officers assigned to the aid for information office were Lt.(jg) J. Oelriche, and Ensigns H. N. Hallett, H. D. Stone, and Merritt Thompson. The intelligence organization was disestablished at the close of the war, and DIO-2ND was not activated during World War II; what would have been its area of responsibility was covered by DIO-1ND and DIO-3ND.[52]

3rd Naval District, New York City

Intelligence activity in the 3rd Naval District dated back to August 1917. During World War I and for a short time thereafter, the officer in charge was known as the aid for information and had a staff of 150. Upon demobilization, the office took on the title of district intelligence office. In addition to handling duties of an investigative nature, it maintained contact with Military Intelligence and local law enforcement agencies. It was also the naval district security office and maintained files on radical, subversive, and ultrapacifist organizations and individuals. After World War I, the 3rd Naval District's intelligence organization also included a public relations section that reviewed local Navy press releases to ensure that they disclosed no confidential information.

LCdr. Paul F. Foster was DIO-3ND from April 1928 to March 1929. Attached to his office were three civilian intelligence experts, James Cumming, Frank Watts, and Robert Peterkin. Close working relationships were maintained with the New York City Police Department, the FBI office in New York City, the Treasury Department's Secret Service, the special agents of the Post Office Department, and New York State law enforcement officers. Particular targets of mutual interest were the undercover operations of the Communist party and foreign nationals suspected of espionage activities.

The three civilian experts, in their surveillance of Japanese tourists arriving in New York City, noted that many were army and navy officers traveling in civilian clothes and that these officers made a practice of calling at, and sometimes delivering packages to, the offices of the Japanese Imperial Railways. They were suspected of turning in cameras or photographic material. Many had arrived in ships that had transited the Panama Canal, a worthwhile photographic target.

Further careful examination of the operations of the Japanese railway office and its personnel uncovered evidence that it might be the central clearinghouse for Japanese espionage activities in North America. After consultations with government officials in Washington and New York City, LCdr. Foster decided to obtain definitive evidence to confirm or negate the suspicions, and he put into effect a coordinated operation that, some months later, acquired a copy of the complete Imperial Japanese Secret Code.[53]

In March 1929, LCdr. Glenn F. Howell succeeded Foster as District Intelligence Officer, 3rd Naval District. His offices were at the corner of South and Whitehall Streets in lower Manhattan. The area assigned to Howell comprised all of New York, Vermont, the northern half of New Jersey, and Connecticut as far east as New Haven. He operated, for the most part, directly under the Office of Naval Intelligence; his work for the commandant of the district, RAdm. Louis R. deSteiguer, was of minor importance compared with what he did for Washington. Howell had a direct telephone line to ONI that he used at least once, and often two or three times, a day. Howell found the naval district commandant too impetuous and excitable to be kept informed of what he was doing for ONI, and he did not confer with deSteiguer other than to sum up the results some time after the conclusion of an operation or case. The summaries were usually given at weekly conferences that Howell had with the commandant and assistant commandant.

Cases that he did not summarize included two spectacularly successful operations. In May 1929, on five successive nights, Howell with two of his operatives, a locksmith, and two photographers entered the offices of the Japanese Inspector of Naval Machinery in New York City. On the last four nights, they were joined by an officer from Washington who could read and understand Japanese. Documents from two old-fashioned safes were selected and photostats made of them, including a 100,000-word secret code that the Japanese had been using for the past two years and was still effective; tables revealing every Japanese aircraft type with all their respective characteristics, and photographs of shells with data on their muzzle velocities. The material was acquired without arousing the slightest suspicion, for the Japanese continued to use the same code for some time thereafter.

On the night of 25 August 1929, a raid was made on the New York headquarters of the Communist party of America. DIO-3ND civilian agent James Cumming managed the operation for Howell, and it was designed to create consternation and confusion at the headquarters. Files and desks were broken open, and records were removed that were believed to be either of security value to the United States or that would disrupt the direction of subversive operations by the headquarters. In the latter category were check books, bank books, and accounting records. Files not taken were strewn about the office to make it difficult to determine what had been taken. The timing of the raid was chosen to coincide with a possible reunion between the leaders of two opposing factions of the Communist party in the United States, William Z. Foster and Jay Lovestone. The USSR had come out in support of the former, and the latter had threatened to seize the very records that DIO-3ND did take. Consequently, the Communist party headquarters blamed the raid on Lovestone and his associates. No one suspected the Navy's involvement in the event.[54]

Howell anticipated trouble at the Brooklyn Navy Yard and at other naval activities in the area on 6 March 1930 when the communists had an assembly scheduled at Union Square. To keep RAdm. deSteiguer and various Navy commands in the area informed of events as they developed, Howell placed a man in a building overlooking the square; he made reports directly to Howell throughout the day of "the worst riots in New York City history."[55]

On 1 April 1930, Director of Naval Intelligence Capt. Alfred W. Johnson came to New York for a conference with deSteiguer and Howell to discuss the triangular relationship between the commandant, ONI, and the DIO. Capt. Johnson also briefed them on the London Naval Conference then being held.[56]

As a result of the "*Florida* incident" (the distribution of subversive handbills on board the battleship *Florida* in New York on 10 May 1929), Howell also began to investigate the methods being used by the Communist party of America to subvert U.S. Navy bluejackets in the New York area. Howell's own account of his experiences as a sailor follows:

> When I took over this duty as Intelligence Officer of the Port of New York, I found myself much hampered by my ignorance of the city and particularly of those regions frequented by the sailors. One of my primary duties here is to act as liaison between the Navy and the police, and at the beginning I found myself completely ignorant of the liberty life of the sailor and the problems he encountered ashore. With Cumming and Watts [two DIO-3 agents] I tried wandering through these parts of the city, but they have "detective" written all over them. . . . [Either] of them might as well have shouted his mission from the housetops. Then my friends in the police force suggested to me that the only way I could ever understand the conditions of the sailor ashore was to be one myself. The results of my expeditions fully justified them. I gained an amazing knowledge that I continually put to good use during my contact with the police. I was able to put a stop to various activities directed against our bluejackets, and I learned just how far the Communists had gone with our men and received first-hand information as to their methods. Most important of all, I got away with these investigations with no one recognizing me or suspecting my true identity. I picked out Benny Friedman, a smart young radioman on duty here in the [District] headquarters, as my companion in these expeditions, and it was a good choice. He is shrewd, bright, quickwitted, and the fact that he is a Jew has been of continual help toward diverting possible suspicion, particularly from the Communist point of view. Then he coached me in my role until I could act my part convincingly. Benny's apartment was up in the Bronx, and here we kept my sailor suit, and I would go there usually in the early afternoon. A little later two radiomen, 1st class, would walk out of the apartment, board the El, and set forth for downtown. From Riverside Drive to Coney Island, Benny and I ranged. We were refused admission to dance halls on account of our uniform, and the next day invariably I would send one of my detectives to interview the managers and threaten Federal suit for discrimination against the uniform. My detectives were hopelessly puzzled as to the source of my minute information, but I never confided [my source], and they never guessed. I brought suit against a roller-skating rink at Coney Island for barring us on one of these expeditions, and I had the police close up several undesirable places.
>
> As soon as I was sure of myself and felt and looked at home in my suit, Benny and I began hanging around Union Square. The Reds let us alone the 1st night, but when we next appeared, about a week later, they began work upon us. Never did anyone have better opportunity to study Communism in the raw than last summer [1929], when I was doing the Square every eight or ten days.
>
> It was fascinating sitting with Benny at a big table at the Coop (The Red's name for the "Workers' Cooperative Cafeteria") on the ground floor of their Union Square headquarters, with half a dozen Reds, listening to their wild arguments, being propagandized constantly in all shades from delicate to downright, Benny and me slowly being converted, arguing back, sometimes amused, sometimes sullen as if suddenly struck with the wrongs of our lot. . . . The Reds got hopeful and believed they were converting us. Twice they insisted on driving us home, and I thanked Providence for our address, which convinced them as to our identity.
>
> A dozen times I played this game last summer. Then came my raid on 25 August on the Communist Headquarters. . . . Some of my necessary advance dope on that place I obtained as a gob, for it is amazing what you can learn from these Reds if you keep your ears open. The most important information I got this way was the fact that after months underground in vaults, the secret files had been again placed in the safes in the Communist headquarters. I knew then when to strike. After the raid, Benny and I made one more appearance on Union Square, found everybody upset and disinclined to talk, and that night, when I got out of my suit, I told Benny to put it away in mothballs.[57]

In the 1930s, DIO-3ND used naval intelligence reserve officers to collect whatever political, economic, social, and naval force information that became available in the district, which was a particularly fruitful source of information on the Far East and Russia. During the late 1930s, DIO-3ND was actively engaged in procuring information about Japanese industry, shipping, ports and harbors, economic conditions, and army and navy installations.[58]

The cosmopolitan population in New York City and its suburbs, where one could find representatives of every race and nationality of the (more or less) civilized world, meant that critical developments anywhere might, and probably would, stimulate some type of reaction, pro or con, in the city. Consequently, DIO-3ND had to be alert to international events and developments and had to anticipate any local reaction that might be detrimental to Navy and Marine Corps interests. The pockets of various nationalities in the district also offered collection opportunities when the DIOs were still responsible for fulfilling requirements for both intelligence and counterintelligence information. Hostile intelligence operatives were also working from, or attempting to exploit, the same sources during both world wars and during the periods of "peace" preceding and following World War II.

Sometimes, prior to, concurrent with, or following a coup or other foreign event, people would come to the DIO-3ND office to volunteer information about the event or the persons involved, particularly for Latin American countries where various groups who had supporters in New York were constantly struggling for power or attempting to take over the governments. The informants' decision to make their information available to the U.S. Navy probably reflected their belief that having the Navy on their side in any future attempt to take over was desirable. That is, they viewed "gunboat diplomacy" as a still useful technique for exerting power.[59]

Naval intelligence received excellent press in the New York area whenever it was needed during the 1930s. Many of the naval intelligence reserve personnel in the area were influential in the legal, shipping, insurance, newspaper, political, and banking professions. The DIO-3ND could give specific assignments to any of the Naval Reserve officers. They were all in a position to give effective and respected response to his requests, and did not accept pay or drill credits to carry out their assignments.

In about 1936, efforts were begun to recruit carefully selected men for commissions as reserve intelligence officers, and on 1 September 1939 some inactive intelligence officers were called to active duty when the national emergency was declared. Some inactive duty reservists conducted investigations without pay, and one reservist, C. J. Gass, prepared the *Manual for Security of Waterfront and Shipping*, issued during 1938.[60]

Representatives of naval intelligence were recognized as "good guys" when the cartoon character Dick Tracy was sworn in as a naval intelligence officer. For a great many people, the cartoon strip was a first exposure for them to naval intelligence.[61]

Beginning in 1935, the Japanese Desk of the Counterintelligence Section of DIO-3ND tabulated a list of "Oriental persons" in the 3rd Naval District. Following Pearl Harbor, it was able to assist in apprehending 235 Japanese aliens of suspected loyalty who were on the list, all by 1100 hours on 8 December 1941. A report received from Postal Censorship resulted in the detention of a Japanese who was familiar with the workings of radar. Information was collected and forwarded to ONI in late 1941 that secret orders had been sent by the Japanese government to all Japanese merchantmen to proceed to Japanese ports. A list of salary and bonus payments to Japanese employees at the Japanese bank in New York was secured, resulting in the collection of $400,000 in unpaid U.S. income taxes.[62]

To make the intelligence service more efficient and capable of expansion in time of emergency, the commandant on 5 October 1938 established seven outlying zone offices. Five more were added by 20 October. Both officers and civilian agents of DIO-3ND developed liaison arrangements within their zones to permit access to confidential official records, credit records, and court files. Useful contacts were also established in business, finance, industrial, labor, and public offices. Informants were developed at defense plants that had Navy contracts in order to assist in obtaining information on security effectiveness and on subversive and enemy alien activities.[63]

The Coastal Information Section was organized to collect, evaluate, and disseminate information on enemy activities and suspicious events along the coast of the 3rd Naval District. It developed liaison with federal, state, and municipal agencies whose personnel might serve as informants. Liaison was also maintained with the Eastern Sea Frontier, the Army, the Army Air Corps and the Coast Guard. Yacht club members, airline employees, and crews of fishing vessels were also developed as informants. Between 1 February 1942 and 15 June 1943, survivors of merchant ships that had been attacked by submarines were interviewed upon their arrival in the district.[64]

The district intelligence office established liaison with the FBI, Army, Coast Guard, Customs, Immigration and Naturalization, and local police and firefighting organizations along the waterfront. As a result of the Delimitation Agreement, ONI directed that the DIOs turn over to the FBI their lists of Nazis, Fascists, Japanese, Communists, and miscellaneous persons classified as "known dangerous" for immediate pick-up or as "suspects" on whom more specific information was needed.

The public relations function was removed from the DIO on 9 May 1941, and in March 1942, censorship operations were transferred to the Office of Censorship.

The DIO was charged with keeping track of labor strikes in the district that would affect war production and with investigating all known or suspected dangerous persons in the Naval Establishment. A major investigation of Communist party activities in New York Navy Yard in 1941 disclosed that a number of navy yard employees were involved.[65]

The Commerce and Travel Section came into being in the DIO-3ND early in September 1939. Initially, its work involved reporting on movements of Japanese vessels, testing the sentiment of shipping companies with respect to possible later collaboration, and establishing liaison with government agencies, particularly Customs and Immigration. Prewar activity also involved limited ship boardings, developing waterfront informants and contacts, and creating a file of merchant marine suspects.

Between 10 March and 7 December 1941, the Commerce and Travel Section organization in DIO-3ND took shape, and its work was divided into three units: ship information (ship boarding, ship search, waterfront investigations), passenger information (identification of all passengers arriving by ship and aircraft from overseas), and waterfront information (working with Customs and the port captain, particularly on port security, file checking in connection with the licensing of all small boats and fishing craft, and the employment of dock guards).

By the summer of 1942, passenger information gave way to travel control, with emphasis on counterintelligence. Close contact was maintained by the Commerce and Travel Section with Immigration, the State Department, the FBI, and the Army's Military Investigative Service.

The Commerce and Travel Section, before and during World War II, also maintained liaison with the following organizations:

American Merchant Marine Institute
Board of Economic Warfare
British Ministry of Transport
British Security Coordinator
Bureau of Immigration and Naturalization
District Coast Guard Office and Port Captain
Department of Commerce
Marine Division of the New York Police Department
National Foreign Trade Council
Netherlands Merchant Marine Commission
Norwegian Shipping and Trade Commission
Port Director
Port Authority of New York
U.S. Army Transport Service, N.Y. Port of Embarkation
U.S. Maritime Commission
War Shipping Administration

Port security was the responsibility of the Coast Guard, but DIO worked closely with the port captain to tighten security. Prior to the war, groups of naval intelligence personnel from the passenger information unit offices and, on occasion, Army intelligence officers began meeting important ships and planes from abroad. The first efforts were with the Pan American Clippers and with those American merchant ships still trading with Lisbon. The activity was under the direction of LCdr. Bernard M. Baruch, Jr. Most information at that time was of a positive intelligence nature, but there was also some counterintelligence information developed. After 7 December 1941, the clippers were among the few passenger carriers to the United States from Europe, and their passengers provided considerable information on conditions in German-occupied territories. Since clippers also touched at Africa, information was developed on Vichy French territory, Nazi agents working along the coast in cooperation with submarines, and the attitudes of native populations. Ships' crews were interviewed on possible sightings during crossings, and stewards provided information on what they saw ashore while arranging provisioning details. Groups of ship boarders interviewed travelers returning from Burma and "Flying Tigers" personnel returning for incorporation into the Army Air Corps about native population loyalties, possible locations for landing fields, and flying conditions over "the Hump" in southwest China. Passengers from Argentina sometimes provided names and descriptions of Nazi agents in that country as well as details on their methods and the propaganda being distributed there.[66]

In October 1940, Lts. Roland W. Kenney and T. Connors, both former FBI agents and recently called-up naval reservists, were sent by Cdr. Jack Hughes, Assistant DIO-3ND, to the Brooklyn Navy Yard Zone Office to check out a tip on a suspected espionage agent frequenting the bars just outside the navy yard gate. The suspect was soon in custody of the DIO-3ND officers, and interrogation revealed that he was Richard J. H. Krebs, a German national and an ex-convict who had served three years in San Quentin for armed robbery before being deported in 1929. Krebs was also a former professional Communist revolutionary and former Nazi secret agent. He was in the United States illegally, having jumped ship at Norfolk, Virginia, in March 1938.

Krebs claimed he had become disenchanted with both the Communists and Nazis, that he had a book about to be published which would expose the operations of both the Communists and the Gestapo, and that, if he could remain in the United States for a year or so, he would provide ONI with the names of all those whom he knew to be members of the Communist and Nazi parties, particularly those in the seamen and longshoremen unions

Kenney and Connors checked out Krebs's story, found it to be true, photographed and fingerprinted him, and arranged with the Immigration and Naturalization Service to have him arrested if he didn't fulfill his part of the bargain.

Krebs's book, *Out of the Night*, published under the pen-name Jan Valtin, was on the best-seller's list in 1941 and was the Book-of-the-Month selection for January 1941. Krebs also kept his word with ONI and provided some very valuable counterintelligence information before he was finally picked up by the Immigration and Naturalization Service in March 1941.[67]

One of the zone offices under District Intelligence Officer, 3rd Naval District was located at the Submarine Base, New London, Connecticut. The officer in charge during the period from October 1940 to May 1944 was the aforementioned Lt. Roland W. Kenney, USNR. His staff included Lts. Keating, Learned, and Bradus, a Chief Warrant Officer Starks, and five Civil Service clerical personnel. New London was of considerable interest to ONI as a potential target for foreign espionage, not only because of the submarine base there, but also because of the Electric Boat Company's submarine building program and the sonar development work ongoing at the Underwater Sound Laboratory.

Two noteworthy espionage cases in which the New London Zone Office became involved were those of Count Anastase A. Vonsiatsky and William C. Colepaugh (who later infiltrated the United States after coming ashore from a German submarine in 1944, as related above). Vonsiatsky and Colepaugh had lived in eastern Connecticut in towns not far from New London, and much of the early evidence of their pro-Nazi inclinations and anti-U.S. activities was developed from surveillance carried out under ZIO New London supervision. Both cases were successfully prosecuted and led to prison terms for both men.[68]

Fishing vessels were used as observation craft by DIO-3ND beginning in December 1941. Since the boats normally operated up to 500 miles from the coast, they made good advance scouts. A fishing fleet unit was set up in the DIO-3ND organization to maintain contact with the fishermen, and Navy personnel made daily visits to major ports for that purpose. Minor ports were visited two to four times a week. Vessels were equipped with Navy-owned A3 (voice) radio-transceivers, and a civilian technician with a mobile laboratory maintained them.

Since the Navy couldn't pay the fishermen, the Fishing Fleet Unit acted as the fishing industry's "sponsor" in facilitating clearances through Coast Guard, Army, and Navy restrictions affecting their operations. The unit also arranged for the adjustment of the fishermen's compasses and had information passed to them on the location of schools of fish spotted from blimps. As a collateral duty, the Fishing Fleet Unit also became the central agency for the receipt and evaluation of flotsam.[69]

Shortly after the United States entered World War II, the State Department arranged to charter Swedish vessels to return Axis internees in exchange for Americans detained in Germany. Only ONI and the FBI were sufficiently security-minded about the possibility of foreign agents using forged documents to enter the United States with the returnees. ONI took the lead and, together with the Army MIS and FBI, inaugurated "travel control."

The departure of the internee repatriation vessel SS *Drottningholm* from New York on 7 May 1942 was badly handled, too much attention being paid to "diplomatic immunity" and too little to persons and luggage *not* under diplomatic immunity. At first, the State Department didn't want any searches made, but when its representatives relented, many papers and documents were taken from nondiplomatic personnel. The free access granted to visitors to the ship may have negated some of the work, however.

In spite of efforts by the Navy, Army, and FBI to improve the organization for the intelligence interrogation of the passengers, when *Drottningholm* returned from Europe on 30 June 1942, confusion was almost as great as before. A complete lack of authority among the government agencies resulted in the passengers being held incommunicado for eight days. Naval intelligence officers remained on board to see that no one interfered with regulations.

Despite the unpopularity of the insistence on the government agents' right to interrogate *Drottningholm*'s passengers, the procedure resulted in the exposure of one Herbert Bahr. Travel Control officers became suspicious because the passenger list indicated that the U.S. consul in Zurich had been hesitant to give Bahr a visa. Bahr had been picked up by Swiss authorities for failure to declare $200 he was carrying. After questioning by DIO-3ND officers, Bahr admitted he had in his possession three volumes that might be of interest to the Navy; two dealt with reaction blading in steam turbines, and the other contained turbine performance data. Bahr claimed that he hadn't been searched when he left Germany. ONI felt he might be a German agent. Customs had received tips that Bahr might be trying to smuggle money into the country, and when ONI's search of his luggage produced $1,500 concealed under a cigar box label, Bahr admitted receiving it from the Gestapo. He was convicted under the Espionage Act.

After the Bahr case and other such incidents, ONI, Army MIS, and the FBI met and agreed on the principles required in future joint examinations of passengers. A later meeting in Washington ironed out details. Representatives of the following departments and agencies in the New York area conferred to finalize plans: the State Department, Coast Guard, MIS, FBI, ONI, the Department of Justice, the Customs Service, and the Navy Office of Public Relations and Office of Censorship. The plan called for passengers to be interviewed after baggage inspection and within 24 hours of their arrival in the United States. The new procedures were first tried out with the arrival of the repatriation ship SS *Gripsholm* on 25 August 1942, when intelligence officers were brought in from the district intelligence offices of the 1st, 4th, 5th, and 8th Naval Districts to help DIO-3ND.

The DIO-3ND section responsible for traffic control initiated an investigation that led to the arrest of espionage suspects on board SS *St. Cergue*, and it recovered highly confidential blueprints that had been lost in a shipwreck. The Traffic Control Section also furnished information that assisted in the successful prosecution of an espionage case on 25 May 1945.[70]

The Travel Control Section took on its final organizational form in July 1942 with increased emphasis on internal security and less emphasis on passenger information. From 1 January to 30 June 1944, its duties dwindled, and so did the number of personnel assigned to the section. Interviews of passengers were limited to persons of interest to the Navy, and only selected ships were boarded. Ship searches were limited to liaison with the U.S. Customs Service Enforcement Division.

The Crew Security Unit checked crew lists against the Merchant Marine Suspect List, and the resulting information was evaluated in light of the ship's itinerary and the cargo carried. If the crew list contained names appearing on the Barred List put out by ONI (under a 26 February 1943 directive), the Coast Guard could remove the crewmen thus identified. Routine checks of crew lists were stopped on 4 October 1944, and the Director of Naval Intelligence ordered the Barred List and the Merchant Marine Suspect List be destroyed on 27 June 1944.[71]

The name of the Coastal Information Section of DIO-3ND was changed to Operational Intelligence on 4 June 1943. Under the new name, it became involved in processing prisoners from enemy submarines.

During World War II, DIO-3ND, as a routine part of the coverage of waterfront activities, arranged for contact with certain underworld characters, chief among whom was Charles (Lucky) Luciano, the New York City vice king. He had been successfully prosecuted by the State of New York in 1936 and had been serving time in Dannemora prison.

Reserve officer LCdr. Charles Radcliff "Rad" Haffenden, then on active duty at DIO-3ND as the head of the Investigative Section (B-3), was the principal contact man with Luciano and requested the gangster's help to the extent of having Luciano pass the word to his underworld contacts that it was desirable to talk to Navy representatives.

In a letter written in 1945 to assist the parole board in its deliberations, then-Cdr. Haffenden extolled Luciano's contributions, but he later testified before the Kefauver Committee in 1951 that his claims regarding Luciano's value had perhaps been exaggerated. The Herlands Commission, established by New York Governor Thomas E. Dewey in January 1954, was able to establish, however, that naval intelligence officers from DIO-3ND did meet with various major criminals in addition to Luciano.[72]

Following World War II, the DIO-3ND was radically and rapidly reduced in size to a peacetime organization. Some of the personnel who had been on active duty were converted to Civil Service status or became contract special agents.

From 1946 on, the primary concern of DIO-3ND was keeping itself informed on Communist activities against U.S. maritime capabilities. Suspected Communist activities included the infiltration of seagoing and dock workers' unions, and DIO-3ND attempted to collect information on reported Communist plans to interrupt the operation of the port of New York. There were struggles between the anti-Communist American Federation of Labor and the Communist-infiltrated and -controlled Congress of Industrial Organization unions. The port was vulnerable to a complete close-down whenever subversive elements desired one.

DIO-3ND had to develop reliable contacts in the unions, on the waterfront, and at various levels throughout the merchant marine industry in order to be assured of advance information on subversive actions and corruption that might be prejudicial to the Navy's operational interests and responsibilities. In due course, DIO-3ND became the channel for communications for other government agencies and departments. For example, when a waterfront strike stopped the movement of the U.S. mail, the Postmaster General would contact the Secretary of the Navy (via the Department of Defense after that department was established) and the Secretary would have ONI advise DIO-3ND to arrange for unloading the mail.

In another instance, a tugboat strike closed the port of New York completely, and the electric companies were running out of fuel, causing blackouts. DIO-3ND was able to arrange through its contact, to have Navy tugs move vital oil barges to the electric company docks to get their generators going again.[73]

DIO-3ND was extensively involved in the clearance of personnel employed at industrial plants having Navy classified contracts during the Korean War. The 3rd Naval District in the early 1950s contained plants and industrial home offices receiving about 50 percent of the Navy's contract dollars, and background investigations were needed for many personnel who required access to classified information. Close collaboration was maintained with Navy bureau representatives and inspectors of naval material to assure that investigations were made only on those persons with bona fide requirements for a security clearance.

Sources on board ship who were potentially able to determine subversive elements in the U.S. merchant marine required briefing and debriefing around the clock, seven days a week. These sources were also used in the collection of foreign intelligence information. Sources in the Military Sealift Command were so extensive that a separate office was established at that command, with an officer and at least one agent assigned.

DIO-3ND sources within the public news media were still helpful during the Korean War. Labor editors, who were experts in the area of subversive operations, kept the office informed on anything of interest that might be going on, making it possible to take neutralizing actions against any operations that might be potentially detrimental to the Navy. Through one such contact, DIO-3ND obtained a copy of the Communist party newspaper, the *Daily Worker*, that identified Harry Bridges as the West Coast Communist leader. To admit that Bridges was a Communist in such an authoritative documentary source, when Bridges had denied it, was a gross error by the Communists, and extensive efforts were made by the local Communists to retrieve all copies. Having been alerted to the incriminating press run, DIO-3ND was able to get a copy. When the Immigration and Naturalization Service started deportation proceedings against Bridges, an Australian national, the newspaper was made available to Mario Noto, the INS attorney prosecuting the case.

As indicated from the above, in the early 1950s DIO-3ND had a very close-knit organization of collectors. Nothing happened in the district that someone in the chain was not aware of or was not in a position to resolve. If the Navy wanted to obtain information about an individual, a newspaper source was sent out to interview the person. For example, the Navy was interested in what Joseph Curran, head of the National Maritime Union, thought on certain subjects. On one occasion, Curran was scheduled to appear on a moderator-led television interview program. One of DIO-3ND's sources was able to talk to Curran while he was waiting to go on the program and got the necessary answers.

The use of newspaper sources could often be helpful to more than one party. For example, DIO-3ND learned via other sources that Harry Bridges was coming to New York for a meeting where he would attempt to organize the East Coast maritime unions and tie them to his West Coast organizations. One of the newspaper sources was alerted to get pictures of Bridges and other attendees entering and leaving the meeting. From the pictures, the DIO was able to identify known Communist party members among the attendees. Subsequent publicity on Communist participation in the union meeting let the public in on the situation for the first time.

One of the very important contributions made by DIO-3ND during the early 1950s was to neutralize the actions of the Communist party and other subversive organizations against the Navy and Marine Corps. With a valid database filled with good evidence on the membership of the Communist party and a skilled analytical team to use it when reviewing objective reports from an equally skilled investigative team, the DIO provided high-quality documentation when it was needed for effective counterintelligence action. The personnel clearance investigations conducted by Naval Intelligence were better than those from other investigative organizations, and its recommendations for denial or termination of security clearances generally stood up because they were based on supporting documentation submitted by the DIO and the Naval Intelligence organization as a whole.

Contract employees were particular targets of Communist espionage during the early 1950s. Soviet intelligence was especially interested in research and technical information on any new developments still on the drawing board. Thus, it was a real challenge for DIO-3ND to know the hostile collectors, their contacts, and their operational methods. The security of contractor plants had to be checked thoroughly and frequently, and contractor employees had to be forewarned of the espionage danger so that they could insulate themselves against being duped into security violations.[74]

District Intelligence Officers
3rd Naval District
1935–1945

Cdr. Wallace B. Phillips	4 Nov. 1935–25 Jun 1937
Cdr. Howard F. Kingman	25 Jun 1937–10 Apr 1939
Cdr. James J. Graham (Acting)	10 Apr 1939–13 Nov 1939
Cdr. Delavan B. Downer	28 Nov 1939–10 June 1940

Capt. Roscoe C. MacFall (Ret)	10 Jun 1940–16 Dec 1943
Capt. William B. Howe (Ret)	8 Jan 1944–4 May 1944
Capt. Elliott B. Nixon	4 May 1944–6 Jan 1945
Cdr. J. I. Coddington (Acting)	6 Jan 1945–30 Jan 1945
Capt. T. S. King	30 Jan 1945–10 May 1945
Cdr. J. I. Coddington (Acting)	10 May 1945–5 Jul 1945
Capt. Wallace B. Phillips	5 Jul 1945–

4th Naval District, Philadelphia

From its establishment around 1932 until 1937, District Intelligence Office, 4th Naval District (DIO-4ND) consisted of one officer, who served as a member of the commandant's staff. He occupied small quarters on the ground floor of Building No. 1 near the center of Philadelphia Navy Yard. In 1937, a civilian was added to the office, and, beginning in 1939, the staff began to grow steadily until mid-1943, when it reached its peak of 202 officers and 118 enlisted men.

In March 1941, on orders from the Secretary of the Navy, the office moved to the 14th floor of the Bankers' Securities Building at Juniper and Walnut Streets in the center of Philadelphia. The move was designed to facilitate the acquisition of intelligence information from all sources and to locate DIO-4ND near other government agencies. In the fall of 1945, DIO-4ND moved back to the Navy Yard to Building No. 734, where it was still located in 1966 when the DIO system ceased to exist.

In March 1960, the district intelligence office was placed under an officer in charge who had additional duty as intelligence officer for the Commandant 4th Naval District.[75]

Records for the activities of DIO-4ND for most of the period of its existence have not yet been researched by the author, but the basic functions and variations in organization were similar to those of the other district intelligence offices.

Naval activities served by DIO-4ND included the Naval Base and Shipyard, Philadelphia; the Naval Air Station, Philadelphia; the Naval Air Station and Naval Air Testing Center, Willow Grove; and the Naval Supply Depot, Philadelphia. Resident agents were maintained at Philadelphia, Camden, Pittsburgh, Cleveland, Columbus, Cincinnati, and Harrisburg.

In the postwar period, Reserve Intelligence Units supervised by DIO-4ND were located in Philadephia, Scranton, York, Camden, Cleveland, Columbus, and Cincinnati. Naval Air Intelligence Reserve Units drilled at Willow Grove Naval Air Station, to the north of Philadelphia.[76]

District Intelligence Officers
4th Naval District
1932–1966

Cdr. W. D. Kilduff	1932–1933
LCdr. (?) Morris	1933–1933
Capt. L.F.S. Moran, USMC	1934–1934
Maj. P. C. Marmion, USMC	1934–1935
LCdr. R. W. Wuest	1935–1937
Cdr. H. F. Newton	1937–1939
LCdr. T. A. Thornton	1939–1939
LCdr. T. H. Tonseth	1939–1939
Cdr. R. W. Wuest	1939–1941
Capt. W. T. Smith	1941–1945
Capt. T. M. Dell, Jr.	1945–1949
Capt. A. M. Kowalzyk, Jr.	1949–1951
Capt. H. P. Wright, Jr.	1951–1952
Capt. M. M. Riker	1952–1957
Capt. P. D. Williams	1957–1958
Capt. M. E. Vandera	1958–1960
Cdr. G. W. Roberts	1960–1962
Cdr. J. A. Meyertholen	1962–1964
Capt. P. J. Foley	1964–1966

6th Naval District, Charleston, South Carolina

In 1936, the duties of the district operations officer in the 6th Naval District were separated from those of the war plans officer. An officer assigned to the district as DIO was given the additional billets of district operations officer and district personnel officer. The policy of assigning one officer to the combined billets of Intelligence, Operations, and Personnel continued through 1939. In October 1939, an officer reported as port director for the Charleston area and was given additional duty as operations officer. The DIO was then relieved of all duties other than intelligence.[77]

In February 1942, the District Intelligence Officer, 6th Naval District was Capt. Sidney W. Souers. He was relieved in October 1942 by Capt. J. Lloyd Abbot. The offices of DIO-6ND were located in an old residence on East Battery in Charleston, South Carolina, except for the Operational Intelligence Section, which was co-located with the District Operations staff at the Fort Sumter Hotel. The DIO-6ND staff was composed of approximately fifty officers plus numerous enlisted yeomen and clerks who were organized broadly into Operational Intelligence, Investigations, and Security Sections.

The Operational Intelligence Section generally provided support to the antisubmarine operations of 6th Naval District surface craft and airplanes, and collected, evaluated, and disseminated information about actual or potential enemy activity. This task mostly concerned submarine activity and, to a lesser extent, the possible landing of saboteurs or espionage agents by submarine.

The Investigative Section was involved primarily with directing the investigative activities in the district, including the investigations conducted by the zone offices in Brunswick, Atlanta, and Savannah, Georgia, in addition to Charleston. Intelligence officers were also attached to various operating bases and to major shore commands like the Charleston Navy Yard, but they were under the supervision of the district intelligence officer for their investigative and other intelligence activities.

The investigations conducted in DIO-6ND were concerned mostly with the backgrounds of applicants for naval intelligence commissions and of naval personnel proposed for assignment to billets requiring access to classified information. DIO-6ND was also responsible for the investigation of suspected sabotage, espionage, or subversion involving naval personnel or facilities. In some cases, the latter investigations required the use of such extraordinary surveillance techniques as mail covers and wire taps.

Close liaison was maintained, both in the field and at the district intelligence office, with the regional offices of the FBI and Army Military Intelligence located within the 6th Naval District area. All personnel investigations included a file check with the FBI.

The Counterintelligence Section, also known as the B-7 Section, was concerned with the evaluation, collation, and dissemination, as appropriate, of information on cases involving possible or suspected espionage, sabotage, or subversion. Among its other functions was reviewing all investigative reports from the field and, where appropriate, developing and initiating further investigative leads within the 6th Naval District or other districts or commands. It also evaluated, collated, and disseminated relevant data received from other investigative agencies such as the FBI. Summaries of information from the FBI included the identification of all individuals in the 6th Naval District known or suspected of being involved in possible sabotage, espionage, or subversive activities.

An extensive card index file was maintained on the names of all persons, naval or otherwise, identified in any investigative reports, so that any potentially adverse information reflected in any report could be easily located. For example, each port director in the 6th Naval District, working in cooperation with the Operational Intelligence Section of the DIO, would furnish a crew list for each merchant ship entering any of the ports of the district. The crew lists would be checked against the card index file in B-7 for derogatory information on any crew member. If any was found, it was promptly sent to the appropriate field office, which might conduct limited investigations, including an interview with the captain of the ship and, in some instances, with the crew member himself. A close surveillance of the crew member's activities also might prove desirable.

A special counterintelligence effort was made at the zone office in Savannah because a significant segment of the city's population was of German origin. The officer in charge at Savannah Zone Office in early 1943 was Lt. Morgan Cantey, and he was followed later in the year by Lt. William Lassiter. Lt.(jg) James Newsom was assigned to the Savannah office from 28 April to 18 August 1943 to assist with the effort to screen Savannah's population.[78]

7th Naval District, Jacksonville and Miami

In May 1941, the Assistant DIO-6ND at Jacksonville, Florida, was designated DIO-7ND, and the office was subsequently moved to Miami.

As of April 1943, DIO-7ND at Miami consisted of zone offices 1 through 5 located at Daytona Beach, Tampa, Miami, Key West, and Tallahassee. It had outlying branches at Nassau in the Bahamas and at Grand Cayman Island.[79]

In 1943, DIO-7ND Capt. Elliott B. Nixon was also the Gulf Sea Frontier (GSF) Intelligence Officer. Elements of his office supported, helped staff, and supervised the GSF combat intelligence organization. They also maintained close liaison with GSF operations, DIO-8ND at New Orleans, and the army intelligence officer of the 26th Anti-Submarine Wing; maintained a plot of enemy and friendly submarines and neutral surface ships, advising GSF Operations of the estimated positions of the vessels in the frontier and contiguous areas; obtained information from the GSF radio-intelligence officer and his radio-direction finder organization and passed operational intelligence based on information from all sources to GSF Operations; and supported the GSF air combat intelligence officers attached to each operational air unit in the GSF area by briefing and debriefing pilots and crews on matters of operational and combat interest.[80]

On 14 June 1946, DIO-7ND moved from the DuPont Building in Miami to Building 907 at the Naval Air Station, Jacksonville. The DIO at that time was Capt. Laurence E. Kelly.[81]

8th Naval District, New Orleans

The District Intelligence Office for the 8th Naval District was established in New Orleans in July 1938. The Gulf area had previously been covered by DIO-6ND.

One of the difficulties in the early part of World War II was the lack of cooperation in the investigative field between DIO and FBI and Army MID representatives. There was very little exchange of information and considerable conflict in their activities. On the other hand, relations with local officials of Customs and Immigration were reportedly good.[82]

9th Naval District, Chicago

Records of the activities of the District Intelligence Office, 9th Naval District in Chicago could not be located for the period prior to 1957. In the late 1950s, the offices of DIO-9ND were located in the Customs House in Chicago and were thus separated from the district commandant and the rest of his staff at the Naval Training Station, Great Lakes, north of the city. The district intelligence officer, as a consequence, did not attend staff meetings and had difficulty convincing other staff members that the DIO was indeed a member of the naval district staff. As a result, logistical support for the DIO was difficult to obtain.

The work load at DIO-9ND consisted primarily of background investigations, with a few criminal cases. Naval activities served included the Naval Training Station, Great Lakes, and the naval air stations at Grosse Ile, Michigan; Olathe, Kansas; and Glenview, Illinois. Resident agents were maintained at Great Lakes, Milwaukee, Detroit, Indianapolis, St. Louis, Kansas City, Des Moines, Minneapolis-St. Paul, Omaha, and Denver. Supervision of Naval Reserve Intelligence Units and Naval Air Intelligence Reserve Units was a function of the district intelligence office and was handled mostly by the Reserve Intelligence Program officer and the air intelligence officer assigned to the district intelligence office.

The opening of the St. Lawrence Seaway in 1959 expanded to some extent the intelligence collection opportunities in the 9th Naval District. Masters of merchant ships that had recently called at Communist country ports were debriefed.

Cdr. Joseph A. Meyertholen was the Assistant DIO-9ND from June 1957 to June 1962. Until August 1958, Capt. A. K. Kaplan was DIO. Capt. Sam C. Loomis, Jr., served from June 1959 to April 1960, and Capt. George Raser was DIO from April 1961 to August 1961 (all dates approximate). At all other times during that period, Cdr. Meyertholen served as acting DIO until his detachment. In June 1962, after the arrival of Capt. N. D. Zimmerman as the DIO, Meyertholen was detached. During Meyertholen's tenure, the DIO and local representatives of the FBI and Army and Air Force intelligence held weekly meetings in Chicago, each hosting one meeting in four.[83]

10th Naval District, San Juan, Puerto Rico

The zone office of the District Intelligence Office, 10th Naval District, San Juan was established on 23 June 1941 in rented offices in Port of Spain, Trinidad. It was staffed by two officers. On 23 August 1941, the office was moved to the British naval base at Port of Spain, since most of its work concerned liaison with the Royal Navy. The zone office moved back to town, however, on 15 January 1942.

On 21 May 1942, a U.S. Naval Liaison Office for Trinidad was established and reported directly to ONI. Although the office was separate from the zone intelligence office (which reported directly to DIO-10ND), both functioned as a single agency. The liaison office was disestablished on 31 August 1942.

Zone intelligence officers provided briefings for departing naval ships and maintained a submarine plot. They were also charged with censorship, boarding neutral merchant ships, examining prisoners of war, and questioning alleged enemy sympathizers.[84]

Capt. Sidney W. Souers, USNR, reported as DIO-10ND in San Juan, Puerto Rico, and intelligence officer on the staff of Commander Caribbean Sea Frontier in October 1942, and he served in both positions until July 1944. A Legion of Merit was subsequently awarded to Souers for his work during the period; the citation noted his resourcefulness in gathering and transmitting valuable information for planning and executing countermeasures against enemy submarine warfare.[85]

Following the end of World War II, all sub-offices of DIO-10ND were closed except those at Trinidad and Guantanamo.

As of August 1946, the main office of the DIO-1ND in San Juan was processing very few investigative cases and performing no foreign intelligence collection. Yet, it was in the latter field that the district had great potential.

A productive Contact Register program was started whereby the office of the governor of Puerto Rico would notify the DIO every time a U.S. citizen applied through that office for a passport. Information provided to the DIO included the name and address of the applicant, the countries to be visited, the expected travel duration, and the individual's profession. If a name-check indicated "suitability," the applicant would be interviewed upon return. If the initial interview indicated an observant, worthwhile source, the subject would be added to the Contact Register.

Arrangements were made to receive passenger manifests of surface and air carriers passing through the Caribbean. The effort proved to be of particular value in keeping track of Soviet nationals and other persons of interest.

The State Department cooperated by issuing an instruction to U.S. consuls in the Caribbean area authorizing them to deal directly with DIO-10ND, who needed their help in reporting ship movements and other matters of naval interest. A special observer program was inaugurated to obtain current information on the Dominican Republic and on revolutionary activities there.

Contacts were maintained with government officials and agencies in Curacao, Venezuela; Trinidad, Jamaica; and the Dominican Republic. Dutch Naval Intelligence and Immigration in Curacao and the British Defense Security in Jamaica and Trinidad were particularly helpful.[86]

U.S. naval facilities at the islands of Eleuthera, San Salvador, and Grand Turk were not within any naval district, and their coordination control had been assigned to the Commandant 10th Naval District by 1962. Based largely on easier access to the area, investigative coverage at the island localities had been assigned to the officer in charge, DIO-6ND. In 1962, it became practicable for the officer in charge of DIO-10ND to resume the responsibility at San Salvador and Grand Turk; DIO-6ND continued to cover the Eleuthera facility.[87]

11th Naval District, San Diego

The World War II activities of two zone intelligence offices of the DIO-11ND in San Diego are of interest for the variety of exploits of personnel assigned. The parent organization itself was organized along lines similar to those of other DIOs in major coastal naval districts.

The Coastal Intelligence Section (B-8) of the 11th Naval District Zone Intelligence Office at the Federal Building in San Pedro was established in January 1941. It consisted of newly reported Lt.(jg) Frank Klaveness, USNR, and one chief quartermaster. Shortly thereafter, the two were also responsible for setting up a Commerce and Travel Section (B-5).

The functions of the B-5 section were developed as boarding commercial ships and fishing boats to get voyage reports from masters, the names of any "troublemakers" in the crew and any other information of value; maintaining liaison with Customs, Immigration, steamship companies, shipyards, customs brokers, etc.; inspecting export declarations at the Customs House; maintaining track charts of merchant ships in the Pacific based on *Lloyd's Register*, *Pacific Shipper*, and other trade journals; keeping track of fishing boat movements and activities; and maintaining a plot of wrecks in local waters. For the B-8 section, the duties were investigating reports of flashing lights along the coast; maintaining contact with civilians living on or doing business along the coast; recruiting and instructing coastwatchers; maintaining liaison with local law enforcement officials, forest rangers, operators of fish canneries, etc.; debriefing survivors of marine disasters, enemy attacks, etc.; and manning and operating the inspection barge at the San Pedro harbor entrance.

In connection with Commerce and Travel functions, a large number of foreign merchant ships were boarded, especially the Japanese vessels that continued to call at San Pedro until August or early September 1941. The Japanese ships were inspected for reinforced positions for gunmounts or any unusually heavy deck-strengthening. For those inspections, Lt. Klaveness joined the Customs Service party looking for narcotics. As a search party member, he had entry to all spaces. Klaveness found and reported many Japanese ships that had deck-strengthening for potential gunmounts.

On large passenger ships, Klaveness boarded as an immigration officer to look for individuals reported to be of suspicious intent. When they were spotted, Klaveness would signal the Customs official on the dock, who would give the individual and accompanying baggage a more than usually thorough examination.

Also during the prewar months of 1941, the Coastal Intelligence Section was instructed to locate and card index all alien Japanese living in the San Pedro area. They were listed by name, home address, name of fishing boat or place of employment, and, if appropriate, the name of the cannery that purchase their fish. The card index, completed only about a month before the war started, was the most complete and accurate listing of aliens held by any federal agency in the San Pedro area.

In early July 1941, LtCol. Stratton, Royal Marines, a member of British Internal Security Service, was in the United States to survey security in American ports. He reportedly informed DIO-11ND that the Commerce and Travel organization in San Pedro was better than at any other American port he had visited.

By November of 1941, the Commerce and Travel Section had two more officers assigned. After Pearl Harbor, there were ten to twelve officers in this section and five or six in the Coastal Intelligence Section.

In late December 1941, another officer, Lt. Howard Wickersham, who had considerable shipping experience, reported to ZIO San Pedro and was assigned to take charge of the Commerce and Travel Section. Klaveness retained Coastal Intelligence. With the start of the war, Klaveness set up three small Coastal Intelligence suboffices, one each at Catalina Island, Newport Beach, and the Naval Section Base at Santa Barbara.

The Coastal Intelligence Section maintained a barge just inside the entrance to San Pedro harbor. All commercial fishing boats and other small craft that entered the harbor were required to come alongside the barge for a document check, inspection of radios (which were supposed to be sealed), and any other search deemed desirable. The procedure started in November 1941 when the submarine net for the harbor was completed. The barge was operated during the hours when the gate in the net was open. Communications between the barge and the Coastal Intelligence Office were by radio.

On Pearl Harbor day, officers of the Commerce and Travel Section stopped the Catalina Island steamer in mid-channel, searched it for enemy aliens (none were found), and confiscated exposed film. Fishing boats were also searched, and Japanese aliens were removed from the boats and sent to the Federal prison on Terminal Island, where they were turned over to immigration officials.

Immediately after Pearl Harbor, the FBI was provided with ZIO San Pedro's list of Japanese aliens living in the San Pedro area. The Japanese were then apprehended by three-man teams consisting of one FBI agent, one San Pedro policeman, and a ZIO representative.[88]

The Los Angeles zone intelligence office on the fourth and fifth floors at 7th and Hill Streets, like the one in San Pedro, kept a file of potential German and Japanese subversives during the pre–Pearl Harbor period. In Los Angeles, the information was summarized on 3-by-5-inch cards. Much of the information came from the Los Angeles County sheriff's office. Sheriff Biscalous was a naval reservist, and his office was completely cooperative.

One of the yeomen stationed at ZIO Los Angeles learned about the *Schlarafia,* a German "social club" meeting in his neighborhood. After the auto license numbers of club attendees had been checked and many of the owners found to be already listed on the subversive cards at the ZIO, a covert raid was made in July 1942 on the *Schlarafia* headquarters. The national membership list was copied and replaced, and more copies were made and sent back to ONI and to interested district intelligence offices and other agencies. When Pearl Harbor was attacked, the FBI used the ZIO Los Angeles files in their roundup of Japanese in the Los Angeles area.[89]

Capt. Ellis M. Zacharias was DIO-11ND at San Diego at the time of the Pearl Harbor attack. The officer in charge of ZIO Los Angeles was Cdr. P. Rhodes.[90]

The first several weeks following Pearl Harbor were hectic for DIO-11ND and ZIO San Pedro, but the routine gradually settled down. At evening twilight on 23 February 1942, a Japanese submarine surfaced off Santa Barbara and shelled the oil installation on the coast at Elwood. Early in the morning on 24 February, Lt. Klaveness and a photographer's mate drove to Elwood and interviewed witnesses to the shelling, obtained as detailed a description of the submarine as possible, and took pictures of the damage. They arrived back at ZIO San Pedro at about 1700 and started working up their report.

At about midnight on 24 February, a Red Alert came over the office teletype from Senior Officer Present, Long Beach, signifying an air raid. Klaveness checked outside to see what was going on and observed antiaircraft searchlights illuminating the sky to the northwest over Santa Monica. As the "target" came closer, more searchlights came on and guns could be heard. Eventually, the object passed over the San Pedro-Long Beach harbor under fire from antiaircraft guns at Fort MacArthur and Long Beach. Then it disappeared. The press played up the event, making it a raid by eight to twenty aircraft, with one shot down. However, no bombs were dropped, and no debris from the allegedly downed plane was found. Klaveness's investigation, which included running down many of the "observer" reports, concluded that no aircraft at all had been involved in the "air raid."

In April 1942, the Commandant 11th Naval District ordered LCdr. Stanley A. Wheeler, USNR, and Lt.(jg) I. A. Camp, USNR, from the district intelligence office in San Diego, and Lt. Klaveness to conduct a survey of the Channel Islands off the coast of southern California. There had been many reports of flashing lights and other sightings and rumors that caves on some islands were being used as bases for replenishing Japanese submarines. The commandant wanted action taken to substantiate or refute the rumors. The survey was conducted mostly on foot and by boat to check out the caves on the various islands. On Santa Cruz and Santa Rosa Islands, Klaveness covered much of the area on horseback, he being the only one of the three who had had any equestrian experience. Nothing whatever of a suspicious nature was found.[91]

In early 1942, ZIO San Pedro was organized as follows: Administration Section, Investigations Section, Commerce and Travel Section, and Coastal Intelligence Section. There were approximately fifteen officers assigned to the zone office, which was located in the Federal Building, San Pedro, overlooking the main shipping channel to Los Angeles harbor. In late 1943, the office reached its peak staff level of forty-nine officers, had outgrown its space in the Federal Building, and had moved to the top floor of the YMCA building, one block to the south.

Officers of Commerce and Travel boarded each merchant ship entering the Los Angeles–San Pedro and Long Beach harbors from both U.S. and foreign ports. The master, the chief engineer, the Armed Guard officer (on U.S. ships), and any other ship's personnel of known source potential were interviewed for intelligence purposes. Crew and passenger lists were checked against FBI, Army, Navy, and Censorship files, and interrogations, as appropriate, were conducted by travel control panels made up of ONI, Army G-2, and FBI representatives. The procedure of interviewing all passengers and crew was carried out only for selected neutral flagships.[92]

Boarding reports were prepared and included information on any enemy actions, the characteristics of foreign ships, the identity of merchant seamen suspected of being enemy sympathizers or of being involved in subversive activities, disciplinary cases for the attention of the Coast Guard or the Navy port director, and matters of interest to FBI, Customs, or Immigration officials. Some time in 1943, the practice of writing reports on all boardings was terminated, and reports were prepared on only those boardings that produced information of interest to the Navy. Copies of boarding reports were distributed to the district intelligence offices in the 11th, 12th, and 13th Naval Districts, to ZIO Los Angeles, and to Commander Western Sea Frontier.[93]

Regular informants were developed on ships of Latin American registry that normally ran between South American and U.S. West Coast ports. For example, a bartender on one ship tipped off a ZIO San Pedro contact concerning narcotics being smuggled into the United States; he also reported that platinum picked up in Colombia was being carried to Argentina for transfer to Spanish ships and eventually flown from Spain to Germany, where it was used in electronic equipment.[94]

The shore activities of officers and crew members of Soviet-flag merchant ships became of interest in the summer of 1943, and selective surveillance was maintained until the resultant reports indicated that the job was of primary concern to the FBI, which then took over the surveillance effort.[95]

Several hundred former Polish prisoners taken by Soviet forces when they invaded Poland in September 1939 were sent out of Eastern Europe via the Persian Gulf and Pacific Ocean, arriving eventually at Los Angeles harbor in October 1943 for transfer to northern Mexico. They were accommodated at the Santa Anita race track during their stay in the United States. While there, Army and Navy intelligence officers interrogated about ninety persons from the group who had been selected primarily on the basis of their occupational backgrounds. ZIO San Pedro supplied the Navy interrogators. Most of the information obtained was political and economic, concerning living conditions and how the Soviet Union treated its subjugated peoples and prisoners. Polish linguists were identified from 11th Naval District personnel files in San Diego and were detailed to help the interrogators in the operation. Aside from a few instances like these, however, most boardings in the San Pedro-Los Angeles harbor area did not produce much information of military value.[96]

During the summer of 1944, the commanding officer of an aircraft carrier that arrived at San Diego reported his inability to curb a long series of thefts and other petty crimes that had prevailed for weeks among his officers and enlisted personnel. He requested and was granted the services of naval intelligence. Two days later, eight persons were in the brig, and signed confessions were on file. A crime ring that had affected the morale of the entire ship's company had been removed. The Commandant 11th Naval District considered intelligence officers properly employed when they were assigned to cases involving the morale and welfare of entire Navy units.[97]

There were other instances in which DIO-11ND instituted investigations based on leads of apparently marginal intelligence interest that developed into solutions to cases of war fraud. In the spring of 1944, a report was received that an officer at the Naval Supply Depot, San Pedro, was purchasing film for his own use under the misrepresentation that he was delivering it to the ship's service store aboard a Navy vessel. An ONI investigation disclosed that the officer was using the same procedure in the procurement of other commodities, ordering large quantities of goods from manufacturers and suppliers throughout the country that the Navy did not carry in stock and for which it had no need. The investigation led to the recovery of $76,000 for the government and a conviction by general court martial for the officer involved. Meager information on a petty theft from a general mess at the Naval Auxiliary Air Station, Los Alamitos, developed into a major conspiracy case involving numerous naval personnel and local produce merchants. Truckloads of government-purchased supplies were being diverted from regular channels into the black market. An investigation of a report of minor irregularities in charges made for repairs to government vehicles led to the discovery that the transportation officer, Naval Operating Base, Terminal Island, had removed and transported tools, lumber, and heavy machinery valued at many thousands of dollars to his ranch, which was 50 miles away. All of the above had been moved by Navy trucks, and, over a period of several months, working parties of as many as thirty enlisted men had worked intermittently on construction at the private site.[98]

12th Naval District, San Francisco

The earliest available records indicate that an intelligence officer reported to the Commandant 12th Naval District in 1921, but the organization is believed also to have been active during World War I. From 1921 to 1927, the office in San Francisco maintained normal contacts with representatives of the Department of Justice and other government agencies in connection with its work. Undercover investigations, particularly those of the Japanese espionage system, were conducted by paid confidential informants. Japanese espionage in the United States operated through various fronts, including Japanese commercial companies. It was also active in South America. In September 1922, Cdr. Royal E. Ingersoll of ONI wrote in a personal letter to DIO-12ND that the membership of the Japanese espionage organizations in South America was largely German and Latin American, with a sprinkling of Japanese. If hostilities involving Japan caused the Japanese members to depart, an organization would be left to function in their absence.

In February 1926, ONI began to organize a special group of Reserve Intelligence Volunteer (I-V[S]) officers. A list of qualified individuals in his area was submitted by DIO-12ND, and a reserve unit was established in March 1927. By December, it had sixteen officers who could fill key billets in time of emergency.

In 1935, DIO-12ND was authorized to employ a special, confidential, full-time investigator, and in 1936 a second was employed part-time. The two agents did the bulk of investigations until July 1940. Both were members of the Naval Reserve unit. In January 1941, a number of I-V(S) officers were ordered to active duty in the 12th Naval District.[99]

The intelligence information collected by DIO-12ND during World War II was mainly on military, naval, and economic developments in Japanese-dominated areas, with emphasis on coastal areas that might be scenes of amphibious landings. Major oil companies provided large-scale maps of the Dutch East Indies. Large mining companies were sources of considerable information on the Dutch East Indies, the Philippines, China, Malaya, Borneo, and the Celebes. Heavy machinery companies, however, were sources of information located outside of the 12th Naval District's area of responsibility. Importing and exporting firms, for the most part, provided information of little value, and some sources were of questionable trust. Some valuable information came from small steamship lines, but most information of potential military value was held in company offices overseas and was unavailable. Large marketing firms like the Singer Sewing Machine and the British-American Tobacco Company furnished excellent leads on individuals, but most company data of interest had been held overseas and, therefore, had been lost. Scientists proved to be valuable sources within their specialties. Local businessmen, in general, were believed not to be good information sources. Medical missionaries provided excellent information because of their wide range of travel. Many former residents of the Orient provided useful information, and national societies like the Philippine Society and the Old China Hands kept lists of all former residents and made them available to DIO-12ND. "World travelers" were of little interest, but university map libraries and professors provided much valuable information. ONI instructions prohibited interviewing foreign diplomats and consular officials, but patient, tactful liaison with British, Australian, and Dutch officials resulted in much valuable information being given voluntarily.[100]

In the operational intelligence area, DIO-12ND's Coastal Information Section was established on 4 April 1941 and was headed by Cdr. F. L. Austin, USCG (Ret). There were problems with developing an effective organization prior to 7 December, and staffing with trained officers was not achieved until the fall of 1942. Headquarters for the section was at Treasure Island, and initially it was staffed with non-Intelligence Officers. Branches for liaison with radio-equipped commercial fishing boats were established at Eureka on 28 February 1942, Morro Bay on 11 March 1942 (taken over by the Coast Guard on 11 August 1943), and Fort Point (the Presidio) on 22 March 1943.[101]

The Commerce and Travel Section of DIO-12ND began functioning in June 1941 when one officer was assigned to develop its program. He used World War I-era pamphlets for guidance. The section set up liaison with steamship lines, railroad companies, bus lines, airlines, U.S. Customs, the Immigration and Naturalization Service, the U.S. Department of Agriculture, the Maritime Commission, the Civil Aeronautics Administration, the Federal Communication Commission, the FBI, and Army Military Intelligence. A formal set of instructions on how to organize and operate the Commerce and Travel Section was received from ONI in October 1941.

Much of the information gathered by Commerce and Travel during boardings of merchant vessels was available from other sources, and in January

1945 ONI indicated that the section was disseminating too many boarding reports. Commerce and Travel learned many other useful lessons from its wartime collection efforts and modified its methods and procedures as the war progressed. Checking merchant ship crew lists for subversives was carried out until November 1944, when the duty was transferred to the Coast Guard. Boarding Pan American Clipper airliners proved unproductive because it usually duplicated similar work by DIO-14ND, and the practice was therefore terminated. Setting up observers on merchant ships didn't work well because of the frequent shifting of personnel to different ships. Determining whether entry of particular travelers or merchant ship crew members into the United States would be prejudicial to U.S. interests was a cooperative venture with the FBI and Army Intelligence, but the FBI eventually assumed the entire responsibility and informed Navy only when someone of interest to the Navy arrived.

The forms to be filled out during the processing of survivors of enemy attacks on merchant ships assumed that the sinkings would be caused by mines or submarines rather than by aircraft. Many weeks usually passed before survivors were interviewed, and there was no way to tell if they had been interviewed previously. In order to effectively process captured enemy equipment being brought into the United States, souvenirs were collected from returning servicemen unless the servicemen had a certificate indicating that the souvenirs had been inspected and released by the theater commander. Customs officials would seize and turn over to naval intelligence all uncleared souvenirs.[102]

By the last year of World War II, DIO-12ND had established a number of outlying offices. Branch B (Investigation, Counterintelligence, and Commerce and Travel) was located at 717 Market Street, San Francisco; zone intelligence offices were located at Denver and Salt Lake City; and branch intelligence offices were located at Sacramento and Vallejo. In addition, intelligence units reporting to Branch B of DIO-12ND were located at the Naval Ammunition Depot, Hawthorne, Nevada; the Naval Air Station, Alameda; the Naval Supply Depot, Oakland; the Naval Air Station, Moffett Field; and the Naval Drydock, Hunter's Point. Branch O (Operational Intelligence) was located at Treasure Island and had subordinate units at Fort Point (the Presidio), Eureka, Monterey, and Morro Bay. The Hawthorne, Nevada unit office under Branch B closed on 28 August 1945, and the Moffett Field, Denver, Salt Lake City, and Sacramento sites were closed on 12 September; all the O Branch unit offices were closed on 4 September 1945.[103]

During the Korean War, the Investigative Section of DIO-12ND usually consisted of two officers and ten to fifteen civilian agents under the district intelligence officer, who in 1952–1953 was Capt. Richard S. Andrews. The investigation backlog was between 2,000 and 3,000 cases, most of them background investigations, some as many as two years old. LCdr. D. F. Miller was directed in November 1952 to take charge of investigations and to correct the situation. By organizing the case backlog by type, he was able to clear up 1,600 delinquent cases in two months.[104]

14th Naval District, Honolulu, Hawaii

On 12 July 1921, Chief of Naval Operations Adm. Robert E. Coontz and Secretary of the Navy Edwin Denby approved Commandant 14th Naval District's "Organization and Operation Plan for the 14th Naval District in War." The section of the plan that dealt with intelligence stated:

> (1) Object: (a) To gather information, primarily of a naval character; secondarily, of a military character, and forward such information to "Intelligence" at District HQ and from there to "Operations." The former object is of greater importance for naval purposes than the latter; (b) To disseminate information from "Operations" to the various District activities, as may be necessary. To accomplish these objects without unnecessary duplication, a joint intelligence plan for complete cooperation is to be drawn up with the Military Intelligence Service.
>
> (2) To accomplish both objects, two organizations in the District are established; (a) An Open District Organization, under [the] Section Commander, composed of members of the Naval Coast Defense Reserve (Class 4). Its function is to cover the Section thoroughly, but especially the coast and shipping interests, so that the coast line and all approaches, together with the shipping operating there, shall always be under surveillance by persons living in the vicinity; (b) A Secret Intelligence organization, organized, directed by and reporting to the District Intelligence Officer.
>
> (3) The Open District Organization: (a) This organization is organized, and functions in peacetime, in order that in sudden emergency its services may be at once available. (b) Careful training of assistants is necessary in what to look for and report. Lines of communication from each assistant to his Section Commander are arranged by telephone in most cases; (c) The open organization has the additional advantage of affording a well-known line of communication by which well-disposed persons may report suspicious occurrences.
>
> (4) The Secret Intelligence Service: (a) The head of this Service is the Intelligence Officer of the 14th Naval District. He reports to the Director

of Naval Intelligence, Navy Department, via [the] Commandant of the District; (b) The Naval Intelligence Office is in the Old Naval Station building, Honolulu. This officer is in communication with the Army Intelligence Officer, to whose reports he has access. These include day-to-day information or emergency messages that reach Military Intelligence from all Army Posts or Military Agencies on the Island of Oahu. Close cooperation exists between Military and Naval Intelligence Offices, joint plans being developed to avoid duplication of work. Telephone exchange, Army HQ, Honolulu, is open day and night; (c) Under present organization, the Naval Intelligence Officer has 23 Sectional Aides for information distributed throughout the Hawaiian Islands. During war these Aides frank important dispatches at radio-stations, the Navy code, or other designated code, being used when necessary. These Sectional Aides report only to the Navy Intelligence Officer of the District. When reporting important movements of aliens or enemies, or information where time is an object, they report at once to Naval District HQ, Pearl Harbor, and Naval Intelligence, Army HQ. Effort shall be made to extend intelligence to include one or more agents on each important plantation in the islands.[105]

During the months of growing tension before the Pearl Harbor attack, Capt. Irving H. Mayfield, DIO-14ND, had been trying to circumvent the Federal Communications Act of 1934, which provided penalties of a $10,000 fine and a year in jail for anyone who violated the security of messages filed with a U.S. communications agency. Capt. Mayfield recognized the importance of getting copies of the coded communications between the Japanese consul general in Hawaii and the Japanese Foreign Ministry in Tokyo. He had tapped the consulate's telephone line, but the tap yielded no useful information. The Japanese messages remained sacrosanct until RCA President David Sarnoff visited Honolulu. During Sarnoff's visit, Adm. Claude C. Bloch, Commandant 14th Naval District, persuaded him to release to Mayfield the RCA file copies of the Japanese messages. The first batch was passed to Fleet Radio Unit, Pacific (FRUPAC) less than a week before the Pearl Harbor attack, but being in an unfamiliar diplomatic code, the messages could not be deciphered by the FRUPAC cryptologists.

After the Pearl Harbor attack, DIO-14ND and the Honolulu police made a joint raid on the Japanese consulate before all its files could be burned. Among the messages of interest were some from one of the consular officers, an ex-ensign in the Imperial Japanese Navy, who had made frequent reports on the arrivals, departures, and berthing assignments of ships at Pearl Harbor. One of the Japanese official's messages reported on air defenses at Pearl Harbor and noted the absence of barrage balloons. Another message, sent by Consul General Kita on 6 December 1941, revealed that a German national, Otto Kuehn, who had resided in Honolulu for several years, was a Japanese undercover agent. In anticipation of the loss of communications between Honolulu and Tokyo, Kuehn had devised a system to communicate with Japanese submarines off Hawaii, and it was explained in a 6 December message. He was accordingly taken into custody.

Operational intelligence support to the Hawaiian Sea Frontier by DIO-14ND was provided through an operational center with three or four operational intelligence officers assigned to the staff of Commander Hawaiian Sea Frontier. The officers maintained a 24-hour watch and kept a plot of patrols and contacts. They also maintained the monthly War Diary and submitted daily and monthly intelligence reports. Capt. James C. Landstreet was officer in charge of the operational center during the 1943–1945 period.

Prior to the invasion of Okinawa, gaps in intelligence data, particularly information about the soil trafficability of the island's beaches, led to a request to DIO-14ND to find anyone in Hawaii who might have first-hand knowledge about the target island. LCdr. N. C. Vanzant found an 80-year-old conchologist, D. D. Thaanum, who had collected seashells there. Thaanum's notes, maps, and photographs were valuable to the Joint Intelligence Center, Pacific Ocean Areas (JICPOA) in its preparation of preliminary intelligence reports on Okinawa. Thaanum also supplied the name of another conchologist, Daniel B. Langford, who had been a colleague on his shell-gathering expedition to Okinawa. Langford was located in Arizona, and Vanzant flew there to interview him. Langford proved to be a walking encyclopedia of information about Okinawa, including its beaches and reefs. At the instigation of Adm. Richmond K. Turner, Commander Fifth Amphibious Force, Langford was employed as a civilian expert to accompany Turner's staff during the Okinawa campaign, throughout which he rendered valuable service.[106]

15th Naval District, Canal Zone

The District Intelligence Office, 15th Naval District was established in 1936, but until 1938 it was staffed by only one part-time officer. By 1940, it had three officers, two enlisted personnel, and four civilians. As of December 1941, it had expanded to fifty-four officers, twenty-six enlisted personnel, and fifty-one civilians. The total of 131 included seventy-one persons assigned to censorship duties who, in 1942, were placed under the Director of

Censorship in Washington, D.C. In 1943, DIO-15ND was reportedly staffed by thirty-two officers, twenty-nine enlisted personnel, and ten civilians.

Because the U.S. Embassy, Panama, had no naval attaché accredited to it, the district intelligence officer was charged with certain attaché responsibilities and maintained close liaison with the U.S. naval attachés in other Latin American countries.[107]

The DIO-15ND main office was at 15th Naval District Headquarters at Balboa in the Canal Zone. In addition, there was a Branch Office at Cristobal on the Atlantic side, plus three intelligence units, one each at Cristobal; the Naval Base, Balboa; and the Naval Ammunition Depot, Balboa. The intelligence units were concerned primarily with the security of the facilities to which they were assigned, with the Cristobal unit serving the submarine base and the air station at Coco Solo.

The Branch Office, Cristobal, briefed and debriefed ships leaving and returning on the Atlantic side, including submarines from Coco Solo. Merchant ship masters were also debriefed as warranted. On the Pacific side, headquarters personnel performed the intelligence collection and briefing functions. The DIO-15ND in 1943 was Capt. Ralph Noisat, and the officer in charge of the Cristobal office from January to May 1944 was Lt. Joseph A. Meyertholen.[108]

As an example of the work unique to DIO-15ND, in the summer of 1941, an ONI requirement was received to collect monograph material on the Darien Province of Panama. DIO Capt. Roscoe F. Dillen, USN (Ret.), detailed Lt.(jg) E. Lawrence Adams, USNR, to perform the task. The small naval yacht *Jade* (PY-17) was assigned to convey him. Adams went upriver to get to the interior, thereafter using dugout canoes and going on foot. There were no roads. The main objective was to record what was seen and to talk with local inhabitants to determine if any foreigners had passed through the area recently. "Foreigners," of course, would be suspect as possible enemy agents, although some were known to be involved in smuggling. The report of Adams's reconnaissance was in narrative form and the information about the wild and wet area was mostly negative.[109]

16th Naval District, Cavite, the Philippines

The records of DIO-16ND were lost during the Japanese occupation of the Philippines, and the author has not been able to research any surviving documentation about the organization. A written request from a scholar to the Director of the Naval Historical Center in 1990 brought forth the information that a reserve officer assigned to DIO-16ND, Lt. George W. Greene, USNR, was believed to have been the only DIO-16ND intelligence officer captured by the Japanese (in May 1942 in the Cebu area) during World War II. Greene survived the conflict and was liberated on 1 February 1945.[110]

Chapter Notes

1. CNO (OP-14) ser 24514–39, 14 Oct 1916.
2. Testimony of Josephus Daniels, 20 May 1920, Senate, *Investigation of the Conduct of the Navy in World War I: Hearings Before the Subcommittee of the Committee on Naval Affairs*, 66th Cong., 2d sess., 1921, vol. 2, pt. 2, 2710.
3. District Intelligence Officer, 6th Naval District," History of DIO-6ND in World War II," MS, 356, OA.
4. Department of the Navy, "Administrative History of the Office of Naval Intelligence in World War II," 10 Jul 1946, unpublished MS, Appendix H, hereafter *ONI WWII Admin History*.
5. Naval Investigative Service (NIS) History File, OA.
6. *ONI WWII Admin History,* Appendix H.
7. "Administrative History of Commandant, 1st Naval District in World War II," 6:21–22, hereafter *Admin History of COM1ND*.
8. Ibid., 46–47.
9. Ibid., 8.
10. Ibid., 13–15.
11. Ibid., 40, encl. 67.
12. ONI-T-8-10, 10 Jan 1941, 9, 11.
13. Ibid., 20.
14. *Admin History of COM1ND,* 6:18.
15. Ibid., 21–23.
16. Ibid., 35.
17. Ibid., 37.
18. Ibid., 49–50, encl. 81.
19. Ibid., 55, encl. 87.
20. OP-32Y3 ltr, ser 05719P32, 11 Sep 1946.
21. OP-32D4 ltr, ser 07772P32, 18 Nov 1946, box 2, Job 3770, OA.
22. ONI-19(A), para. 4004.
23. ONI-70-1, 20 Jun 1956, 33.
24. *ONI Quarterly*, vol. 9, no. 3, 23.
25. ONI-70-1, 20 Jun 1956, 67.
26. ONI Instruction 5450.3 of 7 Sep 1956.
27. Marine Corps Order 3850.1A of 12 Mar 1957.
28. ONI Instruction 5520.59 of 27 Feb 1958.
29. ONI Instruction 12000.11 of 21 Jul 1958.
30. ONI Notice 5450 of 29 Apr 1960.
31. SECNAV Notice 5450 of 3 Aug 1960.
32. ONI Notice 12000 of 24 Sep 1962.
33. CNO (NIC-32) ltr, ser 003808P92, 7 Feb 1968.
34. *Admin History of COM1ND*, 6:2.
35. Ibid., 4–7.
36. Ibid., 15–16.
37. Ibid., 39–40.

38. Ibid., 17.

39. Ibid., 18, 21.

40. Ibid., 22, 36.

41. Ibid., 22–25.

42. Douglas F. Miller, interview, 7 Mar 1976; and ltr to author, 16 Jun 1976.

43. *Admin History of COM1ND*, 6:49–50.

44. Ibid., 55.

45. Ibid., 30.

46. Ibid., 41b.

47. Ibid., 42.

48. Ibid., 44.

49. Ibid., 45–46.

50. Ibid., 20–21, 26–28.

51. Ibid., Appendix A.

52. *Navy Directory* for 1 Sep 1918.

53. VAdm. Paul F. Foster, Oral History, Columbia University, New York, NY, 5 Jan 1967, 385–87.

54. LCdr. Glenn F. Howell, diary, 15405–12, OA. The material confiscated by Howell and his associates remains in U.S. Government custody in File 119, RG 38, NA.

55. Ibid., entry for 6 Mar 1930.

56. Ibid., entry for 1 Apr 1930.

57. Howell diary, 15871–86.

58. "History of DIO-3ND in World War II," MS 1–11, OA, hereafter *WWII History of DIO-3ND*. *Interesting Cases*, a general anecdotal history of DIO-3ND compiled by Hollis H. Hunnewell (a "Voluntary Aide"), was prepared at the end of hostilities and was privately printed and distributed; it provides a list of all DIO-3ND personnel, offices, and position titles, as well as a short administrative history written by LCdr. Spencer Eddy, USNRF, the officer in charge.

59. Capt. Martin Randisi, USNR, transcript of interview, 23 Jul 1975, 30–32.

60. *WWII History of DIO-3ND*, 1–11.

61. Randisi interview, 3–4.

62. *WWII History of DIO-3ND*, 160–63.

63. Ibid., 22–33.

64. Ibid., 61–63.

65. Ibid., 12–22.

66. Ibid., 38–49.

67. Capt. Roland N.W. Kenney, USNR, interviews, 16 Aug 1975 and 15 Jan 1976, transcript Exhibit 3.

68. Ibid., Exhibits 1 and 4.

69. *WWII History of DIO-3ND*, 184–95.

70. Ibid., 97–117.

71. Ibid., 85–97.

72. Briefing for the Director of Naval Intelligence, 1958, OA. Rodney Campbell, *The Luciano Project* (New York: McGraw-Hill, 1977) provides ample and convincing evidence of the ONI-Luciano connection and of the postwar attempts to limit disclosure of the association.

73. Randisi interview, 15–17.

74. Ibid., 18–27.

75. NIS History File, OA.

76. Cdr. Joseph A. Meyertholen, interview by author, 28 Mar 1980.

77. "History of DIO-6ND in World War II," MS, 142, OA.

78. James L. Newsom memo to author, 10 Jan 1976, 1–4.

79. DIO-7ND organizational diagram, 28 Apr 1943.

80. Commander Gulf Sea Frontier memo, 10 Jun 1943, "Intelligence Organization and Functions."

81. DIO-7ND ltr, 24 May 1946.

82. Cecil M. Deason, interview, 21 Jul 1976; and ltr, 24 Nov 1976.

83. Meyertholen, interviews by author, 25 Mar and 6 May 1980.

84. Commander Caribbean Sea Frontier, "Administrative History in World War II," MS, vol. 3, 136–37, OA.

85. RAdm. Sidney W. Souers, biographic summary, NHC.

86. DIO-10ND memo to OP-32D, ser 004039a, 31 Oct 1947, TSC #4349, OA.

87. ONI Instruction 5520.68A of 18 Sep 1962.

88. See Capt. Frank A. Klaveness ltr to author, 9 Jan 1975, for all details of ZIO San Pedro activities.

89. Miller interview, 7 Mar 1976.

90. Miller ltr, 16 Jun 1976.

91. Klaveness ltr, 9 Jan 1975.

92. Donald H. Scott, notes on ZIO San Pedro, 13 Apr 1976.

93. Scott, notes on ZIO San Pedro, 9 Sep 1976; and "Summary of Domestic Intelligence," MS, 1947, OA.

94. Donald H. Scott interview, 7 Mar 1976, Port Hueneme, CA, in author's files, OA.

95. Scott, "Summary," 1947.

96. Scott notes, 5 Sep 1976, and interview, 7 Mar 1976.

97. COM11ND ltr to CNO, 5 Mar 1945.

98. NIS History Files, OA.

99. "History of DIO-12ND," MS, No. 131, 1–6.

100. Ibid., 34–45.

101. Ibid., 54–67.

102. Ibid., 13–21.

103. Organizational chart, DIO-12ND, in DIO-12ND "Administrative Branch Manual," vol. A, chap. 3, OA.

104. Miller interview, 7 Mar 1976; and ltr, 16 Jun 1976.

105. Director of War Plans (OP-12E) ltr to CNO, 6 Jul 1921.

106. Capt. W. Jasper Holmes, "History of JICPOA," MS, 55–56, 317–18, OA.

107. "Administrative History of Commander, 15th Naval District and Panama Sea Frontier," MS, 1:70–71, OA.

108. Cdr. Joseph A. Meyertholen, interviews by author, 24 Mar and 30 Apr 1980.

109. Cdr. E. Lawrence Adams, USNR (Ret.) ltrs to author, 27 Jul and 29 Sep 1988.

110. Chris Gilleland ltr, 11 Jul 1990; Rescued Prisoner-of-War Debrief, 2 Feb 1945, attachment to Director NHC ltr to Gilleland, 2 Jul 1990.

CHAPTER 23

Censorship

On 21 December 1904, the General Board rejected a request by Lt. Albert P. Niblack, who would later serve as the Director of Naval Intelligence (DNI), to publish an article on colliers and coaling stations in the United States Naval Institute *Proceedings*. The board's letter stated: "It cannot be necessary to publish to the world . . . precise statements of military supplies on hand, of the state of work in government shops, of contracts made and of prospective deliveries under them." This rejection was considered to be merely good common sense not censorship.[1]

General Order No. 139, dated 16 December 1911, put the Office of Naval Intelligence in the censorship business:

> No person belonging to the Navy, or employed under the Navy Department, shall convey, or disclose by oral or written communications, publications, or any other means, except as may be required by his official duties, any information whatever concerning the naval or military establishment or forces, or concerning any person, thing, plan or measure pertaining thereto, or any information that might be of possible assistance to a foreign power, without the expressed approval of the Navy Department, Division of Operations of the Fleet, Office of Naval Intelligence, for scrutiny.
>
> /s/ George von L. Meyer
> Secretary of the Navy

In 1912, as a result of General Order 139, ONI became responsible for the censorship of articles written by naval officers. In the 1913 *Navy Regulations*, ONI was made additionally responsible for the censorship of photographs.[2]

Since ONI expected its naval attachés overseas to collect intelligence information commensurate with the information given to foreign representatives in Washington, ONI became the contact point for foreign representatives and became the Navy office for the control and classification of U.S. information. That position led to the establishment of a censorship section within ONI. When World War I began, ONI was the logical activity within the Navy to take on an expanded censorship function.[3]

President Woodrow Wilson, in pursuance of his neutrality proclamation, entrusted to the Navy Department the duty of acting as censor for radio communications. The President's Executive Order of 5 August 1913 was deemed "necessary to prevent radio stations under the jurisdiction of the U.S. becoming the instrument through which unneutral messages were sent to vessels at sea or to other radio stations."[4]

Naval officers were detailed to various commercial stations to act as censors, and the two high-power radio stations at Tuckerton, New Jersey, and Sayville, Long Island, which were capable of trans-Atlantic communication, were taken over by the Navy in accordance with President Wilson's order.[5]

Officers were also designated as special censors in other, more or less, high-powered and important stations, particularly at South Wellfleet and Siasconset, Massachusetts; Belmar, New Jersey; and Miami, Florida. Instructions were issued to the various naval district commandants to detail suitable officers to visit all commercial or private radio stations within their districts. The officers were to explain to the stations that they would be subject to special censorship inspections at intervals and that, at those times, copies of all messages transmitted and received would have to be submitted to the censors. At certain radio facility locations, hydrographic officers, recruiting officers, and officers on special duty in the area were directed to act as censors.[6]

The Navy's wartime action in censorship stemmed from recommendations made by the Joint Army and Navy Board in 1916 when a presidential proclamation prohibited the publication of news in time of emergency that would be of value to the enemy. From that point forward, the Navy developed its interests in cable and radio censorship, and the Army specialized in postal censorship—a division of labor that continued in effect through both world wars.[7]

Commercial radio messages to Germany were stopped if the wording did not make sense or appeared to be in code. There were raids on unauthorized radio stations. Cable censors picked up a message from Chile to Berlin concerning Chilean gold in storage in Berlin, and there were other messages concerning German nitrate in Chile awaiting shipment. ONI gave copies of the messages to Bernard Baruch, then head of the War Industries Board. Baruch did some telephoning, and soon there was American gold on the way to Chile and German nitrate on the way to the United States.[8]

In conformity with Executive Order 2604 of 28 April 1917, censorship of cablegrams passing into and out of the United States and its insular possessions over the submarine cables was established by the Navy on 30 April. The Director of Naval Communications was appointed Chief Cable Censor. Censorship of telephone and telegraph lines was a responsibility of the War Department.[9]

The censorship undertaken by ONI consisted of an exchange of information with the Chief Cable Censor; investigation of suspicious firms and individuals found to be originating, receiving, or mentioned in cable messages; and examination and disposition of the censorship material received from censorship stations. When censoring, the origin and destination of the message or article in question often determined whether or not it would be passed for transmission. From its investigations and card file of suspects, ONI had the best available body of information for evaluating the origin and destination and, therefore, played a prominent part in censorship even though it was the duty of another division in the Office of Naval Operations.[10]

The first censorship organization in the Navy is shown on an ONI organization chart of November 1917, which listed "Censorship of Press Articles" as one of the functions under the cognizance of the Filing and Distributing Section (Section D).[11]

Division I of Section D of ONI censored manuscripts submitted for scrutiny in accordance with Article 1534, paragraph 3, of *Navy Regulations, 1913*. After examination of a manuscript, it was referred to the bureau responsible for the subject it treated with a request that the bureau state whether there were objections to the publication of any statements made in the article. Statements to which the bureau objected were cut out. Additionally, all statements were deleted that were considered objectionable by the Director of Naval Intelligence or that were forbidden to be published by orders received from the Office of Naval Operations. A letter was then prepared for the Secretary of the Navy's signature authorizing publication of the manuscript as excised, and a brief that gave the Office of Naval Operations a precis of the manuscript's contents was prepared for the Director of Naval Intelligence's signature.[12]

Censorship during World War I was expected to prevent leaks of sensitive information via postal, telegraph, telephone, cable, and radio communications. It was also, however, a source for detecting spies and their methods, location, and activities; unmasking illicit trading with the enemy; uncovering fraudulent sales; locating enemy goods in storage; uncovering artificial price-fixing; providing tips on sabotage and propaganda efforts; and helping to provide the War Trade Board with a picture of where stocks of critical materials were located and how those materials were moving.[13]

Upon the general military demobilization at the end of World War I, ONI censorship of magazine articles written by naval officers was terminated, but ONI retained control over the release of Navy photographs.[14]

The Joint Army and Navy Board considered censorship again in 1935, and Joint Board No. 319, serial 566 of 11 September 1935, submitted plans for censorship to the Secretaries of War and Navy. On 18 September 1935, the Chief of Naval Operations advised ONI that Joint Board No. 319 had been approved by the Acting Secretaries of War and Navy, and the decision memorandum was quoted for information and compliance.

A plans manual, *Instruction for the Control and Operation of U.S. Cable and Radio Censorship* (ONI-35), was prepared and mimeographed in 1935 by Cdr. Frank D. Pryor, ONI's planning officer. ONI-35 became the basic naval authority for cable censorship planning and organization and for procurement and training of personnel. It proved an invaluable contribution to the operation of cable and radio censorship during World War II.

On about 2 September 1939, Cdr. Herbert K. Fenn was assigned to duty in ONI and was given the task of completing arrangements for cable and radio censorship, procuring and training Naval Reserve personnel, and effecting tentative mobilization plans for contemplated censorship stations. He was designated OP-16-D and worked immediately under the Director of Naval Intelligence.[15]

The first Cable and Radio Censorship class was begun at the Navy Department under the direction of Cdr. Fenn on 15 November 1939. The class was followed by four additional sessions, each with twenty Naval Reserve officers enrolled and lasting four weeks. Subsequent classes were held in naval educational centers in New York and San Francisco. By the time of the Pearl Harbor attack, about 400 Naval Reserve officers had received basic training in censorship duties. Enlisted personnel were also enrolled to handle minor administrative duties, filing, and yeoman work.[16]

At the beginning of World War II, Navy telecommunication censorship teams moved into communications companies and into jurisdictive censorship stations at international communication gateways at New York and San Francisco; they were operating under direction from Washington in a matter of a few hours after Pearl Harbor. The rapid response was achieved as the result of two years of planning and preparation. Smaller stations were brought under censorship control soon after, and within a week a comprehensive operating radio censorship system was in effect that included continental and overseas stations. Cdr. Fenn was promoted to captain and designated Chief Cable Censor.[17]

On 19 December 1941, the President appointed Byron J. Price as Director of Censorship for the United States as a whole. On 23 December, the Cabinet Committee on Censorship (on which the DNI was the Navy representative) met with Price. He stated that he expected to retain personnel then engaged in censorship duty in Washington and to continue the Operating Board on which Capt. Fenn was the Navy member.

On 25 February 1942, the Chief of Naval Operations directed the Bureau of Navigation to transfer naval personnel on duty in Cable and Radio Censorship from Naval Intelligence to duty under the national Director of Censorship. The actual transfer was effected about 1 March. Chief Cable Censor Fenn was ordered to report to the Secretary of the Navy, who then ordered him to report to Director of Censorship Byron Price. All other naval cable and radio censorship personnel were ordered to report to the Chief Cable Censor.

Only the Naval Censorship Unit remained in the Navy Department, and it was designated OP-16-B-2. No additional personnel were assigned to OP-16-D, and as far as the Navy Department was concerned, it had become inactive.[18]

In the Pacific Fleet area in 1941, the fleet intelligence officer was responsible for handling censorship, with the assistance of the public relations officer. It was the former who acted for Commander in Chief, Pacific in distributing to the fleet any directives on censorship that were received from the Navy Department while the public relations officer supervised the actual censorship of press and radio material. Although the fleet intelligence officer was charged with censorship duties, he was only actively concerned in those cases that aroused suspicion because of some serious violation that required investigation. With the arrival of Capt. Lloyd J. Wiltse as Assistant Chief of Staff for Administration, both the flag secretary and the public relations officer reported to him, and censorship matters were thereafter absorbed into the Administration Division.[19]

During World War II, national censorship was controlled at "the absolute discretion" of the civil director, Byron Price. Telecommunication censorship was a branch of national censorship, the censorship applied to the civilian population. The broad mission of national censorship was to deprive the enemy of information and tangibles that might be of value and to collect intelligence for use against the enemy.

Printed copies of cable and radio messages to be censored by the Navy in New York City during World War II were sent to censorship stations via pneumatic tubes. This procedure was possible because the communications companies were all located very close together.[20]

In October 1942, an officer attached to the Special Activities Branch of ONI was ordered to Canada for indoctrination in the intelligence aspects of naval prisoner-of-war mail. Upon the conclusion of his indoctrination in January 1943, he was assigned to the Postal Censorship Station in New York, where, by arrangement with the Office of Censorship, he organized a section for the review of mail to and from naval prisoners of war in the United States. A considerable volume of intelligence was collected; some, of direct operational interest, dealt with escape plans and possible rendezvous with submarines on the Atlantic coast, and the balance dealt with the movements of German naval ships and personnel. [21]

On 14 September 1943 in a letter to the Navy Department, Price called attention to the question of providing censorship in the Asiatic-Pacific area. The question was referred to the Combined Chiefs of Staff. The resulting JCS memo 873/3 of 12 November 1944 directed the establishment of a censorship organization to handle civilian communications in liberated or occupied areas in the Pacific and Asiatic theaters.

Because there was no desk in the Navy Department in 1944 that had cognizance over censorship, efforts were made to reactivate OP-16-D. Capt. Vernon Huber, who had had some previous censorship experience and was on duty on the staff of the Chief of Naval Operations, unofficially took on the additional duty to cover censorship matters. On 9 December 1944 Capt. F. H. Creech was assigned to duty as OP-16-D, and in February 1945 Capt. James W. Whitfield, who had been in Honolulu to assist in initiating a censorship program on Adm. Nimitz's staff, returned to Washington to relieve Capt. Creech as head of OP-16-D.

On or before 15 August 1945, all active censoring ceased, and on 16 August the voluminous records of the Office of Censorship (including the records of its field stations) were burned on orders of Byron Price, despite a request of 3 August from the Vice Chief of Naval Operations to the director that when the Office of Censorship demobilized, all records and files amassed by that office be transferred to the Navy Department.

On 1 November 1945, Capt. Creech was again ordered to duty in D Branch, and as OP-23-D5, he completed assembling the available civil censorship records in the Navy Department, revised the training manual and ONI-35, and prepared a *War History of Cable Censorship* for ONI.[22]

Several useful lessons were learned from the effort on cable and radio censorship during World War II. To get full value from censorship traffic, censors should be thoroughly indoctrinated in intelligence. If the censors fail to recognize the information conveyed when a message passes over their desks, they lose an opportunity to be of service. The value of censorship as a military weapon was manifest throughout World War II. Its possibilities were only partially explored, and full use was not made of censorship facilities, partly because only a limited number of naval personnel really understood censorship's functions and had no occasion or opportunity to appraise its worth. The Navy censorship group was a unit on detached duty under a civilian head, and therefore a considerable gap between the censorship operation and customary naval activity existed. From a military standpoint, carefully selected and indoctrinated naval and military personnel make the best censors; they will know more about security, be more aware of the needs of naval and military intelligence, and be less susceptible to outside pressure or political inclination.

The categories of censorship in the early 1950s were described in the May 1951 *ONI Review* as being:

> *National Censorship* may be declared by the President in the event of war, when authorized by Congress. There are three channels of National Censorship which are commonly referred to as (1) Press and Radio Censorship, (2) Postal and Travelers Censorship, and (3) Telecommunications Censorship.
>
> *Military Censorship* is a function of military command and is operated when necessary by the military departments. There are four categories of Military Censorship: (1) Armed Forces Censorship, which includes censorship of all personal communications of persons serving with or accompanying the Armed Forces of the United States; (2) Civil Censorship, which includes censorship of communications to or from civilians not attached to the Armed Forces, in territory occupied, controlled, or liberated by the Armed Forces of the United States; (3) Prisoner-of-War Censorship, which includes censorship of communications to or from POWs and similar persons held by the U.S. Armed Forces; (4) Press Censorship, which includes censorship of information and material prepared by persons serving with or accompanying the U.S. Armed Forces and intended for publication or broadcast either within or outside the United States.[23]

In 1950, the National Security Resources Board (NSRB) had overall responsibility for national censorship mobilization planning. It had two basic documents: *Censorship Mobilization Planning—Basic Policies and Principles*, NSRB document 119 of 28 July 1949; and *Censorship Mobilization Planning—Readiness Measures*, NSRB document 119/1 of 4 November 1949. The latter required the Navy to prepare for telecommunication censorship by earmarking, training, and assigning Reserve personnel to mobilization billets. It also proposed censorship station locations and estimated the numbers of personnel that would be required to staff the stations

A joint NSRB-Department of Defense committee known as the National Censorship Readiness Measures Committee was formed with representatives from the National Security Resources Board, Department of Defense, Army, Navy, and Air Force. The first monthly meeting was held in February 1950.

During the early 1950s, there were no naval officers assigned full-time to censorship. The head of the Security Branch in ONI served as the Navy representative on the National Censorship Readiness Measures Committee and did as much exploratory work as he could, using the extracurricular services of certain Bureau of Personnel officers and the occasional services of Naval Reserves during training-duty periods.

On 31 August 1950, the Navy took the first step since World War II toward concrete preparations, as distinguished from pure planning, for telecommunications censorship. On that date, the Chief of Naval Operations issued a directive to the Chief of Naval Personnel and to the commandants of all naval districts and river commands stating that district commandants (i.e., the district intelligence officers) would be responsible for telecommunications censorship within their districts. The directive also set forth the locations for censorship stations and the numbers of officers and enlisted personnel required to staff each. One Naval Reserve officer, who was to work full-time on telecommunications censorship preparations, was to be brought to active duty in each naval district having one or more censorship stations. Two officers were to be called to active duty in the 3rd and 12th Naval Districts. Additionally, Naval Reserve volunteer telecommunications censorship units were to be activated in all naval districts. Telecommunications censorship training classes were to be established in New York and San Francisco, and naval district commandants were to include in their war plans the number of telecommunications personnel required for initial operations.

About the time the telecommunications censorship directive was issued, the Telecommunications Censorship Section (OP-321C) of ONI was partially activated for planning on 22 September 1950, and Capt. S. E. Jones, USNR, was assigned as section head. The censorship programs under the cognizance of OP-321C for planning were Armed Forces Censorship and Telcom Censorship.[24]

The peacetime mission and objectives of OP-321C were to achieve, as early as was practicable, a state of acceptable readiness in each of the censorship planning programs. In 1952, in addition to seven officers on duty in OP-321C, Telecommunications Censorship planning officers were attached to the staffs of each district intelligence office, one in each naval district, except none for the 8th and 9th Districts, and two each on the DIO staffs in the 3rd and 12th Districts. The size and complexity of the Telecommunications Censorship planning problem in the early 1950s were immense. Some 563 officers and 2,804 enlisted naval reservist billets were approved in mobilization planning. The planned wartime operations were to have been conducted through a Washington headquarters, eight continental main telecommunications censoring stations, and four overseas main stations, each with a series of subordinate facilities. The planned total of main stations and other censoring activities was about 110. Continental sections would have been tied in with the headquarters by the Telecommunications Censorship National Teletype Network and would have been operated with $750,000 worth of teletype equipment owned by the Navy and already on hand in the appropriate naval districts. Each main station would have had its own local communications network with its own field monitoring stations.[25]

In the latter part of the 1950s, the Office of Civil and Defense Mobilization (OCDM, later redesignated the Office of Emergency Preparedness) was responsible for coordinating all aspects of national censorship planning. On 1 October 1959, in a letter to the Secretary of Defense, the Director of OCDM recommended that a reserve Division of Special Analysis (DSA) be established for the purpose of allocating the censorship product to user agencies, establishing and maintaining systems under which a national watch-list could be prepared and published in the event of the imposition of censorship, and undertaking technical operations (involving the technical examination of communications for clandestine use such as secret writing or hidden codes).

On 7 December 1959, the Department of Defense responded to the OCDM letter and accepted planning responsibility for the Division of Special Analysis, and the first meeting of DSA was held on 19 March 1960 with sixteen reserve officers present (six Army and five each from the Navy and the Air Force). The reservists were designated as the Joint Planning Staff of DSA, which was to be one of five divisions in the Office of Censorship. The others were Postal and Travelers (operated by the Army and Air Force), Telecommunications (Navy operated), Press (civilian), and Broadcast (civilian). DSA was distinct from the other military reserve activities in that it had the additional function of prescribing policy guidance.

Throughout the next two and a half years, the Division of Special Analysis took steps to define the operational, planning, and organizational scope of its activities. On 14 September 1962, in response to recommendations from the Office of Emergency Planning, the Assistant Secretary of Defense accepted recommendations that the Department of Defense assume the same responsibilities for the DSA as it had already assumed for Postal and Travelers Censorship and for Telecommunications Censorship and that DSA should be converted to a joint service reserve unit. The Assistant Secretary also suggested that DSA be renamed the Special Analysis Division (SAD) so that it would not be confused with the Defense Supply Agency with the same acronym.

In June 1972, a former member of SAD, LtCol. James W. McCord, Jr., USAFR, was among those apprehended and eventually convicted of breaking into the Watergate Office of the Democratic Party National Headquarters. The ensuing publicity touched on SAD and its functions. Late in 1973, at the initiative of the House Appropriations Committee, all Defense Department funds for military participation in the Office of Emergency Preparedness National Wartime Information Security Program were deleted from the Defense budget. On 31 March 1974, this participation by the Defense Department was terminated, and Telecommunications Censorship was disestablished as a Navy function in September 1974.[26]

Chapter Notes

1. General Board ltrs, 3:312, ltr no. 441, OA.

2. James R. Green, "The First Sixty Years of the Office of Naval Intelligence" (M.A. thesis, The American University, 1963), 47.

3. Ibid., 50.

4. *SECNAV Annual Report, 1914*, 23.

5. *SECNAV Annual Report, 1915*, 206.

6. Ibid., 271–72.

7. *ONI Review*, Jun 1946.

8. *New York Times*, 26 Apr 1916, 5; 7 Feb 1917, 5; and 30 Dec 1917, 4; and Green, "First Sixty Years of ONI," 60.

9. *SECNAV Annual Report, 1917*, 45–46.

10. Capt. Herbert E. Cocke, USN, "History of ONI," MS, Office of Naval Intelligence, 1931, 28.

11. Department of the Navy, "Administrative History of the Office of Naval Intelligence in World War II," 10 Jul 1946, unpublished MS, 1076, hereafter *ONI WWII Admin History*.

12. ONI, *Office Organization Manual*, 1 Sep 1918, 21B.

13. ONI-22, *Notes on Espionage, Counterespionage, and Passport Control*, 1935, 5–14.

14. *SECNAV Annual Report, 1921*.

15. *ONI WWII Admin History*, 1078–79.

16. Ibid., 1081.

17. *ONI Review*, May 1951, 102–3; and *ONI WWII Admin History*, 1082.

18. Ibid., 1085–86, 1090.

19. Commander in Chief, Pacific Fleet/Pacific Ocean Areas *Command History*, 402.

20. *ONI Review*, May 1951, 104.

21. OP-16-Z History File, "History 1940–43," box 16, Job 3679, 16, FRC/WNRC.

22. *ONI WWII Admin History*, 1091–92, 1096.

23. *ONI Review*, May 1951, 102.\

24. Ibid., 103.

25. Review of ONI Functions, Feb–Mar 1952, Appendix IV, 75, Naval Investigative Service (NIS) History Files, OA.

26. "History of the Special Analysis Division, 1959–1974," MS, OA.

CHAPTER 24

Commerce and Travel

During World War I, the intelligence services of the Allied countries regarded the checking of crews, passengers and cargoes as one of the most important means for preventing the sabotage of vessels or cargoes, and for uncovering enemy agents and spies. Carrying clandestine correspondence and contraband by travelers became a regular business, and large sums of money were paid for smuggling persons, correspondence, and material to or from the belligerent countries. Examination of baggage, ships, and cargoes, without a concurrent search of the persons of the ships' officers, crews, and passengers, was futile. The seagoing public quickly learned that only baggage and ships were being searched, and this loophole was exploited by hiding contraband, letters, and other matter on the person and within clothing. Letters were even written on the skin.

The importance of maintaining rigid control and supervision over all persons and merchant shipping entering or departing U.S. ports was apparent to ONI immediately after America's entry into the war. No definite plan to deal with the problem was considered, however, until U.S. naval attachés abroad drew attention to the rapid increase in smuggling enemy agents traveling to and from the United States on neutral ships and the letter-carrying by ships' crews.[1]

Under the terms of the Espionage Act of 15 June 1917, the Secretary of the Treasury was to make,

> subject to the approval of the President, rules and regulations governing the anchorage and movement of any vessel . . . in the territorial waters of the United States, [and could] inspect such vessels at any time, place guards thereon, and, if necessary in his opinion in order to secure such vessel from damage or injury, or to prevent damage or injury to any harbor or waters of the United States, or to secure the observance of the rights and obligations of the United States, may take, by and with the consent of the President, for such purposes, full possession and control of such vessel, and remove therefrom the officers and crew thereof and all other persons not specially authorized by him to go or remain on board thereof.

Section 4 of the act provided that "the President may employ such part of the land or naval forces of the U.S. as he may deem necessary to carry out the purpose of this title."

Following a Presidential Proclamation invoking the authority of the Espionage Act, the Secretary of the Treasury issued certain general rules and immediately called upon the Navy for aid in protecting shipping.

Imprecise rules and regulations and lax enforcement by civilian departments immediately caused friction and misunderstanding. To obviate the difficulties, various conferences were held and supplementary orders and instructions were issued from time to time. Until ONI formulated, issued, and executed a plan for the protection of shipping in U.S. ports, for the most part, vessels were unprotected.[2]

The civilian personnel of the Immigration Service, as represented by its Inspection Division, were not sufficiently experienced to handle properly a thorough intelligence investigation. Immigration inspectors were trained for, and familiar with, only Immigration Service requirements. They had no concept of an intelligence inquiry. War experience proved that the examination had to be performed by intelligence personnel, and naval intelligence learned that only lawyers (and good ones) could conduct a proper examination. It was doubtful if Customs or the Immigration Service could have recruited a properly trained, equipped, and experienced force for inspection work, as such a force had to be made up largely of men with legal training and the ability to examine and cross-examine. This type of person was easily secured by the Navy but not by Customs or Immigration because those with the necessary qualifications preferred service with the armed forces rather than with civilian departments in time of war. During World War I, the Customs and Immigration services failed to recruit qualified individuals, but the Navy had an unlimited number of volunteers from which to pick.[3]

In July 1917, ONI instituted steps to establish proper control over all seaborne traffic at U.S. ports. It was then realized that all such powers were, by law, assigned to the Treasury Department. ONI, nevertheless, took measures to stop all illegal traffic at American ports on its own responsibility, regardless of the lack of legal sanction.

The final intelligence examination, as conducted by ship inspection units in World War I, as well as the responsibility for final clearance of passengers by naval personnel, was without actual authorization except as it was covered in the Supplemental Order of 21 October 1918 issued by the Department of State. The Navy activities, nevertheless, were permitted by the other interested government departments because the Navy was recognized as being the best equipped to handle the situation. Its lawyers and other experienced personnel were capable of making competent inquiries through oral examinations and responsible reviews of extensive suspect files prior to any examination.

Although the State Department instruction for shipping control had contemplated a board composed of representatives from the Immigration Service, the Navy, and Army Military Intelligence in the case of suspects, the Ship Inspection Board was actually composed only of Navy men, and only the Navy Department took preparatory action before a vessel's arrival. The Immigration Service detained suspects at the request of the Navy, and further investigation of the suspects and their testimony was conducted by Navy Ship Inspection personnel.[4]

In November 1917, the Treasury Department issued its first set of regulations under the new Espionage Act of 15 June 1917. The Customs Service thereupon began to build up an organization in New York that came into conflict with the Navy Ship Inspection Force, which had already been functioning efficiently. The Navy's forces were thereupon withdrawn. But the Customs Service, with no force at any port but New York, requested that the Navy conduct ship searches and personnel examinations at all ports except New York, where Navy personnel simply accompanied Customs Inspectors on board all vessels, stood by, and watched.

Finally, the passage of the Passport Act on 22 May 1918 gave the Navy representatives a legal status in Ship Inspection Service, and thereafter a great part of the friction with the Customs Service was eliminated.[5]

Under an agreement arranged between representatives of the interested offices and bureaus of the Departments of State, War, Justice, Commerce, Labor, and the Navy, ONI was charged with keeping strict surveillance over all passengers arriving in or departing from U.S. ports. All information received by a government department about the arrival or attempted departure of suspects was referred to ONI and transmitted by it to the district intelligence office in the relevant naval district. It was then the duty of the district intelligence officer, through his Ship Inspection Section, to locate the suspect and have him detained by Immigration Service officers, pending investigation. ONI collected any evidence, and, if sufficient proof of guilt was established, it took action through the Department of Justice, leading to prosecution and incarceration. Later, the surveillance by ONI was extended to include ships, their officers, crews, and cargoes, in addition to the passengers.[6]

The ultimate level of efficiency finally attained by ONI in its passenger and crew control at U.S. ports during World War I was made possible largely through the hearty and complete cooperation of the State Department. From the moment the United States entered the war, the State Department exerted every effort to assist the War and Navy Departments in checking enemy traffic moving via American ports.

In all cases of persons seeking permission to depart or enter the United States, State Department instructions provided that

> the Secretary of State, upon receipt of an application for a passport [visa, or identity card], shall forthwith submit notice thereof to the Military Intelligence Division, ONI, and the Bureau of Investigation of the Department of Justice for such information as said respective intelligence services may have available upon the applicant and for such comments as they desire to make.

The four government services operated in all phases of travel control with mutual confidence and interchange of information. Requests for passports, visas, or identity cards were denied by the State Department on the recommendation of any of the three advisory services.[7]

Two preliminary steps were taken to control incoming passenger travel. First, all persons coming from neutral European countries were required to apply to an American consul for a visa for their passports, subject to approval by the local U.S. naval and military attachés. In addition to the investigations made by the attachés and their agents in the home territory of the applicant for a visa, the names of the applicant and their references were submitted by cable to the State Department, ONI, and the Army's Military Intelligence Division (MID), where searches of their suspect files were made. If no adverse information developed from any of the available sources, the applicant's passport visa was approved, and he or she was permitted to sail. Upon arrival at a U.S. port, the passenger was required to appear before an examining board consisting of representatives of the Customs Service, the Bureau of Immigration (Department of Labor), the Bureau of Investigation (Department of Justice), ONI, and MID. If no information was developed against the passenger, he or she was permitted to land. Searches of the persons, if desired, and a search of all the baggage of the passengers were conducted. The baggage of outward-bound passengers was also subject to search.[8]

Before the departure of every neutral vessel from a foreign port for the United States, the entire crew list was furnished to the U.S. consulate at the port of departure, and the consul in turn submitted the list to the local naval and military attachés. If a subsequent investigation found any crew member who might endanger the safety of the United States, the consul refused a bill of health until the person had been removed from the ship. Whenever any information was discovered against a ship or the ship's crew after her departure for the United States, the attachés had only to cable ONI, and proper action would be taken by the Ship Inspection Service upon the vessel's arrival in the United States.

When a ship arrived at a U.S. port, the crew was mustered and checked against the crew list, and the individual seamen were checked against their respective identity cards and photographs. The muster was conducted by Immigration inspectors, with Navy representatives present. If any shortages in the crew list were discovered, or if seamen were found on board in excess of the crew list, the vessel was specially guarded against persons attempting to slip ashore. In some cases, ships were placed on the "black list" for future voyages until officers, and even the masters, had been removed. During the searches, the Customs inspectors also examined the ships' papers and cargo manifests.

Owing to the enormous number of deserters from neutral merchant ships early in the war, it was necessary to issue an order keeping the crews of all neutral vessels on board during their ships' stays in U.S. ports. The prohibition had the added advantage of obviating the need for any examination of crews of outbound ships, except in the cases of new crew members who had been signed on at U.S. ports. A search of these vessels for enemy letters was also considered unnecessary.

A representative of the Navy's Ship Inspection Section was present at the checking and inspection of cargoes conducted by the Customs Service. If a shipment was reported as suspicious, the Navy representative could request an examination, and if, in his judgment, he thought the shipment should be sequestered, the Customs inspector was requested to hold up the shipment.

During the last year of the war, the Commerce and Travel Section of ONI, which had charge of shipping and travel control, increased in scope and effectiveness until it also became an important factor in the nation's efforts to combat enemy trade. Close relations were established between ONI and the intelligence service of the War Trade Board. Valuable information on trading with the enemy, obtained largely by U.S. naval attachés in Scandinavia, Argentina, Brazil, and The Netherlands, was disseminated to the War Trade Board, the Department of Commerce, Army MID, and the State Department.

The War Trade Board cooperated with the Navy to the fullest extent, holding up undesirable imports and exports and otherwise exercising its legal powers wherever those of the Navy were lacking. Firms suspected of trading with the enemy, particularly in neutral countries, were investigated by the Navy for the War Trade Board and the Department of Commerce, as were the transfers of ships to new owners under U.S., Allied, or neutral registry. The War Trade Board was empowered by presidential proclamation to forbid all transfers of shipping until the prospective purchasers had been investigated and found satisfactory.

The War Trade Board expressed the wish that all examinations and searches conducted on behalf of the board be made by representatives of ONI. The searches were necessary because merchant vessels, foreign and domestic, were found carrying enormous quantities of spare machinery and electrical parts, far beyond the normal levels. The spares were taken on in U.S. ports, having been purchased in small quantities from various dealers as "stores" for the vessel and not as cargo. Thus, they were obtained without export license from the War Trade Board and were not entered or declared on cargo manifests. When the vessel returned from a voyage, all the stores would be found to have been "expended," although the quantity was entirely too great to have been used up during the voyage.

In addition, radio equipment was frequently purchased by vessels while lying in U.S. ports and installed on board in the guise of repairs. When the vessels returned, the new apparatus would be missing. Also, apparatus installed in one port might be found missing and then duplicated at another port of call. The same happened with other spare parts and shipboard consumables. Thus, a vessel touching at Boston, New York, Philadelphia, Baltimore, and Norfolk could, and no doubt often did, take on a full line of stores and radio parts in each port for disposal abroad. In addition to the supplies mentioned above, the most important item taken out illegally in quantity was lubricating oil. Ships left New York for Holland with enough lubricants to carry them around the world but, upon returning to New York, were found to have barely enough remaining to have made port.[9]

The Commerce and Travel sections of ONI and of the naval district intelligence offices were deactivated at the conclusion of World War I.

ONI-16 (Revision 1938), entitled *Instructions for the Control and Operation of Commerce and Travel Sections of the Naval Intelligence Service*, was a confidential, registered publication, signed by Chief of Naval Operations Adm. William D. Leahy. It superseded ONI-16, *Instructions and Orders for Port Guards and Naval Ship Inspectors*, which had been issued in October 1932. Both documents supplied the organizational and procedural guidance, based on World War I experiences, that would be needed in any future mobilization of ONI Commerce and Travel activities.

The Commerce and Travel Section (OP-16-B-5) was reactivated in the autumn of 1939, primarily because of the Navy's increasing responsibility growing out of the enforcement of the newly instituted U.S. neutrality laws. In October, a few reserve officers were appointed to prepare a Commerce and Travel training manual. They assumed that, as had happened in 1917–1918, the civilian government agencies would fail to pick up the ball, and the Navy would once again find the whole problem of ocean travel control under its jurisdiction. Reserve officers were not aware that a higher authority had already taken action early in 1939 to formulate a joint plan with the civilian agencies. In a letter of 17 April 1939, Director of Naval Intelligence RAdm. Ralston S. Holmes had stated that the plan was about to be concluded; actually, the joint plan was not accepted until 1940.[10]

In the plan (later promulgated in ONI-19, *Naval Intelligence Manual*), the Commerce and Travel Section was described as "the center to which flows information concerning travellers, crews, ships and shipping." The joint task, in which the naval intelligence function was limited to cooperation with the appropriate action agencies, was defined as preventing loss or injury to merchant ships and waterfront facilities by subversive acts of passengers and crews, compromise of shipping information, arrival or departure of enemy agents or enemy sympathizers, and loss of cargoes essential to the war effort. The effort was also intended to obtain information about foreign areas, especially ports, and to gather reports from the survivors of enemy actions.[11]

The organization of the newly reestablished Commerce and Travel Section consisted of a Commerce Unit, with a Boarding Subunit, a Shipping Information Subunit, and a Casualty Subunit. The Travel Unit had a Travel Authorization Control Subunit, a Passenger Security Control Subunit, a Crew Security Control Subunit, and a Special Studies Unit.[12] Chapter 22 contains additional information on the activities of the Commerce and Travel units at the ONI and naval district levels during World War II.

By 1957, the scope of activities at the Commerce and Travel sections in the district intelligence offices had evolved to include collection of foreign intelligence from sources other than travelers and waterborne personnel. Accordingly, the name Commerce and Travel Section in each district was changed to Foreign Section.

Among the sources used by the foreign sections to obtain information were foreign naval personnel attending U.S. schools, local press reports about foreign naval personnel and shipping in the United States, shipyard officials involved in building or overhauling foreign naval or merchant ships, U.S. Naval Reserve personnel, and other naval district intelligence office security and counterintelligence sources having foreign information. The intelligence collection and security efforts of the foreign sections continued until the DIO system was disestablished in 1966.[13]

Chapter Notes

1. ONI-16, *Instructions and Orders for Port Guards and Naval Ship Inspectors*, Oct 1932, Art. 1001.

2. Ibid., Art. 1003.

3. Ibid., Art. 1004.

4. Ibid., Art. 1005.

5. Ibid., Art. 1006.

6. Ibid., Art. 1007.

7. Ibid., Art. 1008.

8. Ibid., Art. 1011.

9. Ibid., Art. 1013.

10. Department of the Navy, "Administrative History of the Office of Naval Intelligence in World War II," 10 Jul 1946, unpublished MS, 374–75.

11. Ibid., 376.

12. Ibid., 392.

13. ONI Notice 05550 of 5 Aug 1957.

Chapter 25

Foreign Disclosure

Overt collection of foreign technical information by the Office of Naval Intelligence received impetus with the inception of the military attaché system in the late nineteenth century. Open requests for foreign military information was the normal collection method and became the responsibility of military attachés, passing out of the hands of the State Department's diplomatic corps. Attachés were accredited to the military services of the host countries so that government-to-government exchanges of classified military information could be made at the service level.

During the Spanish-American War, ONI was assigned the task of receiving and processing requests for information about the U.S. Navy from foreign representatives in Washington accredited to the U.S. Navy. As navies converted from wood and sail to steel and steam, intelligence collection activities increased among the world's naval powers, primarily through naval attaché systems.[1]

In 1899, the policy of nearly all governments was to be liberal in matters of information on hardware, but giving such information was carried out by most foreign nations on a basis of reciprocity. Heretofore, ONI and the Navy generally had an open door to all foreign inquirers, and the result was that our naval attachés were frequently without the bargaining power necessary to obtain similar information abroad. Consequently, Navy Department Order No. 22 was issued to require that formal official information that was to be supplied to foreign attachés or other foreign officials in Washington or abroad be furnished through ONI to assure that it was commensurate with the information made available to U.S. naval attachés. In part, the requirement for quid pro quo remedied the prior inequity and brought about an understanding between foreign attachés and the Navy Department that proved mutually advantageous.[2]

Contacts with the Navy Department by foreign attachés as of 1921 were all made through ONI. They were not allowed to go anywhere in the department except through ONI.[3]

In the late 1930s, interest in the collection of technical information on foreign naval developments revived as rapid advances in naval technology again created the impetus for navies to keep abreast of events abroad. Up to the beginning of World War II, foreign naval attachés retained their responsibilities as primary collectors of U.S. Navy information, and the responsibility for controlling the release of information to them remained with ONI. No formal disclosure policy existed, but the clear definition of extant military objectives made questions of release and denial relatively clear-cut. While no administrative changes were made in the Department of the Navy that affected the processing of foreign requests for information, the importance of disclosure as an instrument of national policy gained recognition.

In 1934, on the basis of an informal exchange of letters, the Secretaries of State, War, and the Navy formed a committee to provide guidance to the Army and Navy for considering any disclosure matters that might affect national policy. The committee, named the State-War-Navy Military Information Committee (SWNMIC), was composed of a member and an alternate from each of the three departments, and it met on an informal schedule. In its considerations during World War II, the committee looked not only into the security aspects of any proposed disclosures to the many Allied countries, but also weighed the political and economic factors. The committee continued in existence throughout World War II and achieved formal official status at the end of the conflict when it became a subcommittee of the State-War-Navy Coordinating Committee (SWNCC).

In 1945, the SWNMIC subcommittee produced a disclosure policy in broad terms known as SWNCC 206/9 (MIC-1). It received limited distribution within the service departments, and was used as a guide by their releasing agencies. When problems arose that involved exceptions to policy, the case was referred to SWNMIC. In 1948, ONI issued *Security Directives for Classified Matter, No. 1–48*, to implement SWNCC 206/9 (MIC-1).

The national policy governing disclosure of classified military information in 1947 permitted the Navy Department to release practically any category of information, including top secret, to the United Kingdom. Excepted was information about atomic energy, cryptography, Latin America, and the Philippines; information on sources of intelligence; any information derogatory to the United States; and a few other specific categories. The British military agreed that any U.S. military information released to them would be adequately safeguarded, used only for governmental military purposes, and not be released to a third government without the specific approval of the Navy Department. The various British Dominions and colonies were considered separate nations for purposes of the proviso.[4]

In 1949, the SWNCC was abolished, but, with the establishment of the North Atlantic Treaty Organization (NATO) and the intelligence information release policies that were concurrently developed, the Military Information subcommittee survived and assumed the new title State-Defense Military Information Control Committee (S-DMICC).[5]

A foreign disclosure problem arose in 1950 that was related to the procedure for the release of intelligence information to NATO commands involved in the Korean War. Application of the national release policy, the coordination between the Armed Services in the execution of the policy, and the fact that the policies for release of intelligence to each of the various NATO member nations were at considerable variance with each other and with the existing NATO release policy, placed a major burden on ONI's Collection and Dissemination Section. It was necessary for the cognizant evaluating desk, the producing desk, and the release desk to coordinate their views as to the release of specific intelligence materials. Intelligence produced jointly with another U.S. service required further coordination for approval prior to release.[6]

To eliminate the difficulty, S-DMICC issued a revised and updated disclosure policy in MIC 206/29, and the Army, Navy, and Air Force revised their directives accordingly. ONI issued ONI-321, *Disclosure of Classified Military Information*.

Late in 1955, S-DMICC revised the basic policy again in MIC 206/29 (Rev) *Policy Governing the Disclosure of Classified Military Information to Foreign Governments* (commonly known as the U.S. National Disclosure Policy). The revised document reflected more accurately the increased responsibility assumed by the United States in the cause for world peace and enhanced the ability of the individual friendly nations to contribute toward the effort. In the Navy, ONI-321 was superseded by OPNAV Instruction 005510.48. Under the Navy instruction, information was exchanged between the U.S. Navy and its foreign counterparts through U.S. naval attachés abroad and through foreign naval attachés in Washington on a reciprocal basis under security controls established and maintained by ONI.[7]

Final responsibility within the U.S. Navy for the decision in each case of disclosure was assigned to the Director of Naval Intelligence. While certain delegations of authority had been made for limited and clearly defined purposes, the promulgation of the U.S. Navy disclosure policy was principally intended for guidance purposes. All disclosures or releases of classified material to be made by any command, activity, or individual to a foreign representative had to have the prior approval of ONI.

In July 1956, Commander in Chief, Atlantic (CINCLANT) sent representatives to a joint U.S.-United Kingdom intelligence conference on counterintelligence matters in the Caribbean area. These conferences were thereafter discontinued by order of the Joint Chiefs of Staff, the required intelligence exchange being accomplished through staff liaison between Commander Caribbean Sea Frontier and British commands in the area.

CINCLANT released U.S. classified information to foreign nationals in NATO organizations such as Supreme Allied Commander Atlantic in accordance with authority granted for the purpose.

In 1956, CINCLANT requested the Office of the Chief of Naval Operations to re-evaluate the Foreign Officer Integration-Exchange Program with particular regard to the security aspects of assigning foreign officers to sensitive operational billets, where they were inevitably being exposed to highly classified material not authorized for release to the nations from which they came.[8]

Chapter Notes

1. *ONI Review*, Dec 1956, 509.
2. Capt. Herbert E. Cocke, USN, "History of ONI," MS, Office of Naval Intelligence, 1931, 12–13.
3. Adm. Royal E. Ingersoll, Oral History, Columbia University, New York, NY, 1965, 46, copy in OA.
4. DNI memo to "Mr. Kenney", 24 Nov 1947, in OP-321K Top Secret Security Correspondence files, OA.
5. *ONI Review*, Dec 1956, 509–10.
6. Commander in Chief, Pacific Fleet, "History of Administrative Problems, Korean War," 2:16, OA.
7. *ONI Review*, Dec 1956, 510; and ONI-70-1, Change 1 of 4 Nov 1957, 56.
8. Commander in Chief, Atlantic Fleet, *Annual Report, FY 1957*, 17

CHAPTER 26

Plant Protection

World War I

The World War I activities of the Office of Naval Intelligence in plant protection, a manufacturing field not previously recognized as a Navy function, required a large number of persons whose qualifications were fitted for that type of work. ONI

> first made a list of all materials, including agricultural or manufactured, in storage at the port of New York. It then became the business of the Office to be informed as to all facts of a suspicious nature relating to firms, individuals, and companies doing business under contract with the Navy Department; as to firms whose business may be in contravention of the "Trading with the Enemy Act"; as to alien suspects and their activities; as to the apprehending of deserters and absconders; as to evidences of German propaganda; as to information relating to the enemy, sympathizers, or suspected spies taking passage on vessels sailing to or from the U.S.; and as to suspicious ships and shipments.[1]

The first step taken by ONI to ensure the adequate protection of private manufacturing plants engaged in Navy work during World War I was the adoption and insertion in all Navy contracts of the following clause:

> In addition to the ordinary precautions, heretofore adopted by the contractor for the guarding and protection of its plant and work, the contractor shall provide additional watchmen and devices for the Navy Department against espionage, acts of war, and enemy aliens as may be required by the Secretary of the Navy. The contractor shall, when required, report to the Secretary of the Navy the citizenship, country of birth, or alien status of any and all employees. When required by the Secretary of the Navy, he shall refuse to employ, or if already employed, forthwith discharge from employment and exclude from his works, any person or persons designated by the Secretary of the Navy as undesirable for employment or work for the Navy Department.

The Office of the Secretary of the Navy then sent forms to each contractor, to be filled out in duplicate, giving information on the extent and nature of the plant's contracts with the Navy Department, a nationality census of its employees, and a general description of the precautions adopted by the company to protect its plant and the materials used to manufacture products for the Navy Department. The form required miscellaneous facts that would enable ONI to determine promptly the relative importance of the various plants.

Approximately 5,000 plant security forms were filled out and returned to ONI from prime contractors, their subsidiaries, and their subcontractors. Duplicates of the executed forms were forwarded to appropriate ONI branch offices and naval district aides for information. Through receipt of the plant security forms, the Plant Division (Section A, Division VI) of ONI was also in a position to keep Secretary of the Navy Josephus Daniels and the appropriate Navy bureau chiefs informed about the progress of work on Navy contracts.

In general, the Plant Division confined its security recommendations to physical improvements at the plants and to the enforcement of law regarding the employment of enemy aliens. The Plant Division also participated in the investigation of prospective employees. In the vast majority of facilities engaged in Navy work, extensive and badly needed improvements were made, involving, in the aggregate, the expenditure of hundreds of thousands of dollars by the contractors themselves, with but a single instance where the government paid the cost. Frank discussions between the contractors and ONI representatives gave the contractors a more thorough appreciation of the need for extensive precautions in the protection of their facilities and in the selection of employees. They cooperated with the ONI representatives and were anxious to take actions that might prevent sabotage and improve fire prevention. As a result, there were few explosions and little damage by fire or sabotage in plants having Navy contracts during World War I.[2]

In protecting industrial plants having Navy contracts, ONI investigated both strikes and threatened strikes, and frequently its representatives were able to disabuse the minds of the labor leaders, settle difficulties, and get production resumed. The most dangerous element disclosed by plant investigations was the rampant propaganda of the International Workers of the World unions, Bolsheviki, etc. To counteract the effect of propaganda, enterprising ONI representatives addressed the employees, put up posters, and took other steps to stimulate the loyalty of the workers. Because the collapse of Russia afforded German agents the opportunity to gain access to the United States by way of Siberia, ONI instructed U.S. naval attachés in China and Japan to devote special efforts to obtaining information on any suspects attempting to take passage from the Orient to the United States.[3]

The plant protection sections of ONI and naval districts were deactivated at the end of World War I.

World War II

Near the end of 1939, the FBI proposed that it take part in the internal security of private plants involved in the production of material for the armed services and that FBI agents make plant surveys and inspections and perform follow-up "check ups" to ensure that adequate steps had been taken by plant owners to correct any deficiencies. The Army and Navy accepted the offer.[4]

During the period of growing international tension in the 1930s and until the FBI offer had been accepted, the Planning Branch (OP-16-X) and the Investigations (OP-16-B-3) and Security (OP-16-B-4) Sections of the Domestic Intelligence Branch (OP-16-B) of ONI had performed plant protection functions intermittently, and usually independently, without much coordination. In October 1939, however, a new ONI section was established (OP-16-B-6) to "maintain cognizance on behalf of CNO [Chief of Naval Operations] as to security in plants of importance to the Naval Procurement Program." OP-16-B-6 was given the title Plant Protection, and initially one officer was assigned to it.

ONI considered Plant Protection as primarily an inspection service to be operated from naval district headquarters and conducted by selected personnel specifically qualified by their civilian training and experience. While OP-16-B-6 favored increasing ONI's responsibilities in plant security, the Planning Branch urged that ONI's involvement be held to narrow limits and that the inspection duties be assigned to the "various agencies and bureaus placing contracts with the firms in question." Consequently, when the FBI offered to take responsibility for plant surveys and inspections, the Planning Branch view prevailed.[5]

The basic directive on Plant Protection for private commercial facilities was a Secretary of the Navy (OP-16-B-6) confidential letter dated 20 November 1939. It stated that the FBI would make surveys to embrace "physical properties," "personnel methods," and all other aspects bearing on the security of government property and material being used to carry out Navy contracts. The letter further reported that the FBI had been given a list of plants for which security had been designated as being important, as determined by the cognizant Navy bureaus. A copy of the plant list was furnished to each naval district commandant concerned. The commandants were to be notified by the local FBI offices of any contemplated surveys and were to be provided with copies of survey reports and recommendations.

Throughout 1940, the FBI made plant inspections.The Navy bureau interested, by reason of a contract, in the security of a particular private plant would recommend that a survey be made, and the Navy Department would request that the FBI send an agent to the plant. After the survey, the agent would draw up a report, including recommendations for any necessary improvements. The report would be discussed with the plant management and then be sent to ONI for forwarding to the cognizant Navy bureau. Thirty days after the initial survey, the FBI agent would revisit the plant to see if recommendations had been carried out. There is no evidence, however, that the Navy Department took any action if a plant failed to act.

By November 1940, the volume of reports on plant surveys coming in from the FBI had become so great that additional officers had to be assigned to handle the paperwork in the Plant Protection Section of ONI.[6]

Although the Navy was supposed to limit its requests for surveys to private plants that had Navy contracts, the FBI in October 1940 was obliged to write to ONI protesting the inclusion of plants "that have no government contracts and which have never had Government contracts." The protest was the first sign that the FBI was having difficulty in fulfilling its responsibilities in the plant protection program.[7]

By December 1940, the Plant Protection Section had drafted procedures for the acquisition, evaluation, and dissemination of information received from the FBI. ONI was also giving attention to the question of internal security in the naval shore establishment. Secretary of the Navy letter serial 0273816 of 25 October 1940 (drafted by OP-16-B-6) requested that all district commandants make recommendations for improvements in the internal security of the activities under their commands. The OP-16-B-6 letters committing ONI to an extensive internal security program were obviously in conflict with the earlier Planning Branch concept.[8]

The problem of centralized responsibility for plant protection was not unique to ONI. Even within the Navy Department itself, the naval districts and the fleet maintenance divisions of the Office of the Chief of Naval Operations (OPNAV) were interested in industrial security, and their activities sometimes bypassed ONI. Also, the technical bureaus occasionally initiated independent security actions. The FBI, the Army's Military Intelligence Service, and the Federal Power Commission also had varying responsibilities for plant protection, the last being concerned about the security of the nation's electrical utilities.

The problem of coordinating industrial plant protection security activities of the various government agencies was never completely resolved. For example, if a navy yard depended on a municipal power plant for its electricity and a security weakness was apparent, ONI, the FBI, the Bureau of Yards and Docks, the Naval Districts Division of OPNAV, the navy yard itself, and the local naval district commandant could all find the situation within their cognizance. As a general rule, the agency discovering a potential danger would deal with the problem and would seek assistance as necessary. The practice, however, inevitably led to some duplications of effort.[9]

ONI infrequently investigated personnel at a few private plants, even though such activity was not provided for in existing directives. The unauthorized practice, no doubt, occasionally resulted in duplication of FBI actions.[10]

Fear of sabotage dominated ONI's thinking on plant protection. Domestic "subversive" elements were looked upon with more concern than were potential German or Japanese agents. ONI's expectations were that the primary danger lay in attempts by American Communists to obstruct U.S. preparations for war. The view seems to have survived even after Germany's attack on Russia on 22 June 1941.[11]

As of 1 February 1941, there were 312 private plants on the Navy's list of facilities needing FBI surveys. Of those, 133 were regarded as "very vital facilities," 63 as "other important facilities," and 116 as "relatively less important."[12]

By the end of March 1941, J. Edgar Hoover advised ONI that he was thinking of asking the President to relieve the FBI of plant inspection work. Accordingly, on 26 March, the Domestic Intelligence Branch suggested to OP-16-B-6 that a corps of plant survey personnel be established.

On 17 April, Hoover made a direct request to terminate FBI inspections of private plants of interest to the Navy. Besides pointing out the ongoing duplication of effort, he based his request on the fact that the FBI could not exert financial pressure on the private plants to improve their internal security, whereas the Navy bureaus could.[13]

At the time the FBI decided to turn over the responsibility for private plant surveys to ONI, it had completed investigations on 706 of the 1,021 plants that the Navy had requested be surveyed. In addition, 150 surveys were in progress and would be completed by the FBI. The 165 facilities for which surveys had not been initiated, plus all future requests, were to be taken over by ONI.

The details of the transfer of responsibility for plant surveys were worked out in the ensuing weeks. ONI had previously (14 April 1941) requested authority to assign thirty-eight officers for all the naval districts to cooperate with the FBI and the inspectors of naval material in advising plant owners on security measures. As a result of its new industrial security responsibilities, ONI estimated on 1 May 1941 that fifty-one additional personnel, either officers or civilian agents, would be required to be furnished by the naval districts immediately. It was proposed that persons assigned to plant security duties be ordered to Washington to attend a three-week course in plant inspection at a school which opened on 12 May at 416 5th St., NW. The instructors were officers attached to Domestic Intelligence.

The transfer was initially scheduled for mid-May 1941, but it had to be postponed until shortly after 1 June. A directive issued on 21 May gave detailed instructions to the new plant inspection officers. Personnel security was emphasized almost exclusively. Of the seven main headings in the instructions, five dealt exhaustively with various aspects of personnel security: Selection of Personnel, Personnel Morale, Fingerprinting, Supervision (Identification) of Employees and Visitors, and Guard Force (Patrolling Entrances). The other two headings were brief outlines on Material and Fencing and Floodlighting.

Although the transfer of plant survey responsibility from the FBI to ONI gave the latter a more active role in industrial security, plant security still remained a staff function. ONI had no direct responsibility for plant protection and could only advise the organizations that did have that function. The security of commercial plants working on classified and/or aeronautical material for the Navy was the responsibility of the plant owners. The security of Navy classified material kept at the plants rested jointly with the local naval district commandant, the inspector of naval material, the technical bureau concerned, and the plant owner.[14]

Concern for the security of the naval shore establishment, in which ONI had already had an interest before it took over private plant inspections, stimulated an increased possibility that U.S. shipyards would be used, under Lend-Lease, for the repair of battle-damaged British ships. On 17 March 1941, an ONI administration (OP-16-A) letter to the commandants of all the naval districts warned that "it is not unreasonable to expect that enemy agents will attempt to sabotage them."

In mid-1941, a committee of British experts made an inspection of all U.S. continental naval districts and found that the security of U.S. naval shore facilities was "generally good," although waterfront patrols and identification procedures were inadequate. Security of commercial shipyards was "good to poor," but that of commercial docks and piers was "universally poor." The British report of 4 September 1941 emphasized that steps needed to be taken to coordinate the efforts of the Army, Navy, Coast Guard, Customs, Immigration, federal, state, municipal, and private interests in planning for the security of coastal areas and that wartime responsibility for security should be vested in one authority. Additionally, the report recommended that a set of security rules similar to the British Defense Regulations be prepared for immediate promulgation in case of war. The British group also supplied copies of British manuals and instructions, such as *Anti-Sabotage Precautions and Port Regulations*, which were circulated by ONI to all naval district commandants.[15]

On 12 October 1941, Plant Protection was split into two sections. OP-16-B-6 retained the responsibility for physical surveys and inspections of industrial plants. The new section, Personnel Security (OP-16-B-10), assumed the responsibility for personnel security in the naval shore establishment. Also in October 1941, at the suggestion of the Naval Districts Division (OP-30) of OPNAV, the CNO established a billet for a district security officer in each naval district, with the officer to be responsible for all security measures in the district and to serve on the commandant's staff. No clear line was drawn, however, between the duties of the district intelligence officer, who supervised the Plant Protection Force, and the district security officer.

Suddenly, in December 1941, the headquarters structure for Plant Protection was changed as the result of two conversations held on the afternoon of 12 December. The Assistant Secretary of the Navy at 1530 phoned OP-16-B-10 and asked one of the officers of the section to come over to his office and explain the situation. At the conclusion of their discussion, the head of OP-16-B-6 was similarly summoned. Prior to the two conversations, the Assistant Secretary had known little about internal security, but afterwards he apparently felt he had acquired enough information to recognize that a change was needed. He asked the head of OP-16-B-6 to make recommendations for improving the situation.

Accordingly, a letter to the Assistant Secretary was prepared by OP-16-B-6, containing a primary recommendation that the Plant Protection Section of ONI be moved into the Assistant Secretary's office where it could have direct authority over both the district commandants and the bureau chiefs for making the security improvements found necessary as a result of plant inspections. If there was concurrence with the recommendation, the field Plant Protection Force, then under the district intelligence officers, should be placed organizationally under the district security officers. The letter was undated and unsigned, but it was approved in the margin by the Assistant Director of Naval Intelligence. There is no evidence, however, that the letter was seen by OP-16-B, the head of the Domestic Intelligence Branch. On about 14 December, the Assistant Secretary phoned OP-16-B-6 to ascertain the progress being made on his requested recommendations. He was given the substance of the letter over the phone. The next day, he forwarded the recommendations to the Secretary of the Navy, and on 16 December, the transfer was made. The former OP-16-B-6 with its two lieutenants (junior grade) and five civilians became the Security Division of the Assistant Secretary of the Navy's office. Additionally, the Plant Protection force in each naval district was assigned to the district security officer.[16]

On 17 December 1941, the Assistant Secretary of the Navy, in letter serial 0202, defined internal security as

> safeguarding industrial facilities from sabotage, espionage, fire, robbery, and involves plant protection, traffic control, care in selection of employees, alarm systems, and security of essential services (water, electricity, etc.). . . . In brief, internal security is the safeguarding of an activity from any incident, except open attack, which might disrupt its normal efficacy.

This definition became the guide for the responsibilities of the Security Division of the Office of the Assistant Secretary of the Navy. By including sabotage and espionage, which were personnel security matters, the Assistant Secretary had assigned to his Security Division responsibilities still retained by OP-16-B-10, thus ensuring a bureaucratic conflict between ONI and his office.[17]

Numerous discussions took place between the new Security Division and the Army over which organization should be responsible for surveys and inspections of plants of common interest and with the Coast Guard over who had responsibility for ships being built or fitted out at private shipyards. On 25 February 1942, Presidential Executive Order 9074 stated, "The Secretary of the Navy shall be responsible for measures to safeguard against destruction, loss, or injury from sabotage or other subversive acts, accidents or other similar cause, all waterfront facilities except [those] operated by the War Department." "Waterfront facilities" was interpreted as including private shipyards. The executive order terminated ongoing Navy negotiations with the Army to have the latter take over the security surveys for private shipyards.[18]

On 15 June 1942, the Security Division was administratively transferred back to OPNAV in order to place it in the line of military command with the district commandants. The Security Division was retitled the Internal Security Section (OP-30-S) of the Base Maintenance Division now designated OP-30.

The all-inclusive definition of internal security, promulgated at the time the Security Division was established in the Assistant Secretary's office, was continued when the division became OP-30-S. Although OP-30-S had a direct responsibility over personnel security matters, OP-16 continued its active functions in personnel security activities in naval establishments. In addition, there remained in the Assistant Secretary's office a Personnel Section that took action on OP-16's findings in the field of personnel security. Thus, confusion continued, primarily within OP-30-S, as to which Navy organization had authority over personnel security.

OP-30-S continued to write proposals recommending a consolidation of the organizations involved in internal security. On 28 January 1943, it proposed setting up a Security Division under the Vice Chief of Naval Operations to include representatives from ONI, the Coast Guard, and all Navy bureaus. The proposal was vetoed, but it did lead to moving the Personnel Section of the Assistant Secretary's office to OP-30-S in February. A similar proposal made on 16 February included the merging of the naval district intelligence office and the district security officer organizations. This idea also met with little success, but the discussions it stimulated brought forth a letter promulgated on 27 March 1943 that defined the respective duties of the district intelligence officers and the district security officers. ONI and its DIOs won the bureaucratic battle and retained their dominant interest in the personnel security aspects of plant protection throughout the war.[19]

Korean War

In December 1950, the billet of district security officer on the staff of the naval district commandant was reactivated. Plant protection was one aspect of industrial security and was assigned to the Logistics Division of OPNAV. The other aspect was the protection of classified materials in the hands of industry, which was the responsibility of ONI. The district intelligence officers were to advise district security officers of any information concerning physical security or plant protection or of related matters pertaining to naval shore facilities or commercial plants under the security cognizance of the Navy.[20]

Chapter Notes

1. Testimony of Josephus Daniels, 20 May 1920, Senate, *Investigation of the Conduct of the Navy in World War I: Hearings Before the Subcommittee of the Committee on Naval Affairs*, 66th Cong., 2d sess., 1921, vol. 2, pt. 2, 2710.

2. RAdm. Albert P. Niblack, USN, *The History and Aims of the Office of Naval Intelligence* (Washington: GPO, 1920), 19–21.

3. ONI, *Office Organization of ONI*, 1 Sep 1918, 11, 15.

4. "Administrative History of Internal Security (OP-30-S) in World War II," MSS, 5, OA.

5. Department of the Navy, "Administrative History of the Office of Naval Intelligence in World War II," 10 Jul 1946, unpublished mss, 214–16, hereafter *ONI WWII Admin History*.

6. "Administrative History of Internal Security," 6.

7. *ONI WWII Admin History*, 232.

8. CNO (OP-16-B-6) to all Naval District Commandants, ser 0288116, 2 Dec 1940; and "Administrative History of Internal Security," 7.

9. *ONI WWII Admin History*, 217–18.

10. "Administrative History of Internal Security," 7.

11. *ONI WWII Admin History*, 220.

12. Ibid., 231.

13. "Administrative History of Internal Security," 7–8.

14. *ONI WWII Admin History*, 232–35.

15. Ibid., 258, 260–61.

16. "Administrative History of Internal Security," 12, 16–19. The transfer was effected by Secretary of the Navy ltr ser 411216, 16 Dec 1941, to all bureaus and naval district commandants.

17. Ibid., 2, 19.

18. Ibid., 24–25.

19. Ibid., 30, 33–36.

20. ONI Instruction 5540.1 of 9 Mar 1951.

CHAPTER 27

Mission, Functions, and Responsibilities of ONI

Through extensive quotation of official organizational mission and function descriptions, this chapter provides a documentary history of the mission, functions, and responsibilities of the Office of Naval Intelligence.

Establishment

When ONI was established by Secretary of the Navy William H. Hunt, his General Order 292 defined the purpose of the office as "collecting and recording such naval information as may be useful to the Department in time of war, as well as in peace." This first statement of ONI's field of concern, being very general and completely lacking in specifics, made the scope of its information responsibilities almost limitless. ONI's limited capacity to receive and process information, however, kept the nascent organization within the bounds prescribed by the department's most urgent requirements.[1]

Soon after Lt. Theodorus B. M. Mason reported as the Chief Intelligence Officer, the Secretary gave him fairly specific guidance on what ONI's responsibilities should be in the letter of 25 July 1882 (quoted in full in Chapter 1). The unique aspect of Hunt's guidance was the requirement that ONI keep informed on our own Navy, mercantile marine, and coast defense. The same letter also reiterated Lt. Mason's pre-ONI proposal that intelligence collection should be accomplished by naval attachés assigned to U.S. foreign legations, but then added that collection should also be performed by forces afloat. The Navy bureaus were not told they couldn't collect, but ONI was given the task of collating what they did collect, and the bureaus were required to make any information they acquired available to ONI.

Publishing a monthly bulletin was another function assigned to ONI in Hunt's directive, and the Office of Naval Intelligence continued the practice until April 1963 when the *Defense Intelligence Digest* superseded the intelligence periodicals of the three military services. When the need and volume of material so dictated, such as during World War II, ONI's periodical was published weekly. Conversely, it was published quarterly when funds were scarce. In its early version, when ONI was part of the Bureau of Navigation, the monthly bulletin included information on U.S. ship movements and orders to officers, assuring reader interest.[2]

Formal Codification of ONI's Mission, Functions, and Responsibilities

The first General Instructions in Regard to Intelligence Duty appeared in a Navy Department order of 31 March 1885. The order could not be found in the files, but it is referenced in the department's order of 28 October 1892, which reissued the same order in modified form, and included this statement:

> The duty of the Office shall be to collect and classify information upon all subjects connected with war, or which can have a bearing upon naval action, and to prepare plans of campaign covering all contingencies of active naval operations.

From this statement, it can be inferred that ONI was responsible for the Navy's war planning, beginning in 1885.[3]

In 1896, ONI received another nonintelligence task when Secretary of the Navy Hilary A. Herbert directed it to test various schemes proposed by Congress and others for relieving the promotion stagnation then occurring in the Navy. There is, unfortunately, no known report on the test results. It is of possible interest, however, that of the ten officers (all lieutenants or below) on duty in ONI in mid-1896 when the tests were made, five ultimately attained flag rank.[4]

Mission Modified: 1903

In 1903, the Navy Department's order of 28 October 1892, Instructions in Regard to Intelligence Duty, was superseded by a modified order:

> The duty of the Office shall be to collect and classify information upon all subjects connected with war or which can have a bearing upon naval action or plans of campaign and to cooperate with the General Board in the preparation of detailed plans covering all contingencies of active operations by naval forces afloat or on shore.

Under the heading "Office Work" is the following:

> Matter which is of educational value and professional interest alone does not come within the scope of the office; only matter which is of military importance or of material convenience to U.S. Men-of-War cruising should be retained. Exceptions to this rule may be made in the case of important inventions and discoveries, which may later become of military value through adoption by some foreign services.

The order also instructed ONI

> to bring to the notice of the Bureaus and certain Offices of the Navy Department all matter which may seem to be of value to them whether it be carded [indexing and cross-indexing by subject for a filed report] or not.[5]

The General Board of the Navy, in its letter No. 435 of 29 September 1904 to Secretary of the Navy Victor I. Metcalf, recommended the adoption of a revised form for the inspection and classification of merchant ships proposed by the Chief Intelligence Officer, subject to adding details on operational history, repairs, etc., considered by the General Board to be important to the determination of the actual condition of each merchant ship. The General Board also concurred in the request of the Chief Intelligence Officer for a ship draftsman in his office.[6]

World War I Mission, Functions, and Responsibilities

In May 1916, the General Board, in its mobilization plan of that date, assigned the following duties to ONI:

> The Office of Naval Intelligence will investigate and prepare for war in the Atlantic as follows:
>
> In general:
>
> a. Arrange for securing full information of the state of preparation of the enemy's Navy, having especial regard to the length of time it will take the enemy to mobilize his entire fleet.
>
> b. Arrange for securing prompt information from abroad of the movement of the enemy's forces, particularly from and after the time war becomes imminent. For this purpose, select reliable agents at various points and ports in and near enemy country and in probable field of operations; and keep the list of such dependable persons corrected to date.
>
> c. Prepare a complete system of secret service and cipher codes to be used in communicating with such agents.
>
> d. Keep a complete list of merchant vessels under U.S. registry that may be desirable for use during war, corrected to date. (List [to be] kept by War Plans Division).
>
> e. Keep track of orders and shipments of munitions of war to the enemy.
>
> f. Make such plans and arrangements now as will reduce paper work and other work of investigation and preparation to a minimum on the eve of, and during war.[7]

Counterintelligence was added to ONI's responsibilities in 1916; for details on the subject, see Chapter 21. As described by Secretary of the Navy Josephus Daniels in 1918, the function of ONI in time of peace was to keep in touch with naval problems, naval expansion, and subjects of naval interest in all the countries of the world so that the President, the naval administration, and the responsible naval leaders in Congress could have access to the latest information with reference to naval affairs abroad. ONI's reports from its attachés in the leading maritime countries, and from other sources, furnished the General Board, the Chief of Naval Operations (CNO), and the Navy bureaus with data that gave them the benefit of what had been learned, or put into practice, in other countries.

In time of war, Daniels continued, ONI's activities included observation, investigation, and reporting on all subjects affecting the Navy and the prosecution of the war from a naval point of view. ONI was required to report on naval operations at sea and on land, and on the status, changes, and progress of the material and personnel of foreign navies; it was also required to conduct a close counterespionage watch at home. The latter duty included the investigation of unauthorized radio stations, of enemy aliens and suspects, and of matters connected with cable and mail censorship that affected the Navy. ONI was also charged with the protection of waterfronts and vessels, and of industrial plants having contracts with the Navy Department to safeguard them against sabotage. Guarding U.S. merchant ships while in port and guarding against danger from enemy agents among the passengers and crews on both trans-Atlantic and coastal ships was primarily performed by ONI personnel.[8]

An undated, unsigned document, bearing a pencil notation, "approximate date 1919," describes in general terms ONI's functions immediately after the close of World War I:

(1) Collection and classification of information from abroad and at home for the Navy Department and other naval activities that may require it.

(2) Publication and dissemination of such information [to] the Navy and government officials requiring it.

(3) Direction of the Navy Department Library and the Historical Section.

(4) Official channel of communications for all foreign Naval Attachés in the U.S. with the Navy Department.

(5) Direction of Naval Missions of an educational nature to foreign countries.

(6) Cooperation with other executive departments of the Government in discovering and bringing to justice persons engaged in activities against the U.S.

(7) Direction of Intelligence Officers attached to the Offices of the Commandants of the various Naval Districts.

(8) Maintenance of close touch with all Naval activities both in and out of the Navy Department.

(9) Censorship of cables and radio in time of war.

(10) [also, in time of war,] plant protection, ship inspection, espionage, counter-espionage, and passport review.[9]

Return to Peacetime Activities: 1920

In a memorandum dated 16 January 1931 to Capt. Harold R. Stark, Aide to the Secretary of the Navy, Capt. William Baggaley, as Acting Director of Naval Intelligence (DNI), stated that the 15 April 1920 organization of the Office of Naval Operations (OPNAV) provided for an Intelligence Division to perform the following duties:

(a) [It is responsible for] the collection of all classes of information concerning foreign countries but especially that affecting naval and maritime matters, the evaluation of this information, and its dissemination as intelligence to the proper officials of the Navy and to other interested Government Departments. It is particularly concerned with the strength, disposition and probable intentions of possible enemy forces. It maintains naval attachés abroad. It maintains liaison with other Government Departments for the exchange of information from abroad. It is the Navy Department's channel of communication with U.S. naval attachés and naval missions and with foreign officials to U.S. naval establishments and to private manufacturing plants executing naval contracts.

(b) It maintains a liaison between the Navy Department and the investigating services of other Government Departments with a view to apprehending and bringing to justice persons guilty of offenses against the United States. It is charged with espionage. . . .

(c) It exercises such censorship as may be called for by the Navy Regulations and approved war plans.

(d) It has charge of the Navy Department Library and of Navy Records of historical value. It selects, indexes and files the most important of these records, so that they may be available for printing and future historical use.

(e) Through the U.S. Navy Information Section it is the point of contact for press representatives with the Navy Department. In cooperation with the Bureaus and other Divisions of the Department, it supplies the public with pertinent information regarding the U.S. Navy. It answers inquiries addressed to the Department by civilians for such information. It provides protection against foreign propaganda [This latter duty was not actually assigned to ONI until February 1922].[10]

In the first statement of an Information Policy of the Navy, prepared by the General Board and approved by Secretary of the Navy Edwin Denby on 29 March 1922, the following duties applied to intelligence:

To acquire accurate information pertaining to the political, military, naval, economic and industrial policies of our own and foreign countries.

To select, analyze, classify, summarize and make available all information acquired for the purpose of reference and dissemination.

To disseminate appropriate information systematically throughout the naval service.

To preserve for ready reference and for historical purposes information collected and arranged systematically.

To issue analytical studies of important historical incidents with a view to indoctrination.

To cooperate closely with other departments of the government in the collection, preservation, and dissemination of information.

To provide for protection against foreign espionage and propaganda.

To acquire and disseminate appropriate information of the enemy in time of war.[11]

The Navy Department Information Section was established in ONI by a Secretary of the Navy directive of 21 February 1922; for further details, see Chapter 33.

In a memorandum of 5 March 1923, the revised OPNAV organization orders were formally approved by CNO Adm. Robert E. Coontz. The functions of the Naval Intelligence Division under these orders were as follows:

> The Office of Naval Intelligence shall constitute the Intelligence Division of OPNAV. The Intelligence Division is charged with the collection of information for the department and for other naval activities which require it. It publishes and disseminates such information to the Navy and to Government officials requiring it. It cooperates with the other executive departments of the Government in discovering and bringing to justice persons engaged in activities against the United States. It directs all naval attachés abroad, and is the official channel of communication for all foreign naval attachés in the United States.
>
> It is the duty of the Office of Naval Intelligence to keep in close touch with all U.S. naval activities, both in and out of the Navy Department. The office handles all classes of information excepting purely operating information, which is usually and should always be forwarded by the senior officer present direct to the Chief of Naval Operations. Such information is under cognizance of the Ships' Movement Division.
>
> It is contemplated that upon the outbreak of war, a National Director of Censorship will be appointed with various assistants. The DNI will be Chief Naval Censor and will prepare, as far as is practicable, for the taking over of all U.S. Cable and Radio Censorship. It is contemplated that U.S. mail, telephone and telegraph censorship will be under the Army.
>
> The Navy Information Office, established under the Director of Naval Intelligence, collects and disseminates information regarding the naval service. All Divisions are directed to cooperate with the Navy Information Office by furnishing such information as may be pertinent.

In a description of Naval Operations approved by CNO Adm. Edward W. Eberle and dated 15 June 1926, the functions of the Intelligence Division were identical to those issued on 5 March 1923, with the exception of an additional paragraph:

> With the view to the probable establishment by executing order of a national office having cognizance of counter propaganda, the Director of Naval Intelligence will anticipate such work and organization as may be needed to efficiently and immediately cooperate with that office.

In an organization of Naval Operations approved by CNO Adm. Charles F. Hughes and dated 15 April 1929, changes in the functions of the Intelligence Division were noted, and the duties of the division were given as follows:

> The Director of Naval Intelligence is ex-officio a member of the General Board [of the Navy].
>
> The primary duty of Naval Intelligence is the collection of all classes of information concerning foreign countries, but especially that affecting naval and maritime matters, the evaluation of this information, and its dissemination as intelligence to the proper officials of the Navy and to other interested Government Departments. It is particularly concerned with the strength, disposition and probable intentions of possible enemy naval forces. It maintains naval attachés abroad. It maintains liaison with other Government Departments for the exchange of information from abroad. It is the Navy Department's channel of communication with U.S. Naval Attachés and Naval Missions and with foreign Attachés in the United States. It regulates the visits of foreign officials to U.S. Naval establishments and to private manufacturing plants executing naval contracts.
>
> It maintains a liaison between the Navy Department and the investigating services of other Government Departments with a view to apprehending and bringing to justice persons guilty of offenses against the United States. It is charged with espionage and counterespionage. It supervises the Intelligence Volunteer Service of the Naval Reserve.
>
> It exercises such censorship as may be called for by the Navy Regulations and approved war plans.
>
> It has charge of the Navy Department Library and of Navy records of historical value. It selects, indexes, and files the most important of these records, so that they may be made available for printing and future historical use.
>
> Through the U.S. Navy Information Section, it is the point of contact for press representatives with the Navy Department. In cooperation with the Bureaus and other Divisions of the Department, it supplies the public with pertinent information regarding the U.S. Navy. It answers inquiries addressed to the Department by civilians for such information. It provides protection against foreign propaganda.

The Far East Section of ONI was responsible for monograph information on the following areas between 1932 and 1940: The Japanese Empire (including its League of Nations Mandates), China, Siam, Manchukuo, the Philippine Islands, Guam, Hawaii, Samoa, the Aleutian Islands, and the strategic harbors of the Pacific. In a memo dated 22 September 1933, Lt. Arthur H. McCollum of the Far East Section drew up the organization's duties, which were typical of the responsibilities of other analytical sections within ONI during that period:

1. To maintain monographs on (a) the Japanese Empire, (b) China, (c) Manchukuo, and (d) Siam; and to prepare studies from time to time for inclusion in the proper monograph.

2. To prepare monograph material on Japan for issue to the Service and to send out corrections as necessary.

3. To keep the Naval Attachés in Tokyo and Peiping informed as to information desired.

4. To keep the Chief of Naval Operations and the Secretary of the Navy informed on all developments, particularly those affecting U.S. policy and interests, in areas under the cognizance of this Section.

5. To maintain a flow of pertinent information to and from the General Board, War Plans, The War College, the Military Intelligence Division of the War Department, the State and Commerce Departments, the Technical Bureaus of the Navy Department, and the Dissemination Section of ONI.

6. To maintain close liaison with other sections of ONI on matters concerning the geographic limits for which responsible.[12]

A memorandum of 13 January 1933 from DNI Capt. Hayne Ellis to the Chief of Naval Operations forwarded, as enclosures (A) and (B), an "Outline of Functions of [the] Intelligence Division and a Diagram of Organization of the Intelligence Division." They were submitted in compliance with a CNO request in connection with a proposed revision of the office orders. Certain changes in the wording of the functions over those of 1929 are apparent, and the duties, as described in the enclosures, were set forth:

> The primary duty of Naval Intelligence is the collection of all classes of information concerning foreign countries, but especially that affecting naval and maritime matters, the evaluation of this information, and its dissemination as intelligence to the proper officials of the Navy, and to other interested Government Departments. It is particularly concerned with the strength, disposition, and probable intentions of possible enemy naval forces. It maintains naval attachés abroad. It maintains liaison with other Government Departments for the exchange of information from abroad. It is the Navy Department's channel of communication with U.S. Naval Attachés and Naval Missions and with foreign Attachés in the United States. It regulates the visits of foreign officials to U.S. naval establishments and to private manufacturing plants executing naval contracts.
>
> It maintains a liaison between the Navy Department and the investigating services of other Government Departments. It is charged with the protection of the Naval Establishment against espionage. It supervises the Intelligence Volunteer Service of the Naval Reserve.
>
> Except as otherwise specifically provided for, it is charged with general supervision over the security of naval secret and confidential matter.
>
> It exercises such censorship as may be called for by the Navy Regulations and approved war plans.
>
> It has charge of the Navy Department Library and of Navy Records of historical value. It selects, indexes and files the most important of these records, so that they may be made available for printing and future historical use.
>
> The Public Relations Branch of the Office of Naval Intelligence is the point of contact for press representatives with the Navy Department. In cooperation with the Bureaus and other Divisions of the Department, it supplies the public with pertinent information regarding the U.S. Navy. It answers inquiries addressed to the Department by civilians for such information. It provides protection against foreign propaganda.[13]

The Information Policy prescribed by the Secretary of the Navy on 10 May 1933, insofar as it pertained to intelligence, was quite similar to that of 1922:

> To acquire through naval and other agencies, accurate information concerning the political, military, naval, economic and industrial policies and activities of all countries.
>
> To acquire and disseminate expeditiously appropriate information in time of war.
>
> To provide protection against espionage and propaganda.
>
> To keep the public informed of the activities of the Navy, compatible with military secrecy.[14]

The *ONI Intelligence Manual, 1933* (ONI-19), signed out on 5 October 1933 by Capt. Ellis and approved by CNO Adm. William H. Standley, was based on the above Information Policy of the U.S. Navy and prescribed the mission of the Intelligence Service:

> To acquire through naval agencies, and in cooperation with other departments of the Government, accurate information pertaining to the political, military, naval, economic, and industrial policies of our own and foreign countries.
>
> To acquire and rapidly disseminate appropriate information of the enemy in time of war, recognizing the psychological value of information and its vital requirements for sound decision and action.
>
> To select, analyze, and systematically preserve information for historical and indoctrinal purposes.
>
> To disseminate appropriate information systematically throughout the Naval Service.

To provide for protection against foreign espionage and propaganda.[15]

Note that the public relations function, which nonetheless continued to be exercised by ONI, was not included.

The responsibilities of the Director of Naval Intelligence in carrying out the above mission were:

(a) The organization, administration, and operation of all naval intelligence activities, except those delegated to the Commanders of Forces Afloat and the Commandants of Naval Districts with which close liaison [is] maintained.

(b) The preparation of regulations and plans to insure uniformity within the Naval Intelligence Service.

(c) The preparation of standard methods of handling the information and intelligence comprised within the field of Naval Intelligence.

(d) The general supervision of regulations to maintain the security of naval secret and confidential matter, and the censorship of Cable and Radio Traffic during hostilities.[16]

According to the *Naval Intelligence Manual, 1933*, the Office of Naval Intelligence, as a division of the Office of Naval Operations, had the following tasks:

(a) To obtain and evaluate information of foreign nations, particularly of foreign navies, naval, and maritime matters, and to disseminate the derived intelligence; by special intelligence publications, to disseminate technical developments, political affairs, or other important current topics concerning the naval establishment.

(b) To prepare for the CNO conclusions drawn from careful analysis of available intelligence as to the international sitution, the possibility or probability of U.S. becoming involved in war, and the probable intentions and operations of the enemy forces in cases of war.

(c) To prepare and maintain up-to-date statistical data of all the major navies and matters pertaining thereto.

(d) To provide for and maintain cordial relations with the public, the press, and other news agencies, with a view of proper dissemination of naval matter and pertinent information.

(e) To maintain close liaison with the Military Intelligence Divisions and other Government agencies collecting foreign information.

(f) To supervise and advise the Naval Attachés in the performance of their paramount duty, the collection of naval intelligence.

(g) To control personnel of Naval Reserves assigned and of volunteers for intelligence service in so far as pertains to their naval intelligence duties.[17]

In an unsigned CNO organization memorandum dated 1 July 1934, the duties of the Intelligence Division were, for the first time in one of these documents, presented in outline form under what might be construed as branch headings.

A. Foreign Intelligence:

(1) The collection of all classes of pertinent information concerning foreign countries, especially that affecting naval and maritime matters, with particular attention to the strength, disposition and probable intentions of possible enemy naval forces.

(2) Evaluation and disseminating the above.

(3) Directs the activities of the Naval Attachés abroad.

(4) Handles all communications from foreign attachés accredited to the United States.

(5) It is the Navy Department's channel of communications with the Naval Missions abroad.

(6) Maintains liaison with other government departments for the exchange of information from abroad.

B. Domestic Intelligence:

(1) Maintains a liaison between the Navy Department and the investigating services of other Government Departments.

(2) It is charged with the protection of naval establishment against espionage and other subversive activities.

(3) It supervises the Intelligence Service of the Volunteer Naval Reserve.

(4) It exercises such censorship as may be called for by the Navy Regulations and approved War Plans.

(5) Except as otherwise specifically provided for, it is charged with general supervision over the security of naval secret and confidential matter.

(6) Regulates the visits of foreign officials to U.S. Naval establishments and to private plants executing naval contracts.

C. Historical Records:

(1) It has charge of the Navy Department Library and of naval records of historical value.

(2) It selects, indexes, and files the most important of these records, so that they may be made available for printing and future historical use.

D. Public Relations:

(1) Maintains a point of contact for the press representatives with the Navy Department.

(2) In cooperation with the Bureaus and other offices of the Department, it supplies the

public with pertinent information regarding the U.S. Navy. It answers inquiries addressed to the Department by civilians for information in regard to the Navy.

(3) It provides protection against foreign progaganda in the U.S.[18]

The first objective of Naval Intelligence, as expressed by the *Naval Intelligence Manual* revision of 1936, was "to procure, evaluate, and disseminate information (on world conditions) to our naval forces in time to be of service. Security of information on our own forces, as important as obtaining information, is the second objective of Naval Intelligence."

The manual also admonished: "To keep the mouth shut, and the eyes and ears open, is a cardinal principle of intelligence."[19]

In 1938, ONI was actively engaged in the administration of the naval attaché organization; the collection, evaluation and recording of intelligence information of foreign countries, especially their sea coasts and navies; the public relations of the Navy Department; operation of the Navy's historical records and library; preparation and dissemination of data on our own and foreign navies; internal administration; counterespionage; security measures; planning; and relations with foreign naval and air attachés accredited to the United States.[20]

In War Plan-1 (the "Orange Plan" for winning a Pacific Ocean conflict with Japan) of September 1939, ONI's duties were again described:

> The Naval Intelligence Service is the organization, under the Chief of Naval Operations, charged with the duty of collecting, evaluating and disseminating information required by the Naval Establishment, of defending the Naval Establishment against subversive agents and propaganda, and such censorship, national and military, as may be required.
>
> It is operated by the Director of Naval Intelligence, who will operate it in war.
>
> Standard of Readiness NIS-1: The maintenance and operation of a Naval Intelligence Service, (a) which will provide an adequate Intelligence Service, foreign and domestic, in peace, and (b) which is capable of rapid expansion into a War Organization. This standard includes:
>
> (1) The collection and evaluation of information, foreign and domestic, and the timely dissemination of all information of value to the naval establishment and to other government agencies concerned.
>
> (2) Espionage and counterespionage, in cooperation with other agencies of the government. Counterespionage will embrace among its activities:
>
> (a) Assistance in matters of Plant Protection for Plants doing Naval work.
>
> (b) Assistance in the control of such travelers, crews, ships and cargoes as are of concern to the Navy for the prevention of travel or action (sabotage) inimical to the National Defense.
>
> (3) Preparation and dissemination of pertinent information, not incompatible with military secrecy, regarding the U.S. Navy, and cooperation in the preparation and dissemination of propaganda as necessary, for public consumption and to combat enemy propaganda.
>
> (4) Supervision over the security of Naval Secret and Confidential matter.
>
> (5) Censorship of radio and cable communications of the general public in accordance with such Presidential Proclamation, Executive Orders and laws as may be promulgated, and in cooperation with other government agencies concerned within the field laid down in Joint Action of the Army and the Navy, and until a national censorship may be established. In wartime, when a national censorship is organized and established, the naval organization operating for the censorship of public traffic will be transferred to the control of the Public Relations Administration, maintaining, nevertheless, their naval ranks and discipline.
>
> (6) Direction and control of military censorship within the Naval Establishment. In wartime, military censorship will *NOT* be transferred to the Public Relations Administration. Military censorship will *NOT* extend to official communications.
>
> (7) Consultation with the Bureau of Navigation concerning the induction, training, assignment and separation of Intelligence Service personnel with the view of maintaining the efficiency of the Intelligence Service, to achieve accurate, adequate and efficient performance of the tasks contained in the Navy Basic War Plans (COLOR) [war plans were given color codenames].[21]

Expansion During World War II

In 1941, the jurisdiction of the Far East Section of ONI was expanded to include strategic coverage of areas east of 60° east longitude, adding India and the Indian Ocean, Siberia, the Dutch East Indies, Australia, New Zealand, Malaya, French Indochina, Burma, and numerous additional islands in the Pacific to the section's jurisdiction. Formerly, monograph coverage of all these areas had been assigned to the sections responsible for the countries of which they were colonies, dominions, or political components.[22]

According to a memo drafted in February 1941 by Capt. Howard D. Bode, head of ONI's Foreign Section, the DNI's mission was to obtain an evaluated knowledge of foreign nations, primarily of their warmaking capacity, and especially as that capacity existed in naval and maritime strength or potentiality. Specifically, Bode wrote that information on the following was required: military and naval strengths, the composition and location of forces, their equipment, and their disposition and movement; naval and military bases, coast defenses, economic resources, logistics, lines of communications, tactical practices; personalities of leaders, national and racial characteristics, psychological tendencies, thought, religion (fanaticism, etc.), troop morale, and physical condition (especially of combat forces), mercantile marine (personnel and equipment); geographic situation, weather, meteorology, and hydrographic and physiological features; and national policies and politics (so as to indicate probable military action or tendencies).[23]

Capt. Alan G. Kirk, the Director of Naval Intelligence in 1941, contended that ONI was responsible for interpreting possible enemy intentions after evaluating information received from whatever source and also for preparing that section of the formal planning estimates known as "Enemy Intentions." On the other hand, RAdm. Richmond Kelly Turner, Director of the War Plans Division of OPNAV, maintained that his division should prepare the intentions section and interpret and evaluate all information concerning possible hostile nations from whatever source received. Turner contended that ONI was solely a collection and distribution agency and was not charged with sending out any information that would initiate U.S. Navy operations anywhere.

The controversy between ONI and Turner came to a head in April 1941. Capt. Kirk and RAdm. Turner discussed their differing points of view before RAdm. Royal E. Ingersoll, Assistant Chief of Naval Operations, and then the three of them entered the office of CNO Stark to have the matter resolved. The points at issue were reviewed, and Adm. Stark approved the position taken by Turner. Accordingly, Kirk advised his principal branch heads at ONI of the decision and, in due course, also told he would be relieved as DNI by RAdm. Theodore S. Wilkinson, in October 1941. No change was made, however, to the previously approved ONI mission to evaluate intelligence information and produce naval intelligence estimates.[24]

The immense scope of the Navy's involvement in World War II expanded ONI in numbers of personnel and in the variety of duties assigned to Naval Intelligence. In January 1942, the Identification and Characteristics Section was established to collate data on appearances and characteristics of U.S. and foreign naval and merchant vessels and to disseminate identification manuals and other intelligence materials.

On 10 February, the Protocol and Reception Center was set up to help in handling the increasing numbers of foreign military and naval officials visiting the Navy Department. In addition, the center was given general supervision over U.S. naval officers preparing for intelligence duty abroad.

On 5 August 1942, the Special Intelligence Section, which was responsible for the interrogation and handling of prisoners of war, was moved from the Foreign Intelligence Branch and established as the Special Activities Branch (OP-16-Z), with functions to include obtaining, training, and administering of secret agents. In June, the Special Activities Branch also had assumed responsibility for information on captured enemy naval equipment. In connection with its work in developing a secret undercover intelligence service, the branch maintained liaison with the Office of the Coordinator of Information and subsequently with the Office of Strategic Services when the former was absorbed by the latter.

On 7 October, the Special Warfare Branch (OP-16-W) was established to take over the Navy's participation in psychological warfare and bacteriological warfare. Among its tasks was processing naval intelligence for the confidential guidance of the Psychological Warfare Planning Board of the Overseas Branch of the Office of War Information and supplying naval information to that organization.[25]

Post–World War II Missions, Functions, and Responsibilities

The responsibility and authority of the Office of Naval Intelligence in execution of its mission was set forth in 1946 in Article 425 of *Navy Regulations*: "To keep responsible naval authorities informed of the war capabilities and intentions of foreign nations and of situations or trends threatening the security of the Naval Establishment."

That article and several others were included in the *Changes to Navy Regulations* approved by President Harry Truman on 14 June 1946. Until the changes became effective, there had been no previous specific delineation of ONI's duties in relation to the rest of the Navy in *Navy Regulations*.[26]

The same changes to *Navy Regulations* also provided for "an efficient intelligence organization" within the command of a Commander in Chief, or commander of any force or unit of the operating forces not operating under the Commander in Chief. It also required that "the staff shall include an intelligence section headed by a line officer designated as flag (intelligence) officer."

The Sabotage, Espionage and Countersubversion (SEC) Section of ONI in 1946 was responsible for collecting, evaluating, and disseminating information that would enable naval authorities to guard against sabotage, espionage, and subversion. Some of the tasks of SEC were to maintain suspect files on naval personnel and civilians employed by the Navy; maintain files containing reference and background material on subversive organizations and subversive trends of naval interest; collaborate with the Operational Intelligence and Foreign Intelligence Branches and other parts of ONI, and coordinate, within ONI, those activities pertaining to espionage, counterespionage, and subversion; effect liaison with other government agencies; and disseminate counterintelligence information to naval authorities and other action agencies.

The files maintained by SEC were categorized into (1) naval personnel suspected of subversive activity or background; (2) civilian personnel employed by the Navy, suspected of subversive activity or background; (3) general subversive and espionage suspects of definite naval interest; and (4) general subversive and espionage suspects of possible naval interest.[27]

The mission of the Chief of Naval Intelligence, as approved by the CNO, was derived from the Information Section of the *Statement of Naval Policy*, approved by the Secretary of the Navy on 23 January 1947. Pertinent parts of the Information Section required ONI

> 1. To acquire accurate intelligence, in cooperation with other agencies of the government, both civil and military, concerning the political, military, naval, economic, health, technical, and industrial policies and activities of all foreign countries.
>
> 2. To analyze and preserve military information for ready reference and for historical purposes.
>
> 3. To disseminate useful military information systematically throughout the naval service and to other government departments and agencies.
>
> 4. To provide proper security for military information and protection against espionage and sabotage, in cooperation with other departments and agencies.

The mission of ONI was also expressed in the *Naval Intelligence Manual–1947*:

> To administer, operate and maintain Naval Intelligence in order to provide intelligence necessary for operations and war plans and in order to keep responsible naval authorities informed of the war capabilities and intentions of foreign nations; and of situations or trends threatening the security of the Naval Establishment.

The execution of ONI's mission required the coordination of all intelligence activities in the Navy. ONI, a unit of OPNAV, was the headquarters of the Chief of Naval Intelligence, whose function was to assist the Chief of Naval Operations as his authorized representative in the execution of the CNO's mission. The organizational title "Naval Intelligence" encompassed ONI itself, intelligence posts in foreign countries (naval attachés, naval observers, and naval liaison officers), intelligence organizations in naval districts and river commands, the intelligence sections of operating forces and advanced bases, and the naval sections of intelligence activities sponsored jointly by the Army and Navy.

Within ONI, the missions of the intelligence-producing branches were expressed in the *Naval Intelligence Manual–1947*:

> Domestic Branch: To administer, operate, and maintain the Domestic Branch; to collect information within the Naval Districts on foreign powers; to collect and process information from all sources, and to disseminate intelligence derived therefrom, revealing situations or trends threatening the security of the Naval Establishment; to coordinate Naval District and River Command intelligence activities; and to support Naval District and River Command officers and personnel.
>
> Foreign Branch: To administer, operate, and maintain the Foreign Branch; to collect and process information and to disseminate the intelligence derived therefrom, in order to keep responsible naval authorities informed of the war potentials and intentions of foreign nations and of the characteristics of actual or potential theatres of operations; to coordinate intelligence activities on foreign posts; and to support the Naval Attachés and other Naval Intelligence personnel on foreign posts.
>
> Air Branch: To administer, operate and maintain the Air Branch in order to provide intelligence on air power through participation in the Air Intelligence Division and from other sources; to support Air Intelligence Officers serving with the Operating Forces; to have cognizance over the assignment of Naval Officers to the Air Intelligence Division.
>
> Air Intelligence Division: To prepare air estimates on (a) alien capabilities to employ offensively, and to defend against, air weapons and other weapons designed to serve similar purposes; (b) alien and domestic vulnerability to air attack; (c) facilities effecting [sic] the employment of air power; to maintain liaison with, and to supply all air intelligence required by, the Air Branch of ONI.
>
> Operational Intelligence Branch: To administer, operate and maintain the Operational Intelligence Branch in order to insure dissemination of all necessary intelligence, including operational intelligence, to the Operations Division of OPNAV and to the Commanders of the Operating Forces; to coordinate covert intelligence activities of ONI as directed; to control material above the classification of Secret received and needed by ONI; to carry out such other highly classified activities as

> the CNI [Chief of Naval Intelligence] may assign or direct.[28]

Navy Department General Order No. 247, signed by the Secretary of the Navy on 10 February 1947, defined the Policies and Principles Governing the Distribution of Authority and Responsibility for the Administration of the Naval Establishment. Among the duties assigned to the Chief of Naval Operations was the responsibility to promulgate to the bureaus, boards and offices of the Navy Department and to the shore establishment such directives as he deemed necessary with respect to matters of security and intelligence.[29]

Cold War Missions, Functions, and Responsibilities

On 1 February 1952, an OPNAV internal instruction directed immediate establishment of a continuous review of functions and work load to ensure maximum use of military and civilian personnel. The aims of the review were

> to insure that only those activities [were] located at the seat of government that [had to] be so located because of the nature of activities, the source of material or necessary collaboration with other activities; to eliminate non-essential functions; to streamline administrative and operating procedures; [and] to insure that personnel skills [were] used to the maximum.

Incident to the 1 February 1952 instruction, ONI established a Board of Review of Functions and Workload to compile lists of officer, enlisted, and civilian billets by organization and in order of priority for reassignment. The lists were to be based upon consideration of workload, functions performed, and streamlining of procedures, to the end that, if reassignments became necessary, the functions of ONI would be minimally impaired.[30]

In 1953, the mission of the Commerce and Travel Section of ONI was "to exploit all available maritime commerce and travel human resources to obtain intelligence and information of naval interest, and to coordinate those counterintelligence activities of the Naval Establishment which are concerned with international commerce and travel."[31]

The mission of the Director of Naval Intelligence (OP-92) in 1955 was expressed as follows:

> Under the authority and direction of the CNO, to administer, operate and maintain an intelligence service to fulfill the intelligence and counterintelligence requirements of the Department of the Navy, in order to:
>
> 1. Inform the Naval Establishment of the war making capabilities and intentions of foreign nations;
>
> 2. Provide the Naval Establishment with the intelligence needed for plans and operations;
>
> 3. Warn naval authority of threats to the security of the Naval Establishment;
>
> 4. Provide the naval elements of joint, national and international intelligence;
>
> 5. Promote the maximum intelligence readiness of the Operating Forces and other components of the Naval Establishment;
>
> 6. Coordinate the intelligence effort of the Naval Establishment;
>
> 7. Develop and promulgate, subject to approval of the Secretary of the Navy, policies for the protection of classified matter, including such policies applicable to industrial security;
>
> 8. Advise the CNO concerning all matters relating to naval intelligence and security policies for the protection of classified matter.[32]

Navy Department General Order No. 5 issued during 1957 sets forth four principal tasks for the executive administration of the Navy Department. The naval command task required promulgation to the Navy Department of directives embracing, among other matters, those pertaining to intelligence.

In Article 0204 of *U.S. Navy Regulations, 1948*, the duties prescribed for the CNO include the collection, evaluation, and dissemination of all types of intelligence information required within the Navy Department and the supervision of naval attachés. In Article 1502.1, the CNO was required to supplement those regulations with appropriate publications, including the *Security Manual* and such other detailed instructions as were necessary to ensure proper control of classified matter. The effective edition of SECNAV Instruction 5403.13 defined and delimited the investigative jurisdiction and responsibilities of Naval Intelligence. Additionally, Secretary of Defense letter of 21 February 1955 to Director, Office of Defense Mobilization stated that the Navy Department would operate telecommunication censorship in time of war. The Director of Naval Intelligence's mission, as of 1957, derived from the above-cited basic directives.[33]

The Defense Intelligence Agency (DIA) Production Center was activated at Arlington Hall on 1 January 1963. The Air Force Intelligence Center, including its naval component (OP-922V4), located at Arlington Hall, had already been transferred as one of the initial increments of the DIA Production Center. Other units of ONI and their functions were similarly transferred to the newly established DIA on 1 February 1963:

> (1) OP-922G: Basic Intelligence Branch—all personnel and production functions;

(2) OP-922N4: Non-Bloc Navies, Cold War Section—all personnel, except as separately notified, and production functions;

(3) OP-922V: Targeting Requirements and Support Branch—specified personnel and limited functions;

(4) OP-922Y2E: Military Intelligence Unit—all personnel, except as separately notified, and production functions;

(5) OP-922Y2F: Air/Missile Support Unit—specified personnel but no functions;

(6) OP-922B3: Special Assistant for Systems Development—billet but no functions;

(7) STIC-2-G: Medical Unit—personnel and production functions;

(8) OP-922Y1B3: Weather Unit—personnel and production functions; and

(9) OP-922W: Special Assistant for Editorial and Publication Matters—personnel and functions.[34]

In December 1963, Secretary of the Navy Paul H. Nitze advised all ships and stations that instructions issued by the Director of DIA, under authority granted by the Secretary of Defense, had directive force through operational command lines with respect to naval forces assigned to unified and specified commands. It was also noted that, since its inception, the DIA had assumed an increasingly large portion of the intelligence functions of the Navy Department and had established requirements and procedures applicable to elements of the Navy other than those assigned to unified and specified commands. Accordingly, the Secretary of the Navy designated the Director of Naval Intelligence as his representative to review directives of the Director of DIA to determine their applicability to elements of the Navy Department, and to ensure that such elements were aware of, and responsive to, such directives.[35]

As of 2 July 1964, the mission of the Director of Naval Intelligence was

> to collect, process and disseminate intelligence of naval interest (both positive and counterintelligence) in order to support and advise the Secretary of the Navy and the Chief of Naval Operations in the discharge of their responsibilities; to exercise technical direction over the intelligence effort of the Department of the Navy; and to appraise the intelligence programs under his cognizance.[36]

Secretary of the Navy Notice 5450 of 19 April 1967 established, effective 1 July 1967, the Naval Intelligence Command (NAVINTCOM) "to insure the fulfillment of the intelligence, counterintelligence, investigative, and security requirements and responsibilities of the Navy Department."

OPNAV Notice 5450 of 29 June 1967 delegated to NAVINTCOM command and support responsibilities for the Naval Investigative Service and its offices throughout the world, and for other naval intelligence activities including the Naval Scientific and Technical Intelligence Center, the Naval Reconnaissance Technical Support Center, the Navy Field Operational Intelligence Office, the Naval Intelligence Processing System Support Activity, and the Applications Department.

With the establishment of the Naval Intelligence Command on 1 July 1967, the Director of Naval Intelligence had two hats: Assistant Chief of Naval Operations for Intelligence (OP-92) and Commander NAVINTCOM. Under the first, he was the Staff Intelligence Officer for the CNO, with this mission:

> To serve as the principal staff advisor to the Secretary of the Navy and the Chief of Naval Operations on intelligence and security plans and policy matters and to provide staff intelligence support; to implement the responsibilities of the CNO to develop, coordinate and promulgate policies, plans and programs for intelligence and security activities of the Department of the Navy and to advise and assist officials of the Department of the Navy in matters of protocol and liaison with foreign officials.

Under the second hat, the director's mission was "to direct and manage the activities of the Naval Intelligence Command in order to fulfill the intelligence, counterintelligence, investigative and security requirements and responsibilities of the Department of the Navy."[37]

After the promulgation of OPNAV Notice 5450 of 29 June 1967, the OP-92 staff elements performed the following functions:

> OP-92C: Developed and interpreted policy in the counterintelligence, security, and investigation areas; developed plans for armed forces censorship and for telecommunication censorship, and for the training of naval reservists to be assigned to those functions upon mobilization.
>
> OP-92D: Involved principally in the release of U.S. information to foreign governments in accordance with national regulations governing such release, as prescribed by the National Military Information Disclosure Policy Committee, of which OP-92 was a member.
>
> OP-92F: Divided into two parts, the first provided support to OP-92 on USIB [United States Intelligence Board] matters; on National, Joint and Naval Estimates and on Joint Chiefs of Staff plans. The second was Intelligence Plot, where a round-the-clock cognizance was maintained of current international developments having Navy implications. Located side-by-side with the Chief of Naval Operations' Flag Plot, it provided intelli-

gence support to the CNO Duty Captain outside of normal working hours.

OP-92L: Served as point of contact for all foreign naval attachés accredited to the Navy Department, and as social secretary for OP-92.

OP-92P: Concerned with the Department of Defense programming and budgeting system, it served as OP-92 staff outlet for the larger programming office in the NAVINTCOM headquarters organization.

OP-92S: Served as the OP-92 administrative support unit in the Pentagon.

The Naval Intelligence Command Headquarters comprised four elements:

NIC-1 (Administration and Services): Provided administrative support to the Command in matters pertaining to personnel, training, and Naval Reserves; correspondence management and files; communications and Top Secret control; supply and fiscal; and translation.

NIC-2 (Systems Development): Coordinated the development of intelligence collection and processing systems.

NIC-3 (Intelligence Operations Coordination): Coordinated the intelligence collection, production, and dissemination of the Naval Intelligence Command and its six Field Activities and monitored the requirements and guidance for their fulfillment.

NIC-4 (Resources Management): Coordinated program and budget planning and presentation.[38]

On 10 October 1970, the mission and functions of the Commander Naval Intelligence Command were promulgated in OPNAV Instruction 5450.181:

Mission: Under the command of the Chief of Naval Operations, to direct and manage the activities of the Naval Intelligence Command to insure the fulfillment of the intelligence, counterintelligence, investigative, and security requirements, and responsibilities of the Department of the Navy.

Functions: To accomplish this mission, the Commander, Naval Intelligence Command shall:

1. Command the Headquarters, Naval Intelligence Command and assigned shore (field) activities.

2. Direct and coordinate intelligence collection, production, and dissemination to satisfy Department of the Navy intelligence information requirements and Defense Intelligence Agency (DIA) tasking.

3. Direct the development, management, and operation of ocean surveillance information management system.

4. Direct Department of the Navy participation in the DOD Intelligence Data Handling System (IDHS) and in the development, management, and support of Navy intelligence information systems.

5. Fulfill the investigative and counterintelligence responsibilities of the Department of the Navy (less those combat related counterintelligence matters within the functional responsibilities of the Marine Corps).

6. Participate, as appropriate, in matters pertaining to Undersea Warfare intelligence.

7. Direct intelligence support of the participation in Special Warfare and other designated activities as appropriate.

8. Act as Navy Special Security Officer (SSO) and manage the systems for the protective handling and dissemination of Special Intelligence and Special Activities materials within the Department of the Navy.

9. Provide a translation service for the Department of the Navy.

10. Participate in and provide intelligence inputs to naval, Joint, and national plans and policies and Navy studies and analyses.

11. Determine requirements for research, development, test, and evaluation of new and improved intelligence equipment and techniques.

12. Prepare and submit intelligence inputs to Navy Program Objective Memoranda, Office of the Secretary of Defense functional area reviews, and other pertinent planning and programming documents in collaboration with other appropriate commands and agencies.

13. Serve as major claimant for the Naval Intelligence Command budget, authorize Expense Operating Budgets to assigned shore activities of the Naval Intelligence Command, and provide guidance to these activities relative to the allocation and utilization of funds provided for intelligence, counterintelligence, investigative, and security programs.

14. Inspect and appraise the components of the Naval Intelligence Command to insure the maintenance of efficiency, discipline, readiness, effectiveness and economy in utilization of assigned resources.

15. Collaborate in matters relating to the training of personnel and the organizing, training, and equipping of units for assignment to intelligence, counterintelligence, investigative, and security duties.

16. Sponsor requirements for manpower and personnel for designated Navy intelligence activities and functions as sponsor for the Naval Reserve Intelligence and Air Intelligence programs; function as the Designator Sponsor for 135X (AI), 163X, 662X, and 762X officers and NEC [Naval Enlisted Code] Sponsor for YN-2505 and 9592; and act as Career Program

Manager for the Intelligence and Counterintelligence Career Development Program.

17. Maintain liaison with the Defense Intelligence Agency, and with other organizations on intelligence matters as required.

18. Function as the Navy point of contact for Department of the Navy personnel assigned to the Defense Attaché System.

19. Perform such other functions as may be assigned by higher authority.

Chapter Notes

Throughout much of ONI's history, general guidance for Intelligence within the Navy has been provided by *Naval Intelligence Manuals*. Those that have been located and reviewed by the author include: ONI-8, *Instructions for Intelligence Officers* (1923); ONI-19, *Naval Intelligence Manual* (1933); ONI-19, *Naval Intelligence Manual* (1936 revision); ONI-19(A), *Naval Intelligence Manual–1947*; ONI-19(B), *Naval Intelligence Manual–1949*; and ONI-70-1, *U.S. Naval Intelligence Manual* (1956).

As an example of the content of the manuals, chapter titles of ONI-19(A) were:

1. Basic Authority
2. Definitions
3. ONI—Mission, Objectives, Organization
4. Naval Districts and River Commands
5. Foreign Posts
6. Operating Forces
7. Administration—Personnel, Finance, Correspondence, Reports
8. Collection
9. Processing
10. Dissemination
11. Security
12. Joint Activities and Liaison

1. Department of the Navy General Order 292 of 23 Mar 1882.

2. *Defense Intelligence Digest*, vol. 1, no. 1, May 1963.

3. Capt. Herbert E. Cocke, USN, "History of ONI," Office of Naval Intelligence, 1931, 6.

4. Ibid., 16.

5. Ibid., 16–17.

6. *General Board Letters*, 3:240–42. General Board documents are maintained by the National Archives; the surviving material is stored at the Federal Records Center, Suitland, MD.

7. Cocke, "History of ONI," 19.

8. *SECNAV Annual Report, 1918*, 24–25.

9. OP-16-A-3-d file A3-1/EN 3-10, folder 2, copy in author's files, OA.

10. Ibid., for Baggaley memo.

11. *U.S. Naval Policy GB #420–2*, ser 1108, 29 Mar 1922; and Department of the Navy, "Administrative History of the Office of Naval Intelligence in World War II," 10 Jul 1946, unpublished MS, 5, hereafter *ONI WWII Admin History*.

12. OP-23F14, "Outline History of Far East Section," 11 Jan 1946, MS, box 2, Job 14236, OA.

13. Naval Investigative Service (NIS) History Files, OA.

14. ONI-19, *Naval Intelligence Manual, 1933,* para. 1001.

15. Ibid.

16. Ibid., para. 2018.

17. Ibid., para. 2019.

18. Memo in author's files, OA.

19. ONI-19 (1936 revision), III-IV.

20. *ONI WWII Admin History*, Appendix H.

21. WPL-1 (War Plan 1, the "Orange Plan") of Sep 1939, para. 0541, OA.

22. OP-23F14, "Outline History."

23. Bode memo, 20 Mar 1941, File A8–2, box 16, Job 3679, FRC/WNRC.

24. VAdm. Alan G. Kirk ltr to VAdm. Theodore Wilkinson, 19 Dec 1945.

25. *ONI WWII Admin History*, 65–66.

26. *ONI Review*, Sep 1946, 35.

27. ONI-D-4-T, *Naval Intelligence Training Manual* in NIS History Files, OA.

28. ONI-19(A), *Naval Intelligence Manual*, May 1947, paras. 1001–3, 3101–8.

29. Bureau of Personnel, NAVPERS 16047: *Naval Intelligence*, 1948, 8–9.

30. OP-32 ltr, ser 3353P32, 20 Feb 1952, box 1: "Incoming Correspondence File, 1952-53," Job 12876, FRC/WNRC.

31. ONI Instruction 05550.4 of 14 May 1953, NIS History Files, OA.

32. ONI Internal Instruction 05430.12 of 22 Mar 1955.

33. ONI-70-1, *U.S. Naval Intelligence Manual, 1947*, Change 1 of 4 Nov 1957, 10.

34. ONI Notice 05400 of 29 Jan 1963.

35. SECNAV Instruction 5410.114 of 4 Dec 1963.

36. ONI Instruction 05430.2B of 2 Jul 1964.

37. OPNAV Notice 5450 of 29 Jun 1967.

38. ONI Briefing for Assistant Secretary of the Navy, 6 Oct 1967, copy in author's files, OA.

CHAPTER 28

Organization and Personnel

This chapter relates closely to Chapter 27 on mission, functions, and responsibilities. In the absence of other considerations, an organization should be designed to fulfill its mission, and the mission should be based on a legitimate requirement for a service or product. Ideally, if the organization is well designed, directed, and adequately staffed, it will manage itself.

The organization of ONI throughout its early years was designed primarily to collect and produce intelligence and to provide for the administrative and training needs of its personnel. Very few of ONI's staff were devoted to management or compiling statistics to justify the need for ONI to some other department or agency. This situation was equally true of the district intelligence and naval attaché organizations covered in Chapters 3 and 22.

1882

Lt. Theodorus B. M. Mason, on 15 June 1882, was assigned to the Bureau of Navigation to organize the new "Office of Intelligence" and to serve as its first "Chief Intelligence Officer." Three officers were designated to assist him: Lt. M. Fisher Wright, who reported to the Bureau of Navigation for special duty on 1 July 1882; Lt. Albert G. Berry, who had been on duty at the Signal Office, Washington; and Ens. Templin M. Potts, who had reported to the Navy Department and had not yet otherwise been assigned. Clerks were borrowed from other offices as needed and as available because no funds had been appropriated for direct hiring of civilians by ONI.[1]

Lt. Mason organized his new office along functional lines (rather than geographic) in order to facilitate the correlation of intelligence material according to its anticipated usefulness to ONI's primary customers, the bureaus of the Naval Establishment. An index of subjects was set up that was organized parallel to interests and intelligence requirements of the Secretary of the Navy and or each bureau customer.[2]

By the end of 1885, the arrival and detachment of officers steadied ONI's roster at ten.

1889

Capt. Alfred Thayer Mahan was assigned special duty in the Bureau of Navigation on 30 September 1889 to prepare contingency plans for the Navy. He was assigned there until July 1892 and used ONI's files extensively.

1893

By 1893, the files on ONI were beginning to be organized along geographic lines as information was assembled on the navies of the world. The data accumulated, however, were still primarily on naval hardware and weapons and their capabilities. The limited staff of ONI continued to be engaged totally in production and was organized functionally according to customer requirements.

1897

As newly built ships were commissioned and more officers were required at sea, most shore stations had to reduce their staff levels. As of 1 July 1897, ONI officers were reduced from eight to five. To alleviate the personnel shortage, LCdr. Wainwright successfully induced the assignment of two replacement officers by the end of 1897 and borrowed one clerk, one draftsman, and one laborer from other bureaus and obtained the services of four copyists, charging them to the Naval Appropriation Act account "Increase of the Navy."[3]

1898

The outbreak of the Spanish-American War in April 1898 prompted all seven ONI officers to seek sea duty, and by 1 July all had been detached. Two retired officers were ordered to duty in ONI as their replacements and served in the office throughout the war: Capt. John R. Bartlett (Ret.), Chief Intelligence Officer, and Ens. Edward E. Hayden (Ret.). Capt. Bartlett was also Superintendent of the Coast Signal Service, and on 1 July 1898 he was assigned additional duty as Chief of the U.S. Auxiliary Naval Forces.[4] Following the war, some of the previously assigned officers were returned to ONI, building up to five by 1904 and seven by 1911.

1914

In 1914, the five desks of ONI were still organized by subjects: Ships, Communications, Ordnance, Personnel, and Engineering. The personnel assigned to the desks spent their time primarily filing incoming information obtained for the most part from newspapers and dealing with subjects pertinent to their assigned fields of interest. The assistant director was in charge of all official correspondence including that with naval attachés abroad and with foreign attachés in the United States.[5]

1916

In the reorganization prepared by Maj. John H. Russell, USMC, and Cdr. Dudley W. Knox, approved by Secretary of the Navy Josephus Daniels and put into effect on 1 October 1916, ONI was divided into four parts: Division A, Organization and Control of Agencies for the Collecting of Information; Division B, Coding and Decoding; Division C, Collating All Information for Statistical Study; and Division D, Dissemination and Archives. Great emphasis was placed on domestic intelligence, and counterespionage and secret service activities in the United States were established under the cognizance of Division A. During World War I, counterespionage grew to become the major activity of ONI.

1917

When the United States entered the war in April 1917, the number of naval attaché posts expanded from six to eighteen, and the volume of reports from them and the naval district and branch offices of ONI increased markedly. Since most of the reports, including those from naval attachés, were devoted to counterintelligence information, the ONI sections responsible for processing, filing, and taking action on such reports were expanded accordingly and became the predominant activity. Toward the end of the war, ONI's staff was organized functionally, except that some sections were further divided geographically within their functions.[6]

1918

As of 1 July 1918, ONI had expanded to ninety-two officers, of whom sixty-five were listed as being assigned to an "Aviation Section." Included in this section were three officers from the British Royal Navy. Curiously, however, the 1 September 1918 *Naval Intelligence Office Organization* does not show any aviation section, and the existence of such a section within ONI is not substantiated by any other known reports. Indeed, the *Navy Register* entry may have been in error. Capt. Noble E. Irwin was listed as head of the Aviation Section, which included LCdr. John H. Towers who had served as an aviation observer and Assistant Naval Attaché, London, prior to the U.S. entry into the war.[7]

By this time, ONI was divided into eleven sections, and was headed by a director, an assistant director, and an executive aide. The assistant director performed the duties of an executive. The executive aide, in addition to being an assistant to the assistant director, was in direct charge of Sections G through K of the organization described below.

Section A was divided into eleven divisions:

Division I, Collecting, was supervised by the assistant director and was charged with the care of the confidential correspondence of Section A and with the publication of a confidential bulletin on investigative work.

Division II, Legal Matters for Interdepartmental Matters, Internments and Allied Matters, was responsible for general instruction to branch offices and aids for information; conferences with representatives of the Departments of Justice, State, War, Labor, and Treasury concerning needed legislation and regulations; general direction and supervision of all investigations where action under war legislation or the U.S. Penal Code was contemplated or possible; and securing efficient cooperation and coordination with other departments of the government to eliminate duplication of effort in investigations.

Division III, Direction of Aids for Information, Branch Offices of ONI, and Branches in Alaska, was concerned with logistic support, personnel, and providing specific guidance to the aides for information assigned to the naval districts.

Division IV, Trading with Enemy, Ship Inspection, Mail and Cable Censorship, Enemy Goods in Storage, and Unauthorized Radio, was involved in collecting and disseminating information to the War Trade Board and various departments of the government on the economic and trade conditions and activities in all foreign countries, including the investigation of suspicious firms, shipments of goods intended for transfer to the enemy, and suspected shipboard sabotage. It was also responsible for investigating for the State Department all applications for U.S. passports and visas not issued directly by consuls. Censorship included motion picture censorship.

Division V, Investigation, investigated suspects in the Navy, suspicious travelers, employees in plants having Navy contracts, persons suspected of espionage in the vicinity of Navy property, civil employees of the Navy Department, and radio operators.

Division VI, Plants and Contracts, inspected manufacturing plants, recommended changes to improve plant protection, and developed informants in plants to determine which employees were causing delays in the production of naval materials, attempting sabotage, or causing labor disturbances.

Division VII, Intelligence Service in Mexico, Central America, Colombia, Venezuela, Cuba, Puerto Rico, Haiti, Dominican Republic, and Curacao, organized, directed, and handled the U.S. Naval Intelligence effort in those countries, where the work consisted of collecting and transmitting to ONI information on the sea coasts, shipping, potential submarine bases, wireless stations, and all that was adjudged significant and relevant to the war with Germany, including German or pro-German activities.

Division VIII, Intelligence Service in South America, except Venezuela and Colombia, had the same basic responsibilities in its area as did Division VII.

Division IX, Intelligence Service in Europe, attended to the correspondence and cables to and from naval attachés in Europe.

Division X, Intelligence Service in the Far East, in addition to supervising the naval attachés in Tokyo and Peking, also supervised the aids for information at Guam and Samoa (and was to supervise Manila if and when an aid was designated there). Division X acted upon information obtained from the Far East or sent the information to other divisions that had jurisdiction. Much information of value relative to Far Eastern matters was also received from the aid for information at Honolulu and the various agencies of ONI along the U.S. Pacific Coast. The division's interests in substantive matters concerned German activities (espionage, propaganda, intrigue) in the Far East; the movement of suspicious ships and cargoes; naval bases and naval operations in the Far East, including the activities of the German commerce raiders; and political changes or anticipated social disturbances in the Far East.

Division XI, Technical Investigative Methods, was charged in general with the chemical, physical, and photographic examination of mail, printed matter, etc., for the detection of secret writing. It also collected, collated, and compiled information on unauthorized radio sets and the transmission of suspicious messages; cable censorship methods to detect hidden messages; the use of animate and inanimate carriers (carrier pigeons, clothing, toilet articles, etc.); the use of secret inks on letters, printed matter, personal effects, and the body; falsification of documents; agents' operating methods; the use of explosives and poisons; identification methods; and the organization of the German intelligence system.

Section B, Transmitting, handled all cables, telegrams and radiograms coming to, or being sent by, ONI. Incoming messages were paraphrased, references appended, and then routed to the proper section for action; outgoing messages were written up in proper form, given a date/time number, and serialized as necessary. A complete file was kept of all dispatches and arranged chronologically and by locality of originator or addressee. Instruction was given in the use of codes and ciphers to those officers, agents, and others whose duties required this knowledge. Section B also arranged for codes to be used by naval attachés, aids for information, agents, etc., and secured, through the Code and Signal Section, the best channels for communicating with all naval intelligence representatives. A 24-hour watch was maintained in Section B not only to look after dispatches, but also to handle any other important matter. All incoming and outgoing secret mail was handled by Section B.

Section C was divided into three divisions:

Division I, Collating, collated, filed, and disseminated information on the ports of the world, including their repair facilities and availability of fuel and supplies; the war resources of various countries and their naval and military activities; international affairs; commerce and trade; communication facilities; and general data on the progress of the European war. Data on merchant shipping and losses incident to the war were disseminated daily, while matters of less value, but of possible interest to the service, were compiled and issued every two weeks.

Division II, Information on All Navies, Operations, Strategic Subjects, Records of Naval Officers, collated and recorded all obtainable information concerning those subjects and disseminated it to the proper bureaus and offices. Information that was of permanent or historical value was passed to Section D for deposit in its archives. Information of temporary importance, such as the movement of ships and current ship construction, was kept carded on a day-to-day basis for ready reference. "All reports, rumors, and intelligence items of every description" were desired by Division II, whenever they concerned "Navies, or Naval affairs, American or Foreign."

Division III, Mercantile Collations, disseminated collated information on merchant marine activities throughout the world, ships' tonnages, ships under construction, and losses to submarines and from natural causes; shipbuilding facilities; merchant ship routing; new marine machinery, engineering, equipment and fittings; and laws concerning commerce, shipping, and navigation.

Section D, Dissemination, consisted of five divisions:

Division I was charged with registering, carding, and filing all reports on naval and military material, personnel, and operations; receiving and answering all requests for information relative to those subjects; censoring manuscripts submitted by members of the naval service for scrutiny in accordance with *Navy Regulations*; and printing naval intelligence publications. In answering requests for information, Section D obtained, when necessary, special reports from naval attachés or Navy bureaus, collations or copies of previous reports from Section C, or translations from Section E.

Division II was charged with all matters relating to armed guard detachments on merchant vessels. Extracts from the reports of the commanding officers of armed guard detachments were compiled and sent to all ships and stations having an interest. Division II also handled confidential bulletins from the State Department.

Division III compiled the monthly publication *Anti-Submarine Information* and disseminated printed ONI publications to various ships and stations; it was also responsible for war diaries and for the preparation of special papers. The latter included textbooks and information pamphlets on intelligence work.

Division IV compiled a semimonthly bulletin containing all information received that would be of value to principal stations, battleships, cruisers, and transports.

Division V, Camouflage, gathered and disseminated all available information on that subject to the Camouflage Section of the Bureau of Construction and Repair, the U.S. Shipping Board (responsible for administering the acquisition and operation of all U.S. flag merchant ships), and allied foreign naval attachés from countries that were investigating the subject. Much of the collection (reporting and photographing of the camouflaged ships) was accomplished by the district aides for information.

Section E, Translating, translated intelligence documents from French, Italian, Spanish, Portuguese, Russian, Dutch, Japanese, Chinese, and German into English. The section was also charged with filing, clipping, and distributing certain foreign newspapers and periodicals received by ONI.

Section F, Disbursing, audited and disbursed confidential funds, assisted ONI and its branches with their regular Navy Department accounts, and ordered, received, and kept stocks of office supplies and equipment.

Section G was charged with files and with indexing information cards on suspects and personnel.

Section H, Clipping Bureau, received newspaper clippings from ONI branch offices and aids for information throughout the United States and from naval attachés in European and South American capitals. Section H also clipped numerous papers and received material from press clipping services.

Section I, Chief Clerk, supervised civil employees and their records; procured passports; prepared drafts of orders to naval attachés; maintained corrected copies of *Navy Regulations*, general orders, and uniform regulations; supervised printing and binding; and supervised the sale of war savings and thrift stamps.

Section K, Mail, indexed and routed incoming mail. (There was no Section J.)

The Historical Section, to which a letter designator had not been assigned as of 1 September 1918, was established in ONI to collect all material that would be of historical value. RAdm. William W. Kimball (Ret.) was in charge.[8]

1920

By 1 July 1920, ONI's postwar office force had been reduced to eighteen on the statutory rolls and twenty-four former naval reservists, for a total of forty-two. The office had been reorganized back down to basics and consisted of four sections: Section A, Administrative; Section B, Intelligence (or Incoming Information); Section C, Compiling (or Manufacturing Department); Section D, Historical Section (or by-products). By-products included the Navy Department Library; the dead files, which contained war diaries of ships and stations and their correspondence during World War I; statistics; and international law questions and cases that had arisen during the war.[9]

1921

Cdr. Royal E. Ingersoll reported to ONI in June 1921 and was assigned to the Japanese Espionage Desk. ONI at that time, he later recalled, was divided into foreign desks with different officers having different sections for different countries and also responsibility for different subjects like engineering, radio, and gunnery. Cdr. Ingersoll was also in charge of the Domestic Section, responsible for counterespionage. Ingersoll also kept the Japanese monograph up to date, particularly the section on Japan's naval forces.[10]

1922

On 12 January 1922, the Director of the War Plans Division of the Office of Naval Operations (OPNAV), in a memorandum to the Chief of Naval Operations (CNO), recommended the establishment of a press relations office, to be located within ONI. The Director of Naval Intelligence (DNI) concurred with the recommendation on 14 January. The Secretary of the Navy approved the measure and issued a directive to all bureaus and offices of the Navy Department, dated 21 February 1922, that established the Information Section under the DNI. For further information on the Information Section, see Chapter 33.

1926

The Director of Naval Intelligence's annual report, OP-16-A (SC) 212-2 of 10 June 1926, Enclosure A, lists the activities of Section C as the preparation of monographs; the collating and compiling of information of military and naval value concerning foreign countries and the dissemination of the information to our naval services and to other branches of the government; and the furnishing of vast amounts of comparative data on the naval and aviation strength of the Washington Treaty Powers to the committees and individual members of Congress.

Those types of intelligence activities remained at a fairly static level throughout the interwar period.[11]

1929–1930

The Office of Naval Intelligence, late in 1929 and early in 1930, was but a small division of the Navy Department. It had two officers in the Far East Section, one officer on the British Desk, one officer on the European Desk, and one on the Latin American Desk. In addition, there were three or four officers assigned to domestic intelligence or security. Considerable emphasis was being placed on preparations for the London Naval Conference. Capt. Alfred W. Johnson, Director of Naval Intelligence at the time, was swamped with work that included long conferences at the State and War Departments. The Far East Section had a very good filing system, and information that was needed was quickly available, but it did not have enough personnel to prepare special reports.[12]

1931

The ONI organization in 1931 comprised four principal sections: Administrative, including Naval Reserve for intelligence duties; Mail and Translating; Intelligence proper, divided into Domestic and Foreign; Public Relations; and Historical, Library, and Archives. Heavy emphasis was placed on the collection of all classes of information, but particularly that information affecting naval and maritime matters, the evaluation of such information, and its dissemination.[13]

After the reorganization of 1931, the Administrative Branch had seven sections: A-1, Foreign Liaison (under the Assistant DNI); A-2, Personnel; A-3, Mail, Filing, and Archives; A-4, Supply and Accounting; A-5, Legal (inactive); A-6, Translating; and A-7, Photographic (inactive).[14]

The Administrative Branch was staffed by one lieutenant commander, four clerks, two library assistants, and three civilian translators. The nucleus of the branch was within the Chief Clerk's office, which was responsible for handling civilian personnel for all of ONI, all accounts and finances for the office, translations, printing and binding, space allocation, and legal work.[15]

Under the 1931 reorganization, OP-16-B-1 headed the Intelligence Branch. Section B-2 provided Dissemination, and sections B-3 through B-9 handled Domestic Intelligence. The nine Foreign Intelligence sections included B-10, Foreign Intelligence; B-11, British Empire; B-12, the Far East; B-13, Western Europe; B-14, Central Europe; B-15, the Eastern Europe; B-16, the Balkans and Near East; B-17, Latin America; and B-18, Enemy Trade (inactive in peacetime).[16]

1938

Because of personnel limitations resulting from budget problems of the early 1930s, certain ONI branches, sections, and units prescribed in its War Organization were not staffed. There was no peacetime organization chart or other document that set forth the active units of the office to which responsibility had been delegated for various matters assigned to inactive or nonexistent units. Hence, there were certain matters for which no person in the organization, below the assistant director, was responsible to handle administratively. In practice, when such matters arose, they were assigned by the Director of Naval Intelligence to the unit that appeared at the time to be best equipped to handle the particular problem, or they were handled by the DNI himself or the Assistant DNI.[17]

1941–1942

On 28 April 1941, a Secretary of the Navy directive was issued removing the Office of Public Relations from ONI and placing it directly under the Secretary. All the personnel of the Public Relations Branch of ONI were shifted to the Secretary of the Navy's office, with Cdr. H. Raymond Thurber assigned as acting director.[18]

At the outbreak of World War II, the Special Intelligence Section (OP-16-F-9) comprised one retired officer, two Naval Reserve officers, two enlisted sailors, and one Naval Reserve officer undergoing training in London. Instructions were immediately originated by the section and issued by the CNO to all ships and stations as to the conduct of U.S. Navy personnel in the event of their capture by the enemy.

By 30 June 1942, OP-16-F-9 had been augmented by five officers, and three civilians who were awaiting commissions as German interrogators.

On 1 February 1942, the section head was designated to participate in the drafting of recommendations for a Joint Psychological Warfare (PW) Committee for the planning and control of psychological warfare overseas. The committee's recommendations were approved by the Director of Naval Intelligence and by the Assistant Chief of Staff (G-2) U.S. Army, and on 16 February they were submitted to the Joint Chiefs of Staff (JCS) for final approval.[19]

On 7 December 1941, three previously inactive sections of ONI were opened: Commerce and Travel (OP-16-B-5), Plant Protection (OP-16-B-6), and Censorship (OP-16-D).

In January 1942, the Identification and Characteristics Section was established in ONI for the purpose of collating data on the appearances and characteristics of U.S. and foreign naval ships and merchant vessels and to disseminate identification material. On 12 January, the Fleet Intelligence Branch (OP-16-C) was formed. It included the Intelligence Center and the Information Center. In June, the designation was changed to Combat Intelligence Branch, and OP-16-C-2 became the Publication Section.

On 10 February 1942, the Protocol and Reception Center (OP-16-F-12) was set up to help handle matters of protocol and the increasing numbers of foreign military and naval officials visiting the Navy Department. In addition, OP-16-F-12 had general supervision over U.S. naval officers preparing for intelligence duty abroad.

On 5 August 1942, the Special Intelligence Section was removed from the Foreign Intelligence Branch and reestablished as the Special Activities Branch (OP-16-Z). Its functions included obtaining, training, and administering secret agents. In June 1942, it had assumed responsibility for information on captured enemy naval equipment. In connection with its work in developing a secret undercover intelligence service, OP-16-Z maintained liaison with the Office of the Coordinator of Information and subsequently with the Office of Strategic Services (OSS) when the former was absorbed by the latter.[20]

On 29 May 1942, the OP-16-A-2b subsection of the Administrative Branch, which had been created to handle Naval Reserve enlisted personnel, became section OP-16-A-6.

On 7 October 1942, the Special Warfare Branch (OP-16-W) was established to control psychological warfare and bacteriological warfare activities. Among its tasks was the processing of naval intelligence for the confidential guidance of, and the supplying of naval information to, the Psychological Warfare Planning Board of the Overseas Branch of the Office of War Information (OWI).[21]

1943

The Office of Naval Intelligence was reorganized in 1943 to conform as much as possible with the existing structure of the Army's Military Intelligence Service (MIS). The former Assistant Director was retitled Deputy Director of Naval Intelligence (DDNI). The organization as a whole was divided into three main groups, each under an assistant director. The Services Group was composed of the Administrative Branch (OP-16-A) and the Training Branch (OP-16-T); the Intelligence Group was composed of the Intelligence Branch (OP-16-F), the Special Activities Branch (OP-16-Z), and the Publications and Distribution Branch (OP-16-P); and the Counterintelligence Group was composed of five sections: Naval Censorship (B-2); Investigations (B-3); Security of Naval Information (B-4); Commerce and Travel (B-5); Sabotage, Espionage, and Countersubversion (SEC) (B-7); and Coastal Information (B-8).

Special Warfare (OP-16-W) and Naval Records and Library (OP-16-E) continued as branches under the direct supervision of the DDNI. The Protocol and Liaison Branch (OP-16-L) also reported directly to the DDNI.

Two new groups were formed. The first, Planning (OP-16-X), was composed of the DDNI and the three assistant DNIs and had a permanent secretariat. It was intended to formulate plans for the efficient functioning of ONI and to have charge of War Plans for ONI. The second new group, Evaluation and Dissemination (OP-16-ED), functioned as the Navy component of the Joint Evaluation and Dissemination Staff of the Joint Intelligence Agency. It was composed of the heads of geographic sections of F Branch, the head of the Commerce and Travel Section, the head of the Sabotage, Espionage, and Countersubversion Section, and the head of the Special Warfare Branch.

The Intelligence Branch was organized into sections to cover the four main geographic areas (Europe-Africa, Far East, American Republics, and North America) and the Operational Intelligence (OP-16-FO), Foreign Trade (OP-16-FT), and Intelligence Plot (OP-16-FP) sections.[22]

In May 1943, the North American Theater Section (OP-16-FN) was established in the Intelligence Branch to take over foreign intelligence collection in the United States (including Alaska). The Coastal Information activities of OP-16-B-8 were transferred to the new section.[23]

When the Operational Intelligence Section was disestablished on 9 September 1943, the North American Theater Section was divided into FN-1 (North American Intelligence) and FN-2 (Coastal Intelligence). LCdr. Frank A. Klaveness was the head of FN and also served as FN-2 on an interim basis. LCdr. J. H. Black was assigned as the head of FN-1.[24]

OP-16-FN was responsible for obtaining intelligence about foreign places from sources within the United States. In each naval district, FN sections were established and directed to contact importers, exporters, banks, oil companies, etc., as well as private individuals who had traveled extensively.[25]

1944

Cdr. John L. Riheldaffer, USN (Ret.), head of the Special Activities Branch (OP-16-Z), was liaison officer for ONI with the Office of Strategic Services. All requests for information or data, or requests for and transfers of documentary data, made by ONI personnel to OSS were to be sent to him for handling and recording. OSS personnel were under instruction to clear any contacts they wished to make with the Navy Department through the Naval Command, OSS. In implementation of this directive, the Liaison Division was established in the Naval Command, OSS, with the duty to initiate, maintain, and renew contacts between the Navy Department and the OSS to exercise control for security purposes over visits of personnel from each agency visiting the other. LCdr. Daniel Ravenel, Jr., USNR, was officer in charge of the Liaison Division.[26]

1945–1946

In April 1945, ONI, known briefly as the Naval Intelligence Division, formerly under the Vice Chief of Naval Operations (VCNO), resumed the title Office of Naval Intelligence, a division of the Office of the Chief of Naval Operations. At that time the Operational Intelligence Section was made a branch (OP-16-O) under the Assistant DNI, Intelligence Group.[27]

At the conclusion of World War II, in an OPNAV reorganization, ONI was placed under the Deputy Chief of Naval Operations (DCNO) for Administration and was designated OP-23; the Director of Naval Intelligence was given the new title, Chief of Naval Intelligence. The sections and subsections of Domestic Intelligence dealing with naval, cable, and radio censorship, and with security controls relating to commerce and travel, were deactivated.[28]

The postwar abandonment of the wartime combination of CNO and COMINCH (Chief of Naval Operations and Commander in Chief, U.S. Fleet) under FAdm. Ernest J. King caused a merger of the COMINCH staff into the Office of the Chief of Naval Operations. In the postwar reorganization, the Domestic Intelligence Branch absorbed several functions of the COMINCH staff, including publicity security and the security of code designations, the classification of reports, and security control. Those functions were allocated to the Security of Naval Information Section of ONI.

Immediately after the end of World War II, Secretary of the Navy James V. Forrestal directed the CNO to establish a rigorous, centralized control of the disclosure of classified information. The Security of Naval Information Section (OP-23D21) was designated as the agency to perform the function. With the reallocation of duties and functions precipitated by the end of the war, matters pertaining to security, which had been decentralized during the war, were centralized in OP-23D21.[29]

When its designator was changed from OP-16 to OP-23, ONI's organization consisted of seven branches and two staff elements and was approved by the Chief of Naval Intelligence, Commo. Thomas B. Inglis, on 29 October 1945. Branches included 23C, Administrative; 23D, Domestic; 23E, Naval Records and Library; 23F, Foreign; 23V, Air; 23W, Special; and 23Y, Operational. The staff elements included 23L, Liaison, and 23X, Plans. The Administrative Branch consisted of four major sections: C1, Special Publications; C2, Services (which included Personnel, Supplies and Accounts, Reproduction (duplication), and Mail and Files); C3, Training; and C4, Translations. The Domestic Branch consisted of five sections: D1, Investigations; D2, Security; D3, Contact Register; D4, Sabotage, Espionage, and Counterintelligence; and D5, Cable Censorship (inactive). The Naval Records and Library Branch had four sections: E1, Library; E2, Records; E3, Sound Recordings; and E4, Historical Publications. The Foreign Branch consisted of six sections: F1, Collection and Research; F2, Technical; F3, Graphic; F4, Washington Document Center; F5, Specialist Staff; and F6, Dissemination and Administration. The Air Branch had four sections: V1, Collection; V2, Photographic Intelligence; V3, Evaluation; and V4, Dissemination. The Air Branch was also the channel to the Joint Army-Navy Air Intelligence Activities. The Special Branch and the Operational Branch had no separate sections.[30]

The Navy Subsidiary Post-War Plan–Intelligence was promulgated in November 1945. Based on Basic Post-War Plan No. 7.5, it set down in detail the organization and personnel requirements of naval intelligence, incorporating the views of capable and experienced naval intelligence officers.

The greatest differences between the wartime Naval Intelligence organization and the postwar plan were in the reduction in numbers of service personnel and the resultant necessary increases in civilian personnel to carry out postwar naval intelligence functions. Many wartime intelligence operations had ceased, but new and vital functions had commenced, such as opening posts in locations inaccessible during the war, processing the remaining mass of captured documents, collecting intelligence in occupied countries, and keeping pace with the unstable international situation.

Naval Intelligence opined that it would carry out its mission to the maximum extent with whatever funds and personnel were available. However, it was becoming more evident each day that, as the international situation deteriorated, the requirements for intelligence were becoming greater and more urgent. The personnel statistics for naval intelligence billets during the immediate postwar period show the effects of the drastic reductions in personnel.

Table 28.1.
Intelligence Billets, Jul 1945–Aug 1946

Officers

	1 Jul 1945	1 Jul 1946	20 Aug 1946	Postwar Plan
ONI*	599	374	121	195
DIOs†	787	155	21	92
Foreign	253	205	133	120
Total‡	1,639	734	275	407

Enlisted

	1 Jul 1945	1 Jul 1946	20 Aug 1946	Postwar Plan
ONI*	702	265	75	5
DIOs†	1,158	112	98	18
Foreign	337	180	142	81
Total‡	2,197	557	315	104

Civilian

	1 Jul 1945		1 Jul 1946		20 Aug 1946		Postwar Plan	
	C.S.	Cont.	C.S.	Cont.	C.S.	Cont.	C.S.	Cont.
ONI	356	23	305	21	314	31	510	6
DIOs	148	35	94	51	121	72	137	92
Foreign	30	139	52	123	56	111	120	113
Totals	534	197	451	195	491	214	767	211

*Joint Army Navy Air Intelligence Division (JANAID) included.
†District Intelligence Officer, Photo Intelligence Center included.
‡Washington Document Center not included.
Source: Chief of Naval Intelligence, ser 1520P32, 23 Aug 1946, Accession 3770, box 1, ONI Day File, OA.

1946

To a slight extent, the effects of demobilization in ONI were counterbalanced by obtaining authorization for civilian billets to replace released military personnel in key positions. Even such partial replacement, however, was not possible in the naval districts, where the field activities of the Domestic Intelligence Branch were carried on. During demobilization and until the peacetime components of the Naval Reserve could be organized to advantage, the effectiveness of the naval intelligence service in the naval districts was seriously impaired.[31]

The demobilization of military personnel assigned to ONI proceeded rapidly with the close of hostilities as shown in the table below.

Table 28.2.
ONI Personnel, Aug 1945–Jan 1996

Date	Officers	Enlisted	Civilians
1 Aug 1945	563	652	336
1 Sep 1945	506	620	334
31 Jan 1946	421	364	321
30 Jun 1946	267	214	313
30 Sep 1946	165	59	327

Source: OP-32 Quarterly Summary Report, 1 Jul 1946-30 Sep 1946, OA.

On 19 July 1946, VAdm. Forrest Sherman, then Deputy Chief of Naval Operations for Operations (OP-03), in a memorandum to the Vice Chief of Naval Operations, recommended the transfer of ONI from OP-02 to OP-03.

> Prior to the Japanese attack on Pearl Harbor, there was not in the Navy Department adequate coordination between the War Plans Division and the ONI. The War Plans Division undertook to evaluate intelligence by means of a small Op-Intelligence group. During the war, there existed in the Navy Department an intelligence organization in the Headquarters of the Commander-in-Chief, U.S. Fleet, and also the ONI in OPNAV. After the establishment of the present organization [in 1945], there remained in the Operations Division an OPINTEL [Operational Intelligence] Section (OP-32) in addition to the ONI in the Administration Division.
>
> In January 1946, Rear Admiral Inglis (DNI) and Captain Smedberg (OP-32) both advocated consolidating all intelligence activities under DCNO (Operations). For various reasons, I did not consider such action wise at that time but did agree that a consolidation should take place. Accordingly, OP-32 was disestablished as of 11 February.
>
> In order that in the future there may be the closest practicable coordination of intelligence, strategic planning, and operations, it is recom-

mended that the ONI be transferred from the Administration Division to the Operations Division and redesignated as OP-32.[32]

In August, the recommended shift, with its change of designator, was accomplished, except that the Office of Naval Records and Library remained under the DCNO (Administration) and was combined with the Office of Naval History.

1948

The shortage of adequate personnel in ONI continued to be a problem. In February 1948, a plea was made by ONI to the Chief of Naval Operations for an increase in its officer allowance from 96 to 127. It was pointed out that, with the current allowance, insufficient time could be devoted to the production of, and review and comment on, important joint intelligence papers; ONI could not produce the naval phases of strategic intelligence studies on schedules comparable to the Army Intelligence Division's production of its portion; inadequate time was available for thorough processing of information as received; and correspondence could not be answered without unacceptable delays. The Army Intelligence Division had 270 officers on board at that time. A major reorganization of ONI became effective on 1 October 1948.[33]

1950

The organization of ONI early in 1950 did not provide any functional unit for processing Undersea Warfare (USW) intelligence. Coordinating the processing of the elements of Undersea Warfare intelligence was accomplished by a USW panel that included members from the Foreign and Operational Intelligence Sections of ONI. Reorganization along functional lines was under study and was expected to result in better means for processing USW intelligence.[34]

By 21 September 1950, ONI underwent another major reorganization that resulted in three branches and two staff elements.

1952

Essentially, under the Director of Naval Intelligence and the Assistant Director, the functions of ONI were performed through five assistants during the early 1950s. Two assistants were heads of staff sections, one headed foreign liaison, and the other handled JCS papers and the coordination of plans. The two staff sections, with their clerical personnel, constituted the immediate staff group, which was restricted in size. The other three assistants each had a functional division of responsibility encompassing all functions of the ONI except those pertaining to the immediate staff group.

Through the heads of the Security and Intelligence Branches, the DNI performed production functions. Through the head of the Administrative Branch, he exercised "supervision over all matters of administration of the ONI, including budgetary matters, personnel, disbursing and accounting, training, and office services," with an objective, among others, "to relieve the other branches of ONI of as many administrative duties as possible in order that they may devote their activities to intelligence duties."[35]

1953

In 1953, ONI was organized into three branches: Security (including counterintelligence), Intelligence, and Administration, each headed by a senior captain. There was a close working relationship between the Security and Intelligence Branches, and common centralized files were used. Administration supported the other two branches in fiscal, personnel, training (including intelligence Reserve training), and general services matters, and it provided management control of naval attaché posts and the district intelligence offices.

There was an inherent advantage in the ONI organization as compared to that of the Air Force intelligence organization in that all functions concerned with security and counterintelligence were integrated in ONI. This integration was important because material for the counterintelligence files came from both positive intelligence collection and from investigations. Correlation in one organization was more effective. As a consequence, except for the FBI, ONI had the most complete and readily accessible counterintelligence files in the intelligence community.

Another factor in favor of the integration of security functions within ONI was the centralized administration of the district intelligence offices and the naval attachés, with both organizations collecting positive intelligence and carrying out counterintelligence activities.[36]

1954

Office of the Chief of Naval Operations Internal Notice 5430 of 3 May 1954 prescribed a major reorganization of OPNAV, effective 1 June 1954. ONI became an independent office, with its Director reporting directly to the Vice Chief of Naval Operations. The Director of Naval Intelligence was to be responsive to requests for intelligence from the various planning and operational elements of OPNAV.

OPNAV Internal Notice 5430 of 17 May 1954 prescribed the number designators for the offices in the new OPNAV organization. The Assistant Chief of Naval Operations for Intelligence became OP-92.

1955

The Director of Naval Intelligence in the mid-1950s, as OP-92 and ACNO (Intelligence), was a member of the staff of the Chief of Naval Operations and reported directly to the Vice Chief of Naval Operations (OP-09). The DNI also had a direct responsibility to the Secretary of the Navy.

The DNI accomplished his mission through the broad organization known as Naval Intelligence, which was represented within the components of the three principal parts of the Navy Department as follows: (1) the Navy Department (ONI itself); (2) the shore establishment (naval districts and foreign posts of naval intelligence—the naval attachés, navy liaison officers, and naval observers); and (3) the operating forces of the Navy (the fleet and force intelligence staffs, fleet intelligence center, fleet air intelligence augmenting units, mobile intelligence production units, sea frontier intelligence staff, intelligence personnel assigned to ships and air squadrons, intelligence personnel at outlying naval bases, intelligence personnel at advanced naval base, and intelligence officers serving as, or with, naval port control officers). The DNI/ACNO (Intelligence) exercised technical control and carried out assigned functions of management control over Naval Intelligence in the naval districts and over the foreign posts of Naval Intelligence; he also exercised technical control over, and support to, Naval Intelligence elements in the operating forces.[37]

Table 28.3.
Intelligence Billets, May 1958

	Military	Civilian	Alien
ONI	437	420	—
Attaché	313	62	162
District Intelligence Office	288	729	—
Continental U.S. Field Commands (Other than DIO)	88	177	—
Fleet Intelligence	218	—	—
Air Intelligence	681	—	—

Source: OP-92B1B memo, ser 000601P32, 8 May 1958, OA.

1959

On 19 October 1959, it was determined that the CNO intelligence briefers, the editor of the *ONI Bulletin (ONIB)* and the Graphics Unit (an element of OP-922Y2D) would operate more efficiently as a staff function of the ADNI for Foreign Intelligence. Accordingly, these units were removed from the Operational Intelligence Branch and placed under a newly created Current Intelligence Coordinator (OP-922B4). Intelligence Plot and its watch officers remained under Operational Intelligence as OP-922Y2C1.[38]

1961

The establishment and activation of the Defense Intelligence Agency (DIA) resulted in the phased transfer of a number of ONI functions and personnel to the new agency. Reorganization of OP-922 was necessary to adjust for the transfers. The increasing importance of the Flag Plot complex at the Pentagon and the employment of Intelligence Plot more in support of Flag Plot and less in support of the Operational Intelligence Branch led, in June 1961, to the shift of Intelligence Plot and its watch officers from Operational Intelligence to rejoin CNO intelligence briefers in OP-922B4, which was renamed the Flag Plot Intelligence Support Group. The change was primarily inspired by CNO Arleigh Burke's comment that Flag Plot should not be allowed to become a mere message center.[39]

1966

In June 1966, Naval Reserve RAdm. A. Atley Peterson conducted a survey of ONI's organization and activities as an annual active-duty training project. His broad background in the intelligence field (Navy active duty, Naval Reserve training, and civilian occupation) made his observations, based on his survey, particularly worthy of consideration. Among his findings were the following:

> 1. There was no planning group, as such, in ONI. Two or three officers with broad intelligence experience, free of day-to-day requirements, should have been producing an annual ONI operating plan which, in turn, would serve as the basis for ONI's programs and budgets.
>
> 2. ONI should have had a small unit to look to improvements in operating efficiency.
>
> 3. Too few professional intelligence officers fully appreciated the use of machines or the human-machine relationship. More processing should be done at the sensor site to eliminate needless material being sent to the computer center. The data base needed to be purged continuously to eliminate irrelevant detail from the material to be studied by the operational intelligence (OPINTEL) officer or analyst.
>
> 4. Most of the available sophisticated technical systems appeared to have been thrust upon Naval Intelligence by technical offices at the suggestion of industries. There also appeared to be little integration of the different systems, and too few intelligence officers were qualified to appreciate the capabilities and limitations of the technical systems in which they dealt. There was no ONI group giv-

ing management direction to the operations of the many technical systems, which should have been kept constantly under review.[40]

1966–1967

As a result of the Benson Study on how to reduce the number of people assigned to OPNAV, the Vice Chief of Naval Operations directed the ACNO (Intelligence) to take steps necessary to establish the Naval Intelligence Command (NAVINTCOM), remove as many operating functions as feasible from the OPNAV staff, and assign command and support of field activities to NAVINTCOM. An implementing plan was required for the Vice Chief's approval by 1 January 1967. OP-92 submitted the plan by memo serial 09408P92 of 29 December 1966.

Secretary of the Navy Notice 5450 of 19 April 1967 activated NAVINTCOM effective 1 July 1967. The mission of its commander was "to direct and manage the activities of the Naval Intelligence Command to assure the fulfillment of the intelligence, counterintelligence, investigative, and security requirements and responsibilities of the Department of the Navy."

When the 1 July major reorganization of ONI was implemented, all designated ONI billets transferred to NAVINTCOM. The Director of Naval Intelligence became Commander Naval Intelligence Command and also retained the ACNO (Intelligence), or OP-92 title.

ONI was split into two parts, those remaining in the Office of the Chief of Naval Operations under OP-92 and those shifted to the Naval Intelligence Command. The latter was composed of a headquarters and six field activities. The headquarters had four major elements: NIC-1, Administration and Services; NIC-2, Intelligence Systems Requirements and Support; NIC-3, Intelligence Operations Coordination; and NIC-4, Programming and Budget. The six NAVINTCOM field activities were the Naval Investigative Service, the Naval Scientific and Technical Intelligence Center (STIC or NAVSTIC), the Naval Reconnaissance and Technical Support Center (NRTSC), the Naval Intelligence Processing System Support Activity (NIPSSA), the Naval Field Operational Intelligence Office (NFOIO), and the Applications Department. The relocation of NAVINTCOM headquarters away from the Pentagon was to be "phased over a period of time as necessary."

OP-92 staff elements included Counterintelligence Plans and Policy, Foreign Disclosure Policy, Foreign Intelligence, Foreign Liaison and Protocol, Program Planning, and the Secretariat.[41]

The OP-92 plan for establishing the Naval Intelligence Command transferred to that command all intelligence production, processing, and analysis functions; development of intelligence contributions for specific Navy uses; resources management, systems planning, and programming; personnel and fiscal administration; technical guidance; and research and development coordination. The residual OPNAV staff element (OP-92) consisted of the billets and personnel needed to assist the ACNO (Intelligence) in handling those intelligence-related matters requiring direct CNO cognizance or response. It was not intended that the OPNAV staff element would be independently capable of or responsible for instant or total substantive response. All substantive intelligence support would be provided to the CNO by the Naval Intelligence Command, via OP-92.

A total of forty-two intelligence billets remained on the CNO staff. In addition, certain NAVINTCOM operating elements remained in the Pentagon but were not considered a part of the Office of the Chief of Naval Operations. These included Intelligence Plot, the Special Intelligence Communications Center (SPINTCOM), the Special Security Officer (SSO), and the Estimates Group. The forty-eight personnel in the detached NAVINTCOM elements resident in the Pentagon, added to the forty-two CNO staff billets, made a total of ninety Naval Intelligence billets remaining in the Pentagon.

Relocating the balance of current billets out of the Pentagon was expected to generate requirements for additional billets. Many OP-92 personnel had been performing dual functions, but, in some cases, one function was to remain in OPNAV and the other was to transfer to NAVINTCOM and not be reassigned to another billet. Such cases were expected to require a total of ten additional billets. Approximately five billets were expected to make up a small administrative unit in the Pentagon to handle OP-92 message routing, correspondence, and classified material control functions. Maintaining separate classified document control units and SSO detachments at NAVINTCOM, in addition to those elements remaining in the Pentagon, was expected to require twenty-five billets above the current allowance. While ACNO (Intelligence) was to be double-hatted as COMNAVINTCOM, it was expected that a separate deputy would be needed for each hat, thus requiring an additional captain billet. The Deputy COMNAVINTCOM was also to serve as the commanding officer of the Naval Intelligence Support Activity, a new field activity incorporating those intelligence operating functions and NAVINTCOM command and support functions transferred out of OPNAV. Only ten additional billets, however, were approved by the Vice Chief.

Prior to relocation, ONI occupied about 45,000 square feet of space in the Pentagon. The move was expected to make about 33,000 square feet of space in the Pentagon available for reassignment.[42]

When the plan was implemented on 1 July 1967, the actual reorganization had two essential differences: Foreign Disclosure (OP-92D, with twenty-seven billets) and Intelligence Plot and the Estimates organization (with about twenty-four billets between them) remained in OP-92. Other minor changes brought OP-92 strength to ninety-seven, vice the forty-two originally planned under the reorganization.[43]

In his memo serial 04905P92 of 18 December 1967, ACNO (Intelligence) proposed to the Assistant Vice Chief three changes to the original reorganization plan that would return billets to the OP-092 staff. Based on experience gained in the first four months following the establishment of the Naval Intelligence Command, the billets were the three National Indications Center billets in the Pentagon originally transferred to NAVINTCOM; two Human Resources billets that had transferred to NAVINTCOM, but which were involved with Human Resources projects under the cognizance of JCS (SACSA); and the NAVINTCOM Applications Department, which had been set up as a field activity. The reallocation involved thirty billets that were to remain in the Pentagon anyway. The additions increased the OP-92 staff from 97 to 132 but increased intelligence personnel space requirements in the Pentagon by only two.

Following the establishment of the Naval Intelligence Command, a study was made, under the guidance of the Director of Naval Intelligence, by Capt. Donald P. Harvey and L.P.H. Healey to determine the problems that had arisen when NAVINTCOM elements were no longer located in the Pentagon and to propose solutions. They conducted the study between 9 November and 6 December 1967.

Harvey and Healey noted in their report that most of the confusion and apprehension caused by the ONI reorganization had been derived from the following situations:

> The directive to reduce OP-92 to a small staff in a short period of time and to relocate the majority of ONI headquarters personnel outside the Pentagon without adequate assessment of the effect of the dislocation of the real-time support traditionally tendered to the Secretary of the Navy, the Chief of Naval Operations, and their staffs.
>
> The decision to effect, simultaneously with the above, a radical reorganization of ONI itself, superimposing management and coordination staffs at the expense of the substantive quick reaction support formerly carried out by OP-922.
>
> The consequent replacement of former ONI lines of command by initial decision-making lines from the OP-92 staff, NAVINTCOM staffs, and the various NAVINTCOM field activities, all leading directly to OP-92B/NAVINTCOM 01.
>
> The attempt, since 1 July 1967, to wear the garb of the reorganization while at the same time operating and giving support, for the most part, as though the former organization still existed.

The study made the prognosis that ONI would, for an indefinite period (particularly after NAVINTCOM headquarters had moved to a "temporary" location), suffer certain severe disabilities in attempting to support all who traditionally had expected intelligence support while at the same time attempting to expand its former sporadic role of management coordinator of naval intelligence throughout the Navy.

1968

As a result of a reevaluation, ordered by the Navy Inspector General, of the planned move of NAVINTCOM headquarters out of the Pentagon, ACNO (Intelligence) reported his findings to the Chief of Naval Operations.

Because of his second hat as ACNO (Intelligence), the Commander Naval Intelligence Command found that he had to spend the bulk of his time at his Pentagon Office. To determine the effects of his physical separation from his supporting staff at NAVINTCOM headquarters, a one-week survey was conducted to determine the frequency of personal contact between COMNAVINTCOM or his deputy and the key staff personnel engaged in managing NAVINTCOM's operations. Only those contacts that could not have been handled by other means were counted. They included hand-carrying messages and correspondence for review, discussions, and signatures, and presentation briefings. The survey indicated that, on an average day, twenty-five such contacts were required. It was estimated that two to three workdays would be lost each day, even with a careful stockpiling of business to reduce the number of daily visits to the Pentagon from NAVINTCOM's remote location in the Washington suburbs.

The study found that, in addition to, and of much more significance than these personal contacts, were those required between NAVINTCOM headquarters personnel and various other OPNAV offices and other Pentagon activities and agencies. A one-week survey of such contacts conducted between 29 January and 2 February 1968, including visits by persons from those offices and agencies to NAVINTCOM spaces, found that there was an average of 202 visits per day. Again, only those personal contacts were counted that could not have been handled by other available methods. The particular week selected however, did not provide a fair test, because many of the contacts were related to the *Pueblo* (AGER 2) incident. Nevertheless, the remainder were of such magnitude and urgency as to constitute a significant factor in any analysis of the effect of relocating the headquarters outside the Pentagon.[44]

To reflect the increasingly important responsibility for the management of Ocean Surveillance Information, which had been assigned to ACNO (Intelligence), an Office of Assistant for Ocean Surveillance and Intelligence Operations was established within OP-92 on 13 December 1968 and was designated OP-92Y. At the same time, Intelligence Plot (OP-92F6) and the office of the Assistant for Intelligence Operations (OP-92W) were disestablished and incorporated into the new OP-92Y.

OP-92Y (the former OP-92W) was composed of a staff and four major divisions: the Ocean Surveillance Information Center (OP-92Y1, former OP-92F6), the Surface Warfare Division (OP-92Y2), Air/Space Warfare Division (OP-92Y3), and the Undersea Warfare Division (OP-92Y4). All personnel in Ocean Surveillance and Intelligence Operations were assigned to NAVINTCOM (NIC-3), and only OP-92Y and his staff and division heads had additional duty in OPNAV under OP-92. OP-92Y4 was additionally designated as OP-92U with responsibilities for a special Navy program.[45]

1968–1969

On 1 November 1968, the name of the intelligence operations coordination organization (NIC-3) was changed to Ocean Surveillance and Intelligence Operations to reflect a new mission to develop and establish an all-source Navy Ocean Surveillance Information Management System in accordance with OPNAV Notice 5430, serial 1128P00B3 of 13 December 1968.

In 1969, OP-92 became OP-092, to indicate more clearly that the Director of Naval Intelligence was organizationally placed so as to report directly to the Chief and Vice Chief of Naval Operations.

The establishment of an all-source Integrated Ocean Surveillance Center was nearing realization in 1969. Negotiations were underway for a third-generation computer needed for the center, and the target date for activating the NAVINTCOM Automated Data Processing Center at Suitland, Maryland, was 1 September 1970. The consolidation of NAVINTCOM ocean surveillance analytical talent and effort at Suitland was scheduled to take place incrementally as physical arrangements would permit. The Chief of Naval Material had been asked to provide costing estimates for near-term improvements in the Ocean Surveillance Information System (OSIS) that would expedite the flow of perishable information.[46]

Effective 26 May 1969, the office of Assistant for Foreign Disclosure Policy and Control (OP-092D) was disestablished, and the office of Assistant for Security of Military Information (OP-092D) was concurrently established, consisting of a Foreign Disclosure Policy and Control Branch (OP-092D1), a Security Policy Branch (OP-092D2), and a Security Review Branch (OP-092D3). At the same time, the title of OP-092C was changed from Assistant for Counterintelligence Plans and Policy to Assistant for Counterintelligence and Investigations. According to personnel rosters, however, the change of title of OP-092C didn't actually take place until February 1970. All the functions and resources of the disestablished OP-092D and the security policy functions and resources of OP-092C were transferred to the new OP-092D. Additionally, the functions and related resources of the Chief of Naval Information (CHINFO) that were concerned with security review in the area of public affairs were transferred to OP-092D.[47]

OP-092Y memo serial 0407P92 of 21 May 1969 requested a task force designator be assigned to OP-092 "for the purpose of establishing an entity to provide correlated tactical, scientific and technical intelligence data to the fleet and other designated users." The task force was to be related to tactical/technical intelligence programs. The employment of a task force-type organization was chosen to ensure that the resources assigned to it would be completely responsive to unique tactical and technical requirements. Capt. Wallace L. Russell was the first Commander Task Force (CTF) 168, the number assigned to the new organization on 2 June 1969.

TF 168 was established as part of NAVINTCOM's Fleet Support and Intelligence Operations organization (NIC-03). Its purpose was to provide for the comprehensive exploitation of fleet-collected intelligence data on a fully integrated and timely basis. The principal features of TF 168 activities were quick-reaction technical and analytical support to fleet-operated airborne, surface, and subsurface sensors and sensor systems; on-the-scene and real-time intelligence processing; and direct and immediate support of maritime surveillance operations and integrated all-systems/all-source technical intelligence production.

1970–1973

On 21 October 1970, ONI's OP-092 staff elements at the Pentagon were changed: OP-921 became Security of Military Information; OP-922 became Estimates, USIB (U.S. Intelligence Board) Matters and Departmental Support; OP-923 became Fleet Support and Intelligence Operations; OP-924 became Ocean Surveillance Information and Automation Systems; and OP-925 became Undersea

Warfare. The staff elements were OP-092BA, Executive Assistant; OP-092C, Assistant for Counterintelligence Plans and Policy (the OPNAV roster of 1 February 1970 shows the title changed to CI and Investigations); OP-092J, Assistant for Intelligence Personnel and Training; OP-092P, Assistant for Program Planning; OP-092S, Secretariat; OP-092T, Technical Advisor; and OP-092X, Assistant for Security Coordination.

As of 1 November 1970, the organization of the NAVINTCOM headquarters had been modified into the following departments: Administration, Program and Budget, Fleet Support and Automation Systems, Undersea Warfare, and Intelligence Plot. The NAVINTCOM field activities were reduced to five by disestablishing the Applications Department.[48]

Effective 15 March 1971, the Office of Command Support Programs (OP-094) was established, with RAdm. Frederick J. Harlfinger II as its director. OPNAV Notice 5430 of 2 March 1971, which announced the establishment of OP-094, disestablished the offices of the ACNO (Intelligence) and the ACNO (Communications and Cryptology) among others, and made them divisions of OP-094. The Director Intelligence Division was first designated OP-941, but the OP code was changed to OP-942 by OPNAV Notice 5430 of 11 March 1971 in time to be effective when OP-094 was established. In addition, OP-942 was also double-hatted as Assistant for Intelligence (OP-009).

On 31 March 1971, an Assistant for Reconnaissance and Surveillance Programs (OP-094R) was established within the Office of Command Support Programs. Consolidated within OP-094R were certain reconnaissance and surveillance functions previously carried out by OP-335, NIC-33, and elements of the Naval Security Group. The consolidation removed numerous intelligence collection functions from NAVINTCOM cognizance, including the analysis of results to determine the continuing need for collection by reconnaissance and surveillance programs.[49]

In August 1971, a Foreign Operations Division (NIC-32) was established at the Naval Intelligence Command to provide centralized coordination and management of the many navy-to-navy intelligence relationships. NIC-32 provided a focal point within naval intelligence for monitoring the quality and quantity of quid-pro-quo exchanges with friendly navies, a point of contact for U.S. Navy fleet and field sponsors of exchange relationships, and a staff organization to develop initiatives for assistance to foreign navies for mutual gain derived through the improvement of their capabilities.[50]

The general policy for the guidance of NIC–32 was expressed as follows:

> Improve foreign naval intelligence relationships; expand the quantity and quality of intelligence gained through these relationships; improve the timeliness of U.S. analytical responses to foreign intelligence inputs; and assist foreign naval intelligence efforts by staffing initiatives to provide training, reference material, collection equipment, and other means of enhancing the intelligence proficiency of cooperating navies.[51]

The OSIS, which was established in 1972, was made up of the Naval Ocean Surveillance Information Center (NOSIC) at Suitland, Maryland, under the Naval Field Operational Intelligence Office, which had taken over management responsibility from the Naval Intelligence Processing Systems Support Activity; Fleet Ocean Surveillance Information Centers (FOSIC) at each major Fleet Command Headquarters; Fleet Ocean Surveillance Information Facilities (FOSIF) located near numbered fleet commands. FOSIC Hawaii was established in August 1971; and FOSIF Kamiseya, Japan, and FOSIC Norfolk were established in February 1972. Fiscal Year 1972 saw the first full year of operations by NOSIC, FOSIC London, and FOSIF Rota, Spain.[52]

Effective 1 July 1972, the Naval Scientific and Technical Intelligence Center and the Naval Reconnaissance and Technical Support Center were combined to form the Naval Intelligence Support Center (NISC). NAVSTIC and NRTSC had been colocated in Federal Office Building No. 5 at 4301 Suitland Road, Suitland, Maryland, and the new NISC continued to be located there. Command and support for NISC was assigned to Commander Naval Intelligence Command.[53]

On 1 February 1973, the ACNO (Intelligence) was returned to a position directly under the Vice Chief and Chief of Naval Operations and was designated OP-009.[54]

ONI Personnel Lists and Organization Charts

1885

Officers as of 1 February:

Lt. T.B.M. Mason,
Chief Intelligence Officer

Lt. William H. Beehler

Lt. William H. Driggs

Lt. Sidney A. Staunton

Lt.(jg) John C. Colwell

Lt.(jg) Alexander Sharp

Lt.(jg) Templin M. Potts*

Ens. Frank R. Heath

Ens. William L. Rodgers

Passed Assistant Engineer J.P.S. Lawrance

*The *Navy Register* of 1 Aug 1883 shows then-Ens. Potts reporting to the Navy Department for Special Duty on 13 Feb 1883.
Source: *Navy Register*, Feb 1885

1889

Officers as of 1 January:

Lt. Raymond P. Rodgers,
Chief Intelligence Officer

Lt. Frederick Singer

Lt. William H. Beehler

Lt. Charles E. Vreeland

Lt. Sidney A. Staunton

Lt.(jg) John T. Newton

Lt.(jg) Benjamin Tappan

Ens. John M. Ellicott

Ens. John B. Bernadou

Ens. William L. Howard

Passed Assistant Engineer Charles W. Rae

Source: *Navy Register,* 1 Jan 1898

1892

Officers as of 1 January:

Cdr. Charles H. Davis,
Chief Intelligence Officer

Lt. George W. Mentz

Lt. Charles E. Fox

Lt. George H. Peters

Lt. John C. Colwell

Lt. Ridgely Hunt

Lt. Charles C. Rogers

Lt.(jg) Augustus F. Fechteler

Lt.(jg) Charles W. Jungen

Ens. Edward Simpson

Ens. Marbury Johnston

Assistant Engineer W. H. Alderdice

Source: *Navy Register,* 1 Jan 1892.

1895

Officers as of 1 January:

Lt. Frederick Singer,
Chief Intelligence Officer

1stLt. Lincoln Karmany, USMC

Lt. William W. Kimball

Lt. Edward B. Barry

Lt. Edward F. Qualtrough

Lt. John W. Stewart

Lt. Philip V. Lansdale

Lt.(jg) Randolph H. Miner

Lt.(jg) Wiley R.M. Field

Ens. Creighton Churchill

Ens. Clarence M. Stone

Ens. Sumner E. Kittelle

Source: *Navy Register,* 1 Jan 1895.

1897

Officers as of 1 January:

LCdr. Richard Wainwright,
Chief Intelligence Officer

Lt. William W. Kimball

Lt. Herman F. Fickbohm

Lt. John C. Colwell

Lt. Edward B. Barry

Lt. William. S. Hogg

Ens. William. K. Harrison

Ens. Lay H. Everhart

Source: *Navy Register*, 1 Jan 1897.

As of 1 July:

LCdr. Richard Wainwright,
Chief Intelligence Officer

LCdr. Edward B. Barry

Lt. Herman F. Fickbohm

Lt. William S. Hogg

Ens. William D. Brotherton

Source: *Navy Register*, 1 Jul 1897.

1898

Officers as of 1 January:

Cdr. Richardson Clover,
Chief Intelligence Officer

LCdr. William H. Driggs

Lt. Herman F. Fickbohm

Lt. Samuel W.B. Diehl

Lt. William S. Hogg

Lt.(jg) Webster A. Edgar

Ens. Sumner E. Kittelle

Source: *Navy Register,* 1 Jan 1898.

1904

Officers as of 1 January:

Capt. Seaton Schroeder,
Chief Intelligence Officer

LCdr. Charles N. Atwater

LCdr. John B. Bernadou

Lt. Humes H. Whittlesey

Assistant Engineer Robert E. Carney (Ret.)

Source: *Navy Register,* 1 Jan 1904.

1908

Officers as of 1 July:

- Capt. Raymond R. Rodgers, Chief Intelligence Officer
- LCdr. Henry H. Hough
- LCdr. Robert K. Crank
- LCdr. Humes H. Whittlesey (Ret.)
- Lt. Charles H. Fischer
- Lt. Horace P. McIntosh (Ret.)

Source: *Navy Register,* 1 Jul 1908.

1911

Officers as of 1 July:

- Capt. Templin M. Potts, Chief Intelligence Officer
- LCdr. Humes H. Whittlesey (Ret.)
- Maj. Dion Williams, USMC
- LCdr. Powers Symington
- LCdr. John V. Klemann
- Lt. Horace P. McIntosh (Ret.)
- Lt. William N. Jeffers

Source: *Navy Register,* 1 Jul 1911.

1917

Officers as of 1 August:

- Capt. Roger Welles, Director of Naval Intelligence
- Cdr. Edward McCauley, Jr., USNRF*
- LCdr. Humes H. Whittlesey (Ret.) Officer in Charge of Section D
- LCdr. Macgillivray Milne
- LCdr. Orie W. Fowler (Ret.)
- Maj. Dickinson P. Hall, USMC
- LCdr. E. C. Gilpin, USNRF
- LCdr. J. H. Roys, USNRF
- Lt. R. K. Wright, USNRF
- Lt. A. B. Legare, USNRF
- Lt.(jg) E. Menocal, USNRF

*McCauley also spelled his name "Maccauley" on some documents.
Source: *Navy Register,* 1 Aug 1917

1919

An unsigned document, believed to date from 1919, shows the ONI organization as follows:

Director	**Officer in Charge**
Assistant Director	Executive; foreign naval attachés; U.S. naval missions; liaison with foreign officials in the U.S.
Situation Officer	Estimates of different situations
Section A (Administrative)	Personnel; supplies and accounts; maintenance; cleaning and floor space; translating; photostat; photography; draftsmen
Section B (Intelligence)	U.S. naval attachés; DIOs; selection control of agents; liaison with other departments of government; ship inspection; espionage; counterespionage; passport
Section C (Collating and Compiling)	Collating and compiling information
Section D (Censorship and Photographs)	Censorship of cables and radio propaganda at home and abroad
Section E (Information)	Archives; file room; mail room; official publications

Source: ONI organization document [1919] in author's files, OA.

1922

ONI had six sections and two staff elements under the Director and the Assistant Director of Naval Intelligence:

Designation	**Responsibilities**
Section A, Administration	Archives; files; mail; official publications
Section B, Intelligence and Counterintelligence	Attachés; DIOs; plant protection; ship inspection; selection of agents; passport files
Section C, Collection and Compiling	Aviation; hydrographic and navigation equipment; merchant marine; engineering, construction, and ordnance; social conditions; finance
Section D, Censorship and Propaganda	
Section E, Information	
Section F, Naval Records, Library, and Historical	
Staff Elements	Foreign naval attachés' liaison section; U.S. naval missions section

Source: NA, RG 38, File E-9-a, Item 11334B rearranged into ascending order.

1922

Officers in ONI as of 1 July:

- Capt. Luke McNamee, Director of Naval Intelligence
- Capt. Cyrus R. Miller
- Cdr. John P. Jackson

Cdr. William F. Halsey

Cdr. Royal E. Ingersoll

Lt. Robert H. Grayson

Lt. James M. Creighton

Lt. D. M. Collins

Maj. Victor I. Morrison, USMC

Other officers assigned to specific ONI sections:

Cdr. Ralph A. Koch (Information Section)

Lt. R. E. Webb (Information Section)

Lt. John B. Heffernan (Information Section)

Capt. Dudley W. Knox (Ret.) (History Section, Records, and Library)

Col. Harry K. White, USMC (Ret.) (History Section, Records, and Library)

Source: *Navy Directory*, 1 Jul 1922.

1924

Officers in ONI as of 1 July:

Capt. Henry H. Hough, Director of Naval Intelligence

Capt. Frank B. Upham

Cdr. William W. Galbraith, ADNI

Cdr. Claude B. Mayo

LCdr. John W. McClaran

LCdr. Edward K. Lang

LCdr. Robert M. Hinckley

LCdr. Roscoe E. Schuirmann

LCdr. Richard H. Knight

Lt. Richard W. Gruelick

Maj. R. E. Messersmith, USMC

Maj. Harold B. Pratt, USMC

Cdr. Halsey Powell (Information Section)

Lt. John B. Heffernan, (Information Section)

Lt. R. E. Webb (Information Section)

Lt. William F. Dietrich, (Information Section)

Capt. Dudley W. Knox (Ret.) (Historical Section)

Cdr. Jay H. Sypher (Ret.) (Historical Section)

LCdr. Samuel S. Payne (Historical Section)

Source: *Navy Directory*, 1 Jul 1924.

1926

Officers in ONI as of 1 July:

Capt. Arthur J. Hepburn, Director of Naval Intelligence

Cdr. Lloyd W. Townsend

Cdr. David McD. Le Breton, ADNI

Cdr. Leigh Noyes

Cdr. James L. Kauffman

Cdr. Paul H. Bastedo

LCdr. Herbert R. Hein

LCdr. Ames Loder

LCdr. John W. McClaran

LCdr. Robert M. Hinckley

LCdr. Robert H. Grayson

LCdr. Paulus P. Powell

Lt. Charles B. Gary

Cdr. Jerome C. Hunsaker (CC)

LtCol. Robert B. Farquharson, USMC

Maj. Harold B. Pratt, USMC

Cdr. John T.G. Stapler (Information Section)

Lt. Richard W. Gruelick (Information Section)

Lt. Charles G. Moore, Jr. (Information Section)

Lt. H. Raymond Thurber (Information Section)

Capt. Dudley W. Knox (Ret.), (Historical Section)

LCdr. Richard Wainwright, Jr. (Ret.) (Historical Section)

Lt. Robert S. Robertson, Jr. (Historical Section)

Source: *Navy Directory*, 1 Jul 1926.

1928

Officers in ONI as of 1 July:

Capt. Alfred W. Johnson, Director of Naval Intelligence

Cdr. Raymond A. Spruance, ADNI

LtCol. Robert B. Farquharson, USMC

LCdr. John H. Magruder, Jr.

LCdr. Roscoe E. Schuirmann

LCdr. Aaron S. Merrill

LCdr. Paulus P. Powell

LCdr. Richard E. Webb

LCdr. John K. Richards, Jr.

Lt.(jg) David W. Roberts

Cdr. Allan S. Farquhar (Information Section)

LCdr. Francis C. Denebrink (Information Section, also White House Aide)

Lt. Beverley A. Hartt (Information Section)

Lt. Walter R. Jones (Information Section)

Lt. Alfred P. Moran, Jr. (Information Section)

Capt. Dudley W. Knox (Ret.) (Historical Section, Records and Library)

Source: *Navy Directory*, 1 Jul 1928.

1930

Officers in ONI as of 1 July:

Capt. Harry A. Baldridge, Director of Naval Intelligence

Capt. Dudley W. Knox (Ret.)

Capt. Herbert C. Cocke, ADNI

Capt. William Baggaley
Cdr. Cortlandt C. Baughman
Cdr. Charles C. Gill
Cdr. Lucius C. Dunn
LCdr. Scott B. MacFarlane
LCdr. Webb Trammel
LCdr. Herbert R. Hein
LCdr. George D. Murray
LCdr. Ellis M. Zacharias
LCdr. Roscoe E. Schuirmann
Lt. Charles G. Moore, Jr.
Lt. H. Raymond Thurber
Lt. Lyman S. Perry
Lt.(jg) Edward S. Pearce
Maj. William W. Buckley, USMC
Maj. Clark H. Wells, USMC

Source: *Navy Directory*, 1 Jul 1930.

1932

Officers in ONI as of 1 July:

Capt. Harry Ellis,
Director of Naval Intelligence
Capt. Dudley W. Knox (Ret.)
Capt. William Baggaley, ADNI
Capt. Douglas L. Howard
Cdr. Francis D. Pryor
Cdr. Stephen B. McKinney
Cdr. Jonas H. Ingram
Cdr. Walter K. Kilpatrick
Cdr. William R. Munroe
Cdr. Archibald McGlasson
Cdr. Frank Loftin
Cdr. William C. Barnes
LCdr. Aaron S. Merrill
LCdr. Hamilton V. Bryan
LCdr. Hartwell C. Davis
Lt. George F. Mentz
Lt. Charles B. McVay III
Lt. Donald R. Tallman
Lt. Angus M. Cohan
LtCol. Harold B. Pratt, USMC
Maj. R. E. Davis, USMC
lstLt. James M. McHugh, USMC
lstLt. Frank P. Pyzick, USMC

Source: *Navy Directory*, 1 Jul 1932.

1934

Officers in ONI as of 1 July:

Capt. William D. Puleston,
Director of Naval Intelligence
Capt. Dudley W. Knox (Ret.)
Capt. John T.G. Stapler, ADNI
Capt. Augustin T. Beauregard
Cdr. Francis D. Pryor (Plans)
Cdr. William F. Amsden
Cdr. Jonas H. Ingram
Cdr. Samuel A. Clement
Cdr. Clifford E. Van Hook (Head of Security Section)
Cdr. Lucius C. Dunn
Cdr. Frank Loftin
Cdr. Ellis M. Zacharias
LCdr. Ward P. Davis
LCdr. Charles G. Moore, Jr.
LCdr. Ralph C. Alexander
LCdr. George F. Mentz
Lt. Arthur D. Blackledge
Lt. Andrew P. Lawton
Lt. Arthur H. McCollum
Lt. Lucien Ragonnet
Capt. Maurice G. Holmes, USMC
1stLt. Charles C. Brown, USMC

Source: *Navy Directory*, 1 Jul 1934.

1936

Officers in ONI as of 1 July:

Capt. William D. Puleston,
Director of Naval Intelligence
Capt. Dudley W. Knox (Ret.)
Capt. John T.G. Stapler, ADNI
Capt. Fred F. Rogers (under orders)
Cdr. Francis D. Pryor (Ret.)
Cdr. Frank T. Leighton
Cdr. Wallace L. Lind
Cdr. Charles H. Maddox
Cdr. Earle C. Metz
Cdr. Lawrence F. Reifsnider
Cdr. Frederick G. Reinicke
Cdr. Ernest G. Small
Cdr. John M. Creighton
LCdr. Charles G. Moore, Jr.
LCdr. Ralph C. Alexander
LCdr. C. E. Taylor
LCdr. Joseph U. Lademan, Jr.
LCdr. Allen D. Blackledge
Lt. David W. Roberts

Lt. Alfred J. Bolton (also White House Aide)
Lt. Edwin T. Layton
Lt. Robert N. Allen
Maj. Maurice G. Holmes, USMC
Capt. Edward G. Hagen, USMC
1stLt. Harold D. Hansen, USMC
1stLt. Russell Lloyd, USMC

Source: *Navy Directory*, 1 Jul 1936.

1938

Officers in ONI as of 1 July:

RAdm. Ralston S. Holmes,
Director of Naval Intelligence
Capt. Dudley W. Knox (Ret.),
(Head, Historical Branch)
Capt. Allan S. Farquhar, ADNI
Capt. William R. Munroe
(Head, Domestic Intelligence Section)
Capt. Frank T. Leighton (Head, Security Unit)
Cdr. Francis D. Pryor (Ret.) (Plans and Training)
Cdr. Elliott B. Nixon
Cdr. George A. Rood (Head, Administration Branch)
Cdr. Robert B. Simons (Head, Central Europe Unit)
Cdr. Hamilton V. Bryan (under orders)
Cdr. Terry B. Thompson
(Head, Dissemination Section)
Cdr. John M. Creighton (Head, Far East Unit)
Cdr. William S. Popham
(Head, Foreign Intelligence Section)
Cdr. Nathanial M. Pigman
(Head, Western Europe Unit)
Cdr. John S. Phillips
LCdr. Leland P. Lovette
(Head, Public Relations Branch)
LCdr. F. E. Vensel, Jr. (Head, War Records Section)
Lt. John A. Waters, Jr.
Lt. Daniel A. Frost
Lt. Bernard L. Austin
Lt. Redfield Mason
Lt. Alwin D. Kramer (under orders)
Lt. William G. Beecher, Jr. (Public Relations)
Lt. William S. Veeder
Lt. J. H. Armstrong (under orders)
Lt. S. Adams (under orders)
Lt.(jg) Allan B. Roby (under orders)
Lt.(jg) R. W. Germany (under orders)
LtCol. Robert Blake, USMC
(Head, Latin American Unit)
Maj. W. L. Bales, USMC
Capt. Clayton C. Jerome, USMC
Capt. Earl S. Piper, USMC
Capt. Harry C. Lang, USMC

Source: *Navy Directory*, 1 Jul 1938.

1937–1939

An organization chart approved by DNI RAdm. Ralston S. Holmes shows the breakdown of ONI branches by sections during 1937–1939:

OP-16-A, Administrative Branch:
- A-1 Foreign Liaison
- A-2 Personnel
- A-3 Mail, Filing, and Archives
- A-4 Supply and Accounting
- A-5 Legal*
- A-6 Translating
- A-7 Photo and Drafting*

OP-16-B, Intelligence Branch
- B-1 Dissemination
- B-2 Domestic Section
- B-3 Investigating Unit
- B-4 Security Unit
- B-5 Commerce and Travel Unit*
- B-6 Plant Protection Unit*
- B-7 Developments and Patents Unit*
  (B-3 through B-7 were under the Domestic Section.)
- B-8 (Not identified)
- B-9 Foreign Intelligence Section
- B-10 British Empire Unit
- B-11 Far East Unit
- B-12 Western Europe Unit
- B-13 Central Europe Unit
- B-14 Eastern Europe Unit
- B-15 Balkans and Near East Unit
- B-16 Latin American Unit
- B-17 Enemy Trade Unit*
  (B-10 through B-17 were under the B-9 section.)

OP-16-C, Public Relations Branch
- C-1 Public Information
- C-2 Press
- C-3 Propaganda*

OP-16-D, Censorship Branch*

OP-16-E, Historical Branch
- E-1 Library and Archives
- E-2 War Records

OP-16-X, Planning and raining Section

*Inactive
Source: ONI organization documents in author's files, OA.

1939–1940

ONI organization and personnel as of 1 December 1939:

OP	Title	Incumbent
16	Director of Naval Intellligence	RAdm. W. S. Anderson
16-1	Assistant Director	Capt. Jules James
16-A	Head, Administrative Branch	Capt. G. A. Rood
	Assistant, Administrative Branch	Cdr. H. R. Holcomb
	Assistant, Administrative Branch	Lt. D. J. Harkins (USNR)
16-A-3	Mail, Filing and Archives Section	Lt.(jg) D. S. Knox (USNR)
16-A-6	Head, Translation Section	Miss Boernsen
16-A-8	Head, Reserves Section	Cdr. C. C. Miller
	Assistant, Reserves Section	Lt. J. W. Boulware
	Assistant, Reserves Section	Lt. C. N. Walker (USNR)
	Assistant, Reserves Section	Lt.(jg) Nolan (USNR)
16-B-2	Head, Domestic and Special Intelligence Branch	Capt. E. B. Nixon
16-B-3	Head, Investigating Section	LCdr. R. B. Hunt
	Assistant, Investigating Section	LCdr. H. E. Keisker, (USNR)
	Assistant, Investigating Section	Lt. M. J. Perry (USNR)
16-B-4	Head, Security Section	Cdr. J. S. Phillips
	Assistant, Security Section	Lt. H. W. Taylor
16-B-5	Head, Commerce and Travel Section	LCdr. C. J. Gass (USNR)
16-B-6	Head, Plant Protection Section	Lt. A. D. Condon (USNR)
16-B-8	Head, Coastal Intelligence Section	LCdr. E. S. Earnhardt (Ret.)
16-B-9	Head, Foreign Intelligence Branch	Capt. H. D. Bode
	Assistant, Foreign Intelligence Branch	LCdr. H. W. Baltazzi (USNR)
16-B-10	Head, British Empire Section	Cdr. W. S. Popham
16-B-11	Head, Far East Section	LCdr. A. H. McCollum
	Assistant, Far East Section	Maj. Ronald A. Boone, USMC
	Assistant, Far East Section	Lt. A. D. Kramer
	Assistant, Far East Section	Lt. S. A. Carlson
16-B-12	Head, Western European Section	Cdr. N. M. Pigman
16-B-13	Head, Central European, Balkans, and Near East Section	Cdr. R. B. Simons
16-B-16	Head, Latin American Section	LtCol. Robert Blake, USMC
	Assistant, Latin American Section	Capt. Earl S. Piper, USMC
16-C	Head, Public Relations Branch	Cdr. L. P. Lovette
	Assistant, Public Relations Branch	Lt. W. G. Beecher
	Assistant, Public Relations Branch	Lt.(jg) V. F. Blakeslee (Ret.)
16-C-2	Head, Press Section	LCdr. B. L. Austin
	Assistant, Press Section	LCdr. N. W. Sharpe (USNR)
	Assistant, Press Section	Ens. F. B. George (USNR)
16-D	Head, Censorship Section	Cdr. H. K. Fenn
	Assistant, Censorship Section	LCdr. V. Huber
	Assistant, Censorship Section	LCdr. A. H. Oswald
16-E	Head, Historical Branch	Capt. D. W. Knox (Ret.)
16-E-2	Head, War Records Section	LCdr. R. S. Robertson, Jr. (Ret.)
16-S	Chief Clerk	Mr. H. C. Daniels
16-X	Head, Planning Branch	Cdr. F. D. Pryor (Ret.)
16-Z	Head, Dissemination Branch	LCdr. A. T. Emerson (Ret.)
	Aide to DNI	LCdr. C. O. O'Connell (USNR)

*Forty-four officers and forty-eight civilians, plus nineteen officers under instruction on the list were on duty in ONI at that time.

Source: ONI Personnel Roster, 1 Dec 1939, in author's files, OA.

1943

Organization of ONI's Intelligence Branch:

- FA, Eur-African Theater
  - FA-1, British Isles
  - FA-2, Africa-Middle East
  - FA-3, Western European
  - FA-4, Central European
  - FA-5, Eastern European
  - FA-6, Southern European
- FE, Far East Theater
  - FE-1, Japanese Empire (including Manchuria)
  - FE-2, Continental Asia and India
  - FE-3, Pacific Islands (including Dutch East Indies, Aleutians and Philippines)
  - FE-4, Australia and New Zealand
- FL, American Republics Theater
- FN, North American Theater
- FO, Operational Intelligence
- FP, Intelligence Plot
- FT, Foreign Trade

1946

ONI organization as of 1 August:

OP-32	Chief of Naval Intelligence
OP-32B	Deputy Chief of Naval Intelligence
OP-32C	Administrative Branch
OP-32D	Domestic Branch
OP-32E	Classified Operational Records Branch
OP-32F	Foreign Branch
OP-32L	Liaison Branch
OP-32V	Air Branch
OP-32X	Plans Branch
OP-32Y	Operational Intelligence Branch
OP-32JIS	Joint Intelligence Staff
OP-32JISPB	Joint Intelligence Study Publishing Board
OP-32JSC	Joint Security Control
OP-32JIOA	Joint Intelligence Objectives Agency
JANAID	Joint Army Navy Air Intelligence Division (became Air Intelligence Division (AID) on 1 September 1946)

Source: ONI organization diagram of 1 Aug 1946, in author's files, OA.

1948

ONI's organization and prinicpal supervisory personnel:

OP	Title	Incumbent
32	Chief of Naval Intelligence	RAdm. T. B. Inglis
32B	Deputy Chief	Capt. A.C.J. Sabalot
32B1	Plans and Policies	Capt. J. M. Ocker
32B2	Protocol	Capt. F. F. Ferris
321	Assistant Chief, Security	Capt. S. W. DuBois
321A	Executive Assistant	Mr. H. E. Keisker
321D	Investigations	Mr. C. R. Wilson
321E	SEC	Mr. W. Abbot
321K	Policy and Control	Capt. E. P. Hylant
322	Assistant Chief, Intelligence	Capt. E. G. Fullinwider
322N	Intelligence Staff	Capt. F. S. Habecker
322H	Service	Capt. F. R. Duborg
322F	Foreign	Capt. G. A. Lange
322F1	Geographic Unit	Cdr. W. Outerson
322F2	Technical Unit	Cdr. H. C. Lawder
322F3	Navies, Merchant Marine-Shipbuilding Unit	Cdr. W. S. Howell
322F4	Amphibious Unit	LtCol. C. M. DeHority, USMC
322V	Air Subsec.	Capt. D. L. Day
322V1	Coordination with Air Intelligence Div.	LCdr. C. A. Mitchell
322V2	Deputy Director, Air Intelligence Div.	Capt. W. M. Nation
322Y	Operational Subsec.	Capt. W. S. Veeder
322Y1	Special Intelligence	Cdr. R. L. Taylor
322Y2	Fleet Intelligence	Cdr. W. S. Post
322Y3	Fleet Support	Cdr. J. A. Marks
323	Assistant Chief, Administrative Sec.	Capt. G. B. Helmick
323M	Office Services	LCdr. R. W. Rastetter
323P	Personnel	Cdr. H. W. Sadler
323R	Fiscal	CPC P. T. Lane
323T	Training	Cdr. M. W. Graybill

Source: ONI roster, 1 Oct 1948.

1950

ONI's organization and principal supervisory personnel:

OP	Title	Incumbent
32	Director of Naval Intelligence	RAdm. F. L. Johnson
32B	Assistant Director	RAdm. C. F. Espe
32B1	Plans and Policies	Capt. J. M. Ocker
32B2	Foreign Liaison	Capt. F. F. Ferris
321	Security Branch	Capt. E. P. Hylant
321D	Investigations Section	Cdr. A. L. Redon
321E	SEC Section	Cdr. M. Slayton
321K	Policy and Control Section	Capt. M. B. Duffill
322	Intelligence Branch	Capt. R. H. Rodgers

322N	Intelligence Staff	Capt. R. E. Malpass
322H	Collection and Dissemination Section	[Not listed]
322F	Foreign Section	Capt. A. L. Maher
322V	Air Section	Capt. D. L. Day
322Y	Operational Section	Capt. W. S. Post, Jr.
323	Administrative Branch	Capt. C. J. Stuart
323M	Offices Services Section	Capt. L. I. Jones
323P	Personnel Section	Cdr. H. W. Sadler
323R	Fiscal Section	LCdr. H. W. Smith
323T	Training Section	LCdr. H. J. Hulings

Source: ONI roster, 21 Sep 1950.

1955

ONI organization and principal superviors personnel as of 15 September:

OP	Title	Incumbent
92	Director of Naval Intelligence	RAdm. C. F. Espe
92B	Deputy Director	RAdm. R. W. Cavenagh
92B1	Policy and Plans Coordination	Capt. T. Ashcraft
92B2	Foreign Liaison	Capt. F. F. Ferris
92C	Deputy DNI for Security	RAdm. R. H. Rice
921	Assistant Director, Security	Capt. S. E. Jones
921B	Assistant to Assistant DNI, Security	Capt. D. Nash
921C	Censorship Branch	Cdr. F. A. Klaveness
921D	Investigations Branch	Cdr. J. O. Johnson
921E	SEC Branch	Capt. M. S. Schmidling
921G	Commerce and Travel	Cdr. E. L. McIntosh
921K	Security Control	Cdr. O. F. Salvia
922	Assistant DNI, Intelligence Production	Capt. D. T. Eller
922B	Executive Assistant and Production Coordination Staff	Capt. H. G. Moore
922F	Foreign Branch	Capt. R. E. Styles
922H	Collection and Dissemination Branch	Capt. W. H. Packard
922N	Estimates Branch	Capt. B. E. Wiggin
922V	Air Branch	Capt. G. M. Clifford
922Y	Operational Intelligence Branch	Capt. B. S. Weber
923	Assistant DNI, Administration	Capt. C. J. Stuart
923M	General Services Branch	Capt. F. C. Acker
923P	Personnel Branch	Cdr. J. R. Tenanty
923R	Fiscal Branch	Cdr. J. W. Robbins
923T	Training Branch	Cdr. E. A. Crispell, Jr.

Source: ONI roster, 15 Sep 1955.

1959

ONI organization and principal supervisory personnel as of June:

OP	Title	Incumbent
92	Director of Naval Intelligence	RAdm. L. H. Frost
92B	Deputy DNI, Intelligence	RAdm. A. L. Reed
92C	Deputy DNI, Security	Capt. S. B. Frankel
92B1	Assistant for Plans, Estimates, and Joint Matters	Capt. W. G. Corliss
92B2	Assistant for Foreign Liaison Matters	Capt. C. B. Jackson
92B3	Assistant for Research and Requirements	Capt. W. H. Packard
92C2	Foreign Disclosure Policy and Control Staff	Capt. O. B. Parker
92M	Assistant for Marine Corps Matters	Col. Max C. Chapman, USMC
921	Assistant DNI, Security	Cdr. F. M. Murphy (Acting)
921C	Censorship Branch	Cdr. E. B. Martin
921D	Investigations Branch	Cdr. R. A. Klare
921E	Counterintelligence Branch	Cdr. J. C. Lacy
922	Assistant DNI, Production	Capt. C. M. Bertholf
922X	Special Assistant	Mr. George Kidd
922B3	Commanding Officer, Naval Photographic Interpretation Center	Cdr. D. L. Soper
922C	Senior Navy Representative, National Indications Center	Capt. L. E. Johnson
922G	Basic and Technical Intelligence Branch	Capt. T. M. Peterson
922H	Collection and Dissemination Branch	Capt. R. A. Kotrla
922N	Maritime Branch	Capt. O. M. Butler
922V	Air Branch	Capt. L. T. McQuiston
922Y	Operational Intelligence Branch	Capt. W. M. Stevens
922Y1	Officer in Charge (Fort Meade) Section	Cdr. D. M. Showers
922Y2	Operational Intelligence Evaluations and Dissemination Section	Cdr. J. E. Whatton
922Y3	Special Projects	Cdr. J. M. Larsen
922Y4	Electronics Intelligence Section	Cdr. E. G. Hutchinson

923	Assistant DNI, Administration	Capt. C. M. Sugarman
923M	General Services Branch	LCdr. R. R. Clement (Acting)
923P	Personnel Branch	Cdr. R. W. Werthmuller
923R	Comptroller	LCdr. W. A. Novak (SC)
923T	Training Branch	Cdr. R. D. Faubion

Source: ONI roster, Jun 1959.

1961

The first steps in the reorganization of OP-922 took place effective 1 November 1961 and included the following changes:

OP-922B4	Current Intelligence Coordinator disestablished; functions shifted to OP-922Y2.
OP-922C	Name changed to Plans, Coordination and Estimates vice Senior Navy Representative, National Indications Center (NIC); absorbed OP-92B3 (Assistant for Research and Requirements) functions.
OP-922C1	Special Assistant for Estimates, Research and Requirements (vice OP-922X3).
OP-922C2	Special Assistant for Systems Development (vice OP-922X2).
OP-922D	Commanding Officer, Photo Intelligence Center (vice OP-922X3).
OP-922E	Coordinator of S&T Intelligence Matters (vice OP-922B5).
OP-922F	Special Assistant, Editorial and Publications Matters (vice OP-922X1).
OP-922H	Name changed to Attaché, Collateral Support, and Dissemination Branch.
OP-922H1	Graphics and Special Projects disestablished; functions shifted to OP-922Y3.
OP-922H2	Collection Section changed to Attaché and Collateral Support Section.
OP-922H4	Collateral Support Section disestablished, functions shifted to OP-922H2.
OP-922N1	Name changed from Foreign Navies Section to Non-Bloc Navies Section; cognizance over Bloc Navies shifted to OP-922Y2.
OP-922N4	Foreign Air Section disestablished; limited coverage by OP-922Y2.
OP-922V	Named changed from Target Intelligence Branch to Targeting Requirements and Support Branch.
OP-922Y	Name changed from Operational Intelligence Branch to Composite Support Branch.
OP-922Y2	Name changed from OPINTEL Evaluation and Support Section to OPNAV Staff Support Section.
OP-922Y3	New section, retaining name and functions of OP-922H1, Graphics and Special Projects.
OP-922Y4	Name changed from Electronics Intelligence Section to Fleet Support Section.
OP-922Y5	New section: Security Control.

Source: ONI Notice 5400 of 1 Nov 1961.

1968

Senior supervisory personnel assigned to the Naval Intelligence Command:

NIC	Incumbent
00	RAdm. Eugene B. Fluckey
01	Capt. Frank M. Murphy
1	Capt. H. P. Lyon
11	Cdr. C. W. Goins
12	Capt. M. J. Kleczewski
13	Cdr. R. E. Hall
14	LCdr. M. K. Maugans
15	Mr. P. Thomas Koines
2	Capt. L. W. Moffitt
2B	Capt. J. J. Pavelle, Jr.
3	Capt. J. E. Whatton
3F	Capt. Donald P. Harvey
311	Cdr. C. W. Streightiff
312	Cdr. L. C. Boston
32	Capt. H. W. Holschuh
321	Mr. C. J. Oleniacz
322	Mr. L. G. Fleming
323	LCdr. T. W. Welch
4	Capt. W. O. Myers
412	Mr. E. J. Lessinger

Field Activities under the Naval Intelligence Command as of 1968:

Field Activity	Name
Naval Investigative Service (NIS)	Capt. E. G. Rifenburg, Director Capt. T. L. Stevens, Deputy Director
Naval Scientific and Technical Intelligence Center (NAVSTIC)	Capt. W. R. Banks, Officer in Charge
Navy Field Operational Intelligence Office (NFOIO)	Capt. W. R. Quisenberry, Officer in Charge Cdr. W. S. Eckhout, Assistant Officer in Charge
Naval Reconnaissance and Technical Support Center (NRTSC)	Capt. Charles D. Payne, Commanding Officer
Naval Intelligence Processing Systems Support Activity (NIPSSA)	Capt. W. J. Furnas, Commanding Officer
Applications Department (AD)	Capt. J. F. Bradley, Jr., Head

Source: Naval Intelligence Command, NIC Notice 5000 of 8 Jan 1968.

1971

Office of ACNO (Intelligence) (OP-942) organization:

OP	Title
942B	Deputy Director
942C	Assistant for Counterintelligence Plans and Policy
942D	Security of Military Information Branch
942E	Fleet Support and Intelligence Operations Branch
942F	Estimates, USIB Matters, and Departmental Support Branch
942J	Assistant for Plans, Policies and JCS Matters
942L	Assistant for Foreign Liaison and Protocol
942M	Assistant for Intelligence Personnel and Training
942P	Assistant for Program and Budget Planning
942Q	Ocean Surveillance Information and Automation Systems Branch
942S	Secretariat
942T	Technical Advisor for Operational Intelligence
942U	Undersea Warfare Branch
942X	Special Assistant for Security Coordination

Source: OPNAV Notice 5430 of 5 Oct 1971, encl. 1.

1973

Office of the Director of Naval Intelligence (OP-009) organization as of 1 February:

OP-009*	Director of Naval Intelligence
OP-009A	Executive Assistant
OP-009A1	Personal Aide
OP-009B*	Deputy DNI
OP-009C†	Assistant for Counterintelligence Plans/Policy
OP-009D	Director, Security of Military Information Division
OP-009E†	Assistant for Intelligence Operations
OP-009F	Director, Estimates and Departmental Support Division
OP-009J‡	Assistant for Plans/Policy/JCS Matters
OP-009L	Director, Foreign Attaché Affairs and Protocol Division
OP-009M†	Assistant for Intelligence Management
OP-009P†	Assistant for Program Budget and Planning
OP-009Q†	Assistant for Systems and Sensors
OP-009R	Assistant for OEG
OP-009S	Secretariat
OP-009U	Director, Undersea Warfare Division
OP-009X	Assistant for Security Coordination
OP-009Z	Special Advisor to the DNI

*Double-hatted as OP-094Q and OP-094QB (Assistant and Deputy Assistant for Intelligence Support).
†Double-hatted in NAVINTCOM.
‡Double-hatted as OP-094J (Special Assistant for Joint/Allied Matters).
Source: ACNO (OP-009B) ser P009/24, 8 Feb 1973.

Chapter Notes

1. RAdm. Albert G. Berry, "The Beginning of the Office of Naval Intelligence," *USNIP*, Jan 1937.

2. *ONI Review*, Apr–May 1957, 172.

3. *SECNAV Annual Report, 1897*, 45.

4. *Navy Register*, 1 Jul 1898.

5. Capt. Herbert E. Cocke, USN, "History of ONI," MS, Office of Naval Intelligence, 1931, 17.

6. Department of the Navy, "Administrative History of the Office of Naval Intelligence in World War II," 10 Jul 1946, unpublished MS, 11, hereafter *ONI WWII Admin History*.

7. *Navy Directory*, 1 Jul 1918.

8. ONI, *Naval Intelligence Office Organization*, 1 Sep 1918, 1–31.

9. RAdm. Albert P. Niblack, USN, *The History and Aims of the Office of Naval Intelligence* (Washington: GPO, 1920), 23.

10. Adm. Royal E. Ingersoll, Oral History, Columbia University, New York, NY, 1965, 45–56, 102, copy in OA.

11. *ONI WWII Admin History*, 507.

12. Capt. Henri H. Smith-Hutton, Oral History, USNI, Annapolis, MD, 1979, 1:99–100, copy in OA.

13. *ONI WWII Admin History*, 27.

14. ONI, "Annual Report of the Administrative Branch," Jul 1932.

15. *ONI WWII Admin History*, 10, 104–5.

16. Ibid., 536.

17. Ibid., Appendix H.

18. F. Donald Scovel, "Helms A'Lee: History of the Development of the Public Affairs Function in the U.S. Navy, 1861–1941" (MA thesis, University of Wisconsin, 1968), 129.

19. ONI, "History of the Special Activity Branch," file A12-8, box 16, Job 3679, FRC/WNRC.

20. *ONI WWII Admin History*, 63, 66.

21. Ibid., 65–66.

22. Ibid., 97, 99.

23. Ibid., 330.

24. Assistant Director of Naval Intelligence, ADNI Intelligence Group Memo No. 7 of 9 Sep 1943.

25. C. Frank Klaveness, MS papers on OP-16-FN, in author's files, OA.

26. DNI Memo, 8 May 1944, file A3-2, box 16, Job 3679, FRC/WNRC, Suitland, MD.

27. ONI organization diagram, 6 Apr 1945, in author's files, OA.

28. ONI organization diagram, 29 Oct 1945, in author's files, OA

29. OP-23D, "Narrative of Activities, 1 Oct 1945 to 29 May 1946."

30. ONI organization diagram, 29 Oct 1945, in author's files, OA.

31. ONI, "Contribution to CNO Annual Report, 1 Oct 1945–1 Jul 1946," OA.

32. DCNO (Operations) memo to VCNO, OP-03 ser 73P03, 19 Jul 1946, copy in author's files.

33. OP-03 memo, ser 11P32, 18 Feb 1948.

34. VAdm. Francis F. Low, "Study of Undersea Warfare," ser 001P003, 22 Apr 1950, 15, Post 1 Jan 1946 Command File, OA.

35. "Review of Functions and Workload of ONI," 4 Feb 1952, 14, OA.

36. ONI memo to VCNO, 19 Nov 1953, "Study of ONI," OA.

37. Department of the Navy, ONI-70-1, *U.S. Navy Intelligence Manual*, 20 Jun 1956, 14.

38. ONI Notice 5410 of 8 Jun 1961.

39. ONI Notice 5410 of 8 Jun 1961.

40. RAdm. Atley Peterson, USNR, report, 24 Jun 1966, in author's files.

41. SECNAV Notice 5450 of 19 Apr 1967.

42. OP-92 ltr, ser 09408P92, 29 Dec 1966, "Plan for Establishing NAVINTCOM."

43. OP-92/NAVINTCOM roster, May 1968, in author's files.

44. ACNO (Intelligence) memo, ser 01077P92, 26 Feb 1968.

45. OPNAV Notice 5430 of 13 Dec 1968.

46. DNI memo to CNO, "Developments in Naval Intelligence for 1969."

47. OPNAV Notice 5430 of 26 May 1969.

48. OPNAV Notice 5430 of 21 Oct 1970.

49. OPNAV Notice 5430 of 31 Mar 1971.

50. DNI, "Report to the Chief of Naval Operations for Fiscal Year 1972," 30 Nov 1972, 41, in author's files, OA.

51. DNI, "Plan for Foreign Navy Intelligence Cooperation," 2 Apr 1972.

52. DNI, "Report to the Chief of Naval Operations for Fiscal Year 1972," 34.

53. OPNAV Notice 5450 of 3 Aug 1972.

54. OPNAV Instruction 5430.48, *CNO Organizational Manual*, 1 Feb 1973, copy in author's files.

CHAPTER 29

Intelligence Reserves

The part played by naval reservists in intelligence in World War II and the Korean War was of major significance, particularly during World War II, when most intelligence billets, both ashore and with the operating forces, were filled by reservists. Even during World War I, before there was an intelligence component specifically designated within the U.S. Naval Reserve Force, more than three quarters of the officers assigned to ONI as of 1 July 1918 were members of the USNRF. The percentage was even higher in the naval districts and naval attaché offices. This chapter about the Naval Intelligence Reserves is general, mainly because very few histories of specific units have been located. Its relative brevity, therefore, should not be taken as a negative indicator in an evaluation of the significant and vital requirements that the Reserves have successfully fulfilled in wartime. Those parts of all the other chapters that relate to wartime periods could appropriately have been included in any history of the Naval Intelligence Reserves.

Establishment of the Naval Intelligence Volunteer Service

The Naval Intelligence Volunteer Service was created by the Naval Reserve Act of 28 February 1925 (Public Law No. 512, Sixty-eighth Congress, First Session). The object was to provide a nucleus of reserve officers who, by virtue of their education, experience, and training in civilian life, would be immediately available in time of national emergency to assume duties and perform the important functions of intelligence officers at home and abroad.[1]

The functional code designation for reserve intelligence officers was Class I-V(S), or Intelligence Volunteer (Specialized).[2]

Little effort was made initially during the 1920s to procure I-V(S) officers because of the predominantly pacifistic outlook of the general public in the United States at that time. In February 1926, the Office of Naval Intelligence asked the district intelligence officers to submit a list of qualified individuals, "preferably key people in the news and writing world, who in time of peace can keep in touch with this office and in time of national emergency can be actively coordinated with the duties of Naval Intelligence."[3]

A file card record was kept of all naval reservists enrolled, those awaiting appointment, and those proposed. The last included persons who had had previous experience or who had volunteered their assistance. The cards gave the name, rank, date of acceptance, and a brief of their qualifications, together with their mobilization assignments. The cards were filed under four categories of prospective assignment: ONI Intelligence, ONI Public Relations, ONI Censorship, and Naval Districts. In addition to the card record, a separate file on each reservist was maintained in the ONI file room.

Quotas for Intelligence Volunteer Reserve officers for ONI and the naval districts were revised upward during Fiscal Year 1933 by the Chief of Naval Operations (CNO) following a study on the subject by the General Board of the Navy. The training of the Intelligence Reserves for their specific war duties was accomplished by the district intelligence offices, and a correspondence course prepared in the ONI Administrative Branch was issued by the Bureau of Navigation (BUNAV) (Reserve Section) through the Navy education centers.

When funds were available, BUNAV authorized training with pay either in the naval districts or in ONI, depending on where the officer was to be assigned.[4]

One of the early reserve intelligence officers was Sidney W. Souers, who was appointed as a lieutenant commander on 29 April 1929. Souers served in an inactive status as Senior Intelligence Officer in St. Louis, Missouri, and had the responsibility to study the development of the Naval Intelligence organization, its publications, and its officer procurement.[5]

Intelligence Reserves in the 13th Naval District in the Pacific Northwest area in the 1930s were influenced to a great extent by LCdr. (later Cdr.) Luke May, a private detective with an international reputation in scientific crime detection and an expert in the development of informant networks. Monthly or semimonthly meetings were held in Cdr. May's home or in the Washington Athletic Club in downtown Seattle. All training was in crime detection and surveillance. Practical drills in the late 1930s included surveillance of Japanese "suspects" under the indirect guidance of the local FBI and the boarding of Soviet and Japanese merchant ships.

Personnel in the 13th Naval District Naval Reserve group included lawyers, law enforcement officials, shipping and travel specialists, public relations personnel, and foreign language experts (especially in Japanese, Russian, and German). The naval reservists received neither pay nor retirement credits, but they did get promotion credits for drills attended. Active duty training in a nonpay status on board West Coast Navy ships and stations was common. Completion of correspondence courses was required for the reservists to remain in the program.[6]

In 1933, under the guidance of the ONI Planning Division, most of the naval reservists under the cognizance of ONI were given mobilization assignments compatible with their qualifications.[7]

In the mid-1930s, conditions and government policy changed as it became apparent that an international conflict was approaching. The "Estimate of the Situation," beginning in 1935, stressed the need for expanding intelligence duties and personnel.[8]

The allowed number of I-V(S) officers was gradually being increased to meet the needs of the Navy, as determined from its approved war plans. Reports from the naval districts during the late 1930s indicated that progress was being made in the enrollment of desirable persons and in their general instruction, indoctrination, and assignment to specific billet-related training.[9]

The allowance of I-V(S) officers was increased in Fiscal Year 1936 from 459 to 536, and other increases were contemplated as the naval districts prepared their estimated requirements according to their individual war plans.[10]

Difficulties were being encountered in 1937 in finding persons with the necessary qualifications to serve abroad as naval attachés or assistant naval attachés. On 16 March 1937, ONI sent a letter to all naval districts advising of vacancies in forty cities throughout the world and requesting "a careful and confidential survey . . . for the purpose of locating suitable personnel to fill these vacancies."[11]

Preparations for War, 1938–1941

A survey was made in July and August 1938 by the ONI Planning Officer, Cdr. Hamilton V. Bryan, to obtain information on the readiness of the Naval Intelligence Reserves for active duty. He concluded that it was not ready and that the organization was ineffective, possibly because there were no pertinent ONI directives. Bryan reported that the commandants of the naval districts did not appreciate the importance of district intelligence or of intelligence reservists; the morale in the district intelligence offices and of the I-V(S) reservists was low; war plans were in a backward state; I-V(S) personnel had been commissioned without regard for the tasks that they were expected to perform, and their fitness for the tasks had not been determined; no effective liaison had been developed with the government and private agencies that would play an important part in future naval intelligence activities; and the opportunities for enrollment and training through existing civilian activities had been neglected.[12]

In ONI, there had been no real attempt for many years to supervise and coordinate the training and education of the personnel of the Naval Reserve or of the Navy, active or retired, who were slated for intelligence duties in wartime. The training of Naval Reserve personnel had been assigned to the commandants of the naval districts where the records of the personnel were maintained. No attention had been paid to Naval Reserve officers residing abroad; their records were maintained at the Bureau of Navigation.[13]

As a result of Bryan's survey and report of August 1938, and a subsequent directive issued in April 1939, efforts were begun to improve the readiness of the Naval Intelligence Reserves. By June 1939, approximately two-thirds of the naval districts had completed defining their organizational and personnel needs. The approved war mobilization complement for the entire naval intelligence service was 150 retired regular naval officers to be recalled to active duty, 2,023 reserve officers, 505 warrant officers, and 3,934 enlisted personnel. Of the total, 80 percent was authorized to be procured during peacetime.

In January 1939, another step in improving the readiness of the Intelligence Reserves was taken by dividing the I-V(S) officers in the Washington area into boards to produce training manuals on plant protection, commerce and travel, investigation, censorship, general intelligence, espionage, administration, public relations, and coastal intelligence activities.[14]

Article H-2104 of the *BUNAV Manual* authorized the appointment of Naval Reserve officers in warrant grades in peacetime only in exceptional circumstances. CNO (OP-16-X) letter of 25 May 1939 to the Chief of BUNAV requested authorization for the procurement of warrant officers for the I-V(S) category up to 12 percent of the total officer requirement. It was explained that many mobilization billets required a type of person whose economic status precluded his acceptance of an enlisted rating but whose professional status, while not up to commissioned requirements, was still of great value to the Reserves. The Chief of BUNAV, on 29 November 1939, approved making warrant grade appointments in the categories of boatswain, machinist, and carpenter, I-V(S), USNR.[15]

Although the district intelligence officers might have been able to make a fair estimate of their wartime personnel requirements and might have even succeeded in filling their allowances of reserve intelligence officers, they rarely knew how many of those officers would be available to accomplish active duty training or how many would be drawn off for assignment to billets in ONI or outside the continental limits of the United States.

When funds were made available to pay I-V(S) officers under training, they were used as investigators. Quarterly meetings were held for all I-V(S) officers for training and work study assignments. Much of the investigation of officer applicants was performed by I-V(S) officers on inactive duty as spare time training, performed when their civilian occupations permitted.

By directive from ONI in December 1939, each district intelligence office was to designate one or more suitable inactive I-V(S) officers to establish and maintain liaison with local FBI and Army organizations in the naval districts that were far from the district intelligence offices.[16]

It was believed that the value of reserve intelligence officers could be degraded by publicity concerning their reserve status and activities. Furthermore, publicity could prove embarrassing to reservists residing abroad. To help reduce the chance of adverse publicity, naval district commandants were instructed by the Department of the Navy to inform all I-V(S) officers that

> (1) All communications to I-V(S) officers residing abroad will be mailed in plain envelopes, addressed to them as civilians.
>
> (2) All communications to I-V(S) officers residing in the U.S. or in its possessions will be mailed in franked envelopes but shall be addressed to them as civilians.
>
> (3) No publicity will be given to luncheons, meetings, etc. of I-V(S) officers.
>
> (4) To casual inquiries, I-V(S) officers should state that they belong to the Naval Reserve but not mention the branch.
>
> (5) I-V(S) officers must refrain from using their affiliation with the Naval Reserve for political, business, or social purposes.
>
> 6) I-V(S) officers should be indoctrinated in the policy that their status and mission are confidential.
>
> (7) I-V(S) officers will be omitted from District directives.[17]

The problem of obtaining qualified reserve intelligence personnel continued in 1940. In addition to the standard qualifications required to be met by all naval reservists and special-service volunteer reserves, reserve intelligence officer candidates had to meet special standards because of the highly confidential and sophisticated nature of the matters and material with which they would be required to deal. Those special standards included

> broadness of outlook, familiarity with public events and international affairs, social understanding, tact, imagination, reliability, force, loyalty, enterprise and perseverance. In addition, he is required to have intellectual background suitable to Service requirements, versatility, adaptability, clean-cut Americanism, professional ability, sobriety under strain, and lastly, an unimpeachable record.

From the beginning of the Intelligence Reserve Program, the task of procuring intelligence officers had been in the hands of the district commandants, with the processing of applications and commissions coming under the immediate jurisdiction of the district intelligence officers.[18]

On 18 July 1941, the Bureau of Navigation ordered all district commandants to nominate immediately for active duty all officers in Class I-V(S) other than those assigned to censorship billets or those residing abroad.[19]

In December 1941, BUNAV directed the commandants of naval districts to forward all reserve officer applications to Washington without regard for district mobilization billets or previously assigned district quotas.[20]

Naval Intelligence Reserves in World War II

Immediately after the United States entered World War II, the number of applications for intelligence commissions increased. In some naval districts, so many intelligence duty applications were awaiting action that many persons considered it futile to apply. A new, more efficient system was needed for speeding up the mechanics of selecting and processing applicants.

On 17 February 1942, the Secretary of the Navy directed that one or more officers charged with the paramount duty of procuring naval officers be assigned as soon as possible in each continental naval district, under the direct supervision of the BUNAV and separate from the naval district headquarters. In May 1942, CNO Adm. Ernest J. King directed that an intelligence officer be detailed to each of the officer procurement offices to assist in interviewing and determining the qualifications of I-V(S) applicants.[21]

In June 1942, an intelligence officer who had already been assigned to pass on the qualifications of applicants for intelligence appointments in the Division of Naval Intelligence was assigned to the Bureau of Personnel (BUPERS) (previously part of the Bureau of Navigation) to process the applications. The officer was given additional duties as liaison officer to the Division of Naval Intelligence in connection with the ongoing planning for procurement of I-V(S) personnel.[22]

Because the Navy failed to provide intelligence training to its regular officers, most wartime intelligence functions had to be taken on by the reserves. Many of the reserve officers were well qualified for intelligence work through previous civilian experience in fields such as law, engineering, investigations, news reporting, linguistics, and professional writing. With a quick Navy indoctrination, they performed very well and helped fill most of the billets in the Naval Intelligence service during World War II.[23]

Post–World War II Reorganization

Secretary of the Navy letter (Pers-1D2 serial 48) of 27 March 1946 activated the postwar Naval Reserve Program. As part of the implementation of the program, Bureau of Personnel letter (Pers-1D2 serial 505) of 22 May 1946 established the Reserve Component of the Naval Intelligence Program.[24]

Effective 1 July 1946, the Reserve Component activated in the naval districts was divided into two parts: the Organized Reserve, which was required to perform specified work assignments, and the Volunteer Reserve.

The intelligence work of the Reserve Component in each district was under the control of the district intelligence officer. Both the Organized and Volunteer groups were subdivided on the basis of their mobilization assignments, i.e., to ONI, to the naval districts, to foreign posts, and to the operating forces afloat and ashore. Each subgroup was given instruction and training appropriate to their prospective active duty assignments.

Most of the officers in the Reserve Components at that time had had wartime intelligence experience, but provision was made for enrolling NROTC (Naval Reserve Officer Training Corps) graduates and civilians who had the specific qualifications and aptitudes required by Naval Intelligence to fill vacancies in mobilization billets.[25]

The organization and training of the Reserve Component of Naval Intelligence continued to be carried out by the district intelligence officers under the direction of the Chief of Naval Intelligence, according to the BUPERS letter of 22 May. The training included periodic meetings of reserve officers in the various districts at which lectures were to be delivered by the district intelligence officers and other officers with wartime experience.[26]

To alleviate the critical personnel situation in ONI and the district intelligence offices, a program for using officers of the Organized Reserve Component of Naval Intelligence on two-week training duty was inaugurated on 7 August 1946. Each branch in ONI and each district intelligence office was to prepare a schedule of work projects that could be performed by reserve officers during a two-week tour of active duty. Each district intelligence officer was also to canvas Organized Reserve officers for volunteers for two-week tours of active duty.[27]

With the reactivation of the Reserve Component of Naval Intelligence and the establishment of the Organized Reserve, a general information letter was addressed to all district intelligence officers on 23 August 1946 in order to assist in getting the program in operation as soon as practicable. It was recognized that the active duty program would place an additional administrative burden on the district intelligence offices, and the Bureau of Personnel had been requested on 12 August to authorize each district commandant to order to active duty one special duty intelligence officer, S(I), as the former I-V(S) category had been redesignated) and one enlisted reservist who were to establish and administer the Organized Naval Reserve Intelligence Program in each naval district.

Quotas by rank for the Organized Reserve were established on 29 July 1946. The quotas authorized for each district intelligence office did not mean that all S(I) officers assigned would be earmarked to fill mobilization billets in the district concerned nor that they would perform their two-week annual training duties in the district intelligence office. In selecting S(I) officers for the Organized Reserve, the district intelligence officers were to exercise care to nominate only those who would be of the most value and benefit to Naval Intelligence as a whole. For guidance, each district intelligence office was instructed to adhere to a specific breakdown of

Organized Reserve officers by mobilization billets as shown below.

Table 29.1.
Organized Reserve Officer Billets

	Totals	Percentages
Naval Districts	461	31.0%
Operational Intelligence	180	
Domestic Intelligence	257	
Pool	24	
ONI	415	28.0%
Foreign Posts	171	11.6%
Operational Forces	428	29.4%
Totals	1,475	100.0%

Source: OP-32C5 ltr, ser 15390P32, 23 Aug 1946, ONI Day File, OA.

In October 1946, Naval Intelligence became one of the first reserve programs in the 13th Naval District to be reconstituted after World War II. By mid-1947, there were sixty-five officers and chief petty officers drilling in units in Seattle, Spokane, and Portland. Participants were mainly officers who had been in intelligence and related billets ashore and afloat during World War II. Training emphasis was on operational intelligence, especially for amphibious operations.[28]

The facilities for continuing the training of air combat intelligence (ACI) officers for the postwar Naval Air Reserve Program were established by CNO Planning Directive 16-A-46 serial 225P510 of 21 August 1946. To provide information on the new program to all reserve air combat intelligence officers who had returned to inactive status following World War II, and to invite them to participate, the Chief of Naval Air Reserve Training Command at Naval Air Station, Glenview, Illinois, issued a letter describing the program. Officers enrolled in the Naval Air Reserve Program would be assigned to one of the following:

1. The Organized Reserve, which would be composed of officers who would regularly attend drills, receive retainer pay, and take two weeks' active duty for training annually. The quota for the ACI Organized Reserve was 225 officers.

2. The Volunteer Reserve (Associated), which would be composed of officers who would regularly attend drills with the Organized Reserve at their own volition and without pay, while awaiting the availability of a billet in the ACI Organized Reserve. The Volunteer Reserve Associates were to be eligible to request annual active duty for up to two weeks with pay and allowance.

3. The Volunteer Reserve (Inactive), which would include those ACI officers who were unable to take an active part in ACI Reserve training but who would receive routine information sent out to all Reserve officers. The inactive Reservists would be encouraged to organize and hold meetings from time to time in their own communities, and a limited number could request up to two weeks of active duty annually with pay and allowances, depending on the availability of funds. It was recognized that a majority of the former ACI officers would of necessity remain in the inactive category, but the Navy hoped they would retain an active interest in the Naval Intelligence Reserve Program.

At first, the Naval Intelligence Reserve drills and training periods were with the Organized Reserve squadrons and air groups at the 28 Naval Air Reserve stations and Naval Air Reserve Training units. Later it was found that the Air Reserve intelligence officers could provide better support to the stations and squadrons if they drilled together at a station rather than work individually with single squadrons. Annual active duty training, however, continued to be performed with the squadrons to which the reservists were assigned.[29]

The training syllabus for the Organized and the Volunteer (Associated) Reserves was developed cooperatively by the Air Branch of (OP-32V) ONI, the Postgraduate School of Naval Intelligence, and the Office of the Deputy Chief of Naval Operations (DCNO) for Air (OP-55T). The Reserve Air Intelligence Program remained the responsibility primarily of DCNO (Air) and the Chief of Naval Air Reserve Training until 1 June 1950, when ONI took it over.[30]

Naval Intelligence Reserves at the Local Level, 1946–1951

On 15 May 1946, a meeting was convened at the Zone Intelligence Office, Los Angeles, for the purpose of planning the establishment of a Naval Reserve Intelligence Unit (NRIU) to be composed of local S(I) and former I-V(S) officers. As a result of the meeting, the first official drill of NRIU Los Angeles was held on 12 November 1946 at the zone intelligence office in the Van Nuys Building at 210 West Seventh Street. Cdr. Robert Sibert, USNR, was the first officer in charge; Cdr. Beryl E. Burchfiel, USNR, was first Assistant Officer in Charge; and Cdr. William D. Bretz, USNR, was responsible for training and administration.

The initial complement of the Los Angeles unit in 1946 was 26 officers. It expanded until 1951 when it had 33 officers in a pay status, 33 nonpay officers, 82 I–V(S) officers, and 12 enlisted men, for a total of 160 unit members. Unit strength declined thereafter for several reasons, including the establishment in 1951 of a separate Telecommunications Censorship Unit, the elimination in May 1951 of the enlisted allowance for the Organized Naval Reserve

Intelligence Program, the establishment in 1952 of a separate unit at Long Beach, and the establishment of minimum drill attendance requirements.[31]

The Naval Reserve Intelligence Unit, Denver, Colorado, was organized in 1947 under the 12th Naval District. There was no naval investigative office in Denver at that time, and it was expected that the unit would help to fulfill the District Intelligence Office, 12th Naval District requirement for one. The first officer in charge was Lt. B. Palmer King, USNR, and the unit met in his basement until 1952, when it moved into the office of the Senior Resident Agent at the new Customs House in downtown Denver. In 1951, the unit was designated NRIU 9-1-11 when it shifted from the 12th to the 9th Naval District. Membership remained at around twelve during those early years, and the unit functioned primarily in the investigative field.[32]

As of 1948, Reserve Intelligence Division Three in New York City was composed of eight officers; all in a nonpay status. They met one night a week at different offices to discuss foreign affairs, in which they specialized. At first, they did not meet in uniform, although the commanding officer of the division was the local district intelligence officer. In 1949, the Chief of Staff of the 3rd Naval District decided that a naval reservist should be the commanding officer, and Cdr. Robert H. Barnum, USNR, was appointed.

A training program was needed. Barnum discussed the matter with the 3rd Naval District staff to determine what would be expected of a reserve intelligence officer recalled to active duty with the fleet. It was concluded that the preparation of a complete intelligence briefing for an amphibious operation would be an appropriate training exercise. An island near Indonesia was selected as the target area for an imaginary amphibious operation to be executed in the near future. About one year was spent collecting all the necessary information, including maps, charts, and data on climate, beaches, the people, their language, medical problems, the probable opposition, the kinds of reconnaissance needed, and the historical and political background. When the briefing was completed, it was submitted to the Director of Naval Intelligence.[33]

The Durham, North Carolina, Training Unit of the 6th Naval District Reserve Intelligence Division 6-1 was launched in the first week in July 1948 with LCdr. Egbert Haywood as officer in charge and Lt. James Newsom and Ens. John Kerr in attendance. Weekly drills were held at the local U.S. Naval and Marine Corps Reserve Training Center at 724 Foster Street in Durham. Initially, training consisted primarily of studying the *Operational Intelligence Manual*, but the pattern gradually evolved into a lecture and classroom course format. The courses included strategic intelligence, counterintelligence, operational intelligence, international relations, security of classified matter, photo intelligence, intelligence organization and functions, naval orientation, investigations, and data processing. In the late 1950s, as a practical exercise, the Durham unit developed an extensive study and analysis of the port of Wilmington, North Carolina.

Annual Active Duty for Training (ACDUTRA) for naval intelligence reservists of the 6th Naval District in the early years was mainly devoted to attending operational intelligence courses at Little Creek Amphibious Base at Norfolk; counterintelligence and investigations training at Fort Holabird, Maryland; Photo Interpretation School at Anacostia in the District of Columbia, and on-the-job training in investigations at Charleston, South Carolina.[34]

Naval Intelligence Reserves in the 1950s

As of 1951, the Intelligence Reserves were organized into three major groups: Intelligence, Air Intelligence, and Telecommunications Censorship.

The Volunteer Intelligence Reserve (Telecommunications Censorship) units were set up at all naval districts, with an experienced, full-time Telecom Censorship officer assigned to most districts. The first training seminar for Telecom Censorship Reserve officers was completed in New York City on 26 January 1951. The first West Coast seminar was started in San Francisco on 5 February 1951. The Potomac River Naval Command Telecom Censorship unit formed the nucleus headquarters group for the Chief Telecommunication Censor and assisted to a considerable extent in planning the program. See Chapter 23 for more information on the Telecom Reserves in the 1950s.[35]

Promotion in the Intelligence Reserves required not only attendance at weekly drills but also completion of specified correspondence courses, including difficult and time-consuming Naval War College courses in international relations and international law. There were some pay billets, but the number of pay-qualifying drills per year was changed frequently. The Intelligence Reserves in the 3rd Naval District grew, in spite of the pay problem, to about 120 officers. The largest group of officers was in New York City.[36]

The first drill meeting of Naval Reserve Intelligence Unit, Long Beach (NRIU 11-1-3), was held on 5 February 1952 at the Zone Intelligence Office, Long Beach, in the Times Building, at 215 American Avenue. Cdr. Beryl E. Burchfiel, USNR, was the first officer in charge of the unit, and Lt. J. Fred Reeves was the first assistant officer in charge. Prior to February 1952, the reserve intelligence officers in the Long Beach area had been members of the Los Angeles Naval Reserve Intelligence Unit 11-1-1.

Drills were held by NRIU 11-1-3 four times a month from February 1952 to July 1954, three times a month from July 1954 to June 1956 (only two of which were paid drills), and two times a month from July through December 1956. After that, a return to four drills a month, or forty-eight drills per year, was authorized. In due course, weekend drills were adopted and found to provide better use of Navy training facilities, better contact with the regular Navy and with naval reservists with other designators, and more adequate blocks of time for training activities and productive activity. Weekend drills also stimulated better attendance, particularly by those living in outlying areas, and gave time for better field trips. Unit strength increased from its initial complement of about ten to its all-time high of forty-seven officers in Fiscal Year 1968.

Naval Reserve Intelligence Division 11-1 was awarded the Meritorious Unit Commendation in 1968 for its support of operations in Southeast Asia. The Long Beach unit produced a pocket survival manual specifically designed for aircrews downed in combat areas. Much of the work was carried out during off-duty hours.

The membership of NRIU 11-1-3 was drawn from diverse professional backgrounds: law, engineering, teaching, law enforcement, public administration, business, and science. As a result of the talent and high educational level of the officers of the unit, the training program was conducted at what amounted to a graduate school level. An emphasis on the areas of leadership, professional training, appearance, and attendance resulted in a high degree of spirit and pride in the unit.[37]

After the decline in interest in the Naval Reserve in the immediate post-Korean period, reservist activities in the 13th Naval District revived during the late 1950s to the point where the surface and air programs combined reached a total of 100 officers drilling at Helena, Montana; Eugene, Oregon; and Pocatello, Idaho, in addition to the original units at Seattle, Spokane, and Portland. Most recruits during the period were former Navy deck and aviation officers with no previous intelligence experience. Training during the late 1950s was derived from standard ONI syllabi on investigations, security of classified information, counterintelligence, operational intelligence, and intelligence organization.[38]

Naval Reserve Intelligence Unit 9-1-10A was formed in April 1955 in Madison, Wisconsin, with LCdr. John Bruemmer, USNR, as the officer in charge. From its first days, the unit made exceptional use of the talent at the University of Wisconsin, its faculty serving as visiting lecturers on the history, geography, economics, and social systems of the Middle East, with special reference to the Arab-Israeli problem. Several members of the unit also studied Russian during drills with the help of a Russian language instructor from the university.[39]

In 1959, ONI established the Naval Reserve Translation Program in order to use linguistically qualified naval reservists as an additional resource for producing the translations required by the Navy. The Translation Program also enabled the reservists to acquire points toward their eventual retirement from the Naval Reserve. The program proved to be a tremendous asset, both for the reservists and the Navy. The ONI Translation Unit (OP-923M4) was responsible for supervising the effort.[40]

Naval Reserve Technical Intelligence Unit (NRTIU) 6-1-25 was established at Cape Canaveral, Florida, in 1959 as an integral part of the Reserve Intelligence Program for the purpose of using the talents and knowledge of personnel involved in the aerospace technologies being developed in the area. The mission assigned to the Cape Canaveral unit was to provide analysis and produce reports for Fleet Intelligence on missile and space matters as directed by the Naval Scientific and Technical Intelligence Center at the Naval Observatory in Washington. As the only Technical Intelligence Unit in the Naval Reserve, and as the first unit specifically organized and tasked to produce fleet projects, the members developed and maintained a proud tradition of dedication and service to the organizations they supported.[41]

NRIU 9-1-11 in Denver merged in 1959 with the Naval Air Intelligence Reserve Unit (NAIRU) at Naval Air Station, Buckley Field. The NRIU was the absorbing unit, and its membership rose to about forty. Also, in 1959, Cdr. G. Edward Lewis, USNR, relieved Cdr. B. Palmer King as officer in charge, and the unit started meeting in the U.S. Naval and Marine Corps Reserve Training Center in the Denver Federal Center.

Cdr. Arthur E. Abrams became the third officer in charge of NRIU 9-1-11, succeeding Lewis in 1962. Abrams was the first officer in charge to come from the former NAIRU. He worked hard to obtain active duty training for unit members, but with only marginal success. The inability to obtain active duty for training was a problem that was endemic to the Naval Reserve intelligence units of the period. Increasingly, the unit's members tired of hearing one another give training lectures, and outside speakers were recruited from local universities and other sources.[42]

Naval Intelligence Reserves in the 1960s

Every naval district in the United States had a Naval Reserve Intelligence Division in the early 1960s. Each division was composed mostly of officers, plus six to eight yeomen and personnelmen who were assigned to perform administrative functions. The reserve intelligence divisions were organized on a district-wide basis for supervision and administration. Each division was further subdivided into units, and, as of 1960, there were more than 100 units located in the principal cities of the United States, in Puerto Rico, and in the Canal Zone.

The total Naval Reserve Intelligence Division enrollment in 1960 was approximately 1,150 men and women officers in pay status and 400 in nonpay status. Those in pay status were members of the Selected Reserve, with mobilization billets already assigned and precut orders ready for issue, instructing them to report to active duty in case of a national emergency. Most of the nonpay officers were members of the Ready Reserve and were available for mobilization.

The training program for each reserve intelligence division was under the direction of the reserve training section in each naval district. A Reserve Intelligence Program Officer (RIPO) was assigned to the training section on full-time active duty. In cooperation with the district intelligence officer, the RIPO was responsible for the procurement and training programs for the Intelligence Reserves. To prepare the reservists for their prospective mobilization billets, the training program was designed to make them administrators, investigators, and research specialists. It was also intended to give them an understanding of world affairs, U.S. foreign relations, and strategic interests, and to instruct them in logistics and counterintelligence functions. In addition, it was important for the naval intelligence reservists to keep up with new developments and technical progress in naval weaponry, command and control, and the various types of naval warfare.

Classroom courses were conducted during weekly drill periods, and intensive two-week courses were available to naval intelligence reservists at the amphibious bases at Little Creek, Virginia, and Coronado, California, during their periods of annual active duty for training. In addition, opportunities for practical experience were provided by the district intelligence office, which, in those days, was involved in most phases of naval intelligence, counterintelligence, and security investigations.[43]

In 1960, (qualified) reserve officers on inactive duty were selected to translate unclassified Russian language documents for point credit. The documents included books and professional journals on technical subjects. The program was administered by ONI's Translation Section, primarily for the benefit of the other sections of ONI.[44]

During the early 1960s, the Durham, North Carolina, Naval Reserve Intelligence Training Unit participated in interesting practical exercises in surveillance in conjunction with its classroom course in investigations. In cooperation with the Durham Police Department and various individuals in the community, students were afforded the opportunity to learn by doing, which proved to be an effective instructional technique. A changeover to weekend drills took place in the mid-1960s, and in 1965 the Durham unit entered the Fleet Project Program, producing transportation studies for the Atlantic Intelligence Center (later known as FICLANT) at Norfolk.[45]

In 1965, various units of Naval Reserve Intelligence Division (NRID) 8-1 in the 8th Naval District were selected for a Fleet Intelligence Center, Europe (FICEUR) pilot project to determine if reserve intelligence units could successfully produce studies for the active fleet. NRID 8-1 was selected for the experiment because it had won the Naval Reserve Intelligence Trophy for Fiscal Year 1965 (it won again in 1966).

During 1966, twelve Naval Reserve units of NRID 8-1 worked for FICEUR, as did NRIUs from the 1st, 6th, and 9th naval districts. NRIU 8-1-6 at Albuquerque, New Mexico, worked on transportation studies, which it completed in 1970, earning a letter of commendation from the Commanding Officer of FICEUR. On the basis of its top performance in that effort, NRIU 8-1-6 was selected for another pilot project, the conversion of data input from the IBM 1410 to the IBM 360 computer system.[46]

Following the institution in 1966 of the practice of assigning reserve intelligence units to work on production projects, Unit 11-1-3 at Long Beach spent two years producing a *Navy Survivor's Manual* for the Fleet Intelligence Center, Pacific (FICPAC). For that effort, the unit was awarded a Meritorious Unit Commendation.[47]

In 1969, NRIU 6-1-8 moved from Durham to Raleigh, North Carolina, where weekend training was conducted at the U.S. Naval and Marine Corps Reserve Training Center at 2725 Western Boulevard. Also in 1969, responsibility for the support of Naval Reserve intelligence fleet projects was shifted to FICEUR at Jacksonville, Florida, for whom studies were produced about beaches, ports, and harbors. For its noteworthy contribution in intelligence support of the operating forces, NRIU 6-1-8 received a letter of commendation from the Commanding Officer of FICEUR in January 1973.[48]

Naval Intelligence Reserves in the 1970s

In 1970–1971, FICLANT was supported by 580 officers and enlisted personnel from forty naval reserve intelligence units in the 1st, 3rd, 4th, and 5th Naval Districts. The units provided substantial support in the production of amphibious warfare studies and target analysis. During the period from 1 January 1970 to 30 June 1971, 141 reserve officers and enlisted personnel completed their two-week active duty for training at FICLANT.[49]

To provide policy guidance for integrating the Naval Reserve within the Naval Establishment, and to amalgamate responsibilities and procedures into a formal program, an Intelligence Program Contributory Reserve Support Plan (IPCRESS) was developed and promulgated by CNO letter serial 2060P942 of 1 November 1971.

In 1971, Commander Naval Intelligence Command (COMNAVINTCOM) was designated by the Chief of Naval Operations as the program sponsor for the intelligence component of the Naval Reserve. At that time, 3,200 officers were involved in the air intelligence and regular intelligence reserve programs. For years, the intelligence personnel resource potential had not been fully exploited. In recognition of that deficiency and the fact that training could be enhanced by the naval reservists actually working in a "real world" environment, a program was launched during Fiscal Year 1972 to improve the support provided to the Navy by each reserve intelligence unit. The expertise and special qualifications possessed by the individual members of the reserve force were identified, and their special talents were applied to the intelligence requirements identified by the regular components of the Navy. A computerized personnel data file on all officers and enlisted personnel in the Naval Reserve Intelligence Program was established and designated as the NAVINTCOM Reserve Personnel History Record.

To ensure that the Reserve Intelligence Program included enlisted personnel of the YN (yeoman) rating trained in intelligence duties, eligibility requirements for Naval Enlisted Classification (NEC) Code 2505 were promulgated by Naval Intelligence Command Instruction 1221.1. As a follow-on to the establishment of NEC Code YN 2505 requirements, an enlisted personnel allowance total of 400 was established for the Reserve Intelligence Program in order to ensure an adequate level of technical and clerical support to the intelligence components that were providing production support to fleet commanders and other major naval commands.

A Naval Intelligence Reserve Training Program was formally established during Fiscal Year 1972 by Commander Naval Intelligence Command to provide opportunities to reserve intelligence officers to qualify in operational intelligence and in Naval Investigative Service operations. The program emphasized the use of active duty training activities and deemphasized the use of Naval Reserve drill periods for classroom training.[50]

Reserve officers were assigned to the Naval Ocean Surveillance Information Center (NOSIC) in Suitland and its parent organization, the Navy Field Operatonal Intelligence Center (NFOIO) at Fort Meade, providing both organizations with a blend and depth of knowledge that would not otherwise have been readily available.

In the Washington area, reserve intelligence officers contributed significantly to the production and analysis of operational intelligence. Since the assignment of the original six reserve officers to NOSIC in September 1971, the program grew by June 1973 to twenty-four personnel who performed their weekend drills and their two-week active duty training at NOSIC or NFOIO. The reservists completed significant special projects in support of operational intelligence analysis and reporting while, at the same time, augmenting the various analytical desks in the Current Operations Center at NOSIC.[51]

In 1974, after nearly three decades of command association with the naval districts, the Naval Reserve Intelligence Units were reorganized under reserve intelligence area coordinators and were given designators according to their mobilization functions and the type of activities to be supported.[52]

The activities that were designated in 1974 to receive Naval Reserve intelligence support, and the total number of reserve units organized to provide that support are listed in the table below.

Table 29.2.
Naval Intelligence Reserve Reorganization

Activity	Number of NRIUs
Armed Services Document Center	4
Armed Services Photo Interpretation Centers	1
Contingency Support	15
Army Security Center	1
ELINT (Electronics Intelligence) Centers	1
Armed Forces Air Intelligence	3
Defense Intelligence Agency	4
Defense Attaché System	5
Attaché direct support	1
Intelligence Support Center	13

Task Group	1
Intelligence Processing System	1
Field Operational Intelligence	2
Ocean Information Center	1
Fleet Ocean Information	2
Naval Investigative Service support	38
Fleet Intelligence Centers, Area Analysis	30
Data Handling/Special Communications	1
Intelligence Processing Training	1
Naval Intelligence Command support	15*
Naval Investigative Service (NIS)	6†
Fleet Intelligence Training	2

*Twelve assigned to the Naval Intelligence Command, one to intelligence audit, one to dissemination, and one to intelligence collection.
†Five to NIS investigative teams and one to NIS headquarters.
Source: OPNAVNotice 5400 of 15 May 1974.

Chapter Notes

1. CNO, OP-16-4, *Annual Report, 1932*, 2.

2. Department of the Navy, "Administrative History of the Office of Naval Intelligence in World War II," 10 Jul 1946, unpublished MS, 34, hereafter *ONI WWII Admin History*.

3. F. Donald Scovel, "Helms A'Lee: History of the Development of the Public Affairs Function in the U.S. Navy, 1861–1941" (M.A. thesis, University of Wisconsin, 1968), 92.

4. ONI Administrative Branch, *Annual Report, FY 1933*, 3–4.

5. RAdm. Sidney W. Souers, biographical summary, OA.

6. Naval Reserve Intelligence Command (NRIC) 13–2, "History of the Reserve Intelligence Program in the 13th Naval District," OA; and Capt. James V. McCabe ltr to author, 12 Dec 1973.

7. ONI Administrative Branch, *Annual Report, FY 1933*, 3.

8. *ONI WWII Admin History*, 34.

9. *ONI Annual Report, FY 1935*, 2.

10. *ONI Annual Report, FY 1936*, 1.

11. *ONI WWII Admin History*, 34.

12. Ibid., 29 and Appendix A.

13. Ibid., Appendix H.

14. Ibid., 30 and Appendix B.

15. DNI ltr, ser 01640116, 14 Aug 1945, Naval Investigative (NIS) History Files, OA.

16. District Intelligence Office, 1st Naval District, "Administrative History of DIO-1ND in World War II," MS, passim, OA.

17. ONI-19, *Naval Intelligence Manual*, Change 4, Oct 1940, Art. 112.

18. *ONI WWII Admin History*, 49, 51–55.

19. "Administrative History of DIO-1ND," 18.

20. *ONI WWII Admin History*, 51–55.

21. "Administrative History of DIO-1ND."

22. *ONI WWII Admin History*, 51–55.

23. *ONI Review*, Sep 1946, 35–36.

24. Chief of Naval Air Reserve Training (CNARESTRA) ltr, ser 28398, 22 Oct 1946.

25. *ONI Review*, Sep 1946, 36.

26. CNO, OP-32-C3, Memo on Training Activities, 1 Jul 1946–30 Sep 1946, citing Bureau of Personnel ltr, ser 505, 22 May 1946.

27. OP-32X memo, ser 04847P32, 7 Aug 1946, Accession 3770, box 1, ONI Day Files, OA.

28. NRIC 13-2, "History of Reserve Intelligence Program in 13ND," passim; and McCabe ltr, 12 Dec 1973.

29. Capt. Charles S. Melvin, USNR (Ret.) ltr to author, 4 Oct 1976.

30. CNARESTRA ltr, 22 Oct 1946.

31. Naval Reserve Intelligence Unit (NRIU) 11-1-1 (Los Angeles), "History of NRIU 11-1-1," MS, 18 Mar 1974, 2, 5.

32. NRIU 9-1-11 (Denver), "History of NRIU 9-1-11," MS, Sep 1973.

33. RAdm. Robert H. Barnum, USNR, interview by Lt. Robert B. Rackleff, 27 Mar 1976, transcript, 28–30, Barnum folder in author's files, OA.

34. NRIU 6-1-8, (Durham, NC), "History of NRIU 6-1-8," MS, 29 Dec 1973, 1–4.

35. *ONI Review*, May 1951, 104.

36. Barnum interview, 27 Mar 1976, 28, 30.

37. NRIU 11-1-3 (Long Beach), "Historical Log," 1 May 1971.

38. NRIC 13-2, "History of Reserve Intelligence Program in 13ND"; McCabe ltr, 12 Dec 1973.

39. NRIU 9-1-13 (Madison, WI), "History of NRIU 9-1-13," MS, 20 Oct 1973.

40. Intelligence Program Contributory Reserve Support Plan [*IPCRESS Flyer*] No. 80, 24 Feb 1975.

41. NIC-02R History File, OA.

42. "History of NRIU 9-1-11," Sep 1973.

43. Cdr. Donald H. Scott, "Intelligence in Reserve," *All Hands*, Mar 1960, 17–18.

44. ONI Notice 3400 of 3 Jun 1960.

45. "History of NRIU 6-1-8," 29 Dec 1973, 1–3.

46. NRIU 8-1, "History of Naval Intelligence in New Mexico," passim.

47. Cdr. E. Crilly, USNR, interview by author, 6 Jun 1976, transcript in author's file, 3.

48. "History of NRIU 6-1-8," 3.

49. Fleet Intelligence Center, Atlantic (FICLANT), *Command History, 1 Jan 1970–30 Jun 1971*, I-4.

50. *DNI Annual Report to CNO, FY 1972*, 53–54.

51. *DNI Annual Report to CNO, FY 1973*, V-5, VIII-2.

52. NRIU 8-1, "History of Naval Intelligence in New Mexico."

CHAPTER 30

Officer Training in Naval Intelligence

In the early years of U.S. Naval Intelligence, training in intelligence procedures was gained largely through experience. Lists of naval officers with intelligence experience, as well as officers who had acquired a proficiency in foreign languages or who had knowledge through travel of foreign countries, were kept at the Bureau of Navigation (the predecessor to the Bureau of Personnel and today's Naval Military Personnel Command). The lists were used when selecting officers for assignment to intelligence billets in the Office of Naval Intelligence and to naval attaché posts.

Not until World War II did training for intelligence work achieve recognition as an essential preliminary step toward providing effective intelligence support to naval planning and operations. One persistent and serious gap in intelligence training has been the inadequate indoctrination of prospective commanders in the use of intelligence.

This chapter on intelligence training begins with the first assignment of officers to language training in Japan and China. Chapters 13, 21, 22, 29, and 31 also contain information on training.

Foreign Language Students

In 1910, the first group of oriental language students was sent to Japan. Among them were Lt.(jg) Fred F. Rogers (who served as Naval Attaché, Tokyo from 1933 to 1936); Lt.(jg) George E. Lake; 1stLt. William T. Hoadley, USMC; and 1stLt. Ralph S. Keyser, USMC. The officers assigned to Peking, China, for language study were Marine Corps officers Capt. Thomas Holcomb and 1stLt. E. L. Bigler. All students had diplomatic status and were assigned for administrative purposes to the Naval Attaché, Tokyo and Peking, Capt. John H. Shipley and later LCdr. Lyman A. Cotton. The language training program was terminated in 1913 with the establishment by President Woodrow Wilson and Secretary of the Navy Josephus Daniels of a policy of having as few naval officers as possible on shore duty.[1]

The post–World War I period saw the revival of the language training program; LCdr. Ellis M. Zacharias was sent to Japan during Fiscal Year 1921 and spent three years learning the language and customs. He was followed by others, among them Ens. Thomas Ryan, who was awarded the Medal of Honor for rescuing a woman from the burning Grand Hotel in Yokohama during the earthquake of 1923; Ens. Arthur H. McCollum, later the head of the Far East Section of ONI prior to the Pearl Harbor attack; Lt. Joseph J. Rochefort, who had charge of the Radio Intelligence Unit at the 14th Naval District at the time of Pearl Harbor; Lt.(jg) Edwin T. Layton, the Pacific Fleet Intelligence Officer at the time of Pearl Harbor and throughout World War II; and Lt.(jg) (later Capt.) Henri H. Smith-Hutton, Naval Attaché, Tokyo, from 1939 to the time of the Pearl Harbor attack. There were other language programs: Chinese and Russian were taught in China, and in 1924 Lt. Boyd R. Alexander was a French language student in Paris.[2]

No intelligence collection assignments were given to the Japanese language students in Japan. Their primary job was to learn the language, and they were not trained in collection techniques. On the other hand, they were directed to report to the naval attaché anything of naval interest that they inadvertently learned. When nearing the end of their courses, they were sometimes employed in translation work at the embassy in Tokyo. For example, Lt.(jg) Smith-Hutton was given the task of translating part of the 1926 revision of the *Japanese Coast Pilot* in response to a request to the Naval Attaché, Tokyo, by the U.S. Hydrographic Office.[3] Lt.(jg) William J. Sebald, one of the Japanese language students in Japan, was also given the job of translating part of the *Japanese Coast Pilot* for the naval attaché. Sebald spent several months doing nothing but working on the translation.

The language students were also assigned tasks of writing intellience reports. In March 1927, LCdr. Franz B. Melendy was working on "Comparative Gun Power of American and Japanese Fleets, Its Effect Upon Tactical Handling of the Different Classes of Ships"; Lt.(jg) Sebald was preparing "The Shipbuilding Industry in Japan, Number and Capacity of Shipyards, Government and Private; Possibilities for Expansion of Plants; Developments in Merchant Shipbuilding"; Lt.(jg) David W. Roberts was assigned to report on "The Steel Industry of Japan, Location and Capacity of Steel Mills, including Blast Furnaces, Sources of Raw Ores, Coke, Coal, etc., Data on Yearly Output, Same for Imports"; and Lt.(jg) Smith-Hutton was researching and writing "The Chemical Industry of Japan, Particularly Those Factories Directly Pertaining to War." The sources of information for the reports were the files of the naval attaché, military attaché, commercial attaché and the consul general, plus trade journals, official Japanese government reports, and American and foreign businessmen.[4]

In the early 1930s the U.S. Army and Navy were sending selected officers to countries sharing borders with the Soviet Union for Russian-language training. Among the most popular of these locations were Harbin, Manchuria, and Riga, Latvia. Until World War II, Navy interest in the Soviet Union was primarily confined to ensuring that there were always a certain number of officers on active duty who could understand the Russian language.

There were many international considerations in assigning officers to Russian language training duty. In April 1930, when the question arose of stationing a Marine Corps officer in Harbin for the purpose, the State Department was concerned over the effect that this would have on relations with Japan. After a discussion between LCdr. Zacharias, who was then assigned to ONI, and representatives of the State Department, it was determined that the Harbin assignment would not be taken as showing any special U.S. interest in the affairs of northern Manchuria, or the Chinese Eastern Railroad. Neither was it felt that any harm could result from the association of the language student-officers with the Russian Communist officials of the railroad.

The establishment of the puppet state of Manchukuo (formerly Manchuria) by Japan in March 1932 further complicated the Russian language program. Previously, the students had been given diplomatic passports, were accredited to China (which up to then controlled Manchuria), and were then assigned to residencies in Harbin. The United States refused to recognize the creation of Manchukuo, and the Navy Department withdrew its two students on 7 December 1932, permitting one to remain another three months to complete his training.[5]

In 1933, there were six Navy lieutenants (junior grade) and one Marine first lieutenant language students assigned to the naval attaché office in Tokyo. There were also nine language students at Peking, China: three Navy lieutenants (junior grade), four Marine captains and two first lieutenants. Two of the Marine captains in Peking were studying Russian. There were also Marine officers studying Russian in Shanghai under the aegis of the Commanding Officer, 4th Marines.[6]

In August 1934, Capt. David R. Nimmer, USMC, the first assistant U.S. naval attaché to the Soviet Union, who had been a Russian language student in China, visited Tallin, Estonia, to investigate its suitability as a site for stationing Russian language students. He rejected Tallin because he found that the instruction opportunities and housing facilities there were inferior to those at Riga, Latvia. He also felt the presence in Tallin of British officers in a Russian language program would inhibit the use of Russian in off-duty hours. In reference to the British students, Nimmer observed, "There are now six British officers studying Russian in Tallin, and . . . it is not understood why the British Government goes to such extreme efforts and expense to teach Estonian girls to speak English."[7]

In 1935, the Navy began sending Russian language students to Riga. To meet the wishes of the State Department, the language student-officers were enjoined to refrain from any intelligence activities while stationed at Riga.[8]

Lt.(jg) Carroll H. Taecker was the first officer sent to Riga as a Russian-language student. He arrived in July 1935 and was assigned to the Naval Attaché, Berlin, for administrative purposes in an arrangement that was approved by the State Department as long as such students were not listed as being on the official staff of the U.S. Embassy in Berlin and did not appear in uniform. The students were also required to keep the American Mission in Latvia informed of their movements and activities while in Riga, and they did not have diplomatic status in Latvia.[9]

Lts.(jg) Samuel B. Frankel and George F. Schultz followed in 1936 and were the first officers to receive a complete two-year course. Frankel and Schultz were ordered to report to the Naval Attaché, Berlin, and were further instructed to go to Riga on detached duty for a two-year stay during which they were expected to become completely proficient in Russian. The two officers lived with a Russian family and engaged White Russian instructors. They also talked with Russian military personnel in Latvia from time to time in order to learn military terms.[10]

In 1935, at a U.S. Government initiative, the dozen or so Navy officers assigned to language training in Japan were removed from the diplomatic list. The move was prompted by the presence of over fifty Japanese assigned as "language officers" in the United States who were enjoying diplomatic immunity while they were deeply involved in espionage.[11]

Other than those already mentioned, naval officers assigned to Riga, Latvia, for instruction in the Russian language were Lts.(jg) Harry E. Seidel, Jr. (1937–1938), Arthur L. Wilson (1938–1940), and Stanley W. Lipski (1939–1940). Lipski continued his Russian instruction in Stockholm in 1940, and in 1941 he was assigned as Assistant Naval Attaché, Stockholm, resident at Helsinki, Finland.[12]

In 1941, there were nine competent young officers assigned to the Naval Attaché, Tokyo, for language training. Before they left for the mountains or seashore for the summer, they were told to have their personal effects ready for departing Japan on a moment's notice. Naval Attaché Capt. Henri Smith-Hutton had exchanged letters with ONI, pointing out that the Japanese language students did not have diplomatic status, and that, in case of war, they would probably be seized by the Japanese. He suggested that they should leave Japan and continue their studies in Hawaii or another place where there were Japanese teachers. Smith-Hutton also recommended that he send to the United States the books needed for an expanded Japanese language program, such as dictionaries, phrase books, and grammars. ONI agreed completely with the plan, and, in late July 1941, Smith-Hutton and his assistant, LCdr. Martin R. Stone, telephoned each of the students and told them (in Japanese) to get back to Tokyo as soon as possible. Most of the students had anticipated the call and had put their household possessions in storage in Yokohama.

After a good deal of negotiating and difficulty with local officials, the students were evacuated, moving by train from Tokyo to Kobe, where they took a ship to Shanghai arriving on Labor Day 1941. The Army did not take similar action, and a number of their students were interned for about six months before being repatriated on *Gripsholm* in 1942.[13]

The Navy language students in Japan at the time of the evacuation were Lt. William R. Wilson; Lts.(jg) Forrest R. Biard, Rufus L. Taylor, John R. Bromley, Allyn Cole, Jr., Ted A. Hilger, Thomas R. Mackie, and Gilven M. Slonim; and Marines Capt. Bankson T. Holcomb, Jr., and 1stLt. Ferdinand W. Bishop.[14]

In addition to ensuring that the student officers safely left Japan, Smith-Hutton also sent back all the Japanese dictionaries he could buy, and several hundred were shipped back in mail bags. They were of great value later when the Navy set up its Japanese language school.[15]

Table 30.1.
U.S. Navy and Marine Corps Language Officers Who Studied in Japan, and Their Subsequent Intelligence Asssignments

Name	Dates	Training and Subsequent Duties
Rogers, Fred F.	1910–1912	Training in Japan, (Lt.[jg])
	1919–1920	ONI (Cdr.)
	1933–1936	Naval Attaché, Tokyo (Capt.)
Lake, George E.	1910–1912	Training in Japan (Ens./Lt.[jg])
Hoadley, William T.	1910–1913	Assistant Naval Attaché, Tokyo Training (1stLt., USMC)
Keyser, Ralph S.	1912–1914	Training in Japan (1stLt., USMC)
Redles, William L.	1915–1918	Training, Assistant Naval Attaché, Tokyo (Capt./Maj., USMC)
	1920–1921	ONI (LtCol.)
Zacharias, Ellis M.	1920–1923	Training in Japan, (LCdr.)
	1923–1925	ONI (LCdr.)
	1925–1928	Fleet Intelligence Officer, Asiatic Fleet (LCdr.)
	1928–1931	ONI, Head FE Section (LCdr.)
	1934–1935	ONI, Head FE Section (Cdr.)
	1938–1940	DIO-11ND
	1942–1943	Assistant Director of Naval Intelligence (Capt.)

Name	Dates	Training and Subsequent Duties
Davis, Hartwell C.	1920–1923	Training in Japan (LCdr.)
	1931–1932	ONI
	1940–1941	ONI (Cdr., Ret.)
McCollum, Arthur H.	1922–1925	Training in Japan (Ens.)
	1928–1930	Assistant Naval Attaché, Tokyo (Lt.[jg])
	1933–1935	ONI, FE Section (Lt.)
	1935–1936	Assistant DIO-11ND, San Diego (Lt.)
	1939–1942	ONI, Head FE Section (LCdr.)
	1942–1945	Fleet Intelligence Officer, 7th Fleet (Cdr./Capt.)
	1944–1945	Additional duty CO SEFIC
	1946–1948	CIA Naval Administrative Command (Capt.)
Ryan, Thomas J., Jr.	1922–1924	Training in Japan (Ens.) Awarded Medal of Honor for action in Yokohama earthquake in 1923.
Sullivan, William A.	1923–1926	Training in Japan (Capt./Maj., USMC)
Hickey, B. F.	1923–1924	Training in Japan (Capt., USMC)
Roberts, David W.	1924–1927	Training in Japan (Ens.)
	1928	ONI (Lt.[jg])
	1932	Assistant Naval Attaché, Tokyo (Lt.)
	1933	Fleet Intelligence Officer, Asiatic Fleet (Lt.)
	1935–1937	ONI (Lt./LCdr.)
Melendy, Franz B.	1924–1927	Assistant Naval Attaché, Tokyo (LCdr.)
	1930–1932	As above
Sebald, William J.	1925–1928	Training in Japan (Lt.[jg])
	1942	ONI, FE Section (LCdr.)
	1943–1945	COMINCH, Pacific Section (LCdr.)
	1945	OP-32P (Cdr.)
Monahan, James S.	1925–1928	Training in Japan, (2dLt./1stLt., USMC)
Smith-Hutton, Henri H.	1926–1929	Training in Japan (Lt.[jg])
	1929–1930	ONI, (Lt.[jg])
	1931–1932	Fleet Intelligence Officer, Asiatic Fleet (Lt.)
	1932–1935	Assistant Naval Attaché, Tokyo (Lt.)
	1937–1939	Fleet Intelligence Officer, Asiatic Fleet (LCdr.)
	1939–1941	Naval Attaché, Tokyo (Cdr.)
	1942–1944	COMINCH Intelligence Officer (Capt.)
	1947–1952	Naval Attaché, Paris (Capt.)
Libenow, Louis D.	1926–1929	Training in Japan (Lt.[jg])
	1930	ONI (temp) (Lt.)
	1935–1936	ONI (Lt.)
Pearce, Edward S.	1927–1930	Training in Japan (Lt.[jg])
	1930	ONI (Lt.[jg])
	1937	ONI (Lt.)
	1943	Commander South Pacific Forces Intelligence Staff (Cdr.)
	1944–1945	ONI, Head Japanese Empire Section (Capt.)
Birtley, Thomas B., Jr.	1927–1930	Training in Japan (Lt.[jg])
	1931	ONI (Lt.[jg])
	1941–1944	Fleet Radio Unit, Pacific
Ringle, Kenneth D.	1928–1931	Training in Japan (Lt.[jg])

Name	Dates	Training and Subsequent Duties
Watts, Ethelbert	1928–1931	Training in Japan, (Lt.[jg])
	1935–1937	Assistant Naval Attaché, Tokyo (Lt.)
	1941–1943	ONI (LCdr.)
	1950–1952	Assistant Naval Attaché, London (Capt.)
	1952–1954	Naval Attaché, Tokyo (Capt.)
Pyzick, Frank P.	1929–1932	Training in Japan (2dLt., USMC)
	1932	ONI (1stLt.)
	1941	Shanghai (POW thereafter)
Layton, Edwin T.	1929–1932	Training in Japan (Lt.[jg])
	1932–1933	Assistant Naval Attaché, Peiping (Lt.[jg])
	1936–1937	ONI (Lt.)
	1937–1939	Assistant Naval Attaché, Tokyo (Lt.)
	1941–1946	CINCPACFLT Combat Intelligence (LCdr./Capt.)
	1948–1950	Director, Naval Intelligence School (Capt.)
	1950	DIO-14ND (TAD Fleet Intelligence Office, NAVFE, 7/50–9/50) (Capt.)
	1951–1953	Fleet Intelligence Officer, CINCPACFLT, and CINCPAC AC/S(I) (Capt.)
	1953–1956	Joint Staff, J-2 (RAdm.)
	1956–1958	CINCPAC AC/S(I) (RAdm.)
	1958–1959	Director, Naval Intelligence School (RAdm.)
Rochefort, Joseph J.	1929–1932	Training in Japan (Lt.)
	1941–1942	Officer in Charge, Fleet Radio Unit, Pacific (Cdr.)
Mason, Redfield	1930–1933	Training in Japan (Lt.[jg])
	1937–1939	ONI (Lt.)
	1940–1941	Fleet Intelligence Officer, Asiatic Fleet (Lt.)
	1942–1945	Navy COMINT (LCdr./Capt.)
McCallum, Daniel J.	1931–1934	Training in Japan (Lt.[jg])
	1934–1935	ONI (Lt.)
	1938	Fleet Intelligence Officer, Asiatic Fleet (LCdr.)
	1939–1941	Assistant Naval Attaché, Tokyo (LCdr.)
	1946–1948	Naval Liaison Officer, Batavia (Capt.)
Kramer, Alwin D.	1931–1934	Training in Japan (Lt.[jg])
	1934–1935	ONI (Lt.[jg])
	1938–1943	ONI FE Section (Lt./LCdr.)
	1944–1945	Joint Intelligence Center, Pacific
	1945	ONI FE Section (Capt.)
Claiborne, Henri deB.	1931–1934	Training in Japan (Lt.[jg])
	1934–1935	ONI (Lt.[jg])
	1943	ONI FE Section (Cdr.)
	1945–1946	Fleet Intelligence Officer, 7th Fleet (Capt.)
Cornell, Kenneth H.	1931–1934	Training in Japan (1stLt., USMC)
	1935	ONI (1stLt., USMC)
	1938	4th Marines, Shanghai
Fullinwider, Ranson	1932–1935	Training in Japan (Lt.[jg])
	1935	ONI (Lt.[jg])
	1941–1945	Fleet Radio Unit, Pacific
	1946–1949	Naval Attaché, Buenos Aires (Capt.)
	1952–1953	Naval Attaché, Karachi (Capt.)

Name	Dates	Training and Subsequent Duties
Carlson, Spencer A.	1932–1935	Training in Japan (Lt.[jg])
	1939	ONI FE Section (Lt.)
	1940	4th Marines, Shanghai
	1941–1942	Cast Unit Corregidor
	1942–1944	Fleet Radio Unit, Melbourne
	1945	Fleet Radio Unit, Pacific (Capt.)
Finnegan, Joseph	1934–1937	Training in Japan (Lt.)
	1937–1938	16ND Cast Unit
	1942–1945	Fleet Radio Unit, Pacific
	1947–1950	CIA (Capt.)
Karrer, Harold E.	1934–1937	Training in Japan (Lt.[jg])
	1938	ONI
	1939–1940	16ND Cast Unit (Lt.), died 22 September 1942
Stone, Martin R.	1934–1937	Training in Japan (Lt.[jg])
	1941	Assistant Naval Attaché, Tokyo (LCdr.)
	1950–1951	N2 COMNAVFE (Capt.)
Jordan, Francis D.	1934–1937	Training in Japan (Lt.[jg]), died as POW, Asiatic area 24 October 1944
Richardson, Gill M.	1935–1938	Training in Japan (Lt.[jg])
	1939	RIO, 14ND
	1940–1942	Assistant Fleet Intelligence Officer Asiatic Fleet and Cast Unit
	1942–1945	Fleet Radio Unit, Melbourne
	1947–1949	Fleet Intelligence Unit, CINCNELM (Capt.)
Lasswell, Alva B.	1935–1938	Training in Japan (1stLt., USMC)
	1938–1939	Cast Unit, 16ND
	1939	4th Marines, Shanghai
	1941–1945	Fleet Radio Unit, Pacific
Hudson, Robert E.	1936–1939	Training in Japan (Lt.[jg])
	1941–1945	Assistant Intelligence Officer, CINCPACFLT (Lt./Cdr.)
	1945–1946	ONI, Operational Intelligence Section (Cdr.)
	1948–1949	N2 CINCPAC/FLT (Cdr.)
	1949–1960	Various DIOs (Capt.)
Ballard, Nixon L.	1937–1939	Training in Japan (1stLt., USMC)
Roenigk, John G.	1938–1941	Training in Japan (Lt.[jg])
	1941–1942	CINCPAC Intelligence Staff (Lt.)
	1942–1945	Fleet Radio Unit, Pacific (Lt./Cdr.)
	1957–1960	Naval Attaché, Stockholm (Capt.)
	1962–1964	Naval Attaché, Tokyo (Capt.)
Benedict, Arthur L.	1938–1941	Training in Japan (Lt.[jg])
	1941–1942	CINCPAC Intelligence Staff (Lt.)
	1942–1945	Fleet Radio Unit, Pacific (Lt./Cdr.)
Cole, Allyn, Jr.	1939–1941	Training in Japan (Lt.[jg])
	1941–1945	Fleet Radio Unit, Pacific (Lt./Cdr.)
	1947–1950	ONI, Operational Intelligence (Cdr.)
Mackie, Thomas R.	1938–1941	Training in Japan (Lt.[jg])
	1941–1942	Cast Unit, 16ND (Lt.[jg]/Lt.)
	1942–1945	Fleet Radio Unit, Melbourne (Lt./Cdr.)
	1946–1947	ONI, Operational Intelligence (Cdr.)

Name	Dates	Training and Subsequent Duties
Taylor, Rufus L.	1938–1941	Training in Japan (Lt.[jg])
	1941–1942	Cast Unit, 16ND (Lt.)
	1942–1943	Fleet Radio Unit, Melbourne (Lt./Cdr.)
	1943–1944	OPNAV (Cdr.)
	1945	Fleet Radio Unit, Pacific (Cdr.)
	1945–1946	OPNAV (OP-20) (Cdr.)
	1948–1951	ONI, Operational Intelligence (Cdr.)
	1951–1953	OP-20 (Cdr.)
	1953–1955	OSD (Cdr./Capt.)
	1955–1956	N2 COMNAVFORJAP (Capt.)
	1956–1959	Fleet Intelligence Officer, CINCPACFLT (Capt.)
	1959–1963	ONI, OP-922/OP-92B (Capt.)
	1963–1966	DNI and ACNO(I) (RAdm.)
	1966	Deputy Director DIA (VAdm.)
	1966–1969	Deputy DCI (VAdm.)
Wilson, William R.	1938–1941	Training in Japan (Lt.[jg])
	1942–1945	POW
	1950–1952	ONI, Plans and Policies (Cdr./Capt.)
Biard, Forrest R.	1939–1941	Training in Japan (Lt.[jg])
	1941–1945	Fleet Radio Unit, Pacific and Joint Intelligence Center, Pacific (Lt./Cdr.)
	1945–1946	ONI, Technical Intelligence (Cdr.)
Bromley, John R.	1939–1941	Training in Japan (Lt.[jg])
	1941–1943	Fleet Radio Unit, Pacific (Lt./LCdr.)
	1943–1945	Fleet Radio Unit, Melbourne (LCdr./Cdr.)
	1946	ONI, Head Japanese Desk (Cdr.)
	1947–1949	N2 COMNAVFE (Cdr.)
	1949–1950	ONI, Dissemination (Cdr.)
	1950–1951	ONI, Intelligence Staff (Cdr.)
	1951–1954	Naval Attaché Lima (Cdr./Capt.)
	1954–1956	DIO-1ND (Capt.)
	1956–1959	ONI, ADNI Security (Capt.)
	1959–1962	Naval Attaché Tokyo (Capt.)
Slonim, Gilven M.	1939–1941	Training in Japan (Lt.[jg])
	1941–1945	Fleet Radio Unit, Pacific (Lt.[jg]/Cdr.)
Holcomb, Bankson T., Jr.	1935–1937	Training in Chinese in Peiping (1stLt., USMC)
	1939–1941	Training in Japan (1stLt.)
	1943–1945	NAVGROUPCHINA
Bishop, Ferdinand W.	1940–1941	Training in Japan (1stLt., USMC)
	1943	SOPAC POW Interrogator, killed in plane crash in North Pacific
Hilger, Ted A.	1941	Training in Japan (Lt.[jg]), killed March 1942

Source: *Navy Directory*, various editions; Capt. Joseph Finnegan, manuscript narrative, OA.

Navy School of Oriental Languages During World War II

The need for more Japanese language officers in the Navy became evident when a check was made in December 1940 of the status of those officers who had received language training in Japan. Of the sixty-five or so officers, only twelve were fully proficient in the use of spoken and written Japanese.

To correct the situation, Cdr. Albert E. Hindmarsh, USNR, a former language professor at Harvard, was instructed in February 1941 to make a nationwide survey of Japanese linguists in and out of the Navy, with a view to establishing a practical course that would produce junior Naval Reserve officers thoroughly trained in writing, reading, and speaking Japanese. The survey, conducted between March and June 1941, found fifty-six persons with sufficient knowledge of Japanese to justify inviting them to become the nucleus of a U.S. Navy Japanese language course.

On 1 August 1941, Cdr. Hindmarsh submitted a plan to establish two training centers, one at Harvard University, the other at the University of California at Berkeley. The Director of Naval Intelligence and the Chief of the Bureau of Navigation approved Hindmarsh's plan.

Arrangements were made to have the Naval Attaché, Tokyo, procure fifty complete sets of the seven-volume Naganuma Japanese Language Course, which had for many years formed the basis of the course given to U.S. Navy language officers in Japan. The books reached the United States in September 1941 and were duplicated for the first language class, which was to be convened on 1 October 1941.

Between 25 August and 22 September, Hindmarsh and Glenn Shaw, the chief Japanese language expert in ONI, interviewed and examined student applicants at various U.S. cities. During the trip, forty-eight students were enrolled in the course.

In January 1942, it was decided to enroll additional students, and on 22 February forty-seven new students were selected on the basis of personal interviews conducted by Hindmarsh and Shaw. The students reported to Berkeley, where the course was being conducted far more smoothly and efficiently than at Harvard. It was decided to let the contract at Harvard expire on 30 September 1942 at the end of the one-year contract period.

Between 18 May and 15 June 1942, Hindmarsh and Shaw visited centers throughout the country and enrolled 153 additional students for the course at Berkeley that was scheduled to begin on 1 July. On 23 June, however, the school was transferred to the University of Colorado at Boulder because of a Western Defense Command order requiring all persons of Japanese ancestry to be evacuated from the West Coast. The school had eleven ethnic Japanese teachers and was expecting twenty more.

Between 5 November and 20 December 1942, 302 additional students were enrolled, and approximately 200 were prepared for enrollment, based on interviews and written examinations. Approximately 80 percent of the applicants were rejected because they failed to meet the high minimum standards, which included a college degree and either previous study in Japanese or Phi Beta Kappa standing.[16]

In December 1943, there were seventy WAVES enrolled in the Navy's Japanese Language School, but it was decided to curtail further enrollment until women officers could be assigned outside the continental United States.[17]

In January 1944, courses in Chinese and Malay were added, and the name of the school at Boulder was changed to the Navy School of Oriental Languages. On 6 March 1944, a course in Russian was approved. The new courses began with a small number of students on 3 April: Russian, 29; Chinese, 16; and Malay, 9. The courses lasted eighteen months for Chinese, six months for Russian, and three months for Malay.

The Navy School of Oriental Languages at Boulder and a second facility that had been established at Oklahoma A&M at Stillwater were closed upon activation of the Language Division of the Naval Intelligence School at Anacostia in the District of Columbia on 1 July 1946 under then-Capt. Hindmarsh.[18]

Naval Intelligence Training in the World War II Era

In the Director of Naval Intelligence's annual report for Fiscal Year 1935, it was reported that

> there is [a] definite need for officers with training in intelligence work including knowledge of the principles underlying investigating work. To this end, a school of instruction has been initiated in the Division of Intelligence, and the first class, consisting of four officers who have completed one year at the Postgraduate school and one Marine officer, reported during June. In this connection, arrangements have been made with the Bureau of Investigation, Department of Justice, for the participation of the officers in question in the special course for investigators of the Bureau.

For additional details, see Chapter 21.

Upon the establishment of the Planning and Training Section (OP-16-X) of ONI on 27 August 1938, it was tasked to organize naval intelligence training programs and courses and to prepare for the inspection of the training and readiness of naval reservists in the naval districts. Although OP-16-X had the responsibility for preparing, disseminating, and correcting ONI training and procedural manuals, much of the actual work was gradually assumed by the Administrative Branch because of the small number of people assigned to OP-16-X.[19]

As a result of his study in England in early 1941 of the British methods of extracting information from photographs taken over enemy territory, LCdr. Robert S. Quackenbush, Jr., recommended the establishment of a naval school to train officers in the science of photo interpretation. On 12 September 1941, the Chief of Naval Operations authorized the creation of a photo interpretation school under the Bureau of Aeronautics to be located at the Naval Air Station (NAS), Anacostia.

The first class of the School of Photographic Interpretation convened on 5 January 1942 with a faculty composed of Navy and Marine Corps officers. The school was intended to teach its students how to extract intelligence data from factual evidence contained in photographs of enemy holdings and to present the information obtained in a useful and readily understandable form for operational commands.[20] For more information on the school, see Chapter 13.

On 2 May 1941, a Training Section (first OP-16-A-9, then A-8) was set up within ONI to conduct a three-week course for the indoctrination of officers destined for domestic and foreign intelligence duties. On 1 February 1942, the indoctrination classes were moved to Frederick, Maryland, where the Basic Naval Intelligence School was set up in the Francis Scott Key Hotel. The school was closed, however, on 4 September 1943.

On 1 February 1943, a school for advanced intelligence training was established at the Henry Hudson Hotel in New York City. Its two principal courses were Operational Intelligence and Commerce and Travel. The length of each term was eight weeks, later lengthened to ten.[21]

The purpose of the Operational Intelligence course was to train officers for duty with advance bases, staffs, and forces afloat in foreign theaters. The curriculum was modified several times during its 2½ years of operation. The basic course included photo intelligence, ship and aircraft recognition, communications, navigation, amphibious warfare, and organization and strength of enemy forces. A mock-up of a shipboard combat information center (CIC) equipment and layout was used.[22]

In April 1942, it was determined that specially trained air combat intelligence (ACI) officers were needed to brief pilots on their missions and to ensure a flow of information from combat reports. The Naval Air Combat Intelligence Officers School was set up on 15 April 1942 at Quonset Point, Rhode Island, under the supervision of the Aviation Intelligence Branch of the Bureau of Aeronautics. The Aviation Intelligence Branch selected the students, worked out the curriculum, and made recommendations on the assignment of ACI officers upon completion of their training.[23]

Virtually all ACI students were selected from graduating classes of the A-V(S) (Aviation Volunteer Specialist) Indoctrination School at Quonset Point, until that school closed in January 1944. Thereafter, they came from active duty in the field. In addition, a number of officers from the Marine Corps and ONI attended the school. A major proportion had backgrounds in law, journalism, teaching, or advertising, or extensive administrative experience in the business world. The number of students per class during the first year averaged about 150. After 1 May 1943, classes were limited to 100 students per class.

At first, the curriculum took advantage of the valuable experiences of the British Royal Air Force, particularly with regard to the briefing and debriefing of pilots. Principal courses included intelligence briefing and debriefing procedures, maps and charts, elements of photo interpretation, air tactics, economic geography, aircraft and ship recognition, aerial navigation, elements of aerology, naval communications, performance characteristics and armament of the principal air forces of the world, antisubmarine warfare, radar, flak analysis, amphibious warfare, and air support doctrine. Originally, the course lasted eight weeks, but it was lengthened to ten weeks in 1943. The last class graduated in September 1945, after the Naval Air Combat Intelligence Officers School had trained over 1,800 officers during its three-year existence.[24]

Post–World War II Naval Intelligence Training

By 1945, the Advanced Naval Intelligence School (ANIS) was conducting two-week refresher courses in New York City for officers returning to the United States before they were reassigned. The curricula were tailored to meet the needs of each individual's next assignment, when known. The officers repeated courses whose content had not been used during their previous duty assignments, and they also took new courses that had not been available during their previous attendance at the school. Approximately seventeen officers attended each of the refresher sessions, and they were billeted at the Henry Hudson Hotel. Frequently, the refresher course students were asked to serve as temporary instructors in regular operational intelligence (OPINTEL) classes, where they could speak on subjects with which they had had personal experience.

The ANIS staff had consisted of six officers when it was established in January 1943. By 1945, the staff consisted of fifteen officers, thirteen of whom had served in operational intelligence billets outside of the United States during World War II. At this time the ten-week regular OPINTEL curriculum consisted of the following courses: Naval Staff Procedure, 52 hours; Amphibious Warfare, 27 hours; Operational Intelligence Procedure, 18 hours; Navigation, 24 hours; Means Available and Opposed (i.e., U.S. and Japanese order-of-battle, capabilities, and tactics), 34 hours; Communications, 18 hours; Identification of Ships and Aircraft, 36 hours; Theater Areas, 24 hours; Aerology, 6 hours; Antisubmarine Warfare, 9 hours; Mine Warfare, 5 hours; Photo Intelligence, 38 hours; CIC and Radar, 25 hours; and miscellaneous lectures, 34 hours.

Beginning in November 1944, ANIS conducted three special military government classes, graduating fifty-three officers who reported to the Naval Civil Affairs Staging Area at Monterey, California, for assignment as intelligence officers with military government teams.[25]

World War II experience gave rise to the belief within ONI that a substantial group of officers, both regular and reserve, should be available at all times for intelligence duty. Director of Naval Intelligence Commo. Thomas B. Inglis, in a directive dated 26 September 1945, established the Intelligence Training Unit (OP-16-X-A), a special activity under the administrative control of the ONI Planning Branch that was charged with preparing courses of instruction, including a textbook on naval intelligence, for use in the curricula of the Naval War College, the U.S. Naval Academy Postgraduate School, Naval Reserve Officer Training Centers (NROTC), and for intelligence reserve officers on inactive duty.

FAdm. Ernest J. King, Commander in Chief, U.S. Fleet, in a memo dated 9 October 1945, concurred in the need for intelligence training and stated:

> Experience gained in this war demonstrated conclusively that a Naval Intelligence Service, such as we finally developed during the last months of the war, is a prime essential of modern warfare. . . . The Commander-in-Chief, U.S. Fleet and the Chief of Naval Operations desires that a continuing program of specialized instruction in intelligence and the appreciation of good intelligence be undertaken without delay.[26]

OP-16-X-A completed its assignment and was disestablished in March 1946.

A comprehensive program of training and instruction of officers of the regular Navy in naval intelligence was carried out at the Naval Academy Postgraduate School and at the Naval War College during the immediate post–World War II period. A short lecture series was given to the members of the graduating class at the Naval Academy to acquaint them with the basic principles of intelligence.

Officers in the General Line and Applied Communications courses at the postgraduate school were given a text and lecture course based on the textbook *Naval Intelligence*, prepared by ONI and issued by the Bureau of Personnel under the short title NAVPERS 16047. Intensive instruction in naval intelligence in relation to staff and command requirements was given in the junior and senior courses at the Naval War College. The War College also conducted a correspondence course in naval intelligence that was available to all officers in the regulary Navy and in the Reserves on active duty.[27]

Secretary of the Navy James V. Forrestal in March 1946 authorized the Navy Intelligence School to train officers of the regular Navy and the Naval Reserve in naval intelligence in accordance with the program developed by OP-16-X-A and approved by Chief of Naval Intelligence RAdm. Thomas Inglis. The school began operations on 1 July 1946 as a subordinate activity of the Navy Postgraduate School of the U.S. Naval Academy.

The Intelligence School was initially under the command of Capt. Albert Hindmarsh and occupied quarters recently taken over at the Naval Receiving Station, Anacostia, D.C. The first class included fifty naval officers and five Marine officers. Half of the naval officers were reservists in the process of transferring to the regular Navy as intelligence specialists.

Instruction was in the basic fields of operational, strategic, amphibious, and air intelligence. The subjects were covered in thirty courses, many of a highly technical nature, and scheduled over a period of seven months. The classwork was followed by ten weeks of practical instruction under Commander in Chief, U.S. Atlantic Fleet or Commander in Chief, U.S. Pacific Fleet, mostly at sea. On completion of the practical work, the students returned for study in Spanish, French, German, Italian, Portuguese, Russian, Chinese, or Japanese. The length of the language course varied from four months for Spanish to eighteen months for Chinese and required approximately fourteen hours of study per day (five hours of classwork and nine hours of homework). Finally, following the language course, five weeks were spent studying the geography, history, government, economics, politics, customs, and other aspects of the area of language specialization.

Language instructors for the Intelligence School program were selected from applicants who had had a minimum of twenty years' residence in a country using the language they were to teach. The language courses were available for students from other government departments, but the intelligence courses were not. Both the Army and the State Department had language students in the school from the outset.[28]

As of 1947, officers of the regular Navy who successfully completed the naval intelligence course of the postgraduate school could, when there was a vacancy, become qualified as special duty intelligence (SDI) officers and serve continuously thereafter in intelligence billets.[29]

A two-week training course for a limited number of Organized and Volunteer Reserve air combat intelligence officers (ACIO) was held at the Naval Intelligence School at the Naval Receiving Station, Anacostia, commencing on 7 June 1948. The course included instruction in the latest developments in air intelligence and related fields. Each naval air station and naval air reserve training unit nominated two ACIOs to attend the course. The students, in turn, were expected to pass on what they had learned in the course to other ACIOs attached to their groups, squadrons, and units in training at their respective stations and naval air reserve training units.[30]

The first revision of the textbook, *Naval Intelligence* (NAVPERS 16047), was prepared by the Naval Intelligence School for ONI, and it was issued in 1948 over the signature of RAdm. Inglis. The book was used both at the Naval Intelligence School and in the correspondence course on naval intelligence by regular and reserve officers who had access to secure classified storage facilities.

Intelligence Officer Training During the Korean War Period

When the Korean War broke out on 25 June 1950, there were no Quonset-trained, combat-experienced air combat intelligence officers on active duty in the Navy, and there was no ACI school. The air intelligence function in fleet aviation squadrons was being handled as a collateral duty by assigned pilots and was consequently accorded secondary priority.[31]

Efforts were made to locate former ACIOs, and fifty-two were successfully induced to return to active duty by October 1950. Of those, thirty-five went to the fleet without refresher training, and seventeen went to the Pentagon, with part of the latter group detailed to set up the Air Intelligence Section of the Naval Intelligence School. The first class convened on 21 August 1950, with Cdr. A. Simontacchi, USNR, as the senior member of the facility. The course took nine weeks. Students came from among Officer Candidate School (OCS) graduates and from among recruiters in the district intelligence offices. Likely recruits went to Officer Candidate School and, if they passed OCS and if their clearances for access to classified information were obtained while there, they automatically went to Air Intelligence and Photo Intelligence schools (AI/PI). The first class consisted of nineteen students.

LCdr. Charles S. Melvin, USNR, one of the fifty-two former ACIOs, was assigned in December 1950 to set up OP-323P3D (Air Intelligence Reserve Personnel) in ONI. He helped select the instructors and students for the school, gave one or two lectures to each class, interviewed the students regarding their preferences as to billet assignments, and then tried to match their desires with the Navy's requirements at graduation time.

Both Commanders Air Forces Pacific and Atlantic set up AI/PI schools at NAS Alameda and NAS Norfolk, respectively, to train additional officers for the fleet and to give brief area indoctrination courses to graduates of the Washington AI school who were en route to fleet assignments.[32]

In the early phase of the Korean War, it was found that the training of air intelligence officers (AIO) needed to be revised to provide a basic concept of general intelligence and a better grasp of the duties of staff, ship, air groups, and squadron AIOs in combat and in preparation for combat.[33]

On 27 July 1950, the Deputy Chief of Naval Operations for Operations, in a memo to the Deputy Chief of Naval Operations for Personnel, had requested the immediate establishment of a training program for Organized Reserve air intelligence officers who had no previous intelligence experience. The step was necessary to meet the then-current need for air intelligence officers. A sixty day course consisting of thirty days of instruction in basic intelligence and thirty days in the specialized aspects of air intelligence was determined to be adequate. Twenty-five students per class could be accommodated at existing Naval Intelligence School facilities, where the staff was augmented by six specially qualified AIOs of the Organized Reserve.[34]

On 9 August 1950, the Naval Intelligence School was included on a priority list of Navy organizations that could readily be removed from the Washington area in the event of war. The list, prepared by OP-213, was sent by the Secretary of the Navy to the chairman of the Joint Military Decentralization Board, apparently without the approval of the Director of Naval Intelligence.

A reclama was initiated by ONI, pointing out the dependence of the school upon all federal intelligence agencies at the headquarters level for major curriculum support. Close proximity to ONI was considered essential to the continued excellence of its output, not only because of the experienced lecturers readily available to the school, but also because the school could respond quickly to the need for refresher courses or modified curricula as anticipated and facilitated by direct and personal liaison with ONI. Additionally, having the school in the Washington area afforded an opportunity for ONI to select students for especially sensitive assignments by personal interview.[35]

Accordingly, Chief of Naval Operations Adm. Forrest P. Sherman requested that the Naval Intelligence School be retained in the Washington area. Secretary of the Navy Francis P. Matthews concurred and modified his previous list by stating to the chairman of the Joint Military Decentralization Board that "in the event of the relocation of the Naval Intelligence School, a site be selected which is within a maximum of one hour's driving distance from the Capitol."[36]

The Air Intelligence Section of the Naval Intelligence School at Anacostia graduated 459 Navy and Marine Corps officers during the Korean War period, 1950–1953. Fifteen classes were held. Classes 1-50 through 4-51 received nine weeks of instruction, while classes 5-51 through 4-53 received eleven weeks of instruction, including two weeks of photographic interpretation training.

Table 30.2.
Air Intelligence Section Classes
1950–1953

Class #	Date Convened	Date Graduated	Students Graduated
1-50	21 Aug 1950	27 Oct 1950	19
2-50	30 Oct 1950	29 Dec 1950	19
1-51	29 Jan 1951	30 Mar 1951	24
2-51	9 Apr 1951	8 Jun 1951	28
3-51	18 Jun 1951	17 Aug 1951	33
4-51	27 Aug 1951	26 Oct 1951	31
5-51	5 Nov 1951	18 Jan 1952	34
1-52	28 Jan 1952	11 Apr 1952	28
2-52	21 Apr 1952	3 Jul 1952	35
3-52	14 Jul 1952	26 Sep 1952	35
4-52	6 Oct 1952	19 Dec 1952	36
1-53	12 Jan 1953	27 Mar 1953	35
2-53	6 Apr 1953	19 Jun 1953	35
3-53	13 Jul 1953	25 Sep 1953	35
4-53	5 Oct 1953	18 Dec 1953	32

Source: Senior Member AI Section, Naval Intelligence School ltr, 18 Jan 1954 and encl. 2.

On 24 June 1952, Class VIII of the Naval Intelligence School graduated 62 students, consisting of 39 reserve officers, one Marine officer, 21 regular officers, and one CIA student. Seven previous classes and one special class had a total of 452 graduates.

Prior to the Korean conflict, the Naval Intelligence School course had been nine months long. With the outbreak of the struggle, it became necessary to accelerate the output of graduates to meet the demand for trained intelligence officers, and the course was shortened to six months by eliminating field trips and condensing selected subjects. There were twenty-five officers assigned to the school staff during the Korean War.[37]

Cold War Period

ONI's Air Intelligence School was located at the Naval Observatory in Washington, D.C., from 1954 to 1957, when it returned to Anacostia. The course had been lengthened to about nine months. Many of the students came from Officer Candidate School via other preliminary training schools such as the Aviation Ground Officers School.

The Hoover Commission on Intelligence Activities recommended "that the Navy continue and expand its efforts to improve the intelligence consciousness at all ranks and levels of the Department [of the Navy]." As of March 1956, the current directives pertinent to the intelligence curriculum at Navy schools were being reviewed. Also, the course of instruction at the Naval Intelligence School that provided postgraduate training in intelligence had been lengthened from six to nine months.[38]

Training facilities of the armed forces and appropriate civilian agencies of the government were employed to provide training for personnel assigned to duty within ONI. Courses of instruction varying from public speaking to a detailed technical study of Navy missile components were available.

The courses of instruction formerly offered at the Air Intelligence Division of the Naval Intelligence School and the Photo Interpretation Center were integrated at the Naval Intelligence School during 1958. The resultant course was designed to produce an officer qualified in air intelligence, photo intelligence, and radar analysis in a period of thirty-two weeks.

A training program for intelligence specialist (1630/1635) officers on duty in ONI was established in October 1958; the objective was to train such specialists in the several fields of naval intelligence. It had been found that too many active duty and Naval Reserve intelligence officers were serving for extended periods in one type of intelligence duty and thus becoming specialists within a specialty. It was the policy of DNI RAdm. Laurence H. Frost that intelligence officers gain as broad a knowledge as possible of the various phases and fields of naval intelligence through training and rotation of duty.

The training program within ONI included self study (the correspondence course, *Naval Intelligence*, NAVPERS 10774), organized classroom training, and training opportunities in the Washington area.[39]

In 1961, it was stated that basic preparation for a career specialization in intelligence should begin at approximately the sixth year of an officer's commissioned service and include an appropriate period of postgraduate study in the principles, methods, and tools of naval intelligence. The only available source for the necessary education and training for intelligence officers at that time was the nine-month course at the Naval Intelligence School.

At the intermediate (lieutenant commander) level, it was believed there should be a period of study at a Group I service college (preferably the

Naval War College or the Armed Forces Staff College) in the theory of warfare and the military planning process. Functional training for specific assignments was to continue for officers at the intermediate level.

At the senior career level, those officers who displayed a potential for the most senior and responsible assignments in intelligence were to be sent to a Group I service college, preferably the Naval War College. At this level, except for intensive area briefing and language refresher courses for specific assignments, there would normally be no need for further functional training.

The foregoing represented an ideal career training pattern. Unforeseen and changing requirements continually caused deviations in the careers of intelligence officers. Regular Navy ensigns and lieutenants (junior grade) were rarely commissioned as intelligence specialists. Changes of designators usually occurred when the officers became senior-level lieutenants or junior lieutenant commanders.

A secondary source of personnel for filling intelligence billets in the early 1960s was the line officer (11XX and 13XX) subspecialist. Those officers were primarily oriented toward a general-line career. During their fourth to sixth year of service, they were selected from a volunteer list to attend the Naval Intelligence School. Upon graduation, they were assigned to the same types of billets as intelligence career specialists. The subspecialist officers normally were reassigned to a second tour in an intelligence billet later in their careers on a when-available basis.[40]

In 1962, the Naval Intelligence School was transformed into the Defense Intelligence School under the newly formed Defense Intelligence Agency. It continued to occupy the same dilapidated, wartime temporary buildings in Anacostia.[41]

Graduate-Level Training

In 1972, steps were initiated to establish a graduate-level course in naval intelligence. An objective of establishing a master's degree program was to fulfill the educational needs of naval intelligence for persons capable of developing systems analysis and computer techniques for intelligence research; having a broad understanding of technology and its defense applications, based on a non-engineering approach; and familiar with the national security structure and policy of both the USSR and the United States, with special emphasis on the Soviet ocean strategy and the Soviet navy. Another objective of equal importance was to attract talented young line officers to the intelligence subspecialty. The Defense Intelligence School was unable at that time to upgrade its level of instruction to meet either the standards for a master's degree or the educational requirements of Naval Intelligence. These conclusions were enunciated in DNI RAdm. Earl F. Rectanus's memo of 14 September 1972 to the Chief of Naval Personnel, and the recommendation was made that the course be set up at the Naval Postgraduate School at Monterey, California.[42]

After much correspondence and numerous conferences with the Director of Naval Education and Training, the Bureau of Personnel, the Naval Postgraduate School, and the Defense Intelligence School, approval for the course was obtained and ONI provided the funding. The first class convened in September 1973, the students having been selected administratively, since there hadn't been time to circulate a request for applicants.[43]

Chapter Notes

1. James R. Green, "The First Sixty Years of the Office of Naval Intelligence" (M.A. thesis, The American University, 1963), 47.

2. Ibid., 78; and *Navy Directory*, various editions.

3. Capt. Henri H. Smith-Hutton, Oral History, USNI, Annapolis, MD, 1979, 1:67, 175, copy in OA.

4. Ambassador William J. Sebald (Capt., USNR[R]), Oral History, USNI, Annapolis, MD, 1:120–23.

5. Capt. Henry S. Shields, USAF, "A Historical Survey of U.S. Naval Attachés in Russia: 1904–1941," MS, Defense Intelligence School, Mar 1970, 31–32.

6. ONI Administrative Officer (OP-16-A), *Annual Report to DNI*, 6 Jun 1933, 6.

7. Capt. David C. Nimmer, USMC, Report of 30 Aug 1934 to ONI, entry 195, RG 38, NA.

8. Department of the Navy, "Administrative History of the Office of Naval Intelligence in World War II," 10 Jul 1946, unpublished MS, 5a, hereafter *ONI WWII Admin History*.

9. Shields, "Naval Attachés in Russia," 33-34.

10. RAdm. Samuel B. Frankel, Oral History, USNI, Annapolis, MD, 1972, 47–48, copy in OA.

11. Capt. Joseph Finnegan, manuscript narrative, OA.

12. *Navy Directory*, various editions.

13. Smith-Hutton oral history, USNI, 1:318–19; and VAdm. Rufus L. Taylor, interview with author, 11 Oct 1975, tape recording in author's possession.

14. *Navy Directory*, 1 Apr 1941.

15. Smith-Hutton oral history, 1:319.

16. Navy School of Oriental Languages, "Administrative History," MS, first draft, 1–15, OA.

17. *ONI WWII Admin History*, 714.

18. CNO ltr, ser 11788P23, 1 Jul 1946, Accession 3770, box 1, ONI Day File, OA.

19. *ONI WWII Admin History*, 1394, 1398.

20. Bureau of Aeronautics (BUAER), " Administrative History of the Bureau of Aeronautics in World War II," unpublished MS, 1946, 19:24–25, hereafter *BUAER WWII Admin History*.

21. *ONI WWII Admin History*, 143–44.

22. *ONI Review*, Nov 1945, 39–40.

23. *ONI WWII Admin History*, 1208; and *BUAER WWII Admin History*, 19–20.

24. Ibid., 36–37.

25. ONI, "OPINTEL (Operational Intelligence) Notes," Jun 1945, OA.

26. *ONI WWII Admin History*, 1399–1400.

27. *ONI Review*, Sep 1946, 35.

28. Commo. Charles J. Rend (Deputy Chief of Naval Intelligence),

press conference of 21 Jun 1946, copy in author's files.

29. ONI-19(A), *Naval Intelligence Manual*, May 1947, para. 7104.

30. Chief of Naval Reserve Training (CNARESTRA) ltr, 11 Mar 1948.

31. Capt. Charles S. Melvin, USNR (Ret.), ltr to author, 4 Oct 1976. Melvin was an ACIO during World War II and was Air Intelligence Reserve Personnel Officer in ONI during the Korean War.

32. Ibid.; and Capt. Charles S. Melvin, interview by author, 11 Jul 1976.

33. Commander in Chief, Pacific Fleet, *Interim Evaluation Report No. 2, Korean War Naval Operations,* 213.

34. Deputy Chief of Naval Operations (OP-922Y3) memo, ser 000347P32, 27 Jul 1950.

35. OP-322Y3 memo to SECNAV, ser 000480P32, 11 Sep 1950.

36. SECNAV memo, ser 000481P32, 13 Sep 1950.

37. OP-322N7 memo, ser 724P32, 2 Jun 1952.

38. OP-922 ltr, ser 00063P92, 12 Mar 1956.

39. ONI Instruction 3870.26 of 17 Oct 1958.

40. OP-923P1C memo to OSD (SO), ser 000124P32, 8 Mar 1961, ONI Green Day File, OA.

41. *DNI Report to CNO, FY 1972*, 43.

42. OP-009M ltr, ser 1543P009, 14 Sep 1972.

43. Summary of copies of correspondence in author's files, OA.

CHAPTER 31

ONI and the Naval War College

There was much in common between the Office of Naval Intelligence and the Naval War College (NWC) in their purpose, origin, interests, and problems. Consequently, the two worked closely together during their formative years, and there was even a period of several years during which considerable thought and some pressure was applied to combining the two organizations to form a Navy general staff.

Though there is no known documentary evidence that Commo. Stephen B. Luce helped Lt. Theodorus B. M. Mason sell the idea of setting up an Office of Naval Intelligence, it is apparent that they were friends, and that each had an active interest in the U.S. Naval Institute and contributed to the early editions of its periodical, the *Proceedings*. They both recognized the need for and value to the Navy of information on foreign progress in naval science. It seems reasonable to assume that Mason may have at least had Luce's verbal support for his project.

The General Staff Concept

On 3 May 1884, Commo. Luce was designated the president of a board that was directed to consider and report on the subject of a postgraduate school to be established by the Navy Department. Cdr. William T. Sampson and LCdr. Casper F. Goodrich were the other members of the board.

In its report to Secretary of the Navy William E. Chandler dated 13 June 1884, the Luce Board expressed the hope "that every officer's useful attainments, such as foreign languages, sketching, photography, draughting, surveying, painting, naval architecture, etc., may form a part of his record at the Navy Department [so] that his fitness for any special work may be known and utilized." Similar guidance was given in the Secretary of the Navy's first directive to Lt. Mason on establishing ONI.

The Luce Board also recommended that optional courses in modern languages, watercolor painting, and photography be added at the prospective postgraduate school, as they would bring their own reward in foreign service as well as in military and naval reconnaissance.

The Naval War College was established by General Order No. 325, signed by Secretary Chandler on 6 October 1884, and its first class was convened during September 1885, with Commo. Luce as president and one of the principal lecturers. As in the case of ONI, many people opposed the establishment of the War College and contrived to impede its progress. With mutual assistance, however, ONI and the Naval War College were both successful in attaining their common objective of aiding naval officers to acquire a better understanding of naval science.

Beginning in 1887, ONI staff officers served as lecturers at the Naval War College courses. Although not on ONI's staff at the time, Lt. Charles C. Rogers gave four lectures in three successive years on the functioning of a general staff. The subjects covered in his lectures included Intelligence Branch, Intelligence Systems of Foreign Armies, General Consideration of Naval Intelligence Departments at Home and Abroad, The Meaning of Naval Intelligence in Detail, Reconnaissance, Reasons for a General Staff, and Essence of Intelligence Work in the Preparation for War. As a result of Rogers's presentations, the Naval War College understood the need for a general staff and, until 1916, advocated the establishment of one for the Navy.[1]

Other lecturers from ONI included Lts. Carlos G. Calkins, Washington I. Chambers, and John M. Ellicott, and LCdr. Richard Wainwright. ONI usually sent one or two staff officers as students to each year's course. They were able to add up-to-date information from ONI's studies that was pertinent to the courses being presented at the War College.

In 1889 LCdr. French E. Chadwick, Naval Attaché, London, sent back to ONI the first *Kriegspiel*, a German concept for wargaming as a method for teaching military tactics, and strongly advocated its use at the War College.

Professor James R. Soley, who was in charge of the Navy Department Library and was closely associated with ONI, was a frequent lecturer at the Naval War College. When Soley was appointed Assistant Secretary of the Navy in 1890, friends of the NWC were encouraged by the fact that the college was being placed under his supervision. ONI and NWC were organizationally together under the Assistant Secretary, as they had been previously under the Chief of the Bureau of Navigation.[2]

In the Secretary of the Navy's *Annual Report for 1896*, Secretary Hilary A. Herbert stated:

> A close union should be maintained between ONI and the War College, both working to the end of meeting all possible naval problems that may arise from any international difficulty, keeping all the time abreast with the actual facts and existing conditions of naval warfare.

NWC President Capt. Henry C. Taylor also stated in the annual report:

> By order of the Department, Lieutenant Commander Wainwright, Chief of the Office of Naval Intelligence, attended a portion of the college session. The desirability of close relations between these two institutions leads me to hope that each year we shall have one or two officers from the Office of Naval Intelligence in attendance upon the college sessions.[3]

In 1897, NWC President Cdr. Casper F. Goodrich reported in his annual report:

> It having been recommended to the Department by the President of the College and the Chief Intelligence Officer that officers of the Office of Naval Intelligence and the Naval War College should interchange at stated periods, Ensign J.V. Chase, of the college staff, was ordered to report to the Chief Intelligence Officer for duty in his office for one month from January 4 [1897]. At the completion of this duty, Ensign Chase returned to the college. When the college staff is recruited to its normal strength, it would be well to order Lieutenant [Joseph B.] Murdock to Washington for a month to repeat what Ensign Chase did last January.[4]

Another area of close association between ONI and NWC in the 1890s was in the field of war planning. ONI prepared the Navy's war plans in collaboration with the War College, and the latter would test them in wargaming exercises.[5]

On October 2, 1898, former NWC President Capt. Henry Taylor wrote to RAdm. Luce recommending that the War College and the Office of Naval Intelligence be gradually drawn together to form a general staff, but only if ONI was not hostile to the idea.

In February 1900, Taylor sent to Luce an undated and unsigned copy of a ten-page memo that he had prepared as a reply to an inquiry from Secretary of the Navy John D. Long on what should be done to develop a general staff. Taylor's memo stated:

> That in the development of the Intelligence Office and the War College, the Navy has been unconsciously forming the elements of a General Staff. . . . That the Secretary issue an order to the War College and the Office of Naval Intelligence that their work shall be regarded as directly connected and interdependent, and that the chiefs of the two institutions and their first assistants and the Chief of the Bureau of Navigation shall constitute a permanent board of five members, who shall meet frequently and consult as to war plans and information.
>
> That one half of the Intelligence Officer force shall pass four months of each year at the Naval War College, and one half of the Naval War College Force, four months at the Intelligence Office. That the combined work of the College and Intelligence Office [shall] be under the Chief of the Bureau of Navigation's general direction and orders.

Possibly as a result of the above guidance, and obviously in conformity with some of it, Secretary Long established the General Board in March 1900. Admiral of the Navy George Dewey was senior officer, and among the other eight members were the Chief Intelligence Officer (Capt. Charles D. Sigsbee), the NWC President (Capt. Charles H. Stockton), and the Chief of the Bureau of Navigation (Capt. Arent S. Crowninshield), with the last serving as chairman of the Executive Committee.[6]

In the first direct participation of the General Board in the work of the Naval War College, board members were present at the summer conference that met from 1 June to 30 September 1909. From the time of its establishment in 1900, the General Board had referred questions on strategic and tactical matters in numerous areas to the NWC staff for consideration and opinion.[7]

In 1901, Capt. Sigsbee stated:

> It is believed that still greater efficiency (in ONI work) would result if the natural relations existing among the General Board, War College and ONI were recognized by legislative action, enabling the Department to organize and adjust work on the systematic basis of a General Staff.[8]

Even when ONI had only five officers assigned (in June 1903), one was detailed on temporary duty at the Naval War College.

The Modern Era

Following World War II, intensive instruction in naval intelligence in relation to staff and command requirements was given in the junior and senior courses at the Naval War College. The college also conducted a correspondence course in naval intelligence. Officers of the regular Navy and Naval Reserve officers on active duty were allowed to take the course.

On 18 August 1947, an ONI team gave a series of talks at the War College as an "Introduction to Naval Intelligence." The team included Director of Naval Intelligence RAdm. Thomas B. Inglis who spoke on "The Organization of the Naval Intelligence Service"; Capt. Carl Espe, head of OP-32Y, who spoke on "Strategic and Operational Intelligence"; LtCol. T. L. Ridge, USMC, who discussed "Operational Intelligence Support to the Amphibious Problems"; and Capt. P. Henry, head of OP-32V, who covered "Air Intelligence."

From 1947 on, ONI has almost uninterruptedly provided lecturers to the college on an annual basis, usually the Director of Naval Intelligence and senior ONI officers. The NWC has also been on the distribution list for most ONI products. In due course, after the Navy adopted the practice of designating officers for intelligence duty only, at least one designated intelligence officer has been assigned as a student at the War College each year. When the availability of intelligence specialists permitted, an officer completing a year as a student would be extended for a year as a member of the college staff.

Commencing in 1958, at least one, and sometimes two, ONI civilian analysts attended the college, except for the courses commencing in 1969 and 1970.

Capt. Arthur F. Newell, Jr., USN (Ret.), was brought back on active duty on 1 April 1969 and became the Staff Intelligence Officer of the college. At the same time, NWC President VAdm. Richard G. Colbert was actively advocating setting up a series of military "chairs" for the various areas of naval warfare. Some of the first to be established dealt with air strike warfare, submarine warfare, and naval strategy.

In early 1971, Capt. Newell was instructed to prepare the paperwork to establish a military chair of intelligence. On 19 March 1971, Adm. Colbert signed a letter to RAdm. Frederick J. Harlfinger II, Assistant Chief of Naval Operations for Intelligence, converting the position of Staff Intelligence Officer to the Military Chair of Intelligence, requesting that ONI (OP-92) sponsor the chair, and proposing that the chair be named for RAdm. Edwin T. Layton, USN (Ret.). Others considered for the honor had included RAdm. Ellis M. Zacharias, VAdm. Rufus L. Taylor, RAdm. Roscoe H. Hillenkoetter, and RAdm. Samuel B. Frankel. Layton was judged to be most deserving of the honor, based on his success as Adm. Nimitz's Fleet Intelligence Officer throughout World War II, his becoming the first intelligence specialist to achieve flag rank on active duty, and his service as intelligence officer (J-2) of the Joint Staff.

The duties of the person who held the intelligence chair involved the same close liaison with ONI as had been the case for many years. The establishment of military chairs merely formalized the relationship.

On 22 April 1971, RAdm. Harlfinger, as ACNO (Intelligence), accepted the sponsorship of the Intelligence Chair and concurred with the choice of RAdm. Layton as the officer for whom the chair should be named. Capt. Newell became the first occupant of the Intelligence Chair and served until he retired on 30 December 1971. He was succeeded by Capt. Lewis Connell, USN.[9]

Chapter Notes

1. *SECNAV Annual Report, 1888*; and Adm. Austin M. Knight and William D. Puleston, *History of the Naval War College* (Newport: Naval War College, 1916), 1887 section, 2.

2. Knight and Puleston, 1890 section, 1.

3. *SECNAV Annual Report, 1896*, 46.

4. *SECNAV Annual Report, 1897*, 135.

5. Capt. Damon E. Cummings, *Admiral Richard Wainwright and the United States Fleet* (Washington: GPO, 1962), 74.

6. RAdm. Stephen B. Luce Papers, Manuscript Division, Library of Congress.

7. Knight and Puleston, *History of the Naval War College*, 1909 section, 1; and 1900 section, 1.

8. *SECNAV Annual Report, 1901*, 584.

9. Richard G. Colbert ltr to RADM Frederick J. Harlfinger, 19 Mar 1971; and Harlfinger ltr to Gilbert, 22 Apr 1971, Naval War College file, author's files, OA.

CHAPTER 32

Naval Intelligence and the Spanish-American War

In 1895, when the rebels in Cuba started their effort to achieve independence from Spain, the Office of Naval Intelligence began keeping its files as up-to-date as possible on Spain's navy and on Spanish ports throughout the world. During 1896–1897, ONI's working files were cleared of obsolete material and reorganized so that a limited number of temporary civilian and naval personnel who had had no familiarity with them could quickly find information requested by the Navy Department and the operating forces. To the extent that it was able, with the limited personnel and facilities at its disposal and no appropriated funds, ONI was ready for the Spanish-American War.

At the time of the sinking of the *Maine* on 15 February 1898, several ONI staff officers had already been ordered to sea duty without replacement. When Congress declared war on Spain on 25 April, there remained in ONI four active officers, plus a retired ensign, Edward E. Hayden, who had been recalled to active duty on 23 April. Also on 25 April, Capt. John R. Bartlett, USN (Ret.), reported as the relief for Chief Intelligence Officer Cdr. Richardson Clover. Bartlett relieved Clover on 1 May. By 1 July, only Bartlett and Hayden, plus the usual borrowed clerks and messengers, remained in ONI.

The Naval War Board was established on the day that war was declared. It was an outgrowth of an informal advisory board for the Secretary of the Navy that had been in existence for some time. RAdm. Montgomery Sicard was president of the board, and the other members were Capt. Alfred Thayer Mahan, USN (Ret.), who had been ordered back to active duty on 25 April; Capt. Arent S. Crowninshield, Chief of the Bureau of Navigation; Cdr. Clover, until he left for sea duty; and Lt.(jg) Alphonso H. Cobb, who acted as secretary of the board. Assistant Secretary of the Navy Theodore Roosevelt was a member of the board until 9 May, when he severed his connection in preparation for resigning as Assistant Secretary on 10 May.

The Naval War Board was in session daily throughout the period of hostilities, and the Bureau of Navigation, to which ONI was again subordinated, was open day and night to receive and act on information, keeping the board members advised. Most of the cables to and from naval attachés were in cypher, and it was ONI's responsibility to cypher and decipher this message traffic. Such additional duty led Capt. Bartlett to recommend that officers be added to the staff of ONI to perform only cypher duties in future periods of hostility. During the five months of the war, approximately 300 outgoing and 800 incoming cable messages were processed, aggregating approximately 34,500 words.[1]

As in most wars, the prime question for the Navy was, where are the enemy's naval forces and what are their capabilities and intentions? At the start of the Spanish-American War, the Spanish had three major forces: one in the Philippines under RAdm. Montojo, one in Spain under RAdm. Camara, and one assembling in the Cape Verde Islands under RAdm. Cervera. The first force was expected to remain in the Philippines, but the question was, where? Camara's force could go east or west or remain in Spain, and Cervera was apparently coming west—but what was his intended destination?

The naval resources immediately available to collect information to answer some of these urgent questions included the operating forces and the naval attachés, particularly Lt. William S. Sims in Paris and Cdr. Francis M. Barber, USN (Ret.), in Berlin. Sims had a few previously established sources in Spain, and Barber had an agent, an American named Dr. Breck, who traveled through Spain masquerading as a German. There were also U.S. consuls in various foreign ports. Dr. Breck sent in several good reports. At the end of the war, he wrote up his experiences and sold the story to *Cos-*

mopolitan magazine, where it was published anonymously in the November and December 1898 issues.

When Commo. George Dewey was ordered in 1897 to take command of the Asiatic Squadron, he visited ONI for information about the Philippines and found that the most recent material available was dated 1876. No U.S. Navy ship had been to the Philippines in several years, and there were no naval attachés posted to the Far East. Accordingly, Dewey bought every book he could find on the Philippines and took them with him when he departed the United States on 7 December 1897.[2]

Upon relieving RAdm. Frederick V. McNair as commander of the Navy's forces in the Far East at Nagasaki, Japan, on 3 January 1898, Dewey found that the files of the Asiatic Squadron were also devoid of recent information on the Philippines, other than some press accounts of the rebellion in progress. There was no official information relating to the rebellion or indicating to what degree American interests were involved.

After completing protocol and diplomatic formalities, including an audience with the Emperor of Japan, Dewey took his flagship, the protected cruiser *Olympia*, to Hong Kong in order to be at the nearest port and cable station external to the Philippines. He made the move on his own initiative, in anticipation of an imminent requirement to attack the Spanish naval force based in the islands. Hong Kong was also the most promising point from which to gather recent information on the Philippines and on Montojo's naval force.[3]

Dewey's forces arrived in Hong Kong on 17 February, and, on 25 February, he received a telegram from Assistant Secretary of the Navy Theodore Roosevelt ordering his squadron, except the side-wheeler river gunboat *Monocacy*, to Hong Kong and instructing him to prepare to prevent the Spanish squadron from leaving the Asian coast in the event of war with Spain.

Dewey telegraphed the U.S. Consul at Manila, Oscar F. Williams, to obtain all possible information on Spanish fortifications, underwater mines, and other defenses of Manila, and to advise on any movements of the Spanish fleet. Williams performed his intelligence collection duties well and, in spite of threats on his life, stuck to his post and advised Dewey concerning six new guns on Corregidor, submarine mines and their cable connections to Corregidor, the number of naval ships and other vessels in Manila Bay, and feverish activities in preparing fortifications.

Other intelligence sources exploited from Hong Kong included an American businessman employed by a firm in Hong Kong who made periodic trips to Manila and patriotically checked on specific matters as requested. Commo. Dewey's aide, Ens. Frank B. Upham, impersonating an inquisitive traveler, picked up valuable items of information by going on board steamers in Hong Kong as they arrived from the Philippines. From these reports and those from the consul in Manila, Dewey felt he had a valid estimation of the Spanish defensive resources for the port and the strength of their squadron.

On 31 March, Dewey sent to the Navy Department his first letter since assuming command, declaring his readiness for combat and giving an estimate of the Manila situation. As he found out one month later, he had considerably underestimated both the number of Spanish ships and the number and caliber of guns in the shore defenses. He was correct, however, in the most important point: "I believe I am not overconfident in stating that with the squadron now under my command, the vessels could be taken and the defenses of Manila reduced in one day."

The local Hong Kong papers kept proclaiming the impregnability of Manila, the vast extent of its mine fields, and the great strength of its forts. If the morale of Dewey's squadron had not been unshakeable, it could have been adversely influenced by the psychological barrage. Large bets were offered at Hong Kong clubs that the squadron would not return from Manila; its impending mission was an open secret.

Beginning on 15 April, Consul Williams was advised repeatedly to come to Hong Kong, but it was not until 23 April that the British consul in Manila cabled that Williams had safely left. When Dewey received the news, he notified the Navy Department that he would move to Mirs Bay, 30 miles from Hong Kong, to await Consul Williams if war was declared and the squadron was required to leave Hong Kong for neutrality reasons before the consul arrived.

On 23 April, the acting governor of Hong Kong issued a neutrality proclamation requiring the ships of belligerents to leave the waters of the colony by 4:00 p.m. on 25 April. In consequence, Dewey's ships duly shifted to Mirs Bay, some of them on the 24th and the remainder on the 25th. Communications between Hong Kong and Mirs Bay were maintained by commercial tug.[4]

On 25 April at 1900 local time, Dewey received Secretary Long's message of 24 April:

> War has commenced between the United States and Spain. Proceed at once to Philippine Islands. Commence operations particularly against the Spanish fleet. You must capture vessels or destroy. Use utmost endeavor.[5]

Consul Williams boarded Dewey's flagship, *Olympia*, in Mirs Bay, Hong Kong, at 1300 local time, 27 April, and the fleet was underway for

Manila at 1400. On 30 April, having been advised that the Spanish fleet might be at Subic Bay, Dewey sent the cruisers *Boston* and *Concord* to reconnoiter Port Subic, but no fleet was found there. Dewey was also advised that the entrances to Manila Bay were obstructed by mines, torpedoes, and land defenses. The squadron, however, transited the south channel that night undamaged. In fact, Dewey's fleet went undetected until it was almost beyond the range of the Corregidor defenses and other heavy guns at the entrances to Manila Bay.[6]

It should be pointed out that, aside from the intelligence material about the naval ships of Spain and other navies that ONI had forwarded in its normal distribution to all U.S. Navy ships and squadron commanders on a regular and routine basis during the previous years and the information collected while his squadron was at Hong Kong, Dewey received no further intelligence support after leaving Hong Kong. Any other reports would have had to come by cable to Hong Kong and then by dispatch boats to Dewey, wherever he might be. Needless to say, none reached him until after the Battle of Manila Bay.

RAdm. Montojo, on the other hand, had access to cables reaching Manila, including information from the Spanish consul in Hong Kong, who reported accurately that the "Enemy's Squadron sailed at 1400 [27 April], from Bay of Mirs, and according to reliable accounts they sailed for Subic and then will go to Manila." Montojo was probably also alerted when Subic Bay was reconnoitered on 30 April. Thus, he shouldn't have been surprised when he found the U.S. squadron in Manila Bay at daybreak on 1 May.[7]

The Spanish force was thoroughly defeated by lunchtime, but it took until 7 May for Dewey's report of victory to reach Washington via dispatch boat to the Hong Kong cable office.

The Far East naval action in the Spanish-American War showed the importance of providing basic intelligence to the operating forces and for keeping it up to date during peacetime so that it is ready for immediate use upon the commencement of hostilities. Dewey demonstrated an appreciation for the need of having the best possible current intelligence in hand before going into battle, and he also displayed the proper regard for intelligence of unknown reliability. In his operational decisions, he assumed each intelligence report might be true or false, whichever was worse for his objectives, until he could prove otherwise. From the intelligence that he had been able to gather while in Hong Kong, and from ONI reports on the identification and capabilities of the ships of the Spanish fleet, Dewey thought he had the superior force. But when he left Hong Kong, Dewey didn't know whether or not Montojo had remained at Manila during the week since Consul Williams had collected his latest information.

Beginning on 20 May, Dewey received a series of cabled reports from Secretary Long on the movement of RAdm. Camara's squadron from Spain, possibly towards the east. On 22 June, Camara's fleet was reported to have progressed as far as Cartagena, with its ultimate destination still uncertain. But on 25 June it was reported that the Spanish fleet had passed Cape Bon, Tunis, was heading east at 1500 on 22 June, and consisted of fifteen ships. On 27 June, Secretary Long reported to Dewey the arrival of the Spanish fleet at Port Said on the previous day, giving a much more complete report on the identity of the ships, their condition, and their armament. Much of Long's report probably came from Ens. Buck (whose exploits are discussed later), although it was also supplemented by reports from other sources. On 5 July, the Spanish fleet was reported transiting the Suez Canal after having coaled in the Mediterranean. On 8 July, however, Camara and his fleet were recalled to Spain because of a belief in Madrid that the United States was going to send a force to attack Spain, and Secretary Long's series of reports to Dewey ended.[8]

In the Atlantic theater of the war, public pressure, whipped up by the press, made it necessary for the Navy to assume, until intelligence information proved otherwise, that both Camara and Cervera would attack the U.S. East Coast. To get the information it required, the Navy Department took two separate and specific actions, placing ships on a scouting line across the most logical Spanish approach route and dispatching covert observers to Europe to follow the Spanish naval forces, wherever they might go. The latter effort was a backup measure that, from hindsight, was unnecessary but would have been important if other intelligence resources had failed and if the Spanish force under Cervera had delayed its transit west or had backtracked.

On 30 April, two volunteers from the Bureau of Navigation, Ensigns William H. Buck and Henry H. Ward, departed the United States with orders to follow Camara and Cervera, respectively, and to keep the Navy Department informed. On Sunday, 8 May they arrived at Liverpool, where they both assumed the names and identities of British subjects. Two yachts had been purchased and provisioned for them, and, after consultations on 12 May with Lt. John C. Colwell, USN, the U.S. Naval Attaché, London, Ward left Liverpool on "a pleasure cruise." Buck followed the next day in the other yacht.

After leaving Liverpool, Ward stopped briefly at Brest to advise the Navy Department (via a cover address) of his anticipated schedule, his progress, and the few bits of news he had picked up there.

Then he proceeded to Lisbon, where Buck had already arrived. The information they picked up in Lisbon was mainly from newspapers and other unreliable sources, and it conflicted both with itself and with what Ward had learned, also from newspapers, at Brest. The ensigns could not determine whether or not Cervera had departed Cape Verde and, if he had, whether all of the ships that had reportedly assembled there had left with Cervera.

A cable from the Navy Department advised of a report that Camara's force was at Cadiz and about to depart for the United States. The news induced Ward to set out for Cadiz, leaving Buck at Lisbon to follow later. A second cable from the department, in response to Ward's message from Brest, directed that neither Ward nor Buck should go to Cadiz. Ward, however, had already left for that port. Buck canceled his planned call there and proceeded directly to Gibraltar, from where he was able to keep track of Camara's movements. In due course, Buck followed Camara's fleet to Suez and back to Spain, keeping the Navy Department informed by cable of all its movements. With the return of Camara's fleet to its bases, Buck's service was completed, and he left his yacht at Gibraltar and returned to the United States.

Ward was delayed at Cadiz by heavy weather, which gave him time to reconnoiter Cadiz completely, including the dockyard, and to identify all ships present. Having accomplished this task, he proceeded to Gibraltar, where he arrived on 23 May and cabled his information to the Navy Department. Then on 27 May, he headed west via Tangier and Madeira, arriving at St. Thomas on 11 June. Ward inspected Culebra and Crab Islands and then cruised along the north shore of Puerto Rico to San Juan. He was met by an armed tug off the harbor entrance and, after a preliminary examination by the Spanish, was allowed to enter the port, where he was boarded by port officers. After inspecting his papers, they informed him that he must not leave the harbor or communicate with the shore. The Spanish destroyer *Terror* anchored about 100 yards on his port beam and placed a picketboat to starboard.

Ward had requested the Spanish boarding officer to request the British consul to come out, but the acting vice consul came, accompanied by the same Spanish officials who had been on board before. After a half-hour reinspection of his papers and close cross-questioning, the Spanish set a time of departure of only a few hours later, released the yacht, and allowed Ward to go ashore (accompanied, however, by the Spanish officials). A Spanish naval officer with him suggested a call on the naval commandant, which was made and assisted Ward in confirming his adopted British identity.

After a few hours' stay, which enabled Ens. Ward to establish that the only ships present were the gunboats *Isabel II* and *General Concha*, the destroyer *Terror*, and two tugs, each armed with two six-pounders, he left for St. Thomas to send his report. The remainder of Ward's cruise was spent in checking ports eastward as far as Martinique and Barbados. On 26 June, Ward learned that Camara's force had moved eastward in the Mediterranean, resolving the remaining question from among those encompassed in the original purpose for his cruise.[9]

On 9 July, Ward was directed by cable to go to Curacao and Aruba to investigate reports of the assemblage there of Spanish supplies, which he found to be incorrect. On 17 July he proceeded by mail steamer to New York.

To cover the eventuality that sources in Europe, including Ward, would not be able to keep in touch with Cervera and his force, the Navy Department deployed the auxiliary cruisers *Harvard*, *Yale*, and *St. Louis* (formerly the passenger liners *New York*, *Paris*, and *St. Louis*, respectively, all capable of 20 knots and possessed of favorable steaming endurance) to patrol across the most logical route from the Cape Verde Islands to the West Indies. On 29 April, the *Harvard* was directed to proceed from Tompkinsville, New York, to eastward of the Windward Islands to patrol a line extending between 15°38'N, 59°40'W and 14°25'N, 59°30'W. The *St. Louis* was assigned a patrol line between 16°55'N, 59°50'W and 15°38'N, 59°40'W, which was an extension of the line assigned to the *Harvard*. The *Yale* was directed to cruise around Puerto Rico. All of the auxiliary cruisers were to watch for the Spanish fleet, and all were to report any relevant information by telegraph to the Navy Department and to the Commander in Chief, North Atlantic Station, RAdm. William T. Sampson.[10]

Capt. Charles S. Cotton, commanding the *Harvard*, reported on 11 May from St. Pierre, Martinique, that the Spanish torpedo-boat destroyer *Furor*, one of Cervera's ships, had arrived at Fort de France, Martinique, at 1600 and had departed shortly thereafter. In a cable sent on 12 May, Cotton reported that the Spanish destroyer *Terror*, also of Cervera's force, was at Fort-de-France, and that five large ships, hull down to westward, had been seen from Fort-de-France but not identified. Finally, on 13 May, Cotton cabled the department that he was blockaded by the Spanish fleet in St. Pierre.

On 13 May, as a result of Cotton's messages, Secretary of the Navy Long sent orders to the *St. Louis* to take the word to Sampson, who was off San Juan, that the Spanish squadron had been off Martinique on 12 May. Long also ordered Commo. Winfield Scott Schley and the "Flying Squadron," which had been positioned at Hampton Roads, Virginia, where it could be visible protection for the East Coast in case of attack by the Spanish fleet, to proceed to Charleston, South Carolina, and to be ready to reinforce either Sampson or the blockade of Cuba.

More auxiliary cruisers were sent on 13 May to scout the various routes that Cervera might take from Martinique. The *St. Paul* was to patrol between Morant Point, Jamaica, and the west end of Haiti, and the protected cruiser *Minneapolis* was assigned to cruise between Monte Cristi Island, Haiti, and Caicos Bank in the Bahamas. On 14 May, however, the U.S. Consul at Curacao reported Cervera's force arriving there for coal, and the Navy Department also received a report from London, dated 10 May, that British colliers were probably to meet the Spanish fleet off the north coast of Venezuela.[11]

Secretary Long therefore sent a telegram on 14 May to Schley instructing him to continue on to Key West, and Long notified Sampson about the location of the Spanish fleet at Curacao and of Schley's move to Key West. Long also canceled his orders of 13 May to *St. Paul* and directed that the ship proceed to Key West to be available for dispatch service.

The main problem faced by Secretary Long in trying to pass intelligence information to Sampson at sea was that he was never quite sure through which cable station he could reach the fleet commander or whether there would soon be a dispatch ship calling at the cable station to carry telegrams to the battle force at sea. Sampson repeatedly advised the Secretary that he wasn't receiving any information on the Spanish ships. Consequently, the same information was sent to him through several different communication channels, and eventually he received it. Sampson was obviously and understandably impatient to get some authoritative and timely information on the exact location of all of the Spanish ships, but so was Secretary Long.

Even when Cervera took his force into the enclosed harbor at Santiago de Cuba on 19 May, no U.S. Navy ship saw him enter, nor could it be confirmed from seaward that any of Cervera's ships were present. Schley left Key West on 19 May for Cienfuegos and arrived there on 21 May. After the Navy Department received the report on Cervera's probable arrival at Santiago, it became a problem to get the word to Schley and to have him move his force east to Santiago. On 24 May, Schley reported that he had ascertained that the Spanish fleet was not at Cienfuegos and that he would move east on the 25th.

On 25 May, the Navy Department ordered Schley to contact Cuban insurgents 5–6 nautical miles west of the entrance to Santiago harbor and to obtain their help in observing the harbor from the surrounding heights to determine which Spanish ships were present. On 28 May, Schley acknowledged receipt of the order, but, because all his ships were low on coal and the weather would not permit coaling at sea from colliers, he could not remain off Santiago.[12]

It was not until 10 June that Sampson directed *Suwanee* to carry out the surveillance of Santiago harbor. The insurgent headquarters was found to be at Aceraderos, 19 miles west of the entrance to the harbor. Lt. Victor Blue of *Suwanee* contacted the headquarters to obtain a guide. In consultation with several scouts who had recently returned to the encampment, Blue was informed that the best place to get through the Spanish lines was to the north of Santiago and that it would take two or three days to make the trip.

Several insurgent officers volunteered to serve as Blue's guide, but only one was selected. It was insisted that Blue would have to travel as a Cuban. The insurgent general loaned him a mule. Blue and his guide started out on 11 June, and, by following mule paths and mountain streams and avoiding Spanish pickets and troop concentrations, they reached the camp of an insurgent battalion on the Santiago-Manzanillo road by sunset. Blue obtained new guidance there on the best route for reaching his objective. Three insurgent soldiers were added to the party, and it resumed its movement through dense forest and swampy jungle until more open country was reached. The guide then halted the group for rest and to await daylight.

Off again at daybreak, with the help of local Cubans, the scouting party evaded the Spanish sentries and reached the crest of a hill from where the bay could be observed. Some of Cervera's ships could be identified, but some that should have been there could not be seen from this vantage point. Moving to a higher hill and then to another point nearer the sea finally permitted a complete view of the harbor and an accounting of all of Cervera's ships.

The party then made its way back to the insurgent headquarters late on the night of 12 June. In the morning of 13 June, *Suwanee's* whaleboat picked up Blue and his authoritative information. RAdm. Sampson and Secretary Long soon received the accurate and complete intelligence information on the current location of Cervera's ships that they had been seeking since 25 April. Sampson was able to position his fleet properly for the Battle of Santiago, which terminated the Spanish naval threat in the western Atlantic.[13]

Chapter Notes

1. *SECNAV Annual Report, 1898*, 2:33.

2. Laurin H. Healy and Louis Kutner, *The Admiral* (Chicago and New York: Ziff-Davis, 1944), 140; and Nathan Sargent, *Admiral Dewey and the Manila Campaign* (Washington: Naval Historical Foundation, 1947), 7.

3. Sargent, *Admiral Dewey*, 7.

4. Ibid., 9–21.

5. Ibid., 22.

6. *SECNAV Annual Report, 1898* 2:70, 87.

7. Ibid., 89.

8. Ibid., 99–110.

9. RAdm. French E. Chadwick, *The Relations of the United States and Spain: The Spanish American War*, 2 vols. (New York: Scribners, 1911), 2:359–62.

10. *SECNAV Annual Report, 1898*, 2:360.

11. Ibid., 383–86.

12. Ibid., 395, 397.

13. RAdm. Victor Blue, "The Sighting of Cervera's Ships," *USNIP*, Sep 1899, vol. 25, no. 91, 585.

CHAPTER 33

Public Relations

As a result of the unfavorable publicity for the Navy that followed the 1921 aircraft bombing tests by the U.S. Army against unmanned obsolete battleship hulks anchored in Chesapeake Bay, the Navy recognized that it needed a means to counter the possible misinterpretation of information about Navy operations and programs. Secretary of the Navy Edwin Denby issued a directive on 21 February 1922 establishing the Navy Department Information Section within the Office of Naval Intelligence.[1]

ONI had been responsible for press censorship during World War I and, when the Information Section was set up, the Director of Naval Intelligence (DNI) also held the title of Chief Naval Censor. It apparently seemed logical and not inconsistent to make the same organization responsible for both the protection and the release of information.[2]

Cdr. Ralph A. Koch was designated the first head of the Information Section, and he took over the new organization in February 1922. As assistants, he obtained Lts. John B. Heffernan and William F. Dietrich, and one civilian.

On 1 March 1922, the Secretary of the Navy directed all bureaus and offices of the Navy Department to assign an officer and the necessary clerical help to assist the Information Section of ONI. A similar letter directed fleet commanders, naval district commandants, and commanders of overseas stations to designate an officer to collect information and pictures from ships and stations to be forwarded weekly to ONI.[3]

On 15 March 1930, noting the lack of results of service-wide information programs, especially throughout the lower echelons of command, DNI Capt. Alfred W. Johnson wrote the commandants of naval districts to emphasize the importance of good press relations. The letter provided a statement of Navy policy enunciated by Secretary of the Navy Curtis D. Wilbur on 6 October 1928 directing the Navy "to furnish the public with full information on the Navy not incompatible with military secrecy, including its activities at home and abroad, its educational features, and its contributions to science and industry."

On 17 November 1930, Secretary of the Navy Charles F. Adams issued a directive to all bureaus and offices outlining the duties of the Information Section. The Secretary endorsed the Wilbur policy and stated that the public relations effort was to be accomplished by cooperating with radio broadcasting agencies, motion picture, photographic, and newsreel companies and by complying with requests made by such agencies and companies for general information about the Navy.

> The Information Section is governed by the following approved principles, applicable throughout the Navy Department: (a) To avoid any discrimination in dissemination of news; (b) To ensure no statements derogatory to, or critical of, other branches of the government; (c) Neither to enunciate nor to comment upon policies.

According to the directive, the Secretary of the Navy would determine public information policies and that questions involving the supply of naval information would be referred to the Information Section of ONI.[4]

A SECNAV memo to all Navy bureaus and offices, dated 1 August 1935, called attention to the fact that reporters who had been refused information by the "Public Relations Branch" of ONI on advice of the bureau or office concerned had subsequently been scooped by a rival paper when it obtained the desired information from direct contact with an officer or employee of the same bureau or office. (Several internal ONI memoranda refer to the former Information Section of ONI as the Public Relations Branch during the 1932–1935 period.) The memo directed all heads of bureaus and offices to take steps to ensure that information for publication be cleared through the Public Relations Branch (Press Section) of ONI.[5]

In the summer of 1939, ONI issued *Training Instructions for Public Relations Personnel of the Naval Intelligence Service,* a 300-page manual containing monographs dealing with various aspects of public relations organization and practice.[6]

When Cdr. H. Raymond Thurber reported to the Public Relations Branch on 16 July 1940 as officer in charge, he was directed "to build the office up for an emergency." Thurber's office had two other regular Navy officers who had recently reported, two Naval Reserve officers, two experienced civilian assistants, four civilian clerical personnel, and two enlisted Marines.[7]

War plans for the Public Relations Branch, written in 1924 and revised through 1939, called for a rear admiral as director, a deputy director, and press, radio, photographic (stills), motion picture, and general information sections. Provisions were made for liaison with Navy public relations branches in the naval districts and with Navy public relations representatives afloat.

The mission of the Public Relations Branch was derived from official U.S. Navy policy and was intended to make available to the public—through press, radio, pictorial, and other media—all information concerning the Navy that was compatible with military security in order to inform the public of the activities and conditions of the Navy.

To accomplish its mission, the Public Relations Branch under Cdr. Thurber updated its war plans organization to include a director, an assistant director, and sections for administration, plans, press, radio, pictorial, scripts, reference, civic liaison, and naval districts. The first step towards building up the office was to obtain additional experienced personnel. Records of naval reservists slated for wartime duty in public relations were reviewed, and tentative selections were made of those who would head the various sections. Selected candidates were interviewed and asked to enter active service, if permitted by their personal situations. There was no legal requirement that they do so in 1940, but this recruitment procedure was effective nonetheless.

More office space for the Public Relations Branch was acquired in August 1940 outside the regular ONI spaces in an area more accessible to the press.[8]

In December 1940, the Press Section commenced issuing "Navy Radio News" to the fleet and outlying stations. The United Press had been supplying a news digest to the Navy Department communications watch officer, but in October 1940 United Press desired to terminate the service. Cdr. Thurber negotiated a continuation in modified form. The Press Section watch officer received the digest and edited and augmented it with news of particular Navy interest.

The foundations of the Pictorial Section were laid by an art project. A number of artists, illustrators, and etchers were induced to start the shore-based phase. Vernon H. Bailey was obtained in the spring of 1941 to produce a comprehensive record of shipbuilding at the start of what became the Navy Combat Artist Group, whose works have since been displayed throughout the country. The Pictorial Section supplied material for Navy recruiting posters, "spy" posters being prepared by the Society of Illustrators, and a "Think American" series of posters.[9]

On 28 April 1941, a SECNAV directive severed the Public Relations Branch from ONI and placed it directly under the Secretary. All personnel were shifted to the Secretary of the Navy staff; Cdr. Thurber was assigned as acting director.

On 9 May 1941, the Chief of Naval Operations directed naval district commandants to transfer their public relations offices organizationally from the cognizance of the district intelligence officer and to set them up as separate activities directly under the commandant's control.[10]

The growth of Navy public relations during 1940–1941 added to the volume of work placed on the Security Section (OP-16-B-4) of ONI, which was—and continued to be—responsible for the security and clearance for all Public Relations Branch projects. Conveniently located adjacent to the Public Relations Branch, the Security Section and was in the charge of Cdr. John S. Phillips, with LCdr. Edwin S. Earnhardt, USN (Ret.), as his deputy. The cooperation of Phillips and Earnhardt in providing timely clearance of material intended for publication was outstanding, and their suggestions for saving time were considered invaluable to the Public Relations Branch.[11]

Chapter Notes

1. F. Donald Scovel, "Helms A'Lee: History of the Development of the Public Affairs Function in the U.S. Navy, 1861–1941" (M.A. thesis, University of Wisconsin, 1968), 81–85.

2. Capt. Herbert E. Cocke, USN, "History of ONI," MS, Office of Naval Intelligence, 1931, 28; and Department of the Navy, "Administrative History of the Office of Naval Intelligence in World War II," 10 Jul 1946, unpublished MS, 1076.

3. Scovel, "Helms A'Lee," 85.

4. Ibid., 110. Johnson's 15 Mar 1930 letter was OP-16-F ser A8-3/A7-1; Adams's 17 Nov 1930 directive was ser EN 3(13)/A-1(1) (301117).

5. Scovel, "Helms A'Lee," 118.

6. Ibid., 124.

7. Ibid., App. A, 116.

8. Ibid., 167–68; App. A, 167.

9. Ibid., App. A, 171–80.

10. Ibid., 129–30.

11. Ibid., App. A, 183.

Chapter 34

Operating Forces

This chapter deals with the intelligence activities carried on by Navy operating forces until the start of World War II and with certain general directives that governed fleet intelligence activities in the post–World War II era. It is closely linked with Chapter 18, concerning operational intelligence, although the latter relates mainly to intelligence support from the Office of Naval Intelligence to the operating forces. Subsequent chapters deal with fleet intelligence during World War II in specific geographical areas. Together this chapter and these subsequent chapters cover activities by the operating forces to fulfill their own intelligence needs as well as their efforts to fulfill collection requirements placed on the operating forces by ONI and higher authorities to meet national-level requirements.

In addition to the chapter on operational intelligence, Chapters 2, 4, 5, 6, and 32 contain information on intelligence activities of the operating forces.

Beginnings of Fleet Intelligence

Lt. William S. Sims arrived on the China Station in the cruiser *Charleston* in 1894. He had been appointed by the commanding officer to be the ship's intelligence officer even though Sims professed to knowing nothing of intelligence work. His commanding officer brushed the excuse aside with the reply, "Neither do any of us."

The Sino-Japanese War was in progress, and it had drawn an unusually large number of neutral warships to the area. Sims gathered information on each of the neutral naval vessels present and submitted his reports to ONI. The war itself was demonstrating numerous significant lessons on the use of modern weapons in naval warfare. Following the Civil War in the United States, advances in ship design, ordnance, and armor had been made in Europe. The Sino-Japanese War provided the first opportunity to observe the value of many of these innovations.

From a British report about the Yalu River battle that Sims had obtained, he reported to ONI on the ability of the Chinese ironclads, or battleships, to withstand the withering fire of the Japanese cruisers. He also noted that modern shells, especially from secondary batteries, could set woodwork afire very easily. Other reports of special interest related to methods used by the British protected cruiser HMS *Crescent* during 6-inch gun target practice. The ship had a new type of gunsight that permitted continuous aim. One gun fired twenty-four shots in three minutes, obtaining eighteen hits, a record far superior to the target practice results being obtained by the U.S. Navy at that time. The event seems to have provided the impetus for Sims's subsequent efforts to get the U.S. Navy to adopt more effective fire control methods.[1]

In October 1901, Lt. Sims joined the staff of Commander in Chief, Asiatic Fleet Adm. George C. Remey in the armored cruiser *Brooklyn* as aide "with special intelligence duties." Sims had arrived in the Far East on board the new battleship *Kentucky* (BB 6) in 1900, having come directly from duty as Naval Attaché, Paris. While engaged in his attaché duties, Sims's observations further convinced him that American naval gunnery was far less effective than that of the great foreign powers, and he repeatedly reported as much in strong terms. Because of his derogatory reports about the U.S. Navy, Sims was not ordered back to Washington en route to his next assignment. This development was contrary to the usual practice of having naval attachés review and discuss their reports at the ONI offices immediately following their foreign duty. Adm. Remey was in sympathy with Sims's efforts to stimulate improvement in the Navy's gunnery and gave him a free hand in reporting his observations.[2]

In November 1901, *Brooklyn* visited Vladivostok, and Sims submitted reports to ONI on the Russian warship *Gromovoy* and on the defenses of the Russian Far East city. He commented particularly on the ship's unusually heavy armament at the expense of protection to the guns and, especially, to the ammunition supply. Yet he judged *Gromovoy's* inadequate protection to be superior to that in contemporary American cruisers.

To obtain information on the defenses of Vladivostok, Sims selected and briefed two young ensigns, assigning each of them one side of the harbor from which to observe specific points of interest. One of the ensigns made his observations without any difficulty, getting information on coast defense guns and their location, caliber, arcs of train, etc. The other did equally well but was apprehended in a guarded area as he was returning to town at the end of the day. The young officer was questioned at the Russian military headquarters and made to trace his hiking route on a military map (which he was able to study and subsequently report about when he got back to the ship).[3]

RAdm. Frederick Rodgers, Commander in Chief, U.S. Naval Force, Asiatic Station, reported to the Bureau of Navigation for the period 20 March–30 June 1902: "A considerable amount of intelligence duty in connection with naval ordnance and target practice has been performed by the intelligence officer and inspector of target practice [Sims], and some important reports have been made by officers attached to vessels of the fleet."[4]

Fleet Intelligence During World War I

In 1917 Sims, then a rear admiral and Commander, U.S. Naval Forces Operating in European Waters (COMUSNAVFOREUR), selected London as the location for his headquarters because the predominant naval effort in the war was British. Also, the highly efficient Intelligence Division of the British Admiralty received all important naval information, which, in turn, was made available to Adm. Sims and his staff. Consequently, Sims believed it was unnecessary for him to include in his staff of twelve officers an organization for collecting information. A small intelligence section headed by his aide, Cdr. John V. Babcock, however, was maintained for the purpose of collating, digesting, and disseminating intelligence information. The intelligence section kept in close touch with the British Naval Intelligence Division, with one officer detailed to spend most of his time there.

The U.S. naval attachés at Paris and Rome provided communication channels between the U.S. naval force commander and the ministries of marine in France and Italy. In addition, the U.S. naval attachés in Holland and the Scandinavian countries forwarded all information they obtained (see also Chapter 3).

The COMUSNAVFOREUR Intelligence Section was constantly engaged in making summaries of information and in compiling statistical and other data in convenient form for other sections of Sims's staff. It also transmitted all important information received and the results of its own analyses to the Navy Department, Army Headquarters, and the U.S. operating forces.[5]

Fleet Intelligence Between the World Wars

The senior U.S. Naval Officer, Turkey, was also Commander U.S. Naval Forces, Near East, and U.S. High Commissioner, Turkey, from the end of World War I until U.S. diplomatic relations with the newly reorganized Turkish government were reestablished. He flew his flag on the U.S. station ship at Constantinople (the yacht *Scorpion* until July 1919, then the small cruiser *Galveston*) but maintained his offices at the U.S. Embassy. His staff included an operations office, a communications office, and an intelligence office. The principal and most important work of the intelligence office was that of watching, reporting on, and following the political activities of the Allies, as well as those of various other nations represented in the Near East. Ships of the Near East force visited ports throughout the eastern Mediterranean and the Black Sea in support of relief and Red Cross activities in the Near East and in support of RAdm. Newton A. McCully's mission in southern Russia. Information was gathered on the ports of the area, particularly on the availability of coal, water, and other supplies. The Russian situation also demanded the attention of the intelligence office, which watched trends in the effect of Bolshevism on political and economic conditions in the Near East.[6]

The work of the intelligence officer on the staff of Commander in Chief, Asiatic Fleet in 1920 had proven so successful that consideration was being given to the assignment of officers for similar duty with all the fleets.[7]

Aircraft were used by the U.S. Fleet in the 1920s for scouting and spotting gunfire. The airplanes were equipped with radios so that they could send back contact and information reports to the ship or force that they supported.[8]

Also in the 1920s, the need to develop a fleet cryptanalysis capability began to be recognized as essential in the collection of intelligence for the U.S. Fleet.[9]

A study on collecting and disseminating intelligence was made by the four force commanders for Commander in Chief, U.S. Fleet in 1932. The study, with its recommendations, was forwarded to ONI as a basis for establishing good working relationships between ONI and Navy operational units. It was anticipated that ONI would produce a pamphlet on naval combat intelligence similar to one issued by the Army. ONI-19, the *Intelligence Manual*, was issued in 1933 and included a chapter on combat intelligence.[10]

Combat intelligence units were set up in flag commands and in capital ships during fleet problems (operational exercises) in 1932 and 1933. The units were primarily intended to supply information and analyses for use in tactical situations. Instructions about making intelligence information reports on enemy forces were issued in 1933 by Commander in Chief, U.S. Fleet to all fleet units.[11]

During the interwar period, the Asiatic Fleet was unique among the Navy's operating forces in that it had a full-time intelligence officer on the staff of its commander in chief. In 1929–1930, the officer serving in the billet was LCdr. Hartwell C. Davis, who had had previous duty in the Far East as Assistant Naval Attaché, Tokyo.[12]

Soon after Adm. Montgomery Taylor assumed command of the Asiatic Fleet in August 1931, Lt. Henri H. Smith-Hutton was shifted from his billet as Taylor's flag lieutenant to that of fleet intelligence officer. Adm. Taylor was more interested in what would happen tomorrow than in what had happened yesterday, and this attitude, of course, influenced the fleet intelligence officer in the execution of his duties. Taylor read the newspapers and the many reports received from military and diplomatic representatives in his area. He liked to discuss the significance of events and situations with someone, and Smith-Hutton was the logical staff officer for this duty. Few of the local reports were passed to ONI by Smith-Hutton unless the admiral wished to comment on or add to a report, since they were already available in Washington.

The Asiatic Fleet intelligence officer's other duties at the beginning of the 1930s included acting as a confidential secretary to the commander in chief because all classified correspondence, except registered publications, was handled by the intelligence officer and his yeoman.

No agents were employed ashore by the Asiatic Fleet intelligence officer, and no other means of covert collection were employed. Overt collection efforts were adequately carried out by diplomatic and consular officials. From time to time, fleet units were directed to photograph and describe the harbor facilities of a particular port to be visited, but most ports had already been well covered. Close contact was maintained with the officers of foreign navies in the Far East area. France, for example, was responsible for Catholic missions in China. The heads of the missions, many of whom were Jesuits, were extremely well informed and had many sources of information not usually available to other foreigners. French intelligence officers, therefore, had good information on how the Chinese were thinking about local situations.

The Asiatic Fleet had no communications intelligence (COMINT) collection capability in the early 1930s, but it did have one officer from OP-20G detailed to the commander in chief's staff, Lt. Joseph Wenger, who was especially competent in the means of collecting communication intelligence and whose duty it was to prepare plans for an expanded intercept network.[13]

The intelligence components afloat in 1933 consisted of "intelligence officers on the staffs of fleet, force and task group commanders and all personnel, especially or primarily detailed for intelligence duties, either afloat or ashore, operating under such commander," and "the officers assigned intelligence duties on staffs of smaller units or in individual ships." The organization, training, and operation of intelligence personnel afloat, both in peace and war, was a responsibility of the fleet commander in chief.[14]

The intelligence work of the forces afloat was intended to provide information for the following:

1. The commander in chief in carrying out his peacetime mission.
2. The commander in chief (on foreign station), the State Department representative, and the government in formulating U.S. policy.
3. American business in foreign countries: commercial, financial, industrial, and agricultural.
4. Naval and military commanders in time of war.

Officers permanently assigned as unit or ships' intelligence officers were expected to use the services of all available officers in the collection of information.[15]

The sources available to the fleet for strategic information were considered in 1933 to be radio intercept and cryptanalysis; surface, subsurface, and air observation; reconnaissance; merchant vessels; advance forces; scouting (all types); radio direction finder, plotting and tracking; underwater sound bearings; and shore stations that could provide radio tracking, intercepted messages, and data on U.S. and neutral merchant ships.[16]

ONI's requirements for information from the forces afloat in peacetime included information on foreign ports; reports on foreign combat ships and merchant vessels, limited to data not shown in available publications; reports on foreign naval personnel relative to their efficiency, morale, training, etc.; and tactical information about foreign naval formations, tactics, and maneuvering ability and "smartness."[17]

Lt. Smith-Hutton reported for the second time as Fleet Intelligence Officer, Asiatic Fleet on 8 February 1937 and was promoted to lieutenant commander during his tour. The fleet commander in chief was Adm. Harry E. Yarnell, who flew his flag in the heavy cruiser *Augusta* (CA 31). Periodic situation reports were received on board *Augusta* from U.S. consuls in all major Chinese ports, the various military commanders ashore, naval and military attachés in the Far East area, and the Embassies at Peking and Tokyo. Radio intercept transcriptions were also available from Assistant Communications Officer Lt. Jack S. Holtwick, Jr., who brought them to Smith-Hutton for translation of material of interest to the commander in chief and his chief of staff.[18]

In July 1937, the Asiatic Fleet visited Vladivostok. Before the visit, ONI advised that the most recent reports about that area had been made by the Siberian Expedition in 1920. In order not to antagonize the Russians unnecessarily, no collection of information was to be attempted other than to note any new construction. No new ships or installations were seen, but reports were made on some old and obsolete Soviet submarines and ancient gunboats. Assistant Fleet Intelligence Officer Lt. George R. Phelan, an expert photographer, did take many pictures of the harbor and harbor installations, but only from on board *Augusta*.

When conditions heated up between the Japanese and Chinese in early August 1937, *Augusta* moved from her summer port of Tsingtao to Shanghai, arriving on 12 August. Because the 4th Marines was part of the defense force for the International Settlement within that cosmopolitan city, Adm. Yarnell decided that the fleet intelligence officer could follow the action better by being at the 4th Marines' headquarters, and Smith-Hutton was ordered ashore on about 18 August. His daily routine was to spend the night ashore studying reports and situation maps and then, after lunch, to return to *Augusta* and report to the admiral and answer any questions.

Almost every morning, Adm. Yarnell went to the office of the U.S. Consul General, Mr. Gauss, to discuss the local situation. Col. Charles F. B. Price, commander of the 4th Marines, also attended the conferences unless he was otherwise involved with urgent duties. The admiral liked, respected, and had great confidence in both Mr. Gauss and Col. Price.[19]

The intelligence officer with the 4th Marines was Capt. Ronald A. Boone, USMC, a Chinese-language officer, and his assistant was 2dLt. Victor H. Krulak, USMC (who was to become well known and retire as a lieutenant general). Their staff also included a small group of Marine noncommissioned officers. The organization's reports were highly reliable because the Marines had good sources of information in Shanghai. Capt. Boone had been in China a long time and was on good terms with the Shanghai police force, local and international newsmen, and Chinese authorities, including the military. The radio intercept group with the 4th Marines was copying Japanese diplomatic traffic, much of which they were able to decode. The Japanese messages considered to be of importance were translated by Smith-Hutton for Adm. Yarnell and Col. Price. Smith-Hutton also made periodic visits to the Japanese military headquarters in the Hongkew section of the International Settlement, and the Japanese officials talked quite frankly about their operations and intentions. Thus, Capt. Boone, as the expert on the Chinese, and LCdr. Smith-Hutton, as the expert on the Japanese, were able to follow quite well the progress of the fighting and even to forecast some of the events with reasonable accuracy.[20]

In early 1938, when the Soviets sent four fighter squadrons and two bomber squadrons to the Chungking-Hankow area to help the Chinese, U.S. gunboats were still operating that far up the Yangtze River, and the gunboat commanding officers were instructed to learn all they could about the Soviet personnel and their equipment. Capt. Claire Chennault, U.S. Army (later of the Flying Tigers), who was even then operating with the Chinese, also sent in reports from time to time.[21]

"Intelligence" in the various U.S. Navy fleets in 1938, except in the Asiatic Fleet, was still a largely theoretical concept. Press relations and, to a very limited extent, counterintelligence and security were actively engaged in, but other types of intelligence activity were generally ignored. In the Asiatic Fleet, approximately four officers and four enlisted personnel were working full time on intelligence as a primary assignment. And, one officer in each ship and on each staff of the Asiatic Fleet was assigned to additional duty as intelligence officer. In contrast, in the U.S. Fleet in the Atlantic, only one officer was assigned intelligence functions as additional duty on each staff, air base, submarine base, large ship, and in each division of small ships.[22]

In late October 1939, when the first detailed reports of early naval actions of World War II started coming to the Commander Battle Force, his flag secretary, acting as the intelligence officer, started the *Force Intelligence Bulletin*. The initial distribution was 100 copies, but requests for copies started rolling in, and the distribution was soon running over 1,000 copies per week.[23]

In the years immediately prior to U.S. involvement in World War II, commanders afloat were responsible for the organization and administration of the intelligence efforts within their commands. Such organizations were expected to conform to general directives prescribed by the Chief of Naval Operations insofar as they touched upon, or required coordination with, other parts of the Naval Intelligence service.[24]

On 14 May 1942, Adm. Ernest J. King, Commander in Chief, U.S. Fleet (COMINCH), directed the Chief of the Bureau of Aeronautics to train personnel for twelve aviation intelligence units that were to be assigned to the Joint Intelligence Center at Pearl Harbor and to Advanced Joint Intelligence Centers in other locations in the Pacific. The advanced centers were not to be established until the main center at Pearl Harbor had been activated and became well organized. By the end of June 1942, personnel and material planning for the centers was in progress under the coordination of the Director of Naval Intelligence. As of 8 September 1942, the advanced centers were partly constructed or were in the final planning stages for the South Pacific area at Auckland, New Zealand, for the Northern Pacific area at Kodiak, Alaska, and for the Southwest Pacific at Bellconnen, Australia.

Adm. King intended to establish centers for the Atlantic Fleet, with the main center at Norfolk, as soon as the Pacific area intelligence centers were sufficiently advanced. One aviation intelligence officer was ordered by COMINCH to report to Naval Air Station, Norfolk, as early as 28 August 1942 for duty in connection with the establishment of the Atlantic Fleet Air Intelligence Center at the base. Next in priority were intelligence centers for the five sea frontier commanders (Eastern, Western, Panama, Caribbean, and Gulf). All ten components of the air combat intelligence organization were in operation by 16 November 1942.[25]

The wartime activities of the various wartime and postwar fleet intelligence organizations are discussed in Chapters 35 through 40.

Organization of Fleet Intelligence After World War II

Changes to Navy Regulations 1920, published on 21 June 1946, contained a new article, 687-A, in Chapter 18: "The Commander-in-Chief, or commander of any force or unit of the operating forces not operating under the Commander-in-Chief, shall maintain an efficient intelligence organization within his command." *Navy Regulations* also contained subparagraph (2)(c) of Article 786: "The organization of the staff shall include an intelligence section headed by a line officer designated as flag intelligence officer."

The Naval Intelligence Manual–1947, ONI-19(A), prescribed the mission of an intelligence officer assigned to duty with the operating forces to be as follows:

> a. To provide his commander or commanding officer with the strategic and operational intelligence required for the execution of his mission;
>
> b. To deny to the enemy or hostile forces all information of own forces;
>
> c. To combat sabotage and subversion in own forces; and
>
> d. To supply ONI with information and intelligence of value.[26]

Fleet Air Intelligence Augmenting Units (FAIAU) were established and used during the 1950s, primarily to provide the fleets with the capability, in an emergency, to immediately augment trained intelligence personnel for the forces afloat (usually aircraft carriers) in a forward area, and secondarily to assist the Fleet Intelligence Centers (FIC) in the production of intelligence. FAIAUs were attached either directly to the commander in chief or to the Fleet Intelligence Center of the fleet to which they were assigned.[27]

Fleet Intelligence Centers were established in the 1950s to provide the major fleets to which they were assigned with an intelligence production and intelligence personnel augmentation capability (see Chapter 40 about the specific FICs).

Mobile Intelligence Production Units were designed for rapid deployment so they could provide the fleet to which they were assigned with a mobile intelligence production capability.

Each intelligence organization in the operating forces was under the operational and administrative control of the command or commands to which it was assigned. In accordance with General Order No. 19, the Director of Naval Intelligence exercised technical control over intelligence matters through the Department of the Navy, including those relating to the operating forces.

Publications produced by ONI to provide guidance in intelligence activities in the operating forces during the 1950s included *Operational Intelligence*, ONI Y-1; *Operational Intelligence Manual (Air)*, ONI 52-2; *Operational Intelligence Manual (Amphibious)*, ONI-52-6; and *Intelligence Manual for Operating Units*, ONI-52-7.

To fulfill his mission in the Cold War era, the Director of Naval Intelligence required information concerning the current organizations; planned wartime organizations; operating plans; collection requirements; material status, training, techniques, and procedures; and command support for all intelligence activities and elements in the operating forces. The *U.S. Naval Intelligence Manual*, ONI-70-

1, of 20 June 1956, required that this information, together with appropriate comments and recommendations, be included in the intelligence activity reports to be submitted by the operating forces in accordance with the effective edition of OPNAV Instruction 05440.53. An officer in the operating forces who had been assigned to intelligence could be assigned collateral duties only to the extent that such duties did not interfere with his primary duty, according to the same directive.[28]

To increase the amount of intelligence collected by the operating forces, and focus better collection efforts on the highest priority gaps, a system of formalized, yet flexible, programs was established in 1958–1959. The general parameters of the programs were established by the CNO to provide technical guidance and support as well as to monitor the execution of the programs and the processing of the end results. Actual missions under the program were planned and conducted by the fleet commanders. The intelligence collected was initially processed by the fleets for fleet support requirements and was then forwarded to ONI for complete technical analysis and exchange with other intelligence services.

The basic objectives of the fleet intelligence collection programs were, in order of priority, the collection of intelligence to determine the operational characteristics and capabilities of new enemy material and equipment and to support research and development on countermeasures; the operational status and production level of new material and equipment as related to their effects on enemy strategic capabilities; and the current deployment and employment of new material and equipment as related to enemy order-of-battle and tactical capabilities.

Consistent with national intelligence collection requirements, top priority in all programs was assigned to the collection of intelligence on the potential enemy's state of the art in missilery, including information about naval forces that were considered capable of carrying missiles. A slightly lesser priority was the collection of intelligence on the USSR's capabilities in undersea warfare and air defense.

The fleet intelligence collection programs instituted during the late 1950s provided for the use of submarines, surface ships, and air forces, normally operating independently on specific assignments but occasionally participating in a joint effort. Special equipment and collection devices were made available to the designated forces, particularly for the interception and collection of electronic and acoustic emissions. In addition, the best possible photographic equipment was procured, and special detailed briefings and instructions were given to participating personnel to ensure the maximum coordination of visual, photo, and electronic observations.[29]

Chapter Notes

1. Elting E. Morison, *Admiral Sims and the Modern American Navy* (Boston: Houghton Mifflin, 1942), 40–43.

2. Ibid., 50, 98.

3. Ibid., 111–13.

4. *SECNAV Annual Report, 1902*, 134.

5. ONI, *Monthly Information Bulletin*, 15 Apr 1919, 38–42.

6. ONI, *Monthly Information Bulletin*, 15 May 1920, 140–41.

7. *SECNAV Annual Report, 1920*, 126.

8. Commander in Chief, U.S. Fleet (CINCUS), *Annual Report, Fiscal Year 1925*, 19.

9. *CINCUS Annual Reports, FY 1927*, 43; *FY 1930*, 61; and *FY 1931*, 48.

10. *CINCUS Annual Report, FY 1932*, 67.

11. *CINCUS Annual Report, FY 1933*, 55; and *FY 1934*, 14.

12. VAdm. Felix L. Johnson, interview by author, 14 May 1975.

13. Capt. Henri H. Smith-Hutton, Oral History, USNI, Annapolis, MD, 1979, 1:110–13, 122.

14. ONI-19, *Naval Intelligence Manual*, 1933, para. 4001–4002.

15. Ibid., para. 4006, 4009.

16. Ibid., para. 5004.

17. Ibid., para. 4010.

18. Smith-Hutton oral history, USNI, 2:209–10.

19. Ibid., 224–27.

20. Ibid., 237–39.

21. Ibid., 258–59.

22. Department of the Navy, "Administrative History of the Office of Naval Intelligence in World War II," 10 Jul 1946, unpublished MS, Appendix H.

23. James O. Richardson, *On the Treadmill to Pearl Harbor: The Memoirs of Admiral J. O. Richardson as told to Vice Admiral George C. Dyer, USN (Retired)* (Washington: GPO, 1973), 176.

24. ONI-19, *Naval Intelligence Manual*, Oct 1940, Article 110(d).

25. Bureau of Aeronautics (BUAER), "Administrative History of the Bureau of Aeronautics in World War II," unpublished MS, 1946, 19:25–28, 31–34.

26. ONI-19(A), *Naval Intelligence Manual*, May 1947, para. 6004.

27. ONI-70-1, *U.S. Naval Intelligence Manual*, 20 Jun 1956, 40.

28. Ibid., 64–70.

29. OP-092B1 memo, ser 000176P92, 30 Apr 1959, "Annual Report to the President's Board of Consultants on Foreign Intelligence Activities," passim.

Chapter 35

Operating Forces, Pacific

Pearl Harbor and the Aftermath

On 6 January 1940, Adm. James O. Richardson relieved Adm. Claude Bloch as Commander in Chief, U.S. Fleet (CINCUS). As part of Fleet Problem XXI, the fleet deployed to Hawaii on 1 April from its home bases on the U.S. West Coast, but it remained in Hawaii at the conclusion of the exercise, on the order of President Roosevelt.

In October 1940, Adm. Richardson wrote to Chief of Naval Operations (CNO) Adm. Harold R. Stark that Orange Plan 1, for a war with Japan, called for a major military operation to capture Base One in the Japanese Mandated Islands. Such an operation would require detailed knowledge of the area for proper planning, but the necessary information was not then available. The plan anticipated that sufficient information might be obtainable by reconnaissance after hostilities had commenced. The basic concept of delaying the gathering of necessary intelligence until after the opening of hostilities, Richardson said, presaged disaster.[1]

When Adm. Husband E. Kimmel relieved Adm. Richardson on 1 February 1941 as Commander in Chief, Pacific Fleet (CINCPACFLT) and CINCUS, he was supported by a small staff designed for basing on board ship. On 21 March 1941, the staff moved to temporary facilities at the Submarine Base, Pearl Harbor. Between March and September 1942, the staff (less the War Plans Section) accompanied the admiral on board ship only briefly for short cruises. CINCPACFLT decided in early August that "in order successfully to prosecute a campaign in the Pacific, a shore headquarters at the principal base must be available." Commandant 14th Naval District was requested to erect a new headquarters building, and plans for the building were nearly complete at the time of the Pearl Harbor attack.[2]

In addition to the small intelligence section of the Fleet staff, discussed below, intelligence support to the CINCUS staff included the Radio Unit, Pearl Harbor, headed by Cdr. Joseph J. Rochefort. Established in 1936 as an activity of the 14th Naval District, it was known in 1941 as the District Combat Intelligence Unit. Rochefort's organization served as the primary source of tactical intelligence for the fleet intelligence officer.[3]

When Gen. Tojo took over the Japanese government on 16 October 1941, CNO Adm. Stark alerted the Navy by message and directed that due precautions and preparatory deployments be made. Adm. Kimmel took various actions, including putting submarines on "war patrol" off Wake and Midway Islands and sending twelve patrol planes to Midway to conduct daily patrols within 100 miles of the island.[4]

Upon receipt of the CNO's 27 November "war warning" message, Adm. Kimmel ordered a squadron of patrol planes to Wake from Midway with instructions to search the ocean areas en route. Three days later, Kimmel deployed a replacement squadron from Pearl Harbor to Midway. The squadron proceeded to Midway via Johnston Island, making a reconnaissance sweep along its track. The replacement squadron conducted distant search sweeps of not less than a 500-mile radius and of varying sectors from Midway on 3–6 December. The squadron then proceeded to Wake and, on 2 December, searched to a distance of 525 miles from that island. On 7 December, five of the Midway-based patrol planes were searching out to 450 miles from the island in the 120° to 170° sector.[5]

No distant reconnaissance patrols were conducted from Pearl Harbor. Adm. Kimmel had considered doing so, but he had only 49 patrol planes in flyable condition, and he estimated that another 84 planes would have been needed to patrol a full circle to 800 miles. Only such a massive search sector would, he felt, ensure against a surprise attack by fast, carrier-based planes, and if the searches were conducted on a daily basis for a protracted period, 250 patrol planes would have been required.[6]

Kimmel's command at Pearl Harbor has been much maligned because of the Pearl Harbor attack—unjustifiably so in certain respects. Of the various potential targets for Japanese attack, geographic location made Pearl Harbor one of the lesser possibilities after those, such as Malaya, the Philippines, and Borneo, that had been more positively indicated as being probable Japanese targets by various items of intelligence.

Another mitigating factor relates to direct intelligence support. Insofar as communications intelligence (COMINT) was concerned, Rochefort's Pearl Harbor Unit had to rely on what it received from Corregidor and Washington; the unit had no organic intercept capability.

The unscheduled change of call signs and cypher by the Japanese around 1 December 1941 was an additional signal that hostilities were about to be undertaken. The radio intercept unit at Corregidor was the first to detect that major change, and it was duly reported. Also, when the Japanese carrier force left home waters, some of the enlisted intercept operators and traffic analysts at Corregidor suspected that transmitters from the carriers had been put ashore and were continuing to transmit as if exercises were continuing in the home waters. The suspicion was reported but was sufficiently speculative, so it could only be considered an uncertain possibility that the carriers had left Japanese waters. The big question remaining unanswered was, if the Japanese had left, where had they gone?

That the command at Pearl Harbor did entertain the possibility of an attack on Pearl Harbor was indicated by Kimmel's employment of his inadequate reconnaissance resources. But the search sectors selected as most likely didn't include the northwest, from where the attack actually came.[7]

On 6 December 1941, a sighting report from the PBY flying boat reconnaissance effort of Commander in Chief, Asiatic Fleet Adm. Thomas C. Hart was received, stating that a concentration of Japanese transports and naval vessels, including submarines, was south of Camranh Bay, Indochina, and that other ships were headed toward the Gulf of Siam. Kimmel sent Intelligence Officer LCdr. Edwin T. Layton to show Hart's message to Adm. William S. Pye, Commander Battle Force, embarked in the battleship *California* (BB 44), and to get Pye's comments. Adm. Pye and his acting chief of staff both read the message and estimated that the Japanese were probably going to occupy a position in the Gulf of Siam as an advance base from which to operate against the Burma Road. Pye and his staff asked Layton for his thoughts; he told them that he didn't believe the Japanese would stop there, although part of their operations might be against the Burma Road. Layton believed that the Japanese had objectives further south, probably the East Indies oil resources, inasmuch as the United States had stopped its export of oil to Japan. Layton also added that, since the Japanese never left their flanks exposed, he didn't think they would leave the unsecured Philippines on their flank, and that the United States would thus be at war.

Adm. Pye and his chief of staff both said in effect, "Oh, no. The Japanese won't attack us. We're too strong and too powerful." Layton reported their comments back to Kimmel. At lunch that day, several officers of the CINCPACFLT staff asked Layton about the significance of the Japanese troop transports heading toward the Gulf of Siam. He repeated the comments he had made to Adm. Pye and expressed his belief that the United States would be at war the next day. That drew the usual remarks about "Layton and his Saturday crisis." The next morning, during the attack, Capt. William A. Kitts III, who had been in the wardroom the day before and had heard Layton's forecast, acknowledged that Layton's audience should have listened more seriously to him.[8]

During and following the Japanese attack on Pearl Harbor on 7 December 1941, there was considerable uncertainty as to the direction from which the attack had come. Layton had arrived at his office in the submarine base at 0820, and shortly thereafter Lt. Wesley A. (Ham) Wright of Rochefort's unit informed Layton that they had had one bilateral radio direction-finding (D/F) bearing, 353°/183°, on the attack force but that they couldn't communicate with the second D/F station at Wahiawa, which would have given a crossbearing and resolved whether the Japanese force was north or south of Pearl Harbor. (Later it was learned that the Army had taken over the telephone circuit to the Wahiawa station, which explained why the facility was out of communications when it was most needed.)

Layton laid down the reciprocal bearings on a chart in Operations Plot. Adm. Kimmel was rather irked that Intelligence couldn't tell him whether the enemy was to the north or south. To make matters even worse, a garbled message was received at about that time from a Navy ship reporting two carriers south of Pearl Harbor. Actually, this sighting, as originated, was of two U.S. Navy cruisers.

It was not until later in the afternoon on 7 December that positive information was obtained that the attack had come from the north. A "plot board" from one of the Japanese aircraft that had crashed into the seaplane tender *Curtiss* (AV 4) was recovered and delivered to Fleet Intelligence. Layton examined the plot board and the pilot's navigation sheet, which showed the aircraft's course to Pearl Harbor and the intended course back to the Japanese carrier. With the plot board was a temporary callsign card listing the radio calls for all commands and ships in the attack force. The card was passed on to Rochefort's unit.[9]

When the attack had begun, the Army called its troops to operate lookout stations, gun batteries, security guard posts, etc. It was at this time that the plug was pulled on the telephone circuit to the Navy's second D/F station. Beach patrols and other lookouts then started sending in a series of "the damnedest reports you ever heard," many of which were passed to Commander in Chief, Pacific (CINCPAC) by the Intelligence (G-2) organization of the Army's Hawaiian Command. Some examples included: "Enemy ships bombarding . . . beach, landing in progress." "Two carriers south of Oahu." "Paratroopers landing, wearing blue uniforms with red sun insignia on back." "Shells landing on . . . beach; we are taking enemy ships under counterbattery fire." The Army was reminded that the U.S. Marines had been scheduled to conduct exercises that day and that they were probably shooting at the Marines. Furthermore, since the Marines had no ammunition for their exercise, any shells landing behind the Army observers were probably "shorts" from their own counter-battery fire (which turned out later to be the case). Thus a confused and frantic day was made more hectic for the Navy intelligence staff trying to evaluate the true situation. The climax came that night when Army batteries commenced firing at PBY reconnaissance aircraft landing at Pearl Harbor, prompting reports that enemy airborne troops were landing from flying boats.[10]

At the outbreak of hostilities in the Pacific, Adm. Kimmel's intelligence staff consisted of an intelligence officer (LCdr. Layton), one assistant (Lt. Robert E. Hudson), and one enlisted yeoman. According to a CINCPAC Staff Instruction from 1941, the intelligence staff was responsible for assembling, evaluating, and disseminating enemy information; providing information essential for developing current estimates to Fleet Operations Officer Capt. Walter S. Delany and War Plans Officer Capt. Charles H. "Sock" McMorris; directing counterespionage and counterinformation efforts; supervising reconnaissance and photographic activities; and collecting, evaluating, and distributing information on foreign naval vessels and merchantmen. In practice, however, before the war, the Fleet Intelligence Office was mainly occupied with counterespionage and with the analysis of existing information on the strength and location of Japanese fleets and advance bases.[11]

After 7 December 1941, the fleet intelligence officer handled all types of intelligence needed by the fleet and the Pacific area commanders (North, Central, and South). It was almost immediately apparent that all those duties could only be carried out by a much larger organization; instead of the fleet intelligence staff being enlarged, however, other organizations were formed under the administrative control of the Commander 14th Naval District (COM 14) and under the operational control of CINCPAC to serve the fleet and area commanders. By placing the new intelligence organizations under COM 14, the intelligence producers were relieved of having to perform many purely administrative functions.[12]

Development of the Wartime Intelligence System in the Pacific

After Adm. Chester W. Nimitz took command of CINCPACFLT, and when LCdr. Layton had had his first chance to talk with him, Layton asked to be detached. He wanted to go to sea in command of a destroyer, if possible, and kill Japanese. Nimitz told Layton that he wanted him to stay on and that Layton could kill more Japanese by sitting at his desk on the CINCPACFLT staff than he ever could by commanding a destroyer.

Nimitz thereupon expounded on his requirements for intelligence support. He said that good intelligence was vital to a good estimate of the situation and, in turn, to making sound decisions. As he saw it, intelligence support to operations would become of the greatest importance. Nimitz told Layton,

> I want you to be the Adm. Nagumo [the Chief of the Imperial Japanese Navy General Staff] on my staff, where your every thought, every instinct, will be that of Adm. Nagumo's; you are to see the war, their operations, their aims, from the Japanese viewpoint and keep me advised what you [as a Japanese] are thinking about, what you are doing, and what purpose, what strategy, motivates your operations. If you can do this, then I think you will be able to give me the kind of information I need for the prosecution of my mission.[13]

Nimitz wanted Layton to be at his office ready to brief him daily at 0755 (the time was later changed to 0800). Promptly at that time, the intelligence briefing was started in an easy, informal atmosphere. After the briefing, Nimitz would ask questions about various things having to do with intelligence, the war, and enemy reactions. In addition to the daily briefing, Layton would go to Nimitz's office whenever he had additional intelligence reports or specific bits if information deserving priority consideration. Soon, Nimitz informed his aide, Lt. H. A. Lamar, that Layton was not to be required to wait to see him. If Layton said he had something very important, Nimitz was to be informed, even if Nimitz was in conference with important people. In such a case, he would excuse his visitor while he received Layton's report. Layton didn't find it necessary to use the privilege often.[14]

The prewar Orange plans prescribed that the Navy's mission in a war against Japan was, basically, to advance and capture a position in the Marshall Islands at which to establish a forward fleet base. The Japanese success in their attack on Pearl Harbor caused the Navy to postpone that initial step, not only because of insufficient air, surface, and amphibious strength, but also because of a lack of intelligence on Japanese defenses in the Marshalls. Adm. Nimitz conferred with Layton on how best to remedy the intelligence deficiency. Submarine reconnaissance of Japanese strongpoints in the Marshalls and other Mandated Islands was initiated as part of the war patrols that were deployed immediately following Pearl Harbor. The resulting periscope sightings confirmed the Japanese militarization of the islands, and the receipt of the submarine intelligence contributed to Nimitz's decision to order a carrier task force raid on the Marshalls that was carried out on 1 February 1942. Although the orders urged that photographs be taken for intelligence purposes, none of value were obtained. Thereafter, photography for intelligence purposes was made a specific requirement in any carrier attack on enemy positions.

The strategic and operational planners soon realized their need for intelligence, particularly photographic intelligence, before they could plan and execute an amphibious assault on the Marshalls with any hope of success. The use of carriers for such intelligence-gathering missions, however, was out of the question at that stage of the war. The United States had too few carriers to be able to provide aerial reconnaissance on a continuing basis until some future D-Day.

Thus, in order to meet the requirement for vertical and oblique aerial photographs, a base had to be built within range of the first island objectives. At the outbreak of war, the Japanese were confident that the United States would attempt to seize the Marshall Islands. Accordingly, to protect the Marshall Islands southern flank, they seized and strongly fortified Tarawa in the Gilbert Islands. The Japanese action forced the United States to set up its initial intelligence-gathering base in the Ellice Islands to the southeast of the Gilberts.[15]

After Midway, the subject of the need for more people in intelligence came up again, and again Adm. Nimitz stated his determination not to let the size of his staff get out of hand. Layton told Nimitz that he could not give him effective intelligence support unless he had enough people to do the work; that as the war progressed and U.S. Navy operations expanded in scope, the intelligence requirements would increase in magnitude accordingly; and that more people would be needed to do the job. Some time after Layton's plea, Nimitz announced that Layton had been justified in asking for more intelligence personnel. When Layton said that he needed forty to sixty additional people, however, Nimitz said flatly that he wouldn't consider expanding his staff to that degree. Layton then pushed the idea of assigning the necessary personnel for intelligence to the 14th Naval District to work specifically for him. That led to the establishment in July 1942 of the Intelligence Center, which eventually became the Joint Intelligence Center when Adm. Nimitz became Commander in Chief, Pacific Ocean Areas. (See Chapter 19 for more information on the Joint Intelligence Center, Pacific Ocean Areas [JICPOA].)

Intelligence Center, Pacific Ocean Areas (ICPOA) was thus established by Commander 14th Naval District, on 13 July 1942. Initially, part of the Fleet Intelligence organization became the Combat Intelligence Section of ICPOA, and the District Combat Intelligence Unit was assigned in October 1942 to ICPOA as the Radio Intelligence Unit of the Combat Intelligence Section. The fleet intelligence officer and the part of his section remaining with the CINCPAC staff provided a personal advisory unit for CINCPAC himself and was concerned primarily with tactical intelligence and with collecting and collating information on the location and movement of enemy naval, ground, and air units.[16]

Shortly after the war began, the Fleet Intelligence Office had been augmented by Lts. Arthur L. Benedict, John G. Roenigk and H. B. Coleman (who became the fleet security officer). With Lt. Robert Hudson, they were placed on a one-in-four intelligence watch under the direction of Layton. Benedict and Roenigk, both Japanese linguists, assisted in the translation of captured documents, particularly those from the midget submarine that had been captured off Bellows Field after the Pearl Harbor attack. The Fleet Intelligence watch was maintained until about August 1942, shortly after ICPOA was set up. At that time, Benedict and Roenigk were detached and assigned to the Radio Intelligence Unit of ICPOA. The continuous intelligence watch was then terminated until June 1943 when three Naval Reserve officers, Lts. K. A. Brown, A. M. Ellerby, and R. L. Jackson, were assigned, trained, and established as an intelligence watch.[17]

On 6 September 1943, the day before ICPOA became JICPOA, the Radio Intelligence Unit was removed from ICPOA, assigned to CINCPACFLT, and given the name Fleet Radio Unit, Pacific (FRUPAC). The officer in charge of FRUPAC, Capt. William B. Goggins, was given additional duty on the staff of CINCPAC-CINCPOA as Communication Intelligence Liaison Officer. His primary duty was to supply to CINCPAC all information derived by communications intelligence methods, by all units of the U.S. naval communication intelligence organization, and by all similar Allied organizations.

Also on 6 September 1943, the CINCPAC Staff Intelligence Office, headed by then-Cdr. Layton, became the Combat Intelligence Section of the Staff Intelligence Division. It handled all urgent intelligence material and controlled the dissemination of intelligence at the highest classification. In daily conferences with CINCPAC-CINCPOA, Layton presented the special intelligence material, and it was largely through his briefings that Nimitz received the information necessary to make decisions on the employment of his forces.

Specifically, the Combat Intelligence Section assembled, collated, and made appropriate distribution of information on the enemy; made a daily review for CINCPAC of the current enemy situation and apparent intentions; kept a strategic plot of enemy naval and air forces; prepared daily and special intelligence bulletins for distribution to appropriate echelons of Nimitz's command; disseminated combat intelligence to appropriate fleet, area, and task force commanders; analyzed what was known of the current logistic and material condition of the enemy; directed counterespionage, counterintelligence, and counterpropaganda; and carried on other general intelligence duties.

A concise daily message to distribute current information about the enemy was sent out by the Combat Intelligence Section by radio. Although the addressees on the message varied from time to time, they usually included the most important naval commanders in the Pacific, including British, and all important Army commands in the forward areas. In addition, Capt. Layton helped to supervise the Estimate Section of JICPOA and was the key officer responsible for intelligence matters in support of other divisions of CINCPAC's staff, especially Plans and Operations. Of necessity, the work of the Combat Intelligence Section and the Operations Division overlapped where an intelligence function ended and an operational function began. The smooth transition of such functions was achieved by mutual understanding between the heads of the two organizations.[18]

When the Joint Staff was set up in September 1943 (at the insistence of the Army Chief of Staff and at the direction of COMINCH Adm. Ernest J. King, a distinction was drawn between the Fleet and Joint Staffs by the use of F (Fleet) and J (Joint) designators for particular billets. Some officers, including almost all those in the Plans and Operations Divisions, were double-hatted and given both F and J designations. Fleet Intelligence Officer Layton and his assistants were listed as F only.

The staff functioned as one unit, with the separation between Fleet and Joint on paper only and serving no real useful purpose. In December 1943, when the original controversy that had led to the establishment of the Joint Staff had abated, the Fleet-Joint distinction and the F and J designators were discontinued. The CINCPAC-CINCPOA staff remained what it had been all along, a single joint staff organization.[19]

Of the four main divisions of the staff, two were headed by naval officers (War Plans and Operations) and two by Army officers (Intelligence and Logistics). The Assistant Chief of Staff for Intelligence, Col. J. J. Twitty, was also officer-in-charge of JICPOA. The fleet intelligence officer headed the Combat Intelligence Section of the Intelligence Division.[20]

In planning for the attack on the Marshall Islands and its occupation, it was noted that reinforcements of the perimeter islands were continually being made by the Japanese at the expense of Kwajalein. Some troops had even been moved from the Japanese headquarters on Kwajalein to the perimeter islands.

Adm. Nimitz, in late 1943, called a conference with his Marshall Islands assault commanders, Marine MajGen. Holland M. Smith, Adm. Raymond A. Spruance, Adm. Richmond Kelly Turner, and others. The flag and general officers had all been making plans to occupy Mili, Wotje, and Tarawa (Maloelap). Nimitz reviewed the intelligence situation, order-of-battle, etc. with them and asked the senior officers if they still wanted to follow through with their assault planning for the same three islands. After receiving an affirmative reply, Nimitz announced that the assault would be on Kwajalein.

His assault commanders thought Nimitz had lost his mind. They believed that Japanese air strength in the outer islands would make penetration to the central island, Kwajalein, much too hazardous. Nimitz, however, pointed out that heavy strikes by carrier air and surface bombardment, plus close reconnaissance of outer island strong points, would not only reduce the hazards posed by the strong points, but would also falsely confirm to the Japanese what they expected would be the U.S. objectives, leaving the defenders to be surprised by landings on weakly defended Kwajalein. Nimitz's estimate proved to be correct.[21]

Prior to the invasion of Kwajalein in January 1944, small U.S. Army raider units were to seize two small islands astride the southeastern entrance to the lagoon just before the main assault was to begin. Both units landed on the wrong islands; each unit was one island west of its prescribed target. The more western unit found some Japanese naval personnel whose ship had been sunk during the preliminary naval bombardment and aerial bombing. The senior Japanese officer was carrying a roll of red-edged charts. In Turner's flagship, the Japanese material was identified as being top secret charts of all the Japanese Mandated Islands showing areas that were mined and areas that had been cleared of coral heads and wire-dragged to a certain depth. Copies of the captured Kwajalein chart were immediately reproduced and distributed to all ships and commands, and they were used in clearing Kwajalein atoll of mines and guiding the amphibious invasion to safe and sheltered anchorages in the lagoon. The other charts were later used in the operations against Eniwetok, Saipan, Tinian, Ulithi, etc.[22]

To the Japanese, long-range photographic reconnaissance missions by four-engined PB4Y "Liberators" meant that a U.S. carrier attack would soon follow. That had happened at islands in the Marshalls and at Eniwetok before the early raids on Tarawa. Thus, when a Marine photo plane flew out of the Solomons for a photo mission over Truk, Adm. Koga ordered his fleet out of the base. Most of the Japanese combatants (battleships, cruisers, and destroyers) moved out on Koga's order, but the auxiliaries (supply ships, fleet oilers, repair ships, submarine tenders, ammunition ships, etc.) were delayed in their departure and were sunk by a carrier task force raid on 17 February 1944. The raid and the loss of the vital auxiliaries effectively terminated the Japanese navy's capability to carry out overseas offensive operations from Truk or any other forward base.[23]

The lack of good maps and charts of Pacific islands was a problem throughout the war. Even Guam, which had been a U.S. possession since the Spanish-American War, had not been mapped adequately enough for military-amphibious operations. Similarly, when the United States decided to recapture Attu in the Aleutian Islands chain, no satisfactory terrain maps of the former U.S. island were to be found. That deficiency was corrected by frequent aerial photographic missions flown just after the Japanese occupation of Attu. Photo interpretation kept track of the Japanese buildup of defensive installations and order-of-battle on Attu and also enabled the production of accurate terrain maps.[24]

Guam and the other islands of interest in the Marianas were beyond the range of Allied reconnaissance aircraft. A carrier task force under Adm. Marc A. Mitscher was therefore sent to make an offensive sweep of the Marianas but primarily to fly aerial photographic missions. The intelligence information obtained on 23 February 1944 from Mitscher's reconnaissance operations was needed for planning the capture of Guam, Saipan, and Tinian.

In support of the Hollandia landing by forces under Douglas MacArthur, carrier task forces struck Palau in early April 1944 with devastating effect. In anticipation of the raid, Adm. Koga (who had taken over from the late Adm. Yamamoto) and his staff had left Palau for the Philippines in two flying boats. Adm. Koga, in the first plane, was never heard from again. His chief of staff in the other plane ran into a terrific storm and was forced to land near Cebu, where he and his briefcase were captured by guerrillas.

MacArthur's headquarters was informed of the briefcase with its apparently important papers. The Seventh Fleet was directed to send one of its submarines to collect the documents. The gist of a non-Navy translation of the document was received by dispatch at Nimitz's Pearl Harbor headquarters and prompted a request to MacArthur for photostatic copies of the originals. The copies arrived promptly by air and were translated immediately. Copies of the Japanese defense plans were mimeographed and were sent with a cover letter to all unit commanders of the Marianas invasion forces assembling in Eniwetok. In the Japanese defense plans, the Marianas were included among the areas considered vital to the defense of the empire and were designated as areas where a major U.S. assault or invasion would be counterattacked by a concentration of all available Japanese forces.

Consequently, after the invasion of Saipan had started, when Spruance received intelligence that the Japanese navy was concentrating for a counterattack, he decided to remain close to the invasion area and neutralize the enemy airfields there to disrupt any use by the Japanese as staging points for shuttle bombing. (In shuttle bombing, aircraft take off from carriers, drop their bombs, land at a nearby land base for refueling, and then return to the Japanese carriers.)[25]

On 5 January 1945, a billet was established for a radio intelligence officer in the Communications Division of the CINCPAC-CINCPOA staff. In addition to assisting the communications officer, the radio intelligence officer acted as liaison between the fleet communication officer and the fleet combat intelligence officer and between CINCPOA and the top Army and Army Air Force commands in the theater that were involved in communications and intelligence, and between CINCPAC and FRUPAC.

Early in 1945, CINCPAC adopted a policy of furnishing mobile Radio Intelligence Units to the fleets, task forces, and principal task groups afloat. The Radio Intelligence Units were assigned after an initial period of training at FRUPAC. The radio intelligence officer kept in close touch with the program and drafted the necessary directives, assigning units and moving them to the commands for which they were destined.[26]

The establishment of the Combat Intelligence Office at the CINCPAC Advance Headquarters on Guam in January 1945 required a group of officers to serve as watch standers and intelligence analysts in specific fields. Accordingly, Capt. Layton brought with him to the Advance Headquarters section Assistant Combat Intelligence Officer Lt. Donald M. Showers, USNR, a specialist on Japanese naval order-of-battle; Lt. L. H. Mann, USNR, geographic specialist; Lt. L. B. Fowler, USNR, Japanese air specialist; Lt. G. M. Page, USNR, Japanese merchant shipping specialist and photo interpreter; Lt.(jg) J. A. Rutter, USNR, Japanese economics specialist; and 1stLt. H. F. Leathers, U.S. Army, Japanese Army order-of-battle specialist. Lt. Mann was subsequently released for duty with the Advance Intelligence Center when that organization was set up at Guam as a forward echelon of JICPOA.

The need for direct and secure communications between the Advance Headquarters intelligence section and the Pearl Harbor intelligence agencies prompted the installation of a radio teletype circuit for that purpose. Another teletype circuit to the Radio Analysis Group, Forward Area was maintained in the Combat Intelligence Office for handling intelligence material disseminated to the fleet from Advance Headquarters. The location of the communication equipment within the Combat Intelligence Office necessitated the assignment of four communication watch officers to stand 24-hour watches concurrently with the intelligence officer analysts named above. By that arrangement, CINCPAC communications was relieved of handling special intelligence material, and all such material was received, processed, and disseminated by the Combat Intelligence Section at Advance Headquarters, providing the additional benefit of an increase in security.[27]

Advance Headquarters functioned throughout the Iwo Jima and Okinawa operations and the final actions leading to Japan's surrender. It was closed on 19 September 1945.

Korean War Era

The Intelligence Section of the staff of the Commander in Chief, Pacific Fleet, during Fiscal Year 1950 continued to disseminate intelligence information within the Pacific Command by means of its *Weekly Intelligence Digest*. Within the CINCPACFLT staff, intelligence was disseminated through daily and weekly summaries and by oral briefings. It appears that the importance of intelligence to the daily routine of operations at CINCPACFLT had sunk to a low level after the Second World War; intelligence appeared as a sketchy report in the "Miscellaneous" section of the Fiscal Year 1950 CINCPACFLT Annual Report.[28]

Upon the outbreak of Korean hostilities, the primary problem for CINCPACFLT was to secure adequate intelligence personnel for the naval forces deployed to the Western Pacific. The immediate solution to the problem was complicated by the orientation of the U.S. armed forces toward the European theater, and the loss, since demobilization in 1945–1946, of many skilled intelligence specialists, such as photo interpreters and air combat intelligence officers. Through assignment of additional billets, establishment of Fleet Air Intelligence and Photo Interpretation Schools, and the recall to active duty of reserve officers, the intelligence personnel situation was well on the way toward solution as of 20 September 1950, but it was still not satisfactory.

Other problems facing CINCPACFLT were the need to provide timely and adequate dissemination of intelligence to Commander Naval Forces, Far East (COMNAVFE), and receipt of intelligence from the Western Pacific by CINCPACFLT. The difficulties were caused by an overload of traffic at the message-handling facilities. A top secret 24-hour telecommunications circuit between CINCPACFLT and COMNAVFE was established as a solution.

The reproduction and dissemination of photographs presented yet another problem. The need for timely photographs, both in Pearl Harbor and in Washington by intelligence agencies and for publicity purposes, was also extremely pressing. Experience had shown that if the required number of copies was not printed and the film cataloged while events were fresh, exploitation was seldom carried out at a later date due to lack of personnel and the overriding priority of subsequent requirements. To correct the situation, it was suggested that the film be sent to Pearl Harbor for reproduction when theater facilities weren't able to provide the requisite copies, especially when large numbers of copies were needed. The establishment of a courier service provided expeditious transfer of film and prints.[29]

Shortly after the start of the Korean conflict, CINCPACFLT's Intelligence Section was raised to the status of a staff division. The Intelligence Division conducted daily oral briefings for CINCPACFLT and staff and gave special briefings to type, force, group, and unit commanders, commanding officers of individual ships, and intelligence officers of subordinate staffs. Intelligence was also disseminated to the staff and subordinate commands through intelligence annexes to various plans, by means of the *Weekly Intelligence Digest*, and via formal intelligence estimates and staff studies.

A photo interpretation school was initiated at the Naval Air Station, Alameda, under Commander Naval Air Forces, Pacific, and the CINCPACFLT Intelligence Division coordinated the flow of fleet photography from the forward areas to the respective Navy bureaus in Washington, with collateral distribution of prints to other naval commands as required.[30]

The Submarine Evaluation Board was established in 1952 to provide for a systematic and rapid evaluation of submarine contact reports received by CINCPACFLT. The Assistant Chief of Staff for Intelligence served as senior member on the board.[31]

The Joint Operational Intelligence Agency, Pacific Command (JOINPAC) was established during Fiscal Year 1953 by integrating the CINCPACFLT Intelligence Division (N-2) with the CINCPAC Joint Intelligence Division (J-2). The consolidation was made to maximize efficiency and economy of personnel and funds by physically integrating intelligence operations, administration, personnel management, and facilities. The Special Intelligence Production Unit was established and functioned under the operational command of CINCPACFLT for the production of targeting materials. The Submarine Evaluation Board, established at CINCPACFLT during Fiscal Year 1952, was renamed the Submarine Classification and Damage Assessment Board, but it retained the same functions and membership.[32]

Expansion of the Peacetime Intelligence Capability in the Pacific, 1954–1969

In reaction to increasing Chinese Communist support to Viet Minh aggression against the French in Indochina, the United States ordered two aircraft carriers and a squadron of destroyers to the Philippines in February 1954, ostensibly for six weeks of "fair weather training." Commander First Fleet VAdm. William K. Phillips, with a small operational staff, was flown from San Diego to Sangley Point via Honolulu (during the long Washington's Birthday weekend) to assume command of the force. At CINCPACFLT Headquarters, VAdm. Phillips and his staff were briefed on the situation in Indochina and on his mission while deployed to the South China Sea. First Fleet Intelligence Officer Cdr. Wyman H. Packard was also briefed by the CINCPACFLT Intelligence Officer, Capt. Samuel B. Frankel, on the intelligence support that could be expected and the intelligence collection requirements and possibilities in the anticipated operating area.

Dien Bien Phu had been under Viet Minh attack since late 1953, and when the U.S. carrier force arrived in the Philippine area, the situation was becoming critical for the French. VAdm. Phillips's classified mission was to be ready for combat operations in case a decision was made to employ his force in support of the French. Initially, the aircraft carrier *Wasp* (CV 18) was the flagship, and the intelligence officers (air group and ship) on board provided staff support to Flag Intelligence Officer Cdr. Packard. Appropriate maps and charts were obtained from Commander Naval Forces, Philippines for plotting the situation in Indochina and for planning possible air strikes. Daily situation reports were received from CINCPACFLT, and other intelligence reports were received from the Naval Security Group Detachment on board *Wasp*.

While the force maintained a high state of readiness for contingencies, it conducted a wide variety of training exercises. Reconnaissance patrols were flown to identify and photograph shipping in the area of the force when it was at sea. Detailed surveys were made of the Subic Bay, Cubi Point, and Sangley Point facilities in the Philippines to determine their adequacy to serve as fleet bases.[33]

On 19 March 1954, Chief of Naval Operations Adm. Robert B. Carney ordered VAdm. Phillips to maintain a 12-hour alert and to prepare to steam near the entrance to the Gulf of Tonkin, ready to begin operations in support of the French on about three hours' notice. The force was accordingly moved to an operating area about 100 miles south of Hainan Island. On 25 March, CINCPACFLT recommended that carrier aircraft from Phillips's force conduct reconnaissance of nearby Chinese airfields, assembly points for shipment of supplies, and critical roads and trails over which artillery and other military items had been flowing to the Viet Minh for their Dien Bien Phu siege. Lang Son and Caobang across from Kwangsi Province, and Lao Cai south of Yunnan Province were also to be covered in the reconnaissance flights. Adm. Carney concurred with the recommendation on 29 March.

Using photo plane detachments from Phillips's two carriers, and with the force positioned in the Gulf of Tonkin about 125 miles east-southeast of Haiphong, the photo reconnaissance missions were successfully carried out. Flying in pairs at high altitudes, the photo aircraft covered railroads from west of Nanning and south of Kunming to Hanoi. The aircraft also took pictures of port facilities and airfields in the Hanoi and Haiphong areas and on

Hainan Island. No Chinese Communist reaction was noted.[34]

On 7 April, the Chief of Naval Operations advised CINCPACFLT to complete his reconnaissance by 12 April, including coverage of the Dien Bien Phu battlefield and the Viet Minh supply routes leading to that position. The reconnaissance effort was carried out with marginal success due to unfavorable cloud conditions, and the force then returned to the Philippines. The decision had been made that there would be no unilateral military action by the United States to save Dien Bien Phu.[35]

On 28 April, VAdm. Phillips was informed that he was to be designated Commander Southeast Asia Defense Command, a subordinate unified commander under CINCPAC. Phillips was to be prepared to establish his headquarters in Saigon. He detached his deputy chief of staff and his intelligence officer from his operating staff to return to San Diego to prepare for the possibility. The flagship's intelligence officer was assigned the added responsibility to act as flag intelligence officer. Although the Southeast Asia Defense Command was established for planning purposes in San Diego, with Army, Air Force, and CIA personnel assigned, it was never deployed to Saigon.[36]

While the "fair weather training" did not deter the Viet Cong operations against Dien Bien Phu, it did provide the opportunity to gather intelligence information and photographs that would be of value for comparative purposes when the United States became more directly involved in the area ten years later.

The separation of CINCPAC and CINCPACFLT staffs on 1 July 1956 took place as Capt. Rufus L. Taylor relieved Capt. Samuel B. Frankel as Fleet Intelligence Officer. Frankel's assistant, Col. Robert Lawson, USAF, relieved him as J-2 until RAdm. Edwin Layton arrived in September 1956. Adm. Felix Stump continued to be double-hatted as CINCPAC and CINCPACFLT, heading up both staffs, but he acknowledged that he could not remember which intelligence officer, Layton or Taylor, was on which staff. When Stump had a question in the intelligence field, he would ask the first officer who came to mind.

There was no serious conflict between the two intelligence staffs as to the delineation of their duties. Sometimes, when N-2 started a new project to provide improved support to the fleet, J-2 would object if it appeared to be a service that all three components could use and was consequently a project that J-2 should take over.[37]

When JOINPAC was disestablished and CINCPACFLT's Intelligence Division resumed functioning as a completely separate entity, it was able to fulfill more adequately the intelligence needs of CINCPACFLT and provide a greater responsiveness to fleet needs. At the same time, the Intelligence Division provided support and services in terms of substantive naval intelligence and special communications facilities to the CINCPAC Joint Intelligence Division.

Current intelligence was regularly furnished to CINCPACFLT and his staff through briefings and through a written *Daily Intelligence Summary*. Other commands received current intelligence by means of the weekly PACFLTINTSUM (Pacific Fleet Intelligence Summary) message. Special reports and daily INTSUMs were provided fleet forces during critical periods, such as the Middle East crisis in November 1956 and the P4M Mercator reconnaissance aircraft shoot-down off the China coast by Chinese fighter aircraft on the night of 22–23 August 1956. Intelligence estimates and annexes to various CINCPACFLT operations plans were prepared, as well as contributions to CINCPAC estimates. The Intelligence Division also coordinated the collection of intelligence information by fleet units, as well as the production of target lists and folders by the Fleet Intelligence Center, Pacific (FICPAC). During the Indonesian crisis in 1957–1958, an additional SITSUM (Situation Summary) was issued daily.[38]

The Intelligence Division conducted weekly briefings for CINCPACFLT and his staff on the current situation and on special subjects as needed. Briefings were also given to selected officers from major combatant ships and forces en route to WESTPAC, and representatives and intelligence officers of returning commands were debriefed when appropriate.

An Intelligence Standard Operating Procedure for the Pacific Fleet was issued in March 1958 to outline functions and responsibilities for the following: intelligence collection; command and communications; counterintelligence; captured and recovered equipment, documents, and personnel; maps, charts, publications and photographs; the Eniwetok Proving Grounds; and interpreters, translators, and specialist teams. During the same month, an instruction was issued on Prepositioned Intelligence Material for Emergency Use.[39]

During 1959–1960, a program for the miniaturization (microfilming) of Tactical Target Materials (TTM) for use on board Pacific Fleet aircraft carriers was approved, and funding was authorized for purchasing the necessary equipment. The miniaturization system, initially proposed by CINCLANTFLT, consisted of 35mm negatives of TTM documents mounted in standard IBM punchcards, known as aperture cards. A 3M viewer-printer permitted review, selection, study, and reproduction of the microfiche material as desired. In addition, an

IBM printout provided an inventory of the information available on all targets by country and an overall target index.[40]

Prior to the adoption of the miniaturized Tactical Targeting Materials system, CINCPACFLT had been forced to remove TTMs from Pacific Fleet carriers due to the volume of the materials required for the extensive geographical area having potential operational interest. The problem of excessive volume had been resolved, in part, by prepositioning complete Tactical Target Dossiers at various Pacific Fleet Air Navigation Offices for emergency distribution. The arrangement, however, had not been satisfactory, and the miniaturization program was adopted in order to return the TTM files to the carriers, thus greatly improving their readiness for action. During Fiscal Year 1961, the TTM program was expanded to include files for Laos, North and South Vietnam, and Cambodia.

Fleet camera pools, under the overall control of FICPAC were established at San Diego, Pearl Harbor, Cubi Point, and Yokosuka to furnish basic camera needs to Pacific Fleet ships not having a photographic equipment allowance, enabling those that were scheduled to visit foreign ports or forward areas to perform photo collection missions.[41]

In September 1960 and in January and March 1961, the USSR began operating missile range tracking ships in the mid-Pacific in connection with Soviet missile and satellite launchings. The ships and their activities were kept under close surface and air surveillance by Pacific Fleet units.[42]

During special surveillance and photo collection efforts against the Soviet missile range instrumentation ships (SMRIS), the use of KB-10A cameras proved to be a marked improvement over the standard shipping surveillance cameras. In response to recommendations by CINCPACFLT, the Bureau of Weapons revised the camera allowance of all reconnaissance (VP) squadrons to include one KB-10A camera for each aircraft. The fast, high-resolution cameras immeasurably improved the quality of shipping surveillance photographs obtained in the Pacific area.

In the latter part of Fiscal Year 1961, a major effort was made to obtain photo coverage of Southeast Asia that greatly improved U.S. knowledge and tactical targeting ability for that area. The reconnaissance was also an extremely valuable exercise in testing the Navy's photography capabilities, limitations, and deficiencies. To improve the amount and quality of photo collection, an Aerial Photo Intelligence Requirements Series (APIRS) was initiated to provide a ready reference list of vertical and oblique aerial photo coverage, by various specified scales, desired for Southeast Asia and the Philippines.

During the same period, an IBM-407 computer was installed at FICPAC, and an IBM-704 was installed at the Fleet Operational Control Center, Pacific at Kunia, Oahu. The two machines provided the Pacific Fleet with an almost unlimited capacity for systematically storing readily retrievable intelligence materials.[43]

The Office of Naval Intelligence monitored the development of the highly mechanized Operational Control Center at Kunia, through representative attendance at the frequent conferences on its progress in Honolulu and Washington and through close coordination with the Operational Research Division of the Applied Mathematics Laboratory of the David Taylor Model Basin (DTMB), where the programs for the control center were being developed. It was estimated that the intelligence input to the center would amount to approximately 30 to 70 percent of the total data input. The ONI Mechanization Committee worked closely with DTMB in the development of CINCPAC's various inputs to the automated intelligence files.[44]

During Fiscal Year 1962, aerial photo coverage of critical areas in Southeast Asia by Pacific Fleet units was completed, resulting in improved tactical targeting material and numerous special photo-intelligence reports needed by planners for contingency operations. Photo mapping of South Vietnam by fleet photo squadrons was nearing completion at the end of Fiscal Year 1962, and the results were intended to be used by the Army Map Service, Far East, to standardize its maps of South Vietnam.

The Class D photo laboratory located at Cubi Point was upgraded to a special fleet lab during Fiscal Year 1962. Additional equipment and personnel were made available to the facility under its new classification, and the Cubi Point facility was geared to support emergency and wartime Seventh Fleet forward photo-intelligence efforts.

To assist forces afloat in emergency and critical situations, augmenting photo interpreter teams of four photo technicians each were assigned to Navy Heavy Photographic Squadron 61, Detachment Alfa, at Cubi Point, during Fiscal Year 1962. At the end of the period, there were two augmenting teams at Cubi Point available to photo interpretation officers on board Seventh Fleet ships and attached to flag staffs.[45]

During the early 1960s, intelligence material contained in CINCPACFLT war plans and operational orders was revised and updated as new information become available. Most revisions were contained in the intelligence annexes of various CINCPACFLT contingency plans, although periodic naval intelligence contributions were also made to Pacific Command Intelligence Estimates produced by CINCPAC.

The annual Pacific Fleet Collection Plan, derived from requirements for intelligence to support CINCPACFLT General War Plan 1-63, was revised during Fiscal Year 1963 to show known gaps in intelligence holdings on Far East subjects and to assign collection tasks to principal fleet collectors.

As a result of the constantly changing political and military situation in Southeast Asia, a contingency collection plan was issued. It was based on the intelligence required to support CINCPACFLT OPLAN 32-63 and to fill gaps in the intelligence needed for potential amphibious landings and other military operations in Southeast Asia.

A daily intelligence briefing was included in the morning briefing to CINCPACFLT, local type commanders, and selected members of their staffs. The intelligence briefings were developed from message traffic received during the past twenty-four hours and from information received from CINCPAC and other component commanders over secure phone lines.

Pacific Fleet units were kept informed of world events through the CINCPACFLT Intelligence Summary, disseminated daily by message to more than 100 addressees. A mailed version of the INTSUM was prepared daily for delivery to naval and other military commands in the local Hawaiian area. To provide certain authorized recipients with additional information required for special operations, an intelligence summary at a higher classification was produced and disseminated on a weekly basis or more often, as required.

All fleet units rotating to the Western Pacific were given intelligence briefings upon arrival at Pearl Harbor. Emphasis was placed on Soviet electronic warfare techniques and on the equipment used by the Soviets in their efforts to detect, locate, and identify Pacific Fleet units. Unit commanders were made aware of Soviet air reconnaissance efforts.

An extensive intelligence collection program was carried out throughout the Pacific area during the early to mid-1960s. It involved numerous aircraft and ships operating from the Chukchi Sea in the north to the Indonesian archipelago in the south.

Although aircraft played an important role in collection efforts against Soviet ICBM (intercontinental ballistic missile) impacts in the mid-Pacific, destroyer escort radar pickets (DER), especially configured for intelligence collection and rapid communications, provided an equally valuable contribution to the national collection program. The DERs were to maintain surveillance over the Soviet missile range instrumentation ships and to provide timely operational intelligence so that other collection platforms could monitor missile reentry and impact collection operations.

To fill important amphibious intelligence requirements in Southeast Asia, two major survey operations were carried on for CINCPACFLT during Fiscal Year 1963. A fast amphibious transport (APD) of the Seventh Fleet continued coastal and landing beach surveys in the Republic of Vietnam area that had been started during the previous year. Also, a Marine Coastal Survey Team, supported by a hydrographic survey ship, made an extensive beach survey of the east coast of the Isthmus of Kra of Thailand.

An operational intelligence plot was maintained and staffed on a 24-hour-a-day basis in support of the CINCPACFLT OPCON (Operational Control) Center. Detailed situation, order-of-battle, and special intelligence plots were updated daily and were available to the OPCON Center on instant notice.

Soviet Tu-16 Badger and Tu-95 Bear reconnaissance aircraft overflights of Pacific Fleet carrier forces provided the operating forces with considerable insight on Soviet bomber tactics and operating procedures. Soviet fighter reaction to peripheral reconnaissance flights by Pacific Fleet aircraft continued at the same high level begun in 1962 and provided information on Soviet offshore detection procedures and capabilities.[46]

The CINCPACFLT INTSUM messages continued to be issued on a daily basis throughout Fiscal Year 1964, in one version based on all-source material and another on collateral-source material only. The number of addressees for the collateral-source summary increased to 175. The contents of the summaries were tailored to meet the needs of CINCPACFLT subordinate commands. The mailed version of the collateral summary, previously sent to various commands of the three military services in the local area, was discontinued shortly after 1 January 1964 to reduce costs and extra paperwork. Special summaries of interest to certain commands involved in specific and sensitive operations were prepared as need dictated. For example, the "Concord Squadron," composed of the carrier *Bon Homme Richard* (CVA 31), fleet oiler *Hassayampa* (AO 145), and destroyers *Shelton* (DD 790), *Blue* (DD 744), and *Frank Knox* (DD 742), which deployed to the Indian Ocean from 4 April to 10 May 1964, was kept informed on unusual events in the Indian Ocean and the Arabian Sea throughout the period.

An extensive collection program continued during Fiscal Year 1964 throughout the Pacific, as well as in the Indian Ocean, the Gulf of Aden, and the Arabian Sea. In the last two areas, reconnaissance by a carrier-based aircraft squadron satisfied a large number of national, Pacific Command, and fleet intelligence collection requirements on the Indian Ocean and contiguous areas.

Port data collection in the Pacific Fleet area during Fiscal Year 1964 was highlighted by the visit of the survey ship *Rehoboth* (AGS 50) to Nakhodka, the commercial port for Vladivostok, from 9 to 13 November 1963. The visit was in reciprocity for a previous visit by the Soviet research nonmagnetic sail research ship *Zarya* to Honolulu and San Francisco. The *Rehoboth* had also conducted an oceanographic and hydrographic survey in the Sea of Okhotsk from 7 to 26 October, prior to the Nakhodka visit.[47]

Chapter Notes

1. James O. Richardson, *On the Treadmill to Pearl Harbor: The Memoirs of Admiral J.O. Richardson as told to George C. Dyer, USN (Retired)* (Washington: GPO, 1973), 290.

2. Commander in Chief, Pacific Fleet/Commander in Chief, Pacific Ocean Area (CINCPACFLT/CINCPOA) "Command History, World War II," MS, 44–45, OA.

3. Ibid., 327.

4. U.S. Congress, Joint Select Committee, *Pearl Harbor Attack: Hearings Before the Joint Committee on the Investigation of the Pearl Harbor Attack*, 39 vols. 1946, 39 (summary volume): 104.

5. Ibid., 106.

6. Ibid., 111.

7. VAdm. Rufus L. Taylor, interview by author, 29 Oct 1975.

8. RAdm. Edwin T. Layton, Oral History, USNI, Annapolis, MD, 1970, 74–75.

9. Ibid., 92–93.

10. RAdm. Edwin T. Layton ltr to author, 29 Jun 1976.

11. CINCPAC/CINCPOA, "Command History, World War II," 319.

12. Ibid., 320.

13. Layton oral history, USNI, 79.

14. Ibid., 108.

15. Ibid., 114–16.

16. Ibid., 127–28, 319–22.

17. "History of the CINCPACFLT Combat Intelligence Section, 7 Dec 1941–Sep 1945," MS, 3, OA; and Layton ltr, 29 Jun 1976.

18. CINCPACFLT/CINCPOA, "Command History, World War II," 322–24, 327.

19. Ibid., 57–58.

20. Ibid., 67.

21. Layton oral history, USNI, 145–48.

22. Ibid., 162.

23. Ibid., 154–55.

24. Ibid., 151.

25. Ibid., 158–61.

26. CINCPACFLT/CINCPOA, "Command History, World War II," 354–55.

27. "History of CINCPACFLT Combat Intelligence Center," 5.

28. *CINCPACFLT Annual Report, 1 Oct 1949–30 Jun 1950*.

29. CINCPACFLT ltr, ser 00052, 20 Sep 1950, box 5, Job 11146, TSC 8691, OA.

30. *CINCPACFLT Annual Report, FY 1951*, 65.

31. *CINCPACFLT Annual Report, FY 1952*, 69.

32. *CINCPACFLT Annual Report, FY 1953*, 79.

33. VAdm. Edwin Hooper, Dean C. Allard, and Oscar Fitzgerald, *The U.S. Navy and the Vietnam Conflict*, vol. 1, *Setting the Stage to 1959* (Washington: Naval History Division, 1976), 235–37.

34. Ibid., 252–53.

35. Ibid., 255.

36. Ibid., 262.

37. VAdm. Rufus L. Taylor, interview by author, 23 Nov 1975.

38. *CINCPACFLT Annual Report, FY 1957*, 61; and *CINCPACFLT Annual Report, 1 Feb 1958–30 Jun 1958*, 72.

39. Ibid.

40. *CINCPACFLT Annual Reports, FY 1960*, 60; and *FY 1961*, 28.

41. *CINCPACFLT Annual Report, FY 1960*, 60.

42. *CINCPACFLT Annual Report, FY 1961*, 8.

43. Ibid., 27–28.

44. OP–923M, "Review of ONI's Efforts in Intelligence Mechanization," 25 Jan 1961, passim.

45. *CINCPACFLT Annual Report, FY 1962*, 37–38.

46. *CINCPACFLT Annual Report, FY 1963*, 40, 42–45.

47. *CINCPACFLT Annual Report, 1 Oct 1963–26 Jun 1964*, 13, 15, 49–50, 52.

CHAPTER 36

Operating Forces, South and Southwest Pacific Area

The South Pacific and Southwest Pacific areas are considered together in this chapter, partly because many of the sources of intelligence information were common to each and also because, initially, most of the current intelligence support to Commander South Pacific Area and Forces (COMSOPAC), came from various Southwest Pacific (SWPAC) commands and activities.

The amount of detailed information located on naval intelligence activities in the two areas has been disappointingly small, possibly because, as commands in these areas completed their operations, they moved on without summarizing or preserving their records. Lt. James A. Michener, USNR, was Historical Officer in 1945 on the staff of COMSOPAC, and he reportedly had access to some of the intelligence files. COMSOPAC's official history manuscript has not been located, although certain elements may appear in Michener's *Tales of the South Pacific*.

Commander South Pacific

In preparation for the landings at Tulagi and Guadalcanal in 1942, all immediately available information on those islands was collected. In addition, Gen. Alexander A. Vandegrift, USMC, Commanding General, 1st Marine Division, sent LtCol. Frank B. Goettge to Australia, where he obtained valuable information on the southern Solomons from Australian intelligence centers and from individuals familiar with the area.[1]

Following the landings on 7 August 1942, VAdm. Robert L. Ghormley, COMSOPAC, with headquarters in Auckland, New Zealand, was almost completely dependent upon Gen. Douglas MacArthur's Southwest Pacific Area Command (SWPAC) for current intelligence on Japanese reactions to the landings. This was because the line of demarcation between SOPAC and SWPAC ran just west of Guadalcanal and separated it from the approaches that Japanese naval forces would use coming down the "Slot" from Rabaul.

Although reports from SWPAC aerial and submarine reconnaissance indicated that Japanese forces were on the move on 7 and 8 August, their inaccuracies and delayed receipt by the forces covering the Guadalcanal landings contributed to the disaster that befell the U.S. Navy at the Battle of Savo Island on the night of 8–9 August. Furthermore, no intelligence officers were assigned to the covering forces and, consequently, no professional analysis was made of the fragmentary and conflicting reports received before the Japanese attack.[2]

Subsequently, Australian coastwatchers in the northern Solomons, New Ireland, and New Britain provided invaluable information about the concentration and movement of Japanese air and naval forces threatening U.S. forces in the southern Solomons. The coastwatchers established jungle hideouts at vantage points where they could observe and report on Japanese-held ports and bases and on the movements of combatant forces and troop reinforcements en route to the Guadalcanal-Tulagi area.[3]

When VAdm. William F. Halsey relieved VAdm. Ghormley as COMSOPAC on 18 October 1942, he established his headquarters at Noumea, New Caledonia. His Force Intelligence Officer was Col. Julian Brown, USMC. Collaboration between Brown and the rest of the staff, particularly with Halsey's operations officer, was very close. Intelligence information from COMSWPAC, CINCPAC (Commander in Chief, Pacific), aerial and submarine reconnaissance, coastwatchers, and communications intelligence (COMINT) was brought as received into Flag Plot and the Operations Section, and was discussed as to its relevance to current and planned operations. In turn, operation reports were made available to Force Intelligence when they contained information on enemy forces.

Some of the intelligence staff officers assigned to COMSOPAC during the 1942–1943 period included Cdr. Marion C. Cheek, Cdr. Edward S. Pearce (a Japanese-language officer), LCdr. Logan Jenkins, and Lts. Harris Cox and John Goodbody, USNR. Col. Brown was relieved as Force Intelligence Officer early in 1943 by Col. F. P. Munson, U.S. Army.

Photographic interpretation, prisoner-of-war interrogation, and other specialized intelligence units performed support functions appropriate to their specialities. Commander Air Forces, South Pacific and the 1st Marine Aircraft Wing rotated air combat intelligence officers to the various squadrons operating from Noumea, Espiritu Santo, Guadalcanal, Tulagi, and Munda.

During the height of the campaign for Guadalcanal, LCdr. Daniel J. McCallum, a Japanese language officer, was on that island for eight months as part of COMSOPAC's intelligence group, listening to the voice communications of the Japanese army and navy in the area and reporting what he heard to the local U.S. commanders.[4]

In March 1943, the staff of the Chief of Naval Operations in Washington recommended that the Radio Intelligence Unit (RIU) at Auckland be merged with the Advanced Unit at Melbourne. It was believed that the interests of COMSOPAC could best be served by a major center at Melbourne. A small RIU coding board would continue operation at COMSOPAC headquarters in Noumea. CINCPAC and COMSOPAC both concurred in the move, and the latter made the necessary arrangements to carry it out.[5]

As of March 1944, Col. Ronald A. Boone, USMC, who had been the 4th Marines intelligence officer in Shanghai before the war, was COMSOPAC Intelligence Officer. The officer in charge of the Counterintelligence Unit was Capt. Emil Kruger, USMC. Other elements of the intelligence staff included the Operational and Combat Intelligence Sections, the Photo Intelligence Section, and the COMINT Section. The Operational Intelligence Section included a Lt. Byron R. White, USNR, who later became a U.S. Supreme Court justice.

In 1944, air strikes were still being conducted out of bases in SOPAC, particularly against targets such as Rabaul. The strikes required a considerable photo intelligence effort, and a rather large Photographic and Reproduction Unit was maintained on the COMSOPAC staff at Noumea, New Caledonia.

The Counterintelligence Unit of the COMSOPAC staff was concerned with any possible sabotage, espionage, or subversion involving naval facilities or personnel. It maintained close liaison with the Army Counter-Intelligence Corps office in Noumea and with the local French civil police agencies. It also maintained close liaison with the port director at Noumea for the purpose of checking on crew members of merchant ships arriving in port. Occasionally, leads would be received from district intelligence offices back in the United States concerning suspected crew members due to arrive in SOPAC ports. In those instances, the ship involved would be boarded and, if appropriate, surveillances would be conducted. The French population in Noumea, at that time, had a small but active Communist party and a couple of active Communist-front organizations.[6]

Commander Southwest Pacific

A few days after the Battle of Coral Sea in May 1942, RAdm. J. G. Crace, RN, commented that Allied aircraft reporting on enemy ships was very bad and that the recognition of ships by aircraft was completely inadequate. In fact, it was so bad that his flagship had been attacked by U.S. Army Air Corps B-17 bombers. VAdm. Herbert F. Leary, Commander Allied Naval Forces SOWESPAC, replied that efforts would be made to improve Army Air Corps ship recognition. When the Air Corps was approached, however, it was found that the subject was so distasteful that the commanding general prohibited further discussion on the grounds that the problem had been exaggerated.

To study possible cooperation between submarines of the Seventh Fleet and aircraft of the Fifth Air Force, RAdm. James Fife, Jr., rode as a passenger on several Air Corps reconnaissance flights during the spring of 1943. In December 1942, he had been successful in persuading the Air Corps to extend its reconnaissance into areas where more targets for submarines might be found. But from his experience as a passenger on extended reconnaissance flights, Fife concluded that the Army pilots were not properly trained in ship identification or in sea reconnaissance. Sending naval observers on the flights was suggested but was not carried out due to a lack of personnel.[7]

The U.S. Navy established numbered operational fleets on 15 March 1943, and the U.S. naval forces in the Southwest Pacific Area were designated the Seventh Fleet. Its Intelligence Officer, Capt. Arthur H. McCollum, had a dual capacity; he acted as advisor to Commander Seventh Fleet (COM7THFLT) on intelligence matters and also served as commander of the semi-independent Seventh Fleet Intelligence Center (SEFIC). The Seventh Fleet Intelligence Division advised on intelligence policy and handled public relations and censorship; SEFIC gathered intelligence for dissemination to organizations needing intelligence support. Initially, Seventh Fleet headquarters and SEFIC were located at Brisbane, Australia.[8]

The Seventh Fleet Intelligence Center operated from early 1943 to the end of the war. Its staff grew from a few operational intelligence officers to over 200. Approximately sixty officers were ultimately located at Seventh Fleet headquarters to maintain plots, brief the admiral and his staff, process intelligence, and disseminate information to the operating forces. SEFIC also assigned officers and personnel to related intelligence activities in the area, such as the Mobile Explosive Investigation Unit; the Tactical Air Intelligence Unit; the Central Interpretation Unit, which handled photographic intelligence; the Allied Translator and Interpreter Section, which processed all prisoners of war and captured documents; and the Army's MIS-X (Military Intelligence Service, Unit X), which was responsible for developing survival intelligence.[9]

SEFIC was organized functionally and was composed of sections to provide intelligence support material for the various types of naval operations, such as air, amphibious, and submarine. Briefings were prepared and given daily to COM7THFLT and COMSWPAC (Gen. MacArthur).

As U.S. forces fought their way up through New Guinea and the Southwest Pacific islands to the Philippines, SEFIC moved from Brisbane to Hollandia to Leyte. It sent an advance party forward to provide the necessary intelligence support to the naval commands and ships involved in the New Guinea operations. SEFIC Unit No. 1, composed of eight officers and two yeomen, was set up at Manus with LCdr. Cecil M. Deason, USNR, as the officer in charge.[10]

By January 1944, officers from SEFIC had been assigned to Cruiser Division Five, Destroyer Division Five, Commander Task Force 71 (submarines), Commander Motor Torpedo Boat Squadrons, long-range reconnaissance aircraft squadrons, the Fifth Air Force, and Australian Forces. Also, six officers staffed the SEFIC advanced echelon at Port Moresby.

On 1 September 1944, SEFIC moved to Hollandia with the flag, leaving a small rear echelon at Brisbane with the Deputy Commander Seventh Fleet. An advance unit of the Allied Translator and Interpreter Section also moved to Hollandia to handle captured material and prisoners.

SEFIC started publishing a *Weekly Bulletin* while it was located at Brisbane. The *Weekly Bulletin* staff later moved to Hollandia with the rest of the SEFIC staff. Distribution of the SEFIC bulletin was made to all Seventh Fleet cruisers, destroyers, motor torpedo boat squadrons, and other assigned units.

After the Battle of Leyte Gulf, SEFICU No. 4 was established at Tacloban on Leyte, and, in January 1945, SEFIC moved from Hollandia to Tolosa on Leyte. Unit No. 2 was with the Tactical Air Intelligence Unit, and Unit No. 3 was composed of Japanese language officers. Later, a Unit No. 5 was set up at Subic Bay, and Unit No. 4 was moved to Clark Field.[11]

Chapter Notes

1. VAdm. Robert L. Ghormley, "Narrative on SOPAC," 10 Apr 1942, 13, OA.

2. Capt. Wyman H. Packard, USN (Ret.), "Intelligence and the Navy," *Naval Review 1968* (Annapolis: USNI, 1969), 209–11.

3. Naval Forces, Southwest Pacific (NAVFORSWPAC), *Command History, 1946.*

4. Capt. Henri H. Smith-Hutton, Oral History, USNI, Annapolis, MD, 1979, 1:173, copy in OA.

5. Chief of Naval Operations (OPNAV) msg 102152Z Mar 1943; and COMSOPAC msg 121415Z Mar 1943.

6. Cdr. James L. Newson ltr to author, 10 Jan 1976. Newson, then a lieutenant (jg), was assigned to the Counterintelligence Unit of the COMSOPAC staff and also had collateral duty as liaison officer with the French colonial government of New Caledonia.

7. *NAVFORSWPAC Command History, 1946*, 48–50.

8. Ibid., 28; and RAdm. Arthur H. McCollum, Oral History, USNI, Annapolis, MD, 1971, 2:510–11, copy in OA.

9. *ONI Review*, Nov 1945, 40.

10. Cdr. Cecil M. Deason, USNR (Ret.), interview by author, 21 Jul 1976. Deason was Prosecuting Attorney, Jefferson County, Alabama, when called to active duty, 31 March 1941. He was assigned to DIO-8ND in New Orleans until June 1943, attended Advanced Naval Intelligence School, New York (Henry Hudson Hotel), and was then sent to Seventh Fleet Intelligence Command (SEFIC), Brisbane.

11. ONI, "OPINTEL Notes," Jul 1945.

Chapter 37

Operating Forces, Far East Area

This chapter includes information on the intelligence activities under Commander U.S. Naval Group (COMNAVGRP), China during World War II and Commander Naval Forces, Far East (COMNAVFE), and Commander Seventh Fleet (COM7THFLT) after World War II.

Information on Naval Intelligence activities in the Far East is also found in Chapters 4, 5, 12, 13, 15, 32, and 40.

Asian area commands not researched for this book include Commander Naval Forces, Philippines and Commander Naval Forces, Japan. A chapter on the Vietnam War should probably also be written, although, except for support to naval operating forces, most of the Navy's intelligence resources in Vietnam were expended in the joint service intelligence centers in Saigon. The Navy was required to provide personnel to those centers primarily so that they would be "joint," not to fulfill any need for persons specifically qualified in naval intelligence.

Naval Group, China

U.S. Naval Group, China was established in early 1942 by Cdr. Milton "Mary" E. Miles, who acted on oral orders from Commander in Chief, U.S. Fleet (COMINCH) Adm. Ernest J. King to set up a network of weather reporters and coast watchers and to harass the Japanese in China. Cdr. Miles set up his headquarters at Chungking.

Liaison and close collaboration were established with Gen. Tai Li, the Nationalist Chinese intelligence chief, expediting the establishment of a weather observation and coastwatcher net. Initially, the effort was called the Friendship Project, and it was officially designated the SACO (Sino-American Cooperative Organization) Agreement in April 1943. The U.S. Navy part of the organization was assigned the title U.S. Naval Group, China.

Approximately 2,500 U.S. Navy and Marine Corps personnel were assigned to NAVGRP China. Intelligence officers assigned were of both the S(A) (Special Duty, Aviation) and S(I) (Special Duty, Intelligence) classification; the S(A) officers primarily handled intelligence liaison with the air forces, and the operational intelligence officers supervised the coast watcher nets. NAVGRP China also received intelligence information from the Chinese army intelligence organization and from the U.S. Fourteenth Air Force (14th AF). The Chinese supplied about fifty reports a week, and the Fourteenth Air Force provided photo intelligence reports and daily and weekly summaries on Japanese air, shipping, and related operations.

A daily dispatch was sent by COMNAVGRP China to local commands and to COMINCH, CINCPAC, COMSOWESPAC, and the XX Bomber Command. Fleet Liaison Officers at what later became the Sino-American Cooperative Organization headquarters expedited the dissemination of urgent intelligence information to fleet units operating off the China coast. Such intelligence from a coastwatcher enabled *Barb* (SS 220) to attack a Japanese convoy at night in the Chinese harbor of Namkwan, for which the submarine's commanding officer, Eugene B. Fluckey, later received the Medal of Honor.

An Air Ground Aids Section was established at COMNAVGRP China for escape and evasion assistance to downed airmen. In September 1944, NAVGRP intelligence officers made a survey of coastal areas to obtain data needed for survival assistance. Village officials were given guidance in the recognition of U.S. airmen and in procedures to assist them. Approximately 900 Army Air Force and Navy fliers were rescued in China during the war, representing 90 percent of all airmen bailing out or ditching in Japanese-occupied Chinese territory.

In early 1945, in cooperation with the Commander of the Fourteenth Air Force, Miles established the Anti-Shipping Control Center, which was designed to "obtain, evaluate, and disseminate all shipping information from all sources in China and direct the air effort against Japanese shipping." NAVGRP China directed Army Air Force mining efforts along the China coast to force Japanese shipping out into the deeper waters where U.S. submarines were operating.[1]

Following Cdr. Miles's arrival in early May 1942, the first contingent of personnel and equipment, consisting of Lt. Daniel W. "Webb" Heagy III, six enlisted radiomen, and six tons of radio equipment, for COMNAVGRP China reached Chungking in September 1942. They were the first members of Naval Group, China to occupy "Happy Valley," a 200-acre site eight miles outside Chungking that was to become the operational and training center for the group.[2]

The second and third contingents arrived in October and November 1942. They included LCdr. Edward Gilfillan, USNR, an explosives expert, chemical and mechanical engineer, and long-distance swimmer; Maj. John Masters, USMC; and Lt. Raymond Kotrla, an aerologist, multilinguist, and expert photographer.[3] Marines of SACO actively trained Chinese troops for guerrilla operations. One such officer, 2dLt. Robert H. Barrow, became the 27th Commandant of the Marine Corps.[4]

On 22 September 1942, Cdr. Miles had been unexpectedly appointed as coordinator of Office of Strategic Services (OSS) activities in the Far East. Because of the complications and uncertainties introduced by such an appointment, Chiang Kai-shek directed Miles and Gen. Tai Li to work up a written agreement to take the place of the oral gentlemen's agreement that had previously been observed. The new formal agreement, which became the SACO Agreement, was to be signed by the highest available authorities of the two countries.[5]

By terms of the agreement, which was initialed by Dr. T. V. Soong on 31 December 1942 (and eventually by President Roosevelt on 1 April 1943), the United States was to train guerrillas, intelligence groups, weather teams, saboteurs, and raiding squads. The United States was also to set up weather and radio stations using American equipment and, for the most part, Chinese personnel. In addition to personnel, the Chinese were to furnish transportation and material facilities in China, including bases of operations, and they would make available the intelligence facilities already established. The director was to be Chinese (Gen. Tai Li), and the deputy director was to be American (then-Capt. Miles), each with veto power over the operations of SACO.[6]

Ultimately, SACO put ten units at widely scattered locations to train Chinese guerrillas in small arms and demolition, intelligence collection, weather reporting, and the use of portable radios for reporting.[7]

On the recommendation of French Army Gen. Henri Giraud, NAVGRP China made contact in Indochina in 1943 with Cdr. Robert Meynier of the French navy. Meynier had recruited and organized a large number of agents in Indochina, and their communications were routed through COMNAVGRP China. Consequently, during the remainder of the war, reports were received from Indochina about shipping and port information, weather, Japanese aircraft, prisoners, and the status of wounded. The Indochinese reporting continued even after Cdr. Meynier was forced out in 1944 by de Gaullist elements.

In early 1944, Miles was promoted to the one-star rank of commodore. Up to that time, he had technically been attached to the U.S. Embassy at Chungking with the title of Naval Observer. The Navy part of SACO was now made a "Group of the U.S. Fleet," operating directly under COMINCH Adm. King in Washington.[8]

The administrative change in its status had no apparent effect on the logistic support problems SACO had to put up with throughout its existence. SACO's quota for supplies to be airlifted over the "Hump" from India was 150 tons per month, an inadequate amount, considering the number of people being supported and their extensive intelligence-gathering and operational support responsibilities.[9]

When Commo. Miles visited Washington in March 1944, Adm. King instructed him to be ready for fleet landings on the Chinese coast, possibly by December 1944. Upon his return to China, Miles spent two weeks personally surveying the coastal situation, selecting sites for coastwatchers, and recruiting pirates to provide assistance. When Miles returned to Kunming from the survey mission, he found Capt. W. L. Painter who Adm. Chester W. Nimitz (CINCPAC) had sent to survey the China coast also. Miles gave Painter all the data and pictures he had just gathered, provided supplies and guerrillas for his protection, and published Painter's report in book form, using SACO's printing plant in Calcutta. The title of the book was *The Painter Expedition*.[10]

Ultimately, SACO had under its control nearly 100,000 guerrillas, more than fifty weather stations, more than sixty coastal units, and numerous other small intelligence-gathering services. Its staff organization as of late 1944 was as follows: S-1 (Personnel), S-2 (Intelligence), S-3 (Operations), S-4 (Supply), S-5 (Communications), S-6 (Radio Intercept), and S-7 (Aerology). SACO's major assets included two radio stations that were capable of working directly with Washington, San Francisco, and Pearl Harbor. The operations officers in Chungking were Cdrs. Walter G. Ebert and I. Joseph Galantin, both submariners.

The commanding officers of Coastal Intelligence Units in China were supplied with four-letter codes for communication with Chungking. Submarines operating in certain areas off the Chinese coast were given the relevant code and ordered to monitor the appropriate frequencies. The submarines did not know the originator of the messages on the special frequency, and the Coastwatcher Unit did not know that fleet units could decode their transmissions. Operational immediate messages in the code usually related to Japanese ship movements and were, of course, of immediate interest to U.S. Navy submarines in the area.[11]

During World War II, naval personnel were trained for naval guerrilla warfare at Fort Pierce, Florida. The trainees were being prepared to train and organize indigenous Asian peoples into naval guerrilla forces. As an example, a SACO unit in the Amoy-Swatow area consisted of about 2,000 Chinese fishermen and coastal pirates, with their native small craft and about 300 U.S. Navy officers and enlisted personnel. The unit was intended to interdict the Japanese seaborne resupply effort by coastal and river mining operations; attack and destroy Japanese shipping by employing armed junks and sabotage operations; occupy and defend coastal islands; establish a network of cells and routes for the removal of Allied personnel from China; establish communications networks for the surveillance and reporting of Japanese shipping; interdict Japanese overland supply routes by raids and sabotaging bridges and transports; and establish meteorological reporting stations in conjunction with U.S. Navy forces in the area.[12]

Aerology was one of the important original reasons for the Navy being in China; weather reports were vital to the fleet in the Pacific, because the weather conditions coming down from the SACO area affected weather conditions in the Pacific and thus fleet operations. Cdr. Irwin T. Beyerly arrived in August 1943 to take charge of SACO's Weather Central. By early 1944, enough equipment had been assembled to form a SACO weather-reporting net. Chinese personnel attended a ten-week course similar to that given for Navy aerographers. Weather observation stations were set up at SACO camps and at General Claire L. Chennault's airfields. SACO's U.S. Navy Weather Central sent four daily broadcasts to the fleet.

SACO personnel assigned to Chennault's Fourteenth Air Force were soon called Navy Unit 14. In May 1944, when Cdr. Charles J. Odend'hal took over as the detachment's first commanding officer, its responsibilities included photo reconnaissance and interpretation, mining, radio intelligence, air combat intelligence, and some ground-to-air target guidance. Besides Kunming, elements of Navy Unit 14 were also located with air groups at Kweilin and at forward airfields.[13]

Odend'hal's group also worked with Technical Air Intelligence (TAI) and helped organize pilot rescue. TAI was a mixed group responsible to Washington and originally attached to the Naval Attaché, Chungking. The organization was later transferred to SACO and roamed the China-Burma theater looking for downed enemy aircraft, shells, mines, and other military equipment. In late 1944, the Navy's TAI was combined with its Army counterpart in China and placed under Army command. The intelligence office of the 14th AF tried to pick up everything brought in or photographed by 14th AF pilots that might be of help either to the fleet or to SACO. In turn, the detachment furnished the 14th AF intelligence with information from SACO's various sources that might prove useful to its operations.[14]

Air combat intelligence (ACI) officers went to all active fronts to collect information on Japanese aviation that would help fleet pilots. The first ACI officers sent to SACO, LCdr. Sam S. Savage and Lt. Henry F. Shoemaker, arrived in February 1944. LCdr. Marvin Plake arrived in May and was later commended for his writing of the "Tactics" section of the 14th AF weekly intelligence summary. Other ACI officers included Lts. Stanley E. McCaffrey, Frank Balsley, George H. Fiske, Alfred H. Driscoll, John A. MacLellan, and Edward Bolger; all became involved in combat operations.

Lt. Fiske was liaison officer with the 68th Composite Wing at Liuchow, where B-24s were flying ocean patrols in response to fleet requests. One of the patrols sighted and reported the Japanese Northern Carrier Force that diverted Adm. Halsey away from San Bernardino Strait at a crucial time during the Leyte landings.[15]

By the summer of 1945, fifteen American SACO officers were working for the Air Ground Aid Service (AGAS). Lt. Richard C. Scott was the first SACO officer to be assigned to the downed pilot rescue effort along the Chinese coast, and, before the end of the war, sixty-seven Army and Navy downed airmen had been rescued in his area around Nanking and Foochow. Lts. Frank Balsley and Stanley McCaffrey were two other SACO officers loaned to AGAS. Balsley helped twenty downed pilots escape from the Hangchow Bay area. During 1945, SACO coast camps picked up about twenty-five fliers who had been recovered by pirate fishermen loyal to Gen. Tai Li.[16]

In October 1945, Lt. Joseph A. Meyertholen, under Miles's direction in Shanghai, worked with a Japanese lieutenant commander to identify and plot Japanese minefields in the Formosa Strait, Shanghai approaches, Min River (at Foochow), Amoy Harbor, and Swatow.[17]

The SACO organization was formally disbanded on 30 September 1946 under Joint Chiefs of Staff decision 1290/8, although by that time most of the U.S. personnel had long since returned home. Throughout almost four years of war, SACO lost only three Americans as prisoners to the Japanese.[18]

Post–World War II China

Following World War II, U.S. Naval Forces, Western Pacific, maintained an intelligence liaison office in Shanghai, China, in order to stay informed on events transpiring within that unsettled country. In 1948, the liaison officer was Cdr. T. W. Joyce. One of Joyce's responsibilities was to develop capabilities to supply timely intelligence information under emergency conditions if the Chinese central government should lose control of the greater Shanghai area. Accordingly, he took steps to develop contacts that could provide essential basic information to his intelligence liaison office under such conditions. This office closed when the Chinese Communist powers gained control of China.[19]

Operational Intelligence Forces During the Korean War

On 1 June 1950, the staff allowance of the Intelligence Section of Commander Naval Forces, Far East, included one officer, Cdr. Arthur F. Johnson, plus one civilian interpreter and one enlisted yeoman. Following the start of the Korean War on 25 June, the allowance for the Intelligence Section was increased, and by 1 November 1950 the personnel allowance had grown to nineteen officers (one captain, one commander, four general-line lieutenant commanders, two aviation lieutenant commanders, four aviation lieutenants, and seven general-line lieutenants). The actual number of officers on board was twenty-three, headed by Capt. Martin R. Stone. Of these, five were on temporary duty in Korea to collect coastal information for fleet surface forces involved in blockading, shore bombardment, minesweeping, and evacuation.

Based on a COMNAVFE Staff Instruction of June 1948 that was still in effect at the start of the Korean War, the Intelligence Section was charged with procuring, evaluating, and disseminating to interested Navy and Army commands technical and current intelligence bearing on naval operations, counterintelligence data, and political and economic intelligence primarily of naval interest. The organization was also instructed to maintain liaison with the Army and Air Force intelligence organizations in the Far East and with the office of the chief counterintelligence officer in the Army's Far East Command. In addition, the COMNAVFE Intelligence Section had responsibility for writing the intelligence sections of COMNAVFE operation plans and operation orders.

During the early months of the Korean War, commands subordinate to Commander Naval Forces, Far East, that had established intelligence sections included Commander Seventh Fleet (two officers); Commander Amphibious Group One (three officers); Commanding General, 1st Marine Division (as of 15 November, eighty-nine officers and enlisted personnel, including a Combat Information Center team of seventeen and a prisoner-of-war interrogation team—both teams having been supplied by the Army); and Commander Blockade and Escort Force (section established in September, with two officers ordered in and due to report about 1 December).[20]

At the beginning of the Korean War, the flow of intelligence to naval forces was seriously hampered by the inadequate personnel in the COMNAVFE Intelligence Section, a breakdown in the normal incoming intelligence channels, and the overloading of communications channels.[21]

Capt. Edwin T. Layton was one of the officers drawn from the Pearl Harbor area at the start of the Korean War to augment COMNAVFE's intelligence staff. He had reported as District Intelligence Officer, 14th Naval District (DIO-14ND) in Honolulu in June 1950, having just come from duty as the commanding officer of the Naval Intelligence School at Anacostia, D.C. Layton reported to Yokosuka, Japan, on 8 July for temporary duty as COMNAVFE Intelligence Officer. He found four new graduates of the Intelligence School busily working on the Intelligence Annex to COMNAVFE's Operation Plan, an exercise similar to their last school "problem." They worked all day and all night for three or four days to finish the annex, doing a masterful job, for which they received many compliments.

COMNAVFE's original staff was small and not organized for a shooting war. After some delays, an Intelligence Section with thirty personnel billets was approved, but no additional intelligence officers reported for several months. After the Inchon landings (15–29 September), and more than two months of correspondence, Capt. Martin R. Stone relieved Capt. Layton, who was able to return to his regular duties as DIO-14ND, arriving back at Pearl Harbor in October 1950.[22]

On 7 October 1950, VAdm. C. Turner Joy, Commander Naval Forces, Far East, sent a Special Intelligence Team into Korea to gather data on ports, harbors, the coast, and landing beaches; to photograph anything of intelligence interest; to exploit captured documents and enemy equipment; and to interrogate prisoners of war. The team was to make their reports to Commander Seventh Fleet, the commanders of Task Forces 90 and 95, Underwater Demolition Teams 1 and 3, Commander Mine Squadron 3, and to United Nations forces participating in the conflict.

The Special Intelligence Team included Officer in Charge Lt. Chester J. Oleniacz, USNR, with Lt. Horace G. Underwood, USNR, as Korean linguist; Capt. Ronald E. H. King, Royal Marines; and Lt.(jg) Raymond Moley, Jr., USNR, as Russian linguist. In performing their tasks, members of the team accompanied armed reconnaissance patrols into unsecured areas to collect mine and coastal defense information of importance to UN minesweeping forces in the Wonsan-Songjin area, contributing to their successful operations along the Korean east coast. While working through the Korean Military Advisory Group with the Eighth Regiment of the Republic of Korea (ROK) army's Capital Division, the Special Intelligence Team contributed to the Koreans' successful operations in the fighting prior to the capture of Myongchon and the crossing of the Orangchoon River. Their tactical interrogation of prisoners supplied on-the-spot information to ROK field commanders and to air-ground control and naval gunfire support teams. The team also checked the heavily booby-trapped buildings, caves, and snow-covered harbor installations in search of critically needed mine information at Chongjin; took photographs of important installations from Kilchu to Chongjin that proved to be of value in later attacks by UN forces against the enemy in that area; and rendered valuable services and provided advice to responsible UN commands and civil authorities during the orderly and successful withdrawal of personnel and material from the Chongjin-Songjin area during the December Communist counteroffensive. Following the withdrawal, the Special Intelligence Team was dissolved on 13 December 1950.[23]

Organized at the end of 1950 as an integral part of the COMNAVFE Intelligence Section, the Shipping Surveillance Center collated, evaluated, and disseminated reports of sightings received through air, surface, subsurface, and radar searches and reports from coast watchers.[24]

At the end of 1951, air, surface, and Marine forces had a relatively adequate number of qualified intelligence personnel. Shortages of photo interpreters, linguists, technical intelligence personnel, and trained enlisted men continued, however. The Air Intelligence Schools and the Photo Interpretation School at Anacostia were not graduating enough trained personnel to meet combat requirements. The shortage was expected to become more acute as reserve officers who had reported at the start of the Korean War were released to inactive duty.[25]

Intelligence in Support of the Taiwan Straits Patrol and Commander Seventh Fleet, 1950–1969

On 4 August 1950, the antiaircraft cruiser *Juneau* (CLAA 119) and two destroyers were ordered to patrol the waters around Taiwan. The group was designated Task Force (TF) 72 on 24 August 1950 and was the initial detachment of a force that eventually became known as the U.S. Taiwan Patrol Force. TF 72 surface units were supported by aviation patrol units, and, in due course, Commander Fleet Air Wing One was double-hatted as Commander Task Force 72.

On 7 March 1953, RAdm. Thomas B. Williamson hoisted his flag on the seaplane tender *Pine Island* (AV 12) as the first Commander Formosa Patrol Force, under the operational control of Commander in Chief, Pacific Fleet (the title was changed to Taiwan Patrol Force on 1 November 1955, in deference to the Chinese Nationalists' use of the name "Taiwan" for the island). The purpose of the force was to conduct reconnaissance to detect enemy forces capable of invading Taiwan and to assist in training the Chinese Nationalist Navy.

On 11 December 1953, VAdm. Alfred M. Pride was designated Commander Formosa (later Taiwan) Defense Command, a unified command under CINCPAC. The Commander Formosa/Taiwan Patrol Force was the Navy component of the Commander Taiwan Defense Command.[26]

Total control of air and sea areas around Korea made it possible for UN naval forces to use relatively few ships on patrol and blockade duties, to operate major ships with a minimum of escorts because of the negligible submarine threat, and to use only token air cover. That freed the Seventh Fleet for maximum offensive operations in support of UN ground and air forces in Korea, but it also put a crucial demand on intelligence to detect immediately any change in the air, surface and subsurface threat to the fleet.[27]

As of early 1952, intelligence information received from the extensive collection agencies in the Far East and carefully evaluated through the coordinated efforts of the intelligence sections, ashore and afloat, provided adequate intelligence for conducting current operations. Commander Seventh Fleet believed that sufficient reliable information was being supplied to determine with reasonable accuracy the current disposition of North Korea's military forces and their condition of readiness. Due to the short distances involved and the known concentration of enemy air strength along the Antung border, however, it was recognized that early warning capabilities against air attack were not adequate for alerting naval forces in the area.

On the other hand, it was believed that early warning of any serious Chinese Communist amphibious attack on Taiwan or the Pescadores would be received at least a week in advance. However, early warning of a limited air attack by Chinese Communist aircraft deployed within striking distance of Taiwan would be received only a few hours or even minutes in advance.

The potential enemy was judged to have capabilities to launch submarine attacks against UN naval forces and shipping in the Far East. The effort, it was theorized, could be implemented by submarines reportedly operated by Chinese Communist personnel, by covert operations by Soviet submarines, or by Soviet submarines operating under the guise of being Chinese Communist (in point of fact, neither China nor North Korea possessed submarines during the Korean War).[28]

To provide all units of the Seventh Fleet with the basic intelligence necessary to perform their tasks, intelligence annexes were prepared and disseminated for current operation plans and orders. The annexes were kept current by issuing bimonthly changes that incorporated the latest information on the enemy situation.[29]

The intelligence needs of the Taiwan Patrol Force were concerned primarily with an early warning of any hostile actions or intentions by the Chinese Communists that would indicate preparations to invade Formosa or the Pescadores. Possible indications were expected to include a concentration of Chinese Communist naval forces, junks, or merchant shipping; unusual troop and supply buildups in possible staging areas; and/or the presence of unusual aircraft activity over the Formosa Strait or on the mainland opposite Formosa. Sources available to CTF 72 for fulfilling the intelligence collection requirements were limited to aerial and surface reconnaissance patrols. On occasion, special aerial photographic missions were flown along the coastal areas of China or over Chinese Communist-held islands with the approval, or by direction, of Commander in Chief, Pacific Fleet (CINCPACFLT).

TF 72 maintained a continuous patrol of the Formosa Strait, except for periods of forced evacuation to avoid bad weather. The patrol plane squadrons maintained daily surveillance of Chinese coastal waters from Hong Kong north to latitude 32°N, with an average of three flights per day. In addition, surveillance and photo-reconnaissance flights over the islands and main shipping lanes of the South China Sea were flown at irregular intervals. In January 1953, at the request of CINCPACFLT, the frequency of surveillance flights along shipping lanes in the South China Sea was increased to one per day to provide closer coverage of ships entering Hong Kong and Canton.

On numerous occasions, patrol aircraft of TF 72 were fired on by Chinese Communist surface craft in coastal waters. Many such incidents occurred in September 1952 off the Whangpoa Delta, where a Chinese landing ship had been stationed, apparently as a radar picket. On 20 September 1952, a PB4Y aircraft of VP-48 was attacked by two MiG-15 fighters; five firing runs were made on the U.S. Navy aircraft, but it sustained no damage.[30]

With the termination of hostilities in Korea, the amount of intelligence information being received was greatly reduced, due in large measure to the restrictions against overflights of North Korea. The only new information received was from the debriefing of returning prisoners of war.[31]

The most significant intelligence development during the first half of 1954 in the Far East was the shift in intelligence interest from Korea to China, occasioned by both the stalemated situation in Korea and the increased Chinese Communist activity in the Formosa/Southeast China area. Of particular importance was the increased air and naval activity in the area surrounding the Chinese Nationalist-held offshore islands.[32]

VAdm. Stuart H. Ingersoll relieved VAdm. Pride on 19 December 1955 and served concurrently as Commander Seventh Fleet and Commander U.S. Taiwan Defense Command until VAdm. Wallace M. Beakley relieved him as COM7THFLT in February 1957. VAdm. Austin K. Doyle relieved Ingersoll of his Taiwan post on 28 July 1957.[33]

In 1956, the Taiwan Patrol Force consisted of one land-based reconnaissance squadron based at Naha, Okinawa; one seaplane squadron based at Sangley Point in the Philippines; a special electronic warfare aircraft squadron at Iwakuni, Japan; and a destroyer squadron based in Taiwan.

Patrols were flown daily, covering the coast of China from 32°N through the Taiwan and Luzon Straits into the South China Sea. It was routine practice to divert from normal patrol tracks to investigate any large concentrations of junks since a great percentage of the Communist Chinese capability to move troops was in their multitudes of seaworthy civilian smallcraft.

With an increase in Soviet bloc merchant shipping activity in the Far East in 1956, shipping surveillance to determine the movements of potential war materials became a primary mission of the Taiwan Patrol Force.

In February 1956, an air patrol in the South China Sea spotted buildings under construction (and a Communist flag flying) on Woody Island in the Paracel Islands. Because the islands were claimed by seven different countries, the sighting of an apparent takeover by the Communists was a cause for the apprehension that the islands might become another area of international confrontation. Close scrutiny, however, indicated that the local guano deposits apparently induced what proved to be temporary activity on the island group.

On 22 August 1956, a P4M aircraft of Electronic Countermeasures Squadron One was shot down while on a routine patrol over international waters off Chusan.[34]

During Fiscal Year 1956, close working relations were maintained by Seventh Fleet staff intelligence and the intelligence section of the Taiwan Defense Command, which, in turn, had close exchange agreements with the Nationalist Chinese. The arrangements permitted COM7THFLT to receive excellent intelligence on Communist Chinese activities and targets on the mainland in the vicinity of the Taiwan Strait.[35]

Commander Patrol Forces, Seventh Fleet (CTF 72) was also Commander Fleet Air Wing One and Commander U.S. Taiwan Patrol Force. The intelligence mission of CTF 72 was to conduct air and surface surveillance patrols in order to collect and report information of intelligence value regarding enemy units moving on or under the surface of the ocean in Seventh Fleet's area of responsibility. The intelligence collected included visual, photographic, and electronic data on Communist ships (including submarines, naval auxiliary survey ships, naval electronic intelligence collectors, and merchant and fishing vessels).

TF 72 aircraft (P-5 Marlin seaplanes, P-2 Neptune and P-3 Orion land planes) conducted patrols on a regular basis in the Sea of Okhotsk, Sea of Japan, Yellow Sea, East China Sea, South China Sea, Philippine Sea, and the Gulf of Thailand. Surface patrol units also operated in the Taiwan Strait to collect intelligence.[36]

In addition to the usual sightings of Soviet naval combatants and auxiliaries in naval exercises in the Sea of Japan, eighteen positive submarine contacts were prosecuted by TF 72 aircraft from Iwakuni, Japan, during 1966–1967. The most significant contact event resulted in the first photographs of a Soviet Echo-II-class cruise missile submarine in August 1966.[37]

On 12 September 1967, during TF 72 surveillance of two Soviet guided-missile destroyers that were operating in the Philippine Sea, another Echo-II was caught on the surface. It apparently had suffered a casualty and was unable to dive to avoid detection. A Soviet *Kotlin*-class destroyer and a *Prut*-class submarine rescue ship came from Vladivostok and escorted the submarine back to port during one of the more productive intelligence collection opportunities of the period.

Since the 1950s, a major collection effort had been made in the early fall of each year to sight and identify Soviet naval units being transferred from the Baltic and Northern Sea fleets to the Pacific naval fleet via the Northern Sea polar route. TF 72's participation in the effort during the late 1960s consisted of flying surveillance patrols in the Sea of Okhotsk and the northern parts of the Sea of Japan. Between 4 and 25 October 1967, fifty such patrols were flown as part of the coverage of the annual interfleet transfer.

Communist shipping to North Vietnam was monitored by TF 72 to determine the quantity and nature of the cargoes being carried. Soviet shipping originated from both European and Asian ports, while Chinese Communist vessels used coastal routes into Haiphong. Special shipping surveillance flights were also flown on a continuous basis during the "Market Time" operations off the coast of Vietnam. On two occasions in 1967, the patrols discovered Communist steel-hulled trawlers attempting to smuggle arms and supplies into South Vietnam. On 14 March, a P-3 Orion aircraft made its contact known to a surface unit which forced the trawler to beach, where it was then blown up. Similarly, on 11 July, a P-2 Neptune aircraft discovered a suspicious vessel and vectored a surface unit to investigate; this trawler was also forced to beach and was found to be carrying a large supply of weapons and ammunition.[38]

There was an increasing national-level interest in Soviet navy operations in the Indian Ocean in 1969. As a result, CTF 72 was called on increasingly to conduct daily locator flights on Soviet units proceeding to or from the Indian Ocean. During 1969, fourteen different Soviet transits, in which twenty-seven Soviet naval ships were involved, were covered by U.S. Navy patrol aircraft.[39]

Chapter Notes

1. *ONI Review*, Dec 1945, 29–34.

2. VAdm. Milton E. Miles, *A Different Kind of War: The Little-Known Story of the Combined Guerilla Forces Created in China by the U.S. Navy and the Chinese During World War II* (Garden City, NY: Doubleday, 1967), 78, 83, 92.

3. Ibid., 99–100.

4. Gen. Robert H. Barrow, USMC (Ret.), Oral History, Marine Corps Historical Center, Washington, DC, 19.

5. Ibid., 88, 107–8.

6. Ibid., 109, 111, 116.

7. Ibid., 156–57.

8. Ibid., 183, 189, 192.

9. Ibid., 208.

10. Ibid., 244–45, 258–59.

11. Ibid., 267, 290–94; and Cdr. Joseph A. Meyertholen, USNR, ltr to author, 30 Apr 1980. Meyertholen was an intelligence special duty officer who served with Commander Naval Group, China.

12. OP-922H4E ltr, ser 000301P92, 16 Mar 1956 to Commander Amphibious Training, Atlantic.

13. Miles, *A Different Kind of War*, 296–97, 309–10.

14. Ibid., 315, 324.

15. Ibid., 317–18, 320.

16. Ibid., 323.

17. Cdr. Joseph A. Meyertholen, USNR, interview by author, 24 Mar 1980.

18. Miles, *A Different Kind of War*, 417.

19. Intelligence Liaison Officer Shanghai ltr 48-77, 23 Nov 1948, box 1, TSC 6158, Job 10430, OA.

20. Commander in Chief, Pacific Fleet (CINCPACFLT), *Interim Evaluation Report No. 1: Korean War Naval Operations*, 1477–79.

21. CINCPACFLT, *Interim Evaluation Report No. 2: Korean War Naval Operations*, 1746.

22. RAdm. Edwin T. Layton, Oral History, USNI, Annapolis, MD, 1970, 174–75.

23. Capt. Chester J. Oleniacz, USNR, medal citation in author's collection.

24. CINCPACFLT, *Interim Evaluation Report. No. 2*, 1745.

25. CINCPACFLT, *Interim Evaluation Report. No. 3: Korean War Naval Operations*, 19-5.

26. Commander Taiwan Patrol Force (COMTAIWANPATFOR), *Command History*, 10 Aug 1959 and 2 Mar 1960.

27. Commander Seventh Fleet (COM7THFLT), "Report of Operations, Mar 1951–Mar 1952," 7.

28. Ibid., 36–37.

29. Ibid., 24.

30. COM7THFLT, "Report of Operations, May 1952–Jul 1953," 65, 67–68.

31. COM7THFLT, "Report of Operations, Jul–Dec 1953," 15.

32. COM7THFLT, "Report of Operations, Dec 1953–Jun 1954," 17.

33. *COMTAIWANPATFOR Command History*, 10 Aug 1959.

34. *COMTAIWANPATFOR Command History*, 2 Mar 1960, 3, 5.

35. COM7THFLT, "Report of Operations, Fiscal Year 1956," 7.

36. Commander Patrol Force, Seventh Fleet (COMPATFOR7THFLT), *Command History, 1967*, XI-1.

37. *COMPATFOR7THFLT Command History, 1966*, passim.

38. *COMPATFOR7THFLT Command History, 1967*, XI-2–XI-4.

39. *COMPATFOR7THFLT Command History, 1969*, X-3.

CHAPTER 38

Operating Forces, Atlantic

The initiation of the Neutrality Patrol by Commander Atlantic Squadron in September 1939 was the start of U.S. operational involvement in World War II. Primarily an intelligence-gathering action, the Neutrality Patrol initially was required to "observe and report in code on the movements of combatant vessels of nations in a state of war." In December 1939, however, the mission was expanded to include reporting on the movements of German merchant ships, subsequently leading to the capture by the British of several German merchant vessels following their interception by units of the U.S. Navy.[1]

With the entry of the United States as a formally committed participant in World War II and the designation soon thereafter of Adm. Ernest J. King as Commander in Chief, U.S. Fleet (COMINCH), his organization took over operational control of, and combat intelligence support to, the Navy's antisubmarine warfare forces in the western Atlantic. The Commander in Chief, Atlantic Fleet (CINCLANTFLT) organization, when it was reestablished in Norfolk in 1941, was concerned primarily with administration, training, and logistic support to antisubmarine forces and was not involved in combat intelligence support during World War II. See Chapter 16 for further information on COMINCH and on Commander Tenth Fleet's involvement in intelligence support to forces in the Atlantic.

Fleet Intelligence in the South Atlantic During World War II

On 10 May 1941, RAdm. Jonas H. Ingram, Commander Cruiser Division Two, arrived at Recife, Brazil, in his flagship, the light cruiser *Memphis* (CL 13), to initiate a neutrality patrol based at San Juan, Puerto Rico, and Guantanamo Bay, Cuba, with Recife and Bahia as replenishment ports. The area of the patrol included the triangle formed by Trinidad, the Cape Verde Islands, and the shoulder of Brazil. Axis and Vichy French merchant ships were to be kept under surveillance.

The Naval Observer, Recife, LCdr. William A. Hodgman, USN (Ret.), who had arrived on post on 26 February 1941, advised RAdm. Ingram that Recife was superior to Bahia for naval patrol purposes; the latter, although it had a better natural harbor, was too far below the "shoulder" of Brazil for effective patrol work against Axis submarines.

RAdm. Ingram revisited Recife on 3 July, 15 August, and 12 October 1941, and from the contacts made during his visits created an atmosphere of good will and mutual trust preliminary to necessary future cooperation. Brazil, like the United States, was still neutral in October 1941, but Ingram needed support from Brazil in the form of food, fuel oil storage, and improved harbor facilities.

Although LCdr. Hodgman's duties were mainly concerned initially with reporting ship movements and harbor activities to the Office of Naval Intelligence, he became increasingly involved in providing assistance to U.S. Navy ships visiting Recife. The number of visits remained at fewer than six per month until September 1941, when the average increased to ten. From time to time, Hodgman was absent from Recife to collect information that would be needed for planning future naval and air bases. To assist Hodgman, 2dLt. D. J. Kendall, USMCR, was assigned as his assistant in August 1941.

More naval observers arrived in Brazil in 1941: Cdr. M. B. Saben, USN (Ret.), at Bahia on 1 October; LCdr. H. C. Frazer at Natal on 14 October; and LCdr. Edward Breed, USNR, at Belem on 17 November. Collecting intelligence information and cultivating good relations with local Brazilian authorities were the primary duties of the naval observers during the neutrality period.

At the time of the Pearl Harbor attack, U.S. Navy ships were still making only occasional visits to Brazil, mostly to Recife. In December, the U.S. Navy forces were designated Task Force (TF) 3 (redesignated TF 23 of the Atlantic Fleet in early 1942). On 20 December 1941, a section of Patrol Squadron (VP) 52 seaplanes arrived at Natal along with the seaplane tender *Clemson* (AVD 4) and a contingent of Marines. Assisting the new arrivals accounted for the greater part of the activity of the Naval Observer, Natal.

Upon the U.S. entry into World War II, three companies of Marines were ordered to Brazil to guard airfields at Recife, Belem, and Natal. Brazil, however, was not yet at war with the Axis, and it did not want armed Marines arriving uninvited. When Naval Observer, Recife, LCdr. Hodgman learned of the Brazilian objection, he arranged a compromise. The Marines would crate their arms before entering Brazilian territory. They did this during a stop at Trinidad. At first, the Marines in Brazil served as unarmed shore patrols until the populace became accustomed to seeing them. Later, the Marines were allowed to guard U.S. aircraft and to drill with their arms. Finally, all remaining restrictions were lifted in August 1942 when Brazil declared war on Germany and Italy.

In spite of Brazilian neutrality in early 1942, VAdm. Ingram had persuaded the Brazilian admiral in command of the forces between Rio de Janeiro and the Amazon River of the desirability of close cooperation. On the U.S. side, cooperation included training Brazilian ships in U.S. communications procedures and providing to Brazil the results from patrols by the VP-52 detachment at Natal.

In April 1942, VAdm. Ingram visited Rio de Janeiro to meet U.S. Ambassador Jefferson Caffery and Assistant Naval Attaché Capt. Edward Brady. Brazilian President Vargas invited Ingram and Brady to visit him at his resort in the southwest of the country. As a result of Ingram's adroit negotiating during the visit, Vargas "unfroze" Brazilian merchant shipping and named Ingram as his "Sea Lord" and "Naval Advisor." Ingram considered the meeting with Vargas to have been the diplomatic high point of his experiences in Brazil.

Coordination with the British in the South Atlantic was arranged through a conference at Recife in July 1942 that was attended by VAdm. Ingram and his staff, the British and U.S. naval attachés from Rio, the U.S. Naval Observer, Recife (Hodgman), and a Brazilian naval officer. The British officers recommended to their government that Ingram receive the same intelligence information from British offices at Montevideo, Uruguay; Kingston, Jamaica; Freetown, Sierra Leone; and Capetown, South Africa, that was sent to British warships. The British government agreed.

When Brazil entered the war in August 1942, VAdm. Ingram and his CTF 23 staff moved aboard the oiler *Patoka* (AO 9) in the harbor at Recife to improve his contact and exchange of information with the Brazilian navy. In September, arrangements were also made with the British West Africa Command for the establishment of liaison with CTF 23 to coordinate efforts to block the passage of Axis ships along the line between Recife and Takoradi in the British Gold Coast colony.

On 12 September 1942, President Getulio Vargas ordered his naval forces placed under the command of Commander South Atlantic Force (COMSOLANTFOR), Ingram's new designation, thus completing the final steps for full naval cooperation with the U.S. Navy.

RAdm. Augustin T. Beauregard was Naval Attaché, Rio de Janeiro, in 1942, and his office was almost completely involved in operational and logistics matters at the port. The U.S. Naval Mission in Rio de Janeiro, on the other hand, had little connection with COMSOLANTFOR. In mid-November, COMINCH Adm. King directed VAdm. Ingram and RAdm. Beauregard to develop a new organization for U.S. Navy activities in Brazil. They proposed the following, which was approved by COMINCH in December:

> (1) A naval operating base would be established at Rio de Janeiro, to include the operational functions previously performed by the naval attaché. Beauregard would become Commanding Officer, Naval Operating Base, Rio, and would also take command of the Naval Mission in Rio;
>
> (2) All assistant naval attachés in Brazil would report to the Chief of the Naval Mission (Beauregard) as his assistants and would also perform their operational duties under the Naval Operating Base organization;
>
> (3) A suitable officer would be ordered to Brazil as the official naval attaché;
>
> (4) All Naval Mission officers would be assigned additional duty on the staff of COMSOLANTFOR;
>
> (5) All naval observers on the east coast of South America would come under COMSOLANTFOR, but in each naval observer office there would be one or more intelligence officers operating directly under ONI while also maintaining liaison with Naval Attaché, Rio de Janeiro as well as with COMSOLANTFOR; and
>
> (6) Any additional naval activities established in Brazil or Uruguay would come under the jurisdiction of COMSOLANTFOR.

By early 1943, COMSOLANTFOR and his staff had moved into a permanent shore headquarters at Recife, and numerous land bases and airfields had been completed or were under construction. Navy commands attached to COMSOLANTFOR included

Cruiser Division Two with five cruisers, Destroyer Squadron Nine with eight destroyers, and Fleet Air Wing (FAW) 16 with three patrol plane squadrons. COMSOLANTFOR's operations were almost exclusively involved in antisubmarine reconnaissance or in efforts to stop Axis blockade runners as they passed through the south Atlantic.

On 15 March 1943, the U.S. Fleet was divided into numbered fleets for operational purposes, and COMSOLANTFOR was designated Commander Fourth Fleet under the operational command of CINCLANTFLT. In addition to the forces stated above, the Fourth Fleet included five patrol craft, two minesweepers, various auxiliaries, fifteen Brazilian naval ships, and the Brazilian air force.

The naval observers at Recife, Bahia, Natal, and Belem had become commandants or commanding officers of the naval operating facilities at their locations. Although their duties were concerned with convoy routing and control and with the operation of shore facilities, their staffs included intelligence officers to provide intelligence support to local operating forces and to Ingram's Fourth fleet staff. The intelligence officers were also involved in debriefing survivors from sunken merchant ships and in the initial interrogations of German and Italian prisoners of war.

July 1943 was the peak month for Axis submarine activity in the South Atlantic, with fourteen Allied merchant ships and six Axis submarines sunk. As a result of the heightened activity in the South Atlantic, Airship Wing 4 was commissioned at Recife in August 1943, and the first blimp arrived in September to expand Fourth Fleet reconnaissance and convoy escort capabilities. In October, FAW-16 was also augmented to its maximum strength of nine patrol aircraft squadrons operating in Brazil.[3]

Fourth Fleet Intelligence Officer and Chief Censor during 1943–1944 was LCdr. Howard E. Moore. A combat intelligence section at Fourth Fleet headquarters maintained a plot of movements of enemy submarines and blockade runners up to the time they entered the area where Fourth Fleet forces could begin their surveillance and attack efforts. Lt. Harry Offutt, Air Combat Intelligence Officer, FAW-16, worked in the Fourth Fleet Combat Intelligence Section and kept the patrol squadrons informed of developing targets. When enemy ships were within range, the squadrons tracked, plotted, and attacked these targets. Aside from continuing to provide intelligence information, a minimum of supervision was exercised from FAW headquarters at Recife. Each VP squadron had an air combat intelligence officer, who was responsible for briefing and debriefing the air crews.[4]

Fleet Intelligence Under CINCLANTFLT, 1954–1964

The critical Honduran-Guatemalan situation during the latter part of May and most of June 1954 clearly demonstrated that CINCLANTFLT's operational intelligence facilities were not adequate to meet wartime requirements for support to the fleet. Only by working long hours in overcrowded spaces could the CINCLANTFLT intelligence staff meet top priority requirements. Special studies on Honduras and Guatemala were produced and distributed to commands involved in preparation for the evacuation of U.S. nationals.[5]

In the mid-1950s, a current plot of Soviet and Satellite merchant shipping was maintained in CINCLANTFLT's Operational Intelligence (OPINTEL) Plot. The types of cargoes shipped and individual ship movements were under continuous study to detect any trends or indications that would give early warning of hostile intentions.

A Sound Surveillance Control Center (SSCC) was activated in the OPINTEL Plot on 1 February 1956. The Sound Surveillance System (SOSUS) net, through the SSCC, provided a rapid means of acquiring early warning against submarine penetration into the western Atlantic and Caribbean areas. The coordination of submarine contact reports from SOSUS, high-frequency direction-finding (HF/DF), and other sources was accomplished in the OPINTEL Plot.

In 1956–1957, plans were developed to establish a Fleet Intelligence Center in Scotland, to be activated and staffed during the first month of any authorized mobilization. The projected center was to provide intelligence to fleet units operating in forward areas, primarily air intelligence in support of Second Fleet carrier forces. The proposed basic organization for the center had been approved by the Chief of Naval Operations, but the facility never was built.

Commander in Chief, U.S. Forces, Atlantic (CINCLANT) target listings were converted to the IBM machine record and accounting system. An airfield target priority system, adapted to current IBM machine accounting as well as to digital computation, was also established. A series of Evasion and Escape studies covering selected areas of northern Europe and the northwestern part of the USSR of interest to CINCLANTFLT were being produced by the CINCLANTFLT intelligence staff during Fiscal Year 1957.[6]

By the end of Fiscal Year 1957, the personnel of the Oceanographic Unit, Norfolk, had been integrated into the CINCLANTFLT staff. Plans and estimates for combining ocean acoustic barrier plots were submitted, and an extension and expansion of the SOSUS acoustic array network was recommended.[7]

Commander Antisubmarine Defense Force, Atlantic was established during Fiscal Year 1958 to increase overall coordination in ASW and was given the responsibility for conducting ASW surveillance and defensive operations throughout the Atlantic Fleet area. In directing investigations of any reports of submarine contacts, the force commander was to take operational control of all forces assigned, including the SOSUS system.[8]

By 1958, an Operational Intelligence Plot was being maintained in conjunction with the CINCLANT/CINCLANTFLT Operational Control Center. Order-of-battle and situation plots were maintained in the OPINTEL Plot, as was a library of intelligence publications required by the control center. A special intelligence plot was also maintained in a secure area adjacent to the OPINTEL Plot.

The intelligence annexes of both the CINCLANT and CINCLANTFLT basic war plans and operation orders were limited to the collection and dissemination aspects of intelligence. Objective intelligence on potential enemy forces was disseminated through the *Atlantic Command Intelligence Estimate*.

The principal medium through which major CINCLANT subordinate commands were kept informed of significant current intelligence was the CINCLANT Intelligence Summary, normally issued weekly in standard naval message form. During periods of heightened international tension, it was disseminated more frequently. An intelligence summary was prepared daily for written and/or oral presentation to CINCLANTFLT and to appropriate members of his staff.[9]

In 1959, CINCLANTFLT's Operational Intelligence Plot was relocated in the new Operational Control Center building. Closed-circuit television was used for the first time in the presentation of the daily intelligence summary for CINCLANTFLT and his staff members.

During 1959, the CINCLANTFLT intelligence staff prepared an extensive study, including recommended targets and operating areas, for use in the deployment of Polaris nuclear-powered ballistic missile submarines that were about to become operational. That same year, CINCLANTFLT recommended an expansion of the naval attaché system in Africa in view of the Navy's expanding operations in that area and the need for contingency wartime planning.[10]

Commencing on 1 August 1960, a CINCLANT/CINCLANTFLT/CINCWESTLANT intelligence duty officer watch was established in the Operational Intelligence Plot to furnish immediate intelligence support to the headquarters command duty officer.

The *Atlantic Command Intelligence Estimate* was reissued in 1960 as the *Atlantic Command Intelligence Estimate for Planning (1960)*. The new edition was expanded to include required basic intelligence on specific countries subject to contingency planning, including Cuba, Iceland, the Dominican Republic, and Trinidad. A supplement to the estimate detailed changes to Cuban order-of-battle data that had occurred late in the fiscal year.

Also in 1960, revised intelligence annexes covering Iceland, the Azores, Cuba, Haiti, the Dominican Republic, and Trinidad were prepared in support of the Atlantic Cold War Plan. In addition, a separate intelligence annex was prepared and distributed that provided information on fifteen African countries south of the Sahara; the annex was later revised and expanded to include information about sixteen countries.

Three new Port Briefing pamphlets were issued during Fiscal Year 1961, expanding CINCLANTFLT's port information program in support of fleet units to cover fifty-six ports and cities. Recording tactical targeting data on IBM punch cards was introduced that year, greatly improving the speed of distribution of the material to fleet and staff planners.[11]

Significant changes were made in 1962 at CINCLANT/CINCLANTFLT in the handling and processing of operational intelligence. In April, the CINCLANT Indications Center was activated, providing headquarters with a twenty-four-hour intelligence processing and analytical capability and thus fulfilling a requirement for active CINCLANT participation in the worldwide indications system of the newly established Defense Intelligence Agency (DIA). Later in the year, a Current Intelligence Section was created and made responsible for processing and disseminating all current and operational intelligence and for the operation of the CINCLANT Indications Center.

Current intelligence was presented in daily oral briefings to CINCLANT, in periodic staff briefings, and in the Daily Intelligence Summary. During 1962, direct support was provided to commanders involved in Cuban contingency operations and in "quarantine" operations in the Cuban area.

As an adjunct to the Indications Center, the CINCLANT/CINCLANTFLT OPINTEL Plot continued to function in support of the Operational Control Center and CINCLANT subordinate commands. Considerable use was made of the Operational Control Center's 1103A computer. During the Cuban quarantine (Cuban Missile Crisis) operations in October and November, reports of positions and movements of Soviet and Bloc merchant ships were continuously entered into the computer, and a complete machine tabulation of Bloc shipping of interest was disseminated on a need-to-know basis.

CINCLANTFLT patrol assets conducted surveillance of the Soviet fleet exercise in the Norwegian Sea in July 1962. During the Soviet exercise and during the Cuban quarantine operations, a large number of positive submarine contacts were reported, a substantial number in the western Atlantic.

The need for current, complete, and accurate information on Soviet Bloc activity increased in importance as Soviet naval, submarine, and merchant ship operations grew in scope and intensity in the Atlantic. At the same time that the Soviets were holding their northern fleet exercise, an increase in Soviet merchant ship movements to Cuba was noted. A total of 151 Soviet merchant ship arrivals in Cuba occurred during July through September, and during August and September nine Communist-flag passenger ships arrived. An estimated 21,000 Soviet military personnel were in Cuba by October.

New demands for vitally needed intelligence about situations in Latin America and sub-Saharan Africa were generated, largely due to the continuing penetration and subversion efforts on the part of the Sino-Soviet Bloc. Even excluding the period of the October-November Cuban crisis, developments in 1962 resulted in greatly increased requirements for intelligence about Cuba. A substantial collection effort was made against Cuban targets, including an intensive program of aerial reconnaissance.

The CINCLANT intelligence staff prepared a special estimate on possible military operations in Cuba in September 1962, and a Caribbean ground order-of-battle on Cuba, Haiti, and the Dominican Republic had been published in March of that year. An existing Cuban radar order-of-battle database was adapted to electronic accounting machine processing in June 1962, but two complete revisions and six interim changes had to be issued to keep up with the changing situation during the balance of the year.

A CINCLANT Intelligence Acquisition Section was authorized in 1961 and was staffed and activated in February 1962. Action was then taken to convert CINCLANT/CINCLANTFLT collection procedures to the Defense Intelligence Agency system. A CINCLANT Intelligence Collection Plan was promulgated, and the existing intelligence collection instructions were rewritten to conform to the DIA instruction. Machine reports (IBM punch cards) were used to control and monitor collection.[12]

In 1963, the CINCLANT/CINCLANTFLT Current Intelligence and Indications Center (CIIC) operated on a 24-hour-a-day basis to provide intelligence support to senior, adjacent, and subordinate commands. For Atlantic area Polaris submarines, a current intelligence briefing was given to selected members of each crew and an intelligence support folder was provided prior to deployment on each patrol. Attack aircraft carriers were alerted early on Soviet air activity over the Atlantic during their transits to and from Mediterranean and Atlantic operations; the CINCLANT Daily Intelligence Summary continued to be transmitted to subordinate commanders as their primary source of current intelligence and CINCLANT and selected members of his staff were given a daily briefing on current intelligence in the Fleet Operational Control Center.

The responsibility for analyzing, classifying and reporting unidentified submarine contacts was assumed by Commander Antisubmarine Warfare Forces, Atlantic in 1963. The Current Intelligence Section continued to monitor ASW activities and to provide CINCLANTFLT with any additional information that became available.

In 1963, CINCLANT initiated a collection effort to assist in the surveillance of ships of less than 1,000 tons gross registered tonnage in the Caribbean and Gulf of Mexico in conjunction with Cuban and Latin American subversion and arms-running efforts. Patrol aircraft squadrons based in the Caribbean continued to obtain high-quality photos of ships and deck cargo inbound to Cuba. In addition, a special requirement for color photos of outward-bound passenger ship traffic was fulfilled.

A study of the Cuban crisis showed that an increase in intelligence personnel was necessary to support the planning responsibilities of CINCLANT for contingency operations.[13]

The CINCLANT Current Intelligence and Indications Center continued to provide intelligence support in 1964 on a 24-hour-a-day basis to senior, adjacent, and subordinate commands. Of particular note were the requirements for CIIC to support the following: Polaris submarines, which were provided with an all-source briefing and intelligence folder prior to departure on patrol and with operational intelligence by message while they were on station; large combatants and unit commanders, which were provided with operational intelligence on the activities and tactics of Soviet long-range aircraft reconnaissance and intelligence-gathering trawlers or other potential "snoopers"; and fleet forces in the Caribbean during periods of high tension, which were provided with operational intelligence on potentially friendly and enemy forces and suspect shipping.

CIIC monitored the positions and activities of Soviet and Cuban fishing trawlers operating near the continental United States and kept the U.S. Coast Guard promptly informed. CIIC was also the focal point for intelligence on ships of special interest that were approaching Cuba, and it maintained detailed files on incidents occurring between Cuban-based forces and U.S. military and commercial ships and aircraft.

In 1964, CIIC began to keep intelligence files on insurgency activity throughout the Caribbean area, and a collection effort was maintained to determine Cuban subversive activities against South America. During the same year, camera pools were set up at Newport, Norfolk, Charleston, and Mayport to distribute cameras to small combatants for intelligence collection purposes.

Because of indications that Cuba was smuggling agents, arms, and propaganda to Latin America, CINCLANT stationed two destroyer escorts, supported by maritime air, in the southern Caribbean as a surveillance/intercept ready-response force. While on station from 15 January to 13 April 1964, the patrol force sighted 1,269 ships and boats but found none to be on suspicious missions. At the request of the Joint Chiefs of Staff, CIIC performed an analysis of shipping in the approaches to Venezuela for the period of 24–31 January 1964 to determine a shipping density pattern for use in conjunction with ongoing analysis of Cuban arms smuggling.[14]

Chapter Notes

1. James O. Richardson, *On the Treadmill to Pearl Harbor: The Memoirs of Admiral J. O. Richardson as told to George C. Dyer, USN (Retired)* (Washington: GPO, 1973), 156–58.

2. The preceding discussion on Fleet Intelligence in the South Atlantic is taken from "Commander South Atlantic Force," MS command history prepared post–World War II, 11:11–16, in Navy Department Library.

3. Commander U.S. Naval Forces, South Atlantic, "COMSOLANT History," MS, 8–10, OA.

4. Ibid., 33; and Commander Fleet Air Wing 16, "FAIRWING 16 History," 60–63, 84.

5. Commander in Chief, Atlantic Fleet (CINCLANTFLT), *Annual Report, 12 Apr 1954–30 Jun 1954*, Supplement, 12.

6. *CINCLANTFLT Annual Report, FY 1957*, 72.

7. Ibid., 30.

8. *CINCLANTFLT Annual Report, FY 1958*, 36–37.

9. *CINCLANTFLT Annual Report, FY 1959*, 68–71.

10. *CINCLANTFLT Annual Report, FY 1960* (to 29 Feb 1960), 63–66.

11. *CINCLANTFLT Annual Report, FY 1961*, 75–78.

12. Commander in Chief, Atlantic (CINCLANT), *Annual Report, FY 1962*, 5–8, 11.

13. *CINCLANT Annual Report, FY 1963*, 5–6, 10.

14. *CINCLANT Annual Report, FY 1964*, 4–5, 7, 9–10, 14.

CHAPTER 39

Operating Forces, European Area

This chapter contains information on naval intelligence activities of commands in the European and Mediterranean areas during and after World War II. These commands include Commander Naval Forces, Europe (COMNAVEU); Commander in Chief, U.S. Naval Forces, Eastern Atlantic and Mediterranean (CINCNELM), and Commander Naval Forces, Germany (COMNAVFORGER).

Commander Naval Forces, Northwest African Waters (a World War II command) and Commander Sixth Fleet should have been covered in this chapter but are not—the former because of the lack of information and the latter because of lack of research time.

Commander Naval Forces, Europe Organization, 1942–1944

With U.S. entry into World War II, Special Naval Observer (SPENAVO), London, RAdm. Robert L. Ghormley was designated Commander U.S. Naval Forces, Europe. See Chapter 3 for other details on SPENAVO. Chief of Staff COMNAVEU also served as Naval Attaché, London, and the Naval Air Attaché served as the head of COMNAVEU's Air Section. The Technical Section of the naval attaché's office had previously been consolidated with the Material Section of SPENAVO and remained in COMNAVEU's Readiness Division when that organization was established, but its reports were transmitted to the Office of Naval Intelligence via the naval attaché. The Intelligence Division of COMNAVEU took on the operational intelligence reporting function; it also collected data and submitted intelligence information reports about subjects not covered by the office of the naval attaché.

When former Chief of Naval Operations Adm. Harold R. Stark relieved VAdm. Ghormley in March 1942, the many U.S. Navy units in the European area involved in intelligence functions were authorized to be assimulated under COMNAVEU. The transition, although complicated, was facilitated by having the chief of staff of COMNAVEU also assigned as the naval attaché.

ONI did not favor the consolidation of the naval attaché and COMNAVEU Intelligence Division functions, preferring that the latter handle operational intelligence matters and the former handle "strategic intelligence." In practice, it was difficult to make such a distinction. RAdm. Alan G. Kirk (a former Director of Naval Intelligence), as chief of staff to Stark and also as naval attaché, did not deem it advisable to change his situation.[1]

In March 1942, the U.S. Navy and Army in the European theater began discussions about possible joint collection, analysis, and dissemination of intelligence information. It was not until 6 March 1943, however, that representatives from the European theater of operations, U.S. Army, the U.S. Army Air Force, COMNAVEU, the U.S. military and naval attachés, the Office of Strategic Services, and the Office of War Information met under Army BGen. Kroner to consider pooling and centralizing intelligence data from the various U.S. agencies in London. COMNAVEU voiced a need for additional, specially qualified personnel to fulfill its growing intelligence role, especially with regard to any additional interagency support responsibilities. People, logistics, and reporting relationships and procedures were worked out during the summer and fall of 1943. COMNAVEU took on the direct responsibility for the naval intelligence effort and for the personnel assigned to it, although many of them were involved in liaison and joint committee work. On operational intelligence matters, ONI agreed that COMNAVEU could channel the reporting of information as he considered best to meet the current situation. Additional officers were assigned to COMNAVEU's Intelligence Section.[2]

On 25 May 1943, ONI's desire to separate COMNAVEU and the naval attaché was partially satisfied when the naval attaché, Capt. Paul H. Bastedo, was relieved by Capt. Gail B. Wilson as Chief of Staff COMNAVEU, with Bastedo remaining as naval attaché. COMNAVEU Adm. Stark objected, stating that he wanted all officers assigned in his area to be on his staff. Stark thereupon countered the separation by making Capt. Bastedo his Assistant Chief of Staff for Intelligence. The situation was finally resolved on 21 July 1943 when Capt. Bastedo returned to the United States due to illness. Stark requested that his chief of staff, now a rear admiral, be given additional duty as naval attaché, and Wilson was appointed to the position on 17 October.

Col. William T. Clement, USMC, took over Capt. Bastedo's duties as Assistant Chief of Staff for Intelligence on 21 July 1943 and served until 30 November 1943, when Wilson took on the additional duty as COMNAVEU Intelligence Officer. On 8 February 1944, Capt. Tully Shelley reported as COMNAVEU Intelligence Officer, and, on 14 April 1944, he promulgated a program to coordinate naval intelligence activities in Europe that improved the Navy's intelligence situation there.[3]

The conflict between COMNAVEU and ONI over who would control intelligence activities in England probably adversely affected the Navy's intelligence effort in England during 1942–1943. The Army successfully circumvented a similar problem by using the Office of the Military Attaché as its intelligence-gathering organization.

In October 1943, a Civil Affairs Section was established as an element of the Intelligence Division of COMNAVEU. Cdr. William H. Tuck, USNR, was appointed head of the section. Because Tuck was on a trip to Washington at the time of his appointment, Cdr. Tracy B. Kittredge, USNR, head of the Political Warfare Section of the Intelligence Division, also served as acting head of the Civil Affairs Section.

By the spring of 1944, the number of officers in the Civil Affairs Section had grown to seventy-one, and, on 15 April 1944, the section was set up as a separate staff division of COMNAVEU.

Civil affairs on the European continent was primarily an Army responsibility. Naval civil affairs responsibilities related principally to problems arising in ports and in coastal areas where the Navy had units ashore or where the civilian population was engaged in activities related to the sea.[4]

The Combined Intelligence Objectives Subcommittee (CIOS) was an organization created in London under a directive from the Combined Chiefs of Staff to work with Supreme Headquarters, Allied Expeditionary Force (SHAEF), in selecting intelligence targets of importance to the military operations under SHAEF's command. The U.S. Navy was represented on CIOS by COMNAVEU Intelligence Officer Tully Shelley.

Before the Normandy invasion, Anglo-American intelligence organizations had worked together assembling information for use by the Supreme Commander's forces. COMNAVEU and CTF 122 (Commander Naval Forces, France) intelligence sections prepared reports and collected information. For the invasion, the COMNAVEU Intelligence Section organized a Special Intelligence Unit that operated with the First U.S. Army during 6–20 June 1944 and went ashore in France on D-Day. On 20 July, the unit came under the administrative supervision of CTF 125 (Commander U.S. Ports and Bases, France). It was designated Commander Task Group (CTG) 125.8, but it operated under directives from the COMNAVEU Intelligence Section. CTG 125.8's mission was to collect, evaluate, and disseminate naval intelligence information of a long-range nature. It was self-contained and mobile and was ordered to travel close to the area of combat activity in order to exploit intelligence targets wherever found. CTG 125.8 was composed of thirteen officers and four enlisted men, plus three officers and one enlisted man attached to it but serving at the headquarters of the Third Army. In September, CTG 125.8's mission was revised to include carrying out duties in connection with the interrogation of German prisoners of war (POW) and with the exploitation of captured documents. On 5 October, CTG 125.8 was transferred from CTF 125 to CTF 122, its headquarters was moved from Cherbourg to Paris, and its title was changed to COMNAVEU Forward Intelligence Unit (FIU). In early 1945, most of the organization's officers were transferred to the Naval Technical Mission in Europe.[5]

COMNAVEU Intelligence Activities During the Normandy Invasion

COMNAVEU intelligence personnel participating in the Normandy campaign during June and July 1944 and the units to which they were attached were as follows:

LCdr. L. E. Riggins, USNR, G-2 HQ, FUSA
Lt. C. W. Chattaway, USNR, G-2 VII Corps
Lt. R. M. Dubois, USNR, G-2 VII Corps
Lt. J. Kaitz, USNR, G-2 VII Corps
Lt. G. E. Kidd, USNR, G-2 VII Corps
Lt. J. E. Lambie, Jr., USNR, G-2 VII Corps
Lt.(jg) P. Dibble, USNR, G-2 1st Division
Lt.(jg) D. Longmaid, USNR, G-2 1st Division
Lt.(jg) H. T. Hardenburg, USNR, G-2 4th Division
Lt. J. G. Lyons, USNR, G-2 4th Division

LCdr. R. D. Whichard, USNR, G-2 90th Division
Lt.(jg) H. A. Warren, USNR, G-2 90th Division
Lt. J. E. Kearley, USNR, Cherbourg Naval Port Party
Lt. I. C. Kitchin, USNR, Cherbourg Naval Port Party
YN3 R. L. Steere, Cherbourg Naval Port Party[6]

Among the officers assigned to COMNAVEU's Special Intelligence Unit for the forthcoming invasion of France were four German-speaking representatives from ONI's Special Activities Branch (OP-16-Z): Lts. George E. Kidd and John Lambie and Lts.(jg) Philo Dibble and Henry T. Hardenburg. They had arrived at Liverpool on 24 February and had spent the three months prior to the invasion in various courses of instruction, including a two-week commando drill at Roseneath, Scotland. Their drill master, Lt. J. E. Kearley, USNR, a former Marine sergeant, drilled them as stridently as if they were Marine recruits.

In late May, the above four ONI officers were assigned to various Army units and ordered to the staging areas in the south of England. Dibble went to the 1st Division, Hardenburg to the 4th Division, and Kidd and Lambie to VII Corps. It was only then that they were told where they were to land, and they spent the remaining time being briefed and reading all available reports on the Normandy landing areas.

On D-Day, 6 June, Lambie and Kidd spent a good part of the morning interviewing the crews of landing craft returning to their transport from runs to the beach so they could learn of landing conditions, beach defenses and obstructions, and the intensity and type of enemy fire.

At about 1400, Col. King, Intelligence Officer (G-2) of VII Corps, sent Kidd and a German-speaking Army order-of-battle analyst officer ashore to try to get information on the identity of the opposing German units. About a mile inland from the beach, they found the 4th Division Command Post and a forward hospital unit caring for several German wounded. From the wounded Germans, Kidd and the Army officer were able to identify several enemy units, including a division that had appeared unexpectedly in intelligence reports shortly before the invasion. Their information was given to an officer from the advance echelon of VII Corps to pass to Col. King.

On 7 June, Lt. Kidd, after spending the night in the home of a French farmer, resumed his interrogations on the beach, where prisoners of war were awaiting evacuation to England. Kidd particularly sought information on the location of any German S-boats and submarines in the area that might be able to attack Allied ships off the beaches. That evening, Kidd wrote a longhand report on the information he had collected and on the results of his efforts thus far. He sent it via a landing craft to the transport *Bayfield* (APA 33), addressed to LCdr. Robert Thayer, intelligence officer on the staff of Commander Task Force Uncle (RAdm. Don P. Moon), responsible for transporting troops from England to the beachhead.

On 8 June, Lambie and Hardenburg came ashore and, with Kidd, fixed up a German dugout behind the seawall on Uncle Red Beach as a prisoner-of-war interrogation cubicle. Dibble, who had come ashore on 6 June at Omaha Beach, had been pinned down on the beach for a few hours by the tenacious German resistance, and it was several days before he could make contact with others in the OP-16-Z group. Lts. John Lyon and Michel Dubois of the COMNAVEU team, along with Army interrogators, also turned up on 8 June and joined in the interrogation effort. On 10 June, LCdr. Leslie Riggins, an assistant to Capt. Shelley, and Lt. Irwin Kitchin, another OP-16-Z interrogator, along with YN3 Robert Steere, who spoke German and French, came ashore as members of the Cherbourg Port Party. Kitchin and Steere were later attached to CTG 125.8.

For the next three weeks, until Cherbourg was captured, Lambie, Kidd, and Hardenburg worked in loose coordination, checking various prisoner-of-war enclosures for German naval personnel. They estimated that they found and interrogated about 2,000 naval prisoners during the Cherbourg campaign. Some prisoners were quickly screened out as being of little interest. Others were questioned in detail and provided considerable information on German defenses, such as the locations and construction features of strong points along the northeast coast of the Cherbourg Peninsula and details about German beach obstacles, pillboxes, gun emplacements, and a V-1 missile installation. One Frenchman provided details about the naval dockyard at Cherbourg, including a map overlay. Daily handwritten reports were sent to the *Bayfield* via the beach master, who had set up a courier service to the ship.

For several days following the invasion, Lambie, Kidd, and Hardenburg alternated in helping LCdr. Thayer review the captured German documents beginning to arrive in quantity and to interrogate wounded prisoners of war who had been brought on board the *Bayfield*. One bag of documents contained a set of plans for a new type of contact mine designed for use against landing craft.

The German naval prisoners found and interrogated in the Cherbourg area had worked either at the torpedo arsenal or at other naval-related jobs around the port. They were cooperative and seemed relieved that the war was over for them. In late

June, Lt. Kidd found one prisoner who had worked for several years as a civilian at the German Navy Mine Testing Command at Kiel. Interrogation brought out that he had firsthand information on a new German pressure mine for which no countermeasure or safe sweeping technique had yet been devised. Information about the mine was on the most-wanted list. Kidd tried to get the Army to fly the "hot prospect" to England where he could be interrogated thoroughly with mine specialists attending. But this procedure was against Army policy, so Kidd decided to take the prisoner personally across the Channel on board a landing ship to make sure he would be safely delivered to the Interrogation Center at Kempton Park. Two days later, Kidd successfully made his delivery and then went to London to alert Capt. Shelley's office and the COMNAVEU mine experts before returning to Normandy.

Two other OP-16-Z German linguists were working at the prisoner-of-war holding compound at Devizes, England. They were Lts. Ridgely White and Angus McLean Thuermer. Later, Thuermer and Hardenburg exchanged places, and Thuermer remained with TG 125.8/Forward Intelligence Unit throughout the rest of its organized existence.

Lt. Lambie distinguished himself at Cherbourg. As United States troops were closing in on the city in late June, Lambie and Riggins attached themselves to the Army regiment advancing toward the German naval headquarters at Villa Meurice on a hill beyond Cherbourg. Lambie went looking for the underground passage from the villa that he had read about in intelligence reports. At about the same time that Lambie found the tunnel entrance, a German machine-gunner spotted him and lacerated his cheek. Lambie, however, was able to get help from a tank destroyer that fired a few rounds into the mouth of the tunnel and induced the occupants to surrender. They included the two highest-ranking German officers at Cherbourg: Gen. von Schlieben, commander of Fortress Cherbourg, and Adm. Walter Hennecke, commander of the port. Lambie was quickly decorated with the Silver Star and Purple Heart; Riggins received the Bronze Star.

Gen. von Schlieben and Adm. Hennecke were taken to VII Corps headquarters where they were questioned by MajGen. J. Lawton Collins. Lt. Kidd tried to get permission to talk to Adm. Hennecke, but MajGen. Collins not only refused the request but also failed to pass on any naval information, probably because he didn't get any.

There was one fatality among the COMNAVEU intelligence officers: Lt. J. E. Kearley. He was last seen alive on 28 June while searching through the arsenal at Cherbourg harbor after its capture. On 8 July, his body was found in the water near the E-boat pens; he had been shot in the head.[7]

The fall of Cherbourg and the cleaning up of intelligence targets at that port marked the close of the initial phase of activities of COMNAVEU's Special Intelligence Unit. The lull in the fighting before the Army's breakout from the Cherbourg peninsula permitted a review of the unit's activities and an assessment of its strengths and weaknesses. The unit found that it had an urgent need for adequate jeep transportation; during most of July, the unit had had one jeep, a borrowed motorcycle, and three captured enemy vehicles. The unit's leadership also believed that it needed a tighter, more centralized organization and a means for channeling the distribution of collected material and information. On the plus side, however, a great deal of naval intelligence information was available about the French shore, and the officers assigned were qualified to dig it out. Additionally, relations with the U.S. Army at every echelon were considered to be generally excellent. Up to that time, the unit had been a loosely organized group, with its officers assigned to various Army units upon which they were dependent for transportation, billeting, and messing.[8]

Another member of COMNAVEU's Special Intelligence Unit for the Normandy campaign was Lt. Joseph Kaitz, who came from the Investigations Section (B-7) of the District Intelligence Office, 3rd Naval District, in New York City, via various OSS schools. Kaitz landed on Omaha Beach on D-Day with the 1st Army Division and then later shifted to the 47th Regiment, 9th Division. He had been directed to obtain information on enemy ship movements, minefield locations, salvage operations, and the location of hulks in harbors. Information about the locations of sunken ships was especially needed by Commo. William A. Sullivan, the Navy's Supervisor of Salvage, who was responsible for getting the French harbors and ports back in operation as quickly as possible. There was also a strong interest in captured ships that were still operational, such as minesweepers and salvage ships. For Kaitz to fulfill his directive, he had to conduct continuous search, interrogation, and investigation.

Using a nail on the green shutter of a building in the port area of Cherbourg, Lt. Kaitz scratched the words "U.S. Naval Headquarters, Cherbourg," thus opening the first naval intelligence office in the city. Since he spoke neither French nor German, his duties were mainly the retrieval of documents from safes or other secure storage facilities and to conduct investigations.[9]

On 20 July 1944, TG 125.8 was formally established, and LCdr. George T. O'Neill reported on 21 July as the task group commander. He brought with him four jeeps, one yeoman first class, and LCdr. F. Murray Forbes to set up a Cherbourg office from which to distribute the intelligence gathered by Task Group 125.8 and to handle routine administrative matters for the group.[10]

Most of the personnel in the Special Intelligence Unit were incorporated into the newly formed TG 125.8 under the administrative control of CTF 125. TG 125.8's mission was stated in a COMNAVEU directive to CTG 125.8, dated 31 July 1944:

> To collect, evaluate, and disseminate naval intelligence information in advance of, and in connection with, the occupation of ports and bases by the U.S. Navy.
>
> To collect, evaluate, and forward for transmittal to the Navy Department, Washington, and other interested activities or agencies, naval intelligence information of a long range nature.
>
> Task Group 125.8 will operate as a mobile, self-contained naval unit. Its territory will be limited only by the availability of naval intelligence information, and the collection or dissemination thereof. It will establish and maintain its own headquarters in the field, in close contact with Army commands to whom port and coastal targets are progressively assigned. It will move with these commands on to their targets as the latter are opened up and will remain at, or in the vicinity of, such targets after capture, until its mission has been completed.[11]

The personnel assigned to the Mobile Unit of TG 125.8 were as follows:

LCdr. G. T. O'Neill, USNR
LCdr. F. M. Forbes, Jr., USNR
Lt. J. E. Lambie, Jr., USNR
Lt. G. E. Kidd, USNR
Lt. C. W. Reuss, USNR
Lt. J. Kaitz, USNR
Lt. I. C. Kitchin, USNR
Lt. C. W. Chattaway, USNR
Lt. R. M. Dubois, USNR
Lt. J. G. Lyons, USNR
Lt.(jg) C. A. Rocheleau, USNR
Lt.(jg) P. Dibble, USNR
Lt.(jg) A. M. Thuermer, USNR
Yn1 R. L. Thompson, USNR
Sp(P)2c J. E. Billups, USNR
Sp(P)2c C. P. Reckert, USNR
YN3 R. L. Steere, USNR[12]

Also attached to TG 125.8 and working at Third U.S. Army headquarters in cooperation with the Mobile Unit were the following:

LCdr. A. K. Train, USNR
Lt. M. W. Rehor, USNR
Ens. E. H. Taliaferro, USNR
YN2 E. E. Young, USNR

When United States forces broke out of the Cherbourg Peninsula, Gen. George S. Patton and the Third Army turned west into the Brittany Peninsula and headed for Brest. Other Allied forces pushed in the opposite direction toward Paris. TG 125.8's mobile element was divided on 17 August into two units: one, consisting of three officers and a yeoman second class, was attached to Third Army headquarters; the other followed the forces headed for Paris. The latter unit included LCdr. O'Neill; Lts. Kidd, Kaitz, and Chattaway; and two yeomen. They went directly to Chartres to organize the coverage of, and await the move into, Paris, which was 55 miles away. The Paris unit of TG 125.8 was the first U.S. naval organization to enter the city, arriving after dark on 25 August. For several days LCdr. O'Neill was the senior U.S. naval officer in Paris.

During the first days of the occupation of Paris, TG 125.8 personnel searched through the various buildings and installations that the German navy had occupied at a complex on the outskirts of the city. The British 30th Assault Unit, an armed intelligence unit of the Royal Marines, had moved quickly into the complex. Lt. Lambie had been assigned as liaison officer to the British unit when TG 125.8 was activated, and cooperation between the two groups was good.

One joint U.S. Navy operation with the British 30th Assault Unit was a search through the German Torpedo Arsenal, West, located in an extensive maze of underground passages near Versailles and well stocked with torpedoes and related equipment and machinery.[13]

The Investigations Unit of the TG 125.8 mobile element that entered Paris mainly devoted its efforts to finding French contractors who had worked for the German navy and to collecting documents of interest from their offices. For example, drawings were obtained that showed the degaussing equipment installed in the harbors at Le Havre and Cherbourg; and a large amount of information about German construction of defensive installations was also uncovered. The unit turned over any information that it found about land defenses to the Army.

Another Navy investigative unit in Paris operated under LCdr. William Abbott and was involved in locating and interrogating American and Japanese nationals who had remained in France throughout the German occupation. From the Japanese information was sought on the equipment, weapons, and technical developments that they had obtained in Europe and sent back to Japan. The American nationals were checked as to their loyalty.[14]

In the first few days after TG 125.8 personnel arrived in Paris, numerous documents and pieces of ordnance of interest were gathered, and several Frenchmen were located who could provide specialized information about the Germans. Some of the French citizens had already been arrested as collaborators by French authorities, and the French were reluctant to release them to American custody.

The TG 125.8 unit that moved with the front into Brittany found a German sailor whose ship had laid mines in the port of St. Malo. A Dutch merchant shipmaster, who was familiar with minefields in the Channel Islands, charted the fields in detail for the unit. At Brest, they were the first intelligence officers to search the German naval base, the U-boat flotilla headquarters, the Torpedo Kommando Headquarters, and the U-boat pens. Interrogating French civilians also aided the TG 125.8 unit in locating minefields.[15]

Two British naval officers from the Royal Navy's Forward Intelligence Unit, Lts. Ralph W. B. Izzard and Brian Connell, joined the TG 125.8 unit in Brittany to give a helping hand and to observe U.S. methods. The two British officers remained with TG 125.8 throughout most of the Brittany campaign. A few weeks later, a team of TF 125.8 interrogators was sent to the Channel coast to join the Royal Navy's FIU, which was operating with the First Canadian Army. TG 125.8 officers were also sent on intelligence-gathering missions to various other French ports and cities, such as Bordeaux, Vannes, La Rochelle, and Rennes. Lt. Charles A. Rocheleau joined TG 125.8 at Cherbourg about three days after the liberation of Paris at the end of August 1944, but he was in time to participate with the task group in the siege of Brest, working with the United States VIII Army Corps.[16]

Brest had served as a prewar French naval headquarters. The base area of Brest was located on the waterfront and in underground caves and tunnels in the seaside cliffs. It was the last part of the city to surrender. It was important to get to the German Naval Command Headquarters quickly before documents could be destroyed. Rocheleau, a French linguist, found a French naval engineer outside Brest who could tell exactly where the German headquarters was located, thus saving considerable search time.

Rocheleau and Thuermer spent one day rowing around the submarine basin at Brest and found three submerged submarine hulks that appeared to have been sunk in a way that would block the use of the basin and the underground pens. The information was, of course, reported to COMNAVEU.

After Brest fell, TG 125.8 followed the Army south, in November 1944, where the Germans were still holding out at the port of Nantes. Rocheleau, however, was detached at that time and ordered to Paris, where VAdm. Alan Kirk, Commander Naval Forces, France, was setting up his headquarters. Staff Intelligence Officer Capt. Dallas D. Dupre assigned Rocheleau to be his principal assistant.[17]

COMNAVEU Forward Intelligence Unit and Its Successors

When, in October 1944, TG 125.8 was reorganized and placed under the administrative control of Commander Naval Forces, France as COMNAVEU's FIU, it was ordered to prepare for the advance into Germany. TG 125.8 continued to operate under COMNAVEU directives that were fairly broad and gave considerable latitude to the group's work: "to proceed to such places and omit, revisit or vary your itinerary as may be considered necessary in the collection and dissemination of Naval Intelligence on enemy activities on the continent of Europe."[18]

A mobile communications unit was assigned to the FIU and it set up operations in the Cognac area. It consisted of Lt. C. E. Smith III and four radiomen. The unit's officers in Paris lived at the principal U.S. naval billet, the Hotel Royal Monceau, near the Arc de Triomphe.[19]

Personnel attached to the COMNAVEU FIU during 5 October 1944 to 11 February 1945 included the following:

LCdr. G. T. O'Neill, USNR	Commanding Officer and French Interrogator
Lt. G. E. Kidd, USNR	German and French Interrogator
Lt. I. C. Kitchin, USNR	German Interrogator
Lt. J. E. Lambie, Jr., USNR	German and French Interrogator (liaison with 30 AU)
Lt. J. G. Lyons, USNR	French Interrogator
Lt. M. W. Rehor, USNR	Slavic Interrogator (detached during period)
Lt. C. E. Smith III, USNR	Communications Officer (reported 4 Nov 1944)
Lt.(jg) P. Dibble, USNR	German Interrogator
Lt.(jg) A. M. Thuermer, USNR	German Interrogator
Lt.(jg) P. W. Wilkinson, USNR	Torpedo and Mine Specialist (reported 9 Dec 1944)
SP(X)1 R. L. Thompson	
RT1 R. C. Rands	
MOMM2 O. E. Ames	
RM2 C. J. Borgren	

RM2 R. L. Chagnon
RM2 C. W. Dechnik
SP(X)3 R. L. Steere[20]

When Cherbourg fell on 27 June 1944, the occupying forces included salvage and advance base personnel. With the latter was COMNAVEU's FIU, which had landed during the initial phases of the invasion and had contributed valuable assistance during the Cotentin Peninsula campaign by collecting tactical information and maintaining liaison with the French Resistance organization.

Commander U.S. Ports and Bases, France (CTF 125) was activated on 10 July 1944 under the command of RAdm. John Wilkes, and the FIU became TG 125.8, with LCdr. G. T. O'Neill in charge. In October, Task Group 125.8 became Task Unit 122.11.6 under Commander U.S. Naval Forces, France (CTF 122) VAdm. Kirk, whose headquarters was located in Paris. By March 1945, CTF 122 exercised control of U.S. Navy forces in Holland, Belgium, Germany, and Luxembourg, as well as France.

By June 1945, there was no longer any need for a command as large as CTF 122 in Paris, particularly with the reestablishment of the office of the U.S. Naval Attaché, Paris, and its increasing takeover of the normal peacetime functions of an attaché office. Consequently, CTF 122 was disestablished on 1 July 1945, its residual functions being assumed by Commander Naval Group, France (CTG 124.3), under Commander U.S. Naval Forces, Germany (CTF 124).[21]

The Post-Hostilities Committee of COMNAVEU was established in March 1944 to pool views and information from many specialized sources and to advise COMNAVEU on problems relating to the European Advisory Commission (EAC), which was composed of United States, British, and Soviet Ambassadors and was intended to resolve certain post-hostility problems. One, in which COMNAVEU was actively associated as an adviser to the EAC, was in psychological warfare. One of the Navy officers on the committee staff was Cdr. Tracy B. Kittredge, USNR, who was in charge of political intelligence and acted as COMNAVEU representative at psychological warfare meetings. The nature of the Post-Hostilities Committee changed somewhat in August 1944 when VAdm. William Glassford became Deputy COMNAVEU for Post-Hostilities. Among the officers assigned to the revamped committee were Capt. L. Ragonnet, Intelligence Officer; Capt. Donald Frothingham, Naval Attaché to the Ambassador to the Governments-in-Exile; and Capt. Tully Shelley, COMNAVEU Intelligence Officer.

On 12 September 1944, COMNAVEU created European Affairs Division (EAD) that was intended to have cognizance over European post-hostilities activities and related matters. To aid the EAD, Cdr. Kittredge and the Political Warfare Section of COMNAVEU's Intelligence Division were transferred to the EAD, along with the Civil Affairs Division. The Political Warfare Section personnel formed the Political Section of EAD, which functioned until November 1945, when the section was dissolved and the political intelligence functions were returned to the COMNAVEU Intelligence Division.[22]

In late 1944, COMNAVEU assigned the following officers to Commander Naval Forces, France, for duty with the Army "T" Forces: LCdr. K. H. Baarslag and Lts. J. F. MacMahon, E. K. Salls, and V. G. Davey. The "T" Forces were to follow the combat troops closely and secure intelligence targets selected by the Combined Intelligence Objectives Subcommittee until CIOS Field Teams arrived on the scene to exploit the targets.

The naval intelligence officers assigned to "T" Force were to compile data on naval targets of interest in order to prepare themselves to exploit targets when they were captured; to compile data on naval targets in Berlin and Kiel under the supervision of the special "T" Force commanders for those locations; to determine the targets of naval interest and, through their work with the "T" Forces, provide for the seizure and security of the naval targets until they could be examined by naval experts or CIOS field teams; and to seize captured enemy naval documents and provide for their proper disposition in accordance with existing intelligence directives. The four officers were to travel with the "T" Force of the Army group to which they were attached and either confirm the existing intelligence or ascertain the facts about the targets so that the information could be passed back to the naval coordinator at SHAEF for dispatch to CIOS, the Alsos Mission (a Joint Army-Navy team for technical intelligence exploitation in Europe—see Chapter 11), or other interested naval authorities. The four officers were also directed to locate new targets or the new locations of known targets through on-the-spot interrogation of local civilians. They were to make immediate reports of their findings to the naval coordinator at SHAEF in order to assure the protection of U.S. naval interests.[23]

When the COMNAVEU Forward Intelligence Unit was absorbed into the newly organized U.S. Naval Technical Mission in Europe on 11 February 1945, it became the Intelligence Department of the mission. The headquarters for the mission was set up in Paris, where a cadre of representatives of the various Navy Department technical bureaus was maintained. A steady flow of specialists from the bureaus pursued their assigned projects by making field trips and then returning to Washington. As Allied troops advanced into Germany, NAVTECHMISEU established a major forward base at Villa Lilly, an estate near Wiesbaden, and at a smaller one near Munich.

The Intelligence Department provided guides, escorts, interpreters, and translators for NAVTECHMISEU field parties as they ranged over Germany, France, Austria, Italy, and parts of Czechoslovakia.[24] As an example of the legwork performed by mission personnel, Lt. Joseph Kaitz accompanied the famous aviator, Col. Charles Lindbergh, who was attached to NAVTECHMISEU, to Cologne and then to Freiberg, Germany. Lindberg was in search of German jet engines. He found an aircraft that looked as though it had been configured for jet propulsion, but it had no engine. Another search group eventually found numerous German jet engines.[25]

The intelligence components of U.S. Naval Forces, Germany (CTF 124) and U.S. Ports and Bases, Germany (CTF 126) were organized as a section of the COMNAVEU Intelligence Division and remained under the direct control of the COMNAVEU Intelligence Officer, Commo. Tully Shelley, until 1 March 1945. The "G" Section, as it was known, performed the intelligence work for the two task forces and for the Naval Division of the U.S. Group of the newly established Allied Control Council for Germany. The head of the "G" Section, together with his executive officer and planning officer, made up the intelligence nucleus of the Naval Division of the U.S. Group, and the head of the section represented the Naval Intelligence Service in the Joint Intelligence Command of the U.S. Group. On 1 March, the "G" Section was transferred to COMNAVFORGER and became the Intelligence Divisions of TF 124 and TF 126. (For additional information on these task forces, see the section on NAVFORGER, later in this chapter.)[26]

On 30 August 1945, COMNAVEU's staff was reorganized. Intelligence (N-2) was headed by Shelley, now Senior Assistant Naval Attaché. All intelligence units were brought under the Intelligence Division and comprised four sections, each headed by an assistant naval attaché: General Intelligence, Technical Intelligence, Air Intelligence, and Air Technical Intelligence.[27]

The COMNAVEU Intelligence Division became involved in the exploitation of the captured German naval records found at Tambach Castle. The records were brought to London for photocopying by the U.S. Navy. On 26 September 1945, a microfilm unit was established within the General Intelligence Section. In the spring of 1946, a program was started for indexing, arranging by subject, and selective filming and translating under the direction of Cdr. S. R. Sanders, USNR. For more details on the copying of the Tambach Archives, see Chapter 17.[28]

CINCNELM Intelligence Activities, Post–World War II to 1970

With the end of World War II, the Naval Attaché, London and the Commander in Chief, U.S. Naval Forces, Eastern Atlantic and Mediterranean intelligence staffs were organizationally separated. As of 1947, the major part of the ongoing general intelligence work was being carried on by the Naval Attaché's office, but CINCNELM maintained a small intelligence division. Close coordination was essential between the two staffs, and they collaborated in their collection of intelligence from British sources. The continued delicate state of international affairs and the remarkably close liaison with the British made the coordination important for the effective exploitation of, and for maintaining favorable relations with, their British intelligence sources.[29]

About October 1948, RAdm. Thomas B. Inglis and RAdm. Longley-Cook, RN, the U.S. and British Directors of Naval Intelligence, reached an understanding, thereafter known as the Inglis/Longley-Cook Agreement, whereby an active operational intelligence liaison would be maintained between CINCNELM and the Naval Intelligence Division (NID) of the Admiralty. The need for a combined U.S-British operational intelligence plot in wartime was anticipated, and the U.S. Navy was to assign three lieutenant commanders from CINCNELM to NID as a nucleus for the wartime organization.

The first three reinstated U.S. Navy officers assigned to the British NID under the Inglis/Longley-Cook Agreement were LCdrs. Joseph A. Meyertholen and Harvey Peacock and Lt. John H. Gano. LCdr. Meyertholen served as the head of the Merchant Ship Plot, and Lt. Gano served as the head of the Naval Plot and as Top Secret Control Officer. Despite the unique situation, NID seldom placed any restriction on the U.S. officers' access to information.[30]

Meyertholen made daily reports and provided periodic analyses to his branch head in NID, who, in turn, made periodic contributions to the NID weekly staff meeting. Both Meyertholen and Gano briefed Royal Navy officers being assigned to overseas areas, such as prospective naval attachés and NATO staff officers.

Lt. Edward J. Cummings, Jr., relieved LCdr. Peacock in early to mid-1950 and was assigned to NID's Navy Plot. He also took over Lt. Gano's duties when the latter was detached without relief in October 1950 when CINCNELM experienced a personnel cut. Meyertholen, Gano, and Cummings spent their lunch hours at CINCNELM, where they reported on daily developments. Many items in CINCNELM's *Weekly Intelligence Digest* were based on their reports.[31]

Two shipping intelligence officers covered the movements of Soviet and Soviet satellite merchant ships and Soviet-chartered ships, as well as U.S. merchant ships calling at Iron Curtain ports. They also maintained liaison with the British Admiralty to obtain all intelligence of interest. The one operational intelligence officer maintained a plot of foreign naval ships, their characteristics, employment, and new construction; he also maintained liaison with the Admiralty on matters dealing with the Soviet navy.[32]

A Fleet Intelligence Center (FIC) was established at Port Lyautey (modern Kenitra), Morocco, in March 1954 to support CINCNELM. See Chapter 40 for more details on the FICs.

CINCNELM's intelligence-gathering capabilities were expanded in 1954 and 1955 by the assignment of a captain to the staff as a special assistant for Naval Security Group matters. His functions were related to communications intelligence and security and to special electronic research projects. The research projects were subject to policy control by the Chief of Naval Operations and the Director of the National Security Agency.[33]

The CINCNELM ELINT Center was established in December 1955, and major construction was started in 1955 on a joint Navy-Air Force communication intercept facility in the Middle East that became operational on 15 September 1957.[34] Effective 2 July 1956, the ELINT section at Bremerhaven was moved to Todendorf, Germany.[35]

On 29 October 1956, Israel attacked Egypt after the closing of the Suez Canal. Britain and France joined Israel by making landings on 6 November at Suez. In spite of CINCNELM's close contacts with the British Admiralty, it had had no warning of British intentions to initiate operations in the Middle East, and U.S. support of a Soviet call at the United Nations for a cease-fire resulted in strained relations for a period. The Sixth Fleet was involved in an amphibious exercise at Crete at the time of the unexpected Anglo-French invasion.[36]

In June 1957, CINCNELM began conducting an extensive review of the requirements for intelligence materials by attack aircraft carriers with a view toward reducing the volume of materials and providing a master list. The results of the studies and the clear need for standardization led to a CINCNELM/CINCLANTFLT intelligence materials conference convened on 16 April 1958 at CINCNELM headquarters in London, and attended by representatives of the Second and Sixth Fleet commands, the Navy Hydrographic Office, and Fleet Intelligence Center, Eastern Atlantic and Mediterranean. The adoption of the conference's recommendations greatly improved the intelligence readiness of the fleet and led to the development of a Master Intelligence Requirements List. The recommendations also improved the coordination of aerial surveillance and the reporting of Soviet naval movements, and led to action to fill gaps that were apparent in the preparation of amphibious operation plans.[37]

Operations by Soviet submarines in the Mediterranean in 1958–1959, the first time such operations had taken place, indicated the need for additional intelligence collection activities and also provided an opportunity for such activities.[38]

From 15 July to 25 October 1958, a joint U.S. military operation involving 13,000 Marine and Army troops was conducted in the general area of Beirut, Lebanon, to support a friendly Lebanese government against an Egyptian-encouraged uprising by rebel forces. Intelligence was controlled by the Intelligence (J-2) Division of the Commander in Chief Special Command, MidEast staff. The basic intelligence data on coasts and beaches in the *National Intelligence Survey* publications proved to be out-of-date and unreliable, and, of course, information on rebel order-of-battle was essentially nonexistent. Photographic coverage of key targets was available and current, although coverage was deficient on the interior areas of Lebanon, particularly along the Syrian border. For the initial landings and during the first stages of the operation, all aerial reconnaissance was conducted by TF 60 aircraft operating under standard Navy-Marine Corps doctrine. There was a serious deficiency, initially, in the map and chart coverage of the Middle East at the scale of 1:25,000, the scale needed for accurate planning and operations.[39]

The Soviet navy began a massive increase in its naval operations in the Mediterranean in 1967. In July, there were forty-three Soviet naval units in the Mediterranean, including thirteen submarines. Total USSR unit operating days in the Mediterranean in 1967 exceeded 9,000. The United States countered by dispatching two hunter-killer ASW aircraft carrier task groups to the Mediterranean to participate in a coordinated NATO surveillance exercise. The intelligence collection capabilities of Sixth Fleet destroyers were expanded by the assignment of ELINT riders and combat cameramen during the operations.[40]

Advantage was taken of the intelligence collection opportunities presented during an extensive Soviet naval exercise in the Mediterranean in 1969 that included a joint Soviet, Egyptian, and Syrian amphibious landing near Alexandria and a defensive force intercept of a simulated carrier task group. The latter demonstrated the probable Soviet tactics for providing assistance to their Arab clients in any future confrontations with Israel or in the event Western forces should be asked to intervene in a Lebanon-type situation. Seventy-one Soviet ships were involved, plus Soviet aircraft based at Cairo.[41]

The Fleet Ocean Surveillance Information Facility at Rota, Spain, collated intelligence from all sources on Soviet military, naval, and air presence in the Mediterranean and reported the results to the Sixth Fleet. It also published a daily summary of Soviet activities. Soviet naval unit operating days in the Mediterranean reached a total of 17,888 in 1970.[42]

NAVFORGER Intelligence Activities, 1944–1968

During the summer and fall of 1944, it became evident that the U.S. Navy would become involved in disarming and demobilizing the German navy and in collecting the vast amounts of potentially valuable technical intelligence. The Quebec Conference had made the United States responsible for the occupation of southwest Germany as well as for an enclave in the British Zone in northwest Germany that included the ports of Bremen and Bremerhaven on the Weser River.

To carry out the occupation, COMNAVEU established Commander U.S. Naval Forces, Germany (CTF 124), and Commander U.S. Ports and Bases, Germany (CTF 126), with the latter placed organizationally under the former. The Intelligence Branch of the staff of COMNAVFORGER was designated Task Unit 124.1.3.[43]

A training base for the personnel of TF 124 and TF 126 was set up at Base II, Roseneath, Scotland, and operated from December 1944 through the end of March 1945. The training included German language, history, politics, and people; the U.S. principles of military government; and commando tactics.[44]

Reconnaissance and advance parties were set up under CTF 126 to be ready to move into Bremerhaven (Port Party A) and Bremen (Port Party B) as soon as possible on or after the date of safe entry, when called by the British 21st Army Group.

The Task Force 126 Intelligence Section supplied information on the two ports to aid in planning their disarmament and operation and in planning salvage operations to clear the ports of sunken ships and unusable equipment. A detailed list of targets, with their descriptions and directions for finding them, was produced, and an up-to-date chart of wrecks in Bremen and Bremerhaven was kept as reports were received. A Bremen shipping report was prepared, along with a Weser River Enclave chart showing nets and booms, mobile searchlights, heavy and light antiaircraft gun positions, and coastal defenses. Information was also obtained about Allied experiences in France in the rehabilitation and operation of captured ports. Capt. Albert E. Schrader, Acting CTF 126, visited Le Havre in December 1944 to confer with Col. W. F. Way, U.S. Army, who was in charge of harbor salvage there. Schrader visited Le Havre again in January for conferences with Commander U.S. Ports and Bases, France and his staff.[45]

On 1 February 1945, Port Party A was designated Task Group 126.1, and Port Party B became Task Group 126.2.[46]

Before 28 February 1945, the personnel of COMNAVFORGER's Intelligence Division had been attached mainly to the "G" Section of COMNAVEU's Intelligence Division. On that date, the COMNAVFORGER Intelligence Division was activated. The Intelligence Officer Cdr. W. A. Finn, USNR, was initially stationed in Paris in order to keep Commander Naval Forces, France (the senior U.S. Navy officer on the European continent) informed about intelligence plans for the post-hostilities period. In April, an intelligence unit from CTF 124 was established in Paris.

Problems in administering the Intelligence Division arose because CTF 124 could not become active on the European continent until called forward by SHAEF, and intelligence officers had to be sent forward on temporary duty orders. Thus, in addition to Capt. Finn, two officers were assigned as his Forward Intelligence Unit, and six others (increased to eighteen before the end of May) were attached to the Naval Technical Mission to Europe (TF 128), all on temporary duty. Intelligence officers representing all interests of CTF 124 were attached to Royal Navy parties preparing to go to Hamburg, Kiel, and Wilhelmshaven, and one officer was attached to the French Military Mission for German Affairs.[47]

On 29 March, an alerting message was received for the naval reconnaissance and port parties to be ready to move on short notice. Shortly thereafter, the Bremen reconnaissance party was directed to move to the continent by 2 April.

The combined Bremen-Bremerhaven reconnaissance party of 48 officers and 104 men, under Capt. Vincent H. Godfrey, departed Base II in Scotland on 1 April and arrived at Ostend, Belgium, on 4 April. From there, they pushed towards the U.S. enclave as the fighting permitted. By the morning of 8 April, Godfrey and his party had joined up with the reconnaissance party at Kevelaer, Germany. Together, they moved to Hengelo, Holland, the same day and remained there for two weeks waiting to be called forward.

The next move was to Verden, Germany, on 22 April, and then, on 27 April, Capt. Godfrey and a small reconnaissance party departed for Bremen. That night, he was able to report to the Allied Naval Commander that, although Bremen was widely damaged, the docks, quay walls, cranes, and rail lines were apparently in good shape. An amplified report on 28 April was devoted to harbor conditions and indicated that 30 percent of the wharfage was usable; the remainder was obstructed by sunken craft. The river was believed to be heavily mined.

On 7 May, the TG 126.1 reconnaissance party under LCdr. Richard H. Tenney entered Bremerhaven with the advance elements of the 51st Highland Division. They found the German Naval School for Warrant Officers in excellent condition with facilities and accommodations for 4,000 officers and men. It became the headquarters of the Bremerhaven Command. The former liner *Europa*, one of fourteen ships of over 2,000 tons abandoned in Bremerhaven, was also in good shape but without fuel and silted in. The balance of the month of May was spent making a thorough survey of the shipping in Bremerhaven and of the harbor conditions there.[48]

TF 126 was dissolved on 10 November 1945. Continuing staff functions, including intelligence, were assigned to the U.S. Naval Advance Base, Weser River, under Capt. C. R. Jeffe whose headquarters was at Bremerhaven.[49]

As a result of a review of the U.S. Naval Command set up in Europe, all U.S. naval forces assigned to occupation duties or to the support of U.S. armies in the European theater of operations were assigned to COMNAVFORGER (CTF 124) on 1 July 1945. Commander Naval Forces, France became CTG 124.3. On 2 July 1945, COMNAVFORGER VAdm. Robert Ghormley, Chief of Staff Commo. M. C. Robertson, and several other staff officers shifted from London to Frankfurt and established headquarters there on 3 July.[50]

In early 1945, with the establishment of the Control Council for Germany, COMNAVFORGER became the Naval Division of the council's U.S. Group. Its Intelligence Branch was directed to provide intelligence support for the division; provide naval representation on the council's Joint Intelligence Committee and in the Positive Intelligence subsections of the Intelligence Section of the U.S. Group; ensure that interested U.S. agencies received all naval-related intelligence derived from the British, Soviet, and French Occupation Zones; and to represent the Naval Technical Mission to Europe and the Navy's counterintelligence and censorship efforts in the Intelligence Section of the U.S. Group of the Control Council.

The Intelligence Division of TF 124 (formerly known as the "G" Section of the COMNAVEU Intelligence Division) was assigned to provide naval liaison with British, Russian, and French naval authorities in their headquarters and ports; collect, evaluate, coordinate, and disseminate nontechnical naval intelligence that might be of interest to the military or control authorities concerned or to the Navy Department; provide policy direction on all naval counterintelligence activities, including security measures; and to maintain liaison with civil censorship insofar as it concerned naval intelligence.

The Joint Intelligence Committee of the U.S. Office of Military Government (OMGUS) in Germany met weekly and was composed of representatives of the various U.S. activities in Germany, including the Office of the Director of Political Affairs, the Office of the Economic Adviser, the Armed Forces Division, the Finance Division, the Strategic Services Unit, Information Control Services, and the Office of the Naval Adviser. The representative of the latter was Lt. E. G. Riedel who submitted copies of the minutes of the meetings to ONI.[51]

By 1949, the center for COMNAVFORGER's intelligence organization was in Berlin, where Intelligence Officer Capt. Arthur H. Graubart and two assistants, LCdrs. E. G. Riedel and J. T. Ziegweid, were located. Two offices were maintained in Berlin, one in the Director's Building in the HICOG (High Commissioner, Germany, ex-OMGUS) Compound, and the other in the U.S. Consulate building. The intelligence administrative work load, which included the handling of all incoming and outgoing correspondence, the processing and evaluation of all intelligence collected in Germany, the maintenance of central files, the administration of substations, and liaison with other intelligence agencies both in Berlin and in the U.S. zone proper, was carried out at the HICOG compound. The office at the consulate building served as a "retreat" for receiving and interviewing various sources.

The COMNAVFORGER intelligence substations in 1949 included the following:

1) The Frankfurt office in the E.C.I.C. Building at Oberursel also served as a Joint Interrogation Center for all U.S. intelligence agencies in the U.S. Occupation Zone. Assistant Intelligence Officers (AIOs) LCdrs. S. F. Tyler, Jr., and M. P. O'Brien (electronics specialist) were stationed there. The Frankfurt substation maintained close liaison with all other U.S. intelligence organizations in the area (including those at Weisbaden and Heidelberg, where the Air Force and Army Intelligence organi-

zations were concentrated). The Navy office also acted as a subcentral office for the other outlying Naval Intelligence subposts at Munich, Ulm, Hof, and Bad Hersfeld. Its main collection effort was devoted to the exploitation of German prisoners of war repatriated from Russia.

2) The Munich office was located in the S-2 Building of the Munich Military Post. It had one AIO, Lt. V. L. Rychley, whose main tasks were the interrogation of repatriated German prisoners and liaison with other intelligence organizations in the area.

3) Ulm had one AIO, LCdr. John D. H. Kane, Jr., and was located in the administration building at the repatriated pow reception camp. Its activities paralleled those at Munich.

4) Hof had one AIO, LCdr. J. G. Lyman, and was located in the administration building at the POW reception camp. Its activities also paralleled those at Munich.

5) Bad Hersfeld had no assistant intelligence officer, but an ex-German naval officer maintained an office at the POW camp (which was run by the Germans), where he interrogated repatriated prisoners of possible naval interest and forwarded the results to Berlin.

6) The Bremerhaven office was located in the main building of the Commander U.S. Naval Advanced Bases compound. LCdr. M. P. Horn, a naval aviator, served as assistant intelligence officer and as administrator of the Naval Historical Project, an organization composed of five of the most knowledgeable former German navy flag officers, engaged in compiling various naval studies. Their principal product was an assessment of Soviet naval activities during World War II in Europe. A revised and condensed version was eventually published in 1978 as *The Soviets as Opponents at Sea* by the United States Naval Institute.

7) The Vienna office had one AIO, Lt. C. A. Rochelau, who used office space made available by the Chief of the U.S. Navy Mission, U.S. Forces, Austria. Exploitation of the large number of prisoners of war returning from Russia was the chief function of the Vienna office.[52]

As of 15 February 1950, the COMNAVFORGER intelligence organization was located as follows:

Locality	Officers	Enlisted	Civilians (U.S.)
Berlin (HQ)	1 Capt., 3 Cdrs.	3	2
Frankfurt	1 LCdr, 1 Lt.	1	1
Heidelberg	1 LCdr.	0	0
Bremerhaven	1 Lt.	1	0
Ulm	1 LCdr.	0	0
Munich	1 Lt.	0	0
Hof	1 LCdr.	0	0
Vienna, Austria	1 Cdr., 1 Lt.	0	1
Totals	13	5	4

The intelligence mission within COMNAVFORGER in 1950 was stated to be:

> **Primary**: The collection in Germany and Austria of all possible intelligence bearing on the present and future potential of the Soviet (and Satellites) Navy and the dissemination thereof to the Department of the Navy (ONI) and other U.S. agencies in the theatre and elsewhere having an interest therein.
>
> **Secondary**: The maintenance of close liaison with other U.S. intelligence agencies in Germany having an interest in the over-all intelligence situation. The dissemination of other than strictly naval intelligence to cognizant U.S. agencies in the theatre.[53]

Representatives of COMNAVFORGER were also members of the Intelligence Coordinating Committee, Germany (ICC[G]). All United States covert intelligence activities in Germany were coordinated through ICG(G), but the committee did not have direct authority over, or responsibility for, the separate intelligence functions of the Army, Navy or Air Force.[54]

Chief of Naval Operations letter serial 0050P92 of 15 January 1957, "U.S. Naval Intelligence Organization in Germany (FRG and Berlin) after the Disestablishment of U.S. Naval Forces, Germany," prescribed the mission and functions of the resultant Naval Intelligence organization in Germany.

> **Mission**: Within the context of Executive Order 10608, as implemented by US CINCEUR, and as defined in the pertinent provisions and protocols of the NATO Status of Forces Agreement with Germany to:
>
> 1. Collect, evaluate and disseminate information of naval interest, particularly on the Sino-Soviet Bloc;
>
> 2. Conduct those intelligence operations necessary for the security of U.S. forces.
>
> **Functions**:
>
> 1. To obtain as directed by CINCNELM or CNO (DNI) all available information and intelligence of naval interest, particularly on the Sino-Soviet Bloc;
>
> 2. To serve as CINCNELM's representative with U.S. and Allied Forces in Germany, including West German forces, for the coordination of intelligence and security matters.

3. To effect liaison with appropriate Naval Attachés and with ONR (Office of Naval Research), London, for the coordinated exploitation of sources in Western Europe knowledgeable in scientific and technical (S&T) matters of naval interest; and

4. To maintain close liaison with Naval Attaché, Bonn, towards the most effective realization of the U.S. Navy's overall objectives.[55]

U.S. Naval Forces, Germany, was disestablished on 20 June 1957. One of the elements that remained in existence, however, was the Intelligence Division of the staff, for which responsibility was transferred to CINCNELM on 1 July 1957. The former COMNAVFORGER Intelligence Officer became CINCNELM Representative, Germany, with an office in Frankfurt. Subordinate offices were located in Berlin, Frankfurt, and Munich. Capt. R. H. Tenney became the first representative. U.S. Navy representatives in subordinate offices in West Germany were LCdr. G. C. Nowak in Berlin, Cdr. A. R. Czerwonky in Frankfurt, and Cdr. V. L. Rychly in Munich. U.S. civilians in the Scientific and Technical Unit, which was located at Frankfurt, were Dr. O. F. Schuette, Dr. E. G. Kovach, and James W. Grady.[56]

Making the former NAVGER intelligence organization a part of the staff of CINCNELM gave it co-equal status with Commander in Chief, U.S. Army, Europe and Commander in Chief, U.S. Air Force, Europe and permitted the Navy to continue to participate as an equal and parallel member of the U.S. intelligence community in West Germany.

Berlin was the principal point of confluence of refugees, returnees, and defectors from Soviet Bloc countries and, as such, was a prolific source of intelligence information. Berlin was also the major channel for the movement of agents to and from Bloc countries. Sources available to the CINCNELM representative in Berlin provided continuing and authoritative information on East German naval and merchant ship building programs, on the Soviet naval forces based in East Germany, and on research and development (R&D) targets of naval interest in East Germany.[57]

OPNAV Instruction 011150.1A of 22 April 1960 outlined the naval activities in CINCUSNAVEUR (Commander in Chief, U.S. Naval Forces, Europe, formerly CINCNELM) area and listed the Scientific and Technical Unit, Frankfurt as operating under the military control of the CINCUSNAVEUR representative in Frankfurt and under the technical control of the Director of Naval Intelligence.

The Joint Overt Interrogation Center, Berlin (JOIC[B]) was established in 1960 to satisfy the requirements of local and higher headquarters for information obtainable through overt interrogation of refugees and defectors in Berlin. The JOIC(B) was formed from U.S. Army, Navy, and Air Force overt intelligence elements present in the city.

Refugees from East Germany and elsewhere, arriving in West Berlin and desiring help applied at the Processing Center, Marienfelde. A refugee who appeared knowledgeable on subjects of naval interest during an initial screening was given a secondary screening by an interrogator from the REPGER office in Berlin. If it was then determined that the refugee warranted more extensive interrogation, the refugee would be taken from Marienfelde to a Navy "house" or to the JOIC(B) office, because intelligence activities were not permitted at Marienfelde.

A refugee whose knowledge justified a more technical interrogation would be referred to the Defector Reception Center (DRC) or to Camp King, an Army installation staffed by the 513th Military Intelligence Group and devoted to overt interrogation. At the DRC and its subsection, the Scientific Exploitation Section (SES), Navy interrogators were allowed to question the refugees, but at Camp King, Navy requirements were served only through Army interrogators.[58]

The CINCUSNAVEUR Representative, Germany organization in 1961 had offices in three locations. The Munich office conducted official U.S. Navy liaison with the BND (the Federal German Republic intelligence service; the Navy representative also maintained other, less formal contacts within the Naval Evaluation Group of the BND). The representative in Berlin served as the primary Navy source of immediate information on the Berlin situation. His collection activities included the exploitation of refugees (55 percent of the total effort), covert sources (25 percent), publications (15 percent), and other sources (5 percent). A secondary screening was provided by the Berlin office to the refugee center at Marienfelde, and there were also Navy interrogators at the JOIC in Berlin. The Navy representative at Frankfurt (which included the S&T Unit) worked at the Defector Reception Center and its Scientific Exploitation Section; conducted liaison with, and provided guidance to, the Army's 513th Military Intelligence group; and took advantage of other intelligence activities in the area. The SES had been established under a charter dated 20 November 1958.[59]

The NAVEUR/REPGER Navy Liaison Office (NLO) in Hamburg was not involved directly in clandestine operations but did supply valuable intelligence support to such operations. An important aspect of the U.S. Naval Intelligence presence in Hamburg was its location near a refugee debriefing center. About three refugees per month of interest to the Navy were ship-jumpers or escapees from Soviet bloc countries arriving in West Germany by ship via third countries. In December 1968, however, the NLO Hamburg office was closed.[60]

The broad mission of the REPGER S&T Unit at Frankfurt was described in 1968 as the collection of scientific and technical information of naval interest by exploiting all sources of information on S&T developments (primarily of a research and development nature) in Sino-Soviet and European countries. To accomplish its mission, the unit's four scientists, acting as representatives of the U.S. Navy and the S&T Unit, were to develop and maintain professional contacts with scientists and technical management personnel throughout the European R&D community.

The U.S. scientists also acted as consultants to the various U.S. defense attachés in Europe by providing scientific expertise in the collection, field evaluation, and collation of scientific information under certain data exchange agreements between the U.S. and other countries. The scientists represented Department of Defense components in liaison with foreign scientific and technical organizations in Europe. To assist in the exploitation of defectors and refugees, the S&T Unit developed questions for the interrogators who debriefed scientific and technical personnel fleeing Soviet bloc countries. The S&T Unit scientists also prepared technical reports based on such interrogations, and they assisted in evaluating the sources.

The civil service classifications of the four senior civilian positions in the S&T Unit had originally been established in 1956 as GS-14 intelligence research specialists. Two of the positions were downgraded in 1959 to GS-13. A single GS-11 interpreter performed both translating and interpreting for the S&T Unit. Able to read at least two foreign languages, the interpreter translated a wide variety of documents covering subjects such as chemistry, biology, physics, and electronics. The S&T Unit also employed a GS-7 intelligence assistant to maintain the technical library and to perform file research and other intelligence support work.[61]

Chapter Notes

1. "Administrative History, Naval Attaché London, 1939–46," MS, 22–27.

2. Commander U.S. Naval Forces, Europe (COMNAVEU), "Administrative History, 1940–46," 114–25.

3. "Administrative History, Naval Attaché London, 1939–46," 27–39.

4. COMNAVEU, "History of Military Government and Civil Affairs," passim, MS, Accession 68-A-605, box 258, OA.

5. COMNAVEU, "Administrative History, 1940–46," 340–42, 344.

6. COMNAVEU, "History of COMNAVEU Forward Intelligence Unit," Part I, 2, MS copy in author's files.

7. Capt. George E. Kidd, USNR, interview by author, Jan 1976, 4–13.

8. "History of COMNAVEU Forward Intelligence Unit," Part I, 2, 4.

9. Capt. Joseph Kaitz, USNR, interview by author, Sep 1976, 3, 5–6.

10. Kidd interview, 13.

11. "History of COMNAVEU Forward Intelligence Unit," Part I, 4.

12. Ibid., 5.

13. Kidd interview, 13–16.

14. Kaitz interview, 7, 10–11.

15. Kidd interview, 14, 16.

16. Ibid., 14, 17; and Cdr. Charles A. Rocheleau, USNR, interview by author, 1 Oct 1975 (notes).

17. Rocheleau interview.

18. Kidd interview, 17–18.

19. Ibid., 18.

20. "History of COMNAVEU Forward Intelligence Unit, Part I, 10.

21. COMNAVEU ltr, 1 Jun 1946, encl. A, chap. 1, 2–3, 8, 15, 17–18, 20.

22. COMNAVEU, "Administrative History, 1940–46," 383–403.

23. Head, Foreign Intelligence Section, COMNAVEU ltr to COMNAVFOR France, 21 Dec 1944.

24. Kidd interview, 19–20.

25. Kaitz interview, 7–9.

26. COMNAVEU, "Administrative History, 1940–46," 406–7.

27. Ibid., 442–43.

28. COMNAVEU, "Administrative History, Oct 1945–Mar 1947," 79.

29. Commander in Chief, U.S. Naval Forces, Eastern Atlantic and Mediterranean (CINCNELM), "Quarterly Summary, 1 Apr–1 Jul 1947," 3.

30. Cdr. Joseph A. Meyertholen, USNR, interview by author, 24 Mar 1980.

31. Meyertholen interview, 2 May 1980.

32. CINCNELM ltr, ser 000183, 14 Sep 1950, box 5, TSC 8665, Job 11146, OA.

33. CINCNELM, "Report of Operations, FY 1955," Sec. 6.

34. CINCNELM, "Report of Operations, FY 1956," part 4, 11–13.

35. CINCNELM, "Report of Operations, FY 1957," encl. 2.

36. Commander Sixth Fleet (COM6THFLT), "History," MS, 5 Sep 1959, OA.

37. CINCNELM, "Report of Operations, 21 Feb 1958–1 Jul 1958," 5; and CINCNELM, "Report of Operations, 1 Jul 1958–31 Mar 1959," 6.

38. CINCNELM, "Report of Operations, 1 Jul 1958–31 Mar 1959," encl. 7.

39. Commander in Chief, Special Command, Mid-East (CINCSPECOMME), "Operation Blue Bat," report of Dec 1958, passim.

40. *COM6THFLT Command History, 1967*, IIF.

41. *COM6THFLT Command History, 1969*, IIC.

42. *COM6THFLT Command History, 1970*, IID.

43. Commander Naval Forces, Germany (COMNAVFORGER), "History," (no date), chap. 1, 1–2, 6–7.

44. Commander U.S. Ports and Bases, Germany, "History of TF 126," MS, 1946, 13–14.

45. Ibid., 7–9, 15.

46. Ibid., 10.

47. COMNAVFORGER, "History," chap. 3, 49–52.

48. "History of TF 126," 16–20.

49. Ibid., 24.

50. COMNAVFORGER, "History," chap. 3, 48–49.

51. COMNAVEU, CTF 124 NNI 96 (Intelligence Report Form), 9 Apr 1945, discussing the intelligence organization of COMNAVEU.

52. COMNAVFORGER ltr, ser 000421-49, 22 Nov 1949.

53. NAVFORGER IO (Intelligence Officer) ltr, ser 0011-50, 16 Jan 1950.

54. "Intelligence Coordinating Committee, Germany (ICC[G]), in 1957–1959," OP–922H4 files, OA.

55. CNO ltr, ser 0050P92, 15 Jan 1957.

56. Commander in Chief, U.S. Naval Forces, Eastern Atlantic and Mediterranean Representative, Germany (CINCNELM-REPGER), "Intelligence Collection Plan, 10 Aug 1957," sec. 1, 1.

57. Ibid., 27, 29.

58. OP-922H2G memo, 3 Jul 1961, OP-922H4 files.

59. "H4 file" on REPGER, OA.

60. Naval Intelligence Command, NIC-321F ltr, ser 0055-68, 3 Jun 1968, OP-922H4 files; and REPGER msg (no date-time group), 18 Nov 1968.

61. "Position Management Review, S&T Unit, Frankfurt," 1 Nov 1968.

CHAPTER 40

Fleet Intelligence Centers

Fleet Intelligence Centers were established for the three major fleet commands during the 1950s. The *Naval Intelligence Manual* (ONI 70-1) of 1956 defined a Fleet Intelligence Center (FIC) as "an activity which provides a fleet with an intelligence production and intelligence personnel augmentation capability."

This chapter treats the three Fleet Intelligence Centers in Europe (FICEUR), the Atlantic (FICLANT), and the Pacific (FICPAC) separately, tracing their activities and growth chronologically. The section about FICPAC includes information on its satellite facility in the Philippines, which was established to support the forces involved in the Vietnam War.

The World War II organization known as the Seventh Fleet Intelligence Center is covered in Chapter 36. Another primarily naval World War II organization, the Intelligence Center, Pacific Ocean Areas (predecessor of JICPOA [the Joint Intelligence Center, Pacific Ocean Areas]) is discussed in Chapter 19.

Again, there are many gaps in the material presented and, consequently, many possibilities for further research.

Evolution of Fleet Intelligence Center, Europe

For several years until 1950, the U.S. Naval Forces, Eastern Atlantic and Mediterranean (CINCNELM) organization was concerned about the possible adverse effect on the command's operational readiness should it become necessary to evacuate its London headquarters without being able to salvage the existing intelligence files. The first step in the establishment of what became the Fleet Intelligence Center, Europe was therefore taken on 24 July 1950 when CINCNELM sent a letter to the Commanding Officer, Naval Air Activities, Port Lyautey (modern Kinitra in Morocco), directing the establishment of emergency intelligence files there for CINCNELM. While the Port Lyautey office was not, by any means, a Fleet Intelligence Center, it was the predecessor of FICEUR.

As of 14 September 1950, duplicate essential intelligence files (reports, publications, charts, etc.) were being located at Port Lyautey, and a normal flow of material to the officer custodian of the files had commenced by that time.[1]

In June 1951, CINCNELM requested from the Chief of Naval Operations (CNO) the documents he considered necessary for establishing duplicate emergency files at Port Lyautey. Documents requiring handling via communications intelligence (COMINT) channels were assembled by ONI's Operational Section's Special Unit (OP-322Y1) and shipped via officer messenger on 11 July 1951. Subsequent shipments were made on an average of once a month.[2]

Cdr. James A. Marks, CINCNELM's Deputy Intelligence Officer, had been detailed in August 1950 as intelligence officer at the Naval Air Facility (NAF), Port Lyautey, to set up the CINCNELM duplicate intelligence files. As soon as a makeshift vault area had been put together by the public works department, duplicate copies of all information reports, including attaché reports, from all over Europe started flowing to NAF Port Lyautey. Initially, Cdr. Marks had one secretary; later, LCdr. Bart Connolly was assigned as his assistant.

When the communication facilities of Port Lyautey were expanded, the Naval Air Facility became a naval activity (NAVACT), and the intelligence officer's title and the address for duplicate copies of reports were changed accordingly. In the summer of 1952, a new COMNAVACT office building was completed, with greatly increased office space and vault area for the intelligence officer. (The old vault had been about to overflow.) Cdr. Marks, a qualified French interpreter, had an additional duty as liaison officer with the commandant of the French base at Port Lyautey, NAVACT being a tenant on the French base. Marks also served as interpreter when the U.S. Air Force was negotiating the building of Strategic Air Command bases in Morocco. Marks was detached in September 1952, having been relieved by Cdr. Harold Fleck, USNR.[3]

In October 1952, CINCNELM proposed the establishment of a NELM Fleet Intelligence Center based in a naval vessel, pending the availability of a satisfactory location ashore in the Mediterranean area. The Chief of Naval Operations acknowledged the desirability of the proposal in April 1953, but because of recent personnel cuts imposed on the Navy, the personnel requirement for the proposed FIC could not be met without compensatory billet reductions at other activities. A survey of Atlantic Fleet intelligence billets was initiated to determine if there were any over which CINCNELM's requirements should take priority. Concurrently, the Chief of Naval Operations approved the establishment of a FIC in London as an interim measure, using intelligence billets disestablished in the NELM area.[4]

In March 1954, seven officers and twelve enlisted men of the CINCNELM Air Targeting Section in London were transferred to Port Lyautey, where they became the nucleus of the organization officially designated by the Secretary of the Navy (SECNAV) as the Fleet Intelligence Center, Eastern Atlantic and Mediterranean (FICELM), a part of the operating forces assigned to CINCNELM initially established in May 1953 as an organization on paper to be staffed if and when needed. An associated organization, the Fleet Air Intelligence Augmenting Unit, Eastern Atlantic and Mediterranean, had been established by the same SECNAV directive. In December 1959 that unit was integrated into FICELM, which took over the responsibility to provide rapid augmentation of trained intelligence personnel to naval forces in the forward area in support of actual and contingency operations. In March 1960, CINCNELM was renamed CINCUSNAVEUR (Commander in Chief, U.S. Naval Forces, Europe), and in October FICELM became FICEUR. In January 1964, it was moved from Port Lyautey to the Naval Air Station, Jacksonville, Florida.[5]

The relocation of FICEUR to Jacksonville was the result of a series of compromises influenced by international and domestic considerations. In 1963, it had been determined that FICEUR must leave Port Lyautey because of Morocco's failure to agree on the renewal of U.S. base rights. Many possible locations in Europe were considered, and feasibility and engineering studies were actually completed for Rota, Spain. At about that time, the gold flow problem was a major domestic issue in the United States, and Spain's Generalisimo Franco was holding out for increased funding in connection with U.S. base rights. Consequently, it was decided to move FICEUR to the United States. Jacksonville was selected because the Mayport/Jacksonville/Cecil Field complex provided an excellent point of contact with the Navy's aircraft carriers and air groups between their deployments to the Sixth Fleet. Furthermore, a building was available.[6]

FICEUR's on-board staff as of 1 July 1970 consisted of 52 officers, 115 enlisted, and 9 civilians, against an authorized level of 59 officers, 103 enlisted, and 9 civilians. The organization occupied four wings of a five-wing, run-down, termite-infested, highly flammable, vacated wooden mess hall that had been built in 1941 as a "temporary" structure.[7]

Prior to 1972, FICEUR was responsible for maintaining an intelligence database encompassing all of Europe, including Turkey; the USSR west of longitude 100°E ; Jordan, Syria, Israel, and Lebanon in the Middle East; and Egypt, Libya, Tunisia, Algeria, and Morocco in North Africa. In March 1972, when Strike Command was disestablished in the Joint Chiefs of Staff (JCS) Revised Unified Command Plan, the Middle Eastern countries of Iran, Iraq, and the Arabian Peninsula were added to FICEUR's responsibilities.

Commands receiving intelligence support from FICEUR were Commander in Chief, U.S. Forces, Europe; Commander in Chief, U.S. Naval Forces, Europe; Commander Sixth Fleet, both in its U.S. Forces organizational structure and under its NATO "hat" under Commander Strike Force, South; and Commander Middle East Force, the last after March 1972.[8]

The Secretary of the Navy approved the disestablishment of FICEUR as a separate organization effective 31 December 1974, and FICLANT was redesignated Fleet Intelligence Center, Europe and Atlantic, effective 1 July 1974.[9]

The commanding officers of FICELM/FICEUR were as follows:

Name	Reporting Date
Cdr. Robert P. Williams (OIC)*	Apr 1954
Capt. Howard T. E. Anderson	Jul 1954
Cdr. Nelson E. Marris (Acting)	Apr 1955
Capt. Ernest W. Humphrey	Jul 1955
Capt. Garrett S. Coleman	Aug 1956
Capt. William F. Dawson	Jun 1958
Capt. Richard S. Roberts	Jul 1961
Capt. Hartsel F. McCue	Jun 1963
Capt. K. P. Rehnberg, Jr.	Jun 1965
Capt. William V. Moore	Jul 1967
Capt. Emory R. Sourbeer, Jr.	Aug 1970

*Officer in Charge.

Evolution of Fleet Intelligence Center, Atlantic

In August 1950, in connection with a disaster plan, ONI duplicated its operating manuals, instructions, requirements, and policies and furnished them to the Commander in Chief, Atlantic. Subsequent to the promulgation of the disaster plan, CINCLANT was put on the distribution list for all material received and prepared by ONI, including all naval attaché reports, all studies, and copies of ONI's complete production effort except for the contributions to JCS papers.

The directive's originators conceived of an arrangement whereby each operating fleet would be able to act as an "independent ONI" for its operational area under emergency conditions, and the text employed the term "fleet intelligence centers," although the actual establishment of such organizations was still in the future. The replication of ONI's files at the fleet level, of course, imposed a significant duplication of effort.[10]

As directed by SECNAV Notice 5450 of 3 March 1955, the Mobile Intelligence Production Unit (MIPU) was established at the U.S. Naval Base, Norfolk, under the military command of Commander in Chief, Atlantic Fleet (CINCLANTFLT) and the management control of the Bureau of Aeronautics. The MIPU's mission was to develop an intelligence production capability that would be immediately available to support U.S. naval forces as directed by CINCLANTFLT; to provide U.S. naval forces with the intelligence necessary to carry out their respective missions efficiently; and to maintain a high degree of mobility for rapid deployment to such areas as were directed by CINCLANTFLT. MIPULANT was formally established at CINCLANTFLT Headquarters on 13 June 1955 under an officer in charge, LCdr. A. R. Gitlin.[11]

During Fiscal Year 1956, MIPULANT became engaged in the production of target materials in the form of Atlantic Fleet Target Folders for all assigned targets designated in plans for special weapons operations.[12]

By the end of Fiscal Year 1957, MIPULANT had completed production of the target folders (Phase II ATMP) for a high percentage of the airfield targets listed in Annex Whiskey of CINCLANT Operation Plan 200-58. The production schedule provided for the completion of the target folders for all airfields and selected priority static targets (naval bases, ports, etc.) by January 1958, a goal that was achieved. Production efforts were then shifted to the revision of previously produced folders. In addition, pilot model domestic target folders were produced for pilot training, and Regulus cruise missile target folders were produced for Regulus-carrying cruisers and submarines. A sample Regulus folder was forwarded to CINCPACFLT in 1958 for the purpose of developing a standard fleet-wide missile folder.[13]

In 1958, MIPULANT commenced production of Atlantic Fleet Target Folders to supersede the Atlantic Fleet Delivery Folders it had produced. A program for the maintenance and revision of target folders at six-month intervals was also initiated. Six folders on domestic targets were completed for use in predeployment training by AIRLANT attack squadrons.[14]

MIPULANT had completed production of target folders on all targets listed in Annex "W" to Operation Plan 200-59 by 29 February 1960. In addition to the folders covering foreign targets, fifty others had been programmed for production on domestic targets, of which fourteen had been produced as of February 1960. Folders that had previously been produced and distributed in support of the Regulus-I cruise missile operations were removed from operational units, except from Regulus-I ships under the control of Commander Cruisers, Atlantic.[15]

On 1 May 1961, the name MIPULANT was changed to Atlantic Intelligence Center (LANTINTCEN), and its mission was expanded to include intelligence support to forces assigned to CINCLANT and to the maintenance of sufficient personnel to meet peacetime and wartime augmentation requirements for forces in forward areas. The officer-in-charge title was changed to a commanding officer.[16]

In Fiscal Year 1961, LANTINTCEN started producing Atlantic Fleet Delivery Mission Folders for targets assigned to Commander Second Fleet. In conjunction with the War Gaming Department of the Naval War College, some routes used in the mission folders were verified on the Navy Electronic Warfare Simulator.

During the same period, the entire target photo library of the Atlantic Intelligence Center was reformatted to 35mm photography and was distributed to the fleet on Filmsort aperture cards for rapid production using the Filmsort Model 200 reader-printer. The Chief of the Bureau of Weapons provided funds for IBM equipment, and the center started cataloging its entire intelligence library using the revised Intelligence Subject Code approved by the United States Intelligence Board. The center also produced special studies on Caribbean countries.[17]

In November 1962, LANTINTCEN placed emphasis on intelligence support to forces involved in the Cuban Missile Crisis. On 19 August 1963, it began sending teams to brief pilots on board aircraft carriers in the Mediterranean about Second Fleet strategic target responsibilities.[18]

In 1964, the mission of the Atlantic Intelligence Center was (1) to develop an intelligence production capability that would be immediately available for the support of the U.S. naval forces and other forces assigned to CINCLANT; (2) to provide CINCLANT forces with the intelligence necessary to carry out efficiently their respective missions; (3) as a matter of readiness, to provide for the rapid augmentation of intelligence personnel assigned to forces in the forward areas, as directed by CINCLANTFLT; and (4) to maintain sufficient qualified and trained personnel for assignment to fulfill peacetime and wartime augmentation requirements. The center's areas of interest included Latin America, Africa south of the Sahara, and portions of the Middle East. Actual production of tactical targeting intelligence, however, was limited to Central and South America and Cuba.[19]

The original computer installation, put into operation on 1 April 1964 at the Atlantic Intelligence Center, consisted of an IBM 1410 40K (Kilobytes) (40,000 positions of core storage) central processor with six 729 II tape units; an IBM 1403, 600 line-per-minute printer; and an IBM 1402 card reader/puncher. In December 1964, the Central Processing Unit was upgraded to 80K, and a 1201 Model I disk unit was added. In 1967, the disk unit was exchanged for a 1301 Model II unit, which doubled the storage capability to 56 million characters. Also in 1967, an accelerator was added to the central processor to reduce internal processing time. Later, in March 1968, the 729 II tape units were replaced by 729 VI units to give higher operating speeds and greater data density on the tapes.[20]

As of March 1965, fifteen Naval Air Intelligence Reserve Units (NAIRU) were involved in the production of Tactical Targeting Illustrations (TTI) and Target Area Studies (TAS) for the Atlantic Intelligence Center. Over a three-year period, the NAIRUs had produced more than 500 illustrations and 12 studies.

In April 1965, at the request of Commander Carrier Division Four, a complete study of Cuban operational planning targets was made and the necessary associated route planning was conducted. From that date onward, the Atlantic Intelligence Center became thoroughly involved in specific target intelligence projects to support operating forces committed to CINCLANT contingency operation plans, marking the beginning of the production of Mission Oriented Packages, designed to provide all intelligence required from initial landing beaches to the objective area.

The Dominican Republic crisis in May 1965 required maximum intelligence support on a 24-hour-a-day basis. Related products produced by the intelligence center included a large mosaic of Santo Domingo showing airfields and ports. Additional studies were made about Dominican Republic communications, roads, fuel storage, and power plants. Graphics were provided to the CINCLANT Operations Control Center and to forces involved from the Army, Navy, Air Force, and Marines.[21]

A new 30,500 square-foot Atlantic Intelligence Center building was accepted on 8 February 1966, and personnel moved in on 2 May 1966. The development of an automated data processor (ADP) plotter installation was begun in the new building, and the first ADP electronic order-of-battle transmission was received on 25 July 1966 from *Independence* (CVA 62), operating in the western Mediterranean. The plot data was in plotter-machine language and was transmitted through two radio stations and two land lines. Over 4,000 characters were included in the transmission, and there were only thirty easily corrected errors in the data received at LANTINTCEN. Two and one-half hours after the message was sent by the originator, CINCLANTFLT flag officers were being briefed with a graphic display showing Mediterranean-area radar stations and the various ranges of detection for each.[22]

When the Egyptian-Israeli crisis flared on 27 May 1967, all emphasis on intelligence production at LANTINTCEN was shifted to the Middle East. The center's area of responsibility included Iran, Iraq, Sudan, Yemen, and Aden. A review of existing intelligence holdings disclosed that all pertinent information was at least seven years old. Assisted by fifty naval reservists on two weeks' active duty for training, LANTINTCEN updated the seven-year-old intelligence products in seven days. The local NAIRU from Naval Air Station, Norfolk spent a two-day drill period at LANTINTCEN in a massive search of the photographic files for photos required in the development of studies of the Middle Eastern area. The response to the Egyptian-Israeli crisis demonstrated the highly professional contribution that could be made by NAIRUs and NRIUs (Naval Reserve Intelligence Units) on short notice, but it also demonstrated the need for a routine and continual updating of worldwide intelligence databases as new information is received.[23]

OPNAV Notice 5450 of 13 April 1968 changed the name of the Atlantic Intelligence Center, Norfolk to the Fleet Intelligence Center, Atlantic (FICLANT), effective 1 May 1968. The organization's mission remained the same, and its area of concern, based on CINCLANT's areas of responsibility, included the Atlantic Ocean, the Caribbean Sea, and the Indian Ocean east to Burma. Because Atlantic Fleet forces were designated for participation in Commander in Chief, Middle East, Africa and South Asia operational plans, other areas of concern for FICLANT were the Red Sea and Persian Gulf, Africa south of the Sahara Desert, and Southern Asia west of Burma. Close liaison was maintained with FICLANT's counterpart organizations under Commander in Chief, U.S. Army Forces, Atlantic and Commander in Chief, U.S. Air Forces, Atlantic in preparing intelligence on those areas. The intelligence products of the three intelligence centers were standardized to meet the requirements of the Army, Navy, and Air Force operating forces with a minimum of duplication.[24]

FICLANT developed a computerized biographic file that became operational in November 1968. It combined data on ground order-of-battle personalities produced by FICLANT with counterintelligence biographies produced by Fleet Marine Forces, Atlantic. At the end of the year, the automated file contained 2,408 records.

An aircraft markings publication and an aircraft photo file were produced by FICLANT in 1968. Multiformat Tactical Targeting Illustration files were replaced by a single-format system that allowed more effective cataloging procedures and simplified use.[25]

FICLANT production of Supplemental Photographic Interpretation Reports was started in December 1968 as a result of national-level tasking. The production effort involved "second-phase" analysis of regularly scheduled photo reconnaissance missions, and the reports were produced in human-readable, machine-processable format, permitting ease in updating the computerized data file and speed in transmitting the information to recipients via the Automatic Distribution of Intelligence (AUTODIN) network, as well as by normal communications means.[26]

In 1969, FICLANT modified its files to support the projected requirements of the automated intelligence filing and distribution capabilities of the Naval Intelligence Processing System (NIPS) Phase II system and the Integrated Operational Intelligence System (IOIS). The associated Telecommunications File became operational in August 1969, combining accuracy, timeliness, and comprehensiveness to fulfill the requirements of NIPS documentation. Its major application was anticipated to be in the areas of Tactical, Conventional/Unconventional, and Electronic Warfare.

FICLANT's automatic distribution of amphibious intelligence products was replaced in 1969 by the *Amphibious Products Catalog*, permitting consumers to order only those products desired and in the format most suitable to their needs.

A Collection Requirements and NIPS Management Branch was established on 1 July 1969 to provide a centralized location for the collection of intelligence information and the management of NIPS.

The *Air Intelligence Digest* was published monthly by FICLANT, and, in September 1969, a newsletter entitled *Transportation Notes* was initiated for distribution to seventeen Naval Reserve Intelligence Units. As a result of receiving the newsletter, the units were motivated to increase the number of intelligence articles they submitted to the *Air Intelligence Digest*.

During 1969, FICLANT obtained approval from ONI to use Naval Reserve Intelligence Unit assets to assist in the production of information for a geo-political/socio-economic file, with the information to be processed within the Intelligence Data Handling System (IDHS) 1410 Formatted File System (FFS) produced by FICLANT and used by all other Navy automated intelligence centers.

Two new major equipments were added to the FICLANT photo laboratory in 1969: a Niagara Photo Printer provided a color duplication capability, and a Beacon Precision Enlarger gave a photo reproduction magnification capability from 3x to 62.5x.

A *Coordinated Reconnaissance Planning Guide*, which included photographic and textual data on reconnaissance targets, was produced by FICLANT in 1969.[27]

As of 30 June 1970, FICLANT had 53 officers, 128 enlisted personnel, and 27 civilians on board, against an authorized allowance of 56 officers, 133 enlisted, and 34 civilians. The crowded FICLANT facility contained 31,800 square feet of floor space, an estimated deficiency of 13,000 square feet.[28]

Increasing Soviet naval activity in the Indian Ocean and the Caribbean Sea, as well as U.S. projects such as the Diego Garcia facility, generated an increased demand for area background briefings. From 1 January 1970 to 30 June 1971, forty-five briefings were given to the staffs of three Caribbean ready groups, two consecutive UNITAS (the annual U.S. Navy training cruise to South America) exercise task forces, six Middle East-bound destroyer groups, and a variety of deploying Marine Corps battalion landing teams, SEAL teams, patrol aircraft squadrons, and Diego Garcia supply shuttle ships. The briefings emphasized the Soviet presence and the weapon capabilities of the countries in each area.

In response to an expressed CNO interest in keeping all Navy personnel informed about the political-military importance of Navy deployments and operations, FICLANT coordinated an intensified Pilot Shipboard Intelligence Briefing Program in support of Destroyer Division 82's April 1971 Indian Ocean deployment. For the first time, crew briefings were held in addition to the usual staff and wardroom presentations, and over 250 enlisted men attended.

All-hands briefings thereafter became an integral part of the FICLANT predeployment briefing program.

In August 1970, FICLANT started considering the production of a new tactical target graphic, called a Split Tactical Target Illustration (STTI), to replace the existing Tactical Target Illustration, Quick Response Graphic, Automated Intelligence Graphic, the CINCLANT Area Installation DataBase (Volumes I & II), and the Tactical Target Materials Catalog. Concurrently, the Fleet Intelligence Center, Pacific proposed an Automated Tactical Target Graphic (ATTG) concept that was practically identical to the STTI concept. An ATTG that combined the best of FICPAC's ATTG and FICLANT's STTI was adopted by the intelligence community at a Tactical Target Materials Working Group Conference held in May 1971 by the Defense Intelligence Agency.

The *FICLANT Intelligence Brief* (*FIB*) was first published in September 1970 to replace the *Atlantic Intelligence Digest*. It was subsequently issued whenever the need arose to disseminate information of general interest. Each issue treated a single subject and was distributed to the commands most directly concerned with the subject matter.

The *CINCLANTFLT Port Directory*, a three-volume, loose-leaf publication prepared by FICLANT, covered all areas of the Atlantic Fleet Command responsibility and was updated semiannually. In January 1971, quarterly updates were instituted to provide the fleet with more current information, and to permit fleet units to see results of their collection efforts more quickly.[29]

In November 1970, FICLANT replaced its 1410 computer with an IBM 360/40 third-generation computer, which provided for mass on-line storage of intelligence production files and for remote file maintenance and query, thus improving responsiveness to intelligence analysts. Also provided were communications terminals for the transmission of data over the military communications satellite and land-line systems.

At the beginning of the 1970s, FICLANT was producing finished intelligence in a variety of forms, including catalogs, study folders, 35mm slides, aperture cards, computer tapes, and punch cards to meet the needs of the operating forces in CINCLANTFLT's area of responsibility.[30]

In October 1971, at the direction of the Defense Intelligence Agency, FICLANT began converting its digital database from the 1410 Formatted File System to the Machine Independent Data Management System.

The entire six-person branch of FICLANT involved in the production of target graphics was sent to the Philippines in June 1972 to augment the Fleet Intelligence Center, Pacific Facility (FICPACFAC), which was operating in support of stepped-up U.S. Navy participation in the Southeast Asia conflict.

An *Indian Ocean Fact Book* was completed and issued by FICLANT in July 1972. Invaluable assistance in the production of the book was provided by the nine Naval Reserve Intelligence Units assigned to perform country studies in support of the project. Approximately fifteen to twenty-one additional sections of the book were scheduled to be produced during the following year.

A new product, the Naval Forces Intelligence Brief, was also prepared by reserve units and produced by FICLANT in 1972. A major order-of-battle project, it was designed to supplement the Defense Intelligence Agency's Naval Forces Intelligence Program. Another new product, the Emergency and Evacuation Support Folder, was written and produced by reserve units to assist Atlantic Fleet operating forces that might become involved in the evacuation of U.S. nationals from foreign countries in time of civil unrest or serious natural disaster.

FICLANT's participation in the debriefing of American prisoners of war returned from Southeast Asia included coordinating the supply of intelligence debriefing support materials to East Coast hospitals and maintaining qualified intelligence debriefers. In addition to the seven debriefers assigned as intelligence material liaison officers, one to each East Coast naval hospital, other debriefers were assigned to the Portsmouth, Virginia, Naval Hospital as primary and alternate debriefers. Two FICLANT officers were also assigned as Senior Intelligence Debriefing Coordinators at the naval hospitals at Portsmouth and Great Lakes.

In response to increased Southeast Asia activity, eleven FICLANT personnel (four officers and seven enlisted men) augmented FICPACFAC in the Philippines for periods of three to six months during 1972.[31]

Effective 1 July 1974 and preparatory to the disestablishment of FICEUR, FICLANT's title was changed to Fleet Intelligence Center, Europe and Atlantic (FICEURLANT). Its mission statement was:

> To maintain an intelligence production capability which is responsive to CINCLANTFLT and CINCUSNAVEUR tasking in support of U.S. naval forces assigned, and to provide those forces with the intelligence necessary to carry out their respective missions; to respond to CINCLANT and USCINCEUR requirements as directed by CINCLANTFLT and CINCUSNAVEUR; and to provide a source of qualified intelligence personnel for rapid augmentation of naval operating forces for wartime or contingency operations.[32]

The commanding officers of MIPULANT/LANTINTCEN/FICLANT from 1955 to 1975 were the following:

Name	Reporting Date
LCdr. A. R. Gitlin*	13 Jun 1955
Cdr. J. E. Keefe*	25 Feb 1958
LCdr. C. D. Cullison*	11 Apr 1960
Cdr. J. H. Nevby	29 Jul 1960
Cdr. R. A. Schulze	24 Aug 1962
Capt. H. A. Kelly	17 Sep 1962
Capt. R. P. Fuller	20 Jul 1964
Capt. C. D. Hinds	12 Aug 1968
Capt. C. D. Cullison	2 Jul 1970
Capt. T. B. Yount	21 Jul 1972
Capt. A. Bath	May 1975

*The first three officers were officers in charge. Nevby began with this title, but was later designated commanding officer, the title held by all his successors.

Evolution of Fleet Intelligence Center, Pacific

On 1 July 1952, the Special Intelligence Production Unit (SIPU) was established under Commander Fleet Air, Japan at the Naval Air Station, Atsugi to assist the aircraft carriers of Task Force 77 with their Korean War photographic intelligence work load.[33]

When combat operations ended in July 1953, SIPU remained at Atsugi, and in March 1955 it was redesignated the Mobile Intelligence Production Unit, Pacific (MIPUPAC) and placed under the military command of CINCPACFLT. Shortly thereafter, part of the unit was moved to Ford Island in Pearl Harbor, while some of the personnel remained behind at Atsugi into early 1956 to continue the preparation of air target materials to support aircraft carrier operations in the Western Pacific.[34]

The Fleet Intelligence Center, Pacific was established on 27 May 1955 and placed under the military command of CINCPACFLT and under the management control of the Bureau of Aeronautics. FICPAC was to "develop an intelligence production capability which is immediately available for support of U.S. Naval Forces in the Pacific area, and to provide those forces with the intelligence necessary to carry out efficiently their respective mission."[35]

Before FICPAC was established, no Navy organization was involved in the production of basic intelligence in the Pacific Fleet area. The CINCPACFLT intelligence staff was too small and had to rely heavily on ONI's output.

In 1956, the Special Weapons Evaluation Group, which had been set up in 1954 as Detachment Able of the First Fleet staff at Naval Air Station, North Island, in San Diego, California, was also moved to Ford Island as the Special Weapons Evaluation Group, Pacific (SWEGRUPAC), and was co-located with FICPAC in order to provide better service for aircraft carriers as they stopped at Pearl Harbor for their readiness inspections before proceeding to the Western Pacific. It was recognized that SWEGRUPAC and FICPAC could be mutually supporting, but one of the early deterrents to a full merger of the two was the fact that the personnel billets for the evaluation group came from operational staff elements. CINCPAC/CINCPACFLT Operations didn't want to lose the positions to intelligence, where the billets could not be retrieved if they did not support operations adequately.[36]

SWEGRUPAC worked with FICPAC, but the former was under the supervision of the plans officer on the CINCPACFLT staff, and the latter was under the supervision of Capt. Rufus L. Taylor, USN, Fleet Intelligence Officer. As a result of numerous studies, Capt. Taylor believed that time, trouble, money, and effort could be saved by merging within FICPAC all production of basic and photographic intelligence in the Pacific. Taylor's proposed merger included SWEGRUPAC, whose primary function was the production of CINCPACFLT target lists and associated target planning folders based on and containing extensive photo intelligence. Initially, the CINCPACFLT plans officer was against the merger and having the production of target planning folders placed under intelligence control. He believed there would be a conflict over the direction of the production functions. The operations officer and chief of staff both supported the view of the plans officer, and the subject was dropped at the direction of Chief of Staff VAdm. Maurice E. Curts.

Several months later, the subject was raised again at a staff conference, giving Capt. Taylor the opening to argue once more for the merger. The resulting get-together with Plans Officer Capt. Ernest E. Christensen and Operations Officer Capt. Charles Duncan led to an agreement on the merger of SWEGRUPAC and FICPAC in late 1958.[37]

The Special Intelligence Production Unit element that had been moved to Ford Island from Atsugi had completed its transfer prior to mid-January 1955, when Cdr. Joseph A. Meyertholen arrived to be its officer in charge. The original officer in charge had been Cdr. Sterling T. Dibrell, head of ONI's Graphic Section during most of World War II and into the postwar years. Capt. George B. Raser relieved Cdr. Dibrell and served as interim officer in charge for two or three weeks until the arrival of Meyertholen. Administratively, the Ford Island SIPU was under the commanding officer of Fleet

Composite Squadron One at Barbers Point. SIPU was almost exclusively a targeting support activity and was equipped with a photo lab, photo interpreters, and a print shop. The production effort focused primarily on aerial targets in the Far East. Photo squadrons (both aircraft carrier and shore-based) sent their film to SIPU for processing and review by photo interpreters for targeting purposes. Cdr. Meyertholen had been a Naval Group, China coastwatcher in 1944–1945 and knew Chinese coastal targets intimately; he was also a trained photographic intelligence officer.

Targeting materials, based on a CNO prescribed format, were delivered by the Pearl Harbor SIPU to each aircraft carrier as it passed through Pearl Harbor en route to the Far East. SIPU would retrieve the same material upon the carrier's return. Commander Fleet Air, Hawaii assisted with the necessary liaison. Three of the newly established FICPAC photo interpreter officers were given orders to report to the Mobile Intelligence Production Unit at Atsugi without relief; replacements were expected to come from CINCPACFLT. Capt. Samuel B. Frankel, CINCPACFLT Intelligence Officer, however, said that he could not spare any of his intelligence officers for reassignment to FICPAC. Protestations were made to the Office of the Chief of Naval Operations without success until a CINCPACFLT personnel officer and Cdr. Meyertholen were sent to Washington to explain the situation. The loss of the three regularly assigned photo interpretation officers would leave FICPAC with only the two photo interpreters already on loan from CINCPACFLT, plus Meyertholen, who had been assigned as FICPAC Executive Officer. The transfers to Atsugi were canceled.

The first officer in charge of FICPAC was Cdr. Douglas K. English, who arrived in August 1955. His title was changed to commanding officer on 24 December 1956.[38]

FICPAC also housed and employed the personnel assigned to the Fleet Air Intelligence Augmenting Unit. The unit had been established to augment intelligence staffs in aircraft carriers and air groups during deployment. When not deployed, augmenting unit personnel at FICPAC kept up with current intelligence and updated basic intelligence and target folders. In 1960, the unit was merged with FICPAC, and the latter took over the carrier-staff augmenting function.

The concentration of all basic intelligence at FICPAC was an effective and efficient use of Pacific Fleet intelligence personnel resources. Furthermore, it left the small staff intelligence unit at the CINCPACFLT headquarters at Makalapa free to concentrate on strictly operational intelligence and supervisory matters.[39]

The increased tension in Southeast Asia in 1962–1963 required FICPAC to produce Special Photo Intelligence Reports about Laos and Cambodia, and the SPIRs on Thailand and the Republic of Vietnam were withdrawn that year for revision. FICPAC also compiled the best available photos of Soviet intelligence collection vessels (AGI) and issued a publication entitled *SIG-ONE*, a guide for rapid identification of AGIs by aircraft and other reconnaissance units.

A new set of folders, Survival and Evasion Intelligence, was being produced during 1963 to replace the Evasion and Escape (E&E) Folders in the fleet. The title was changed and the contents of the folders were revised to conform with the current Cold War situation in the Far East. In like manner, the E&E newsletter was revised in concept, and its title was changed to *Survival, Evasion, Resistance, and Escape* to place proper emphasis on survival and resistance to interrogation if captured.[40]

FICPAC began producing separate basic intelligence studies for each country in the Far East during Fiscal Year 1964. Non-Communist country studies were given first priority, followed by separate order-of-battle publications and an area intelligence study on each Communist-dominated country within the Pacific Command area.

During Fiscal Year 1964, FICPAC completed and distributed Survival and Evasion Folders on the Shantung Peninsula, North Korea, and North Vietnam; Special Photo Intelligence Reports on Cambodia and Malaysia; and a new *Tactical Targeting Material Catalogue*.

Responsibility for the production of the Integrated Operational Intelligence System database for *Ranger* (CVA 61) and for subsequent carriers using the system had been transferred from the Naval Reconnaissance and Technical Support Center to FICPAC by the Chief of Naval Operations in the latter part of 1963.[41]

In 1963–1964, with the increasing photo reconnaissance flights over the Republic of Vietnam, the initiation of photo reconnaissance over Laos, and the decision by higher authority to use Pacific Fleet photographic facilities in exploitation of the major portion of the material obtained by reconnaissance missions, the small Navy photographic unit at Cubi Point in the Philippines was temporarily augmented by photographic personnel from the fleet, from the U.S. Army, Pacific, and other Pacific Command resources. Recommendations were made and approved to have the temporary laboratory at Cubi Point established as a permanent organization, with increased equipment, enlarged facilities, and additional personnel permanently assigned. As of June 1964, the Cubi Point photographic laboratory was functioning on a 24-hour basis to help meet national requirements for aerial photography from Southeast Asia.[42]

Fleet Intelligence Center, Pacific Facility, located in the Philippines, was the outgrowth of the Joint Cubi Special Processing Facility and was intended to provide a modern, precision photo-processing, interpretation, and repair center for direct support to fleet, national, and theater consumers. The facility was established by SECNAV Notice 5450 of 31 August 1964 under the military command of the commanding officer of FICPAC and under the management control of the Director of Naval Intelligence. It had the formal mission of satisfying "fleet photographic processing, exploitation and research/library requirements in the Western Pacific area." Because the Pacific facility was within carrier on-board delivery (COD) aircraft range of the Gulf of Tonkin, it could receive, interpret, duplicate, and quickly disseminate photographs from carriers operating off Vietnam.[43]

With the initial deployment of U.S. nuclear-powered ballistic missile submarines in the Pacific in December 1964, a Polaris planning document was developed by FICPAC.

In Fiscal Year 1965, FICPAC was putting increased emphasis on support to contingency plans and produced Tactical Target Illustration sheets in lithographic and aperture card form for 491 targets. Research for 350 additional targets was in progress. FICPAC also assisted the Defense Intelligence Agency in determining data about thirty-four North Vietnamese and Laotian targets for inclusion in DIA's *Bombing Encyclopedia*. In addition, 226 Mission Planning Folders were developed in support of the Pacific Command Nuclear Attack Targeting Materials Program. Of the 226 folders, 95 were original and 131 were revisions of folders that had already been distributed to the fleet.[44]

As a result of photo analysis, a variety of products were developed and distributed during Fiscal Year 1965, including detailed port facility and fuel supply and storage (POL) studies, Russian trawler analyses, Soviet missile range instrumentation ship (SMRIS) reports, and a *Chinese Communist Air Facility Photo Interpretation Key*. Sixty order-of-battle files totaling some 17,000 EAM (Electric Accounting Machine) cards and covering fourteen Western Pacific countries were created by FICPAC for the specific use of the Integrated Operational Intelligence Center (IOIC) in *Ranger*. The computer installation and its related equipment at FICPAC was significantly increased in size and capability during the same period, with the memory core storage doubled to 80,000 character positions.[45]

As of 1966, the Fleet Intelligence Center, Pacific was under the command of CINCPACFLT, with the Chief of Naval Operations having primary support responsibility and with area coordination assigned to the Commandant 14th Naval District. FICPAC's mission was officially stated as follows:

> To develop and produce intelligence to support the U.S. Naval Forces in the Pacific Area; to provide those forces with operational intelligence materials necessary to carry out their respective missions; to provide direct support and assistance to CINCPACFLT which will enable him to fulfill his requirements for nuclear weapons employment and planning; [and] to provide a conventional warfare tactical targeting program based on CINCPAC and CINCPACFLT Contingency Plans.[46]

At the beginning of 1966, FICPAC was primarily concerned with the production of hard-copy, long-range intelligence documents. With the ever-rising tempo of operations in Southeast Asia, however, it soon became apparent that an increased demand for intelligence support would be placed on FICPAC.

Keeping abreast of new targets and the rapidly changing status of old targets in Southeast Asia was creating an enormous burden for the FICPAC Contingency Requirements Department, responsible for production in support of the Tactical Target Illustration program. Accordingly, on 27 May 1966, CINCPACFLT discontinued the production of the target illustrations covering Laos, North Vietnam, and South Vietnam. On 4 June, production was similarly halted on Project Isolation Targets and Armed Reconnaissance Targets, lines of communication covering North Vietnam. The same directive of 4 June assigned to FICPAC the new task of performing third-phase, in-depth exploitation of all reconnaissance photographs obtained over Navy areas of responsibility in North Vietnam.

In conjunction with the exploitation of film from Southeast Asia, FICPAC produced a series of Continuing Review Photo Intelligence Reports (CPIR) that included photographic enlargements, bomb damage assessment reports, and special studies as tasked by CINCPACFLT, Commander Seventh Fleet, and Commander Task Force 77. In preparing the CPIRs, the FICPAC photo laboratory produced an average of 27,000 prints per month.

In September 1966, FICPAC began making targeting recommendations and providing all-source intelligence documentation to assist Seventh Fleet forces engaged in interdicting targets in Southeast Asia. Aided by the center's IBM 1410 computer, analysts were able to determine which lines of communications were being used by the North Vietnamese to transport supplies to their combat forces.[47]

At the start of 1967, the Nuclear Warfare Support Department of FICPAC was involved in the planning and preparation of documents in support of the Pacific Fleet's nuclear-powered ballistic missile submarine force. The *Mission Planning Document* and *Far East Photo Aids to Navigation: Submarines* were issued during April.

In February, the Survival, Evasion, Resistance, and Escape (SERE) sections of the existing aircraft Mission Planning Folders were updated, and the nineteenth in a series of *SERE Newsletters* was produced. The *SERE Guide to Southeast Asia* and thirteen North Vietnam Safe Briefs (guides to escape and evasion techniques) were completed and distributed.

On 22 March 1967, at the direction of CINCPAC, FICPAC assumed overall responsibility for the production of the *Pacific Command Contingency Planning Facilities List* for Laos and North Vietnam. The document became the principal data reference for the annotation of graphics showing targets in those two countries.

On 21 April, FICPAC was tasked by CINCPACFLT to produce Special Imagery Interpretation Reports for direct support to CINCPAC's North Vietnam area analysis.

In response to CINCPACFLT tasking, the production of Radar Target Folders was started to support A-6 Intruder all-weather strike aircraft operations in Southeast Asia. From May through August 1967, forty-four folders were completed and distributed to A-6 squadrons. *SERE Newsletters* 21 through 23 were produced between May and October.

On 2 June 1967, FICPAC began the distribution of *Photographic Aids to Surface Navigation* for use by northern and southern area Sea Dragon ships involved in shore bombardment in the Vietnam War. On 20 June, the Special Projects Office of FICPAC began the photographic interpretation readout of Black Shield photography. To accommodate the additional effort, FICPAC was augmented by thirteen Navy and Marine intelligence personnel on temporary duty from various other Pacific Fleet resources.

During the period July through September 1967, *Special Weather Studies* and a *Weather Study Key* were promulgated, and Change B to the *SERE Guide to Southeast Asia* was distributed. The fifth volume of the *Pacific Aerial Radarscope Photography Guidebook* was updated and reproduced for distribution. Also during the period, thirty Mine Target Delivery folders were revised and updated for distribution along with the *Mine Targeting Document*.

On 11 September, FICPAC promulgated the Tactical Installation File (TIF) and shortly thereafter issued the associated database to all Pacific Fleet attack aircraft carriers equipped with the Integrated Operational Intelligence System. The TIF data file consisted of the North Vietnamese, Laotian, and North Korean Installation and Orders-of-Battle File and was intended for use by fleet staffs, IOIC photo interpreters, and air intelligence officers in targeting, imagery interpretation, and intelligence reporting. The TIF represented a new concept for merging all tactical and FIC-produced intelligence into a unified database that could be rapidly queried and quickly updated by using either electric computers or card processing equipment. The TIF had the capability to arrange and present intelligence logically and in a readable format from a multitude of sources. Incident to the production of the TIF, a more responsive update procedure was adopted, using the AUTODIN network to transmit data electrically from Hawaii to FICPACFAC in the Philippines; from here it was delivered by aircraft to IOIS-equipped aircraft carriers on station off Vietnam.

During 1967, FICPAC supplied many officers to Commander Fleet Air, Hawaii to act as Operational Readiness Inspection observers for aircraft carriers. The observers evaluated the carrier air groups' strike pilot readiness, air wing recognition, SERE training, photo interpretation readiness, and IOIC performance. FICPAC personnel also assisted in the development of inspection procedures for electronic warfare, mine warfare, and nuclear weaponeering.[48]

When *Pueblo* (AGER 2) was captured in January 1968, FICPAC quickly updated its intelligence profile on North Korea, producing targeting information for possible contingencies. CINCPACFLT tasked FICPAC to produce and forward a North Korean IOIC Tactical Installation File to *Enterprise* (CVAN 65). The carrier received the TIF three days later. From photo exploitation of a Black Shield photography mission flown after the *Pueblo* hijacking, quick response graphics were produced for support to possible contingency operations against North Korea. Over 180 Tactical Targeting Illustrations were produced for the Army, Air Force, Navy, the Defense Intelligence Agency, and the State Department, and 53,000 lithographic copies were distributed within four weeks of the *Pueblo* seizure. Also, twenty-seven Special Mission Planning Folders covering contingency nuclear targets in North Korea were produced at the request of CINCPACFLT during the three weeks immediately following the *Pueblo* incident.

During the early days of 1968, the Nuclear Warfare Support Department of FICPAC was also heavily engaged in the planning and preparation of various target documentation. Targets were selected and flight planning was processed to produce Force Application Source Data documents for strike aircraft and source data in support of reconnaissance missions. The Flight Plan Analysis computer program was updated to include the A-7 Corsair-II aircraft.

A Miniature Transparency (MITRAN) index file was designed to automate the handling and selective retrieval of the rapidly growing number of FICPAC MITRANs.

On 4 March 1968, production and distribution of the *North Korean Contingency Plans Facilities List* was commenced in accordance with CINCPACFLT tasking. Also in March, FICPAC began conducting Second-Phase Exploitation of GIANT SCALE SR-71 aerial reconnaissance photography in response to CINCPACFLT tasking. Delivery of GIANT SCALE film was made to FICPAC within twenty-four hours after each mission was flown, and through expeditious production of Photo Interpretation Reports, FICPAC was able to provide timely tactical intelligence in support of CTF 77 operations.

Implementation of the Mark III Formatted Filing System also began in March. The FFS had many program bugs that required constant checking for corrections. A Mark II FFS Data Source Change Proposal to provide the Formatted File System with the capability to handle the Binary-Coded Decimal Automated Intelligence File was completed during the same month.

As of 1 April 1968, FICPAC underwent a major reorganization: the Nuclear Warfare Support Department became the Mission Planning Support Branch of the newly named Intelligence Production Department; the Automatic Data Processing Branch of the former Technical Services Department became the Intelligence Data System Department; and the Library and Distribution Branches of the Administration Department became branches of the Services Support Department.

When the bombing halt of April 1968 became effective, FICPAC was tasked to produce a study of military supply storage and transshipment in Haiphong to be used by CINCPAC for briefing the Joint Chiefs of Staff and the National Security Council. The study, delivered in May, showed that the North Vietnamese were using Haiphong harbor more extensively than before the bombing halt to off-load, store, and transship war materials to the insurgents in South Vietnam.

The Amphibious Analysis Section of FICPAC was formed in April 1968 and was tasked to produce and maintain automated data files on landing beaches, helicopter landing areas, parachute drop areas, and coastal defenses. In May 1968, to aid in meeting Naval Intelligence Processing System requirements for the creation of databases for selected countries in the Pacific Command area by Fiscal Year 1970, the Amphibious Analysis Section was augmented by a group of five Marine photo interpreters on temporary assignment from Fleet Marine Force, Pacific.

In May 1968, FICPAC released the Mark III Data Source Change Proposal to the intelligence community, providing recipients with the capability to handle the Binary-Coded Decimal Automated Intelligence File under the Mark III Formatted Filing System—a particularly valuable improvement because it allowed computer analysts direct access to a vast assortment of intelligence material in support of already created and newly developing data files.

During the first half of 1968, Naval Air Intelligence Reserve Units were assigned to assist FICPAC in the production of Tactical Targeting Illustrations. Reserve units T-1 from Los Alamitos, California; D-1 from Dallas, Texas; K-1 from Olathe, Kansas; and E-1 from Minneapolis, Minnesota, participated. Each unit was assigned a particular country or area for TTI production.

In June 1968, FICPAC was tasked to conduct continuing photo analysis in support of a new interdiction in Vietnam. The area assigned to Task Force 77 for ROLLING THUNDER operations was subdivided into three traffic control areas. Traffic choke points were assigned along each major line of communication within the traffic control areas. FICPAC provided a day-to-day status report about the traffic choke points and an analysis of enemy reactions to the interdiction. The reports were of value not only to the CINCPACFLT targeting and operations staffs but also to CTF 77 planners.

On 25 June 1968, the Ports and Harbors Naval Installation Analysis Section of FICPAC was set up to design, produce, and maintain automated data files about ports and harbors in the Navy's area of responsibility. In August, the Transportation Section of FICPAC was formed to design, produce, and maintain an automated intelligence file on roads, railroads, inland waterways, and terrain obstacles.

New targeting was required with the addition of the battleship *New Jersey* (BB 62) to Task Force 77, and FICPAC revised its Intelligence Data Handling System target intelligence file for North Vietnam. A battleship target list was developed and submitted to Commander Seventh Fleet in August 1968. In September, the Urban Area Analysis Section of FICPAC was formed to design, produce, and maintain automated intelligence files on urban areas and their related transportation, communication, and utilities facilities.

In October, FICPAC completed a study that proposed changes in aircraft carrier SIOP (Single Integrated Operation Plan) launch positions for nuclear strike aircraft that would permit recovering all of the strike aircraft back on board rather than at shore bases. The proposal, providing the aircraft carrier commanders with a restrike capability, was enthusiastically approved for implementation.

In November, CINCPACFLT tasked FICPAC to produce Mission Planning Materials for use in nuclear and non-nuclear delivery against important non-SIOP targets. Also in November, CINCPAC assigned an additional 2,081 targets to FICPAC for the production of Tactical Target Illustrations. The production effort was distributed among the Naval Reserve Intelligence Units that were working on the countries affected. By December, the units had submitted a total of 330 TTIs to FICPAC, and forty-eight officer and enlisted naval reservists had each performed two weeks' active duty in conjunction with the effort.

In November 1968, CINCPAC requested CINCPACFLT and FICPAC to provide photographic proof that the North Vietnamese were taking advantage of the 30 October bombing halt to push military supplies southward. Using selected prints provided by the FICPAC Photo Interpretation Section and the analysis prepared by CINCPACFLT from the prints, CINCPAC forwarded a critical study to the Joint Chiefs and other Washington policymakers.

In January 1969, fourteen NRIUs and eight NAIRUs were assigned to FICPAC's Surface Support Branch to augment its efforts to fulfill the Naval Intelligence Processing System's requirements and to meet a deadline to establish a Phase II NIPS database for the Pacific Command area by December 1970. By June 1969, several naval reservists from each reserve unit assigned to FICPAC had completed two-week active duty periods to work on their respective fleet project areas.

In February 1969, FICPAC requested a new building to house its personnel and equipment at Makalapa, next to the CINCPACFLT headquarters building.

In April 1969, when a Navy EC-121M Constellation electronic reconnaissance aircraft was shot down in the Sea of Japan off the coast of North Korea, FICPAC once again revalidated the North Korean photographic database. In the ensuing weeks, over 500 North Korean Quick Response Graphics (QRG) were produced and distributed, and more than 10,000 copies of TTIs were reprinted for the Air Force. It became apparent that there was a critical lack of radar target materials, and, in consequence, a set of radar-applicable QRGs were devised to give offset aim and radar-validated reference points. Production began in May, and by 30 June more than 300 of the QRGs had been produced for use by A-6 squadrons and by other commands.[49]

In August 1969, Contingency Target Materials (CTM) were introduced into the fleet through the Miniature Transparency program. The CTMs took the place of the Contingency Target Folders and the Freedom Drop Folders held by fleet units. The change resulted in a significant reduction in shipboard storage requirements and provided more and higher-resolution photographs for shipboard reproduction, as well as a capability for more rapid reproduction of more comprehensive material in a simplified format. FICPAC also benefited because the CTMs could be more rapidly and more accurately updated when new information was received.

After three months of tests, the Gerber 1000 Series Digital Plotter became operational at FICPAC in August 1969. Using the machine for automated plotting substantially reduced manual drafting requirements and permitted the conversion of nine drafting billets to critically needed photo intelligence billets.

In September 1969, a small-scale UNIVAC 9300 computer was installed at FICPAC to alleviate the time-consuming input/output processing required on the IBM 1410 system.

The first Mine Field Planning Folder Supplements were distributed on 15 September 1969, superseding the old Mine Target Delivery Folders with a streamlined product that provided a great saving in personnel.

On 7 October 1969, CINCPACFLT tasked FICPAC to provide Commander Seventh Fleet, CTF 71, and aircraft carriers operating in the Western Pacific with selected prints of specific North Korean Contingency Targets after each Korean GIANT SCALE mission. To expedite completion of the support requirement, FICPACFAC was included in the distribution of duplicates of the photographs from the GIANT SCALE missions, and, around December 1969, the production effort shifted to FICPACFAC in the Philippines.

A comprehensive revision of FICPAC's contribution (Chapter 1) to the *SERE Guide to Southeast Asia* had been started in September 1969. After four months of work, a completely new chapter, consisting of 106 pages with 104 graphics, was accepted and distributed in January 1970 as Change C to the *SERE Guide*.

In response to CINCPAC tasking, FICPAC set up the Intelligence Data Handling System Communications Branch in the Intelligence Data Handling System Department in December 1969 to operate the Pacific Command IDHS communication switch. The switch had a direct link to the Defense Intelligence Agency switch and provided service by means of on-line communications, for CINCPAC; CINCPACFLT; Commander in Chief, Pacific Air Force; and Commander in Chief, U.S. Army, Pacific

In February 1970, FICPAC received the *NIPS Phase II Producers' Manual* from Commander

Naval Intelligence Command and began immediate implementation of the numerous required changes to the FICPAC database files.

On 21 February 1970, because of losses in personnel augmentation billets, FICPAC recommended reducing the number of some of its periodic reports to permit an annual saving of 2,200 hours in printing and assembling and approximately 200 hours of computer processing time. The recommendation was approved by CINCPAC effective 1 May 1970.

As a result of actions initiated by FICPAC, silver recovery incinerators were procured and installed at FICPAC in April 1970 and at FICPACFAC in July 1971. The incinerators were used to destroy vast quantities of obsolete classified reconnaissance photos and recover valuable silver ash. It was estimated that $50,000 worth of silver would be recovered annually.

In May 1970, CINCPAC tasked the Fleet Intelligence Center with developing a capability to produce a Cambodia Contingency Planning Facilities List to support Pacific Command contingency planning; FICPAC was also tasked to produce a sample listing. Approximately 1,000 known installations were identified from an automated file, and the information was extracted for use in the listing.

Another new tasking, assigned to the Fleet Intelligence Centers on 5 June 1970, was to reissue the IOIC database to aircraft carriers after their major yard overhaul periods. The task, which previously had been performed by the Naval Reconnaissance and Technical Support Center at Suitland, Maryland, required FICPAC to reproduce Pacific database tapes and approximately 500,000 machine data cards for each IOIC-equipped carrier upon completion of its yard period.

The program to provide intelligence support material to the Seventh Fleet's all-weather aerial strike force was expanded during Fiscal Year 1970 to include developments in radar targeting support materials. In addition to the already existing requirements for material about North Korea and North Vietnam, coverage was expanded by CINCPACFLT in July 1969 to include Laos and targets in the SIOP and in CINCPAC operation plans. In December 1969, tasking was further expanded by CINCPACFLT to include domestic U.S. training materials for use in all-weather attack training in the continental United States.

On 14 April 1970, FICPAC, at a Pacific Command Target Materials Review Group meeting, proposed that an Automated Tactical Target Graphic be adopted to replace the Tactical Target Illustrations and Quick Response Graphics. The latter had been authorized by CINCPAC in August 1969 in lieu of TTIs for point targets within the Pacific Command area and could be rapidly produced and distributed but did not include textual data about the targets. The major advantages of ATTGs were that the cost of production and maintenance would be reduced and that the ATTGs would be more responsive to Pacific Command consumer needs. In June 1970, FICPAC was tasked to produce a sample ATTG for distribution and evaluation by Pacific Fleet users and by the other Fleet Intelligence Centers.

During Fiscal Year 1970, four NRIUs and eight NAIRUs were assigned to FICPAC management. The Naval Reserve units were used in the production of Quick Response Graphics and in the automation and miniaturization of specially formatted intelligence studies to support the amphibious command ship automated intelligence center databases that had been created and maintained under the Phase II implementation of the Naval Intelligence Processing System. Approximately 300 naval reservists participated in the program during their drill periods, and 151 officers performed their annual active-duty training at FICPAC.[50]

Intelligence production by FICPAC, as of 1970, was accomplished under the general management and supervision of the CINCPACFLT Assistant Chief of Staff for Intelligence, who also established the relative priorities for its production efforts. FICPAC's personnel as of 30 June 1970 included 75 officers, 214 enlisted personnel, and 24 civilians, against an authorized allowance of 80 officers, 228 enlisted, and 27 civilians. FICPAC occupied three buildings on Ford Island: a converted warehouse with 34,170 square feet of working space; a former aircraft hangar with 12,000 square feet, and a portion of the Headquarters, Commander Antisubmarine Warfare Forces, Pacific with 9,300 square feet.[51]

Thirty-five percent of FICPAC's production during 1970 was generated by operations in Southeast Asia. The remainder was devoted to the analysis of intelligence for use in mission planning, target folders, emergency and contingency war plans, the development of the NIPS database, and other purposes.[52]

Congress had approved design funds, in January 1970, for a new 74,400-square-foot FICPAC building near the CINCPACFLT headquarters at Makalapa. In December 1971, the Military Construction Appropriation Bill, which included funds for the new building, was signed by the President.[53]

In October 1970, the first automated amphibious database retrieval was performed in response to a request from Commander Amphibious Forces, Pacific. The database provided a listing of the numbers of causeway sections that would be required for amphibious landings at beaches in twelve Asian countries and was used in support of planning requirements for the new *Newport* class tank landing ship (LST).

In January 1971, the Defense Intelligence Agency approved the ATTG, and FICPAC began full-scale production. By 30 June 1972, over 500 ATTGs had been produced.

In February 1971, after the introduction of the A-7E version of the Corsair-II light attack aircraft in late 1970, FICPAC began expanding its all-weather attack targeting document material to accommodate the A-7E's need for validated geographic coordinates to ensure its all-weather capability. In addition, Radar Systems Targets Lists for the A-7E target system were produced on North Korea, North Vietnam, and Laos in support of Southeast Asia operations and other contingency plans.

Through joint efforts by FICPAC and Commander Naval Forces, Vietnam (COMNAVFORV), Supplement I to the *North Vietnam Infiltration Trawler Identification Guide* (*INTRIGUE*) was published in August 1971. The publication satisfied a high-priority COMNAVFORV intelligence production requirement issued during August 1970. Due to its loss of control over the national-level collection resources needed to satisfy the trawler recognition guide requirement, CINCPACFLT had assigned it to FICPAC in April 1971.

During Fiscal Years 1971 and 1972, a total of 270 Naval Reserve officers performed their active-duty training at FICPAC with the Reserve Fleet Projects Program, which was intended to provide intelligence production support.

In May 1972, FICPAC was tasked to produce a *Western Pacific-Indian Ocean Port Directory*. Using FICPAC intelligence resources, Naval Reserve Intelligence Division 14-1 at Honolulu began production of the directory as part of the Reserve Fleet Projects Program.

With the increase in operations following the North Vietnamese invasion of South Vietnam on 1 April 1972, FICPAC rapidly reorganized its photographic interpretation assets to provide CINCPACFLT, CINCPAC, Fleet Marine Force, Pacific, and Commander Seventh Fleet with Operation LINEBACKER targeting support, special operations planning for the 8 May mining operations at Haiphong, analysis of North Vietnamese reaction to the mining, and analysis of North Vietnamese lines of communication. Distribution of reconnaissance film from the Southeast Asia area of operations was increased, and exploitation proceeded on an all-source, around-the-clock basis. Film from carrier reconnaissance aircraft, Air Force reconnaissance drone aircraft, and Strategic Air Command SR-71 aircraft was reviewed, exploited, and reported by means of special graphics, briefings, and messages.

By 15 April 1972, Navy and Marine photo interpreters had been organized into a quick response, target analysis team as part of a special contingency branch of FICPAC to produce selected annotated photographic prints for CINCPACFLT target planning. Production of ATTGs and other major exploitation efforts not in direct support of current Southeast Asia operations was tentatively set aside. New documents reproduced to meet current needs included Naval Gunfire Support Lists and Naval Gunfire Support Overlays. Numerous studies and graphic support projects were undertaken, such as producing seven special studies to update Fleet Marine Force, Pacific contingency plans, updating helicopter landing area studies and beach files, and developing new files based on new photographs.

By the end of Fiscal Year 1972, FICPAC was operating a leased IBM 360/50I computer 24 hours a day, 7 days a week,and its government-owned Gerber 1022 Plotter was being used for the graphic display of information. In addition, FICPAC operated and maintained the Pacific Command's IDHS communications computer switch (using an IBM 360/30F) for on-line file access in a remote, multi-user telecommunications environment. The IBM 360/30F equipment had been received on 25 February 1971, and, on 16 April 1971, an IBM 2703 Transmission Control Unit had been received as the final component for the IDHS communications between the Defense Intelligence Agency and Pacific Command's computer switches. On 28 January 1972, the IBM 360/SOI became operational, starting a new era in FICPAC Intelligence Data System processing capabilities.[54]

During July 1972, FICPAC performed a major effort in support of Southeast Asia operations by making a study of bombing damage to dikes in the vicinity of Thai Binh. In addition, the discovery of North Vietnamese minesweepers in Haiphong harbor led to an intensive study of the entire minesweeping problem in the Haiphong area.

A project to validate the existence of 171 suspected prisoner-of war-camps in Laos was undertaken in September 1972. A review of 300 rolls of photographs from thirty reconnaissance missions resulted in the location of 165 of the 171 facilities and the determination that some 130 were, in fact, being used to hold POWs.

The cease-fire in Southeast Asia in January 1973 required FICPAC to send ten officers to Clark Air Force Base in the Philippines to participate in Operation HOMECOMING, the processing of returning American POWs. Four more officers soon followed. The duties of the fourteen FICPAC officers included receiving the returnees at the Clark Air Force Base Hospital, assisting in meeting medical schedules, ensuring the issue and proper fit of uniforms, handling and meeting with visitors and the press, and personally escorting the returnees on their flights to naval hospitals in the continental United States. The FICPAC escorts conducted thorough initial intelligence debriefings of the returnees in an attempt to determine the status of other American POWs and personnel listed as missing or killed in action. Many of the FICPAC debriefers remained with their returnees and conducted debriefings at Navy hospitals in the United States.

The cease-fire also prompted the disestablishment of CINCPACFLT's FICPAC Support Branch in February 1973 and the change of the *Contingency Planning Facilities List* in March from a monthly to a quarterly publication, with monthly updates.

In October 1973, two enlisted photo interpreters were sent on temporary duty to augment the aircraft carrier *Hancock* (CV 19) during her deployment to the Indian Ocean during the Arab-Israeli conflict.[55]

On 19 November 1973, the Navy accepted the new FICPAC building (Building 352) at Makalapa, overlooking Pearl Harbor.

Commanding officers of the Fleet Intelligence Center, Pacific from its inception through July 1975 were as follows:

Name	**Reporting Date**
Cdr. D. K. English*	Aug 1955
Capt. B. J. Moynahan	Aug 1957
Capt. L. T. McQuiston	Jul 1960
Capt. T. T. Guillory	Jun 1962
Capt. K. E. Gulledge	Aug 1964
Capt. E. E. Kerr	Jul 1966
Capt. R. E. Lytle	Jul 1968
Capt. A. R. Jussel	Jul 1970
Capt. L. E. Connell	Aug 1972
Capt. J. N. Ford	Jul 1975

*Initially as officer in charge.
Compiled from various sources.

Chapter Notes

1. CINCNELM ltr, ser 000183, 14 Sep 1950, box 5, TSC 8665, Job 11146, OA.

2. CNO (OP-322Y1B) ltr to CINCNELM, ser 000545P32, 11 Jul 1951.

3. Capt. James A. Marks, USN, ltr to author, 9 Nov 1973.

4. CNO (OP-322H4C) ltr to CINCNELM, ser 000179P32, 10 Apr 1953.

5. Fleet Intelligence Center, Europe (FICEUR), *Consolidation Study*, Jan–Mar 1973 (no serial number).

6. FICEUR, "Shore Establishment Review," 1 Jul 1970, XVI-1.

7. Ibid., XII-1.

8. Commanding Officer FICEUR ltr, 4 Dec 1973.

9. OPNAV Notice 5450 of 28 Jan 1974; and Change 1 of 29 Mar 1974.

10. OP-322 memo to OP-32B, 30 Aug 1950. The memo's use of the term "Fleet Intelligence Center" is believed to be an informal reference to the intelligence staff assigned to the CINCLANTFLT Intelligence Officer, since the "FICs" had not yet been established.

11. Fleet Intelligence Center, Atlantic (FICLANT), *Command History, 3 Mar 1955–1 Jan 1968*, encl. 1, Annex 4.

12. *CINCLANTFLT Annual Report, FY 1956*, 58–59.

13. *CINCLANTFLT Annual Reports, FY 1957*, 73; and *FY 1958*, 64.

14. *CINCLANTFLT Annual Report, FY 1959*, 68–71.

15. CINCLANTFLT, "Report Upon Being Relieved," 29 Feb 1960, 63–66.

16. *FICLANT Command History, 3 Mar 1955–1 Jan 1968*, encl. 1, Annex 3, citing SECNAV Notice 5450 of 4 May 1961.

17. *CINCLANTFLT Annual Report, FY 1961*, 75–78.

18. *FICLANT Command History*, encl. 1.

19. IBM Corp., *Functional Analysis of the Atlantic Intelligence Center*, 18 Dec 1964, 1, 5.

20. *FICLANT Command History*, Annex 9.

21. Ibid. 8.

22. Ibid., 9.

23. Ibid., 11.

24. Ibid., encl. 1, 6, and Annex 1.

25. *FICLANT Command History, 1968*, 5–6.

26. *FICLANT Command History, 3 Mar 1955–1 Jan 1968*, encl. 1, 4.

27. *FICLANT Command History, 1969*, 3–7.

28. FICLANT, "Shore Establishment Review," 9 Apr 1970, 10.

29. *FICLANT Command History, 1 Jan 1970–30 Jun 1971*, B-1–B-4.

30. Ibid., I-2–I-3.

31. *FICLANT Command History, 1972*, B-1–B-4.

32. OPNAV Notice 5450 of 28 Jan 1974.

33. Commander in Chief, Pacific Fleet (CINCPACFLT), *Interim Evaluation Report No. 4: Korean War Naval Operations*, 3–48.

34. SECNAV Notice 5450 of 3 Mar 1955.

35. SECNAV Notice 5450 (OP-213C ser 1929P21) of 27 May 1955.

36. VAdm. Rufus L. Taylor, interview by author, 23 Nov 1975; and Commanding Officer, FICPAC (Capt. Lewis E. Connell) ltr to author, 7 Nov 1974.

37. Taylor interview, 29 Oct 1975.

38. Cdr. Joseph A. Meyertholen, interviews by author, 25 Mar 1980 and 4 May 1980.

39. Taylor interview, 29 Oct 1975.

40. *CINCPACFLT Annual Report, FY 1963*, 41.

41. *CINCPACFLT Annual Report, 1 Oct 1963–26 Jun 1964*, 49.

42. Ibid., 51–52.

43. *CINCPACFLT Annual Report, 26 Jun 1964–30 Mar 1965*, 55.

44. Ibid., 52–53.

45. Ibid., 18, 52.

46. *FICPAC Command History, 1966*, 1–2, citing OPNAV Instruction 05450.134 of 6 Aug 1965.

47. Ibid., Part II, 3–4.

48. *FICPAC Command History, 1967*, 3–8, and Appendix 6.

49. *FICPAC Command History, 1 Jan 1968–30 Jun 1969*, Part II, 2–10.

50. *FICPAC Command History, FY 1970*, 6–14.

51. FICPAC, "Shore Establishment Review," 25 Feb 1970, 3, 32.

52. Ibid., 42.

53. Ibid., 37; and *FICPAC Command Histories, FY 1971*, and *FY 1972*, II-4.

54. *FICPAC Command Histories, FY 1971* and *FY 1972*, II-5–II-8, II-10–II-13.

55. *FICPAC Command History, Jul 1972–Dec 1973*, II-3–II-5, II-8.

CONCLUSION

It should be reiterated and emphasized that this book is intended as an introductory reference work on U.S. Naval Intelligence. There are many gaps in the information presented herein, mainly because of the voluminous official and informal accounts of most naval historic events as well as of the decisions made that influenced U.S. Navy policy, planning, and operations.

Prior to starting the research for this book, an effort was made to summarize and review the intelligence available and presented to major area and operational commands, which they used to prepare for battles and campaigns against hostile forces. It became obvious that an understanding of the intelligence organization, including its methods and activities relating to collection, production, and dissemination procedures was needed first. The review of intelligence available to justify various command decisions and actions still needs to be undertaken for nonpolitical historic preservation purposes and for possible future command guidance.

Everyone in the Navy, whether or not he or she realizes it, is an element of the naval intelligence effort. Each participates in it as a contributor, user, or both. No one will deny that official duties require intelligence: an ability to learn and understand information, correlate it, determine its validity, and anticipate its need, now and in the future. So, it is hoped this book will be helpful especially to active duty personnel in addition to future researchers of naval history.

APPENDIX

Directors of Naval Intelligence

Lt. Theodorus B.M. Mason
Jun 1882–Apr 1885

Lt. Raymond P. Rodgers
Apr 1885–Jul 1889

Cdr. Charles H. Davis
Sep 1889–Aug 1892

Cdr. French E. Chadwick
Sep 1892–Jun 1893

Lt. Frederick Singer
Jun 1893–Apr 1896

LCdr. Richard Wainright
Apr 1896–Nov 1897

Cdr. Richardson Clover
Nov 1897–May 1898

Capt. John R. Bartlett
May 1898–Oct 1898

Cdr. Richardson Clover
Oct 1898–Feb 1900

Capt. Charles D. Sigsbee
Feb 1900–Apr 1903

Cdr. Seaton Schroeder
May 1903–Apr 1906

Capt. Raymond P. Rodgers
Apr 1906–May 1909

Capt. Charles E. Vreeland
May 1909–Dec 1909

Capt. Templin M. Potts
Dec 1909–Jan 1912

Capt. Thomas S. Rodgers
Jan 1912–Dec 1913

Capt. Henry F. Bryan
Dec 1913–Jan 1914

Capt. James H. Oliver
Jan 1914–Mar 1917

RAdm. Roger Welles, Jr.
Apr 1917–Jan 1919

RAdm. Albert P. Niblack
May 1919–Sep 1920

RAdm. Andrew T. Long
Sep 1920–Jun 1921

Capt. Luke McNamee
Sep 1921–Nov 1923

RAdm. Henry H. Hough
Dec 1923–Sep 1925

Capt. William W. Galbraith
Oct 1925–Jun 1926

Capt. Arthur J. Hepburn
Jul 1926–Sep 1927

Capt. Alfred W. Johnson
Dec 1927–Jun 1930

Capt. Harry A. Baldridge
Jun 1930–May 1931

Capt. Hayne Ellis
Jun 1931–May 1934

Capt. William D. Puleston
Jun 1934–Apr 1937

RAdm. Ralston S. Holmes
May 1937–Jun 1939

RAdm. Walter S. Anderson
Jun 1939–Jan 1941

Capt. Jules James
Jan 1941–Feb 1941

RAdm. Alan G. Kirk
Mar 1941–Oct 1941

RAdm. Theodore S. Wilkinson
Oct 1941–Jul 1942

RAdm. Harold C. Train
Jul 1942–Sep 1943

RAdm. Roscoe E. Schuirmann
Sep 1943–Oct 1944

RAdm. Leo H. Thebaud
Oct 1944–Sep 1945

RAdm. Thomas B. Inglis
Sep 1945–Sep 1949

RAdm. Felix L. Johnson
Sep 1949–Jun 1952

RAdm. Richard F. Stout
Jul 1952–Nov 1952

RAdm. Carl F. Espe
Dec 1952–May 1956

RAdm. Laurence H. Frost
Jun 1956–Sep 1960

RAdm. Vernon L. Lowrance
Sep 1960–Jun 1963

RAdm. Rufus L. Taylor
Jun 1963–May 1966

Capt. Maurice H. Rindskopf
May 1966–Jul 1966

RAdm. Eugene B. Fluckey
Jul 1966–Jun 1968

Capt. Frank M. Murphy
Jun 1968–Aug 1968

RAdm. Frederick J. Harlfinger II
Aug 1968–Jul 1971

RAdm. Earl F. Rectanus
Jul 1971–Sep 1974

RAdm. Bobby R. Inman
Sep 1974–Jul 1976

RAdm. Donald P. Harvey
Jul 1976–Aug 1978

RAdm. Sumner Shapiro
Aug 1978–Aug 1982

RAdm. John L. Butts
Aug 1982–Sep 1985

RAdm. William O. Studeman
Sep 1985–Jul 1988

RAdm. Thomas A. Brooks
Jul 1988–Aug 1991

RAdm. Edward D. Sheafer, Jr.
Aug 1991–Sep 1994

RAdm. Michael W. Cramer
Sep 1994–

ABBREVIATIONS

AAF	Army Air Forces
ACINT	Acoustic Intelligence
ACI	Air Combat Intelligence/Information (after 1942) [officers]
ACNO	Assistant Chief of Naval Operations
Admin	Administrative
Adm.	Admiral
AEW	Airborne Early Warning
AFSA	Armed Forces Security Agency
AGC	Amphibious Command Ship
AGER	Environmental Research ship
AGTR	Technical Research Ship
AI/PI	Air Intelligence/Photo Intelligence
AIC	Acoustic Intercept Committee/Advance Intelligence Center
AIDS	Acoustic Intelligence Data System
AIO	Air Intelligence Officers
ALUSLO	American Legation, U.S. Liaison Office
ALUSNA	American Legation, U.S. Naval Attaché
ANEEG	Army/Navy Electronic Evaluation Group
ANIS	Advanced Naval Intelligence School
AOS	Amphibious Objective Studies
APD	Fast Attack Transport
ASR	Submarine Rescue Ship
ASWORG	Antisubmarine Warfare Operational Research Group
ASW	Antisubmarine Warfare
ATF	Fleet Ocean Tug
ATIS	Allied Translator and Interpreter Section
AUTODIN	Automatic Distribution of Intelligence
BUAER	Bureau of Aeronautics
BUDOCKS	Bureau of Yards and Docks
BUNAV	Bureau of Navigation
BUORD	Bureau of Ordnance
BUPERS	Bureau of Personnel
BUSHIPS	Bureau of Ships
Capt.	Captain
CA	Heavy Cruiser

CCRAK	Combined Command for Reconnaissance Activity, Korea
Cdr.	Commander
CEE	Captured Enemy Equipment Program
CFAW	Commander Fleet Air Wing
CIA	Central Intelligence Agency
CIC	Combat Information Center
CIIC	Current Intelligence and Indications Center
CINCAF	Commander in Chief, Asiatic Fleet
CINCFE	Commander in Chief, Far East
CINCLANT/LANTFLT	Commander in Chief, Atlantic/Atlantic Fleet
CINCMED	Commander in Chief, Mediterranean
CINCNELM	Commander in Chief, U.S. Naval Forces, Eastern Atlantic and Mediterranean
CINCNELMREPGER	Commander in Chief, U.S. Naval Forces, Eastern Atlantic and Mediterranean Representative, Germany
CINCNORAD	Commander in Chief, North American Air Defense Command
CINCPAC/PACFLT	Commander in Chief, Pacific/Pacific Fleet
CINCUS	Commander in Chief, U.S. Fleet (before 1942)
CINCUSNAVEUR REPGER	Commander in Chief, U.S. Naval Forces, Europe Representative, Germany
CIO	Confidential Intelligence Officer
CIOS	Combined Intelligence Objectives Subcommitte
CL	Light Cruiser
CMC	Commandant of the Marine Corps
CNO	Chief of Naval Operations
Col.	Colonel
COLOP	Collection Opportunity
COMAIRPAC	Commander Aviation Forces, Pacific
COMALASEAFRON	Commander Alaskan Sea Frontier
COMBASEFOR	Commander Base Force
COMCARIBSEAFRON	Commander Caribbean Sea Frontier
COMCRUDIV	Commander Cruiser Division
COMDESRON	Commander Destroyer Squadron
COMEASTSEAFRON	Commander Eastern Sea Frontier
COMFAIRMED	Commander Fleet Air, Mediterranean
COMINCH	Commander in Chief (title adopted by Adm Ernest J. King for himself and his operational staff when he was CNO in WWII)
COMINT	Communications Intelligence
COMKEYWESTFOR	Commander U.S. Naval Forces, Key West
Commo.	Commodore
COMNAVDESFOREASTPAC	Commander Naval Defense Force, Eastern Pacific
COMNAVEU	Commander Naval Forces, Europe
COMNAVFE	Commander Naval Forces, Far East
COMNAVFORGER	Commander Naval Forces, Germany
COMNAVFORJAP	Commander Naval Forces, Japan

COMNAVFORV	Commander Naval Forces, Vietnam
COMNAVGRP	Commander Naval Group
COMNAVINTCOM	Commander Naval Intelligence Command
COMNAVNAW	Commander Naval Forces, Northwest African Waters
COMPATWING-2	Commander Patrol Wing Two
COMP	Collection Operations Management Plot
COM7THFLT	Commander Seventh Fleet
COMSOLANTFOR	Commander South Atlantic Forces
COMSOLANT	Commander Naval Forces, South Atlantic
COMSOPAC	Commander South Pacific Area and Forces
COMSOWESPAC	Commander Southwest Pacific Force
COMSWPAC	Commander Southwest Pacific Area (MacArthur's command)
COMUSMACV	Commander U.S. Military Assistance Command, Vietnam
CTF	Commander Task Force
CTM	Contingency Target Material
DAS	Defense Attaché System
DCI	Director of Central Intelligence
DATT	Defense Attaché
DCNO	Deputy Chief of Naval Operations
DD	Destroyer
DDNI	Deputy Director of Naval Intelligence
DER	Destroyer Escort Radar Picket
DIA	Defense Intelligence Agency
DIO	District Intelligence Office/Officer
DIRNAVSECGRU	Director Naval Security Group
DIRNSA	Director National Security Agency
DNI	Director of Naval Intelligence
DOD	Department of Defense
DRC	Defector Reception Center
DSA	Division of Special Analysis
DSIO	District Staff Intelligence Officer
ECMRON	Navy Electronics Countermeasures Squadron
ELINT	Electronic Intelligence
Ens.	Ensign
EUCOM	European Command
FAIAU	Fleet Air Intelligence Augmenting Unit
FBI	Federal Bureau of Investigation
FBIS	Foreign Broadcast Information Service
FEAF	Far East Air Force
FICEUR	Fleet Intelligence Center, Europe
FICLANT	Fleet Intelligence Center, Atlantic
FICPAC	Fleet Intelligence Center, Pacific

FIN	*Fleet Intelligence Newsletter*
FIU	Forward Intelligence Unit
FOSIC	Fleet Ocean Surveillance Information Center
FRUFEC	Field Research Unit, Far East Command
FRUPAC	Fleet Radio Unit, Pacific
FS	SmallCoastal Freighter
HF/DF	high-frequency radio direction-finder
HUMINT	Human Intelligence
I&C	Identification and Characteristics [section]
IAIS	Integrated Air Intelligence System
ICBM	Intercontinental Ballistic Missile
ICIS	Interdepartmental Committee on Internal Security
IFF	Identification, friend-or-foe
IFI	Intelligence Files Index
IIC	Interdepartmental Intelligence Conference
IIR	Intelligence Information Report
ILO	Intelligence Liaison Office
INTERPRON	Photographic Interpretation Squadron
INTSUM	Intelligence Summary
IOIC	Integrated Operational Intelligence Center
IOIS	Integrated Operational Intelligence System
IPIR	Initial Photo Interpretation Report
IR	Infrared/Information Report
JANAF	Joint Army/Navy/Air Force
JANAID	Joint Army-Navy Air Intelligence Division
JCS	Joint Chiefs of Staff
JEIC	Joint Electronic Intelligence Center
JIC	Joint Intelligence Committee
JICA/NA/ME	Joint Intelligence Collection Agency, NorthAfrica/Middle East
JICPOA	Joint Intelligence Center, Pacific Ocean Areas
JIOA	Joint Intelligence Objective Agency
JOINPAC	Joint Operational Intelligence Agency, Pacific Command
JRC	Joint Reconnaissance Center
JUSMAAG	Joint U.S. Military Assistance Advisory Group
JOIC(B)	Joint Overt Interrogation Center, Berlin
LCol.	Lieutenant Colonel
LOFAR	Low-Frequency Acquisition and Ranging
Lt.	Lieutenant
Lt.(jg)	Lieutenant (junior grade)
Ltr	Letter
MAAG	Military Assistance Advisory Group

MCAS	Marine Corps Air Station
MID	Military Intelligence Division (Army)
MIPU	Mobile Intelligence Production Unit
MS(S)	Manuscript(s)
MSC	Military Sealift Command
NA	National Archives and Records Administration
NACA	National Advisory Committee for Aeronautics
NACAIN	Naval Collaboration in Air Intelligence
NAF	Naval Air Facility
NAIRU	Naval Air Intelligence Reserve Unit
NAS	Naval Air Station
NATO	North American Treaty Organization
NAVCINTSUPPCEN	Naval Counterintelligence Support Center
NAVINTCOM	Naval Intelligence Command
NAVSECGRUDET	Naval Security Group Detachment
NAVSTIC	Naval Scientific and Technical Intelligence Center
NAVTECHJAP	Naval Technical Mission to Japan
NAVTECHMISEU	Naval Technical Mission in Europe
NCU	Navy Communications Unit
NFOIO	Navy Field Operational Intelligence Office
NFOSG	Naval Field Operations Support Group
NHC	Naval Historical Center
NIA	National Intelligence Authority
NIC	Naval Intelligence Command
NID	Naval Intelligence Division (British)
NIE	National Intelligence Estimates
NIIS	Naval Intelligence Investigative Service
NIPSSA	Naval Intelligence Processing System Support Activity
NIRM	Naval intelligence Requirements Memorandum
NIRPS	*Naval Intelligence Requirements-Periodic Summary*
NIS	Naval Investigative Service
NIS	*National Intelligence Survey*
NISC	Naval Intelligence Support Center
NISG	Naval Investigative Support Group
NISO	Naval Investigative Service Office
NISU	Naval Investigative Support Unit
NLO	Navy Liaison Office
NOB	Naval Operating Base
NOF	Naval Operating Facility
NOL	Naval Ordnance Laboratory
NOSIC	Naval Ocean Surveillance Information Center
NPIC	Naval Photographic Intelligence Center/ Navy Photographic Interpretation Center
NRID	Naval Reserve Intelligence Division
NRIU	Naval Reserve Intelligence Unit
NRL	Naval Research Library

NRTSC	Naval Reconnaissance and Technical Support Center
NSA	National Security Agency
NSC	National Security Council
NTRS	Naval Tactical Reconnaissance Ship
NWC	Naval War College
NWP	Naval Warfare Publication
OA	Operational Archives, Naval Historical Center
OCDM	Office of Civil and Defense Mobilization
OCI	Office of the Coordinator of Information
ONI	Office of Naval Intelligence
ONIB	*ONI Bulletin*
OPCON	Operational Control
OPINTEL	Operational Intelligence
OPLAN	Operations Plan
OPNAV	Office of the Chief of Naval Operations
OSIS	Ocean Surveillance Information System
OSRD	Office of Scientific Research and Development
OSS	Office of Strategic Services
PACMIRS	Pacific Military Intelligence Research Section
PECM	Passive electronic countermeasures
POW	Prisoner of War
PRISIC	Photographic Reconnaissance and Interpretation Section, Intelligence Center
QRG	Quick Response Graphic
RAdm.	Rear Admiral
RG	Record Group
RIN	Royal Italian Navy
RIU	Radio Intelligence Unit
RN	Royal Navy
RVAH	Navy Reconnaissance Attack Squadron
SAC	Strategic Air Command
SACEUR	Supreme Allied Commander, Europe
SACLANT	Supreme Allied Commander, Atlantic
SAD	Special Analysis Division
SACO	Sino-American Cooperative Organization
SCAEF	Supreme Commander, Allied Expeditionary Forces
SCAP	Supreme Commander Allied Powers
S-DMICC	State-Defense Military Information Control Committee
SEATO	Southeast Asia Treaty Organization
SEC	Sabotage, Espionage, and Countersubversion
SECNAV	Secretary of the Navy
SEFIC	Seventh Fleet Intelligence Center

Ser	Serial
SES	Scientific Exploitation Section
SHAEF	Supreme Headquarters, Allied Expeditionary Force
SICR	Special Intelligence Collection Requirements
SIGINT	Signal Intelligence
SIOP	Single Integrated Operational Plan
SITSUM	Situation Summary
SLAR	Side-Looking Radar
SMRIS	Soviet Missile Range Instrumentation Ship
SOMM	Special Observer–Merchant Marine
SOPACIU	South Pacific Photographic Interpretation Unit
SOSUS	Sound Surveillance System
SPENAVO	Special Naval Observer
SSCC	Sound Surveillance Control Center
SSO	Special Security Officer
S&T	Science and Technology
STANCIB	State-Army-Navy Communication Intelligence Board
STIC	Scientific and Technical Intelligence Center
SWEGRUPAC	Special Weapons Evaluation Group, Pacific
SWNCC	State-War-Navy Coordinating Committee
SWNMIC	State-War-Navy Military Information Committee
T-AGOR	Oceanographic Research Ship
T-AG	Miscellaneous Auxiliary Ship
TARS	Tactical Airborne Reconnaissance System
TAS	Target Area Study
TDC	Taiwan Defense Command
TTI	Tactical Targeting Illustration
TIIC	Technical Industrial Intelligence Committee
TSCC	Technical Surveillance Countermeasures Committee
TTM	Tactical Target Material
UDT	Underwater Demolition Team
UFO	Unidentified Flying Object
USAFE	U.S. Air Force, Europe
USAREUR	U.S. Army, Europe
USCIB	United States Communication Intelligence Board
USIB	United States Intelligence Board
USMC	United States Marine Corps
USNR	United States Naval Reserve
USNIP	United States Naval Institute *Proceedings*
USNRF	United States Naval Reserve Force
USW	Undersea Warfare
VAdm.	Vice Admiral
VAH	Navy Heavy Attack Squadron
VAP	Navy Heavy Photographic Squadron

VC	Navy Composite Squadron
VD	Navy Photographic Squadron
VJ	Navy Utility Squadron
VNN	South Vietnamese Navy
VP	Navy Patrol Squadron
VQ	Fleet Air Reconnaissance Squadron
VW	Airborne Early Warning Squadron
WAVES	Women Accepted for Voluntary Emergency Service
WDC	Washington Document Center
WFASC	Washington Field Activities Support Center
WNRC	Washington National Records Center, Suitland, MD
WSEG	Weapons System Evaluation Group
WWII	World War II
ZIO	Zone Intelligence Office

Select Bibliography

Primary Sources

ARCHIVAL COLLECTIONS

Federal Records Center, Washington National Records Center, Suitland, MD.

File A8-2. Box 16, Job 3679.

"History of the Special Activities Branch, 17 June 1940–31 October" (MS). Box 16, Job 3679.

Office of Naval Intelligence. "USN Activity re Armistice of 1918 and Peace Conference of 1919." Section 103 of "German Monograph." Box 1, Job 7631.

OP-16-Z. "War History" (MS). Box 16, Job 3679.

Library of Congress, Manuscript Division, Naval Historical Foundation Papers.

RAdm. Stephen B. Luce Papers.

National Archives and Records Administration, Washington, DC.

LCdr. Glenn F. Howell Diary. Record Group 38, Records of the Office of the Chief of Naval Operations.

Letters to Officers, Ships of War. Record Group 45, Naval Records Collection of the Office of Naval Records and Library.

Naval Historical Center, Washington, DC.

Aviation History Branch

Aviation Squadron Command Histories

Navy Department Library

World War II Administrative Histories Collection

Operational Archives

Department of the Navy. "History of Administrative Problems, Korean War." Vol. 2, "Chief of Naval Operations. 1955. The typescript series is in eight bound volumes.

Naval Investigative Service (NIS) History Files

Officer Biography File

Post 1 January 1946 Action Reports

Post 1 January 1946 Command File (includes annual reports, command histories)

Privileged Manuscript Collection

Record of Proceedings of the Pueblo Court of Inquiry

United Stated Naval Administrative Histories of World War II, Appendices Collection

World War II Action Reports and Command Files

World War II Administrative Histories Appendices Collection

MEMOIRS

Adams, John. *The Works of John Adams, Second President of the United States*. Boston: Little, Brown & Co., 1854.

Gleaves, RAdm. Albert. *Life and Letters of Rear Admiral Stephen B. Luce, U.S. Navy*. New York: Putnam, 1925.

Halsey, Adm. William F., USN. *Admiral Halsey's Story*. New York: McGraw-Hill, 1947.

von Doenhoff, Richard A., ed. *The McCully Report: The Russo–Japanese War, 1904–1905*. Annapolis: Naval Institute Press, 1977.

ORAL HISTORIES AND INTERVIEWS

Author's Interviews and Correspondence Files (Operational Archives)

Cdr. E. Crilly, USNR. 6 June 1976.

Deason, Cecil M. 21 July 1976.

Fabian, Capt. Rudolph J. Tapes.

Floyd, Joseph H. 7 June 1975 and 10 August 1975.

Fuentes, Lt. Rudolf A., USNR. 11 July 1976.

Johnson, VAdm. Felix. 14 May 1975.

Kaitz, Capt. Joseph, USNR. September 1976.

Kenney, Capt. Roland N. W., USNR. 16 August 1975 and 15 January 1976, transcript Exhibit 3

Kidd, Capt. George E., USNR. January 1976.

McCauley, Capt. Edward. 19 November 1943.

Melvin, Capt. Charles S. 11 July 1976.

Meyertholen, Joseph A. 24 March 1980 and 3 May 1980.

Miller, Douglas F. 7 March 1976.

Randisi, Capt. Martin, USNR. 23 July 1975.

Rocheleau, Charles A., USNR. 1 October 1975.

Taylor, VAdm. Rufus L. 19 September 1975 and 29 October 1975. Also tapes.

Columbia University, Oral History Projects Office, New York, NY (copies held in the Operational Archives).

Anderson, VAdm. Walter S. 1963.

Foster, VAdm. Paul F. 5 January 1967.

Johnson, VAdm. Alfred W. 1964.

Kirk, Adm. Alan G. 1965.

U.S. Naval Institute, Oral History Collection, Annapolis, MD (copies held in the Operational Archives).

Frankel, RAdm. Samuel B. 1972.

Jurika, Capt. Stephen, Jr., USN. 1975–76.

Layton, RAdm. Edwin T. 1970.

McCollum, RAdm. Arthur H. 2 vols. 1971.

Sebald, Ambassador William J. (Capt., USNR, Retired). 2 vols. 1977.

Smith-Hutton, Capt. Henri H. 2 vols. 1979.

U.S. Marine Corps Historical Center, Oral History Collection. Washington, DC.

Barrow, Gen. Robert H., USMC (Ret.). 28 January 1986.

PRINTED GOVERNMENT DOCUMENTS

Library of Congress. Federal Research Division. *The Military Attaché System: Its History and Present Status*. ca. 1971.

U.S. Congress. Joint Select Committee. *Report of the Joint Committee on the Investigation of the Pearl Harbor Attack*. 39 vols. 79th Cong., 2d sess., 1946.

———. House. *Estimates Submitted by the Secretary of the Navy: Hearings Before the Committee on Naval Affairs of the House of Representatives*. 66th Cong., 1st sess., 1920.

———. House. *Navy Department Appropriations Bill for 1947: Hearings Before the Subcommittee of the Committee on Appropriations*. 79th Cong., 2d sess., 1946.

———. House. Committee on Naval Affairs. *Estimates Submitted by the Secretary of the Navy: Hearings*. 64th Cong., 1st sess., 16 March 1916.

———. Senate. *Investigation of the Conduct of the Navy in World War I: Hearings Before the Subcommittee of the Committee on Naval Affairs*. 66th Cong., 2d sess., 1921. Vol. 2, pt. 2.

U.S. Department of the Navy, *Annual Report of the Secretary of the Navy*. Various years.

———. *General Orders and Circulars Issued by the Navy Department From 1863 to 1887*. Washington: GPO, 1887.

———. *Navy Directory*. Various years.

———. *Navy Register*. Various years.

———. *U.S. Naval Activity Relative to the Armistice of 1918 and the Peace Conference of 1919*. Washington: GPO, 1919.

———. Bureau of Personnel. *Intelligence for Naval Officers*. NAVPERS 16047. 1948.

———. *Monthly Information Bulletin*, 15 April 1919, 15 May 1920.

———. *Naval Intelligence Manual*. ONI-19. Washington: GPO, 1933.

———. *Naval Intelligence Manual*. ONI-19(A). May 1947.

———. *Naval Intelligence Manual*. ONI-70-1. 4 November 1957.

———. *Notes on Espionage, Counter-Espionage and Passport Control*. ONI-22. 1935.

———. *Office Organization Manual*. 1 September 1918.

———. *Operational Intelligence (OPINTEL) Manual*, ONI-Y-1 1 January 1951.

———. *Operational Intelligence (OPINTEL) Notes*, August 1945.

———. *Report of Functions and Workload*, 1952.

———. *U.S. Navy Intelligence Manual*. ONI-70-A. 20 June 1956.

Secondary Sources

BOOKS

Campbell, Rodney. *The Luciano Project*. New York: McGraw-Hill, 1977.

Chadwick, RAdm. French E. *The Relations of the United States and Spain: The Spanish American War*. 2 vols. New York: Scribner's, 1911.

Clarke, George R., William O. Stevens, Carroll S. Alden, and Herman F. Krafft. *A Short History of the United States Navy*. Revised edition. Philadelphia: J. B. Lippincott, 1939.

Cummings, Capt. Damon E. *Admiral Wainwright and the United States Fleet*. Washington: GPO, 1962.

Daniels, Josephus. *Our Navy at War*. New York: Doran, 1922.

Davis, Vernon E. *History of the Joint Chiefs of Staff in World War II*. Vol. 1, *Organization and Development*. Washington: Joint Chiefs of Staff, 1972.

Dictionary of American Biography. New York: Scribner's, 1927–1964.

Furer, RAdm. Julius A. *Administration of the Navy Department in World War II*. Washington: Naval History Division, 1959.

Gebhard, Louis. *Evolution of Naval Radio Electronics and Contributions of the Naval Research Laboratory*. Washington: Naval Research Laboratory, January 1976.

Gröner, Erich, Dieter Jung, and Martin Maass. *Die deutschen Kriegsschiffe, 1815–1945*. Vol. 3. Koblenz: Bernard & Graefe Verlag, 1985.

Healy, Laurin H., and Louis Kutner. *The Admiral*. Chicago and New York: Ziff-Davis, 1944.

Hooper, VAdm. Edwin B. Hooper, Dean C. Allard, and Oscar P. Fitzgerald. *The United States Navy and the Vietnam Conflict*. Vol. 1, *Setting the Stage to 1959*. Washington: Naval History Division, 1976.

Knight, Adm. Austin M., and Lt. William D. Puleston. *History of the Naval War College*. Newport, RI: Naval War College, 1916.

Marolda, Edward J., and Oscar P. Fitzgerald. *The United States Navy and the Vietnam Conflict*. Vol. 2, *From Military Assistance to Combat, 1959–1963*. Washington: Naval Historical Center, 1987.

Miles, VAdm. Milton E. *A Different Kind of War: The Little-Known Story of the Combined Guerilla Forces Created in China by the U.S. Navy and the Chinese During World War II*. Garden City, NY: Doubleday, 1967.

Morison, Elting E. *Admiral Sims and the Modern American Navy*. Boston: Houghton Mifflin, 1942.

Naval Historical Division. *Civil War Chronology, 1861–1865*. Washington: GPO, 1971.

Robert Murphy. *Diplomat Among Warriors*. Garden City: Doubleday, 1964.

Niblack, RAdm. Albert P., USN. *The History and Aims of the Office of Naval Intelligence*. Washington: GPO, 1920.

Richardson, James O. Richardson. *On the Treadmill to Pearl Harbor: The Memoirs of Admiral J. O. Richardson as Told to Admiral George C. Dyer, USN (Retired)*. Washington: GPO, 1973.

Ruge, Friedrich. *The Soviets as Opponents at Sea*. Annapolis: U.S. Naval Institute, 1977.

Sargent, Nathan. *Admiral Dewey and the Manila Campaign*. Washington: Naval Historical Foundation, 1947.

Tolley, RAdm. Kemp. *The Cruise of the Lanakai*. Annapolis: Naval Institute Press, 1977.

Wohlstetter, Roberta. *Pearl Harbor: Warning and Decision*. Stanford: Stanford University Press, 1962.

Zacharias, Capt. Ellis M. *Secret Missions, The Story of an Intelligence Officer*. New York: Putnam, 1946.

ARTICLES

Berry, RAdm. Albert G., USN (Ret.). "The Beginning of the ONI." U.S. Naval Institute *Proceedings*. January 1937.

Blue, RAdm. Victor. "The Sighting of Cervera's Ships." U.S. Naval Institute *Proceedings*. September 1899.

Ellicott, Capt. John M. "T.B.M. Mason." U.S. Naval Institute *Proceedings*. March 1952.

———. "Three Navy Cranks and What They Turned." U.S. Naval Institute *Proceedings*. October 1924.

Knowles, James A. Knowles, Jr. "Blue Water Monitor." U.S. Naval Institute *Proceedings*. March 1973.

"The Korean Mission to the United States in 1883." Royal Asiatic Society *Journal*. Vol. 18. 1929.

Packard, Capt. Wyman H., USN, "The Naval Attaché." U.S. Naval Institute *Proceedings*. April 1965.

Sachse, Lt. William A. "Our Naval Attaché System." U.S. Naval Institute *Proceedings*. May 1946.

Troy, Thomas F. "The Coordination of Information and British Intelligence." *Studies in Intelligence*. Vol. 18, Spring 1974.

PERIODICALS

All Hands. March 1960.

Defense Intelligence Digest. May 1963.

ONI Quarterly. Various years (copies in World War II Command File, Operational Archives).

ONI Review. Various years (copies in Post 1 January 1946 Command File, Operational Archives).

PI and Radar Target Intelligence Newsletter. January 1962.

PAPERS

Eddy, Cdr. Spencer. *History and Anecdotes of the New York Branch of ONI, 1916–1918*. Privately reproduced manuscript, 1919.

Green, James R. "The First Sixty Years of the Office of Naval Intelligence." M.A. thesis, The American University, 1963.

Lundeburg, Philip K. "American Antisubmarine Operations in the Atlantic, May 1943–May 1945." Ph.D. diss., Harvard University, 1953.

Scovel, F. Donald. "Helms A'Lee: History of the Development of the Public Affairs Function in the U.S. Navy, 1861–1941." M.A. thesis, University of Wisconsin, 1968.

Shields, Capt. Henry S., USAF. "A Historical Survey of U.S. Naval Attachés in Russia: 1904–1941." Defense Intelligence School, March 1970.

INDEX